MATERIALS SCIENCE RESEARCH
Volume 20

TAILORING MULTIPHASE AND COMPOSITE CERAMICS

MATERIALS SCIENCE RESEARCH

Recent volumes in the series:

MATERIALS SCIENCE RESEARCH • Volume 20

TAILORING MULTIPHASE AND COMPOSITE CERAMICS

Edited by

Richard E. Tressler
Gary L. Messing
Carlo G. Pantano

and

Robert E. Newnham

The Pennsylvania State University
University Park, Pennsylvania

PLENUM PRESS • NEW YORK AND LONDON

Library of Congress Cataloging in Publication Data

University Conference on Ceramic Science (21st: 1985: Pennsylvania State University)
  Tailoring multiphase and composite ceramics.

  (Materials science research; v. 20)
  "Proceedings of the twenty-first University Conference on Ceramic Science, held
July 17-19, 1985, at the Pennsylvania State University, University Park,
Pennsylvania" — T.p. verso.
  Includes bibliographical references and index.
  1. Ceramic materials — Congresses. 2. Composite materials — Congresses. I.
Tressler, Richard E. II. Title. III. Title: Multiphase ceramics. IV. Series.
TA455.C43U54 1986   666   86-15145
ISBN-13:978-1-4612-9309-5   e-ISBN-13:978-1-4613-2233-7
DOI:10.1007/978-1-4613-2233-7

Proceedings of the twenty-first University Conference on Ceramic Science,
held July 17–19, 1985, at the Pennsylvania State University,
University Park, Pennsylvania

© 1986 Plenum Press, New York
Softcover reprint of the hardcover 1st edition 1986

A Division of Plenum Publishing Corporation
233 Spring Street, New York, N.Y. 10013

PREFACE

The proceedings of the Twenty-First University Conference on Ceramic Science held at The Pennsylvania State University, University Park, PA on July 17, 18 and 19, 1985 are compiled in this volume "Tailoring Multiphase and Composite Ceramics".

This Conference emphasized the discussion and analysis of the properties of multiphase ceramic materials in which the microstructure is deliberately tailored for specific applications or properties. Internationally recognized authorities presented keynote and invited lectures on topics dealing with processing and fabrication of multiphase and composite electroceramics, fiber reinforced composites and high temperature multiphase ceramics. Results of recent research were presented in oral and poster sessions by leading researchers from several countries. This collection of papers represents the state of the art in our understanding of the processing-structure-property interrelationships for these materials which possess unique and useful electrical, magnetic, optical, mechanical and thermal properties as a result of their multiphase nature.

We are grateful for the financial support of the National Science Foundation, the Office of Naval Research, the Air Force Office of Scientific Research, and the Defense Advanced Research Projects Agency for this conference. We gratefully acknowledge Prof. Robert Davis' leadership role in steering and expanding this university conference series on ceramic science. We thank Ron Avillion and Linda Rose for their expert assistance in planning and coordinating the meeting. Thanks are due to Ms. Marian Reed, Ms. Judy Bell and Ms. Linda Decker for timely secretarial assistance in organizing the program and bringing the proceedings to press.

Richard E. Tressler
Gary L. Messing
Carlo G. Pantano
Robert E. Newnham
University Park PA
October 1985

PREFACE

PROCESSING AND FABRICATION OF MULTIPHASE CERAMICS

STRUCTURE-PROPERTY RELATIONS IN MULTIPHASE CERAMICS

MULTIPHASE ELECTROCERAMICS

FIBER AND WHISKER REINFORCED COMPOSITES

SINTERING OF MULTIPHASE CERAMICS

R.J. Brook

Department of Ceramics
University of Leeds
Leeds LS2 9JT, U.K.

INTRODUCTION

The great majority of ceramics are multiphase, i.e. comprising more
than one solid phase in addition to such porosity as may be retained. In
the traditional ceramics sector, the multiphase nature is a consequence
of the use of raw materials of relatively low purity; more recently it
has become a feature of the engineering ceramics sector as the deliberate
construction of multiphase structures or composites has been attempted in
the quest for enhanced or novel properties.

On the basis that such deliberately constructed materials can be
considered to contain a distribution of one phase within another, a first
classification can be assembled depending on the nature of the distributed
phase and on its geometrical form (Fig. 1). The purpose of this paper is
to consider the four systems outlined in the figure from the point of view
of sintering, i.e. of the microstructural changes that occur during the
heat treatment that forms part of the fabrication cycle. The current
emphasis given to ceramic fabrication confirms the recognition now accorded
to the critical role of processing in the realisation of material properties.
The more detailed and specific the nature of the desired composite
structure, the more the attention that must be directed to processing in
developing the system towards its required form. The topic of composites
is now one where the available level of processing expertise is often the
factor controlling development.

| | Phase | | |
Geometry	Solid	Liquid	Gas
3D	Grain composites		Porous solid
2D	Liquid phase sintered solids		Powder bed
1D	Fibre composites		

Fig. 1.   Types of composite classified according to the nature and
dimensional extent of the distributed phase.

In considering the four systems, an attempt will be made to summarise current thinking in respect to the sintering of each type; in addition, a form of kinetic analysis will be introduced which may offer some benefit in clarifying or at least presenting the complex changes that take place.

POROUS SOLIDS

While not multiphase in the sense of having more than one solid phase, systems comprising porosity and a single crystalline phase are important as models for many of the issues to be considered for the other types. Their role as model systems is further emphasised by the recognition that such systems occur extremely infrequently in practice.

Despite a substantial literature devoted to mechanisms in solid state sintering[1] and to the study of specific systems, it is now recognised that experimental and interpretive difficulties make clear mechanistic conclusions the exception. The problem is shown in Fig. 2 where it can be seen that there are two broad categories of microstructural change which result in reduction of system interfacial energy, namely densification and grain growth or coarsening. Each category has the additional complexity in that

(i) it can occur by a variety of alternative mechanisms (Fig. 3),

(ii) its dominating mechanism can be changed depending on conditions of temperature, particle size, system chemistry, some of which may well themselves change during any given experiment,

(iii) it depends on the extent of change brought about by the operation of the other category.

The existence of multiple, varying and interacting processes has made the interpretation of microstructural change in simplistic terms a somewhat discredited pursuit.

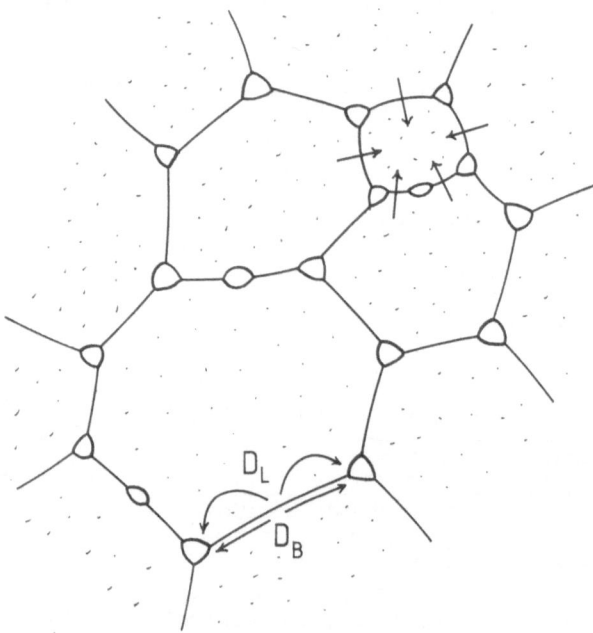

Fig. 2. Schematic illustration of densification (below) and grain growth (top) processes during sintering. The two processes result respectively in pore shrinkage and pore growth (by coalescence).

Densification	Grain Growth	
Lattice diffusion	Surface diffusion	)
		)
Grain boundary diffusion	Lattice diffusion	) Pore drag[2]
		)
	Gas phase transport	)
	Grain boundary diffusion	)
		) Pore ripening[3]
	Lattice diffusion	)
	Impurity drag[4]	

Fig. 3. Atomic mechanisms capable of contributing to the two categories of microstructural change.

The approaches that have been taken in the face of this difficulty have been at two levels. In analytical terms, the ratio[5] of grain growth rate to densification rate has been identified as a parameter helpful in predicting the overall form of sintering behaviour; the avoidance of microstructural heterogeneity[6] (variations in particle size, local density, composition) has likewise been emphasised in analytical treatments. The guidelines for successful sintering emerging from such considerations are now broadly accepted as in Fig. 4 although discussion of relative importance continues. In experimental terms, standards for powder design[7] and for green structure[8] have been strikingly advanced in recent years as the quest for structural homogeneity has been pursued.

Recognising these two aspects of the problem (ratio control; attainment of homogeneity), it is helpful to have experimental means to assess the relative contributions for a given set of conditions. One way to explore the ratio is to employ quantitative stereology on partly sintered structures (density dependence of surface area[9] or of grain boundary area[10]). A second is to use the information in the density vs time behaviour of isothermal sintering runs. As an example, the ratio may be derived[10] at a given density (Fig. 5) from knowledge of the first and second time derivatives of the density during hot pressing. The only assumption is a power law dependence of the densification rate on grain size. A more refined expression[5] of the ratio (Fig. 6) can also be determined[11] from the derivatives.

The second method is applied to the hot pressing of an alum-derived alumina (Criceram, France) in Fig. 7 where the plots show respectively density-time data, and the density dependence of the first derivative ($\dot\rho - \rho$), second derivative ($\ddot\rho - \rho$), and the ratio ($\Gamma - \rho$). For this material, Γ is greater than unity for all pressures used, i.e. pore growth occurs throughout the process.

1. The ratio $\Gamma^* = \dfrac{\text{rate of grain growth}}{\text{rate of densification}}$ must be low

   $\therefore$ Choose temperature)
   powder particle size) to keep Γ^* small.
   additives)

2. The pores must not become separated from grain boundaries.

   $\therefore$ Use pinning additives)
   uniform particle size) to restrain boundary motion.
   uniform pore size)

Fig. 4. Requirements for attainment of dense ceramics by sintering.

$$\dot{\rho} = KG^{-m}$$

$$G = (K/\dot{\rho})^{1/m}$$

$$\dot{G} = -\frac{1}{m} K^{1/m} \dot{\rho}^{-(1+m)/m} \ddot{\rho}$$

(a) $\quad \Gamma^* = \dfrac{\dot{G}}{G}\dfrac{\rho}{\dot{\rho}} = -\dfrac{1}{m}\dfrac{\ddot{\rho}\rho}{\dot{\rho}^2}$

(b) $\quad \dfrac{\dot{G}}{G} = -\dfrac{1}{m}\dfrac{\ddot{\rho}}{\dot{\rho}}$

Fig. 5. The ratio Γ^* of grain growth rate to densification rate (and also the relative grain growth rate) may be obtained directly from density/time data during hot pressing. The constant m is 2 (lattice diffusion controlled densification) or 3 (grain boundary controlled densification). Density and grain size are respectively ρ and G.

$$dr = \frac{\partial r}{\partial G} dG + \frac{\partial r}{\partial \rho} d\rho$$

$$= 0$$

$$\text{for } \Gamma = 1 = -\frac{\dfrac{\partial r}{\partial G}\dot{G}}{\dfrac{\partial r}{\partial \rho}\dot{\rho}}$$

$$= \frac{2}{m}\frac{\ddot{\rho}(1-\rho)}{\dot{\rho}^2} \qquad\qquad \text{hot pressing}$$

$$= \frac{1}{(m+1)}\left[\frac{2(1-\rho)}{\rho} - \frac{2\ddot{\rho}(1-\rho)}{\dot{\rho}^2} + 1\right] \quad \text{sintering}$$

Fig. 6. An alternative form of ratio employs the term Γ which takes the value unity for constant pore size (r) during sintering. Pore growth from coalescence is exactly offset by pore shrinkage as a result of local atom diffusion; the two processes in Fig. 2 counterbalance.

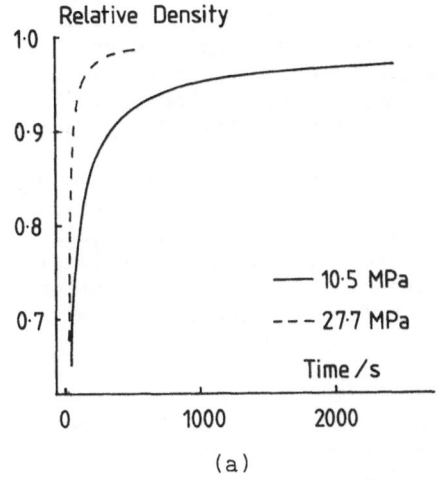

(a)

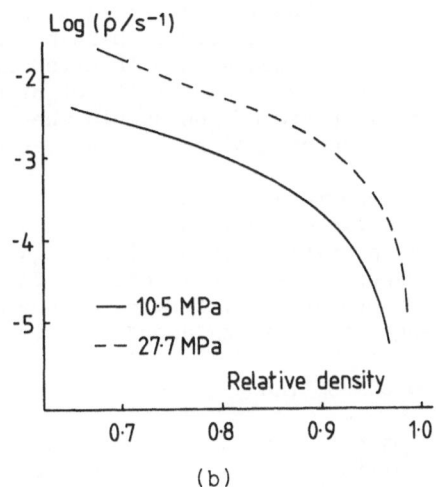

(b)

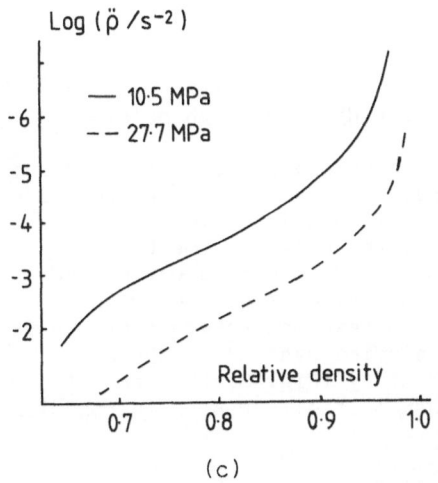

(c)

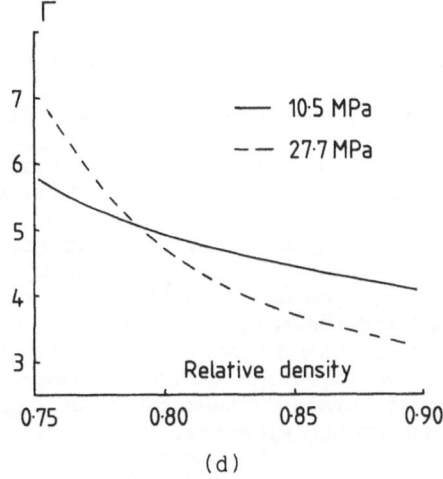

(d)

Fig. 7. Kinetic behaviour of an alum–derived alumina hot pressed at 1450°C. The pressure is effective in enhancing both densification (a and b) and grain growth rates (b and c; see Fig. 5). The effect of pressure on the ratio (d) is beneficial at densities above 80% but is less pronounced than is often assumed.

This method and the stereology method have been used to study[10] the question of the influence of MgO dopant on the sintering of Al_2O_3. Generally the ratio when directly measured is much less affected than has been estimated on the basis of separate measurements[12,13] of the contributing rates. The indication is therefore that the preservation of microstructural homogeneity is perhaps the critical issue for this system, a finding supported by the observed strength of boundary drag in direct measurements[14]. The ability of pinning agents to restrict boundary motion in parts of the sample where rapid densification has occurred[10] as a consequence of packing variations is an important stabilising factor.

In summary, therefore, the behaviour of porous solid model systems is satisfactorily interpreted in the light of Fig. 4. While the ratio can be a critical issue (SiC, Si_3N_4), current emphasis both analytical and experimental is directed to the avoidance of structural heterogeneity. Powder preparation, forming method and choice of pinning additive offer a range of promising approaches in the search for improved processing.

GRAIN COMPOSITES

Materials consisting of two crystalline phases in an equiaxed polycrystalline structure have been prepared in a number of systems but, with the exception of zirconia inclusions, clear property benefits have not generally resulted[15]. Zirconia based systems have accordingly received the bulk of attention[16] and they will be briefly considered here.

There are three ways to prepare such composites:

(i) precipitation from pre-sintered solid solutions (PSZ),

(ii) reaction of pre-sintered reactants such as zircon/alumina which yields zirconia/mullite composites,

(iii) sintering of mixtures of zirconia with the intended matrix phase as in the zirconia/alumina system.

Of these, the second and third involve sintering in multiphase systems.

Since zirconia composites require the zirconia phase to be fine and well distributed for maximum strengthening, an objective in the reaction sintering of such materials is to achieve full density prior to reaction[17] since early reaction followed by the heat treatment for densification risks ripening of the zirconia particle size beyond the optimum. The use of a fine starting particle size is helpful[18] in enhancing the ratio of densification rate to reaction rate; behaviour is, however, system dependent and the zircon/alumina system[17] has been observed to follow a more favourable trajectory (Fig. 8) than the zircon/magnesia system[19] presumably as a consequence of differences in reaction mechanism and kinetics.

There have been a number of studies[16] of sintering and microstructural development[20] in cosintered alumina/zirconia powders. In view of the need to avoid excessive zirconia particle growth, low values of the grain growth/densification ratio are sought; the use of hot-pressing[21] or of fine, chemically coprepared powders[22] or of fine, mixed powders[23] has commonly been adopted.

The interactions that can occur in the cosintering of mixed powders can take a number of forms. These include:

(i) kinetic changes resulting from mutual inter-solubility with

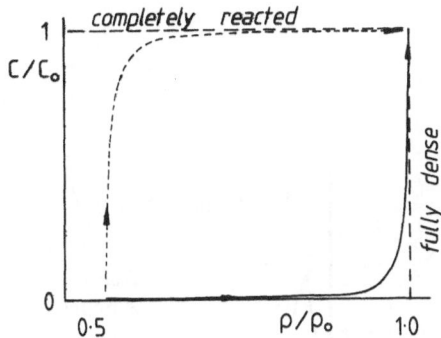

Fig. 8. Possible trajectories for reaction sintering systems. The
favoured path (solid line) has been observed for zircon/alumina;
the path for zircon/magnesia (dotted line) makes exploitation of
transformation toughening difficult to achieve.

consequent changes in defect chemistry[24] and diffusion rates;

    (ii)   structural pinning of matrix phase grain boundaries by
inclusions of the dispersed phase[25] in accord with the Zener criterion;

    (iii)   fault formation arising from local variations in densification
rate as a consequence of poor mixing and green structure heterogeneity;

    (iv)   modification of the distribution of grain boundary phases[26].

While instances of each of these factors have been recognised, the
explanation of kinetic changes on the basis of mechanisms remains complex.
Pinning by zirconia particles has for example been used to explain[23]
the fall in matrix grain size of composites (Fig. 9), the initial rise
at small additions being attributed to abnormal growth in the presence
of a heterogeneous inclusion distribution. Hot pressing and pressureless
sintering studies[27] of poorly and well mixed alumina/zirconia powders all
suggest a clear fall in densification rate (Fig. 10) at small zirconia
additions, a finding most simply explained on defect chemistry arguments[28]
but involving an unexpectedly high solid solution limit. The kinetic
ratio term is more favourably affected (Fig. 11) than the densification
data would at first sight suggest.

    To summarise the position for mixed phase sintering, there is
general recognition of the need for excellent mixing quality and fine
starting powders if favourable microstructures are to result. There is
also good support for the several interaction factors between powders
though the precise contributions of the different factors remain even
for the most studied system unresolved.

LIQUID PHASE SINTERING

    The enhancement of densification by the existence of a continuous
boundary phase, liquid at the processing temperature, is the most common
firing procedure used for engineering ceramics, occurring at times in
systems where not specifically intended. The physical mechanisms of

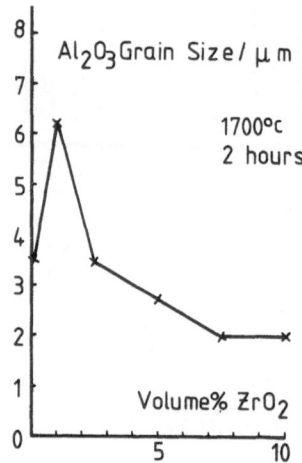

Fig. 9.   Zirconia inclusions influence the matrix grain size during the sintering (1700°C, 2 hrs) of zirconia/alumina composites[23].

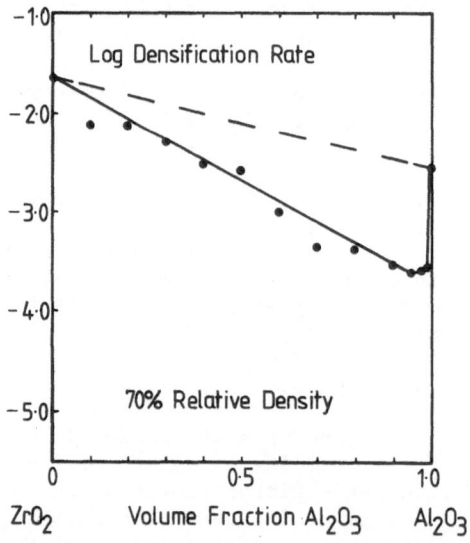

Fig. 10.   Zirconia additions substantially reduce[27] the densification rate of alumina powders during hot pressing (1350°C, 20 MPa). At greater than 5% addition, a rule of mixtures applies. Both powders were of similar particle size (0.5 μm).

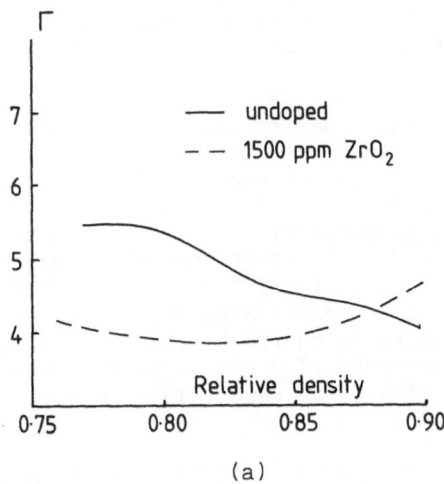

(a)

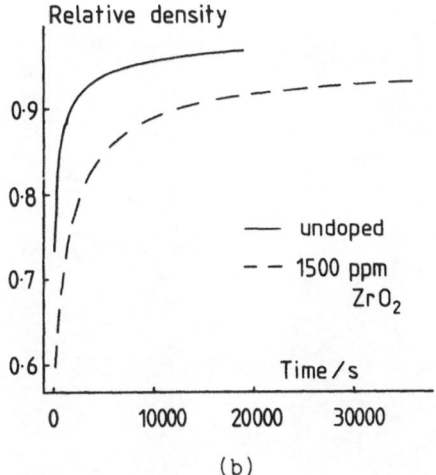

(b)

Fig. 11. Influence of added zirconia (from alcohol solution of
ZrOCl$_2$.8H$_2$O) on the sintering rate (a) of alumina at 1450°C.
The ratio of grain growth to densification (b) is favourably
affected by the additive below 90% density.

solution-diffusion-reprecipitation have long been recognised though argument still revolves around the relative importance of contact flatten-ing[29] (solution from contact points) and of directional Ostwald ripening[30] (solution from small particles in the size distribution). Evidence for the latter in the pressureless sintering of metal systems is now strong[31]. The range of dominance of the two contributions in respect to such factors as applied pressure (in hot pressing) and particle size distribution remains to be determined.

The influence of powder quality is, as in single phase sintering, strong. Factors such as size, size distribution and shape all influence microstructural development and make detailed kinetic interpretation difficult. Taking the example of silicon nitride processed under fixed conditions of additive, temperature and applied pressure, clear variations in behaviour can be seen for different powders, those with wide particle size distribution (Fig. 12) yielding[32] a higher initial density (Fig. 13) but a more rapid grain growth rate (Fig. 14). Although this behaviour is consistent with predictions on the basis of existing packing and ripening models, the existence of other factors[33] (shape, oxygen content, carbon content, α/β ratio) lends uncertainty to attribution of observations to single influences.

The use of liquids during densification brings problems with mechanical performance in subsequent high temperature use; there is corresponding interest in such transient liquid phase methods as densification followed by evaporative removal of the liquid[34] or crystallisation of the boundary phase[35] or solid solution formation[36] with the granular material. Modification of liquid composition[37] has also proved possible and beneficial. Such grain boundary engineering[38] has proved a profitable approach to the problem.

(a)   (b)

Fig. 12. Commercial powders of Si_3N_4 prepared by (a) milling reaction bonded material and (b) reductive nitridation of silica gel. The particle size distribution in the latter is notably narrower.

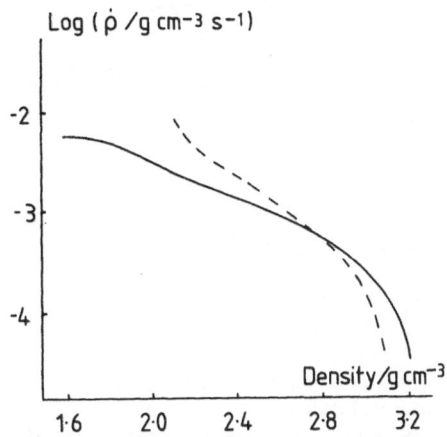

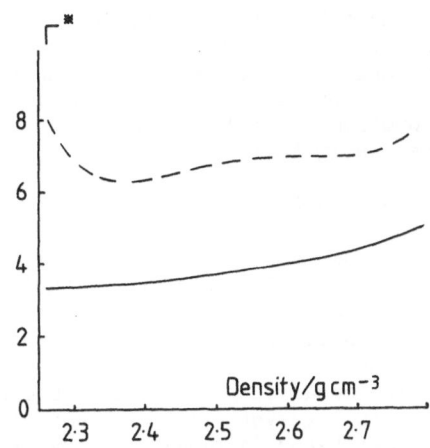

Fig. 13. Hot pressing behaviour
(1650°C, 10 MPa) of milled (dashed
line) and chemically prepared
(solid line) silicon nitrides.
The sintering additive (10^w/o
equimolar MgO/Y$_2$O$_3$) was identical
in both instances.

Fig. 14. The favourable influence of
a narrow particle size distribution
for the hot pressing of silicon
nitride is shown for the powders and
conditions of Figs. 12 and 13.

FIBRE COMPOSITES

    Composites involving the incorporation of short or long fibres
necessarily bring serious inhomogeneity in densification rate between
different components in the microstructure and the fabrication problem
has correspondingly been recognised as severe[39]. Cold pressing and
sintering has yielded cracked composites owing to the shrinkage of the
matrix phase and best results have been produced by hot pressing
methods. These allow the prevention of heterogeneity cracking and result
in satisfactory low porosity composites provided due attention is paid to
thermal expansion matching so that thermal shock cracking does not occur
on cooling from the processing temperature. Whether the matrix phase is
polycrystalline from hot pressed crystalline powder or from devitrification
of hot pressed glass powder, the hot pressing step has generally been
found to be necessary.

    For short fibres, the problem of mixing fibre and matrix powder
brings difficulty with the avoidance of clustering and fibre damage, and
the loading is consequently limited[40] to some 30^v/o. The development of
high aspect ratio grains within a conventionally sintered powder system
provides an alternative approach as in the case of silicon nitride[41].
Long fibre systems lend themselves to prior impregnation of oriented
fibre bundles and thus allow higher fibre loadings but hot pressing is
again commonly required.

The fabrication of fibre composites is an active and largely proprietary research area where the promise of dense systems provides considerable inducement. The theoretical context to the processing of these composites is the recognition of system heterogeneity as a fundamental problem in sintering and the current practical emphasis on hot pressing seems likely to remain.

CLOSING REMARKS

The awareness of the complexity of sintering processes and of the shifts in detailed mechanism likely to result from variation of experimental conditions within an ostensibly constant compositional system has led to a declining emphasis on mechanism identification.

The broad set of physical processes involved in microstructural change is, however, now well established and substantial progress in practical terms is being made by its application. Two major concerns of much current work are powder quality and the attainment of homogeneous green bodies. The latter is by definition a problem with multiphase systems and it remains to be seen how far it will be possible to attain sufficient phase matching to allow pressureless densification. Increasing control of powder quality and design and increasing use of structure stabilising additives perhaps offer the best avenue for future development.

REFERENCES

1. G.C. Kuczynski, Notre Dame Conferences, Mat. Sci. Res. 6 (1973), 10 (1975), 13 (1980), 16 (1984).
2. C.H. Hsueh, A.G. Evans and R.L. Coble, Acta Met. 30:1269 (1982).
3. M.V. Speight, Acta Met. 16:133 (1968).
4. J.W. Cahn, Acta Met. 10:789 (1962).
5. C.A. Handwerker, R.M. Cannon and R.L. Coble, Adv. in Cer. 10:619 (1984).
6. A.G. Evans, J. Am. Ceram. Soc. 65:497 (1982).
7. E.A. Barringer and H.K. Bowen, J. Am. Ceram. Soc. 65:C-199 (1982).
8. I.A. Aksay, Adv. in Cer. 9:94 (1984).
9. J.E. Burke, K.W. Lay and S. Prachozka, Mat. Sci. Res. 13:417 (1980).
10. R.J. Brook, E. Gilbart, N.J. Shaw and U. Eisele, Powd. Met. 28:105 (1985).
11. U. Eisele, personal communication.
12. M.P. Harmer and R.J. Brook, J. Mat. Sci. 15:3017 (1980).
13. C. Monty and J. Le Duigon, High T./High P. 14:709 (1982).
14. S.J. Bennison and M.P. Harmer. J. Am. Ceram. Soc. 68:C-22 (1985).
15. G.W. Groves, pp.9-22 in 'Ceramic Composites for High Temperature Engineering Applications', ed. R.W. Davidge, CEC, Brussels (1985).
16. A.H. Heuer and L.W. Hobbs, Adv. in Cer. 3 (1981); N. Claussen, M. Rühle and A.H. Heuer, Adv. in Cer. 12 (1984).
17. N. Claussen and J. Jahn, J. Am. Ceram. Soc. 63:228 (1980).
18. Y.Y. Shen and R.J. Brook, Sci. of Sint. 17:35 (1985).
19. Y.Y. Shen and R.J. Brook, Ceram. Int. 9:39 (1983).
20. B.W. Kibbel and A.H. Heuer, Adv. in Cer. 12:415 (1984).
21. F.F. Lange, J. Mat. Sci. 17:247 (1982).
22. D.W. Sproson and G.L. Messing, J. Am. Ceram. Soc. 67:C-92 (1984).
23. F.F. Lange and M.M. Hirlinger, J. Am. Ceram. Soc. 67:164 (1984).
24. R.J. Brook and M.P. Harmer, SRC DL/SCI/R15, 80 (1980).
25. D.J. Green, J. Am. Ceram. Soc. 65:610 (1982).
26. E.P. Butler and J. Drennan, J. Am. Ceram. Soc. 65:474 (1982).
27. R. Majumdar, personal communication.
28. F.A. Kroger, J. Am. Ceram. Soc. 67:390 (1984).
29. W.D. Kingery, J. Appl. Phys. 30-301 (1959).
30. W.A. Kaysser, M. Zirkovic and G. Petzow, J. Mat. Sci. 20:578 (1985).

31. D.N. Yoon and W.J. Huppmann, p.55 in 'Contemporary Inorganic Materials 1978', ed. G. Petzow and W.J. Huppmann, Dr. Riederer-Verlag, Stuttgart (1978).
32. H. Pickup, personal communication.
33. G. Wötting and G. Ziegler, Ceram. Int. 10:18 (1984).
34. R.W. Rice, Proc. Brit. Ceram. Soc. 12:99 (1969).
35. M.H. Lewis, A.R. Bhatti, R.J. Lumby and B. North, J. Mat. Sci. 15:103 (1980).
36. R.J. Lumby, B. North and A.J. Taylor, Special Ceramics 6:283 (1975).
37. D.R. Clarke and F.F. Lange, J. Am. Ceram. Soc. 63:586 (1980).
38. R.N. Katz and G.E. Gazza, pp.417-431 in 'Nitrogen Ceramics', ed. F.L. Riley, Noordhoff, Leyden (1977).
39. D.C. Philips, Chapter 10 in 'Handbook of Composite Materials, 4, Fabrication of Composites', ed. A. Kelly and S.T. Mileiko, North Holland, Amsterdam (1983).
40. D.C. Philips, pp. 48-73 in 'Ceramic Composites for High Temperature Engineering Applications', ed. R.W. Davidge, CEC, Brussels (1985).
41. F.F. Lange, Mat. Sci. Res. 11:597 (1978).

THE MORPHOLOGICAL STABILITY OF CONTINUOUS INTERGRANULAR PHASES

W. Craig Carter and Andreas M. Glaeser

Department of Materials Science and Mineral Engineering
and Materials and Molecular Research Division
Lawrence Berkeley Laboratory
University of California
Berkeley, California 94720

INTRODUCTION

The morphological instability of continuous phases has received considerable attention subsequent to the first complete analysis of such phenomena by Rayleigh[1] in 1878. As Rayleigh remarked, "[these] phenomena, interesting not only in themselves, but also as throwing light upon others yet more obscure, depend for their explanation upon the transformations undergone by a [cylindrical body] when slightly displaced from its equilibrium configuration and left to itself."[2] The well known result of the Rayleigh analysis is that any infinitesimal periodic perturbation with a wavelength exceeding the (cylinder) circumference will increase in amplitude, and eventually cause the formation of one discrete particle for each wavelength increment of cylinder.

The Rayleigh analysis has been applied to a number of microstructural phenomena involving capillarity-induced shape changes. Among these are: the stability of lamellar eutectics[3], fibers in composites[4], artificially lengthened precipitates[5,6], shape evolution of field ion emitter tips[7,8], healing of cracks introduced by thermal shock[9], as well as by scoring and welding of bicrystals[10,11], and the stability of the continuous pore phase during sintering of powder compacts.[12,13]

While in some of the aforementioned applications the Rayleigh analysis may be valid, complications arise when the continuous phase is located at a grain boundary. For an intergranular phase, each grain boundary intersection is characterized by some dihedral angle. C. S. Smith was perhaps first to recognize the modifying effect of dihedral angle on the stability of continuous grain boundary phases.[14]

In discussing continuous phases along three grain junctions Smith wrote, "If a second phase forming at a grain edge has a dihedral angle against grain boundaries of nearly 180°, it will behave like a cylinder and will certainly break up. If, however, the interphase tension is low in comparison with the adjacent grain boundary tension, the resulting triangular shape becomes stable at longer and longer lengths until, at a dihedral angle of 60° and below, the phase becomes stable at any length of grain edge."

Continuous phases may be situated at (along) the junctions of an arbitrary number of grains. The analysis presented quantifies the discussion of Smith by extending the method of Rayleigh to geometries of continuous phases surrounded by **n** grains with (variable) dihedral angle ψ. Results indicate the stability condition depends strongly on intergranular phase geometry, and may differ significantly from that of a cylinder.

A complete analysis of morphological instability consists of two parts[15]: a thermodynamic analysis identifying the smallest wavelength (infinitesimal amplitude) perturbation for which the amplitude will increase, and a kinetic analysis determining the particular wavelength for which perturbation growth is most rapid. In this paper, we present a thermodynamic analysis for nonfaceting surfaces with single-valued interfacial tensions. Possible modifications which may result from surface faceting, as well as the implications of the analysis to the kinetics of phase breakdown are qualitatively addressed. A kinetic analysis is forthcoming.[16]

THEORETICAL ANALYSIS: DETERMINATION OF λ_{min}

The ensuing sections describe the assumptions made and procedures used in the analysis. The objectives are the calculation of the surface area and volume of both a perturbed and an unperturbed channel as a function of the number of bounding grains **n**, and the dihedral angle ψ. The results of these calculations are employed to define the condition for thermodynamic stability of a continuous grain boundary phase.

Geometry

Figure 1 illustrates most of the geometrical parameters relevant to the analysis. When the intersection points of the intergranular phase with the n grain boundaries are joined by straight line segments, an n-sided polygon is produced. When the surface energy is isotropic, equilibrium requires the curvature of each bounding interface be constant. If the surface energy of the n interfaces is identical, the polygon will display n-fold symmetry.

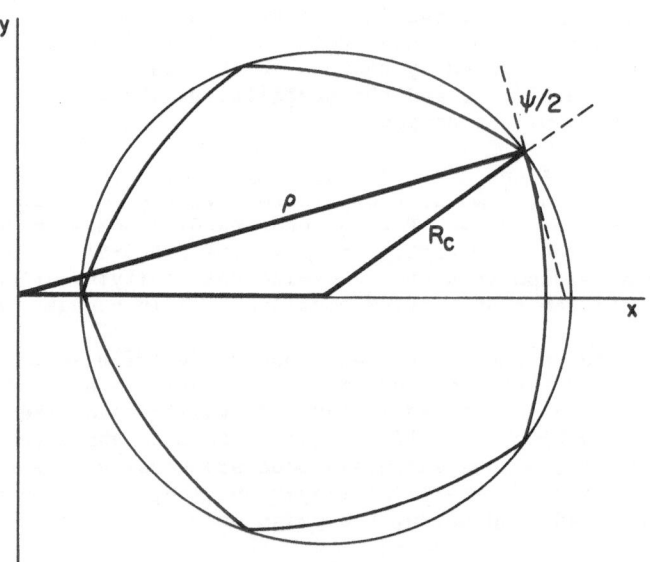

Figure 1   Geometry of intergranular phase, and illustration of parameters ρ, R_c, and dihedral angle ψ.

16

The polygon is circumscribed by a circle of radius R_c. Each interface may be described by a circle of radius ρ. It is always possible to pick one circle of radius ρ such that its center coincides with the Cartesian origin. Requiring the interfaces to have a common dihedral angle ψ leads to

$$\frac{\rho}{\sin (\pi/n)} = \frac{-R_c}{\cos (\psi/2 + \pi/n)} \qquad (1)$$

A perturbation on the radius of curvature can be described as the largest term of some periodic function, i.e.,

$$\rho = \rho_0 + \delta \cos kz \qquad (2)$$

where $k = 2\pi/\lambda$, z is the axial (longitudinal) coordinate of the cylinder, and δ an arbitrarily small amplitude.

Determination of Surface Area

The grain boundary area per wavelength perturbation is

$$A_{gb} = n \int_0^{2\pi/k} (\alpha - R_c) \, dz \qquad (3)$$

$$= n \int_0^{2\pi/k} \alpha + (\rho_0 + \delta \cos kz)\frac{(\cos(\psi/2 + \pi/n))}{\sin (\pi/n)} \, dz \qquad (4)$$

$$= \frac{2n\pi\alpha}{k} + \frac{2n\pi\rho_0}{k} \frac{\cos (\psi/2 + \pi/n)}{\sin (\pi/n)} \qquad (5)$$

where α is some arbitrary length.

The superficial area per wavelength perturbation is (in cylindrical coordinates)

$$A_s = n \int_{-\beta}^{\beta} d\theta \int_0^{2\pi/k} dz \, \rho \, (1 + (d\rho/dz)^2)^{\frac{1}{2}} \qquad (6)$$

where

$$\beta = \psi/2 + \pi/n - \pi/2 \qquad (7)$$

Since $d\rho/dz$ is of order δ and arbitrarily small,

$$A_s \simeq 2n \int_0^{\beta} d\theta \int_0^{2\pi/k} dz \, (\rho_0 + \delta \cos kz) \left(1 + \frac{\delta^2 k^2 \sin^2 kz}{2}\right) \qquad (8)$$

$$= \frac{4n\pi\rho_0}{k} \left(1 + \frac{\delta^2 k^2}{4}\right) \beta \qquad (9)$$

17

Volume Determination and Criterion for Break-up

The cross-sectional area of the intergranular phase (channel) is

$$A_{cs} = n\rho^2(\beta + [\sin^2\beta \cos(\pi/n)/\sin(\pi/n)] - \sin\beta \cos\beta) \quad (10)$$

$$= n\rho^2\chi \quad (11)$$

The channel volume per perturbation wavelength is

$$V = \frac{2n\pi\chi}{k} (\rho_o^2 + \delta^2/2) \quad (12)$$

which yields

$$\rho_o^2 = \xi^2 (1 + (\delta^2/2\xi^2)) \quad (13)$$

where

$$\xi^2 = kV/2n\pi\chi \quad (14)$$

To sufficient approximation

$$\rho_o = \xi (1 - (\delta^2/4\xi^2)) \quad (15)$$

Defining the surface energy $\gamma_s = 1$, then $\gamma_{gb} = 2 \cos \psi/2$. The superficial energy per wavelength is

$$\Gamma = A_s + 2 \cos(\psi/2) A_{gb} \quad (16)$$

$$= \frac{4n\pi\rho_o}{k} (1 + \frac{\delta^2 k^2}{4})\beta + \frac{4n\pi\rho_o}{k} \frac{\cos(\psi/2 + \pi/n) \cos\psi/2}{\sin\pi/n}$$

$$+ \frac{4n\pi a \cos\psi/2}{k} \quad (17)$$

Using Eq. 15 yields

$$\Gamma = \frac{4n\pi\xi}{k} (\beta + \frac{\cos(\psi/2 + \pi/n) \cos\psi/2}{\sin\pi/n}) + \frac{4n\pi a \cos\psi/2}{k}$$

$$+ \frac{n\pi\beta\delta^2}{k\xi} [(k^2\xi^2 - \frac{(1 + \cos(\psi/2 + \pi/n) \cos\psi/2)}{\beta \sin\pi/n}] \quad (18)$$

Eq. 18 is equivalent to an expansion of the surface energy per wavelength about the radius ρ_o

$$\Gamma(\rho) = \Gamma(\rho_o) + \Gamma'(\rho_o) \int_0^{2\pi/k} (\rho - \rho_o)dz$$

$$+ \Gamma''(\rho_o) \int_0^{2\pi/k} \tfrac{1}{2}(\rho - \rho_o)^2 dz \quad (19)$$

Equivalently,

$$\Gamma(\rho) = \Gamma(\rho_o) + \Gamma''(\rho_o)(\pi\delta^2/2k) + \ldots \qquad (20)$$

As in all metastability problems, the stability depends on the sign of Γ''. It is sufficient to approximate

$$\rho^2 = \xi^2 \qquad (21)$$

when establishing the sign of Γ''.

The growth of any perturbation of wavelength $\lambda > \lambda_{min}$ is energetically favorable where

$$\frac{\lambda_{min}}{2\pi\rho} = (1 + \frac{\cos(\psi/2 + \pi/n)\cos\psi/2}{\beta\sin\pi/n})^{-\frac{1}{2}} \qquad (22)$$

which may be written more simply as

$$\frac{\lambda_{min}}{2\pi\rho} = (\frac{\beta}{\chi})^{\frac{1}{2}} \qquad (23)$$

or in terms of the convenient R_c

$$\frac{\lambda_{min}}{2\pi R_c} = (\frac{\beta}{\chi})^{\frac{1}{2}} \frac{\sin\pi/n}{\sin\beta} \qquad (24)$$

DISCUSSION

Results of the analysis for several n are illustrated in Figure 2. For each n, the minimum thermodynamically unstable infinitesimal perturbation wavelength coincides with the Rayleigh result for $\psi = 180°$, increases with decreasing ψ, and tends to infinity as ψ approaches $\pi - (2\pi/n)$, or equivalently, as the interface curvature vanishes. For fixed n, a continuous phase with lower ψ is expected to be more stable than one with higher ψ. The phase is completely stable to perturbations when $\psi \le \pi - (\pi/2n)$. For fixed ψ, the stability increases with n (Table I).

To facilitate comparison between this analysis and that of Rayleigh, a normalization parameter is introduced. Defining R_{eq} as the radius of a

Table I Comparison of $\lambda/2\pi R_c$, $\lambda/2\pi\rho$, and $\lambda/2\pi R_{eq}$ for $\psi = 145°$ and n varying from 2 to 11

n	$\lambda_{min}/2\pi R_c$	$\lambda_{min}/2\pi\rho$	$\lambda_{min}/2\pi R_{eq}$
2	1.19	1.14	1.44
3	1.55	1.21	1.74
4	1.99	1.30	2.16
5	2.62	1.42	2.80
6	3.63	1.57	3.84
7	5.45	1.80	5.71
8	9.44	2.15	9.83
9	22.4	2.87	23.3
10	211	6.05	218
11	∞	∞	∞

19

cylinder having the same volume per unit length as an intergranular phase characterized by a dihedral angle ψ, the ratio $\lambda_{min}/2\pi R_{eq}$ normalizes the actual λ_{min} by the minimum wavelength that would grow in a geometrically similar compact with an equivalent volume fraction of second phase. The dependence of $\lambda_{min}/2\pi R_{eq}$ on n for $\psi = 145°$ is presented in Table I.

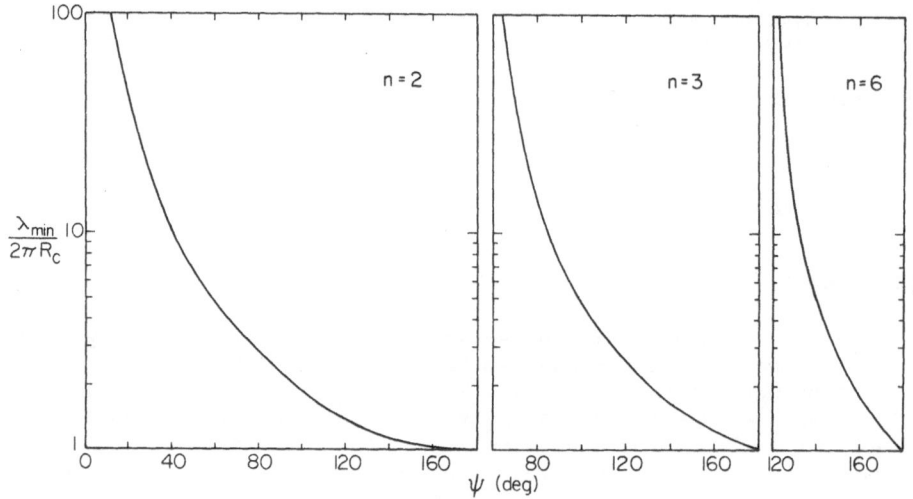

Figure 2   Illustration of effect of varing n and Ψ on the stability of continuous intergranular phases.

General Considerations

The nature of a second phase, its composition, inherent properties, morphology, and distribution within a matrix, can have an important bearing on a material's ultimate properties. Introduction of continuous filaments into polycrystalline matrices may dramatically alter mechanical behavior. High temperature stability and useful lifetimes of such composites will be influenced by the fibers' stability against breakdown. Continuous phases may provide high diffusivity transport paths, or be preferentially leached, thus limiting the utility of a material in storage applications, e.g., containment of nuclear waste. In these cases as well as others, factors influencing the morphological stability of continuous phases may become important elements in materials design. In the following, a number of specific cases are considered, and the extent to which dihedral angle may stabilize an intergranular phase is indicated.

Consideration of n = 2

The dihedral angle, through its effect on λ_{min} will affect both the size and spacing of discrete particles or phases produced by perturbation growth processes. Experiments conducted on bicrystals with a systematic variation in ψ (misorientation) would be expected to reveal a systematic variation in particle size and spacing. In polycrystals with a spectrum of ψ, "scatter" in the sizes and spacings of discrete particles would be expected.

Numerous experiments have investigated the healing of intergranular cracks. High aspect ratio intergranular pore channels have been introduced by scoring and welding bicrystals[10,11], as well as by thermal shock of polycrystals.[9] Micrographic evidence indicates that pore breakup during subsequent annealing is somewhat erratic, i.e., the wavelength of breakup appears to vary from one channel to the next. This may in large part reflect differences in pore channel size. In addition however, results presented by Gupta for alumina[9], indicate the final pore spacing to pore radius **ratio** may vary by as much as factor two.[*]

One expects the morphological evolution of these systems to be affected by the presence of grain boundaries. A factor two variation in pore spacing:pore radius ratio, if entirely due to variations in λ, would require λ to vary by a factor ≈ 3. Although dihedral angles were not measured in the studies cited, recent dihedral angle measurements in alumina by Handwerker[18] have indicated a wide dihedral angle range, 85° to 170°. This range of ψ, would lead to a factor ≈ 2.5 variation in λ_{min}, and must similarly affect the kinetically dominant wavelength. As discussed in Section 3.5, such shifts in pore spacing:pore radius ratios may have an important effect on the magnitude of transport coefficients deduced from such experiments.

---

[*]Although the center to center pore spacing in a 2-D projection permits an assessment of the perturbation **wavelength**, it is not possible to assess the ratios $\lambda_{min}/2\pi R_{eq}$ or $\lambda_{min}/2\pi\rho$ without dihedral angle measurements.

Consideration of n = 3

During intermediate stage sintering, the pore phase is often approximated as a cylindrical channel along three grain junctions. Breakdown of the continuous pore phase, marking the transition from intermediate to final stage sintering has been assumed to occur by perturbation growth processes. Recently, micrographic evidence was obtained suggesting a Rayleigh instability of pore channels along three-grain junctions.[13] The results of the present analysis, as well as inconsistencies between observed behavior and kinetic behavior predicted for cylinders, suggest breakdown of the pore phase may be significantly more complex than previously assumed.

The pore cross-section in an unfired compact comprised of spherical particles is nonuniform. Neck growth associated with densification would be expected to lead to pore channel closure at the point of initial minimum cross-section, the three grain junction midpoint. However, if the initial perturbation is unstable and mass redistribution along the pore channel is rapid in comparison to channel shrinkage, perturbation decay, followed by channel shrinkage and ultimately, pore closure due to the growth of morphological perturbations may become possible.

Dihedral angle will affect the driving force for perturbation growth and densification. As ψ is decreased, the driving force for perturbation growth decreases if $\lambda_{initial} > \lambda_{min}$; the driving force for decay is increased when $\lambda_{initial} < \lambda_{min}$. A decrease in ψ decreases the driving force for densification. Thus, in systems with low ψ and/or high surface diffusivities or vapor pressures, perturbation decay is favored. After sufficient channel shrinkage to support perturbation growth along a facet length channel, a Rayleigh instability may ensue, leading to formation of closed pores along three-grain junctions and at four grain junctions. In contrast, in systems with high ψ and high grain boundary or lattice diffusivities, growth of the initial perturbation may be energetically favorable, or channel shrinkage may occur more rapidly than perturbation decay. Both situations will lead to the development of isolated pores at four-grain junctions only. Resulting differences in pore volume and location will effect the pore mobility, and thus mechanisms and conditions for pore-grain boundary separation.

The dihedral angle distribution may have an important modifying effect on microstructural evolution. In materials with a broad dihedral angle distribution, the processes dominating the pore phase's morphological evolution may vary within a compact, resulting in a spectrum of pore-grain boundary separation conditions. This factor, combined with effects of ψ on pore shrinkage and coarsening behavior, as well as pore mobility[17], may contribute to the development of microstructural inhomogeneities promoting the initiation of abnormal grain growth.

Narrowing the dihedral angle distribution would be expected to lead to more uniform microstructure development. A comparison of dihedral angle measurements in undoped and MgO-doped Al_2O_3[18], has indicated that dopant additions reduce the width of the dihedral angle distribution. Handwerker et al. point out this increases the uniformity of microstructural evolution by reducing the variation in driving forces for densification.[19] The potential benefits of a dopant-induced reduction in boundary mobility have frequently been cited. Dopant effects on the uniformity of the pore structure produced during the transition from intermediate to final stage sintering may also be important.

In addition to the pore phase in powder compacts, second phases at three-grain junctions are commonly found in alloys with a large difference

22

in either the melting points or solubilities of the constituents.[14]   A
residual glassy phase along three-grain junctions may also develop in
liquid-phase sintered materials. Similar stabilizing effects may be of
importance in these cases as well.

Consideration of n ≥ 4

Table I illustrates the pronounced increase in stability to
perturbation growth accompanying an increase in **n** (ψ constant). The
enhanced stability is manifested in two ways. The stabilizing effect
becomes significant at progressively higher ψ as **n** increases, and the
dihedral angle range within which perturbation growth is possible
diminishes (Figure 2). Thus stabilization effects of the type considered
are expected to be extremely important when a continuous phase is bounded
by a large number of grains.

High coordination number stacking faults in wire sintering
experiments would be expected to be extremely stable to perturbation
growth. Similar effects may also be important for stacking faults in
conventional powder compacts. The high temperature stability of fibers in
composites is expected to be affected by the values of ψ and **n**. Subject
to other constraints, it would be advantageous to maximize the number of
coordinating grains. Thus grain size:fiber diameter ratio (**Q**) emerges as
a potentially important parameter in materials design.

Grain growth and fiber coarsening may modify **n**, introducing an
additional time-dependent component to morphological stability. Grain
growth may dramatically decrease fiber stability within certain ranges of
Q. A decrease in **Q** from 20 to 4 may only have a limited effect, whereas
an additional factor of 2-3 increase in grain size would likely have a
profound influence on fiber stability. High temperature stability of
properties may increase from use of grain growth inhibitors during
materials fabrication.

When **n** is sufficiently large to inhibit perturbation growth
processes, other factors inducing mass redistribution along fibers may
assume greater significance. A "perturbation" or variation in the number
of coordinating grains along the fiber axis will produce local curvature
differences which may induce mass transfer from regions of lower n to
higher-n regions. Thus, relatively coarse-grained regions may emerge as
preferential fiber-pinchoff sites. Microstructural homogeneity thus
becomes an issue of considerable importance.

Anisotropic Interfacial Energies

Isotropic surface energies have been assumed ; however, both grain
boundary and surface energies may vary within an individual channel. This
anisotropy may introduce both variable surface curvature and variable
dihedral angles. Geometric modelling of this situation*is tedious but not
difficult: metastable configurations can be determined.   However, each
case is unique, and therefore, the problem is not amenable to
generalization.

----

*It is interesting to note that if all bounding surfaces are
comprised of the same phase, metastability requires that all bounding
segments have nonzero curvature or that all have zero curvature. Thus,
the analysis is not applicable to situations where only one bounding
surface has zero curvature.

Facetting presents an additional complication. Cahn has evaluated the stability of single crystal rods with a specific surface energy isotropic in transverse planes, but a function of $\phi = (\partial R/\partial z)$.[20] Surface energy anisotropy may have either a stabilizing or destabilizing effect, depending on the manner in which γ varies with surface reorientation. If the original surface orientation corresponds to a cusp in the γ versus ϕ plot, the cylinder is stabilized with respect to infinitesimal perturbations, however, may be unstable to finite perturbations. In contrast, when the original surface corresponds to a ridge in the γ - ϕ plot, the cylinder is unstable. More generally, λ_{min} depends on the second derivative of γ with respect to ϕ. If $(\partial^2\gamma/\partial\phi^2)$ is positive (corresponding to minimum in γ at $\phi = 0$), λ_{min} is increased, and conversely.

Although similar behavioral trends might be anticipated for intergranular phases, inclusion of torque terms reflecting the orientation dependence of interfacial energies into an analysis appropriate to intergranular phases would be more difficult, and introduce numerous complications. Even if the continuous phase is a single crystal, there will be n distinct misorientations and hence interphase boundaries for an n-grain coordinated phase at each location along the z-axis. Conceivably, the interface orientation of one interphase boundary lies near a cusp orientation, while the adjacent interphase boundary lies in a position corresponding to a ridge in the γ - ϕ plot. Consequently, quantitative assessment of surface energy anisotropy effects is a formidable task.

Kinetic Implications

The stability condition defined is thermodynamic; if any mass transport systems are operative that permit the development and growth of perturbations, equilibrium will be established by growth of a perturbation with $\lambda > \lambda_{min}$. For intergranular phases, λ_{min}, and thus the kinetically favored wavelength will depend on both ψ and n.

Kinetic models predict the rate at which equilibrium will be established, and the perturbation wavelength which maximizes the driving force:transport distance ratio. For a cylindrical void with isotropic interfacial energy, both the perturbation growth rate and magnitude of the kinetically dominant wavelength are predicted to depend on the dominant mass transport system.[7] An analysis by Nichols and Mullins suggests the most rapidly growing wavelength will vary from $\sqrt{2}\cdot\lambda_{min}$ for surface diffusion dominated growth to $2.1\lambda_{min}$ for volume diffusion dominated growth. Intermediate values are predicted when both mechanisms contribute substantially to breakdown.

Qualitatively similar behavior is anticipated for breakdown of continuous intergranular phases. However, in such cases, ψ and n will have a modifying influence. Since the most rapidly growing wavelength must exceed λ_{min}, the kinetically dominant wavelength must also vary with ψ. One anticipates rates of evolution become vanishingly low as ψ approaches $\pi - (2\pi/n)$, since the chemical potential gradient must vary at least linearly with the inverse of the dominant wavelength.

The dihedral angle and number of bounding grains will affect both the spacing and size of discrete "particles" or phases produced by perturbation growth processes. Since these parameters are used to both identify the dominant transport mechanism and estimate the magnitude of the appropriate transport coefficient, application of kinetic analyses appropriate to cylinders to the breakdown of intergranular phases, without consideration of dihedral angle effects, may introduce systematic errors.[16] Plausibly, transport coefficients determined in this way may be in error, and/or differ considerably from those determined using other methods.

Cautions and Limitations

The stability condition derived in this analysis defines the minimum wavelength necessary for a **sinusoidal** perturbation to increase in amplitude. As detailed in Section 2, sinusoidal perturbations with $\lambda > \lambda_{min}$ decrease the surface area per wavelength perturbation, and thus decrease the surface energy in comparison to that of an unperturbed cylinder having the same volume per wavelength.

The surface area of the perturbed "cylinder" is obviously sensitive to the form of the imposed perturbation. Thus, it is conceivable that a perturbation with an additional radial or rotational component could yield a λ_{min} smaller than that derived here. Hence, although the calculated values for λ_{min} presented serve as a **sufficient** condition for instability, it is necessary to acknowledge the possibility that perturbations of more complex geometry may provide a different and smaller value for λ_{min} as a **necessary** condition for instability.

Acknowledgements

This work was supported by the Office of Energy Research, Office of Basic Energy Sciences, Materials Sciences Division of the U. S. Department of Energy under Contract No. DE-AC03-76SF00098. One of the authors (WCC) was supported in part by an ARCO Foundation Fellowship. Rowland M. Cannon, Lorenzo Sadun and John Salmon are thanked for helpful discussions and comments. Finally, we wish to acknowledge the enthusiastic support of G. O. Bears.

References

1. Lord Rayleigh, "On the Instability of Jets," Proc. London Math. Soc., 10, 4-13 (1879).
2. Lord Rayleigh, Theory of Sound, Vol. 2, 2nd Edition, MacMillan and Co., London, (1929).
3. H. E. Cline, "Shape Instabilities of Eutectic Composites at Elevated Temperatures," Acta Metall., 19, [6], 481-90 (1971).
4. A. J. Stapley and C. J. Beevers, "The stability of sapphire whiskers in nickel at elevated temperatures, Part 2," J. Mater. Sci., 8, 1296-1306 (1973).
5. M. McLean, "The kinetics of spheroidization of lead inclusions in aluminium," Philos. Mag., 27, 1253-66 (1973).
6. D. M. Moon and R. C. Koo, "Mechanism and Kinetics of Bubble Formation in Doped Tungsten," Met. Trans., 2, [8], 2115-22 (1971).
7. F. A. Nichols and W. W. Mullins, "Surface (Interface) and Volume-Diffusion Contributions to Morphological Changes Driven by Capillarity," Trans. A.I.M.E., 233, [10], 1840-48 (1965).
8. F. A. Nichols and W. W. Mullins, "Morphological Changes of a Surface of Revolution due to Capillarity-Induced Surface Diffusion," J. Appl. Phys., 36, [6], 1826-35 (1965).
9. T. K. Gupta, "Instability of Cylindrical Voids in Alumina," J. Am. Ceram. Soc., 61, [5-6], 191-95 (1978).
10. C. F. Yen and R. L. Coble, "Spheroidization of Tubular Voids in Al_2O_3 Crystals at High Temperatures," ibid., 55, [10], 507-09 (1972).
11. O. Maruyama and W. Komatsu, "Observations on the Grain-Boundary of Al_2O_3 Bicrystals," Ceramurgica International, 5, [2], 51-5 (1979).

12. W. C. Carter and A. M. Glaeser, "Dihedral Angle Effects on the Stability of Pore Channels," J. Am. Ceram. Soc., 67, [6], C124-C127 (1984).

13. M. D. Drory and A. M. Glaeser, "The Stability of Pore Channels: Experimental Observations," ibid., 68, [1], C14-C15 (1985).

14. C. S. Smith, "Grains, Phases, and Interfaces: An Interpretation of Microstructure," Trans. A.I.M.E., 175, [1], 15-51 (1948).

15. F. A. Nichols, "On the Spheroidization of Rod-Shaped Particles of Finite Length," J. Mater. Sci., 11, [6], 1077-82 (1976).

16. W. C. Carter and A. M. Glaeser, to be published

17. C. H. Hsueh, A. G. Evans, and R. L. Coble, "Microstructure Development During Final/Intermediate Stage Sintering - I. Pore/Grain Boundary Separation," Acta Metall., 30, [7], 1269-79 (1982).

18. C. A. Handwerker, "Sintering and Grain Growth in MgO"; Sc.D. Thesis, Massachusetts Institute of Technology, Cambridge, MA, February 1983.

19. C. A. Handwerker, R. M. Cannon, and R. L. Coble, "Final Stage Sintering of MgO," pp. 619-43 in Structure and Properties of MgO and Al$_2$O$_3$ Ceramics, Vol. 12 in Advances in Ceramics, W. D. Kingery, Ed., ACS Publication (1984).

20. J. W. Cahn, "Stability of Rods with Anisotropic Surface Free Energy," Scripta Metall., 13, [11], 1069-71 (1979).

ROLE OF SHEAR IN THE SINTERING OF COMPOSITES

Rajendra K. Bordia and Rishi Raj

Department of Materials Science and Engineering
Bard Hall, Cornell University
Ithaca, NY 14853

ABSTRACT

Recent theoretical work[1,2,3] has emphasized the importance of shear deformation in the sintering of composite ceramics. In this report, we will first review the important results obtained in one analysis.[2] Next, an experimental technique will be described to give the key parameters necessary to compare the predicted and the actual densification behavior of composites. Finally, microstructural evidence is given of the processing defects generated in composites.

INTRODUCTION

The presence of a minority phase can lead to an internal stress during sintering due to differential densification.[2,3,4] The internal stress can (a) lead to significant deviations in the shrinkage rate from the rule of mixtures and (b) produce strength degrading internal flaws. The problem is likely to be particularly acute for composite ceramics since the constituents of the composite can have very different densification behavior. The presence of internal defects has been reported in the literature.[5,6,7] Lange et al.[8] report that differential sintering of ZrO_2 agglomerates in Al_2O_3-ZrO_2 composites led to strength degrading crack like voids. The influence of inhomogeneities on the densification response has been reported in earlier studies.[9,10] Recently, DeJonghe et al.[11] have shown that the densification response of ZnO-SiC composites is significantly slower than that predicted from the rule of mixtures.

The problem of composite densification was formulated[2] by considering a spherical region within a matrix, each having a different densification response. The incompatibility stresses and their effect on the

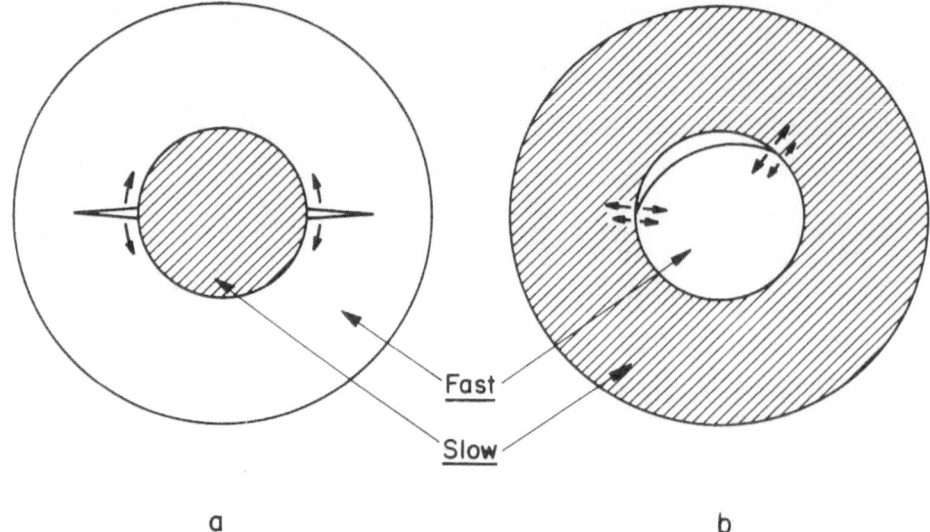

Fig. 1. Schematic of two types of internal defects produced by tensile
stresses when one region (shaded) sinters slower than the other
(from Ref. 2).

densification of the composite were obtained. If the matrix sinters faster
than the spherical inhomogeneity then tensile hoop stresses develop at the
interface which could produce radial cracks [Fig. 1(a)]. On the other
hand, if the inhomogeneity sinters faster than the matrix then tensile
radial stresses develop which could produce circumferential cracks [Fig.
1(b)].

In either case, non uniform densification generates shear stresses
in the matrix which can be relaxed by shear deformation. In (2) the shear
and the densification response are represented by simple viscoelastic
models. In order to obtain simple closed form solutions, the properties
are further assumed to remain constant with time.

In the following sections, the important results from this analysis
are reviewed. The case of a non-sintering spherical inhomogeneity is considered because of its relevance to composites where the dispersed phase,
for example whiskers of SiC in an alumina matrix, is non-sintering. It

will be shown that the interfacial stresses and the composite densification are sensitive to a non dimensional parameter β. It is defined as
the ratio of the shear deformation rate to the densification rate. We
will see that a large value of β is desirable to prevent flaw generation
and to predict the shrinkage rate of the composite. Next, we will demonstrate that sinter forging can be used to measure β. Finally, microstructural evidence for defect formation from our preliminary experiments will
be presented.

REVIEW OF THE PREDICTIONS OF ANALYSIS

For the case of a composite with a non densifying minority phase,
the results of the analysis are rather simple. It was found that the interfacial stresses rise to a plateau value in the very early stages of
sintering and then gradually relax in a period which is of the order of
sintering time. The two quantities of interest are the maximum value of
the hoop stress (which governs the formation of radial cracks) and the
densification behavior of the composite. They can be written in terms
of β as follows:

$$(\sigma_\Theta)_{max} = \frac{2 - 9\beta}{4f + 9\beta} \, p_o \tag{1}$$

$$\frac{\Delta\rho}{\Delta\rho_{max}} = 1 - \exp\left\{-\frac{9\beta}{4f + 9\beta} \frac{t}{\tau}\right\} \tag{2}$$

Where p_o is the intrinsic sintering pressure and τ is the time constant
for densification of the "pure" matrix. "f" is the volume fraction of the
inhomogeneity. Thus, a large value of β reduces the hoop stress. In
particular, if $\beta > 2/9$ then the hoop stresses are compressive. Therefore
radial cracks are likely to form only if $\beta < 2/9$. Also, the probability
of forming radial cracks increases as the size of the inhomogeneity increases.

The predicted densification response of the composite, from Eqn. 2, is
shown in Fig. 2 for a fixed volume fraction of the dispersed phase. It
should be noted that even for 5 volume percent, the composite densification is considerably slower than rule-of-mixtures if β is small. If β
is large the densification response of the composite approaches that of
the matrix.

Since β emerges as an important parameter in the analysis, we will
present an expression for β in terms of experimentally measurable quantities.

The constitutive law used in the analysis for shear deformation is given by the viscoelastic model shown in Fig. 3. G_o is the elastic shear modulus and η_g is the shear viscosity. A rate constant ω_g is defined as the inverse of the shear viscosity:

$$\omega_g = \frac{G_o}{\eta_g} \tag{3}$$

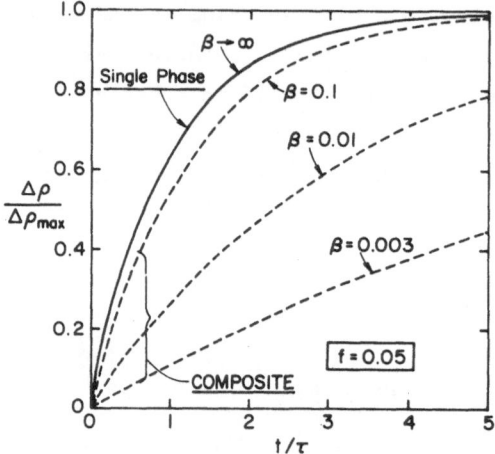

Fig. 2. The solid line represents the assumed sintering behavior without the second phase. The effect of β on the sintering of the composite, containing volume fraction f = 0.05 of the second phase, is shown in dotted lines (from Ref. 15, courtesy The American Ceramics Society).

Thus, if a step shear strain is applied then the shear stress relaxes via a time constant equal to $1/\omega_g$ as shown in Fig. 3(b). The shear strain rate $\dot{\varepsilon}_s$ and the shear stress σ_s are related as follows:

$$\sigma_s = \eta_g \, \dot{\varepsilon}_s \tag{4}$$

The densification behavior is represented by the viscoelastic model shown in Fig. 4(b). The phenomenological densification behavior shown in Fig. 4(a) can, as a first approximation, be simulated by the three element

Kelvin solid. Its response to a step load is shown in Fig. 4(c). For densification the rate constant is defined as:

$$\omega_k = \frac{K_o}{\eta_a} \tag{5}$$

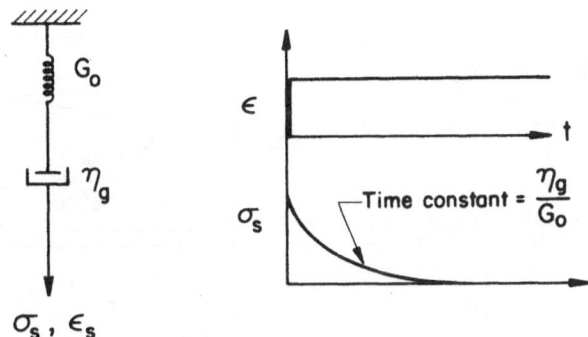

Fig. 3.   (a) The Maxwell element used here as the spring darkspot analog of viscoelastic creep response. (b) Stress relaxation response of a Maxwell element (from Ref. 2).

Then β is given by:

$$\beta = \frac{\omega_g}{\omega_k} \tag{6}$$

The response of the Kelvin model shown in Fig. 4(b) to a step load p_o, the sintering pressure, is given by:

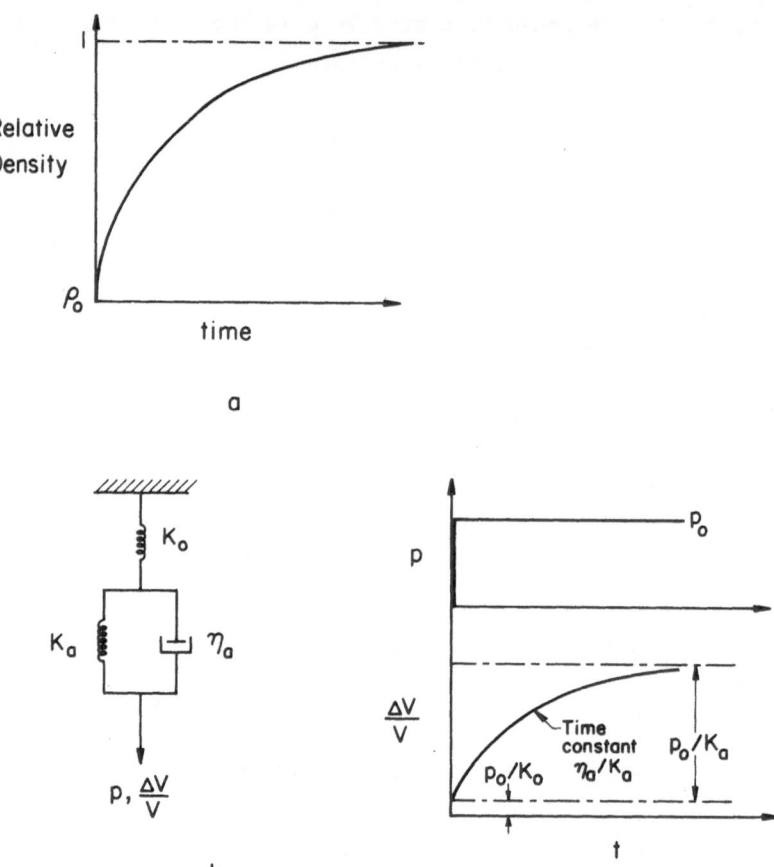

Fig. 4.   (a) Typical densification curves for isothermal sintering.   (b) Single Kelvin-Voigt element to phenomenologically model the densification curve.   (c) The volumetric strain vs time response of (b) subjected to a constant pressure p_o (from Ref. 2).

$$\Delta\rho = \frac{p_o}{K_o} + \frac{p_o}{K_a} \left[1 - \exp(-\frac{K_a t}{\eta_a})\right] \tag{7}$$

Where $\Delta\rho$ is the densification strain.   The phenomenological densification response is assumed to be:

$$\Delta\rho = \Delta\rho_{max} \left[1 - \exp(-\frac{t}{\tau})\right] \tag{8}$$

Where τ is the time constant for densification. Comparing Eqns. (7) and (8), and neglecting the elastic strain, p_o/K_o, we get:

$$\eta_a = \frac{p_o \tau}{\Delta\rho_{max}} \tag{9}$$

Substituting Eqn. (9) in Eqn. (5), ω_k is given by:

$$\omega_k = \frac{K_o \, \Delta\rho_{max}}{\tau \, p_o} \tag{10}$$

Substituting Eqns. (3) and (10) in Eqn. (6), β is given by:

$$\beta = \frac{G_o}{K_o} \frac{\tau p_o}{\eta_g} \frac{1}{\Delta\rho_{max}} \tag{10}$$

This can be further simplified to:

$$\beta = \frac{3(1-2\nu_o)}{2(1+\nu_o)} \frac{\tau p_o}{\eta_g} \frac{1}{\Delta\rho_{max}} \tag{12}$$

where ν_o is the elastic Poisson's ratio.

An Experimental Technique for Obtaining β

Sinter forging can be used to obtain β. In sinter forging, an axial stress, σ_z, is applied on the porous powder compact and both the axial and radial strains, ϵ_z and ϵ_r are measured as a function of time during isothermal sinter forging. For this loading condition, the stress and strain tensors are given by:

$$\underline{\sigma} = \begin{bmatrix} -p_o & 0 & 0 \\ 0 & -p_o & 0 \\ 0 & 0 & -p_o+\sigma_z \end{bmatrix}, \quad \underline{\epsilon} = \begin{bmatrix} \epsilon_r & 0 & 0 \\ 0 & \epsilon_r & 0 \\ 0 & 0 & \epsilon_z \end{bmatrix}$$

Stresses and strains are the true values since sintering involves large deformation. Using the definitions given in Ref. 12, the mean and the deviatoric stresses and strains are given by:

$$\sigma_m = -p_o + \frac{\sigma_z}{3}$$
$$\sigma_s = \sigma_z \tag{13}$$

and

$$\epsilon_m \equiv -\Delta\rho \equiv -\ln(\rho/\rho_g) = \epsilon_z + 2\epsilon_r \tag{14}$$
$$\epsilon_s = \frac{2}{3}[\epsilon_z - \epsilon_r]$$

Where ρ is the density at any time normalized with respect to the theoretical density and ρ_g is the normalized density at the onset of isothermal sinter forging. The stresses and strains are related by the constitutive laws given by Eqns. (4) and (8). Thus, knowing the deviatoric stress and the deviatoric strain rate, η_g can be obtained and knowing the density as a function of time, the time constant for densification, τ and the total densification, $\Delta\rho_{max}$ can be obtained. Finally, the sintering pressure, p_0, can be obtained by noting that at any density, the densification rate is proportional to the hydrostatic pressure i.e.:

$$\frac{1}{\rho} \frac{d\rho}{dt} = K(\frac{\sigma_z}{3} - p_0) \tag{14}$$

Thus, sinter forging experiments done at several different applied stresses, σ_z, give us p_0 as a function of density.

PRELIMINARY RESULTS FOR PROCESSING DEFECTS IN
THE SINTERING OF COMPOSITE CERAMICS

Preliminary experiments on TiO_2-Al_2O_3 composites did indicate the presence of radial and circumferential cracks. The results are in qualitative agreement with the predictions of the analysis.

Three sets of composite powders were made. All of them used alkoxide derived TiO_2 powders. The synthesis technique is similar to the one used in Ref. 13 and the exact details are given in Ref. 14. The freshly prepared powders were washed several times in deionized water followed by drying at 345K for 18 hours and then calcined at 823K for 16 hours. This was then ball milled using Al_2O_3 milling media and finally dried in flash evaporator. The nominal particle size of TiO_2 was 0.3 μm. This powder was divided into three portions.

The first composite powder was made by adding to one portion of TiO_2, 5 volume percent Al_2O_3 agglomerates. The agglomerates were made by freeze-drying Al_2O_3 suspensions at pH = 8.0. The nominal particle size of Al_2O_3 was 0.85 μm. The agglomerates obtained were in the size range 20-50 μm. This composite powder was cold pressed and sintered at 1548K for 90 minutes. Fig. 5(a) is a secondary electron micrograph of a polished and thermally etched section from this composite and Fig. 5(b) shows the same section in the backscattered mode. The presence of large radial cracks around the Al_2O_3 agglomerates is evident.

The next composite powder was made by adding 5 volume percent of as received Al_2O_3 powder of particle size 0.85 μm to TiO_2. In this case the alumina powder had small agglomerates. The preexisting agglomerates were in the size range 1-10 μm. This composite was also sintered at 1548K for 90 minutes. As can be seen from Fig. 6(a) and 6(b), no radial cracks

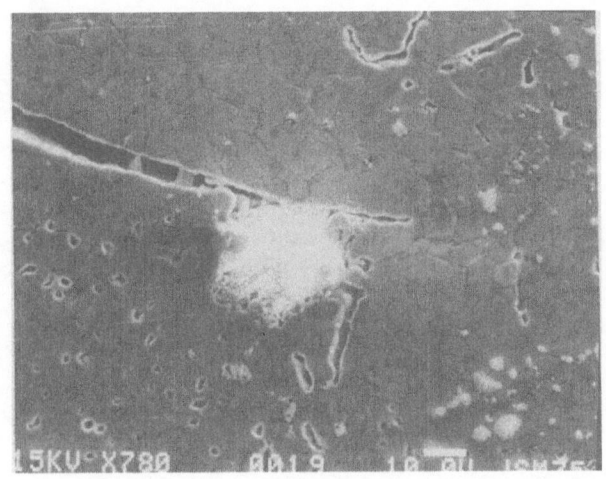

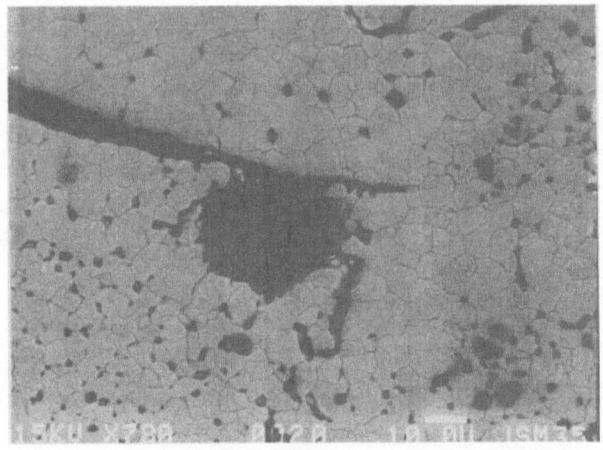

Fig. 5. Polished and thermally etched section of a 95% TiO_2-5% Al_2O_3 composite sintered at 1548K for 90 minutes. The Al_2O_3 agglomerates sinter at a rate slower than the TiO_2 giving rise to radial cracks.

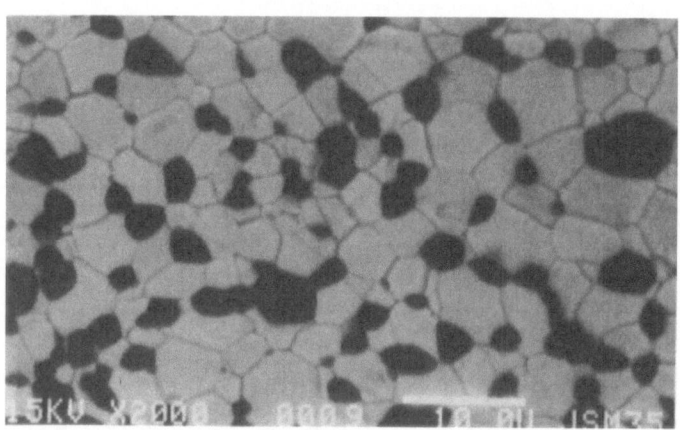

Fig. 6.   Polished and thermally etched sections of a 95% TiO_2-5% Al_2O_3
composite sintered at 1548K for 90 minutes. The Al_2O_3 agglomerates are smaller than the ones used for Fig. 5. The agglomerates sinter at a rate slower than the matrix but do not lead
to radial cracks.

were obtained. Thus, it was confirmed that larger inhomogeneities are more likely to form radial cracks.

Finally, Al_2O_3 agglomerates were made of 0.1–0.3 μm particles by freeze-drying Al_2O_3 suspensions at pH = 8.0 and then lightly presintering at 1373K for 15 minutes to increase their density. Once again 5 volume percent of these agglomerates were added to TiO_2 and the composite sintered at 1548K for 90 minutes. In this case, the Al_2O_3 agglomerates sinter faster than TiO_2. This led to the formation of circumferential cracks as can be seen from Fig. 7(a) and 7(b). It should be pointed out that for this case even small agglomerates, less than 10 μm, led to defect formation as can be seen in the micrographs. This is in agreement with the predictions of the theory. It was predicted in Ref. 2 that radial and circumferential cracks form by different mechanisms and a critical inhomogeneity size exists only for radial defects.

WORK IN PROGRESS

The important results from an analysis published earlier (2) for composite densification have been reviewed. It is shown that the ratio of the shear rate to the densification rate, called β, emerges as an important parameter. A large value of β is desirable for the densification to approach the rule of mixtures and to avoid the formation of internal defects. An expression for β in terms of experimentally measurable quantities had been derived. It is shown that sinter forging with the simultaneous measurement of axial and radial strain can give all parameters necessary to obtain β.

Preliminary experiments on the sintering of TiO_2-Al_2O_3 composites have confirmed the qualitative predictions of the analysis with regard to defect formation. It is found that the tendency to form radial cracks around a slowly densifying inhomogeneity increases as the size of the inhomogeneity increases. On the other hand formation of circumferential cracks around a fast sintering inhomogeneity is independent of the inhomogeneity size. It should be noted that since both radial and circumferential defects are generated in the TiO_2-Al_2O_3 systems, these defects are actually due to differential sintering and not due to thermal expansion mismatch.

Recently, we have carried out sinter forging experiments on alkoxide derived TiO_2 to measure β. We have used wet processing techniques to control the green structure of the compact. In addition, dilatometric measurements have been made on TiO_2-Al_2O_3 composites. In these experiments, we have varied the size and the volume fraction of Al_2O_3. Good agreement has been found between the predicted and the actual response. These results will be published at a later date.

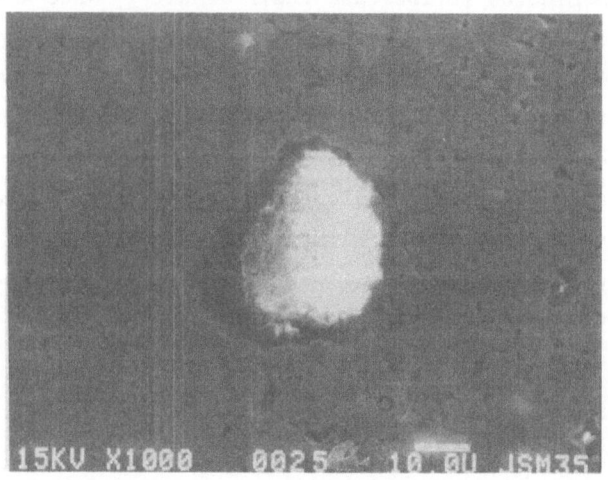

Fig. 7. Polished and thermally etched sections of 95% TiO$_2$-5% Al$_2$O$_3$
composite sintered at 1548K for 90 minutes. The Al$_2$O$_3$ agglomerates were presintered and at the sintering temperature sinter
faster than the matrix giving rise to circumferential cracks.
Note that even very small agglomerate produce this type of
defects.

ACKNOWLEDGMENT

This research was supported by the Department of Energy, under Contract No. DE-AC02-77ER04386-A009. Support was also received from the National Science Foundation through use of the facilities of the Materials Science Center at Cornell University.

REFERENCES

1. A.G. Evans, Consideration of Inhomogeneity Effects in Sintering, J. of Am. Ceram. Soc., 65 (10): 497 (1982).

2. R. Raj and R.K. Bordia, Sintering Behavior of Bi-Modal Powder Compacts, Acta Metall., 32 (7): 1003 (1984).

3. C.H. Hsueh, A.G. Evans, R.M. Cannon and R.J. Brook, Viscoelastic Stresses and Sintering Damage in Heterogeneous Powder Compacts, Acta Metall., to be published.

4. B. Kellett and F.F. Lange, Stresses Induced by Differential Sintering in Power Compacts, J. of Am. Ceram. Soc., 67 (5): 369 (1984).

5. K.D. Reeve, Non-Uniform Shrinkage in Sintering, Am. Ceram. Soc. Bull., 42 (8): 452 (1963).

6. T. Vasilos and W. Rhodes, Fine Particulates to Ultrafine-Grain Ceramics, in: "Ultrafine-Grain Ceramics," J.J. Burke, N.L. Reed and V. Weiss, eds., Syracuse University Press, New York (1970).

7. F.F. Lange and M. Metcalf, Processing-Related Fracture Origins: II, Agglomerate Motion and Cracklike Internal Surfaces Caused by Differential Sintering, J. of Am. Ceram. Soc., 66 (6): 398 (1983).

8. F.F. Lange, B.I. Davis and I.A. Aksay, Processing-Related Fracture Origins: III, Differential Sintering of ZrO_2 Agglomerates in Al_2O_3/ZrO_2 Composites, J. of Am. Ceram. Soc., 66 (6): 407 (1983).

9. F.W. Dynys and J.W. Halloran, Influence of Aggregates on Sintering, J. of Am. Ceram. Soc., 67 (9): 596 (1984).

10. J.P. Smith and G.L. Messing, Sintering of Bimodally Distributed Alumina Powders, J. of Am. Ceram. Soc., 67 (4): 238 (1984).

11. L.C. DeJonghe, M.N. Rahaman and C.H. Hsueh, Transient Stresses in Bimodal Compacts During Sintering, to be published.

12. R. Raj, Separation of Cavitation-Strain and Creep-Strain During Deformation, J. of Am. Ceram. Soc., 65 (3): C46 (1982).

13. E.A. Barringer and H.K. Bowen, Formation, Packing and Sintering of Monodisperse TiO_2 Powders, J. of Am. Ceram. Soc., 65: C199 (1982).

14. R.K. Bordia, Ph.D. Thesis, Cornell University, in preparation.

15. R.K. Bordia and R. Raj, Sintering of a Composite with a Glass or a Ceramic Matrix: An Analysis, J. of Am. Ceram. Soc., in press.

HIP OF LIQUID PHASE SINTERED CERAMIC COMPOSITES

O-H. Kwon and G. L. Messing

Department of Material Science & Engineering
The Pennsylvania State University
University Park, PA 16802

INTRODUCTION

Many ceramics today are multiphasic or composite by design to obtain unique combinations of optical, thermal, electrical and mechanical properties. Because the reliability and the ultimate properties of these materials often require full density and a pore-free, fine-grained microstructure, hot isostatic pressing (HIP) is receiving increased utilization for their fabrication.

HIP differs from uniaxial hot pressing in that pressure is transmitted isostatically to an object via an inert gas phase. The ability to press components isostatically allows two approaches to densification. First, powder compacts can be encapsulated in an impermeable sheath that will deform and conform to the object during pressing. Alternatively, ceramics that have been sintered to the closed pore state can be HIPed without encapsulation. Industrially, HIP is already widely applied for post-sintering densification in the cutting tool[1-3] and ferrite[4] industries. Also, a number of other ceramics such as Si_3N_4[5-8], SiC[9,10] and ZrO_2[11-14] have been demonstrated to yield significantly improved properties after HIPing. While densification theory is well developed for solid state sintering of simple oxides, there has been little fundamental analysis of multiphase and composite ceramics sintering or their subsequent densification behavior during HIP. Furthermore, while liquid phase sintering has been widely examined, HIP densification of liquid phase sintered (LPS) ceramics has been little studied.

In this paper, some of the fundamental phenomena that affect densification during HIPing of liquid phase sintered ceramic composites are discussed. Alumina-glass composites have been chosen as a model system for ceramic composites having a viscous grain boundary phase. Specific issues addressed include the effect of applied pressure on solid solubility in the liquid, liquid phase migration and gas diffusion. These phenomena are presented in relation to microstructural changes and densification during containerless HIPing of closed porosity ceramics.

DENSIFICATION

Densification by solution-precipitation in the final stage of liquid phase sintering is largely a function of the compressive stress at grain contacts. The stress is a complex function of the amount of liquid, the dihedral angle of the liquid and the geometry of the grain assemblage. Densification is obtained mainly by contact point flattening through the dissolution of grain contacts in the liquid phase. During HIP, the total driving force in a closed porosity material has three components of stress as follows:

$$D. F. = \sigma_s + \sigma_a - \sigma_i \qquad (1)$$

where, σ_s is the normal stress at grain contacts resulting from capillarity, σ_a is the applied isostatic pressure, and σ_i is the internal gas pressure of pores. Since σ_s and σ_i are usually much smaller than σ_a in HIPing, the driving force can be approximated by σ_a. However, the applied stress is only supported by particle contacts and the true driving force is given by the effective stress at grain contacts, σ_{eff}, which is a complex function of the packing geometry, the amount of second phase, V_1, and the dihedral angle, $\emptyset$. However, if V_1 and $\emptyset$ are known and a reasonable geometrical model for the grain structure in the final stage of densification derived, the uncertainty in the magnitude of the driving force can be reduced.

If a tetrakaidecahedron grain model is assumed for grains in final stage sintering then the grain boundary area, covered by a second phase, β, can be calculated as a function of V_1 and $\emptyset$ by using the multiphase geometrical model presented by Wray[15].

$$\sigma_{eff} = \frac{\sigma_a}{\{1-\beta(V_1,\emptyset)\}\rho} \qquad (2)$$

where, $1-\beta(V_1,\emptyset)$ is the fractional grain contact area and ρ is the relative density. Fig. 1 shows the increase in the effective stress as a function of relative density. Solid lines 1 to 4 give previous models[16-19] whereas solid line 5 shows the proposed model based on Eqn. (2) at $\emptyset$ = 180°. Dotted lines show the change in σ_{eff} for various dihedral angles. For densification during LPS, material systems often exhibit dihedral angles <60° and sometimes, as small as 0°. Therefore, Fig. 1 predicts that the effective stress is increased by two to three times the applied stress for a 95% dense material when the dihedral angle ranges from 60° to 10°.

According to the Gibbs-Thompson equation the increase in chemical potential or solid solubility at grain contacts is

$$\Delta C = C - C_o = C_o \left[\exp \left(\frac{\sigma_{eff}\Omega}{kT} \right) -1 \right] \qquad (3)$$

where, ΔC is the solubility increase, C is the solubility under σ_{eff}, C_o is the solubility under no applied pressure, Ω is the molecular volume of the solid, k is the Boltzmann constant, and T is temperature. Fig. 2 shows the change in solubility ratio at grain contacts as a function of temperature for various effective stresses as calculated from Eqn. (3) for alumina. Since σ_{eff} can be increased by 2 to 3 times for small dihedral angles, solubility can be enhanced by 40 to 70% at 1500°C for an applied pressure of 100 MPa.

For HIP densification experiments, Al_2O_3 samples with 5 and 10% glass (20.5% MgO, 17.5% Al_2O_3, and 62% SiO_2 in wt%) were prepared to have grain sizes of 2.1, 3.6, and 5.9 μm median diameter. These samples

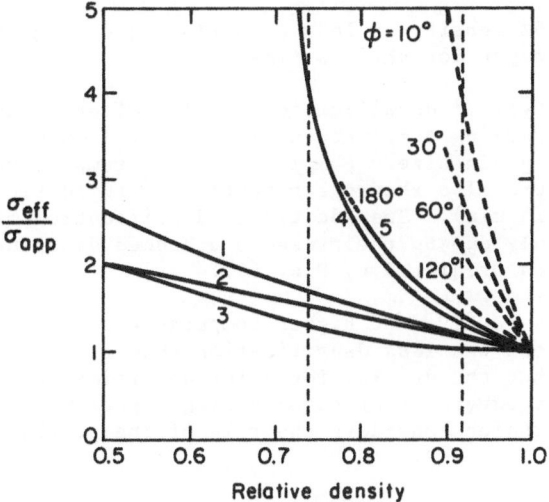

Fig. 1 Effective stress at grain contacts as a function of relative density for various dihedral angles (solid line 1-McClelland[16]; 2-Spriggs and Vasilos[17]; 3-Farnsworth and Coble[18]; 4-Arzt, et al.[19]

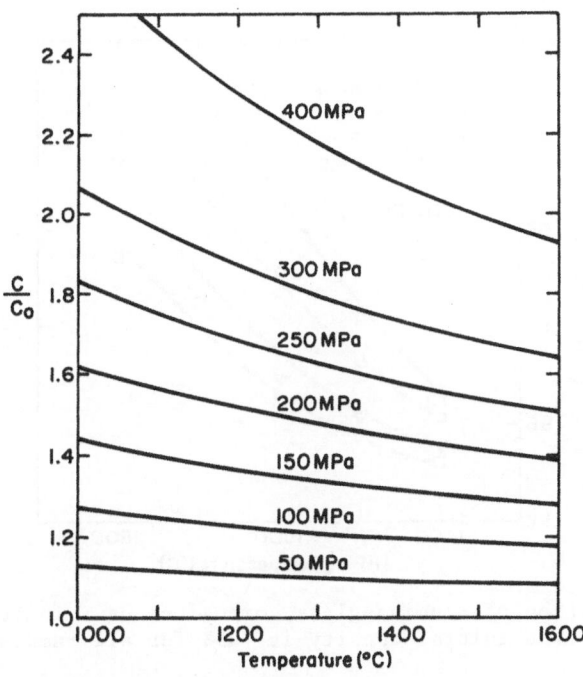

Fig. 2 Relative solubility increase of alumina in glass as a function of temperature for various effective stresses.

were sintered to 95% density at 1600°C in flowing oxygen and HIPed at 100 MPa for 15 min in argon for various times.

Fig. 3 shows typical densification results after HIPing. Density increases with increasing temperature, increasing liquid volume content and decreasing alumina grain size. Samples HIPed at temperatures below the sintering temperature also showed substantial densification in a relatively short time (i.e., 15 min). The additional densification at lower temperature is mainly due to the pressure enhanced dissolution of alumina at grain contacts as predicted by Eqn. (3).

When samples were HIPed at higher temperatures and pressures and for extended times, there was less densification than reported in Fig. 3. This result suggested that the driving force for densification was reduced or that the sample was adversely affected during depressurization and cooling. The following discussion considers the role of the pressurizing gas on these two effects.

EFFECTS OF GAS DIFFUSION AND DISSOLUTION

Examples of gas phase induced problems in materials have been demonstrated in many fields, e.g., voids in metal castings[20], glass refining[21], and fission gas induced swelling of nuclear fuels[22]. Ceramics often have a silicate glass in the grain boundary as a sintering aid or to modify properties. Since the structure of silicate glass is characterized as a random network with a temperature dependent free volume, higher gas diffusivity and solubility are expected in glasses than in their crystalline counterparts[23]. Therefore, ceramics with a glassy grain boundary phase are more susceptible to gas inclusion.

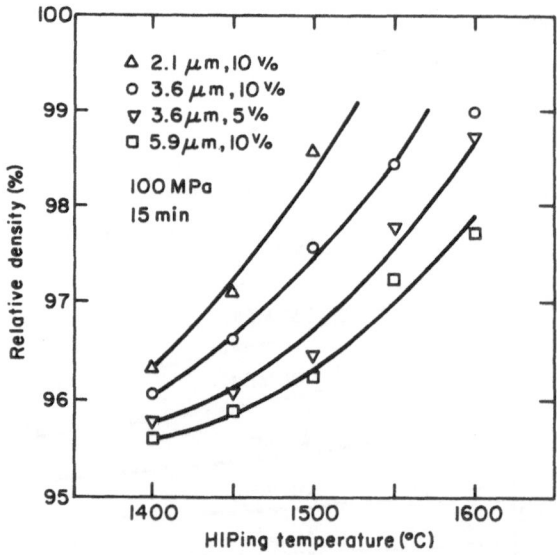

Fig. 3 Densification of alumina-glass composites as a function of HIPing. Note that the initial density is ~95% for all samples.

Diffusivity and solubility data for argon in the silicate glass used in this study at the HIPing conditions studied are not yet available. However, published gas diffusion data in other glasses (Fig. 4) give an idea of relative gas diffusion into silicate glasses for typical HIPing conditions.

The diffusivity of inert gases in fused silica increases with decreasing size of the gas molecule, i.e., He>Ne>Ar>Kr>Xe. In fused silica, diffusivities of Ar, N_2, and O_2, which have similar molecular sizes, are within one order of magnitude of each other. It is noteworthy that diffusivities of gases in silicate glasses at high temperature, which ranged from 10^{-4} to 10^{-6} cm^2/sec at about 1500°C, are approximately the same as the diffusivities of the condensed material in glass[24]. Consequently, the transfer of gas and condensed material through glass channels might represent competitive processes during HIPing. Indeed, if gas diffuses into the glass and then condenses in pore sites it would result in a significant negative driving force (σ_i).

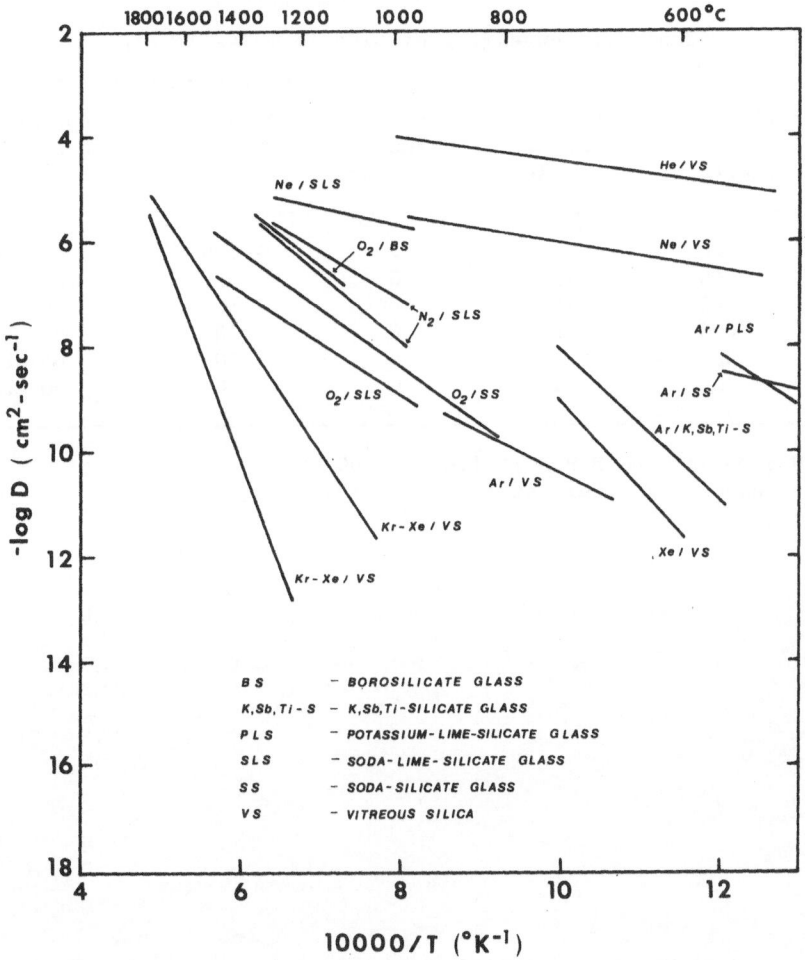

Fig. 4 Diffusivities of gases in glass as a function of temperature.

Faile and Roy[25] showed that gas solubility in glass increases up to 10 mole % with increasing pressure and temperature. Therefore, during containerless HIPing the gas content in the glass can be significantly increased. During cooling and depressurization the gas can precipitate to form voids which can further grow depending on the rate of cooling. This mechanism of pore generation would result in an apparent dedensification.

Since a large amount of Ar diffusion and dissolution in glass during HIP is predicted from the gas diffusion and solubility data, a series of annealing experiments were performed on samples to determine whether there was significant gas dissolved in the glass.

Samples annealed at temperatures $\geq 1200^{\circ}$C showed swelling and cracking (Table 1). The degree of swelling and cracking is high in the samples HIPed for a long time at high temperature and pressure. Fig. 5 shows an extreme example of the annealed sample. Lenticular cracks were formed parallel to the surface because gas diffusion into samples resulted in a pressure gradient parallel to the surface. However, samples annealed at low temperature ($\leq 1100^{\circ}$C) didn't show any change in density. A

Table 1. Annealing of HIPed samples in air

Annealing		No. of samples annealed	No. of samples cracked	Density change (% T.D.)	Appearance after annealing
Temp($^{\circ}$C)	Time(min)				
1600	60	7	7	>-5.0	swelled
1400	60	15	0	-2.7	bloated
1200	60	18	8*	-1.3	no change
1100	600	6	0	0	no change
1000	300	2	0	0	no change
1300**	60	2	0	0	no change

* all 8 samples were HIPed at 150 and 200 MPa.
** second annealing after annealing at 1100°C

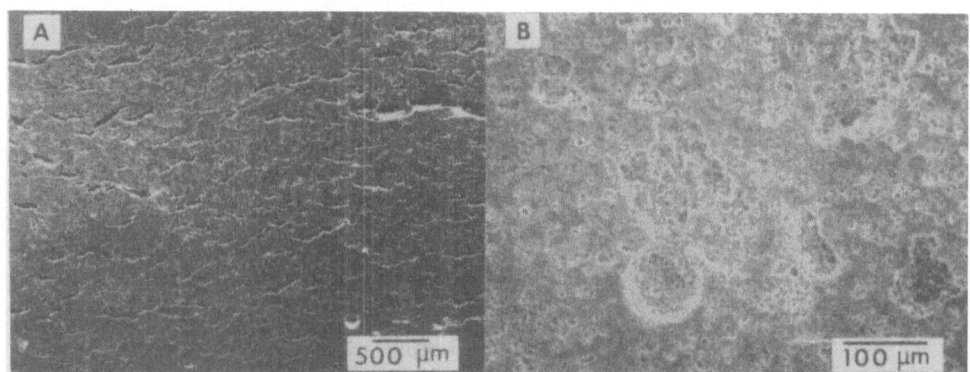

Fig. 5 Microstructure of samples annealed at 1600°C; (a) sample sectioned perpendicular to the surface, (b) sample sectioned parallel to the surface.

subsequent annealing of these samples at $1300^{\circ}C$ for 60 min showed no evidence of swelling or decrease in density. Therefore, the appropriate annealing of HIPed samples at low temperature may prevent any deleterious effects of gas diffusion and dissolution.

CAPILLARY FLOW

Early studies on the capillary flow of liquids into cylindrical capillaries by Washburn[26] and Rideal[27] have shown that the dynamics of capillary flow have certain practical aspects in connection with the movement of liquid through porous materials. Validity of their theory has been proven in many applications.[28] Such an analysis is useful to understand the role of capillary driven liquid redistribution in porous materials during LPS.

Assuming the liquid is Newtonian and under chemical equilibrium with the porous media, Poiseuilli's law applies for laminar flow conditions. The capillary flow rate can be calculated under given driving pressure as follows:

$$\frac{d\ell}{dt} = \frac{\Sigma\sigma}{8r^2\eta\ell} (r^4 + 4Er^3) \tag{4}$$

where, ℓ is the length of the cylindrical capillary at time t, η is the liquid viscosity, r is the radius of capillary, E is the coefficient of slip at the solid-liquid interface, and $\Sigma\sigma$ is the total effective driving pressure which is acting to force the liquid along the capillary.

Powder compacts which have been liquid phase sintered to the closed pore stage can be described as a porous body as schematically shown in Fig. 6(a). If liquid menisci on the surface are balanced by internal liquid menisci, liquid flow in the channels will cease. However, if the liquid covers the surface, capillarity at the external surface is zero because the meniscus radius increases to infinity (Fig. 6(b)). The total driving pressure for capillary flow consists of four separate pressures in this system, the applied pressure, σ_a, the hydrostatic head, σ_h, the capillary pressure, σ_c, and the internal pore pressure, σ_p.

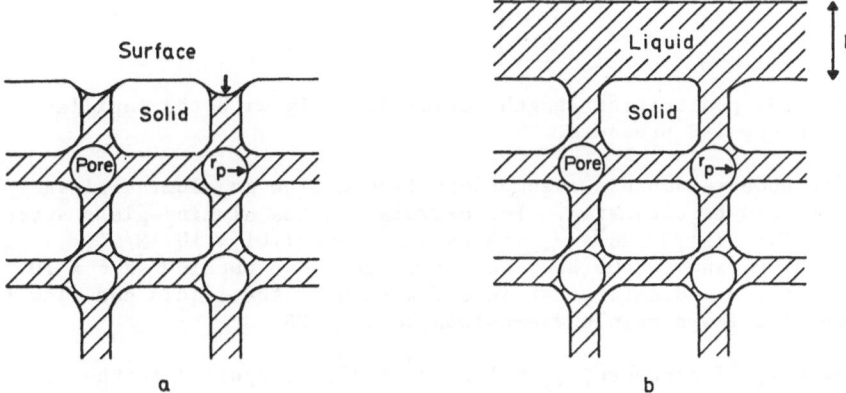

Fig. 6 Schematic diagram of a LPS porous body. Note the liquid menisci on surface, (a), and no liquid meniscus on surface covered with liquid, (b).

$$\Sigma\sigma = \sigma_a + \sigma_h + \sigma_c - \sigma_p \tag{5}$$

where

$$\sigma_h = dgh \tag{6}$$

and, d is the liquid density, g is the gravitational constant, and h is the height of the liquid head.

$$\sigma_c = \frac{2\gamma}{r}\cos\theta \tag{7}$$

where, γ is the surface tension and θ is the contact angle. As discussed earlier, σ_p depends on the solubility and diffusivity of gases entrapped in the liquid.

Summing up and substituting into Eqn. 4 gives the following expression:

$$\frac{d\ell}{dt} = \frac{\sigma_a + dgh + \dfrac{2\ell}{r}\cos\theta - \sigma_p}{8\eta\ell}(r^2 + 4\ Er) \tag{8}$$

If σ_h, σ_p and E are small, then upon integration

$$\ell = \left[\frac{(\sigma_a + \dfrac{2\gamma}{r}\cos\theta)\ r^2 t}{4\eta}\right]^{1/2} \tag{9}$$

Since σ_a is relatively small under atmospheric pressure (1 atm) i.e., $2\gamma\cos\theta/r \gg \sigma_a$, Eqn. 9 can be further reduced for small r to

$$\ell = \left[\frac{\gamma\cos\theta rt}{2\eta}\right]^{1/2} \tag{10}$$

Thus, the length of penetration due to this process is proportional to the square root of time, the capillary radius, and the interfacial energy, and inversely proportional to the square root of the liquid viscosity.

However, if sufficient pressure is applied to the system then the interfacial energy contribution is negligible, i.e., $\sigma_a \gg 2\cos\theta/r$ and Eqn. 9 can be simplified to

$$\ell = \left[\frac{\sigma_a r^2 t}{2\eta}\right]^{1/2} \tag{11}$$

Note that the penetration length varies linearly with the capillary radius under high applied pressure.

With some reasonable assumptions liquid flow in alumina-glass composites can be estimated. For example, in the alumina-glass system at 1500°C, $\gamma = 0.4$ J/m^2, $\theta = 26^\circ$ [29], $\eta = 5$ Pa·sec [30], $\sigma_a = 1.01 \times 10^5$ N/m^2, $r = 0.5 \times 10^{-7}$m, and t=3600 sec, and from Eqn. 9 we obtain $\ell = 2.6 \times 10^{-3}$ m. This estimation indicates that in a few seconds the liquid can flow the distance of a few particle dimensions during LPS.

However, if pressure, $\sigma_a = 1 \times 10^8$ N/m^2, is applied to the liquid and the other values are the same, Eqn. 9 gives $\ell = 7.2 \times 10^{-3}$m, which is three times greater. This demonstrates that liquid flow through a porous body can be significantly enhanced by an applied pressure.

Fig. 7 shows the calculated glass penetration length as a function of capillary radius for various applied pressures at 1500°C for 60 min. If σ_a is small, ℓ increases with $r1/2$, and if σ_a is large, ℓ increases linearly with r. Thus, the effect of pressure on capillary flow has increasing importance as the capillary radius increases.

The above analysis has been restricted to a system in chemical equilibrium. However, in many systems at elevated temperatures, the phases of a solid-liquid-vapor system are often not at chemical equilibrium. Under chemical nonequilibrium conditions, the effect of chemical reactions on the interfacial energies have to be considered. As pointed out by Aksay, Hoge, and Pask[31], a solution reaction results in a possible transient decrease of interfacial energy due to a negative free energy of solution. Also, an instantaneous decrease in contact angle is expected followed by a pull-back to the equilibrium, which is retained while the reaction proceeds to chemical equilibrium. Additional complications for modeling can arise from changes in the initial capillary geometry due to the dissolution and/or the formation of crystalline reaction products at the interface.

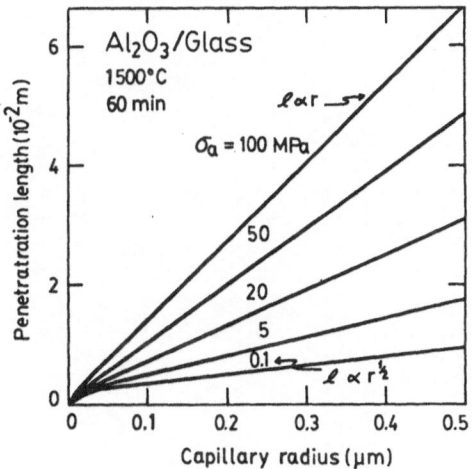

Fig. 7   Glass penetration length in an alumina-glass system as a function of capillary radius for various applied pressures at 1500°C for 60 min.

To physically examine liquid penetration in Al_2O_3-glass composites, alumina pellets with 3 vol% MAS glass (~95% density) were sintered at 1600°C. A small piece of MAS glass of ~0.1 g was placed on the pellet as shown in Fig. 8 (a) and isothermally heated in a furnace at 1500 and 1600°C then quenched in air. Some samples were totally immersed in glass in a Mo or Pt crucible as shown in Fig. 8(b). For HIPing, samples were isostatically pressed at 1500°C/100 MPa in Ar for 30 min. The depth of penetration was measured on optical micrographs of sample cross-sections.

Fig. 9(a) shows the cross section of the entire sample after glass penetration and Fig. 9(b) shows the sample microstructure before glass penetration. After capillary flow of glass, a dense region was formed parallel to the surface as shown in Fig. 9(c). The number and size of

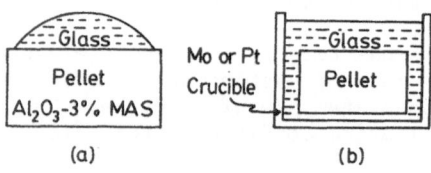

(a)   (b)

Fig. 8 Schematic diagram of the glass penetration experiments showing
(a) glass on a sintered alumina-3% MAS glass pellet of ~95%
density, and (b) a pellet which is immersed in glass in a crucible
for heat treatment or HIPing.

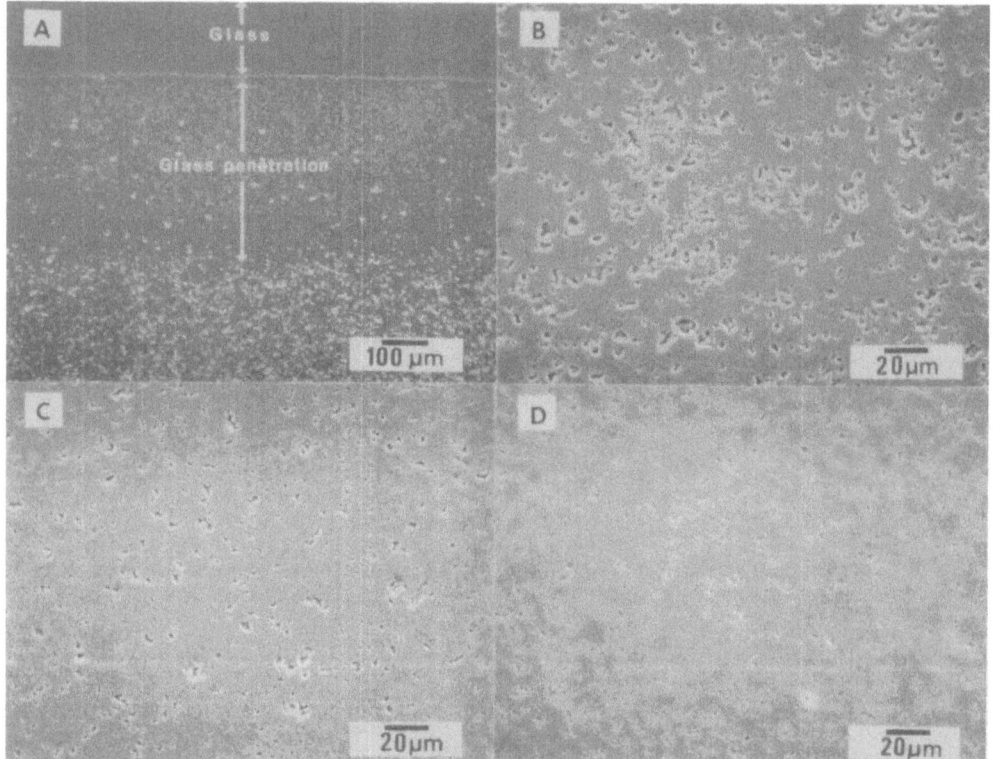

Fig. 9 Microstructure of samples after glass penetration. Since samples
were etched with dilute HF after polishing, the pore sizes are
slightly exaggerated: (a) glass penetration at 1500°C for 10 min.
at atmospheric pressure, (b) initial microstructure of the sample,
(c) microstructure of the glass penetration region, and (d) microstructure of the glass immersed sample HIPed at 1500°C/100 MPa for
30 min.

pores were decreased substantially. However, a small number of reduced pores remained after 60 min. This indicates that pores could not be totally removed by capillarity. Further densification should be obtained by the dissolution and diffusion of the internal gas through the liquid phase. With HIPing the size and number of pores are reduced further as shown in Fig. 9(d) resulting in greater than 99.5% density and only a few small intergranular pores. This suggests that the applied pressure is effective in increasing capillary flow. However, the existence of small pores after HIPing suggests the necessity for gas dissolution and diffusion into the liquid phase for the attainment of full density.

Fig. 10 shows the measured penetration length of glass at 1500 and 1600°C under no applied pressure as a function of time. The same penetration curves were also obtained for the glass immersed samples. Although Eqn. 9 predicts that the penetration length is proportional to $t^{1/2}$, the results show ~$t^{3/4}$ proportionality. This implies that solution assisted penetration is operative simultaneously with capillary flow. This is because the glass composition placed on the surface was not in equilibrium composition in the $MgO-Al_2O_3-SiO_2$ system at these temperatures, thus resulting in the dissolution of alumina at high temperatures and expansion of the capillary. Greater penetration at high temperature (1600°C) is mainly due to the decrease in viscosity of liquid as predicted by Eqn. 9. In contrast with liquid penetration samples, the HIPed sample showed complete glass penetration throughout the sample thickness, ~6mm, resulting in a very dense microstructure.

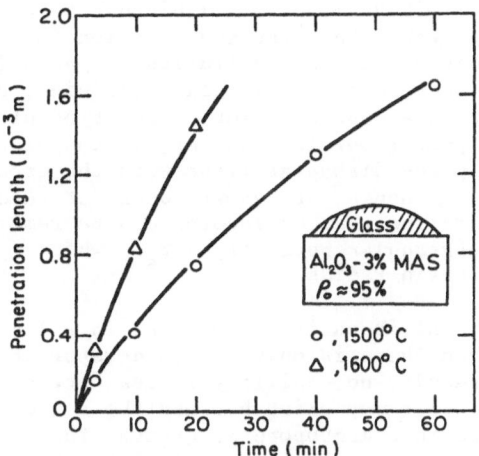

Fig. 10 Measured glass penetration length from the surface as a function of time under no applied pressure.

Raju, Aksay and Pask[32] observed the penetration behavior of silicates in magnesia and forsterite compacts. Penetration occurred when the solid was wet by the liquid and when a continuous or open pore structrue was present. Although they expressed this process as "permeation", this also can be described as capillary flow based on the present model.

Flaitz[33] demonstrated that complete penetration of a liquid in a solid microstructure is not necessarily associated with complete wetting at the solid-liquid-vapor interface and showed that the silicate glass can penetrate into a fully dense, pure polycrystalline alumina at high temperature to reduce the interfacial energy if the thermodynamic requirement, $2\gamma_{sl} < \gamma_{ss}$, is satisfied. The above results demonstrate that the liquid can penetrate rapidly into the liquid phase sintered ceramics which have small, isolated, intergranular pores to minimize the liquid-vapor interfacial energy.

Although the applicability of Eqn. 9 is limited, it gives an order of magnitude estimate of liquid penetration through a porous body and shows the effectiveness of an applied pressure. It also predicts the time required to densify materials by the liquid penetration process and the rate of liquid redistribution. The HIP result shows that pressure assisted infiltration is operative at high temperature in ceramic systems with viscous liquid.

MICROSTRUCTURAL HOMOGENIZATION

Each processing step, starting with preparation of the powder, has the potential for introducing heterogeneities in the microstructure. As these flaws can limit the properties and performance of the ceramic[11], it is important to consider the mechanisms of their removal during HIP. In powder metallurgy, investigators[34-36] have shown that macropores or voids can be filled by liquid flow during LPS. Furthermore, Amberg, et al.[1] and Kwon and Messing[37], showed that macropores can be filled during HIP by liquid flow.

Removal of isolated pores in the liquid during liquid phase densification with a limited amount of liquid can be achieved by two different mechanisms. When the interface reaction for the contact point flattening is fast, dissolution and diffusion of gases in liquid may control pore removal such as occurs in glass refining, but when the interface reaction is slow, the dissolution reaction of the solid at grain contacts may control pore removal. However, if the size of pores in a liquid varies, small pores disappear first with the growth of larger pores by Ostwald ripening. Although pore growth does not result in densification, this implies that macropores can be removed only after removal of small intergranular pores ($r_p < r_s$), where r_p is the pore radius and r_s is the grain radius.

If the bulk material is entirely immersed in a liquid and at the macropore surface, then there is only a liquid-vapor interface in the macropore and, as a result, no capillary forces between grains.[38] However, appreciable viscous forces may exist between the solid grains. Therefore, if the capillary force in a macropore is greater than the viscous forces between grains, the macropore can be filled either by the liquid or by grain-liquid mixture flow depending on the processing conditions. Since the applied pressure can enhance both the interface reaction and capillary flow, it is expected that HIPing will increase pore removal and result in microstructural homogenization.

To examine macroscopic defect removal processes during HIP, macropores were incorporated in the sintered microstructure by mixing 0.5 vol% PMMA spheres in the powder mixture. Fig. 11(a) shows the typical microstructure of a sample sintered to 95% density with 50 and 100 μm diameter induced macropores. The spherical macropores are clearly defined in Fig.11(b) which shows that there is no liquid on the pore surface.

From HIPing experiments it is known that macropore filling strongly depends on the initial sample density. For example, when the HIPed density is lower than 97% of theoretical, macropore filling is not observed which

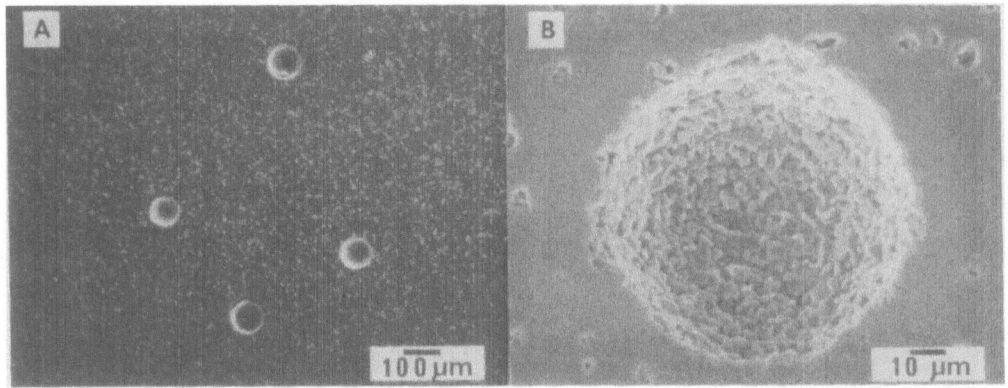

Fig. 11 Microstructures of samples sintered to 95% density at 1600°C
with induced macropores showing (a) distribution of macropores,
and (b) the absence of glass on the surface of a spherical
macropore.

suggests that small pores were preferentially filled. However, when the
HIPed density is greater than 99% of theoretical, all of the observed
macropores are filled to various degrees with the glass. At the higher HIP
densities the macropores have two distinctly different morphologies. Fig.
12(a) shows the typical microstructure of coarse-grained samples HIPed at
low temperature, whereas Fig. 12(b) shows the grain-liquid mixture flow
into the macropore which is characteristic of fine-grained samples HIPed at
high temperature. Figs. 12(c) and 12(d) show completely filled macropores
indicating two concurrent macropore filling processes. Macropore filling
by the second type often cannot be morphologically identified because it
results in increased microstructural homogenization as supported by the
following observations. Since the volume fraction of macropore former is
constant in all samples, samples with 50 μm macropores should have 8 times
as many macropores as the samples with 100 μm macropores. However, there
were <4 times the number of 50 μm macropores than 100 μm macropores counted
on sectioned surfaces. Thus, it is concluded that many 50 μm macropores
were microstructurally homogenized during HIPing by the grain-liquid flow
mechanism of pore filling.

SUMMARY

Hot isostatic pressing studies of sintered alumina-glass composites as
a model system for ceramics having a viscous grain boundary phase have
shown that:

(1) HIP densification is significantly increased as a result of pressure
enhanced dissolution of the solid phase into the liquid at grain
contacts.

(2) Pressure enhanced gas solubility in the liquid phase may be
deleterious but can be controlled by annealing after HIPing.

(3) Applied pressure results in substantial pore removal by
pressure-enhanced capillary flow in closed porosity ceramics.

(4) Microstructural homogenization can occur by pressure driven liquid or
grain/liquid flow into pores resulting in a reduction in the number
and size of flaws.

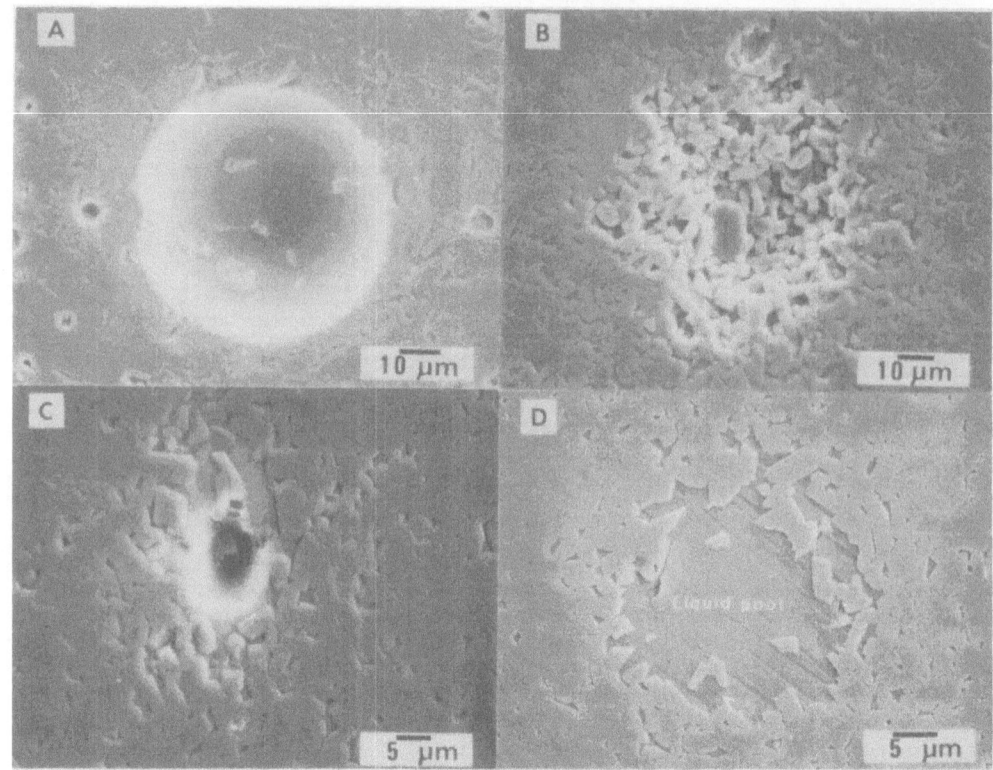

Fig. 12 Various morphologies of macropores after HIPing. (a) partially
filled macropore of 100 μm (1600°C, 100 MPa, 15 min. G.S. =
5.9 μm), liquid advances into macropore forming a hollow liquid
shell, (b) partially filled macropore of 100 μm (1600°C, 100
MPa, 15 min, G.S. = 2.1 μm), the grain-liquid mixture flows
into the macropore, (c) almost filled and homogenized macropore
of 100 μm (1500°C, 100 MPa, 60 min, G.S. = 3.6 μm), and (d)
completely filled macropore of 50 μm (1500°C, 100 MPa,
30 min, G.S. = 2.1 μm).

ACKNOWLEDGEMENT

The authors gratefully acknowledge the financial support of this work
by the Army Research Office under contract No. DAAG 29-82-K-0099.

REFERENCES

1. S. Amberg, E. A. Nylander, and B. Uhrenius, Powder Metall. Intl., 6[4]
    178-180 (1974).
2. E. Lardner, Powder Metall., 18[35] 47-52 (1975).
3. U. Engel and H. Hubner, J. Mater. Sci., 13[9] 2003-2012 (1978).
4. E. Takama and M. Ito, IEEE Trans. on Magnetics, Vol. MAG-151 [6]
    1858-60 (1979). ,
5. H. C. Yeh and P. F. Sikora, Amer. Ceram. Soc. Bull., 58[4] 444-447
    (1979).
6. R. R. Wills, M. C. Brockway, L. G. McCoy, and D. E. Niesz, Ceram. Eng.
    & Sci. Proc., 1[7-8] (B) 534-539 (1980).
7. T. Yamada, M. Shimada, and M. Koizumi, Amer. Ceram. Soc. Bull., 60[11]
    1225-1228 (1981).

8. O. Yeheskel, Y. Gefen, and M. Talianker, J. Mater. Sci., 19[3] 745-752 (1984).

9. K. Homma, T. Tatuno, H. Okada, and T. Fujikawa, Proc. of the 25th Japan Congress on Materials Research, pp. 213-17, Soc. of Mater. Sci., Japan (March 1982).

10. G. K. Watson and T. J. Moore, Automotive Technology Development Meeting, Dearborn, Michigan (October 1981).

11. F. F. Lange, J. Amer. Ceram. Soc., 66[6] 396-398 (1983).

12. S. Hori, M. Yoshimura, S. Somiya and H. Kaji, J. Mater. Sci. Letters, 3 242-44 (1984).

13. K. Tsukuma, K. Ueda, and M. Shimada, J. Am. Ceram. Soc., 68[1] C4-5 (1985).

14. K. Tsukuma, K. Ueda, K. Matsushita, and M. Shimada, J. Am. Ceram. Soc., 68[2] C56-58 (1985).

15. P. J. Wray, Acta Metall., 24 [1] 125-135 (1976).

16. J. D. McClelland, J. Amer. Ceram. Soc., 44[10] 526 (1961).

17. R. M. Spriggs and T. Vasilos, J. Amer. Ceram. Soc., 47 [1] 47 (1964).

18. P. L. Farnsworth and R. L. Coble, J. Amer. Ceram. Soc., 49 [5] 264-268 (1966).

19. E. Arzt, M. F. Ashby and K. E. Easterling, Metall, Trans., 14 A [2] 211-21 (1983).

20. R. L. Coble and M. C. Flemings, Metall, Trans., 2 [2] 409-15 (1971).

21. M. Cable, Glass Tech., 2 [2] 60-70 (1961).

22. C. C. Dollins and F. A. Nichols, J. Nucl. Mater., 66 143-57 (1977).

23. J. E. Shelby, "Molecular Solubility and Diffusion," in: Treatise on Materials Science and Technology, M. Tomozawa and R. H. Doremus (Eds.), Vol. 17: Glass II, Academic Press, New York (1979).

24. H. A. Schaeffer, J. Non-Crystal. Solids 67 [1-3] 19-33 (1984).

25. S. P. Faile and D. M. Roy, J. Amer. Ceram. Soc., 56[1] 12-16 (1973).

26. E. W. Washburn, Physical Rev., 2nd Ser., 18[3] 273-83 (1921).

27. E. K. Rideal, Phil. Mag., 44, 1152-59 (1922).

28. A. M. Schwartz, Ind. and Eng. Chem., 61[1] 10-21 (1969).

29. O-H. Kwon and G. L. Messing, unpublished work.

30. E. F. Riebling, Can. J. Chem., 42, 2811-21 (1964).

31. I. A. Aksay, C. E. Hoge, and J. A. Pask, J. Phys. Chem., 78[12] 1178-83 (1974).

32. A. P. Raju, I. A. Aksay, and J. A. Pask, Amer. Ceram. Soc. Bull., 52[2] 166-69 (1973).

33. P. L. Flaitz, Ph.D. Thesis, University of California, Berkeley (1982).

34. O. J. Kwon and D. N. Yoon, Sintering Processes (Proc. 5th Intl. Conf. on Sintering and Related Phenomena, June 1979, Notre Dame, U.S.A.), ed. G. C. Kuczynski, Plenum Press, New York pp. 203-218 (1980).

35. H. H. Park, S. J. Cho, and D. N. Yoon, Metall. Trans. A 15[6] 1075-1080 (1984).

36. S. J. L. Kang, W. A. Kaysser, G. Petzow, and D. N. Yoon, Powder Metall., 27[2] 97-100 (1984).

37. O-H. Kwon and G. L. Messing, J. Am. Ceram. Soc., 67[3] C43-45 (1984).

38. D. Tabor, in Adhesion, ed. D. D. Eley, Oxford Univ. Press, pp. 115-206 (1961).

SINTERABLE YTTRIA-DOPED ZIRCONIA POWDERS CHEMICALLY COPRECIPITATED IN NON-AQUEOUS MEDIUM

Giulio A. Rossi

Norton Company High Performance Ceramics
1 New Bond Street
Worcester, Massachusetts 01606

ABSTRACT

Yttria-doped zirconia powders were synthesized by coprecipitation, by mixing an alcoholic solution of $ZrCl_4$ and $YCl_3 \cdot 6H_2O$ with an alcoholic solution of NaOH. The purified powders were characterized for physical and chemical properties, pressing behavior and sinterability. The powders are ultrafine, highly pure and presumably chemically homogeneous but strongly agglomerated. TZP (Toughened Zirconia Polycrystals) were pressureless sintered from powders calcined at around 700°C and fired densities around 95-99% were obtained with the appropriate processing. Very uniform microstructures were observed in the sintered samples, with most of the grains in the submicron range. For a 4 w/o Y_2O_3 composition (2.2 m/o) flexural strengths as high as 1157 MPa (3 point) and fracture toughness values (microindentation method) of 5-6 $MPa \cdot m^{\frac{1}{2}}$ were measured.

INTRODUCTION

High performance zirconia ceramics can be produced only through a rigorous control of the microstructure, which, in turn is a function of the physical and chemical properties of the starting powder.

The "ideal" powder is chemically homogeneous at the atomic scale and of very high purity (or with a strictly controlled impurity content) to avoid the formation of intergranular glassy phases which degrade the high temperature properties. It is also submicron in size, spherical and deagglomerated. The ideal particle size distribution is still subject to debate, some researchers[1] preferring monosize powders with others[2] favoring multimodal distributions.

Sub-micron, sinterable powders can be produced with two main routes, i.e., at low temperature starting from chemical precursors, and at high temperature, for example by rapid solidification form the melt, or laser, plasma synthesis, flame oxidation, in the gas phase. Both routes present advantages and drawbacks, in terms of properties obtained and production costs.

This paper describes a new method for the synthesis of yttria doped zirconia powders of very high purity. The powders are coprecipitated in an alcoholic medium in order to avoid the formation of a gel, which inevitably occurs when a doped zirconia is precipitated from aqueous solutions with an alkaline reagent, such as ammonia.

Among the problems associated with gels are: 1) the presence of a salt by-product, which is difficult to remove, and 2) the formation of unsinterable hard granules upon drying.

The first problem (residual salts) may result in the incorporation of impurities hard to remove during the sintering cycle, like gases which may inhibit the densification. The second problem (hard agglomerates) could be eliminated by freeze-drying or supercritically drying the gel, since with both methods the deleterious influence of the liquid meniscus is eliminated. These additional operations, however, are expensive and time consuming.

A third method has been recently used to dry gels to an amorphous ultrafine powder, i.e. azeotropic distillation[3,4]. In so doing, the water is replaced with a non polar organic liquid, generally benzene, which can be subsequently removed without leaving behind undesirable organic residues. Even though the method seems to work well, it is expensive and uses a dangerous chemical.

The method described in this paper, for which a patent has been recently granted[5], does not use aqueous, but alcoholic solutions, with the result that the gel formation is avoided. The details are described in the following section. $ZrCl_4$ powder of very high purity is used as the precursor of ZrO_2 and the precursors of the stabilizing oxides are alcohol soluble salts which give water insoluble hydroxides upon reaction with the precipitating agent (NaOH). For instance, the yttrium and magnesium hydroxides are virtually insoluble in water, whereas calcium hydroxide is appreciably soluble and some of it would be lost during the washing step.

EXPERIMENTAL

Two Y_2O_3-ZrO_2 compositions were synthesized in this study, viz.
2.2 m/o (4 w/o) and 3 m/o (5.3 w/o); both are known to produce strong and
tough TZP ceramics. The reagents used were: $ZrCl_4$ powder, reactor grade
from Teledyne Wah Chang; $YCl_3 \cdot 6H_2O$ from Alfa Products; Baker analyzed
NaOH pellets and anhydrous ethanol from VWR Scientific. A typical coprecipitation procedure is described below.

The $ZrCl_4$ powder is dissolved in ethanol, preferably with cooling,
since the dissolution is strongly exothermic. The color of this solution
slowly changes from pale yellow to deep purple, indicating the formation
of light absorbing chemical species, whose nature has not been identified
(probably chloro-alkoxides). Subsequently $YCl_3 \cdot 6H_2O$ is dissolved in the
appropriate amount.

A second solution of NaOH in ethanol is prepared by dissolving an
amount of caustic soda which will react stoichiometrically with the Zr/Y
salts. The two solutions are mixed, for example by using a peristaltic
pump, or, alternatively, the solution containing the Zr and Y salts is
poured into the other one under vigorous stirring. Adequate cooling is
necessary to prevent the liquid from boiling, being the reaction very
exothermic. Glass containers for the alkaline solution should be avoided
to prevent contamination with silica.

Upon mixing the two solutions an instantaneous coprecipitation of
the Zr and Y hydrous oxides takes place, accompanied by the formation of
NaCl crystals which are sparingly soluble in ethanol. The suspension is
filtered to remove the liquid and the wet cake is dried at low temperature,
preferably a steam bath, in order to avoid the pyrolysis of the alcohol
and discoloration of the powder. (Carbon formation must be avoided,
because the powder coarsens during calcination). Finally, the salt
contained in the dry cake is extracted by washing with water.

Besides ethanol, also methanol can be used, in which NaOH is more
soluble. However, methanol is more dangerous than ethanol and tends to
remain adsorbed on the high surface area of the powder, even after
repeated washings.

It is interesting to note that no gel is formed when the salt
containing cake is washed with water or with an ammonium hydroxide
solution. Boiling with water or water/ammonia does not seem to have any

adverse effect either. On the other hand, washing with acidic solutions causes peptization, since the hydrous oxides are more soluble. Peptization is observed also if the cake is washed with water when it still contains some alcohol. This may be explained remembering that in the perfectly dry cake the hydroxide particles are bound by OH bridges, which prevent the water molecules from penetrating and causing peptization and gelling, whereas in the wet cake the ethanol molecules can still be displaced by water.

The powder dried at 100°C is extremely soft and possesses a very open texture, which is the result of the voids left behind after the dissolution of the NaCl crystals. Such highly open structure may be desirable for other applications, but causes problems during pressing and gives low green densities. However, appropriate treatments can correct this problem.

RESULTS AND DISCUSSION

Powder Characterization

Tables I and II show the results of the physical and chemical characterization of a 4 w/o Y_2O_3-ZrO_2 powder coprecipitated by mixing (at room temperature with a peristaltic pump, using equal flow rates) a 10 w/o solution of NaOH in anhydrous ethanol with an equal volume solution of $ZrCl_4$ and $YCl_3 \cdot 6H_2O$ also in anhydrous ethanol.

Table I

Physical Properties of a 4 w/o Y_2O_3 - ZrO_2 Powder
as Function of the Calcination Temperature

Property	110°C	500°C	700°C	900°C
BET Area (m2/g)	239	65	35	13
XRD Phases	am.	tet.	tet.	tet. + mono.
Average Particle Size (μm)	5.4	3.6	3.1	2.5 (*)
Average Particle Size (μm)	1.5	N.A.	0.2	N.A. (**)

(*) Determined with the Coulter Counter TAII, water only.

(**) Determined with the Nicomp analyzer, 1 w/o ammonium tartrate. N.A. = not available

60

Table I shows how certain physical properties are affected by the
calcination temperature. The BET specific surface area, measured by N_2
adsorption, shows a sudden drop at 500°C, caused by the crystallization
occurring at around 450°C, and then gradually decreases due to particle
sintering. The XRD peaks become sharper at higher temperatures and the
background noise decreases, but the characteristic splitting shown by the
tetragonal phase never appears and the XRD pattern coincides with that of
the cubic polymorph (see Figure).[5] On the other hand, the sintered pieces
obtained from this powder clearly show the tetragonal pattern, as well as
the powder with the same composition obtained by rapid solidification from
the melt.[6] Incidentally, other commercially available powders with the
same composition exhibit XRD patterns of the cubic type. The reason for
the absence of the splitting is not clear; probably it is due to poor
crystallization or to the line broadening that masks the tetragonal
doublets.

The particle size distribution and the average particle size are
strongly influenced by the technique used and by the method of sample
preparation. From the data of Table I it is not clear whether the average
particle size measured with the Coulter Counter as a function of the
calcination temperature decreases because the agglomerates are broken down
or because the size of the agglomerates shrinks. It is clear, on the
other hand, that ammonium tartrate is a good dispersing agent, since
the value of 0.2 microns given by the Nicomp particle size analyzer
agrees with the size shown by the SEM photomicrographs.

Table II

Chemical Analysis of the same powder of Table I, dried at 110°C

Chemical Species	Y_2O_3	Na	Mg	Ca	Al	Si	Cl
Concentration	4.1	<100	<50	<100	<50	~10	~5

Note: Y_2O_3 concentration (w/o) determined by atomic absorption.
The other values are in ppm. Chlorides determined by titration
with $AgNO_3$ (Volhard method).

Table II shows the chemical analysis of the same powder, which was extracted with water for 5 days in a Soxhlet extractor. The Y_2O_3 content is very close to the nominal value, indicating that the hydrous oxide is virtually insoluble in water. The degree of purity is very high and it is believed that this is the purest Y_2O_3/ZrO_2 powder ever prepared. Table II also shows that chlorine can be removed by washing better than sodium. Other batches were washed with a perforated basket centrifuge and the impurity content was a little higher probably because this method compacts the cake or because the amount of water was insufficient.

Figure 1 shows the TGA and DTA curves obtained from the same powder. The weight loss vs. T is gradual and the LOI at $1000^{\circ}C$ is about 15 w/o. The DTA curve shows the crystallization exotherm at $430^{\circ}C$.

Figure 2 shows a SEM picture of this powder calcined at $700^{\circ}C$. The state of agglomeration is visible, as well as the fairly uniform particle size distribution. Figure 3 and 4 show SEM photomicrographs of two commercially available powders with the 5.3 w/o Y_2O_3 composition, i.e. the TZ3Y from TOYO-SODA (Japan) and ZYP from ZIRCAR Corporation (USA). The powder prepared with this method and the TZ3Y powder are coprecipitated from chemical precursors and show similar morphologies. The ZYP powder, on the other hand, exhibits a more "blocky" morphology, probably the result of a different synthesis method.

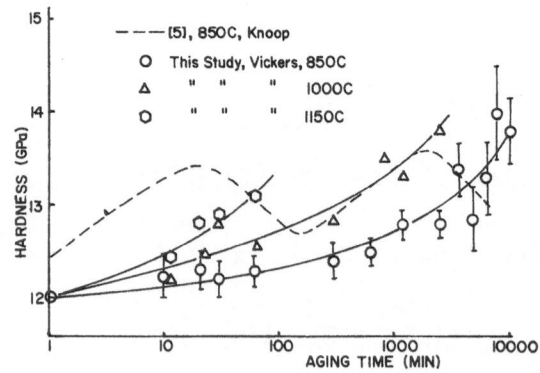

Figure 1

TGA and DTA curves of a 4 w/o Y_2O_3-ZrO_2 powder dried at $110^{\circ}C$.
(Heating rate = $5^{\circ}C/min.$)

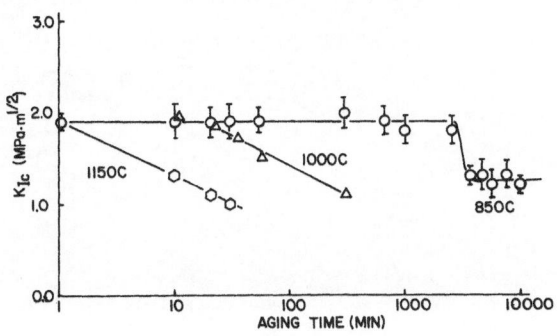

Figure 2

SEM photomicrograph of a 4 w/o Y_2O_3-ZrO_2
powder calcined at $700^{\circ}C$

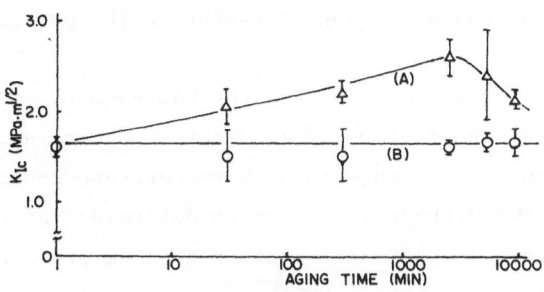

Figure 3

SEM photomicrograph of the TZ 3Y (5.3 w/o Y_2O_3) Powder

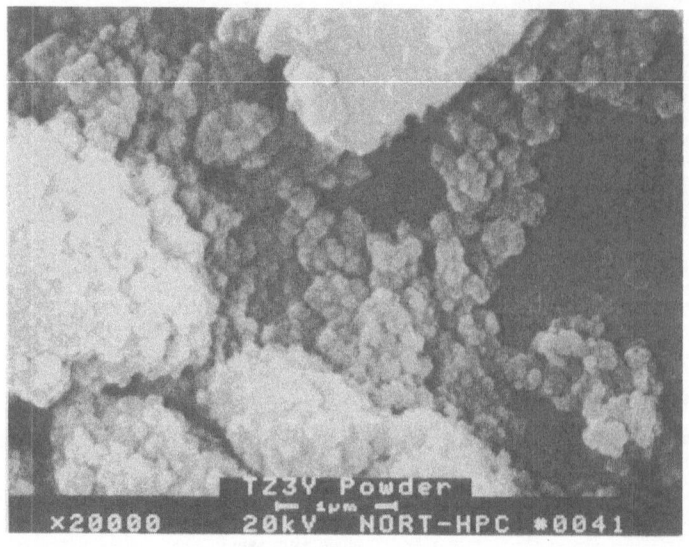

Figure 4

SEM photomicrograph of the ZYP (5.3 w/o Y_2O_3) powder

Figure 5 shows the XRD patterns of a 3 m/o Y_2O_3 powder, calcined at $700^{\circ}C$, and of a sample with the same composition hot pressed at $1470^{\circ}C$. The pattern of the hot pressed sample is virtually identical to that of a 100% tetragonal zirconia reported by Gupta et al[7]. As mentioned previously, the typical splitting is absent in the powder sample.

An interesting question concerns the homogeneity of the yttria distribution in the chemically coprecipitated powders and how it compares with that of powders prepared with high temperature routes, i.e. rapid solidification. Two methods were used to determine the chemical homogeneity, microprobe with EDS and scanning Auger microscopy (SAM). Difficulties were encountered with both methods and a clear answer could not be obtained. Secondary ion mass spectroscopy (SIMS) is probably the best technique for solving this problem.

Sintering

The uncalcined, amorphous powders cannot be pressed and sintered, since the samples disintegrate upon firing, even when a heating rate as low as $50^{\circ}C/hr$ is used. It is not known at what temperature the cracking occurs, but it is believed that the cause is the sudden shrinkage caused by the crystallization around $450^{\circ}C$.

Calcination at about 700°C gave the highest sintered densities, about 95% of TD after firing at 1600°C for 1 hour. The green density after uniaxial pressing at 35 MPa was typically around 28-30% of TD and increased to about 40% of TD after cold isostatic pressing at 207 MPa. Surprisingly the fired densities of the isopressed samples were always lower, which could be explained by the fact that residual gases were not able to escape from a denser green piece. In the most recent experiments appropriate deagglomeration procedures have resulted in fired densities of 98-99% of TD. Figure 6 shows the dilatometer curves of two bars die pressed at 34 MPa, one from the powder described above and the other from the Toyo-Soda TZ3Y powder. It is interesting to note that the bar pressed from the powder prepared with this method was densified to 97% of TD, despite the fact that its green density was much lower than that of the TZ3Y bar. However, a direct comparison cannot be made, since the two samples have different compositions (2.2 and 3.0 m/o Y_2O_3).

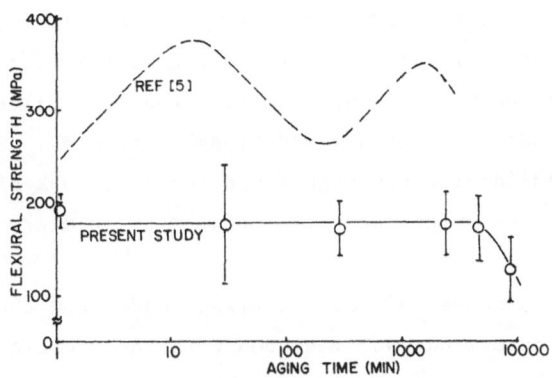

Figure 5

XRD Patterns of (A) 3 m/o Y_2O_3-ZrO_2 powder calcined at 700°C and (B) sample with the same composition, hot pressed at 1470°C.

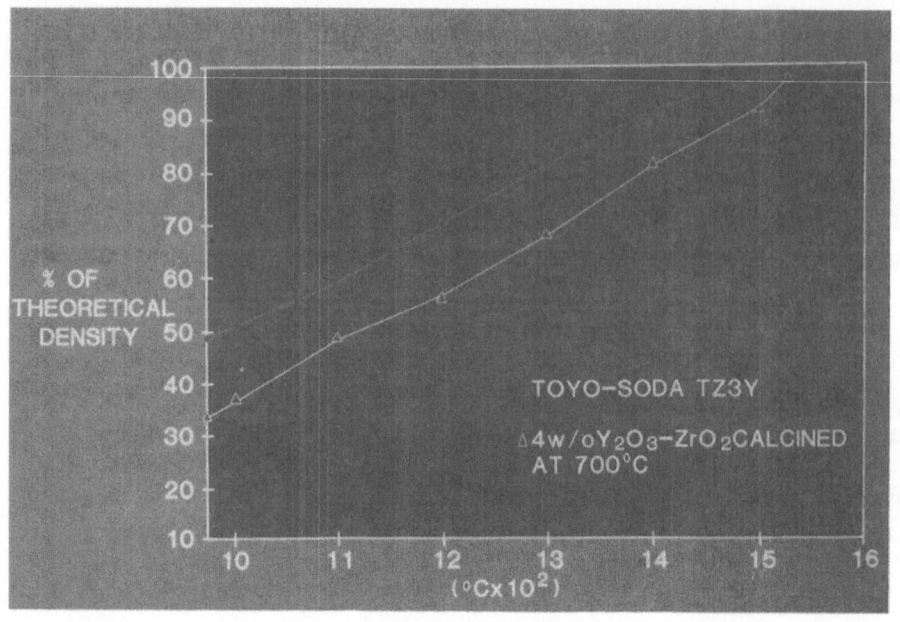

Figure 6

Dilatometer curves for the Toyo-Soda TZ3Y and 4 w/o Y_2O_3-ZrO_2 calcined
at 700°C. (Bars die-pressed at 34 MPa, heating rate 5°C/min.)

Microstructure and Mechanical Properties

Figure 7 shows the microstructure of a polished and thermally etched
section of sintered tile (1600°C, 1 hour, 300°C/hr rate) made from the
4 w/o Y_2O_3-ZrO_2 powder previously described. The tile was die pressed
first at 35 MPa and then cold isopressed at 207 MPa. Figure 8 shows a
similar microstructure exhibited by a TZ3Y tile made in Japan. These tiles
had a MOR (4 points) of 620 and 854 MPa, respectively. Recent improvements
in processing have resulted in MOR values as high as 1157 MPa (3 point).
The obvious importance of the right processing is demonstrated by the low
MOR values (400 to 600 MPa) measured on sintered TZ3Y tiles, pressed using
the as received powder without any additional treatment. Such low values
were unexpected considering the high fired density, 99% of TD, which
confirms once more the importance of a uniform green microstructure.

The fracture toughness of several tiles, made from the 4 w/o Y_2O_3
powder previously described, was measured with the microindentation method
and values in the 5-6 MPa.$m^{\frac{1}{2}}$ range were obtained. Similar values are
exhibited by the best TZP ceramics, i.e. the Japanese Toray, Toyo-Soda,
NGK materials.

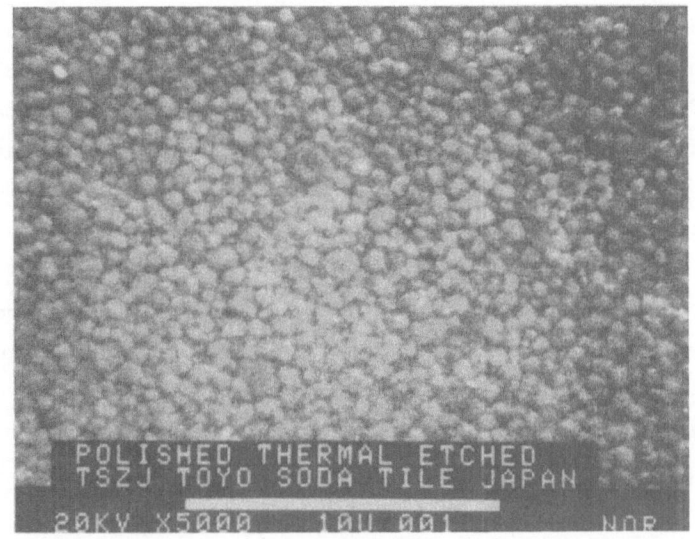

Figure 7

Polished and thermally etched surface of a 5.3 w/o Y_2O_3-ZrO_2 tile
sintered at $1600^{o}C$, 1 hour (5000X)

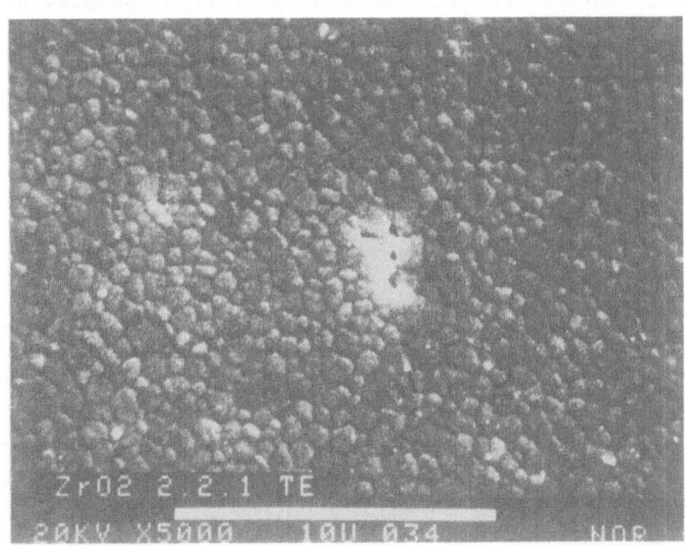

Figure 8

Polished and thermally etched surface of a 5.3 w/o Y_2O_3-ZrO_2 (TZ3Y)
tile from Toyo-Soda (5000X). Sintering conditions not known.

CONCLUSIONS

The method described in this paper can be used to synthesize highly
pure sub-micron, sinterable, doped zirconia powders. The relatively high
cost of the $ZrCl_4$ precursor is compensated by the lack of gel formation,
and the ease of salt removal by washing. Long and expensive operations,
such as azeotropic distillation, freeze-drying, supercritical drying,
hydrothermal treatments, are unnecessary.

The powder agglomeration is a problem, but appropriate treatments
result in higher fired densities and stronger ceramics. Work is underway
to optimize the powder properties, either at the synthesize stage or
after the preparation.

REFERENCES

1. H. K. Bowen and coworkers. (Ceramic Processing Research Laboratory,
   M.I.T. Dept. of Materials Science and Engineering, Cambridge, MA 01239.)
2. I. A. Aksay, University of Washington, Dept. of Materials Science and
   Engineering, Seattle, WA 98195.
3. H. Bernard and J. C. Viguie: U.S. Patent 4,365,011.
4. Information published by Toyo-Soda Manufacturing Company, (Japan).
5. Giulio A. Rossi: U.S. Patent 4,501,818
6. Personal communication from C. E. Knapp, Norton Company, 8001 Daley
   Street, Niagara Falls, Ontario, Canada.
7. T. K. Gupta, J. H. Bechtold, R. C. Kuznicki, L. H. Cadoff, B. R. Rossing,
   J. Mat. Sci. 12 (1977) 2421-2426.

MICROSTRUCTURAL AND CHEMICAL ASPECTS OF A STRONTIA

SINTERING AID ON Mg-PSZ†

R.H.J. Hannink and J. Drennan

CSIRO, Division of Materials Science
Advanced Materials Laboratory
P.O. Box 4331, G.P.O. Melbourne, Victoria 3001, Australia

ABSTRACT

The addition of SrO to magnesia-partially stabilized zirconia (Mg-PSZ) retains the beneficial effects of the contaminant silica as a sintering aid, yet causes the aid to be purged from the sintering body. The advantages of the SrO addition include increased fracture strength and density, a reduction in grain size and slowing down of the decomposition process which accompanies the sub-eutectoid ageing process. This paper describes the microstructural and chemical events leading to the overall improved material.

INTRODUCTION

Magnesia-partially stabilized zirconia (Mg-PSZ) is well established as an advanced engineering ceramic (1-3). The high strength and toughness are derived from the controlled transformation of metastable tetragonal precipitates to the room temperature stable monoclinic form, whilst being constrained in a cubic stabilized zirconia (CSZ) matrix. The transformation is accompanied by a volume dilation and shear strain which results in a compressive 'shield' around a propagating crack or a compressive layer on mechanically worked surfaces. A further advantage of the material is that its microstructure may be tailored by suitable post sintering heat treatment, to suit particular industrial applications (4).

Mg-PSZ is generally fabricated along conventional powder routes, starting with zirconia powder plus the addition of a stabilizer (magnesia in the case of Mg-PSZ). A common contaminant in the zirconia powders, is silica (apart from about 2% hafnia). Although the amount of SiO_2 is kept to a minimum, very few commercial powders can claim a SiO_2 content of less than 0.5wt%. The chemical and structural nature of the silica present in zirconia is not well understood, but it is known to be well dispersed and in a very active state (5).

---

†Subject of Int. Pat. App. No. PCT/AU83/00069, Hughan et al.

A number of studies have been carried out to determine the effects of silica on the structural behaviour of CSZ and PSZ materials (6-11). All the studies found that the presence of silica aided in the consolidation and sinterability of a zirconia body by the formation of a glass phase. The glass phase was found to be rich in the stabilizer oxide. Hence sintering was achieved through a liquid phase sintering process, which is a well known mechanism for achieving improved sintering parameters. Associated with the removal of stabilizer for glass formation are a number of detrimental side effects, all of which are directly reflected in a deterioration of the thermomechanical properties of the ceramic (8,11).

To illustrate the beneficial effects of the SrO addition on the mechanical properties of Mg-PSZ two sets of data are presented in Fig. 1. This figure shows the modulus of rupture (MOR) versus ageing time at 1100°C, for the same materials with and without SrO. From this figure it can be seen that the addition of SrO has the dual effect of: (a) increasing the maximum attainable MOR and (b) slowing the degredation in the mechanical properties which accompanies the sub-eutectoid decomposition process.

In this paper we describe the microstructural and chemical aspects of a small addition of SrO sintering aid to Mg-PSZ and show how the beneficial sintering aid characteristics of the silica contaminant are maintained whilst its detrimental effects are predominantly nullified.

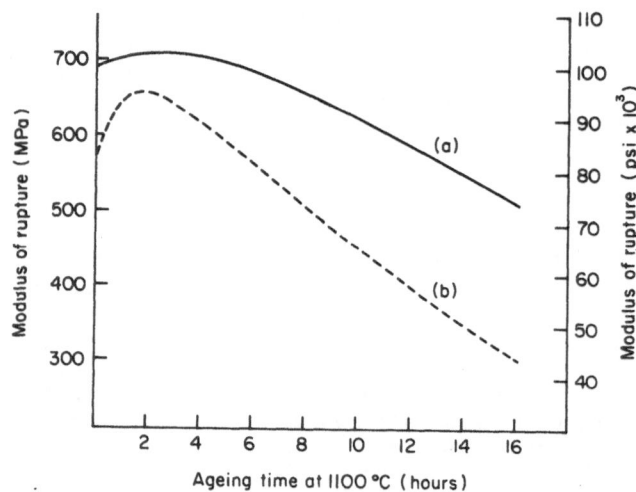

Fig. 1. Modulus of rupture versus ageing time at 1100°C for Mg-PSZ: (a) with SrO; and (b) without SrO additions.

EXPERIMENTAL

Zirconia powder of known silica content (0.05wt%) was mixed with 3.4wt% MgO and 0.25wt% SrO, by thorough ball milling. These powders were fabricated in to test specimen bars by isopressing, sintering at 1700°C and control cooled to room temperature. The bars were further heat treated (aged) at 1100°C, to optimise the mechanical properties (4).

The samples were characterized using optical microscopy and

scanning and transmission electron microscopes fitted with x-ray energy dispersive analytical facilities (EDA).

EDA was used extensively to determine the chemical nature of the grain boundary phases observed in the fabricated Mg-PSZ samples. Because of the close proximity of the Si $K\alpha_{1,2}$ (1.739keV) and the Sr $L\alpha_1$) (1.806keV) peaks, stripping techniques were used to determine the coexistence of Si and Sr in analysed regions. The presence of Sr could be established by observing the Sr $K_{\alpha 1}$ peak at 14.163keV. To aid the peak stripping procedure, standard peak profiles were collected from pure MgO, SrO, SiO_2 and ZrO_2 samples, under identical instrumental conditions.

Trace elemental analysis was performed using photoelectron spectroscopy (ESCA) of polished and freshly cleaved surfaces. Sample analysis was performed in a vacuum of 10^{-8} torr or better at room temperature. Quantitative elemental ratios were determined from known cross sections and the instrumental transmission function.

RESULTS AND DISCUSSION

The Effects of SrO on Grain Boundaries

The effect of the SrO addition, on the consolidation behaviour of Mg-PSZ, is most readily observed by an examination of the central core and surface regions of a test specimen. Fig. 2 shows a series of optical micrographs taken from a region along a tapered section of a test specimen starting from the as-fired surface. As can be seen from the figure the as-fired grain boundaries are coated with a film which is seen to extend into the bulk of the material. Not until approximately 30 µm into the bulk of the ceramic does this grain boundary phase become unresolvable by optical microscopy. Below this depth the usual microstructure of Mg-PSZ is observed ie. that of a dense ceramic with closed porosity, of ~ 5%.

To examine the regions described in Fig. 2 in more detail, transmission electron microscope specimens were prepared from samples cut close to the as-fired surface and from the bulk of the material. The recorded micrographs are shown in Fig. 3 (a & b). Figure 3(a). from the bulk of the ceramic, reveals very clean boundaries with only the occasional particle of isolated second phase being present (arrowed). These isolated particles were in the main, either crystalline regions of forsterite (Mg_2SiO_4) or crystalline regions of an unidentified material rich in Sr, Zr and containing some Mg, the phase may possibly be a strontium-magnesium-zirconate. The boundary phases close to the surface of the ceramic piece were continuous, as can be seen in Fig. 3(b), and consisted primarily of an amorphous/glass phase rich in Sr and Si and containing a small amount of Mg. Intermediate stages between the surface and core forms were observed, i.e. the presence of all three phases. This observation provided evidence that the glass phase formation can be a result of SrO interaction with the forsterite (12), and is important as it demonstrates the affinity of the SrO for SiO_2, whether the SiO_2 occurs as free SiO_2 or as part of a compound.

Independent experiments on the reaction of SrO with forsterite have shown that reactions such as the one below readily occur:

$$2Mg_2SiO_4 + 3SrO \rightarrow (1700°C) \quad Sr_3Si_2MgO_8 + 3MgO$$

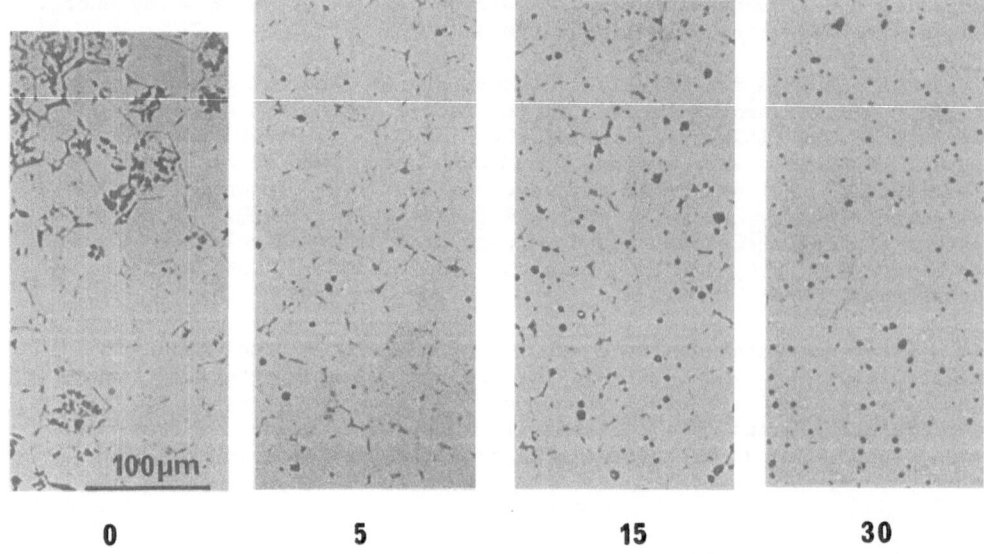

0   **5**   **15**   **30**

Fig. 2. Optical micrographs of ~ 5° tapered section showing the
presence of grain boundary phases at various depths into the samples.
Numbers under images indicate depth below sintered surface, in µm.

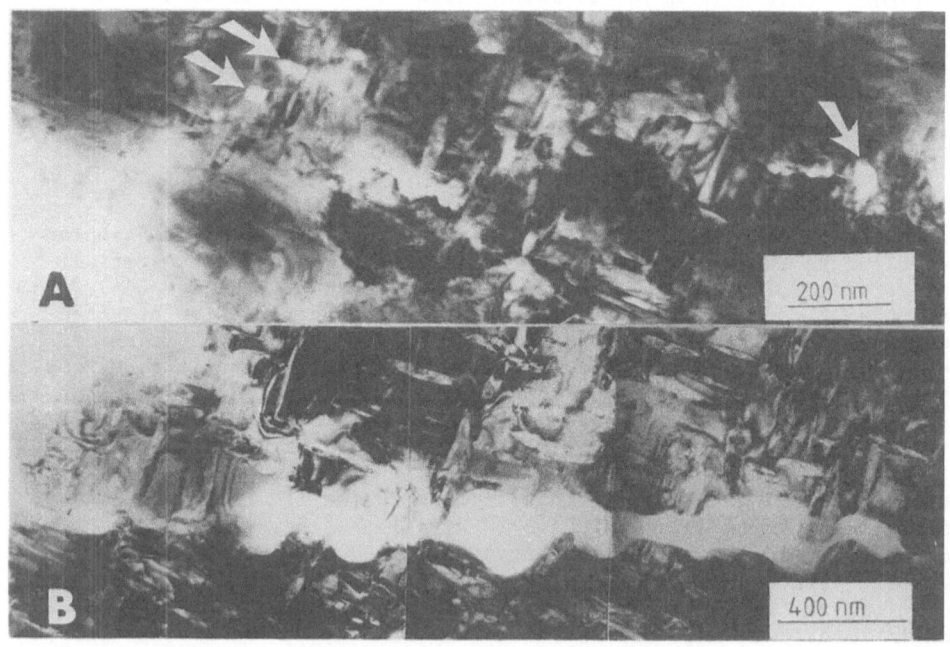

Fig. 3. Transmission electron micrograph (bright field) images of grain
boundaries in a sintered Mg-PSZ material with SrO additions,
(a) from within the bulk of the sample and (b) near the surface.

Figure 4(a to c) summarizes the elemental contents of the phases observed in the TEM, using EDA. Also shown in Fig. 4 are the corresponding selected area electron diffraction patterns. The nature of the small isolated particles in the core of the specimen (Fig. 3 (a)) is not fully understood. EDA analysis indicated a roughly equal Sr/Zr ratio for these particles suggesting possibly $SrZrO_3$. The electron diffraction patterns however, could not be indexed on the basis of the ideal $SrZrO_3$ unit cell, and as mentioned previously may be a $SrO-MgO-ZrO_2$ compound (Fig. 4(a)). EDA analyses of the glass phase near the surface is shown in Fig. 4(b), with its corresponding electron diffraction pattern.

In material to which no SrO has been added, triple point and grain boundaries clearly contained a second phase throughout the material; the extent of which could be directly related to the amount of silica in the starting powder. Analyses of these phases showed them to be almost entirely composed of forsterite, Mg_2SiO_4, (Fig. 4(c)).

Exuded Phases
_____

In addition to the grain boundary phases just described, a supplementary observation provided further evidence of the effects of SrO on Mg-PSZ. At the completion of normal commercial firings, solid glassy droplets containing crystalline phases are almost always observed on the bottom edges of sintered components. One such drip was removed from a large fabricated piece (2-3Kg) and subjected to analysis.

The interface between the drip and Mg-PSZ substrate is shown in Fig. 5. The analysis of the accumulated drip revealed a complex assemblage of crystalline and amorphous phases containing Sr, Si, Mg, P, Na and trace amount of Ti with Zr. The presence of the P was determined from ESCA analysis since in EDA the P $K\alpha_{1,2}$ (2.013keV) and the Zr $L\alpha_{1,2}$ (2.042 KeV) almost overlap. It is the presence of these trace elements which can result in the formation of the $SrO-SiO_2-X$ glass phase.

Powder X-ray diffraction patterns taken from the drip material, confirmed the existence of P by the identification of the compound $\beta-[Sr(Mg)]_3(PO_4)_2$. Minor amounts of two other crystalline phases were also identified these being $Sr_2MgSi_2O_7$ and Mg_2SiO_4.

Chemistry of the SiO_2/SrO reactions
_____

The preferential reaction of SrO with SiO_2 is fundamental to the explanation for its effectiveness as a sintering aid in Mg-PSZ. At the beginning of the results section evidence for the affinity of SrO to SiO_2 was described. To examine this affinity in more detail, a series of samples consisting of MgO, $SrCO_3$ and SiO_2 (2,1, and 1 molar ratio respectively) were mixed aand reacted at various temperatures and the products subjected to x-ray analyses. At 1100°C $Sr_3Si_2MgO_8$ and MgO were the only phases detected. Between 1200 to 1400°C in addition to the two phases already mentioned an additional phase, $Sr_2Si_2MgO_7$, was detected.

As can be seen from these results, x-ray reflections corresponding to those expected from forsterite were not observed within the temperature range examined. It is well known that in Mg-PSZ materials to which no SrO has been added, the major grain boundary phase observed is forsterite. This suggests that without the presence of SrO the reaction of SiO_2 with MgO is favoured. Stability of $SrO-SiO_2$ compounds is much greater than for the $MgO-SiO_2$ compounds (12).

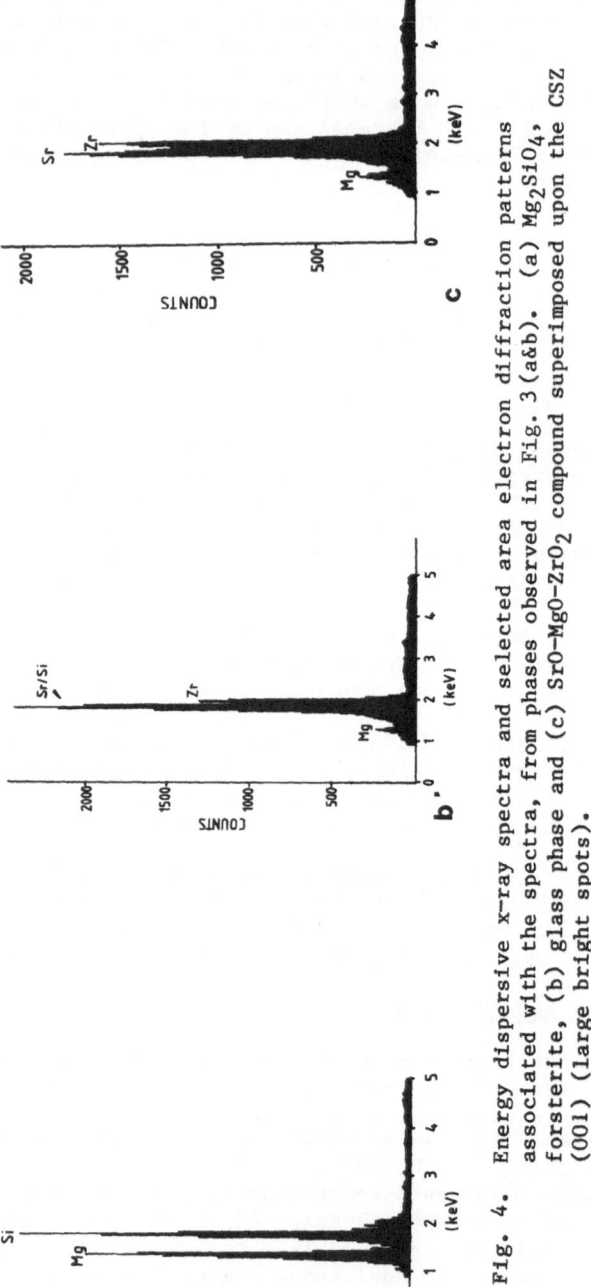

Fig. 4. Energy dispersive x-ray spectra and selected area electron diffraction patterns associated with the spectra, from phases observed in Fig. 3 (a&b). (a) Mg_2SiO_4, forsterite, (b) glass phase and (c) $SrO-MgO-ZrO_2$ compound superimposed upon the CSZ (001) (large bright spots).

The presence of Sr/Si glasses as opposed to crystalline compounds requires some discussion. It is possible that the phases examined by EDA contained some traces of glass forming elements. As mentioned Na and P were detected by ESCA within the glass drip exuded from the fabricated component. Some question could arise as to the occurrence of Na an P. The elements are inherent in the initial starting powders and are not thought to arise as a function of the furnace refractory environment. The exuded product forms irrespective of the type of furnace refractory.

Whilst most of the glass phase has been exuded onto the surface a microscopic film will remain. As Lange (13) has pointed out, within the normal lifetime of a sintering experiment, a glassy phase once formed, will always be present even if the thickness of the film is very small. The actual mechanism by which the glass phase finds its way to the surface is not well understood. A possible driving force for the migration to the surface may be the replacement of volatile species, e.g. Na and P. These may be lost from the surface of the ceramic during the high temperature sintering process and replaced by those present within the bulk.

More detailed examination of the grain boundary chemistry could lead to a much better understanding of the sintering of zirconia alloys and the determination of beneficial sintering aids for other ceramic systems.

Effect of SrO on Decomposition of Mg-PSZ

As stated previously the optimal mechanical properties are obtained in Mg-PSZ following a sub-eutectoid ageing treatment at 1100°C. The beneficial nature of the SrO additions were shown in Fig. 1, in terms of the maximum MOR attainable in two alloys of the same composition with and without SrO. Fig. 6 shows the difference in the optical microstructure of the two materials both aged for 32h at 1100°C.

Fig. 5. Constituent phases in a 'drip' on the surface of a large fabricated component of Mg-PSZ with SrO additions. See text for descriptions.

The white regions delineating the grain boundaries, in Fig. 6 (a & b), are decomposed grain material consisting essentially of monoclinic ZrO_2 containing MgO rods (14). A number of alternate interpretations, as to the effects of SrO on this decomposition behaviour are possible. A full discussion will be presented elsewhere (12). However, the most plausible interpretation is presented here. It has previously been observed that nucleation for the decomposition reaction is favoured at heterogeneous sites, e.g. grain boundaries and pores (14). The effect of SrO is to produce 'clean' grain boundaries thus removing possible heterogeneous nucleation sites. This interpretation may find support from such observations shown in Fig. 7 (a & b). These figures show the decomposition at various sites in two materials, with and without SrO. From such observations it became apparent that in the non-strontiated material, decomposition nucleation occurred more frequently as a result of 'dirty' boundaries, Fig. 7(b). It can be seen from this figure that the boundaries in this material are very porous, in part, the result of phase removal by etching. This second phase contamination provides copious sites for the nucleation of the decompositon reaction. In the strontiated material, decomposition was associated with pores and second phase monoclinic zirconia particles along the boundary. Decomposition nucleation did not appear to occur when the boundaries were obviously clean (as indicated by the arrows in Fig. 7 (a)). These features of the decomposition and their influence on kinetics are further discussed elsewhere (15).

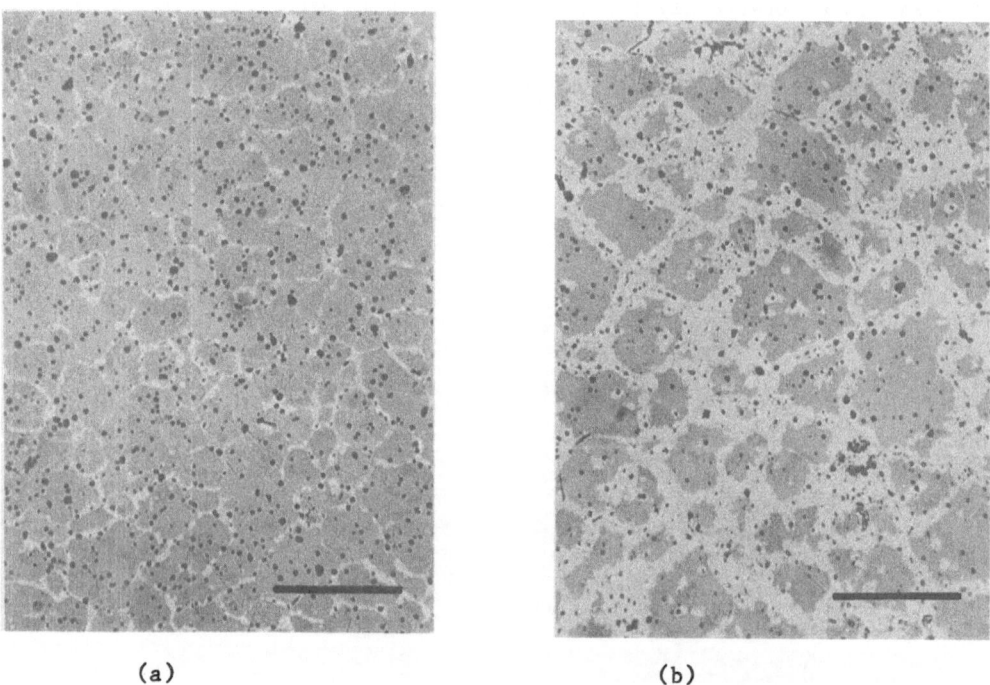

(a)   (b)

Fig. 6. Effect of SrO on the decomposition kinetics of two 9.7 mol %
         Mg-PSZ alloys, (a) with SrO and (b) without SrO. Bar length
         equals 100 µm. Note the difference in grain size between
         (a) & (b). The reduction in grain size is another benefit of
         the SrO additions (12).

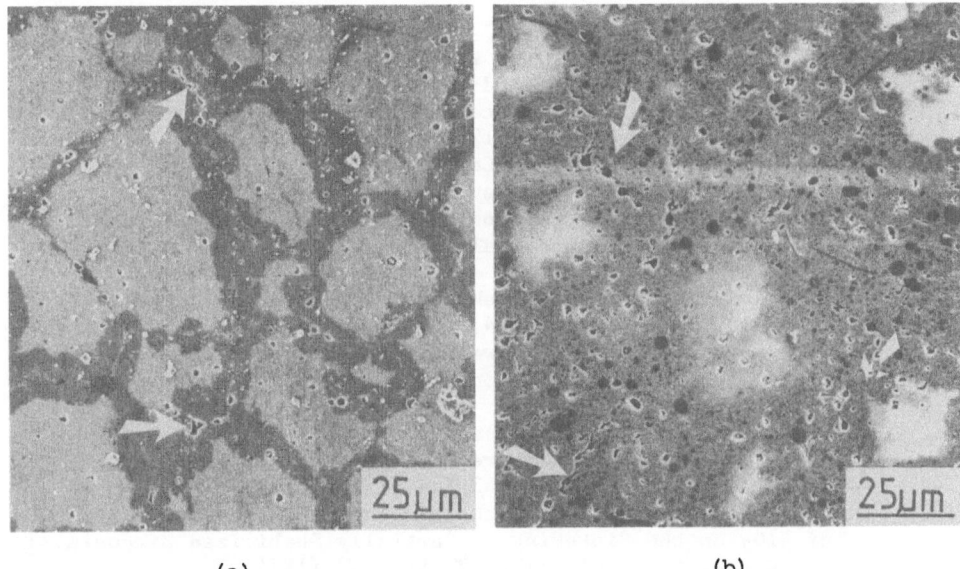

(a)	(b)

Fig. 7. Scanning electron microscope images of the decomposition
nucleation sites (arrowed) in two Mg–PSZ materials, (a) with
SrO, clean boundaries, aged 16 hrs at 1100°C; (b) without SiO,
dirty boundaries, 8 hrs at 1100°C.

SUMMARY AND CONCLUSIONS

The beneficial effects of SrO on the sintering behaviour and final
mechanical properties of PSZ alloys are the result of a preferential
reaction of SrO with SiO_2 contaminant in original ZrO_2 starting
powders. The glassy phase which is formed as a result of the strontia-
silica reaction also scavenges other impurities from the consolidating
body.

The major portion of the glassy phase which is formed is exuded on
to the sintered component surface, and is subsequently removed in the
final mechanical dressing operation. The exudant leaves 'clean' grain
boundaries. This has the effect of lowering the population of possible
nucleation sites for the decomposition reaction, which is a part of the
sub-eutectoid ageing process at 1100°C. The benefit of the clean grain
boundaries appears to result in a reduction in the overall
decomposition kinetics.

The addition of SrO sintering aid the Mg–PSZ induces a marked
improvement in the thermo-mechanical properties.

ACKNOWLEDGEMENTS

The authors take pleasure in acknowledging the assistance of
R.R. Hughan and V. Gross with the sample fabrication and the
experimental programme.

REFERENCES

1. Science and Technology of Zirconia, Advances in Ceramics Vol 3, Eds. A.H. Heuer and L.W. Hobbs, The American Ceramic Society, 1980; also Science and Technology of Zirconia II, Advances in Ceramics Vol 12, Eds. N. Claussen, M. Ruhle and A.H. Heuer, The American Ceramic Society, 1984.

2. R.H.J. Hannink, M.J. Murray and M. Marmach, Magnesia-Partially Stabilized Zirconia (Mg-PSZ) as Wear Resistant Materials, Wear of Materials 1983, Ed. K.C. Ludema, ASME, NY 1983.

3. P.A. Janeway, PSZ – A Breakthrough in Toughness, Ceramic Industry, 122 40-45 (1984).

4. R.H.J. Hannink and M.V. Swain, Particle Toughening in Partially Stabilized Zirconia: Influence of Thermal History, this volume.

5. J. Drennan, unpublished results.

6. T.W. Smoot, D.S. Whittemore, Destabilization of Zirconia, J. Am. Ceram. Soc., 48 163-164 (1965).

7. J.F. Shackelford, P.S. Nicholson and W.W. Sheltzer, Influence of SiO_2 on the Sintering of Partially Stabilized Zirconia, J. Am. Ceram. Soc. Bulletin, 53 865-867 (1974).

8. F.J. Esper and K.H. Friese, The Relationship Between Texture Parameters and Certain Properties of Lime Stabilized Zirconia with Silica Additions, Ber. Dt. Keram. Ges., 55 314-316 (1978).

9. K.C. Radford and R.J. Bratton, Zirconia Electrolyte Cells Part I: Sintering Studies, J. Mat. Sci. 14 59-65 (1979).

10. D. Mallinkrodt, P. Reynen and C. Zorgrafou, The Effect of Impurities on Sintering and Stabilization of ZrO_2 (CaO), INTERCERAM Nr2 125-129 (1982).

11. D. Dou, P.S. Pacey, C.R. Masson and B.R. Marple, Grain Growth in CaO-Stablized ZrO_2 in the Presence of a Liquid Phase, J. Am. Ceram. Soc., 68 C.80-C.81 (1983).

12. J. Drennan and R.H.J. Hannink, The Effect SrO Additions on the Microstructure and Mechanical Properties of Mg-PSZ, submitted J. Am. Ceram. Soc., 1985.

13. F.F. Lange, Liquid Phase Sintering: Are Liquids Squeezed out from Between Compressed Particles? J. Am. Ceram. Soc. – Comm, [2] c-23 (1982).

14. R.H.J. Hannink, Microstructural Development of Sub-eutectoid Aged MgO-ZrO_2 Alloys, J. Mat. Sci., 18, 457-70 (1983).

15. S. Farmer, A.H. Heuer and R.H.J. Hannink, Sub-Eutectoid Behaviour of MgO-ZrO_2 Alloys – A Microstructural Investigation, submitted J. Am. Ceram. Soc., 1985.

PHASE RELATIONSHIPS IN Y-Si-Al-O-N CERAMICS

D.P. Thompson

Wolfson Research Group for High-Strength Materials
Department of Metallurgy and Engineering Materials
University of Newcastle upon Tyne, UK

INTRODUCTION

During the last fifteen years of silicon nitride development, yttrium oxide has emerged as the best sintering additive for producing fully dense materials which combine good room-temperature strength with reasonable high temperature strength at temperatures in excess of 1000°C. β'-sialons, which offer advantages over silicon nitride of easier fabrication, also require an additive for densification and again, yttria has proved to be the most satisfactory. The high temperature properties of these materials are determined mainly by the grain boundary phase assemblage and this in turn is determined by phase relationships in the yttrium sialon system at the sintering temperature, at any post-preparative heat treatment temperature and also at the likely operating temperature of the material. The amount of work involved in determining phase relationships in a five-component system over a wide temperature range is extremely large; researchers have therefore focused their attentions on localised regions of the system which are of particular relevance to commercial materials. A good summary of previous work in the yttrium sialon and other sialon systems has been given by Jack[1]. The present paper summarises the current state of understanding of phase relationships in the yttrium sialon system and includes more recent work carried out at Newcastle on previously unexplored regions of the system.

THE Y-Si-O-N SYSTEM

Early behaviour diagrams for the Si_3N_4-SiO_2-Y_2O_3 system (Tsuge et al.[2], Wills et al.[3], Jack[4], Lange et al.[5]) show preliminary observations based on relatively few experimental results and in some cases there are errors in tie lines and phase compositions. The more detailed work of Gauckler et al.[6] (Figure 1) and the Newcastle group (Figure 2) has confirmed the compositions of the four quaternary oxynitrides and the tie lines between them. In contrast to systems involving larger rare earth cations, there is no yttrium phase of the type $Y_2Si_6O_3N_8$ nor is there a pure oxide apatite of composition $Y_{9.33}(SiO_4)_6O_2$. The Y_2O_3-SiO_2 system shows compounds at only the 1:1 and 1:2 positions but the apparent simplicity of the system is complicated by polymorphism and there are two different forms of Y_2SiO_5 and up to seven modifications of $Y_2Si_2O_7$. The Y_2SiO_5 polymorphs correspond to both types of structure observed in rare earth orthosilicates[7] and the four characterized modifications of $Y_2Si_2O_7$ ($\alpha,\beta,\gamma,\delta$) correspond to the four structure types

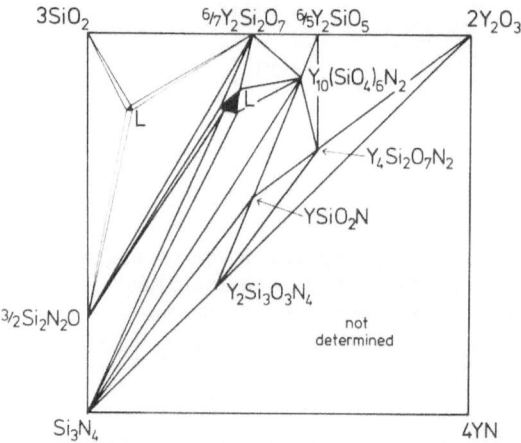

Fig. 1. Phase relationships in the Si_3N_4-SiO_2-Y_2O_3 system at 1550°C (after Gauckler et al.[6])

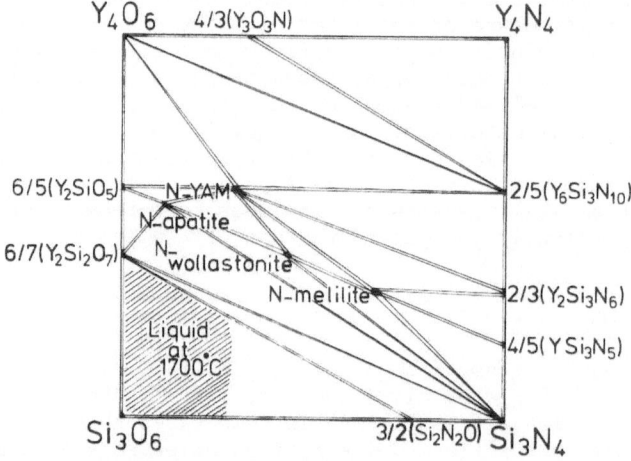

Fig. 2. Phase relationships in the Si_3N_4-SiO_2-Y_2O_3-YN system at 1700°C.

observed in the high atomic number rare earth disilicates[7]. Further characterisation is needed for the remaining three forms of $Y_2Si_2O_7$ (y[8], z[9] and c[8]).
The effect of transformations between the various polymorphs of $Y_2Si_2O_7$ is
important for low z-value β'-sialons and O'-sialons[11] where this phase occurs
after devitrification of grain boundary glass.

No attempt has been made to map out accurately the region of liquid
phase in Figure 2. Gauckler et al.[6] determined the eutectic temperature
along the Si_3N_4-$Y_2Si_2O_7$ join to be 1520°C and the eutectic in the SiO_2-Si_2N_2O-
$Y_2Si_2O_7$ triangle is probably similar. Some authors[6,12] have specifically
determined "sub-solidus" phase relationships by firing compositions above the
solidus temperature and then heat-treating at sub-solidus temperatures to
devitrify any glassy phase. This procedure does not ensure equilibrium at
the subsolidus temperature and may merely produce a mixture of the crystalline
products after firing plus the crystalline products of glass devitrification.
In many cases however there is very little difference, and the subsolidus
relationships reported by Gauckler et al.[6] are identical to their 1550°C
results. Cao et al.[12] consider that the Si_3N_4-$Y_2Si_2O_7$ tie line should
instead be joining Si_2N_2O and $Y_{10}(SiO_4)_6N_2$. This may be due to the different
devitrification conditions they employed and is a good argument for authors
not to assume that sub-solidus phase relationships are temperature invariant.
Results described later in the present work emphasise this point.

Recent work at Newcastle has extended the system described above to
include the Y_2O_3-YN-Si_3N_4 region. Investigation of this part of the system
involves the use of yttrium nitride as a starting material which in powder
form readily hydrolyses and at moderate temperatures oxidizes to yttrium
oxide. Particular care is therefore needed in specimen preparation. The results of firing compositions in nitrogen at 1750°C are shown in Figure 2 and
indicate the existence of three yttrium silicon nitrides and one yttrium oxynitride. No tie line was observed between N-melilite and yttrium oxide but
instead tie lines extend across the system from $Y_4Si_2O_7N_2$ to $Y_6Si_3N_{10}$ and
$Y_2Si_3N_6$. This is in good agreement with the results of Lange et al.[5] who reported additional uncharacterized phases in this area. Table 1 gives X-ray
data for these new phases, two of which have been characterized.

All phases shown on Figure 2 are stable to at least 1600°C with the exception of N-α-wollastonite which decomposes above ~1400°C to give mainly
$Y_{10}(SiO_4)_6N_2$ and $Y_2Si_3O_3N_4$. Its occurrence in high temperature firings is as
a result of either crystallization on cooling from a liquid phase or devitrification from a glass at sub-solidus temperatures.

THE YN-Si_3N_4-AlN SYSTEM

Reaction between the covalently bonded silicon and aluminium nitrides
and the interstitial yttrium nitride proceeds exceedingly slowly even at
1750°C. It is therefore desirable to have some liquid phase present to aid
reaction and the surface oxides plus additional oxidation of YN provide more
than enough liquid for this purpose. Apart from the ternary yttrium silicon
nitrides reported in the previous section, Figure 3 shows that the only other
phase occurring in this plane is α'-sialon which at 1750°C has a range of
composition extending from y=1.8-3.4 in the general formula $Y_{y/3}Si_{12-y}Al_yN_{16}$.
Further discussion of the homogeneity range of α'-sialon is given in a later
section.

CRYSTALLINE PHASES IN THE 2M:3X PLANE

Several phases which occur either as secondary crystalline phases or as
devitrification products of sialon glasses have metal:non-metal ratios of

Table 1.   X-ray data for Y_3O_3N, YSi_3N_5, $Y_2Si_3N_6$ and $Y_6Si_3N_{10}$

YSi_3N_5

Y_3O_3N

Hexagonal, $\underline{a}$: 9.814, $\underline{c}$: 10.621 Å

d_{obs}	I_{obs}	hkl	d_{calc}	d_{obs}	I_{obs}
6.711	mw	100	8.491	8.511	m
5.892	vvw	101	6.627	6.637	s
5.168	w	002	5.300	5.305	m
5.124	vw	110	4.902	4.905	m
3.882	mw	102	4.496	4.501	m
3.313	vw	111	4.450	4.446	m
3.109	s	200	4.245	4.247	w
2.976	vvs	201	3.941	3.939	s
2.944	s	112	3.599	3.602	w
2.749	s	003	3.540	3.541	w
2.656	w	202	3.313	3.323	w
2.584	w	103	3.262	3.263	s
2.560	w	210	3.209	3.216	w
2.426	mw	211	3.071	3.071	s
.	.	.	.	.	.
.	.	.	.	.	.
.	.	.	.	.	.

$Y_6Si_3N_{10}$

$Y_2Si_3N_6$

Orthorhombic, $\underline{a}$: 5.978, $\underline{b}$: 10.288, $\underline{c}$: 9.854 Å

d_{obs}	I_{obs}	hkl	d_{calc}	d_{obs}	I_{obs}
4.974	m	110	5.168	5.175	w
4.755	ms	020	5.144	5.144	w
4.444	m	111	4.577	4.584	m
4.037	m	021	4.560	4.558	w
3.935	m	112	3.566	3.570	vw
3.651	ms	022	3.558	3.588	vw
3.544	w	200	2.989	2.987	ms
3.451	m	130	2.975	2.970	s
3.244	ms	113	2.772	2.774	vs
3.212	m	023	2.768	2.763	w
3.005	s	220	2.584	2.592	w
2.841	vs	040	2.572	2.580	w
2.717	ms	202	2.555	2.555	s
2.699	m	132	2.547	2.544	s
.	.	.	.	.	.
.	.	.	.	.	.
.	.	.	.	.	.

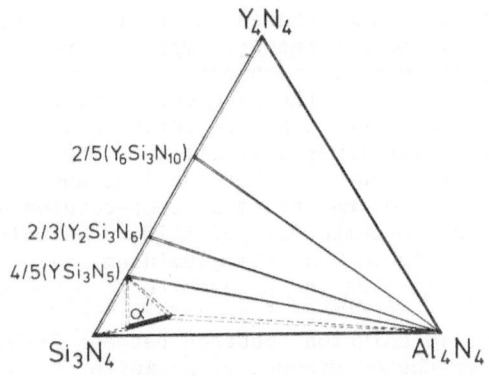

Fig. 3. The YN–Si_3N_4–AlN plane at 1750°C

2:3. In the present section details of composition, thermal stability and homogeneity range of these phases are discussed. Figure 4 shows the plane concerned.

Yttrium aluminium garnet (YAG) is an important secondary phase in sialon ceramics. A range of composition which would offer useful preparational flexibility was originally reported[13] but subsequent Newcastle work has found a much smaller range with a maximum of 0.4 (Si+N) replacing (Al+O) in $Y_3Al_5O_{12}$. YAG offers possibilities of accommodating impurity cations and in fact a wide range of garnet solid solution occurs in nature. However, attempts to incorporate nitrogen into grossular garnet ($Ca_3Al_2Si_3O_{12}$) were equally unsuccessful and approximately the same nitrogen solubility was observed as for YAG.

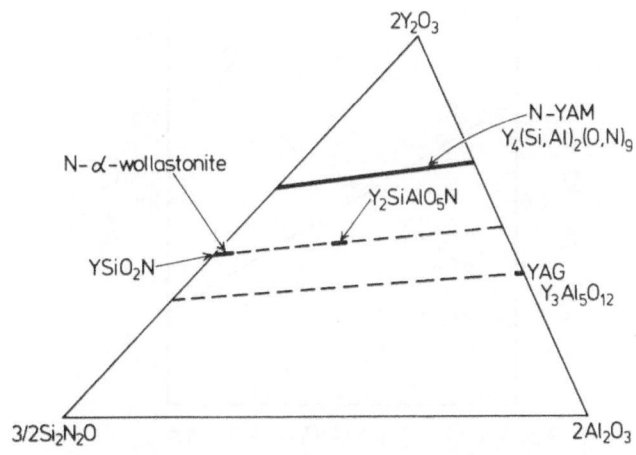

Fig. 4. 2M:3X phases in the Y_2O_3–Si_2N_2O–Al_2O_3 plane

Along the $YAlO_3$–$YSiO_2N$ join, phases occur with the structures of α–wollastonite and perovskite. Both structure types are very stable when the yttrium is replaced by large rare earth cations (La,Ce) but their stability decreases with decreasing size of the rare earth cation. Yttrium, which is equivalent in chemical behaviour to a rare earth cation of intermediate size, forms both compounds but with limited stability. The perovskite form of $YAlO_3$ has been reported to occur only between 1835 and 1875°C[14] but in fact it can occur in yttrium sialon reactions at temperatures as low as 1700°C. The low temperature α–wollastonite form of $YAlO_3$[15] is only stable up to temperatures of ~1050°C. As soon as the aluminium and oxygen in $YAlO_3$ are replaced by silicon and nitrogen the perovskite form ceases to occur but the stability of the α–wollastonite form increases. Rae[16] believed that a complete α–wollastonite solid solution occurred between $YAlO_3$ and $YSiO_2N$; this observation was based on the occurrence of a similar phase intermediate in unit cell dimensions at the composition Y_2SiAlO_5N. Recent work on the devitrification of yttrium sialon glasses has shown that there is virtually no solid solubility of Y_2SiAlO_5N in $YSiO_2N$ or vice versa but that Y_2SiAlO_5N does have some range of homogeneity extending towards $YAlO_3$ as shown in Figure 5. (In this diagram the equivalent hexagonal unit cell dimensions have been used for ease of comparison.) This observation is in good agreement with the crystal structures of these phases, because $YSiO_2N$ has four layers of $[Si_3O_6N_3]$ rings per c repeat whereas Y_2SiAlO_5N and $YAlO_3$ are isostructural with only two layers of $[(Si,Al)_3(O,N)_9]$ rings per repeat.

The remaining series of compounds in the 2M:3X plane occurs between $Y_4Si_2O_7N_2$ and $Y_4Al_2O_9$. It is generally believed that solid solution occurs all the way between these two end-members even though the variation of unit cell dimensions with composition is not linear (Figure 6). Indeed some compositions near the centre of the range and prepared by devitrification of yttrium-rich yttrium sialon glasses (have X-ray patterns which index) on a tetragonal unit cell (see Table 2) related to the monoclinic cell (a:7.56, b:10.45, c:10.81 Å; β:111°) by the relationships:

$$a_T = b_M = c_M \sin\beta \sim 10.5 \text{ Å}; \qquad c_T = \tfrac{1}{2}a_M = c_M \cos\beta \sim 3.8 \text{ Å}.$$

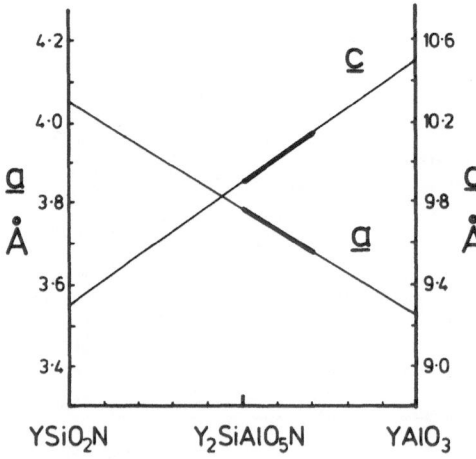

Fig. 5. Unit cell dimensions of yttrium N-α-wollastonites

Table 2. X-ray data for "tetragonal" YAM,
$\underline{a}$: 10.363, $\underline{c}$: 3.738 Å

hkl	d_{obs}	d_{calc}	I_{obs}
110	7.338	7.328	m
200	5.183	5.182	w
210	4.636	4.635	s
220	3.662	3.664	mw
101	3.519	3.517	w
111	3.337	3.330	w
310	3.276	3.277	s
201 ·	3.033	3.032	vs
211	2.910	2.910	s
320	2.875	2.874	s
221	2.615	2.617	w
400	2.593	2.591	m
301	2.535	2.537	m
410	2.516	2.513	mw
.	.	.	.
.	.	.	.
.	.	.	.

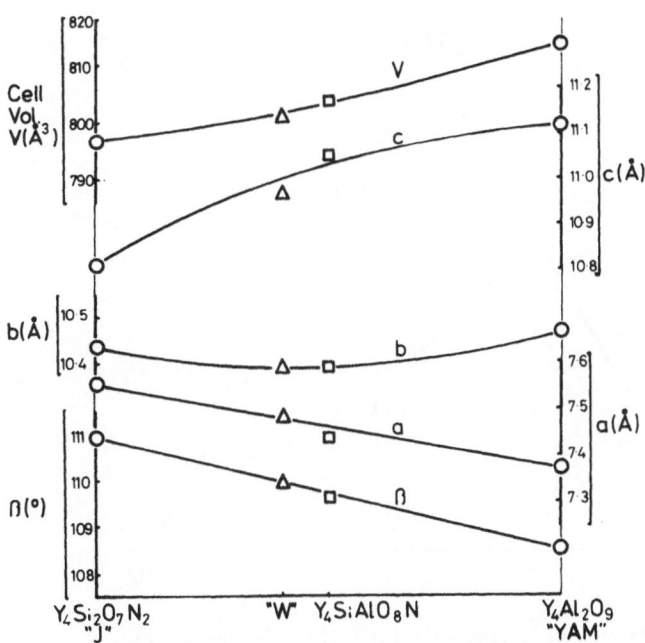

Fig. 6. Unit cell dimensions of YAM-type phases

Further work is needed to establish whether this tetragonal cell is a high-temperature modification or whether the true cell is still monoclinic but with a chance relationship between unit cell dimensions.

The 2M:3X phases described above normally occur as secondary crystalline phases or as devitrification products in β'-sialon ceramics. However, in recent years renewed interest has been expressed in silicon oxynitride, Si_2N_2O, which has a small range of homogeneity extending towards alumina referred to as O' sialon. O' offers the same advantages of easier preparation and fabrication compared with silicon oxynitride as β'-sialon does with respect to silicon nitride. Preliminary results show promise[11] and a range of materials can be expected which show improved oxidation resistance as compared with equivalent β'-sialons. The only detailed study of O' phase relationships is that of Cao et al.[12] who observed an O' homogeneity range of 15m/oAl$_2$O$_3$ which is twice the Newcastle value of 7.6m/o[11]. In the Newcastle work[11], $Y_2Si_2O_7$ was used as the secondary crystalline phase, but Cao et al.[12] have shown that O' is also in equilibrium with YAG and this may offer a more refractory alternative grain boundary phase.

PHASE RELATIONSHIPS IN β'-SIALON CERAMICS

Previous investigations of the yttrium sialon system have been directed towards an understanding of phase equilibria involving β'-sialon. The β' plus liquid region of the Si-Al-O-N square extends into the five component system as a large volume which offers wide compositional flexibility in the preparation of dense β'-sialons. For many applications, the ease of fabrication combined with good low temperature strength retained to temperatures in excess of 1000°C is a powerful combination and applications such as cutting tools, welding shrouds, extrusion dies, seals, and bearings[17] provide examples of this. For strength retention to higher temperatures it is necessary to remove the glass by devitrification, and for this purpose it is important to know which crystalline compounds adjacent to the glass forming region are in

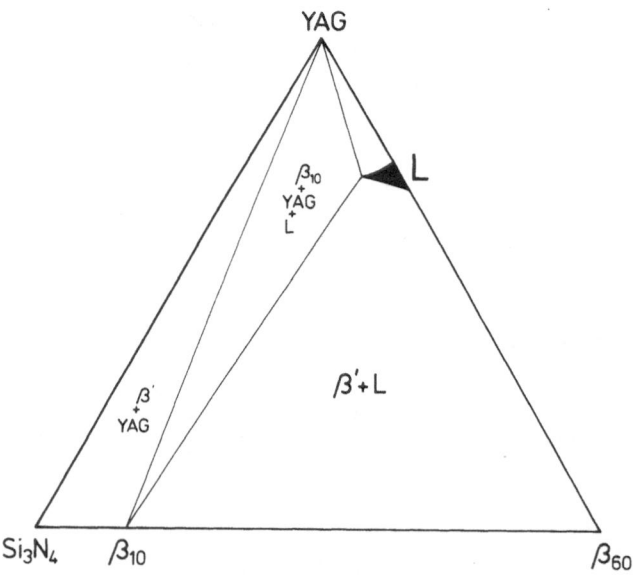

Fig. 7. Phase relationships in the β'-YAG plane at 1550°C (after Bosković[19] and Hohnke and Tien[20])

equilibrium with β'-sialon at devitrification temperatures. Naik & Tien[18] showed that of all the available yttrium-containing phases, β'-sialon was only in equilibrium with $Y_2Si_2O_7$ and $Y_3Al_5O_{12}$. YAG requires less yttria additive in the starting powder mix and is more refractory than the disilicate. In practice, β'-YAG materials show excellent mechanical properties at temperatures up to 1400°C so much so that $β'-Y_2Si_2O_7$ composites have not been seriously evaluated.

For good high-temperature creep resistance it is important to remove as much glassy phase as possible from the fired ceramic. For β'-YAG materials, the starting composition must lie on the triangular plane joining the β' line to YAG and at the firing temperature the liquid in equilibrium with β' must also lie on the plane and devitrify to give β' plus YAG. Bosković[19] and Hohnke & Tien[20] explored phase relationships in the β'-YAG plane (see Figure 7) which showed a large β' plus liquid region for all β' compositions containing more than 10 e/oAl, the liquid occurring along the YAG-β'$_{z=4}$ join at an approximate composition $Y_3SiAl_7O_{14}N_2$. This is quite surprising, because such an aluminium rich eutectic would be expected to have a higher eutectic temperature and would be unlikely to form a glass on cooling. Work carried out at Newcastle showed different results for this system. For convenience, the β'-YAG plane was abstracted from the Y-Si-Al-O-N Jänecke prism and Figure 8 shows that the geometrical shape is significantly different from the equilateral triangle shown in Figure 7. More importantly the complete intersection of the β'-YAG plane with the Jänecke prism includes additional areas extending up to the hypothetical composition $Y_{12}Si_5O_{28}$ on the Y_2O_3-SiO_2 join and down to Al_3O_3N on the base of the prism. During investigations on the glass-forming region in the system, Drew et al.[21] mapped out their results on triangular sections of constant e/o nitrogen. These sections intersect the β'-YAG plane, to give a region of glass formation as shown on Figure 8. The region of liquid formation at 1700°C is obviously larger than the glass-forming region and at this temperature extends as far as the Si_3N_4-YAG join and then spreads rapidly across the plane at slightly higher temperatures as shown in Figure 9. At lower temperatures, complete reaction of the starting nitrides takes much longer and the composition of the liquid phase is more silica rich than can be represented on Figure 9. The diagram illustrates the difficulty of producing β'-YAG ceramics free from other crystalline phases because for this to be achieved the liquid phase must have cooled to form a glass of composition somewhere inside the triangle defined by the β' line and the YAG point composition. This will only be the case for higher z-value β'

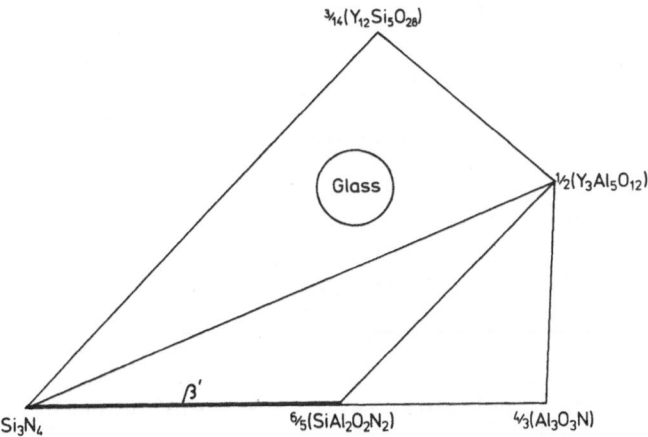

Fig. 8. The glass forming region in the β'-YAG plane (after Drew et al.[21])

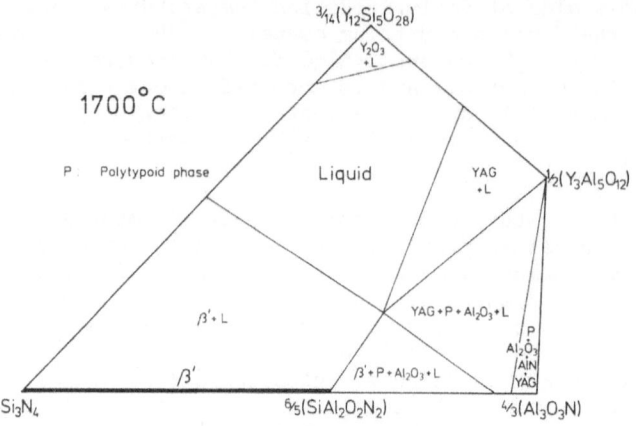

Fig. 9. Phase relationships in the β'-YAG plane at 1700°C

compositions and in general the liquid phase in equilibrium with lower z-value β'-sialons will lie on the silica side of this triangle. On devitrification at lower temperatures additional yttrium silicon oxynitride phases will therefore occur. Figure 10 shows the results of devitrifying compositions in the β'-liquid region of Figure 9 at 1350°C, and in fact significant proportions of N-α-wollastonite do occur along with some nitrogen apatite. The result of devitrifying the same compositions at 1050°C is shown in Figure 11, where different results are now achieved because of the appearance of the phase labelled B, which is the wollastonite phase of composition Y_2SiAlO_5N. In fact this composition is not quite in the plane but is so close that essentially no other phases are introduced. Y_2SiAlO_5N devitrifies from the glass at 1050°C because it has a composition within the glass-forming region and very little structural reorganization is needed to change to a crystalline phase. Diffusion can occur at these very low temperatures

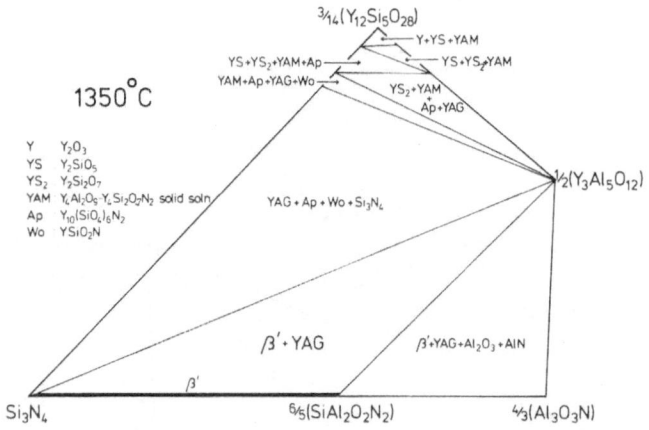

Fig. 10. Phase relationships in the β'-YAG plane at 1350°C

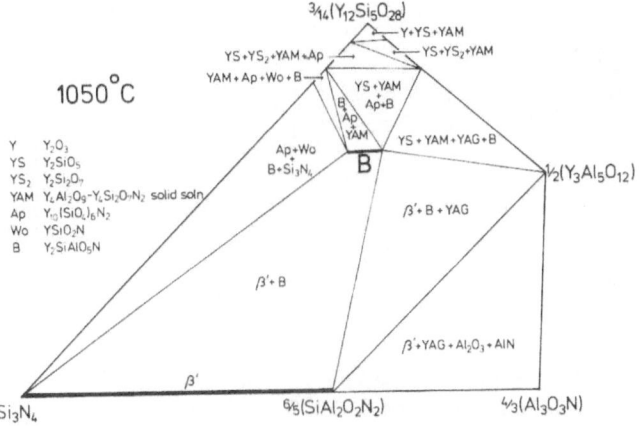

Fig. 11. Phase relationships in the β'-YAG plane at
1050°C

because the glass transition temperature for yttrium sialon glasses is in the range 930–1000°C[21]. These results are in good agreement with the later work of Bošković et al.[22] who observed diffraction patterns corresponding to Y_2SiAlO_5N (labelled N-phase) in β'-YAG specimens prepared from Y_2O_3, Al_2O_3, AlN, Si_3N_4 mixtures. This phase has probably occurred by devitrification of the glass on cooling but provides additional evidence that the glass composition is more silicon-rich than originally suggested by Bošković[19] and Hohnke & Tien[20].

These results emphasise the point made earlier that sub-solidus phase relationships may change depending on the devitrification temperature. They also illustrate the difficulty of preparing β'-YAG composites free from other crystalline phases. However by careful experimentation the levels of phases such as $YSiO_2N$ and Y_2SiAlO_5N can be minimized and the resulting ceramic will still retain far better high-temperature mechanical properties than the as-fired β'-glass materials.

PHASE RELATIONSHIPS IN α'-SIALON CERAMICS

Recent observations on the hardness of β'-sialons have shown that significant improvements can be obtained if α'-sialon is also present[23]. α'-sialons are not currently manufactured commercially, mainly because of the difficulty of preparing dense materials free from secondary crystalline phases. Preliminary explorations show that their potential as high-temperature engineering ceramics is at least as good as β' sialons, and the possibility of incorporating the densifying additive into the crystal structure to form an essentially single-phase material is very attractive.

α'-sialons are represented by the formula[24]:

$$M_xSi_{12-(m+n)}Al_{(m+n)}O_nN_{16-n}$$

where m(Al-N) and n (Al-O) bonds replace m+n (Si-N) bonds in $\alpha-Si_{12}N_{16}$, and x large metal atoms (M=Li,Ca,Y or rare earths with Z>60) statistically occupy the two large interstices per unit cell. In the Y-Si-Al-O-N system, α' com-

89

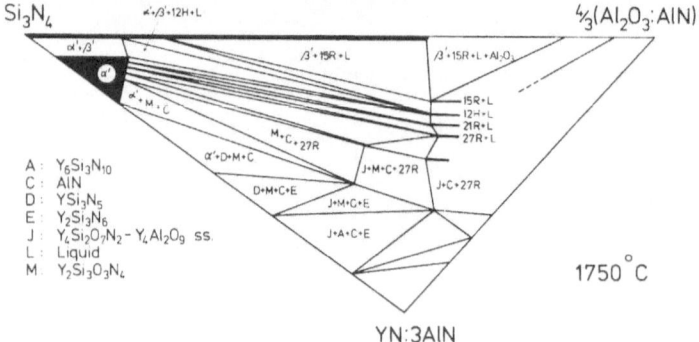

Fig. 12. Phase relationships in the α'-yttrium
sialon plane at 1750°C

positions occur in the triangular plane bounded by Si_3N_4, Al_3O_3N and YAl_3N_4
and phase relationships in this plane at 1750°C are shown in Figure 12. The
single phase α' region only just intersects the $Si_3N_4-Y_2O_3.9AlN$ line which
defines the limit of compositions accessible using yttria rather than yttrium
nitride as the starting source of yttrium. In contrast to β'-sialon, yttrium
α' sialon is in equilibrium with liquid only along the line of compositions
facing β' and the range of liquid compositions is also very restricted. The
preparation of single phase, dense α' plus glass materials therefore requires
precise compositional control. Since no other crystalline phases form tie
lines with α' which intersect this α' plus liquid region, it is impossible to
prepare α' sialon ceramics containing a single refractory crystalline secondary phase by devitrification of grain boundary glass. Polytypoid phases
readily occur as additional phases.

The limit of β'-sialon in equilibrium with α' occurs at a z value slightly less than unity[23]. α'-β'-glass materials can be easily prepared and devitrification to give YAG as a grain boundary phase is possible but the same
difficulties arise as experienced with β' sialon ceramics.

CONCLUSIONS

Many regions of the yttrium sialon system remain inadequately explored,
but these are generally away from the areas where commercial materials are
being developed. Phase equilibria involving β'-sialon are reasonably well
understood but relatively little work has been carried out on α'-materials.
The liquid region in the system and its variation with temperature have not
been adequately explored and there is also scope for further devitrification
studies particularly at lower temperatures. In view of the increasing interest in strong high-temperature materials, further work along these lines
would be beneficial to tailor more precisely the grain-boundary phase assemblages and hence optimise material performance.

REFERENCES

1. K. H. Jack, "The relationship of phase diagrams to research and development of sialons", in: "Phase Diagrams: Materials Science and
   Technology", vol 5, A. M. Alper, ed., Academic Press, New York, 241
   (1978).

2. A. Tsuge, H. Kudo and K. Komeya, Reaction of Si_3N_4 and Y_2O_3 in Hot-pressing, J. Am. Ceram. Soc., 57:269 (1974).

3. R. R. Wills, S. Holmquist, J. M. Wimmer and J. A. Cunningham, Phase relations in the system $Si_3N_4-Y_2O_3-SiO_2$, J. Mater. Sci., 11:1305 (1976).

4. K. H. Jack, Sialons and related nitrogen ceramics, J. Mater. Sci., 11:1135 (1976).

5. F. F. Lange, S.C. Singhal and R. C. Kuznicki, Phase relations and stability studies of the $Si_3N_4-SiO_2-Y_2O_3$ pseudo ternary system, Westinghouse Research Report No. 76-9D4-POWDR-R1 (1976).

6. L. J. Gauckler, H. Hohnke and T. Y. Tien, The system $Si_3N_4-SiO_2-Y_2O_3$, J. Amer. Ceram. Soc., 63:35 (1980).

7. J. Felsche, The crystal chemistry of the rare-earth silicates, Structure and Bonding, 13:100 (1973).

8. A.W.J.M. Rae, Unpublished work, University of Newcastle upon Tyne, (1976).

9. J. Ito and H. Johnson, JCPDS Index card No. 21-1459

10. E. Butler, R.J. Lumby, A. Szweda and M. H. Lewis, Syalon ceramics for high temperature engines; an illustration of grain boundary engineering, in Proc. Int. Symp: "Ceramic components for Engine", S. Somiya, E. Kanai and K. Ando, eds., Hakone, (1983).

11. M. B. Trigg and K. H. Jack, Silicon oxynitride and O'-sialon ceramics, Proc. Int. Symp: "Ceramic components for Engine", S. Somiya, E. Kanai and K. Ando, eds., Hakone, (1983).

12. G. Z. Cao, Z. K. Huang, X. R. Fu and D. S. Yan, Phase equilibrium studies in Si_2N_2O-containing systems, I: Phase relations in the $Si_2N_2O-Al_2O_3-Y_2O_3$ system, (to be published).

13. M. H. Lewis, A. R. Bhatti, R. J. Lumby and B. North, The microstructure of sintered Si-Al-O-N ceramics, J. Mater. Sci., 15:103 (1980).

14. N. A. Toropov, I. A. Bondar, F. Ya. Galakhov, X. S. Nikogosyan and N. V. Vinogradova in: "Phase Diagrams for Ceramists", E. M. Levin, C. R. Robbins and H. F. McMurdie, eds., 2:2344 (1969).

15. E. F. Bertaut and J. Mareschal, Un Nouveau type de structure hexagonale, $AlTO_3$ (T=Y,Eu,Gd,Tb,Dy,Ho,Er), C. R. Acad. Sci. Paris C, 257:867 (1963).

16. A.W.J.M. Rae, Ph.D. Thesis, University of Newcastle upon Tyne, (1976).

17. N. E. Cother and P. Hodgson, The development of syalon ceramics and their engineering applications, Trans. J. Brit. Ceram. Soc., 81:141 (1982).

18. I. K. Naik and T. Y. Tien, Subsolidus phase relations in part of the system Si,Al,Y/N,O, J. Amer. Ceram. Soc., 62:642 (1979).

19. S. Bosković, Densification in the system Si_3N_4-YAG-sialon, Science of Ceramics, 11:225 (1981).

20. H. Hohnke and T. Y. Tien, Solid-liquid reactions in part of the system Si,Al,Y/N,O, Proceedings of the NATO Advanced Study Institute: "Progress in nitrogen ceramics", F. L. Riley, Ed., 101 (1981).

21. R. A. L. Drew, S. Hampshire and K. H. Jack, The preparation and properties of oxynitride glasses, Proceedings of the NATO Advanced Study Institute: "Progress in nitrogen ceramics", F. L. Riley, ed., 323 (1981).

22. S. Bosković and E. Kostic, Formation of sialons in systems Si_3N_4-$3Y_2O_3.5Al_2O_3-Si_2Al_4O_4N_4$ and $Si_3N_4-3Dy_2O_3.5Al_2O_3-Si_2Al_4O_4N_4$, Science of Ceramics 12:391 (1983).

23. C. Chatfield, T. Ekstrom and M. Mikus, Microstructural investigation of SiAlON materials, (to be published).

24. S. Hampshire, H. K. Park, D. P. Thompson and K. H. Jack, α'-sialon ceramics, Nature 274:880 (1978).

THE FABRICATION OF COMPOSITE O'-β' SIALON CERAMICS

W. Y. Sun, D. P. Thompson and K. H. Jack

Wolfson Research Group for High-Strength Materials
Department of Metallurgy and Engineering Materials
University of Newcastle upon Tyne, UK

ABSTRACT

Dense sialon ceramics with the structure of silicon oxynitride (O') are readily prepared by pressureless sintering powder mixtures of Si_3N_4, SiO_2, Al_2O_3 and Y_2O_3. By suitable heat-treatment, the glassy phase in the product can be devitrified to give yttrium disilicate ($Y_2Si_2O_7$), and the resulting O'-$Y_2Si_2O_7$ material has excellent oxidation resistance at high temperatures. Since both these phases occur in equilibrium with β'-sialon - already established as a good high-temperature engineering ceramic - compositions within the O'-β'-$Y_2Si_2O_7$ compatibility region offer good prospects for development as ceramic materials, combining the mechanical properties of β' with the good oxidation resistance of the O' and $Y_2Si_2O_7$ phases. The degradation of the high-temperature properties observed when a grain-boundary glass is present is thereby avoided and, by varying the proportions of the two completely compatible phases in the composite, its microstructure can be more readily tailored than in a single-phase matrix material.

Dense O'-β' materials of varying O'-β' ratio containing one or more of glass, $Y_2Si_2O_7$ and YAG as intergranular phases have been prepared by pressureless sintering. With Si_3N_4, SiO_2, Al_2O_3 and Y_2O_3 as the starting materials, O' starts to form at ~1400°C increasing in amount until at ~1600°C the formation of β' begins and the mix then equilibrates to its expected O':β' ratio. The oxidation resistance of these materials in both the as-fired and devitrified conditions shows improvements with increasing O' content and increasing extent of devitrification.

INTRODUCTION

It has previously been shown[1] that dense O'-sialon, $Si_{2-x}Al_xO_{1+x}N_{2-x}$ with a limiting value of x=0.2, can be prepared by pressureless sintering powder mixtures of silicon nitride, silica and alumina using yttria as a densifying additive. The products have good oxidation resistance up to 1350°C. Phase relationships in the Y-Si-Al-O-N system[2] show that O', β', and $Y_2Si_2O_7$ form a compatibility region, and sialon compositions within this region provide advantages as engineering ceramics since they combine the good mechanical properties of β' with the oxidation resistance of O'. The yttrium-sialon liquid necessary for the high-temperature reaction and densification cools to give an intergranular glass, but subsequent devitrifying heat-treatment im-

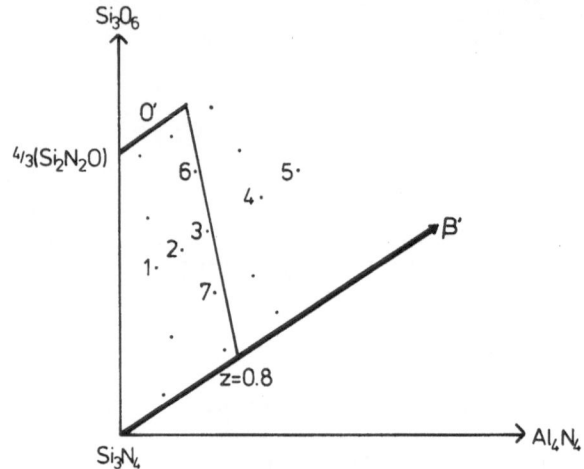

Fig. 1. O'-β' region of Si-Al-O-N system showing
compositions explored

proves the refractoriness of the material. The present work attempts to optimise the processing variables to produce dense O'-β' composite ceramics and
the oxidation resistance of some of these is reported.

EXPERIMENTAL

Appropriate powder mixtures of silicon nitride (grade LC10, Starck-
Berlin), aluminium nitride (Starck-Berlin), crushed quartz crystal (Thermal
Syndicate plc), alumina (Grade A17, Alcoa), and yttria (Rare Earth Products
Ltd.), all >99.9% purity, were wet-milled in isopropyl alcohol for 25h using
alumina media. Powder mixes were dried in vacuum at 115°C to give seventeen
compositions within or near the O'-β'-$Y_2Si_2O_7$ sub-system. In calculating
compositions, allowance was made for surface oxides on the nitrides and for
the additional alumina introduced by milling. It was necessary to use alumi-

Table 1. The effect of intermediate heat-treatment on density of O':β' composites

Firing Conditions	Bulk density, g cm^{-3}		
	3	4	5
1h at 1600°C	2.88	3.06	3.09
1h at 1700°C	2.98	3.08	2.10[*]
0.5h at 1600°C + 1h at 1700°C	2.98	3.09	3.11
1h at 1800°C	3.10	2.00[*]	1.68[*]
0.5h at 1600°C + 1h at 1800°C	3.08	3.13	3.19

[*]bloated

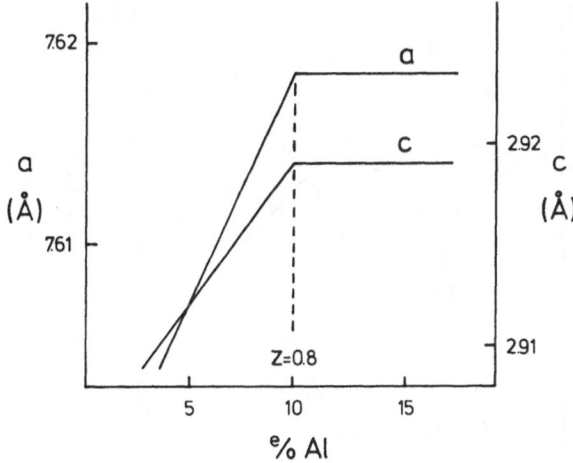

Fig. 2. Variation with total Al concentration (e/o)
of cell dimensions of β' coexisting with
O'; limit of β', z=0.8

nium nitride in the powder mix for only three compositions near the β'
single-phase boundary. In all other runs the required compositions were obtained with silicon nitride, silica, alumina, and yttria mixes. Pellets of each
mix were uniaxially pressed in a steel die, then cold isostatically pressed
at 200MPa and, after embedding in $SiO_2:Si_3N_4$ powder to suppress volatilisa-
tion of silicon monoxide and nitrogen, fired in nitrogen at 1600°-1800°C. In
some cases intermediate heat-treatments at 1550°-1650°C for 0.5-1h were used

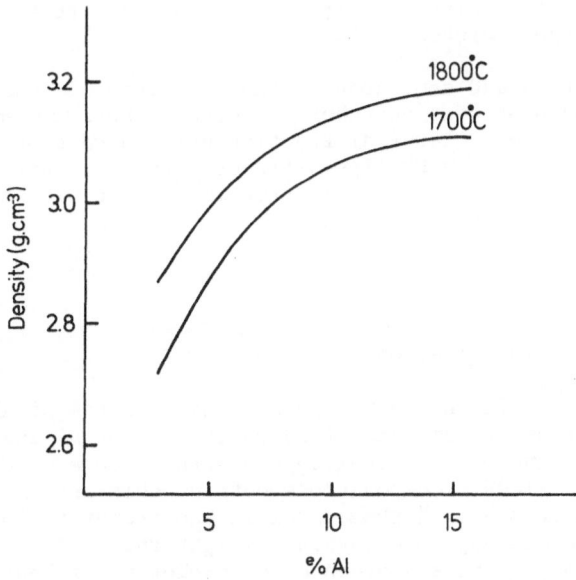

Fig. 3. Variation of density with total Al concentration (e/o) in compositions 1-5

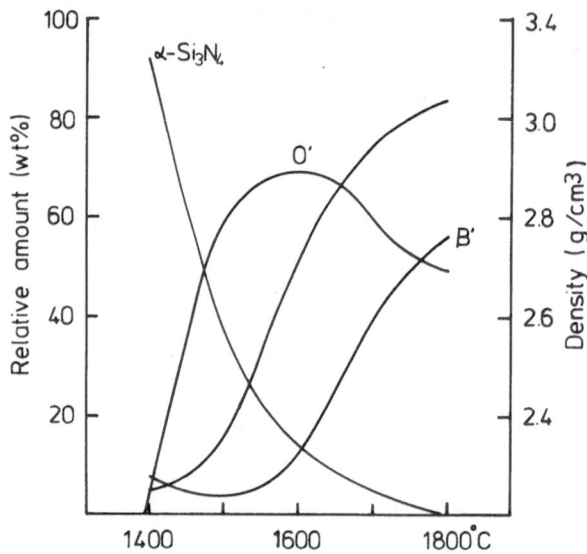

Fig. 4. O'-β' composite sialon formation and densification for composition 3. Different
specimens were fired for 0.5h at 100°C
intervals in the range 1400°–1800°C.

to minimise bloating at higher firing temperatures. Post-preparative heat-
treatment was carried out at 1100°–1400°C in nitrogen to devitrify the grain-
boundary glass.

Oxidation behaviour was followed at 1000°–1400°C by measuring weight
changes with an analytical balance after successive heat-treatments for 48h
in air in a vertical tube furnace.

Phase identification and semi-quantitative analysis of reaction, devitrification and oxidation products, and where necessary their unit-cell dimensional measurements, were made by X-ray diffraction using a Hägg-Guinier focusing
camera for powder samples and a Philips diffractometer for surface examination
of dense, bulk specimens. A JEM-100C electron microscope with EDAX analytical
facilities was used for SEM observations.

RESULTS AND DISCUSSION

Sample preparation and densification

Matrix compositions within the two-phase region of the Si-Al-O-N system
(see Fig. 1) with appropriate amounts of added yttria and silica were expected
to give products contaning O', β', $Y_2Si_2O_7$, and perhaps some undevitrified
intergranular Y-sialon glass. Compositions outside this region were expected
to contain more glass or sialon X-phase. As in the preparation of O'-sialons[1],
the mixed powders were made up to maintain a weight ratio of $Y_2Si_2O_7$:(O'+β')=
0.15. The compositions of the seventeen mixes explored, without the additional 15wt.%$Y_2Si_2O_7$, are given in Fig. 1.

Cell dimensions (see Fig. 2) show that the upper limit of β'-sialon coexisting with O'-sialon is z=0.8 at both 1700°C and 1800°C where the general
β' composition is $Si_{6-z}Al_zO_zN_{8-z}$. With overall compositions outside the

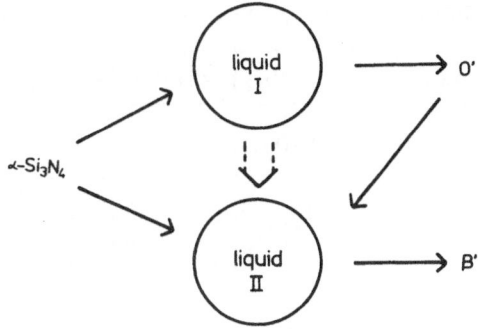

Fig. 5. Reaction sequence for O':β' formation

O'-β' region, bloating occurred above 1700°C and became more severe at higher temperatures and with increasing aluminium concentrations. However, with a modified firing programme in which the specimen was held at 1600°-1650°C for 0.5-1h before the final soak for 1h at 1700°C or 1800°C, bloating was reduced or in some cases completely eliminated. Densities of three compositions (3, 4 and 5) each after five different firing treatments, are given in Table 1. The liquid phase formed by rapid heating to the highest temperatures, 1700° and 1800°C, seems less stable than that at lower temperatures (~1600°C) and more readily generates volatile vapours, probably silicon monoxide and nitrogen. Fig. 3 shows that compositions 1-5 with similar O':β' ratios increase in density with increasing alumina contents and also with increasing firing temperature.

Hot-pressed specimens were assumed to be fully dense and so pressureless sintered compositions 3, 6 and 7 near the limiting O'-β' tie-line, and which approach full density, were used for oxidation resistance measurements.

The reaction sequence

The typical reaction sequence shown by Fig. 4 for composition 3 was determined by quantitative measurement of phase compositions and densities after firing different specimens for 0.5h at 100°C intervals in the range 1400°-1800°C. O' formation starts at about 1400°C and increases with increasing temperature up to about 1600°C. β' then increases rapidly at the expense of unreacted α-Si_3N_4 and some O'-phase. The maximum amount of transient O' depends on the starting composition and, for example, is 80% for run 5 and 20-25% for compositions near the β' single-phase field. A possible reaction mechanism is suggested in Fig. 5. A relatively oxygen-rich yttrium-sialon liquid (I) formed at low temperatures allows dissolved silica to react with silicon nitride and alumina to precipitate O'. At higher temperatures, increasing amounts of silicon nitride dissolve to give a nitrogen-enriched liquid (II) from which β' is precipitated. Depending on the overall composition, some of the first-formed O'-sialon dissolves in liquid (II) and is precipitated as β' to achieve the final equilibrium phase distribution.

Devitrification

The liquid necessary for reaction and densification cools to give an intergranular glass and in order to crystallise this, and hence to improve creep resistance, a variety of devitrifying heat-treatments was explored.

Table 2. Devitrification treatment for composition 3

Treatment	Temperature, °C	Packing powder	Phases present in addition to O'+β'
(1)	1300–1050	$1SiO_2:1Si_3N_4$	Surface: γ-Y2S,s; α-SiO_2,s Bulk: Y2S(β,δ),m; YAG,w
(2)	1300–1050	AlN	Surface: YAG,s Bulk: YAG,m; Y2S(β,δ),w
(3)		AlN	Bulk:
(a)	1300		YAG,w
(b)	1250		YAG,mw; Y2S(δ,y"),vw
(c)	1200		YAG,s; Y2S(δ,y"), w
(d)	1150		⎫
(e)	1100		⎬ No further change from (c)
(f)	1050		⎭
(4)	1300–1200	AlN	Bulk: YAG,s

s, strong; ms, medium strong; m, medium; mw, medium weak; w, weak; vw, very weak.

Single-temperature treatments in the range 1100°–1300°C showed that all compositions at temperatures near 1200°C gave $Y_2Si_2O_7$ in one or more of its crystalline modifications β, δ, and y"[3]. It was clear, however, that complete devitrification could not be achieved by a single-temperature treatment and so the following four modifications were explored:

(1) Holding for 24h at each of six temperatures decreasing at intervals of 50°C in the range 1300°–1050°C (total time, 6d). The specimen was embedded in a packing powder of $1SiO_2:1Si_3N_4$.

(2) As for (1) except that the packing powder was AlN.

(3) A batch of six samples was heat-treated for 24h at 1300°C and cooled, after which one sample was removed and the remainder reheated and maintained at 1250°C for 24h. The procedure was repeated at intervals of 50°C down to 1050°C.

(4) As for (2) except that the range was limited to 1300°–1200°C.

The packing powder in modifications (2), (3) and (4) was AlN, and although this was inactive at 1200°C it reacted with O' at 1300°C to give surface layers enriched in yttrium-aluminium garnet (YAG). However, all treatments starting at 1300°C, even with $1SiO_2:1Si_3N_4$ as packing powder, gave YAG as a product in the bulk of the specimen, thus suggesting that devitrification at 1100°–1200°C is far from complete and that the residual glass at this temperature is aluminium-rich. The phases observed, other than O' and β', after the above devitrification treatments of composition 3 are listed in Table 2. Apart from the surface cristobalite and surface YAG when $1SiO_2:1Si_3N_4$ and AlN respectively are used as packing powders, the devitrification product in the bulk of the specimen is always a mixture of $Y_2Si_2O_7$ and YAG when the temperature of any part of the treatment exceeds 1250°C. At 1200°C and lower, only $Y_2Si_2O_7$ is observed but devitrification is then incomplete. The silicate rejected by treatment at lower temperatures leaves a more stable and more alunium-rich glass that can be crystallised only at higher temperatures.

Oxidation

Typical plots of weight gain against temperature for compositions 3, 6

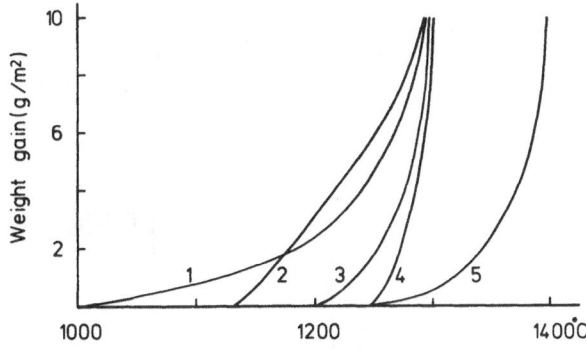

Fig. 6. Oxidation after 48h at each temperature
         in the range 1000°–1400°C

Composition 3:   curve 1, without devitrification
                 curve 3, devitrification at
                          1200°C for 16h
                 curve 5, devitrification with mod-
                          ified treatment (1); see
                          Table 2
Composition 6:   curve 4, devitrified at 1200°C for
                          16h
Composition 7:   curve 2, devitrified at 1200°C for
                          16h

and 7 after 48h oxidation in air are given in Fig. 6. Accompanying oxidation
there is invariably devitrification of the intergranular glass and so the
interpretation of observations must take this into account.

At 1100°C and also at 1400°C the major oxidation product is α–cristobalite, while at intermediate temperatures both cristobalite and γ-$Y_2Si_2O_7$ are

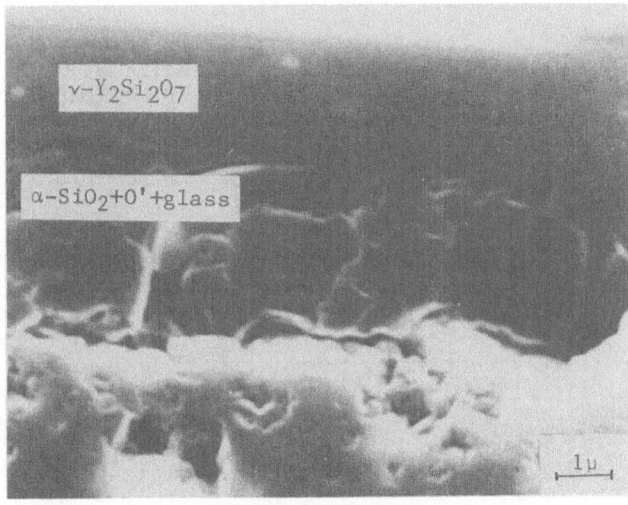

Fig. 7. SEM showing the oxidised layer of the
        O':β' composite (Composition 3) after
        oxidation for 24h at 1300°C.

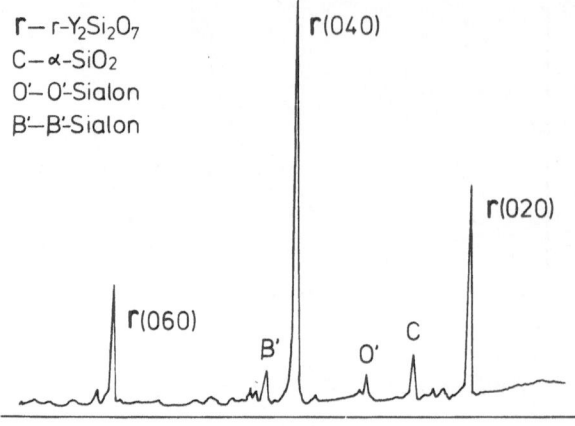

Fig. 8. Surface X-ray diffraction pattern for composition 3
after oxidation at 1300°C for 24h

observed. The ratio γ–$Y_2Si_2O_7$:α–SiO_2 is determined by the oxidation temperature and is a maximum between 1200°C and 1300°C. The monoclinic γ–$Y_2Si_2O_7$ is
formed as perfectly oriented, large crystal plates parallel to the specimen
surface and with the crystalline b-axis normal to the surfaces of both
crystal and specimen. This is shown in Figs. 7 and 8 and is clearly a result
of crystallisation from a surface liquid.

Comparison of curves 1, 3 and 5 of Fig. 6 shows that prior heat-treatment
of composition 3 to devitrify the intergranular glass improves the oxidation
resistance. The comparison of curves 3 and 5 shows further that heat-
treatment to give complete devitrification is more effective in improving
oxidation resistance than only partial devitrification.

SEM with EDAX analysis shows that the oxidation scale consists of two
layers, the outermost layer of oriented, plate-like γ–$Y_2Si_2O_7$ crystals and
then a mixed layer of α–SiO_2 and O'-phase randomly distributed in a glassy
matrix containing Y, Si, and Al as metallic elements; see Figs. 7 and 9.

CONCLUSIONS

Composite O'–β sialon ceramics are obtained with higher than 98% theoretical density by pressureless sintering powder mixes of silicon nitride,
silica, alumina, and yttria at temperatures up to 1800°C. Most of the intergranular glass is devitrified to produce $Y_2Si_2O_7$ and YAG by sequential heat-
treatments at successively decreasing temperatures. The oxidation resistance
of O':β' composites with $Y_2Si_2O_7$ and YAG intergranular phases is good up to
above 1300°C for compositions containing approximately equal weights of the
O' and β' phases. Oxidation products are α–cristobalite and a strongly
oriented surface layer of γ–$Y_2Si_2O_7$ formed by crystallisation from a liquid.

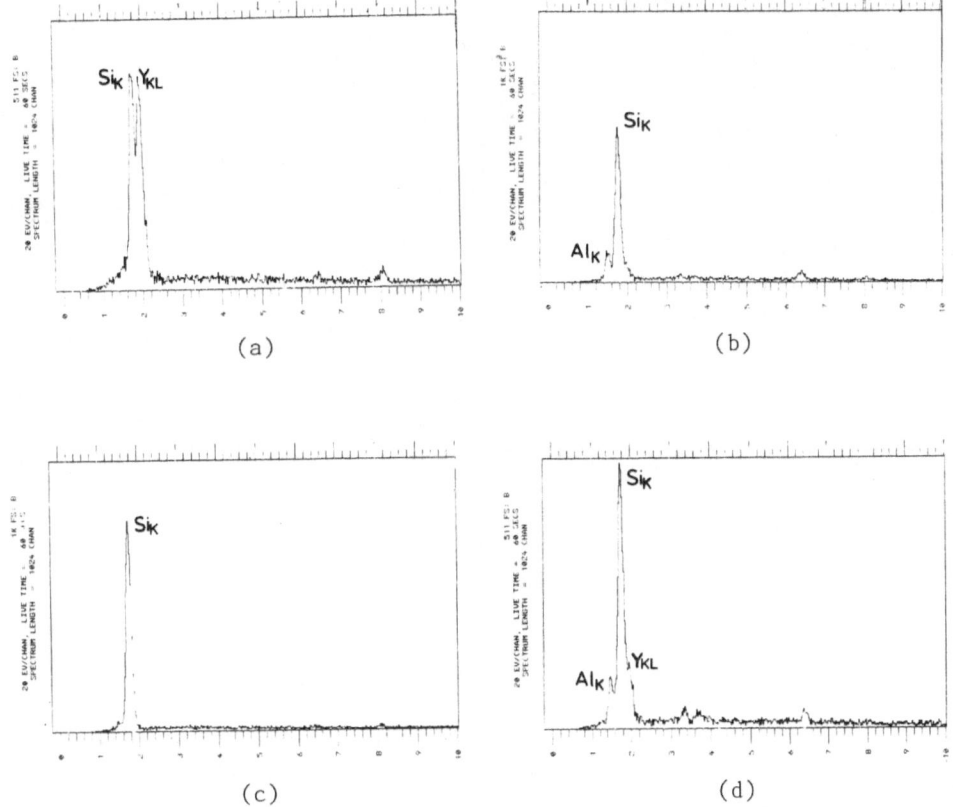

Fig. 9. EDAX of the oxidised layer shown by Fig. 7: (a) plate-like crystals of γ-$Y_2Si_2O_7$ from the outermost surface; (b) O' from the mixed layer; (c) α-SiO_2 from the mixed layer; (d) glass from the mixed layer

ACKNOWLEDGEMENT

One of us (W.Y.S.) thanks Lucas Cookson Syalon for a maintenance grant during part of the period in which the work was carried out.

REFERENCES

1. M. B. Trigg and K. H. Jack, "Silicon Oxynitride and O'-sialon Ceramics". In Proc. "First International Symposium on Ceramic Components for Engine", 1983 Hakone, Japan. Eds. S. Somiya, E. Kanai and K. Ando 1984 Tokyo: KTK Scientific Publishers, pp.343-349.

2. I. K. Naik and T. Y. Tien, "Subsolidus Phase Relationships in Part of the System Si,Al,Y/N,O", J. Am. Ceram. Soc., 62 (11-12), 642-643 (1979).

3. A.W.J.M. Rae, unpublished. Wolfson Research Group for High-Strength Materials, University of Newcastle upon Tyne.

CERAMIC EUTECTIC COMPOSITES

V. S. Stubican, R. C. Bradt, F. L. Kennard,
W. J. Minford and C. C. Sorrel

Department of Materials Science and Engineering
The Pennsylvania State University
University Park, PA   16802

ABSTRACT

Methods of growing ceramic eutectics are reviewed. The basic
microstructure of the obtained oxide eutectics can be generalized from the
relative interfacial area/unit volume and the volume fraction of the minor
phase for the fibrous and lamellar ideals. Crystallographic investigation
of oxide eutectics show the interfacial planes and solidification
directions are all low Miller indices regardless of the crystal structures
of the phases. The orientation relations of directionally solidified oxide
eutectics appear to result from two factors: (i) the minimization of the
lattice misfit and strain (ii) the ionic charge balance at the interface
between two phases. The oxide eutectics without colony boundaries can be
expected to possess superior mechanical properties. The lamellar $ZrC-ZrB_2$,
$ZrC-TiB_2$, and $SiC-B_4C$ eutectics were directionally solidified. All of the
investigated carbide-boride eutectics exhibit superior mechanical
properties when compared with the individual constituents.

INTRODUCTION

Since it was shown by Herring and Galt[1] in 1952 that whiskers could
attain fracture stresses approaching the theoretical strengths of perfect
crystals, greater than 10^6 psi, the properties of composites have been
actively researched in order to exploit these strengths. Due to the fact
that aligned in situ composites possess higher strengths than the
individual components or randomly oriented composites formed by traditional
methods, these materials are of interest from a scientific as well as an
engineering standpoint. These anisotropic materials have been referred to
as directionally solidified composites, in situ composites, and
directionally solidified eutectics.

The directional solidification of metal alloys has been extensively
investigated over the past century. The impetus has been chiefly to
control and improve the microstructure for application at elevated
temperatures. This has led to the directional solidification of melts of
eutectic composition. These have been found to yield oriented lamellar or
fibrous composite microstructures, possessing excellent thermal stability
and a coherent interphase bonding which is impossible to achieve in
composites formed by conventional methods.

Recent interest in materials which can sustain even higher temperatures has directed considerable research emphasis toward ceramic materials. One method of producing a ceramic composite is, as in metallic systems, the directional solidification of a eutectic melt which offers the additional advantage of growing a high melting ceramic composite, in-situ. Although the objectives of the limited amount of work in directional solidification of ceramic eutectics has been to maximize a particular mechanical, electrical, optical or magnetic property, an understanding of the microstructure and the crystallographic relationship between the phases is imperative in directing any improvements.[2] The directionally solidified ceramic eutectics may then find further use in electronic devices, optical imaging, light transmission devices, piezoelectric and ferromagnetic applications.

METHODS OF GROWING CERAMIC EUTECTICS

Several methods can be used to solidify ceramic eutectics directionally in the laboratory. Hurt and Viechnicki[3] used a gradient furnace and vapor-deposited tungsten crucibles. A graphite resistance furnace was used as a heat source, and heating was done in argon or in vacuum. Projecting into the hot zone of the furnace was a tungsten heat exchanger, which supported the crucible and melt. Helium was forced into the heat exchanger to draw heat away from the crucible and melt, thereby producing solidification from the bottom of the crucible. The temperature gradient in the material depended on the rate of flow of helium into the heat exchanger and the temperature of the melt. Viechnicki and Schmid[4] used a Bridgman-type crystal-growing furnace for directional solidification of oxide eutectics.

Kennard et al.[5] used a modified Bridgman-type furnace for their studies. The melting furnace used for a commercial crystal-growing furnace (Arthur D. Little Co.) was powered by a 25-kw induction generator. The maximum melting temperatures used were ~50°C above the liquidus. Ingots were directionally solidified by lowering the loaded crucible through the susceptor at the desired rate. Runs were made in an argon atmosphere of 50 psi. The furnace's temperature profile was measured using a W/3% Re-W/25%Re thermocouple.

Hulse and Batt[6] used a floating molten zone technique for the directional solidification of a number of oxide eutectics. A zone refiner, designed for use with metal systems included a larger chamber for work in inert gas or vacuum atmospheres. A 50-kw power generator operating a ~500 kHz. supplied RF power to the work coil through a step-down transformer. The heating was accomplished by radiation from a small carbon ring susceptor about 1/8-inch thick. The cracking in ingots prepared at high solidification speeds was reduced by using additional susceptors or more insulation just below the heater.

Fragneau and Revcolevschy[7] and Fragneau et al.[8] used a vertical floating zone, arc image furnace to grow NiO-CaO and NiO-Y$_2$O$_3$ eutectics respectively.

Sorrell et al.[9] used the floating molten zone technique to directionally solidify carbide and boride eutectics. The sintered rod of eutectic composites was placed in a 25-kW, 450-kHz Czochralski crystal growth furnace. The sintered rod was lowered through a doubly nested copper induction heating coil. Because most carbides and borides are good electrical conductors, the induction coil and sintered rod were directly coupled.

OXIDE EUTECTICS

Microstructure

Viechnicki and Schmid[4,10] studied the eutectics $Al_2O_3-Y_3Al_5O_{12}$ and
$Al_2O_3-ZrO_2$. Using a Bridgman-type furnace with molybdenum crucibles and a
thermal gradient of approximately 200°C/cm, only colony-type microstructures were obtained in the $Al_2O_3-Y_3Al_5O_{12}$ system; however, certain
colonies did exhibit oriented regions consisting of rods and plates of
$Y_3Al_5O_{12}$ in a matrix of Al_2O_3. Again, in the $Al_2O_3-ZrO_2$ system only colony
microstructures were obtained with the Bridgman-type furnace at solidification rates between 1.3 and 15.6 cm/hr. The eutectic microstructure
consisted of rods of monoclinic ZrO_2 in an alumina matrix. The volume
percent of ZrO_2 at the eutectic composition was approximately 34. However,
with a gradient furnace and a 1200°C/cm gradient the cellular growth was
completely eliminated so that no colonies resulted. Solidification rates
of 0.75 and 1.0 cm/hr were used. Under these conditions the ZrO_2 was no
longer rod shaped but was lamellar or plate-like instead.

Rowcliffe et al.[11] directionally solidified the two eutectics in the
$Al_2O_3-TiO_2$ system using an electron-beam heating technique. The microstructure of the eutectic between Al_2TiO_5 and TiO_2 consisted of lamellae of
Al_2TiO_5 in a TiO_2 matrix. Samples rich in Al_2TiO_5 were also solidified
giving plate-like primary Al_2TiO_5 surrounded by the eutectic. The reported
eutectic composition between Al_2TiO_5 and Al_2O_3 was also solidified.
However, the ingot contained considerable primary dendritic Al_2O_3.

Hulse and Batt[6,12], using a zone melting apparatus with a thermal
gradient of up to 1000°C/cm, investigated a number of oxide eutectics,
including 29 different eutectics with liquidus temperatures <1500°C and
several eutectics of possible interest for structural applications, namely
$Al_2O_3-ZrO_2$ (Y_2O_3), $Al_2O_3-ZrO_2$ (CaO), and $ZrO_2-Y_2O_3$. In the $Al_2O_3-ZrO_2$
(Y_2O_3) system cellular growth was obtained at all solidification rates >0.8
cm/hr with the colony size increasing with an increasing solidification
rate. At solidification rates below 0.8 cm/hr, plane front growth
occurred. At solidification rates above 10 cm/hr, the ingots were cracked,
which Hulse and Batt attributed to thermal shock.

Figure 1. ZrO_2-MgO eutectic
            directionally solidi-
            fied at 1.0 cm/hr.
            Transverse section.

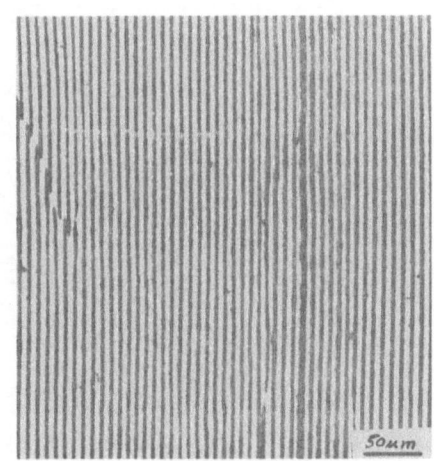

Figure 2. $SrZrO_3$-ZrO_2 eutectic
            directionally solidi-
            fied at 0.5 cm/hr.
            Longitudinal section.

Galasso et al.[13] investigated the $BaFe_{12}O_{19}$–$BaFe_2O_4$ eutectic in an attempt to improve the magnetic properties of $BaFe_{12}O_{19}$ by aligning the c-axis of this permanent magnetic material. The directionally solidified eutectic consisted of 32.5 vol% $BaFe_{12}O_{19}$ blades embedded in a matrix of $BaFe_2O_4$. No improvements in the magnetic properties were obtained since the c-axis of the $BaFe_{12}O_{19}$ plates were preferentially oriented perpendicular to both the growth direction and the plate face and the plates were randomly rotated about the growth axis.

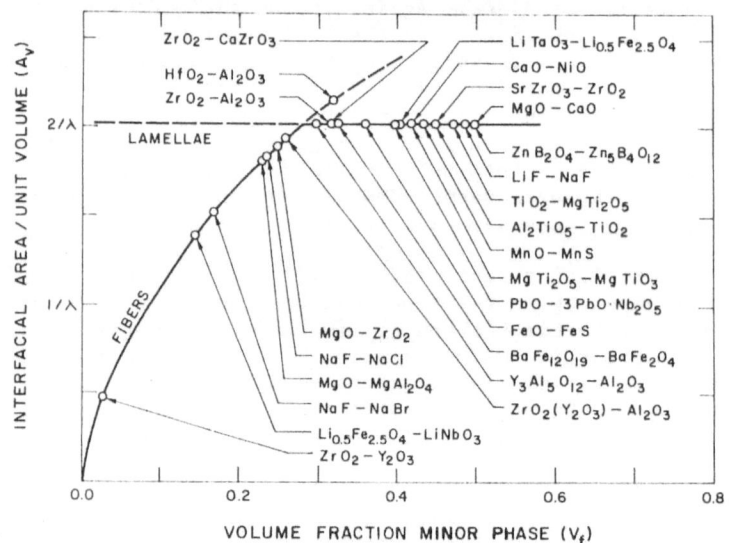

Figure 3. Correlation between the volume fraction of observed nonmetallic eutectic microstructures and interfacial area/unit volume.

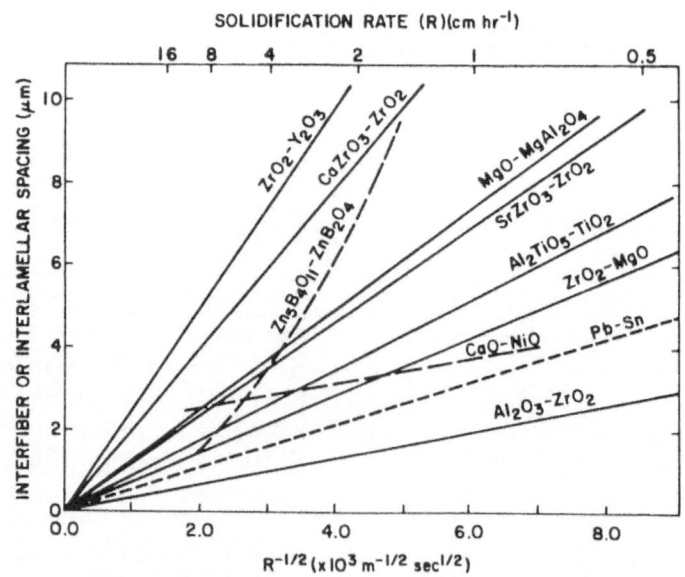

Figure 4. Variation of interlamellar or interfiber spacing with reverse square root of solidification rate for directionally solidified oxide eutectics.

A lamellar $Zn_5B_4O_{11}$/glass matrix structure was produced by Harrison[14] when the $ZnB_4O_{11}-ZnB_2O_4$ eutectic was directionally solidified.

The eutectic in the NiO-CaO system was directionally solidified and produced a lamellar microstructure. Kennard et al.[5,15,16] studied microstructure of directionally solidified eutectics in the systems $MgO-MgAl_2O_4$ and ZrO_2-MgO. In both cases well-oriented microstructures were obtained if the rate of solidification was < 2cm/hr. When grown at 1 cm/hr the microstructure of ZrO_2-MgO eutectic is fully aligned with MgO rods in the cubic ZrO_2 matrix as shown in Figure 1.

Minford et al.[17] studied the microstructures of several directionally solidified oxide eutectics that melt at high temperature. Their study included $ZrO_2-CaZrO_3$, $ZrO_2-SrZrO_3$, $ZrO_2-Al_2O_3$, MgO-CaO, and $MgTi_2O_5-TiO_2$ systems. In all of these systems at solidification rates less than approximately 2 cm/hr, well-aligned eutectic microstructures were obtained as shown in Figure 2.

The basic microstructures of ceramic eutectics can be generalized from the relative interfacial surface area per unit volume (A_v) for the fibrous and lamellar ideals. For fibers, $A_v = 2/\lambda(\pi v/0.866)^{-1/2}$; for lamellae, $A_v = 2/\lambda$, where λ is the fiber/lamellae spacing. Previously reported eutectic microstructures were compiled and are plotted as a function of the volume percent minor phase in Figure 3. Eutectics that solidify in a fibrous morphology are shown as points on the "fibers" curve at the volume percent of the fibrous components; those with a lamellar microstructure are points on the "lamellae" horizontal line. Excellent agreement exists. Similar agreement exists for many metal systems. The basis for this representation is the concept originally advanced by Cooksey et al.[18] for minimization of the interfacial surface area in the development of a lamellar or fibrous eutectic microstructure. In spite of the excellent correlation of microstructures with the volume percent of the minor phase, it is not uncommon to find that eutectics predicted to exhibit a highly oriented lamellar microstructure can occasionally solidify in a fibrous microstructure, as has been observed in both the $SrZrO_3-ZrO_2$[17] and the MgO-CaO[6] eutectics.

The variation of the interlamellar or interfiber spacing with the inverse square root of the solidification rate for oxide eutectics is shown in Figure 4. The predicted relation, $\lambda^2 R = C$[19], is generally followed in both the fibrous and lamellar microstructures with a substantial variation in the proportionality constant, C, from $1 \times 10^{-17} m^3/s$ for $Al_2O_3-ZrO_2$ to 45 $\times 10^{-17} m^3/s$ for $CaZrO_3-ZrO_2$.

Crystallography

Minford et al.[17] investigated the crystallographic features of a number of directionally solidified oxide eutectics by the Buerger precession x-ray method. A zero-level Buerger precession pattern of the $SrZrO_3$ $-ZrO_2$ eutectic whose solution reveals that solidification direction $||[100]_{SrZrO_3}|| \sim[112]_{ZrO_2(m)}$ and lamellar interface $||(001)_{SrZrO_3}||$ $(111)_{ZrO_2(m)}$. Similar Buerger precession patterns were taken and analyzed for the eutectics listed in Table 1. The interfacial planes and solidification directions are all low Miller indices regardless of the crystal structures of the phases. The orientation relations of directionally solidified oxide eutectics appear to result from two factors: (a) the minimization of the lattice misfit or strains between the two phases, and (b) the ionic charge balance at the interface between the two phases. Frank and van der Merwe[20] proposed that the misfit between phases should be <16% for a semicoherent interface. With the exception of the $MgO-ZrO_2$ eutectic, all are well within this requirement.

Table 1. Crystallography of Directionally Solidified Oxide Eutectics

Constituents of eutectic	Crystal structure	Eutectic microstructure	Solidification direction	Orientation relation	Charge density at interface	Lattice misfit (%)
CaZrO₃	Orthorhombic perovskite	Lamellar I	[100]CaZrO₃ ~[112]ZrO₂(c)	Lamellar interface (001)CaZrO₃ (111)ZrO₂(c)	$0.089 +/Å^2$ $0.078 -/Å^2$	9.8
ZrO₂	Cubic	Lamellar II	[011]CaZrO₃ [011]ZrO₂(c)	Lamellar interface (100)CaZrO₃ (100)ZrO₂(c)	$0.089 +/Å^2$ $0.135 -/Å^2$	8.9
SrZrO₃ ZrO₂	Orthorhombic perovskite Monoclinic	Lamellar	[100]SrZrO₃ ~[112]ZrO₂(m)	Lamellar interface (001)SrZrO₃ (111)ZrO₂(m)	$0.085 +/Å^2$ $0.076 -/Å^2$	11.0
MgO MgAl₂O₄	Cubic Cubic	Fibrous (MgO)	[111]MgO [111]MgAl₂O₄	$(hkl)_{MgO} \| (hkl)_{MgAl_2O_4}$	No net charge	4.0
MgO ZrO₂	Cubic Cubic	Fibrous (MgO)	[111]MgO [111]ZrO₂(c)	$(hkl)_{MgO} \| (hkl)_{ZrO_2(c)}$	No net charge	18.8
MgO CaO	Cubic Cubic	Lamellar	[111]MgO [111]CaO	$(hkl)_{MgO} \| (hkl)_{CaO}$	No net charge	12.6
ZrO₂ Al₂O₃	Monoclinic Hexagonal	Fibrous (ZrO₂)	[100]Al₂O₃	(001)Al₂O₃∥{100}ZrO₂(m) (110)Al₂O₃∥<100>ZrO₂(m)	$0.154 +/Å^2$ for (100)ZrO₂(r) $0.165 -/Å^2$ for (001)Al₂O₃	5.9 for [001]ZrO₂(r)∥[110]Al₂O₃ 6.7 for [010]ZrO₂(r)∥[100]Al₂O₃
TiO₂ MgTi₂O₅	Tetragonal Orthorhombic	Platelets (TiO₂)	[001]TiO₂ [100]MgTi₂O₅	(010)TiO₂∥(010)MgTi₂O₅		

The importance of maintaining a charge balance across the lamellar interface is an extension of Pauling's second rule of ionic structures. Fragneau and Revcolevschi[7] discussed crystallographic relations between two solid solutions forming eutectic in the system NiO–CaO. The lamellar interfacial planes between NiO and CaO phases were (111) planes and the preferred growth direction for both phases [110]. For the NiO–Y₂O₃ eutectic Fragneau et al.[8] found the exitaxial relations to be $(1\bar{1}1)_{NiO} \| (110)_{Y_2O_3}$ and $[110]_{NiO} \| [001]_{Y_2O_3}$.

Mechanical Properties

A relation between lamellar spacing and strength somewhat like that which Petch[21] determined in relating yield strength to grain size has been observed for metallic eutectics. One would expect strength to increase with the decrease in λ or with the increase in the solidification rate. No such relationship has been found in oxide eutectic systems. Spacing does not effect the strength of MgO–MgAl₂O₄[15], ZrO₂–MgO[16], or ZrO₂–CaZrO₃[12] eutectics, or else the flexural strength of these eutectics remains approximately constant with the increase in the solidification rate. The strength-controlling microstructural parameter appears to be colony size or more likely, the width of the colony boundaries. The colony size remains essentially constant over the wide range of solidification rates studied. To observe any effect of interphase spacing on strength, it would be necessary to investigate only colony-free specimens.

Microhardness measurements of the MgO–MgAl₂O₄ eutectic[15] at room temperature show that the solidification rate affects the microhardness. The results shown in Figure 5 when plotted vs the inverse square root of the interfiber spacing, yield and approximately straight line, indicating Petch-type behavior. The indentations were made near the cell centers on the surface cut perpendicular to the growth direction. The smaller the interfiber spacing, the greater the hardness. The microhardness of the eutectic is greater than that of either MgO or MgAl₂O₄. Direct comparisons of microhardness are difficult; however, reliable data suggest values of ~850 for MgO and 1175 for MgAl₂O₄.

Minford et al.[2] measured the creep properties of the MgO–MgAl₂O₄ directionally solidified eutectic with colony and grain microstructures and a nearly stoichiometric MgAl₂O₄ single crystal. Eutectic specimens solidified at 8 cm/hr possessed colony microstructure, while specimens solidified at 1 cm/hr were free of colonies but were composed of large columnar grains. The creep resistance at 1600°C of the grain structure was greater than that of the colony structure and of MgAl$_2$O$_4$ single crystal (Figure 6). Voids occur along the grain and colony boundaries in region of creep of specimens subjected to tensile stress, which suggests that a void nucleation and growth process may best describe the deformation mechanism in this eutectic.

It is further evident from these results that the colony boundaries are the reason for the decrease in the mechanical properties of eutectics. Specimens without colony boundaries can be expected to possess superior mechanical properties. Because of the great potential of directionally solidified oxide eutectics for high temperature applications, investigation of specimens free of colonies and columnar grains should be pursued.

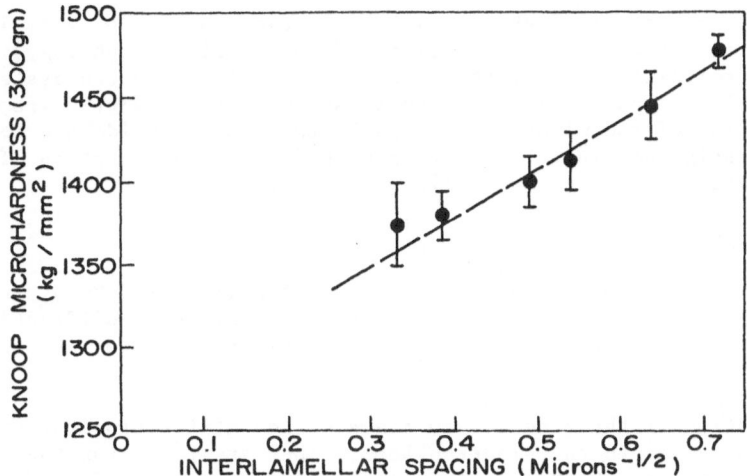

Figure 5. Variation in Knoop microhardness with fiber spacing for MgO-Mg$_2$Al$_2$O$_4$ eutectic.

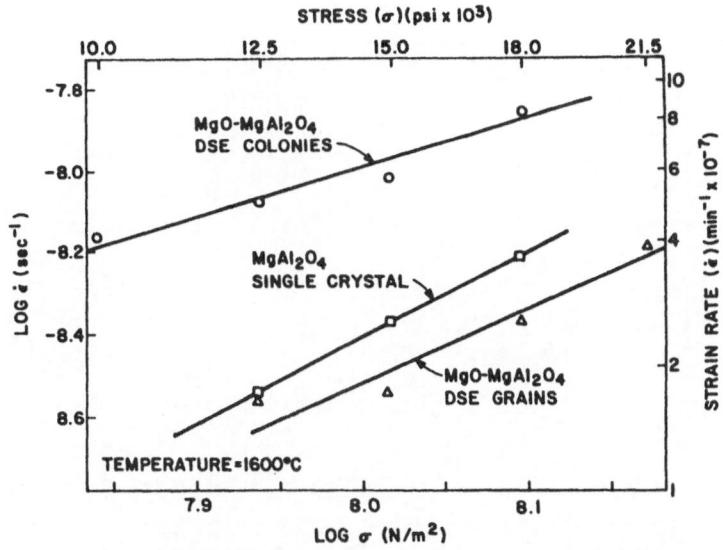

Figure 6. Variation of the mean strain rates with outer fiber stress at 1600°C for the directionally solidified MgO-MgAl$_2$O$_4$ eutectic with a colony and grain structure and single crystal MgAl$_2$O$_4$.

Microstructure

Sorrell et al.[9] studied the directional solidification of both the ZrC-ZrB$_2$ eutectic and the ZrC-TiB$_2$ eutectic. Both were directionally solidified by using the floating zone technique. The ZrC-ZrB$_2$ eutectic consisted of pore-free columnar grains of variable size with parallel lamellae within the grains. The typical microstructure of this eutectic is shown in Figure 7. Attempts to determine the shape of the interface by rapid quenching were unsuccessful, but it may be assumed that planar growth was attained because there were no colonies at growth rates less than ~3 cm/hr. The interlamellar spacings obeyed the $\lambda^2 R = C$ law over the limited growth rates investigated.

The microstructure of the ZrC-TiB$_2$ eutectic consists of very large grains of broken and deformed lamellae (Chinese script morphology). The directional solidification of the eutectic composition in the system SiC-B$_4$C gives a lamellar type of microstructure, as could be predicted from the volume fraction of the minor phase. As in the other carbide and boride systems investigated, interlamellar spacing is a linear function of the inverse square root of the solidification rate.

Crystallography

Crystallography of the eutectic systems TiC-TiB$_2$, TiC-ZrB$_2$, and ZrC-ZrB$_2$ was investigated by using selected area electron diffraction. In each of these systems the carbide phase is cubic and the boride phase is hexagonal.

Application of the zone law to the electron diffraction patterns gives directions parallel to the electron beam. They are [01$\bar{1}$] for the cubic phase (TiC, ZrC) and [01.0] for the hexagonal phase. Therefore, these directions are parallel. It is also observed that the (111) and (00.1) planes of the cubic and hexagonal phases are parallel, and consequently so

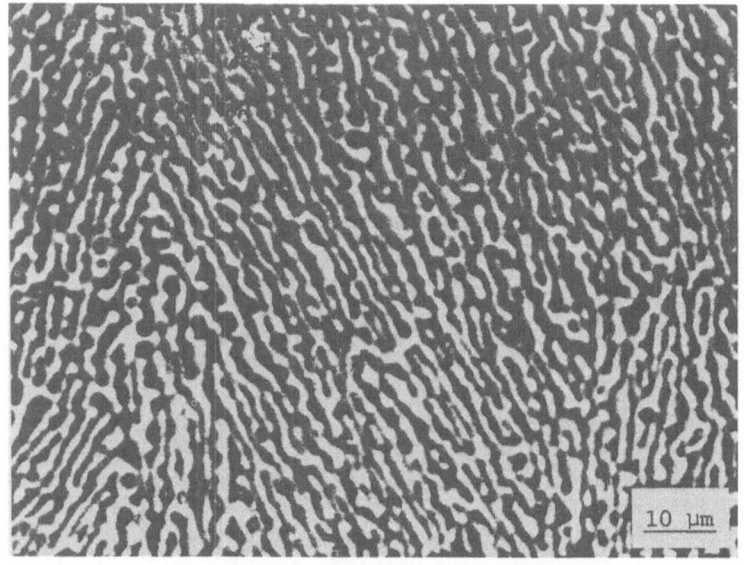

Figure 7. ZrC-ZrB$_2$ eutectic solidified at 1.7 cm/hr.
Transverse section.

are the [111] cubic and [00.1] hexagonal directions. Combining the selected area electron diffraction patterns and the micrographs of the microstructure defines the interfacial planes. In these systems they are the (111) cubic and (00.1) hexagonal planes. This plane pair is a sensible combination because both planes have low Miller indices and they are close packed (Figure 8) with alternating planes consisting of only one type of atom. The misfit between the two phases is only 1.6% for the $TiC-TiB_2$ eutectic, 4.6% for the $ZrC-ZrB_2$ eutectic, and ~9.1% for $ZrC-TiB_2$ eutectic. These are well below the border value of 16% misfit for a semicoherent interface.

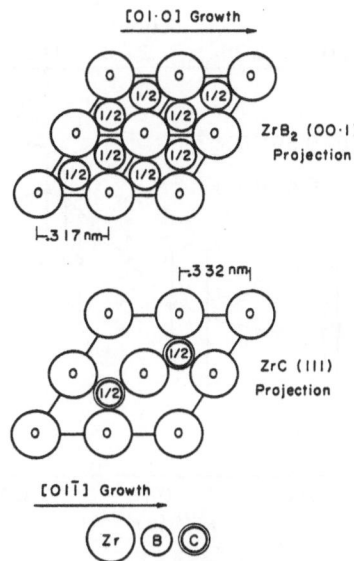

Figure 8. Interfacial planes and growth directions of the $ZrC-ZrB_2$ eutectic.

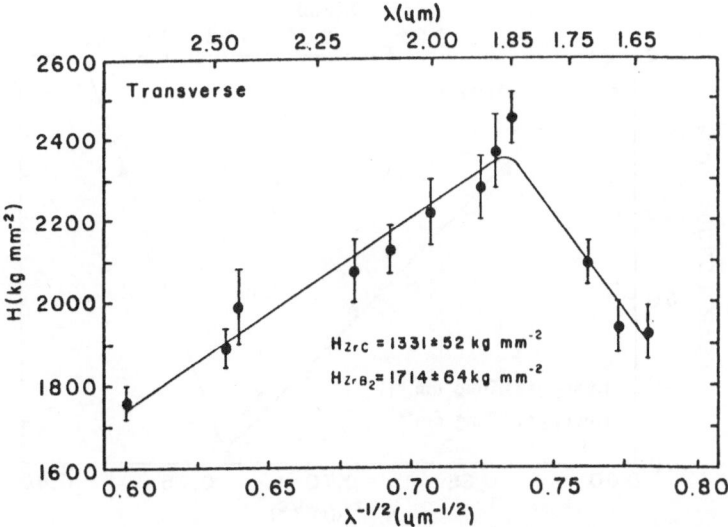

Figure 9. The Knoop microhardness (500 g load, transverse section) vs. $\lambda-1/2$ (λ is the interlamellar spacing) for $ZrC-ZrB_2$ eutectic.

Mechanical Properties

Microhardness is a property that should be strongly influenced by the local microstructure. Hardness has been most closely related to the ease with which dislocations move. Dislocations move easily through perfect crystals, but are impeded by any obstruction. Theoretically, the stress needed to move dislocations is inversely related to the interphase spacing in a two-phase system. Thus a fine, coherent eutectic may be expected to have a drastic effect on the hardness of the material.

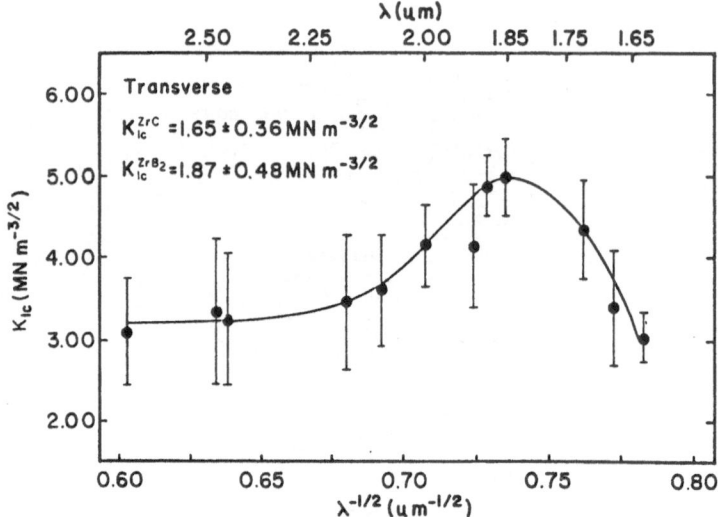

Figure 10.   The Vickers fracture toughness (1500 g load, transverse section) vs. $\lambda^{-1/2}$ for ZrC–ZrB$_2$ eutectic.

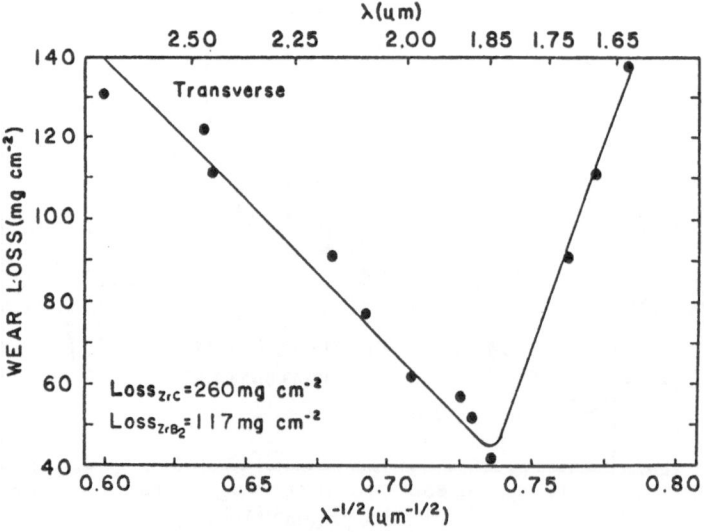

Figure 11.   The wear loss (transverse face) of the ZrC–ZrB$_2$ eutectic (250 cycles, 250 g load).

The transverse microhardness for $ZrC-ZrB_2$ eutectic sample is shown in Figure 9. The hardness of nearly all directionally solidified samples are greater than those of the individual components.[23] Particularly notable is the maximum hardness corresponding to $\lambda = 1.85$ μm, its hardness approaches nearly twice that of the single components. The hardness exhibits Hall-Petch behavior from $\lambda = 2.75$ to 1.85 μm, which corresponds to growth rates of $R = 0.4$ to 2.9 cm/hr; at this point, the hardness decreases. The expected leveling off of the hardness has not been observed in aligned eutectics; only the former portion of the behavior is usually reported. It is probably more than coincidence that the microstructure breaks down and develops misorientation at this point. Similar behavior was observed with the longitudinal microhardness.

Evans and Wilshaw[24] also stated that the microhardness and the indentation fracture toughness of brittle materials may be related to each other. Therefore, it is expected that the fracture toughness might show the same trend as the hardness; Figure 10 bear out the expectations. The fracture toughnesses of the transverse and longitudinal sections of the directionally solidified $ZrC-ZrB_2$ eutectic samples increase with the decreasing interlamellar spacing, as did the hardness, and the K_{Ic} values are generally higher than those of individual components.

Wear loss data for the transverse faces of the ZrC-ZrB eutectic samples are shown in Figure 11. Since the wear loss was approximately linear with grinding time, the weight loss after 250 cycles with a 250-g load is plotted as function of $\lambda^{-1/2}$. As expected, the wear resistance data correlate well with those for the hardness and fracture toughness. A maximum in wear resistance was observed at $\lambda = 1.85$ μm ($R = 2.9$ cm/hr) and the wear resistance of nearly all the directionally solidified specimens is superior to that of the individual components.

It was also found that the directionally solidified $SiC-B_4C$ eutectic has much much better wear resistance than either B_4C or SiC.[22]

ACKNOWLEDGMENTS

This research was supported by the US Research Office, Durham, North Carolina, under grants No. 8115-MC and No. DAA29-70-G-0020.

REFERENCES

1. C. Herring and J. K. Galt, Elastic and Plastic Properties of Very Small Metal Specimen, Phys. Rev., 85:1060 (1952).
2. V. S. Stubican and R. C. Bradt, Eutectic Solidification in Ceramic Systems, Ann. Rev. Mater. Sci., 11:267 (1981).
3. Hurt and D. Viechnicki, in Ultrafine Grain Ceramics, ed. J. J. Burke, N. L. Reed and V. Weiss, Syracuse University Press, p. 273 (1970).
4. D. Viechnicki and F. Schmid, Eutectic Solidification in the System $Al_2O_3/Y_3Al_5O_{12}$, J. Mat. Sci., 4:84 (1969).
5. F. L. Kennard, R. C. Bradt and V. S. Stubican, Eutectic Solidification of $MgO-MgAl_2O_4$, J. Am. Ceram. Soc., 56:566 (1973).
6. C. O. Hulse and J. A. Batt, The Effect of Eutectic Microstructures on the Mechanical Properties of Ceramic Oxides, United Aircraft Research Laboratory Final Report, N910803-10 (1974).
7. M. Fragneau and A. Revcolevschi, Crystallography of Directionally Solidified NiO-CaO Eutectic, J. Am. Ceram. Soc., 66:C162 (1983).
8. M. Fragneau, A. Revcolevschi and D. Michel, Crystallography of Directionally Solidified $NiO-Y_2O_3$ Eutectic, J. Am. Ceram. Soc., 65:C102 (1982).

9. C. C. Sorrel, H. R. Beratan, R. C. Bradt and V. S. Stubican, Directional Solidification of (Ti,Zr) Carbide-(Ti,Zr) Boride Eutectics, J. Am. Ceram. Soc., 67:190 (1984).

10. F. Schmid and D. Viechnicki, Oriented Eutectic Microstructures in the system Al_2O_3/ZrO_2, J. Mater. Sci., 5:470 (1970).

11. D. J. Rowcliffe, V. J. Warren, A. G. Elliot and W. S. Rothwell, The Growth of Oriented Ceramic Eutectics, J. Mater. Sci., 4:902 (1969).

12. C. O. Hulse and J. A. Batt, Fracture of directionally Solidified $CaO.ZrO_2-ZrO_2$ Eutectic, in Fracture Mechanics of Ceramics, Vol. 2, ed. R. C. Bradt, D.P.H. Hasselman and F. F. Lange, Plenum Press, New York, NY, p. 483 (1974).

13. F. S. Galasso, W. L. Darby, F. C. Douglas and J. A. Batt, Unidirectional Solidification of the $BaFe_{12}O_{19}-BaFe_2O_4$ Eutectic, J. Am. Ceram. Soc., 50:333 (1967).

14. D. E. Harrison, Lamellar Glass-Crystal Structures in the System $ZnO-B_2O_3$, J. Crystal Growth, 3-4:674 (1968).

15. F. L. Kennard, R. C. Bradt and V. S. Stubican, Mechanical Properties of the Directionally Solidified $MgO-MgAl_2O_4$ Eutectic, J. Am. Ceram. Soc., 56:160 (1976).

16. F. L. Kennard, R. C. Bradt and V. S. Stubican, Directional Solidification of the ZrO_2-MgO Eutectic, J. Am. Ceram. Soc., 57:428 (1974).

17. W. J. Minford, R. C. Bradt and V. S. Stubican, Crystallography and Microstructure of Directionally Solidified Oxide Eutectics, J. Am. Ceram. Soc., 62:154 (1979).

18. D.J.S. Cooksey, D. Munson, M. P. Wilkinson and A. Hellawell, The Freezing of Some Continuous Binary Eutectic Mixtures, Phil. Mag., 10:745 (1964).

19. H. W. Weart and D. J. Mack, Trans. AIME, 212:664 (1958).

20. F. C. Frank and van der Merwe, One Dimensional Dislocations II, Proc. Royal Soc., A198:216 (1949).

21. N. J. Petch, The Cleavage Strength of Polycrystals, J. Iron Steel List., 174:25 (1953).

22. J. D. Hong, K. E. Spear and V. S. Stubican, Directional Solidification of $SiC-B_4C$ Eutectic: Growth and Some Properties, Mat. Res. Bull., 14:775 (1979).

23. C. C. Sorrel, V. S. Stubican and R. C. Bradt, Mechanical Properties of the $ZrC-ZrB_2$ and $ZrC-TiB_2$ Directionally Solidified Eutectics, to be published in J. Am. Ceram. Soc.

24. A. G. Evans and T. R. Wilshaw, Quasi Static Solid Particle Damage in Brittle Solids I: Observations, Analysis and Implications, Acta Met., 24:939 (1976).

NICKEL OXIDE - BASED ALIGNED EUTECTICS

Alexandre Revcolevschi

Laboratoire de Chimie Appliquée
Université Paris-Sud, Bâtiment 414
91405 Orsay Cédex, France

INTRODUCTION

Because of their potentially interesting anisotropic properties, many directionally solidified eutectic materials have been studied in the past two decades. Mostly metallic systems have been investigated, the main purpose being the improvement of mechanical properties, particularly in the field of turbine blade alloys. More recently, various ceramic eutectic systems have been studied[1] and some physical properties have also been examined[1-3].

The object of the present article is to review in particular the work carried out in the author's laboratory, on several nickel oxide-based aligned eutectics. The choice of NiO as a common oxide to several binary systems, and the selection of particular binary systems was guided by the purpose of studying interfaces associating the simple sodium chloride-type structure of NiO with phases corresponding to other structure types. Furthermore, NiO was selected among other oxides exhibiting a NaCl-type structure (FeO, MnO, CoO,..) because of its stability in air from the melting point ($T_F \sim 1970°C$) to room temperature. Another reason for the choice of nickel oxide was our experience of crystal growth of mono and bicrystals of this oxide[4] and the available data on the structure and properties of grain boundaries in NiO[5-7].

The systems which have been studied to date and which will be treated are : NiO-CaO (sodium chloride) ; $NiO-NiAl_2O_4$ (spinel) ; NiO-stabilized cubic ZrO_2 (fluorite) ; $NiO-Y_2O_3$ (fluorite-related bixbyite) ; $NiO-Gd_2O_3$ (monoclinic B structure of the rare earth sesquioxides) ; $NiO-La_2NiO_4$ (K_2NiF_4 structure).

We shall firstly consider the techniques which were used for the growth of aligned eutectics and discuss the resulting microstructures. We shall then examine the crystallographic relationships between the phases formed during eutectic solidification, discuss the crystallographic nature of the interface, and analyze the transition of one phase to the other. Finally, we shall report some very recent results relative to the chemical reduction of these composite materials.

METHODS OF GROWING OXIDE EUTECTICS

Most of the eutectic temperatures of oxide systems are very high
(> 1500°C) and therefore require special equipment. Several methods have
been reported in the literature [3] : they involve either refractory metal
crucibles heated by graphite or floating zone techniques associated with
R.F.-heated carbon susceptors. The disadvantages of these high-temperature
methods are (i) a risk of contamination of the oxides by the crucible,
(ii) a possible non-stoichiometry of the resulting oxide material, due to
the necessity of operating under vacuum or controlled atmosphere, to
prevent carbon oxidation.

The two methods which were used to grow the nickel oxide-based
eutectics avoid these drawbacks, since they involved crucible-free techniques and were carried out in air.

The first method was a floating zone technique associated with a double
ellipsoid image furnace [8,9] in which a molten zone was established between two
cylindrical rods prepared by compression and sintering (Fig. 1) : the rods
were rotated at about 60 rpm in opposite directions to enhance mixing in the
liquid and translated at a rate of typically 0.5 to 2 cm.h^{-1}. The starting
rods for the floating zone experiments were prepared by pressing (1 kbar)
of finely ground optimum mixtures of 99.999 % pure NiO and > 99.9 % pure
oxides or carbonates. The pressed rods, 10 to 12 mm in diameter, were sintered in air at 1200°C for 24 h.

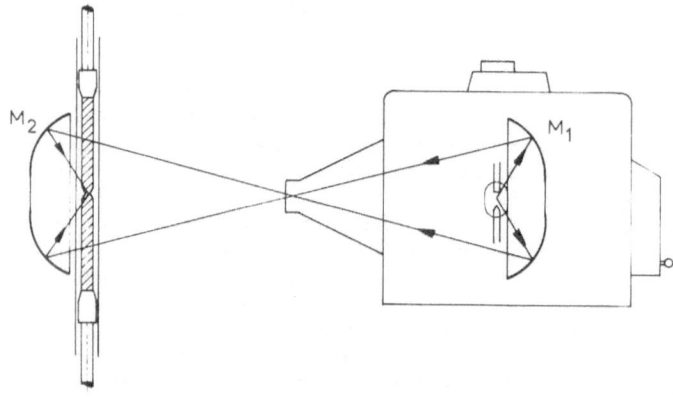

Fig. 1. Floating-zone crystal-growth device associated with an image furnace.

The other technique was the "skull method" , currently used in industry
for the fabrication of single crystalline stabilized zirconia : it consists
of R.F. direct induction-heating of batches of a few hundred grams contained
in a water-cooled copper crucible, about 5 cm in diameter and 6 cm high.
Preheating of the powder mixtures was achieved by using chunks of metal corresponding to one of the oxides. An alternative method is to use a carbon
rod placed inside the powder until direct R. F. coupling occurs. After
melting, the crucible content was slowly cooled, yielding directional solidification at rates of the order of a few mm per hour.

Both methods apparently correspond to plane-front growth conditions
as evidenced by the regular microstructures which were obtained even at
off-eutectic compositions.

Well aligned structures were obtained in all the systems discussed here.

The NiO-CaO phase diagram consists of an eutectic at ∿ 1720°C and two terminal solid solutions[10]. The eutectic structure is lamellar and consists of NiO and CaO solid solutions which have lattice parameters of 4.203 A and 4.754 Å respectively as opposed to 4.168 Å and 4.810 Å for pure oxides. The composition for which the best microstructures were obtained (absence of primary phases and regular structures) was 60 mol % NiO, i.e., slightly different from that indicated for the eutectic in the phase diagram[11].

The phase diagram of the NiO-Y_2O_3 system indicates a eutectic at ∿ 1700°C and a composition of 22 mol % Y_2O_3, and two terminal phases with no detectable solubility of one oxide into the other[12]. Very regularly aligned lamellar structures were obtained at 26 mol % Y_2O_3 (Fig. 2) in spite of a recent redetermination, by solar furnace thermoanalysis, of the eutectic composition at 31 mol % Y_2O_3[13].

Fig. 2. Longitudinal section of NiO-Y_2O_3 eutectic obtained by directional solidification at 2.6 cm h^{-1}. Scale bar, 15 μm.

It seems that, as mentioned earlier, the high temperature gradient available in the techniques which were used, particularly that involving thermal imaging, makes possible off-eutectic growth, so that knowing the exact position of the eutectic in the phase diagram is of secondary importance.

In the NiO-Al_2O_3 system, there is a eutectic between spinel-type NiAl$_2$O$_4$ and a NiO solid solution. The eutectic is near 1900°C and at a composition which is about 14 mol % Al_2O_3. The aligned eutectic structure is made of fibers of NiO solid solution (a = 4.19 Å) embedded in a spinel type matrix (a = 8.071 Å)[13].

In the absence of any available phase diagram for the NiO-ZrO_2 system a systematic study of various mixtures of NiO, ZrO_2, and CaO (the latter being used to stabilize ZrO_2 in the cubic form) was carried out, the proportions of zirconia and calcia being chosen in such a manner that a ZrO_2-15 mol % CaO stabilized zirconia should result in the structure. The best results in terms of absence of primary phase and regularity of the structure were obtained for 70 mol % NiO - 25.5 % NiO - 4.5 % CaO. The lattice parameter of the zirconium oxide phase is a = 5.13 Å which is that of 15 % calcium oxide-stabilized zirconia, indicating that all the calcium oxide has gone into that phase. This is confirmed by the lattice parameter of the nickel oxide phase a_{NiO} = 4.178 Å, which is that of pure NiO. The very regular eutectic structure which was obtained was lamellar (Fig. 3)[14].

Fig. 3. Transverse section of the nickel oxide/calcium oxide-stabilized zirconia eutectic (70 mol % NiO - 25.5 mol % ZrO_2 - 4.5 mol % CaO) solidified at 1.5 cm h^{-1}. Scale bar, 50 μm.

Very recent studies were carried out in the NiO-Gd_2O_3 and NiO-La_2O_3 systems. Phase diagrams were available from the work of B. Willer[15] and J. Cassedane[16]. In the gadolinium sesquioxide-based system, one again is concerned with a very simple phase diagram which makes possible bringing together at eutectic interfaces the two terminal phases of the diagram. The eutectic temperature and composition are T ∿ 1540°C and 55 mol % NiO, respectively. No reciprocal solubility is noted for either phase. The structures are lamellar and the lattice parameters are those of the pure oxides. As for NiO-La_2O_3, growth was carried out at compositions close to 70 mol % NiO, i.e., near the eutectic indicated at T ∿ 1650°C between pure NiO and La_2NiO_4. The structure was found to be fiber-type and indeed consisting of the pure oxides: cubic NiO (a = 4.17 Å) and tetragonal La_2NiO_4 (a = 3.88 Å ; c = 12.65 Å), (Fig. 4).

The occurrence of either lamellar or rod-type structures in the several systems discussed above is in good agreement with the calculations relative to lamellar-to-rod transitions in a eutectic, which established that lamellar structures should be stable when the minor component has a volume

a   b

Fig. 4. Longitudinal (a) and transverse (b) sections of the eutectic NiO-
    La_2NiO_4 solidified at 1.5 cm h^{-1}. Scale bar, 10 μm.

fraction higher than 0.3[17,18]. The only exception to this rule was found in
the case of the $NiO-Al_2O_3$ system for which, in spite of volume fractions of
0.42 for NiO and 0.58 for $NiAl_2O_4$, fiber type aligned structures were
obtained[3]. A few anomalies of this kind have already been observed.

    The effect of the growth rate R on the interlamellar spacing λ of the
eutectic was studied in several of these NiO-based systems and it was found
that the variation of λ with R follows a relation relatively close to the
λ^2R = constant law proposed by Tiller[19] and verified in many eutectic systems
for both fibrous and lamellar structures (Fig. 5). An exception was found
for NiO-CaO where a law of the type λ^5R = constant seemed best to describe
the results. No explanation of this discrepancy was found.

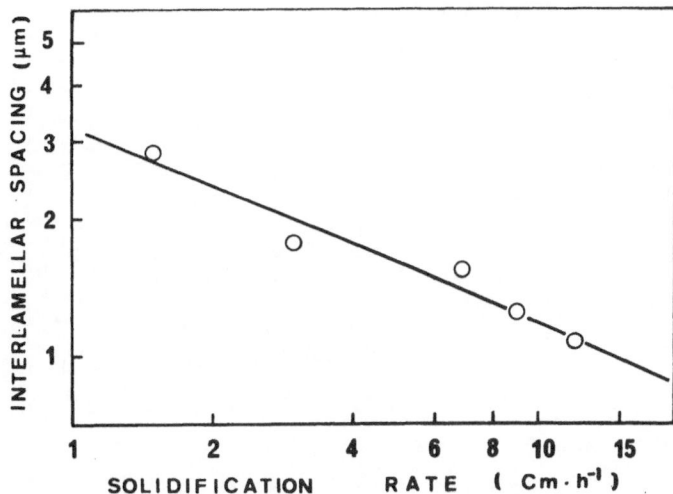

Fig. 5. Variation of interlamellar spacing with solidification rate for
    the $NiO-ZrO_2$ eutectic.

As mentioned earlier, aligned eutectics, with the large interfacial areas they supply, are ideal material for a better understanding of two-phase solid interfaces. X-ray diffraction was used as the main technique to study the crystallography of the eutectics.

Microfragments of the directionally solidified samples oriented along the growth direction were studied by rotating crystal, Weissenberg and Buerger precession methods to determine preferred growth directions and epitaxial relations (Fig. 6-9). The patterns revealed that growth directions have low Miller indices and that precise and simple orientation relations exist between the two phases (Table I). The information concerning growth directions of the eutectic phases obtained from these microfragments was confirmed by X-ray diffractometry experiments carried out on large transverse sections of solidified ingots : high intensity peaks corresponding to planes normal to growth directions were observed.

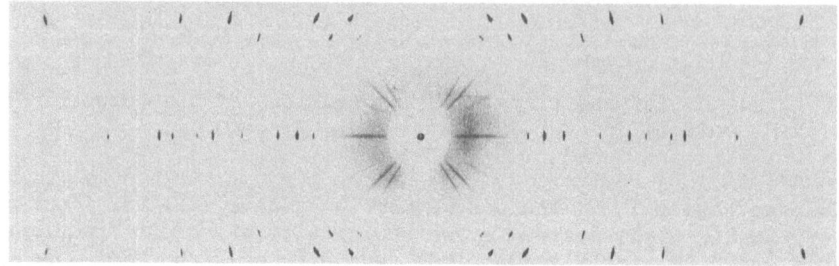

Fig. 6. Rotating crystal photograph, showing <110> axis rotation for both phases of the NiO-CaO eutectic (Cu K$_\alpha$).

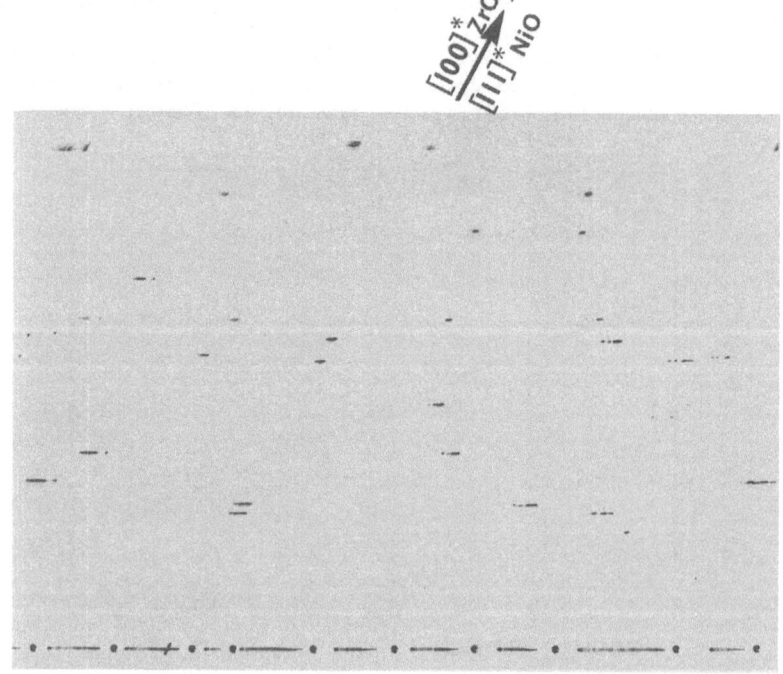

Fig. 7. Weissenberg zero level photograph normal to [001] of ZrO$_2$ and [1$\bar{1}$0] of NiO showing the parallelism of [100]$^*_{ZrO_2}$ and [111]$^*_{NiO}$.

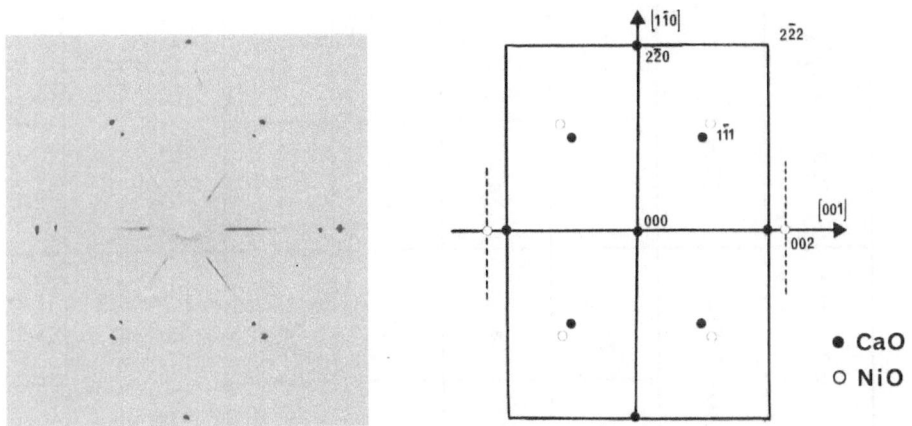

Fig. 8. Buerger precession photograph and corresponding reciprocal plane indexation for NiO-CaO (zero level, <110> axis rotation, Cu K_α).

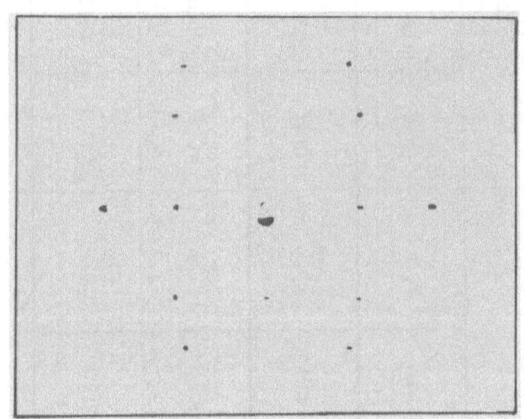

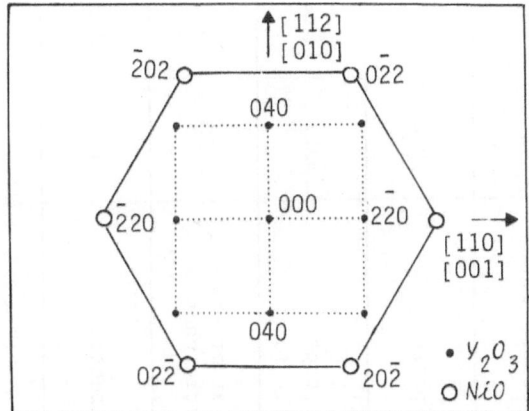

Fig. 9. Buerger pattern and corresponding reciprocal plane indexation for NiO-Y_2O_3 (sample oriented along growth directions $[110]_{NiO}$ and $[001]_{Y_2O_3}$ - Mo K_α).

Table I. Crystallographic relations at the lamellar interfaces

	NiO – CaO		NiO – Y$_2$O$_3$		NiO – Al$_2$O$_3$		NiO – ZrO$_2$(CaO)		NiO – Gd$_2$O$_3$	
Eutectic composition	40 mol % CaO		31 mol % Y$_2$O$_3$		14 mol % Al$_2$O$_3$		70 mol % NiO		55 mol % NiO	
Eutectic phases	NiO	CaO	NiO	Y$_2$O$_3$	NiO	NiAl$_2$O$_4$	NiO	ZrO$_2$	NiO	Gd$_2$O$_3$
Structure type	NaCl	NaCl	NaCl	pseudo CaF$_2$	NaCl	MgAl$_2$O$_4$	NaCl	CaF$_2$	NaCl	B form of Ln$_2$O$_3$
Lattice parameters of the eutectic phases (Å)	4.20	4.75	4.18	10.60	4.19	8.07	4.18	5.12	4.18	a =14.07 b = 3.57 c = 8.73 α =99.9°
Growth direction	<110>	<110>	<110>	<001>	<111>	<111>	<110>	<001>	<110>	<020>
Orientation relations	(hkl) // (hkl)		[1$\bar{1}$0] // [001] (111) // (100)		(hkl) // (hkl)		[10$\bar{1}$] // [001] (111) // (100)		[1$\bar{1}$0] // [010] (111) // (20$\bar{1}$)	
Interface plane	(111)	(111)	(111)	(100)	fibers		(111)	(100)	(111)	(20$\bar{1}$)

To specify the Miller indices of the interphase planes in the case of lamellar structures, a special procedure was applied. Fragments of the directionally solidified samples having a morphology which was previously characterized by optical observation and X-ray diffraction, were placed on a goniometer head and observed by S.E.M. The orientation of the lammellae was estimated micrographically from their intersection with the surfaces of the fragments (Fig. 10) : interfaces are (111) for both NiO and CaO phases in the NiO-CaO system[11]. They are $(111)_{NiO}$ and $(100)_{Y_2O_3}$ for the NiO-Y_2O_3 system[20], $(111)_{NiO}$ and $(100)_{ZrO_2}$ for NiO-ZrO_2, and $(111)_{NiO}$ and $(20\bar{1})_{Gd_2O_3}$ for the NiO-Gd_2O_3 eutectic[21].

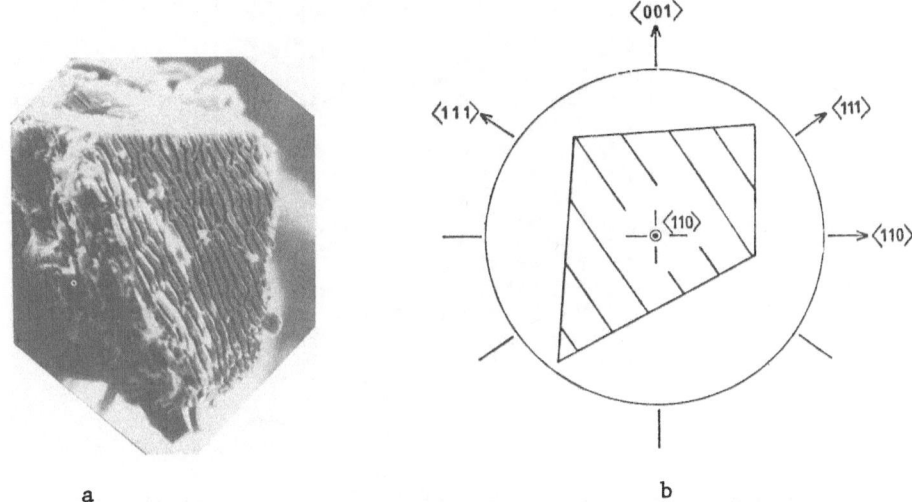

Fig. 10. a) Scanning electron micrograph of fractured elongated microfragment of NiO-CaO eutectic studied by X-ray diffraction (view along fragment axis).

   b) Schematic representation of microfragment habit of fig. 10.a and corresponding crystallographic indices.

Fig. 11 represents a computer simulation of the atomic arrangements corresponding to the epitaxy between four-fold and three-fold symmetry surfaces in the particular case of NiO-Y_2O_3 : the structure of yttria (bixbyite-type) may be described as a fluorite superstructure obtained by ordering of oxygen vacancies on the anionic sublattice and is represented here by a (100) fluorite-type atomic distribution from which one of every four atoms of oxygen is removed.

INTERFACE STRUCTURES

Considering the data presented in table I, it is interesting to represent the structure of the oxides involved in lamellar eutectics, perpendicularly to growth directions, in order to describe atomic arrangements near the interfaces (Fig. 12).

In the case of the NiO-CaO eutectic, made of nickel oxide and calcium oxide solid solutions, the interface can be easily described by (111) alternate planes of metal and close packed oxygen : the metal planes are richer in nickel on one side of the interface and in calcium on the other side, since one is concerned here with solid solutions (Fig. 12.) The distances between anionic planes in each phase differ only slightly.

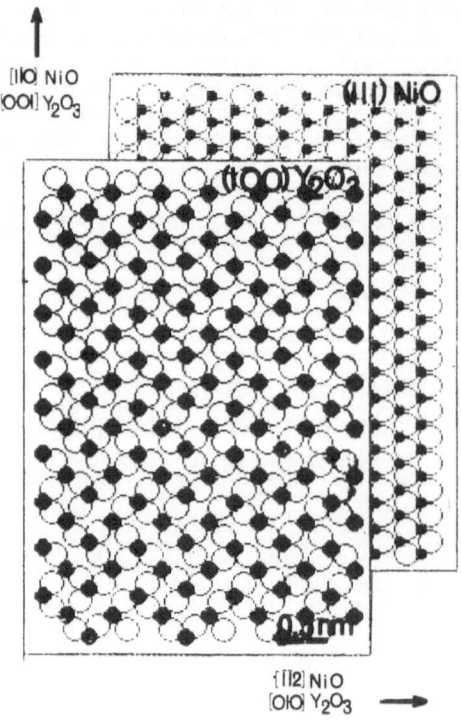

Fig. 11. Computer simulation of the atomic arrangement at the NiO-Y₂O₃ interface, viewed perpendicularly to the growth direction and showing
the mating of three-fold and four-fold symmetry surfaces. The same
crystallographic relationship is observed for the NiO-ZrO₂ eutectic.

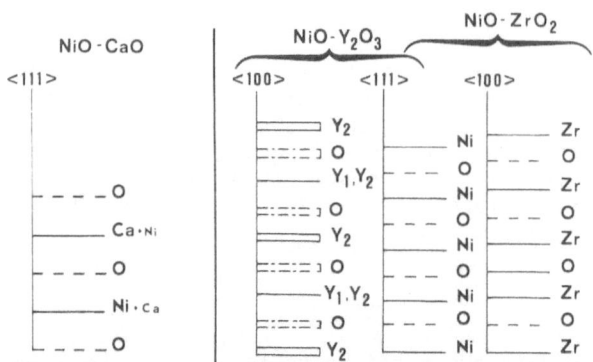

Fig. 12. Representation of the NiO-CaO interface and of the atomic layers
of Y_2O_3, NiO, and ZrO_2, viewed perpendicularly to growth directions
(i.e. parallel to interfaces).

In the case of the NiO-Y₂O₃ eutectic where there is no detectable solubility of one oxide in the other, a description of the interface in terms
of oxygen and metal planes may be also considered (Fig. 12). This description is complicated by the fact that in bixbyite the oxygen vacancy distribution defines two different crystallographic positions for yttrium atoms
and one set of oxygen sites in the cubic structure[22] : along a <100> direction, atoms are not strictly coplanar and define either planes or "sheets"

(Fig. 12.)[20]. Because of the existence of two different types of cationic layers in Y_2O_3, three different alternative forms of epitaxy can be considered in order to satisfy the observed $(100)_{Y_2O_3}$ // $(111)_{NiO}$ relation at the interface. It was shown that the most probable interface structure would be that in which a nickel oxide oxygen plane would be followed by an yttrium plane, itself followed by an yttria type oxygen sheet[20] : this structure is the only one in which the six-fold coordination observed for Y in Y_2O_3 is maintained. Note also that in both NiO and Y_2O_3 phases, the distances between two consecutive metal-oxygen layers, along the direction perpendicular to the growth direction, are very near one another.

The analysis of the $NiO-ZrO_2$ interface seems easier because of the simpler fluorite structure of ZrO_2, viewed perpendicularly to <100> in Fig. 12.: one would have at the interface, successively, zirconium, oxygen and nickel planes. Of course this rough description does not take into account the small distortions associated with the vacancies introduced in the ZrO_2 oxygen sublattice by the calcium oxide which stabilizes the cubic modification of zirconia. Here again, note that in both NiO and ZrO_2 phases the distances between two consecutive metal-oxygen layers along the direction perpendicular to the growth direction, are not very different.

In the case of the $NiO-Gd_2O_3$ eutectic, the analysis of the transition from one phase to the other at the interface, requires a close examination of the structure of Gd_2O_3. The structure of the B form of rare earth sesquioxides is made of layers of composition Ln_2O_3 at y = 0 and y = 1/2. Parallel to the xy plane, layers of oxygen alternate with layers of rare earth atoms. Each rare earth atom has seven oxygen neighbors[23,24]. It is worthwhile giving here also the original description proposed by Caro for the A and B forms of the rare earth sesquioxides Ln_2O_3, the B form corresponding to a slight distortion of the A hexagonal form[25]. The structure is made of a two-dimensional packing of (Ln_4O) tetrahedra, sharing edges and forming an infinite polymeric complex cation $(LnO)_n^{n+}$; these layers are separated and bounded by oxygen planes (Fig. 13). This description involves a predominantly covalent Ln-O bond inside the layers, and an ionic type bond between the layers and the oxygen ion planes. The plane of these layers is (00.1) in the hexagonal A form of Ln_2O_3 and $(20\bar{1})$ in the monoclinic B form. This layered character of the structure is confirmed by the easy cleavage of rare earth oxide single crystals along these planes. If now the structures of

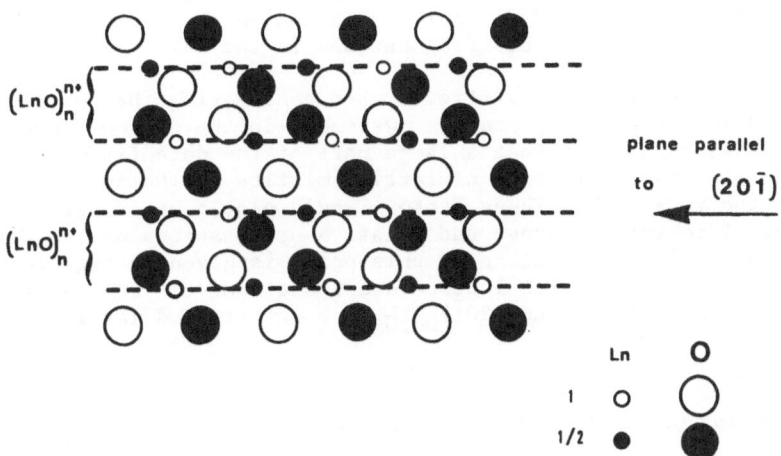

Fig. 13. Projection of the monoclinic B-structure of the rare earth sesquioxides Ln_2O_3 down the b axis and indication of the lamellar character of the structure.

125

NiO and Gd_2O_3 are examined perpendicularly to $<111>_{NiO}$ and to a direction normal to $(20\bar{1})$, it is possible to visualize the atomic arrangement at the interface : the transition from NiO to Gd_2O_3 would occur through a succession of layers "...oxygen - nickel - oxygen - $(LnO)_n^{n+}$ - oxygen - ...".

DISCUSSION OF THE CRYSTALLOGRAPHY OF EUTECTICS

These results bring additional data to the already large variety of crystallographic characteristics determined for non-metallic eutectics. Obviously it appears that in binary structures involving phases of cubic symmetry, the $<110>$, $<111>$, and $<100>$ growth directions prevail. This is true for the systems mentioned in this review but also for $MgO-MgAl_2O_4$, MgO-cubic ZrO_2, and MgO-CaO where growth directions are $<111>$ for all phases[2] . This was also observed in sodium chloride-type halide eutectics such as NaCl-NaF and LiF-NaF[26,27]. This holds also for the fluorite-type phases of the ZrO_2-Ln_2O_3 (Ln = Nd, Sm, Dy) eutectics[28] and for the cubic ZrO_2 phase of the ZrO -$CaZrO_3$ eutectic[2] , which all grow along $<110>$.

When comparing growth directions in structurally homologous systems, we find it hard to make generalizations :(if we note a good agreement between data relative to the NaCl-spinel type systems $NiO-NiAl_2O_4$ and MgO-$MgAl_2O_4$[29], on the other hand, in the case of NaCl-fluorite type systems, we find that the similar results concerning $NiO-Y_2O_3$ and $NiO-ZrO_2$ differ from those proposed by Kennard et al[30] for $MgO-ZrO_2$ where $<111>_{MgO}$ // $<111>_{ZrO_2}$.) Discrepancies exist also among the NaCl-NaCl type systems where $<100>_I$ // $<100>_{II}$, $<111>_I$ // $<111>_{II}$, and $<110>_I$ // $<110>_{II}$ relations have been found[2,27] . These discrepancies are not surprising : indeed, if one considers that in lamellar oriented solidification, interfaces are parallel to growth directions, growth axes should be directions of strong atomic bonding or more precisely of "periodic bond chains"[31] contained in the interface plane. Hence, in one such plane a very limited number of growth directions might be possible, among which, as seen so far in most systems, only one would prevail ; other would possibly be stabilized by local perturbations during the growth process, or by impurities. This seems to be confirmed by observations made for example in the case of the $NiO-ZrO_2$ eutectic for which, besides the major growth axes indicated in Table I, was found, on microfragments oriented along the growth direction and studied by the Weissenberg method, a growth direction $[21\bar{3}]_{NiO}$ // $[103]_{ZrO_2}$. This relationship is equivalent, within a few minutes precision, to $<110>_{NiO}$ // $<001>_{ZrO_2}$ proposed in Table I and indicates that relative lattice orientations are the same in this fragment and in the bulk.

Generalizations also seem difficult when considering the epitaxy relations reported so far for non-metallic systems[3], in the case of lamellar structures. It seems, however, that a large part of the data fit with the well accepted concepts of minimization of lattice misfits and balance of charge densities at the interfaces. These factors would play a major part in the establishment of interface planes and relative orientations of the lattices of the two phases. An illustration of this point is given in fig. 14 which represents the superposition of lattice cross-sections corresponding to interfacial planes $(111)_{NiO}$ and $(20\bar{1})_{Gd_2O_3}$: a rather good lattice matching may be noted.

CHEMICAL REDUCTION OF OXIDE - OXIDE EUTECTICS

Chemical reduction experiments were carried out recently on lamellar NiO-based eutectic structures. The studies were conducted under conditions where only one of the oxides, namely NiO, is reduced into metal. The purpose of the experiments was twofold : (i) evaluating the possibility of

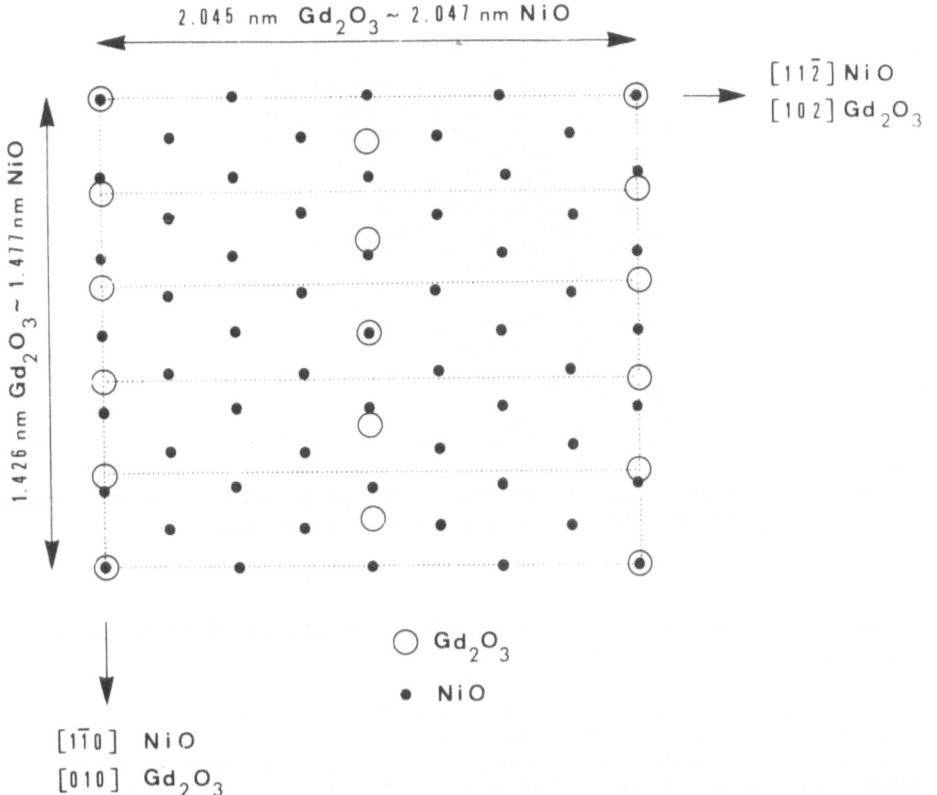

2.045 nm $Gd_2O_3 \sim 2.047$ nm NiO

$[11\bar{2}]$ NiO
$[10\,2]$ Gd_2O_3

1.426 nm $Gd_2O_3 \sim 1.417$ nm NiO

$[1\bar{1}0]$ NiO
$[010]$ Gd_2O_3

○ Gd_2O_3

• NiO

Fig. 14. Superposition of lattice cross-sections corresponding to interfacial planes $(111)_{NiO}$ and $(20\bar{1})_{Gd_2O_3}$.

fabricating aligned metal – oxide lamellar structures, which could not
be grown directly from the melt ; (ii) providing a new type of metal-oxide
interfaces, the crystallography of which could be related to that of the
original oxide-oxide duplex structure.

Studies were performed in particular on $NiO-ZrO_2$ samples which
exhibit a very regular lamellar structure and from which it was expected to
fabricate a fine scale structure made of alternate lamellae of ionically
conducting (ZrO_2) and electronically conducting (Ni) phases.

Directionally solidified samples, 3 mm thick, were submitted for 70
hours to a chemical reduction treatment at 1075°C under a CO/CO_2 mixture
($CO/CO_2 = 0.1$) corresponding to $P_{O_2} = 1.6\ 10^{-11}$ atm., i.e., conditions where
NiO should be totally reduced to nickel[32].

The microstructure resulting from reduction of the samples is presented
in fig. 15 which indicates that the aligned nature of the original eutectic
structure and the cohesion between the two phases are maintained. Powder
X-ray diffraction experiments performed on crushed samples show total reduction of NiO, the only observable lines being those of pure nickel (a = 3.52 Å)
and of unaltered calcia-stabilized zirconia, the lattice parameter of which
is the same as that found in the original eutectic structure. In some areas
of the samples, elongated cavities, presumably caused by the elimination of
oxygen from the nickel oxide phase, can be observed in the nickel phase. A
few cracks can also be noted, the region adjacent to the cracks being
depleted in cavities, indicating cavity coalescence. Observation at higher

Fig. 15. Well-aligned area of Ni-ZrO$_2$ sample prepared by reduction, under CO/CO$_2$, of NiO-ZrO$_2$ eutectic. Scale bar, 15 μm.

magnification by scanning electron microscopy confirms these features and particularly the boundary coherence.

X-ray diffractometry experiments performed on transverse sections of samples of oxide-oxide eutectic structure submitted to reduction by CO/CO$_2$ mixtures show, in comparison with the powder diffractograms of the same material, very strong (220) nickel and (200) and (400) zirconia peaks (Fig. 16). Considering the results concerning the orientation of the two phases of the NiO-ZrO$_2$ samples, indicated in Table I, it appears that the oriented nature of the ZrO$_2$ phase remains unchanged ; the nickel phase seems also to be well oriented, the (110) orientation of the nickel lamellae being that of the original nickel oxide phase. Taking into account the fcc structure

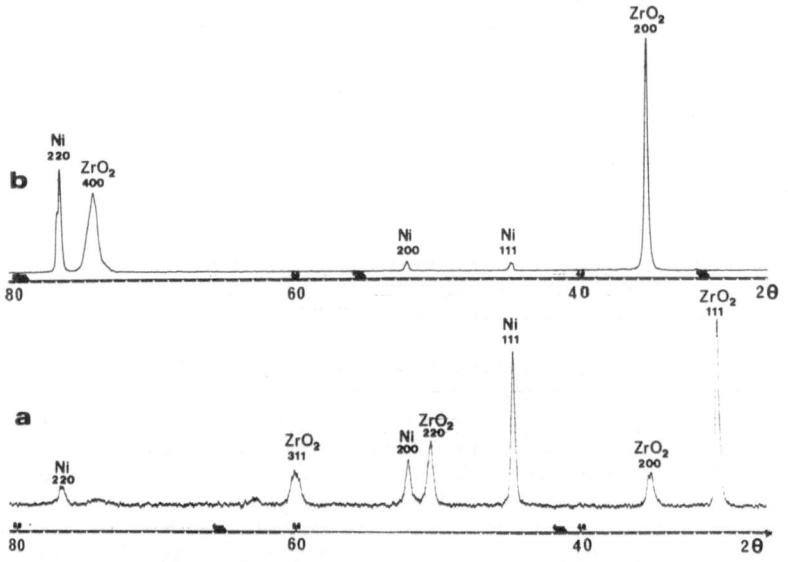

Fig. 16. X-ray diffractometry of the material resulting from reduction of eutectic NiO-ZrO$_2$. a) Powder pattern.
b) Transverse section of eutectic structure.

of nickel and the NaCl-type structure of nickel oxide, it appears that the reduction takes place by a topotactical process in which the departure of oxygen from the NiO phase leaves intact the fcc nickel sublattice which contracts, preserving the monocrystalline nature of the nickel-based phase[33]. Work is currently in progress to examine at a finer scale, i.e., on microfragments of the crystallographic relationships between metal and oxide and their evolution with temperature.

CONCLUSION

The present review relative to crystallographic relations and interface planes in NiO-based eutectics covers a very limited number of systems, so that generalizations can be hardly made. It appears however that, in comparison with metallic systems, for which a large variety of growth directions or interface orientations have been found for one specific system (e.g. Pb-Sn or Al-Al₃Ni), one finds less dispersed results for oxide-oxide eutectics. A unifying element seems to be the presence of densely packed oxygen planes which obviously control crystallographic relationships and interface orientations. Use of T.E.M. and introduction of coincidence site lattice theory should improve our description of oxide-oxide interfaces.

REFERENCES

1. R. L. Ashbrook, "Directionally Solidified Ceramic Eutectics", J. Am. Ceram. Soc., 60 [9-10]:428 (1977).
2. W. J. Minford, R.C. Bradt and V.S. Stubican, "Crystallography and Microstructure of Directionally Solidified Oxide Eutectics", J. Am. Ceram. Soc., 62 [3-4]:154 (1979).
3. V. S. Stubican and R. C. Bradt, "Eutectic Solidification in Ceramic Systems", Ann. Rev. Mater. Sci., 11:267 (1981).
4. G. Dhalenne, A. Revcolevschi and A. Gervais, "Growth of Oriented Nickel Oxide Bicrystals", J. Cryst. Growth, 44 [3]:297 (1978).
5. G. Dhalenne, A. Revcolevschi and A. Gervais, "Grain Boundaries in Nickel Oxide (I)", Phys. Status Solidi A, 56 [1]:267 (1979).
6. G. Dhalenne, A. Revcolevschi and C. Monty, "Grain Boundaries in Nickel Oxide (II)", Phys. Status Solidi A, 56 [2]:623 (1979).
7. G. Dhalenne, M. Déchamps and A. Revcolevschi, "Relative Energy of <011> Tilt Boundaries in NiO", J. Am. Ceram. Soc., 65 [1]:C-11 (1982).
8. A. Revcolevschi, "Arc Image Furnace for X-Ray Diffraction Studies up to 3000°C and High Temperature Crystal Growth", Rev. Int. Hautes Temp. Réfract., 7 [1]:78 (1970).
9. R. T. Cox, A. Revcolevschi and R. Collongues, "Growth of O¹⁷ Enriched Al₂O₃ Crystal by a floating Zone Technique", J. Cryst. Growth, 15 [4]:301 (1972).
10. D. E. Smith, T. Y. Tiem and L. H. Van Vlack, "The System NiO-CaO", J. Am. Ceram. Soc., 52 [8]:459 (1969).
11. M. Fragneau and A. Revcolevschi, "Crystallography of the Directionally Solidified NiO-CaO Eutectic", J. Am. Ceram. Soc., 66 [9]:C-12 (1983).
12. E. N. Timofeeva, N. I. Timofeeva, L. N. Drozdova and O. A. Mordovin, "Reaction of Nickel (II) Oxide with Yttrium Oxide", Izv. Akad. Nauk SSR, Neorg. Mater., 5 [6]:1155 (1969).
13. M. Fragneau, Thesis, Paris XI (1983).
14. G. Dhalenne and A. Revcolevschi, "Directional Solidification in the NiO-ZrO₂ System", J. Cryst. Growth, 69:616 (1984).
15. B. Willer, Thesis, Strasbourg (1970).
16. J. Cassedane, Anais da Acad. Brasileira de Ciencias, 36:13 (1964).
17. D. J. S. Cooksey, D. Munson, M.P. Wilkinson and A. Hellawell, "Freezing of Some Continuous Binary Eutectic Mixtures", Phil. Mag, 10 [107]: 745 (1964).

18. J. D. Hunt and K. A. Jackson, "Lamellar and Rod Eutectic Growth", Trans AIME, 236 [6]:1129 (1966).
19. W. A. Tiller, "Polyphase Solidification", Liquid Metals and Solidification, in Am. Soc. For Metals, Cleveland, OH, p. 276 (1958).
20. M. Fragneau, A. Revcolevschi and D. Michel, "Crystallographic Study of the Lamellar Interface of the Y_2O_3- NiO Eutectic" in "Advances in Ceramics", vol. 6, The American Ceramic Society, Columbus, OH (1983).
21. B. Dubois, G. Dhalenne, F. d'Yvoire and A. Revcolevschi, "Crystallography of the Directionally Solidified $NiO-Gd_2O_3$ Eutectic", submitted to J. Am. Ceram. Soc.
22. B. O'Connor and T. M. Valentine, "Neutron Diffraction Study of the Crystal Structure of the C-form of Yttrium Sesquioxide", Acta Crystall. Section B, B25 [10]:2140 (1969).
23. D. T. Cromer, "The Crystal Structure of Monoclinic Sm_2O_3", J. Phys. Chem., 61:753 (1957).
24. H. L. Yakel, "A Refinement of the Crystal Structure of Monoclinic Europium Sesquioxide", Acta Cryst., B [35]:564 (1969).
25. P. E. Caro, "OM_4 Tetrahedra Linkages and the Cationic Group $(MO)_n^{n+}$ in Rare Earth Oxides and Oxysalts", J. Less-Common Metals, 16:367 (1968).
26. J. W. Moore and L. H. Van Vlack, "Preferred Orientation in Microstructure of Eutectics between Compounds", J. Am. Ceram. Soc., 51 [8]:428 (1968).
27. D. Penfold and A. Hellawell, "Microstructures of Alkali Halide Eutectics LiF-NaF and NaF-NaCl", J. Am. Ceram. Soc., 48 [3]:133 (1965).
28. D. Michel, Y. Rouaux, M. Perez-y-Jorba, "Ceramic Eutectics in the System $ZrO_2-Ln_2O_3$ (Ln = Lanthanide) : Unidirectional Solidification. Microstructural and Crystallographic Characterization", J. of Mat. Science, 15 [1]:161 (1980).
29. F. L. Kennard, R. C. Bradt and V. S. Stubican, "Eutectic Solidification of $MgO-MgAl_2O_4$", J. Am. Ceram. Soc., 56 [11]:566 (1973).
30. F. L. Kennard, R. C. Bradt and V. S. Stubican, "Directional Solidification of the ZrO_2-MgO Eutectic", J. Am. Ceram. Soc., 57 [10]:428 (1974).
31. P. Hartman, Z. Krist, 119:65 (1963).
32. A. Dominguez-Rodriquez and J. Castaing, "Deformation plastique de l'oxyde de nickel monocristallin", Revue Phys. Appl., 11:387 (1976).
33. A. Revcolevschi and G. Dhalenne, "Crystallographically Aligned Metal-Oxide Composite Structure made by Reduction of a Directionally Solidified Oxide-Oxide Eutectic", Nature, 316:335 (1985).

IMPERFECTIONS IN THE DIRECTIONALLY SOLIDIFIED
STRUCTURE OF NiO-CaO EUTECTIC

B. J. Pletka

Department of Metallurgical Engineering
Michigan Technological University
Houghton, MI 49931

ABSTRACT

Imperfections in the microstructure of the directionally solidified NiO-CaO lamellar eutectic were examined by optical, scanning electron, and transmission electron microscopy. Lamellar terminations and mismatch boundaries were observed in all eutectic grains on transverse sections. The lamellar interfacial planes between the NiO and CaO phases were found to be approximately $(111)_{NiO}$ parallel to $(111)_{CaO}$ in agreement with previous work. However, the orientation of the lamellar habit plane was determined to vary from an exact {111} plane from electron diffraction studies. Analysis of Kikuchi line patterns in electron diffraction patterns also showed that random variations in orientation up to 2° occurred between adjacent lamellae about the growth direction. In order to accommodate such variations in orientation, sub-boundaries formed within individual lamellae.

INTRODUCTION

The idealized microstructure of a directionally solidified lamellar eutectic would consist of alternate parallel lamellae of the two phases, free of defects, arranged so that a constant crystallographic orientation was maintained by each phase, i.e., each eutectic grain would be composed of two interpenetrating single crystals of the two component phases. Furthermore, a unique crystallographic orientation relationship would exist between the two phases such that a specific plane in each phase was parallel to the lamellar habit plane and a particular direction in each phase was parallel to the growth direction. In reality, these idealized conditions are not obtained and a variety of imperfections exist in the directionally solidified structure. For example, it has been shown in several metallic eutectic alloys that the orientation of the lamellar habit plane can vary up to 10° from being parallel to low index planes in the two phases and that random variations of 0.5 to 2° exist between adjacent lamellae (see references 1, 2, and 3 for reviews of this general topic). In addition, fault-free grains have been observed in only one metallic eutectic system[4].

The microstructures, crystallography and mechanical properties in a number of ceramic eutectic systems have been investigated (see references

5 and 6 for reviews). However, there has not been a recognition of the faults present in the eutectic grains. In addition, although the crystallography of the orientation relationships between the two phases has been examined in ceramic eutectic systems, these studies utilized x-ray diffraction techniques to characterize the orientation relationships. Since the x-ray beam will encompass hundreds of individual lamellae and usually a number of eutectic grains, it is the average orientation relationship that is determined in these studies. Details of the variation in crystallography of the two phases were only elucidated in metallic systems when transmission electron microscopy and in particular, precise electron diffraction techniques, were used to investigate the crystallography[7,8].

The purpose of the present paper is to demonstrate that faults do exist in the grains of directionally solidified lamellar eutectic ceramics and that variations do occur in the crystallographic orientation relationships between the phases comprising the lamellar eutectic. The specific system investigated was the NiO-CaO eutectic.

EXPERIMENTAL PROCEDURE

The air isobar for the Ni-Ca-O phase diagram is a simple binary eutectic with a reported eutectic reaction isotherm at approximately 1973K and eutectic composition at 42 mole % CaO[9]. Work in our laboratory has suggested the eutectic composition is closer to 37 mole % CaO. Both oxide phases have the rocksalt structure and exhibit rather limited solid solubility.

Lamellar eutectic microstructures were obtained by directional solidification in air in a vertical, floating zone, arc image furnace at growth rates between 20 and 128 mm·h^{-1}. The feed rods were isostatically pressed oxide powders (derived from the carbonates of nickel and calcium using the liquid mix technique[10,11]) that were sintered at 1723K for three hours. The arc image furnace was similar in design to that described by Cox et al[12] and additional details of the apparatus can be found in reference 13.

As-grown ingots contained a polycrystalline lamellar microstructure in which the eutectic grain boundaries were elongated parallel to the growth direction. Transverse sections were cut from the middle of the ingots to insure the lamellar interfaces were aligned nearly parallel to the growth direction; some curvature of the solid/liquid interface during directional solidification has been observed near the periphery of the ingot[13]. Specimens for optical, scanning electron, and transmission electron microscopy observations were prepared using standard techniques. Transmission electron microscopy was performed using a Philips 301 electron microscope operating at 100 kV or a JEOL 100 CX electron microscope operating at 120 kV.

RESULTS AND DISCUSSION

The analysis of imperfections in the directionally solidified structure of the NiO-CaO lamellar eutectic will be divided into two sections. The first will be concerned with the faults that exist within lamellar eutectic grains and the influence these faults may exert on the properties of the lamellar eutectic, in particular, the mechanical properties. The second section will deal with the nature and extent of the variance in crystallography that can exist between the two phases that comprise the eutectic structure.

Faulted Lamellar Microstructures

The faults observed within eutectic grains in metallic systems take several forms. The optical micrograph in Figure 1 shows a number of so-called mismatch boundaries that seem to divide the grain into numerous subgrains. Small changes in crystallographic orientation on the order of 1 to 4° and dislocation sub-boundary structures have been observed across the mismatch boundaries in metallic systems, consistent with the notion these regions can be considered as subgrains[1-3,8].

It should be recognized that the view of the mismatch boundaries in Figure 1 is just a section through a series of defects which lie roughly parallel to the growth direction and perpendicular to the lamellae planes; the entire defect has been defined as a mismatch surface[1]. In a transverse section, the defect appears as an irregularity in the packing of the lamellae on either side of the trace of the mismatch surface. In reference 1, the intersection of the mismatch surface with a transverse section was called a trace line; in this paper, we prefer and use the term, mismatch boundary. Mismatch boundaries exist either with a net difference in the number of lamellae across the boundary, a 'net fault', or no difference in the number of lamellae across the boundary, a 'no-net fault'. An example of a no-net fault is shown in Figure 2.

Within the microstructure, isolated lamellar terminations occasionally occur as shown in Figure 3. The presence of lamellar terminations in disrupting the regular lamellar microstructure is analogous to the disruption an edge dislocation creates in the atomic lattice. Since the presence of a lamellar termination causes a localized disturbance across several lamellae (Figure 3), a distinction between lamellar terminations and mismatch boundaries as different defects is somewhat arbitrary. However, the termination may be thought of as a linear "unit defect" which may or may not be contained along a mismatch boundary. (See reference 1 for additional details.)

Mismatch boundaries and lamellar terminations originate during solidification, and several models have been advanced for their formation[1,14-17]. While there is disagreement over which model is appropriate, it is generally agreed the presence of the mismatch boundaries during growth is to affect changes in orientation so that a preferred orientation relationship exists between the two phases. It has been firmly established that a preferred interfacial plane for each phase exists in both metallic and ceramic systems[1,6]. In NiO-CaO, the orientation relationship has been defined as[18]:

$$\text{lamellar plane} \quad || \; (111)_{NiO} \; || \; (111)_{CaO}$$

$$\text{growth direction} \quad || \; <110>_{NiO} \; || \; <110>_{CaO}.$$

The fact that a preferred orientation relationship exists between the phases in a lamellar eutectic has been interpreted as indicating that a low energy interface has been established, i.e., the energy resulting from the interphase area per unit volume of the lamellar microstructure can be minimized if the phases grew so as to select interfaces of lowest possible energy. A schematic representation of the interfacial energy versus orientation (phase-phase or interfacial planes) in one dimension is shown in Figure 4; a deep cusp is found at one particular orientation. When growth begins in a directionally solidified eutectic, the microstructure will adjust its orientation parameters at the growth front to conform to the desired orientation relationship. Grains unfavorably orientated will

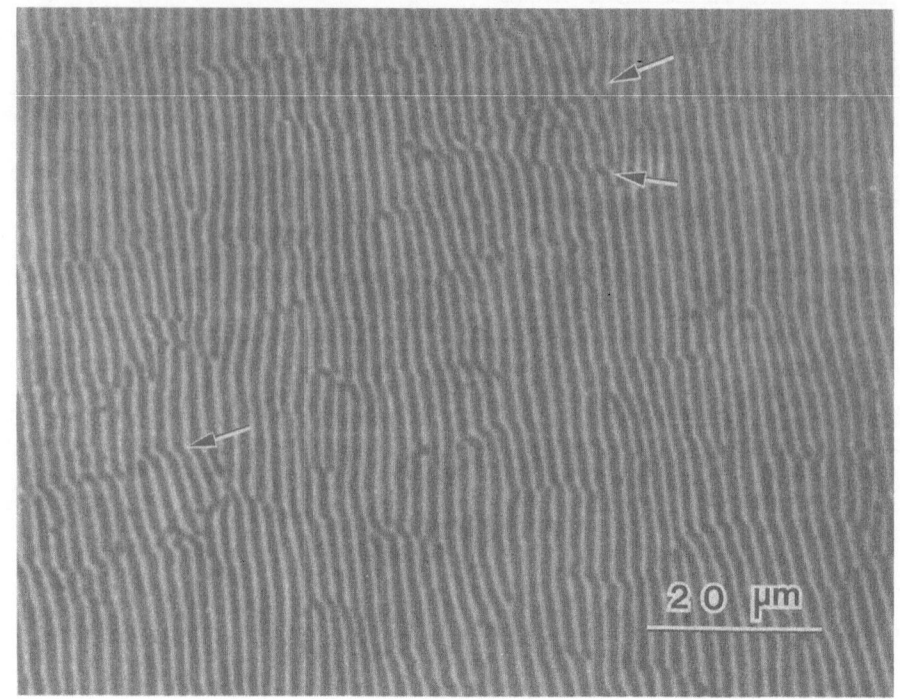

Fig. 1.   Optical micrograph of a transverse section with
          several mismatch boundaries indicated by arrows.

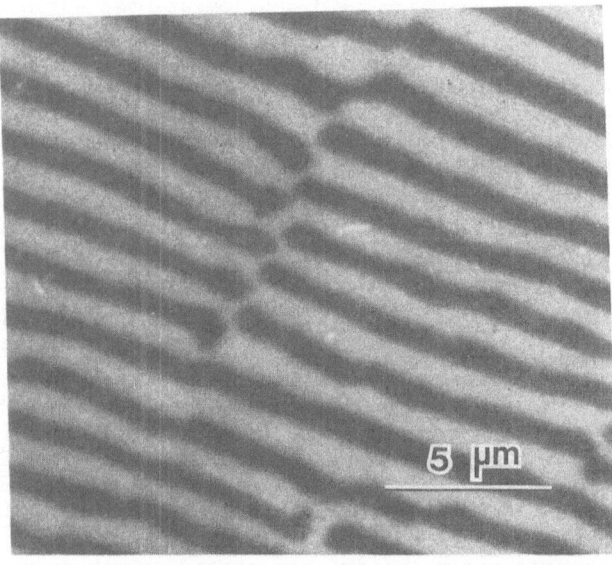

Fig. 2.   Example of a no-net fault; scanning
          electron micrography of a transverse
          section.

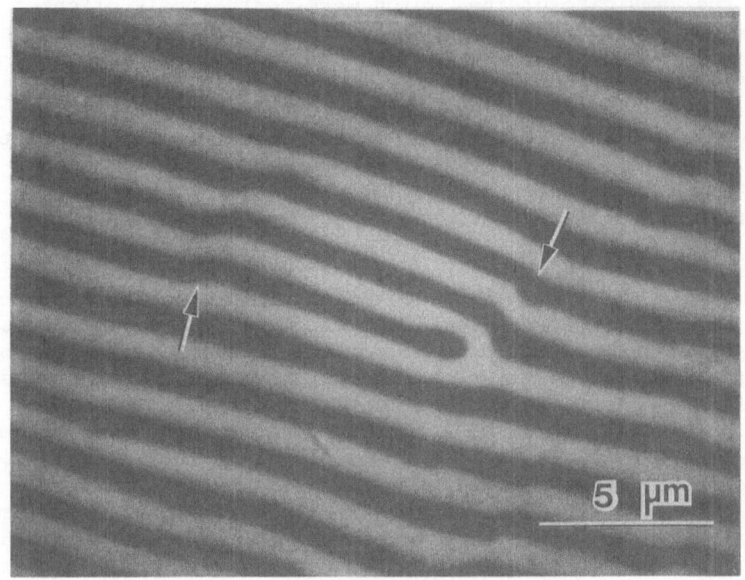

Fig. 3.   Scanning electron micrograph illustrating
           a lamellar termination and the localized
           disturbance it creates in the lamellae
           adjacent to it.

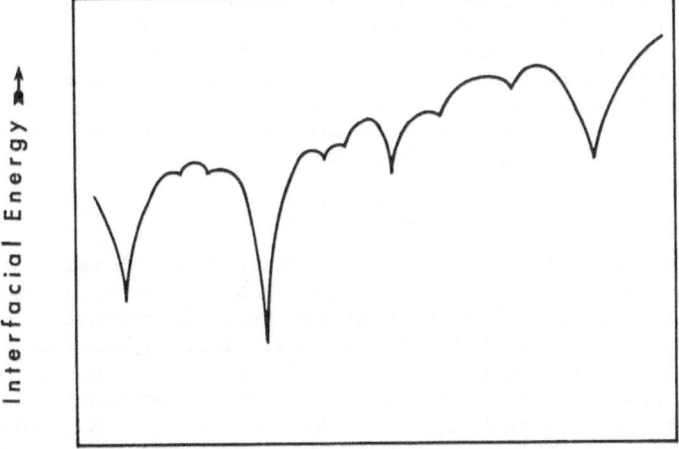

Fig. 4.   Schematic representation of interfacial energy
           versus orientation for a one-dimensional inter-
           face. The preferred orientation occurs at the
           deep cusp in the plot. After Fletcher[19].

grow out and the remaining grains, which were fortuitously oriented close
to the low energy configuration, will continue to adjust their orientation.
The mechanism by which these minor adjustments in orientation takes place
is the formation of mismatch boundaries. For example, the "kinks"
indicated by the arrows in Figure 3 in the vicinity of the lamellar
termination cause a change in lamellar direction with the possibility of
lowering the interfacial energy[1]. For a detailed discussion of how the
formation of mismatch surfaces can affect changes in orientation in order
to develop preferred orientations, see references 1 and 17.

Having established that faults in the form of mismatch boundaries and
lamellar terminations exist in NiO-CaO and other lamellar eutectic ceramic
systems (see Figure 5 in reference 6, for example), the obvious question
to be addressed is whether the presence of these faults affect any of the
properties of the eutectic system? Since an orientation change exists
across the mismatch boundary, degradation of electrical or optical
properties, for example, would be anticipated. Although no experimental
evidence apparently exists to verify or dispute this last statement, it
has been demonstrated recently in a metallic system that lamellar
terminations can affect the yield strength of a directionally solidified
single crystal eutectic[20]; single crystals, in which no eutectic grain
boundaries are present, have been grown in metallic systems. Figure 5
shows the engineering stress-strain curves for a series of Co-CoAl
specimens cut from a single ingot grown at the same solidification rate so
the interlamellar spacing was approximately constant among the specimens.
It can be seen that a large variation in yield strength exists between the
various Co-CoAl specimens. Examination of transverse sections cut from
the specimens revealed that a large variation existed in the termination
density between the specimens. The variation in termination density was
attributed to a region or regions of the ingot possessing a transitional
microstructure between the lamellar and rod microstructures which can both
exist in the Co-CoAl eutectic system. The termination density for each
specimen was measured and when plotted versus the yield strength, the
yield strength was found to decrease with increasing transverse
termination density (Figure 6). The observed dependence of yield strength
on lamellar termination density was attributed to a relaxation in
constraints on deformation of the two phases in that the lamellar
terminations act as localized perturbations in the eutectic
microstructure. It was proposed that intense deformation initiated in the
vicinity of the terminations (consistent with surface slip-line
observations) as yielding began. As the number of terminations increased,
slip could more easily spread throughout the specimen with a resultant
decrease in the macroscopic yield strength.

Crystallographic Variance in Eutectic Crystals

It was discussed earlier that a preferred orientation relationship
develops between the phases in order to achieve a low energy interface.
This would seem to indicate that an invariant crystallographic
relationship must exist between the two phases. This hypothesis appeared
to be confirmed by initial crystallographic determinations on metallic
eutectic systems using x-ray diffraction techniques. Subsequent work,
using electron diffraction techniques on the lamellar Al-CuAl$_2$ eutectic
system[7,8], demonstrated that variations up to ±10° occurred in the
relationship of the lamellar habit plane being parallel to {111} Al and
{211} CuAl$_2$. The variations in the lamellar plane orientation were
attributed to the fact that the boundary energy may not be the only
variable that controls the growth crystallography. It was suggested that
anisotropic growth kinetics may be important or that local growth

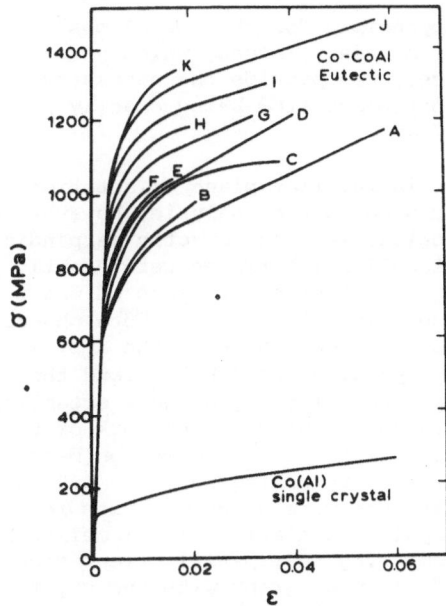

Fig. 5.   Engineering stress-strain curves for Co-CoAl
eutectic specimens deformed in compression at
an initial strain rate of 0.9 x 10^{-4} s^{-1} at room
temperature. The curve marked Co(Al) single
crystal represents the flow curve for a Co
specimen having the same approximate composition
(Co-6.4 wt.% Al) as the Co phase in the eutectic.

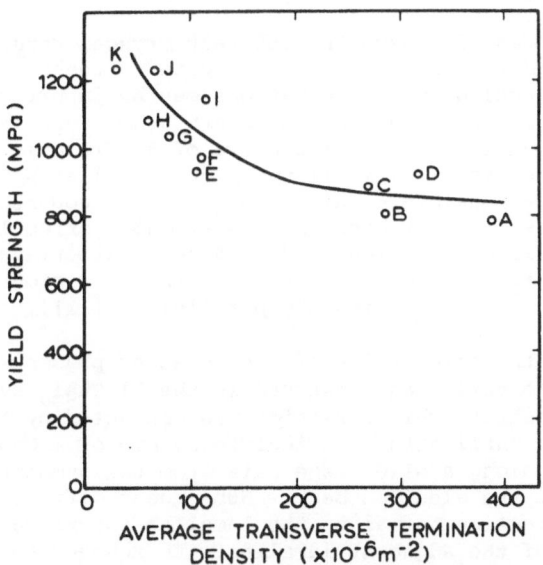

Fig. 6.   Yield strength versus transverse lamellar
termination density for the specimens
identified in Fig. 5.

fluctuations may be responsible for the variations[8]. It has been also proposed that arrays of boundary ledges, which have been shown to exist along eutectic interfaces, can provide the necessary misorientation to permit the lamellar habit plane from being exactly parallel to a low-energy orientation[21-25].

Similar variations in lamellar plane orientation were found in the NiO-CaO eutectic. A problem in the analysis, however, was the uncertainty that the transverse sections were cut exactly perpendicular to the growth direction so that the lamellar interfaces were exactly parallel to the electron beam. Therefore, the procedure adopted was to tilt a eutectic grain until the <110> zone (growth) axes of both phases in a specific region were parallel to the electron beam; the approximate epitaxial relationship of $(hkl)_{NiO}$ parallel to $(hkl)_{CaO}$ and the preferred growth direction approximately parallel to <110> were established from electron diffraction patterns, in agreement with the work of Fragneau and Revcolevschi[18]. The condition that the electron beam was parallel to the <110> zone axis was verified by making sure the Kikuchi lines in the electron diffraction patterns were symmetric with respect to the electron beam. The crystallographic orientation of a lamellar habit plane was then determined by simply comparing the spatial orientation of the lamellar interface in a bright field micrograph with the crystallographic orientation of each phase established from the analysis of the electron diffraction pattern.

Two examples of the analysis are shown in Figure 7. For both examples, the trace of a {111} plane is drawn on the micrograph. It can be seen that the trace is approximately parallel to the lamellar habit plane in Figure 7a but is approximately 13° away from the lamellar habit plane in Figure 7b. Although a more detailed analysis of the growth crystallography of NiO-CaO eutectics will be presented elsewhere, the example contained in the present paper clearly demonstrates that an invariant crystallographic habit plane is not established in this system.

Since both NiO and CaO have the rocksalt crystal structure, the idea that anisotropic growth kinetics may contribute to these variations is unlikely although local growth fluctuations may be important. An additional consideration is that if the sharp cusp drawn in the schematic interfacial energy versus orientation curve of Figure 4 is actually a shallow minimum, additional flexibility in the lamellar habit plane orientation would be possible. This hypothesis is supported by the observation that the lamellar habit planes are not perfectly straight along the lamellae length in Figures 7a and 7b. Rather, variations in habit plane orientation can be observed even in the region where the lamellar habit plane was approximately parallel to {111}.

Variations in the orientation of the adjacent phases were also observed, similar to variations observed in the Al-CuAl$_2$ system[7,8]. The experiment to demonstrate this variation was conducted by tilting a region of a eutectic grain until either an individual NiO or a CaO lamellae was orientated exactly along a <110> zone axis from the symmetry of the Kikuchi lines about the electron beam. Subsequent diffraction patterns were taken in a direction normal to the lamellar interface and the zone axis orientations of the adjacent lamellae with respect to the reference lamellae were established by analyzing the Kikuchi line patterns; changes in orientation of 0.1° can be detected from the analysis of Kikuchi line patterns[26].

Several examples of the analysis are shown in Figure 8. In Figure 8a, the misorientations of the <110> zone axes for the first four lamellae with respect to the reference CaO lamellae are all about 0.8°.

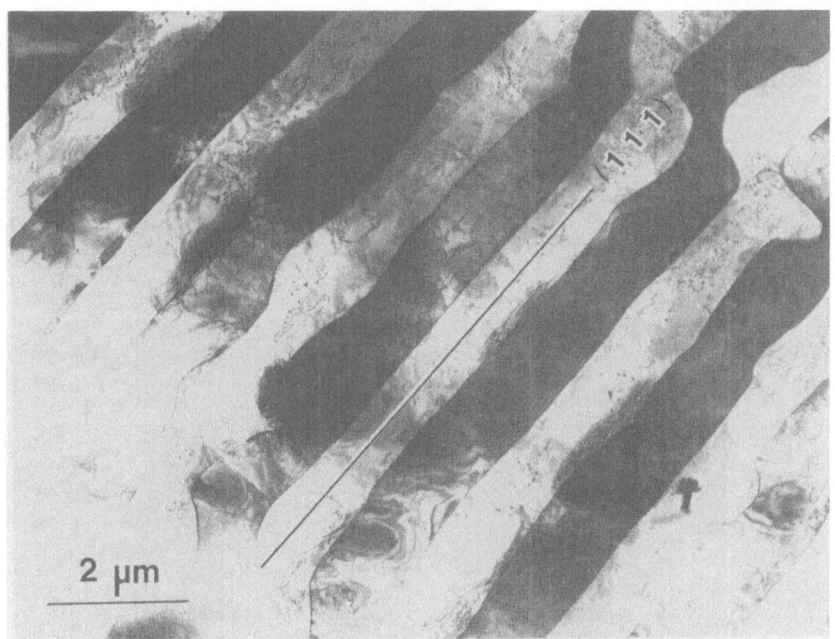

Fig. 7a. Transmission electron micrograph of a transverse
section; trace of a {111} plane approximately
parallel to the lamellar habit plane is shown.

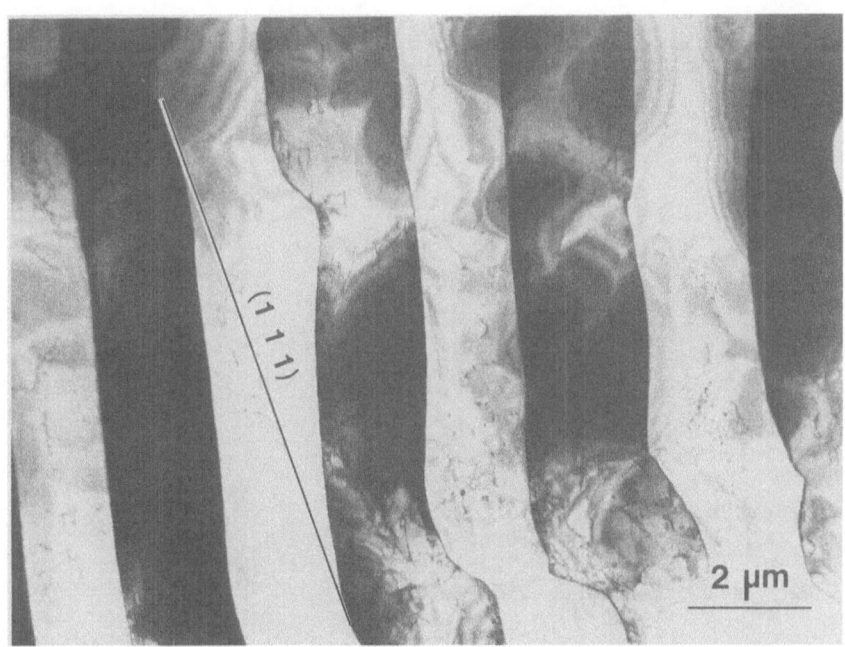

Fig. 7b. Transmission electron micrograph of a transverse
section; trace of a {111} plane that is approximately
13° away from the lamellar habit plane is shown.

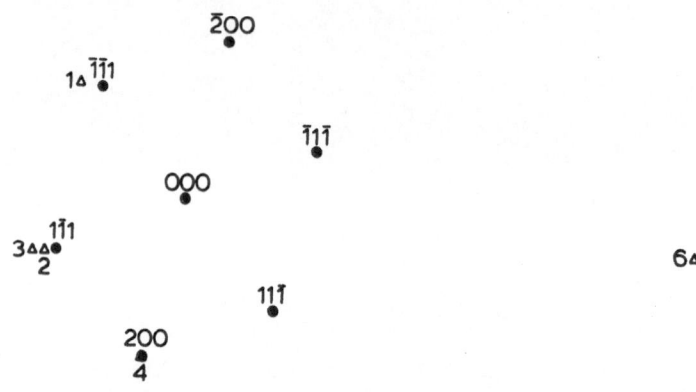

$Z = [011]$

Fig. 8a. Schematic [011] zone axis diffraction
pattern showing the relative orientations
of adjacent lamellae with respect to a CaO
reference lamellae.

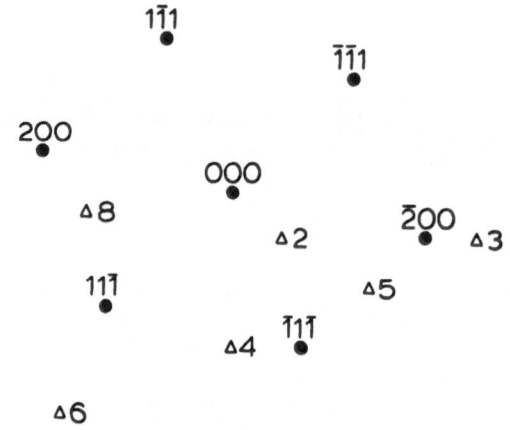

$Z = [011]$

Fig. 8b. Schematic [011] zone axis diffraction
pattern showing the relative orientations
of adjacent lamellae with respect to a NiO
reference lamellae.

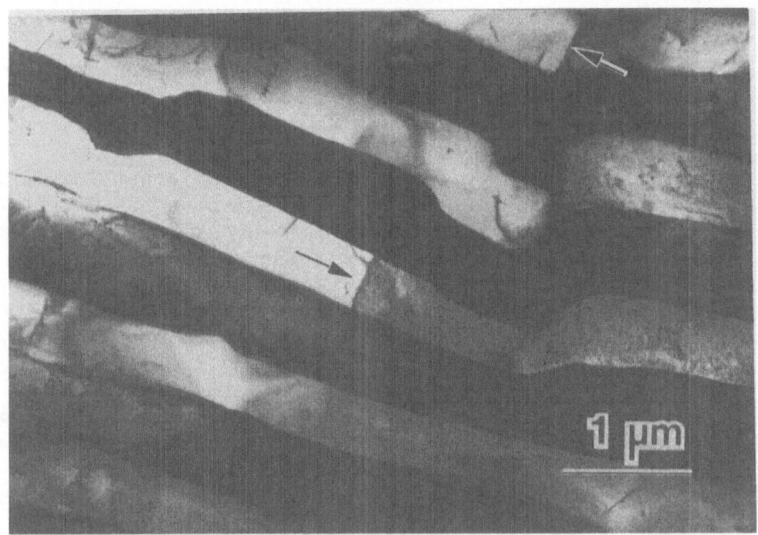

Fig. 9a. A change in contrast (indicated by arrows) occurs along two lamellae due to the misorientation which exists across dislocation sub-boundaries.

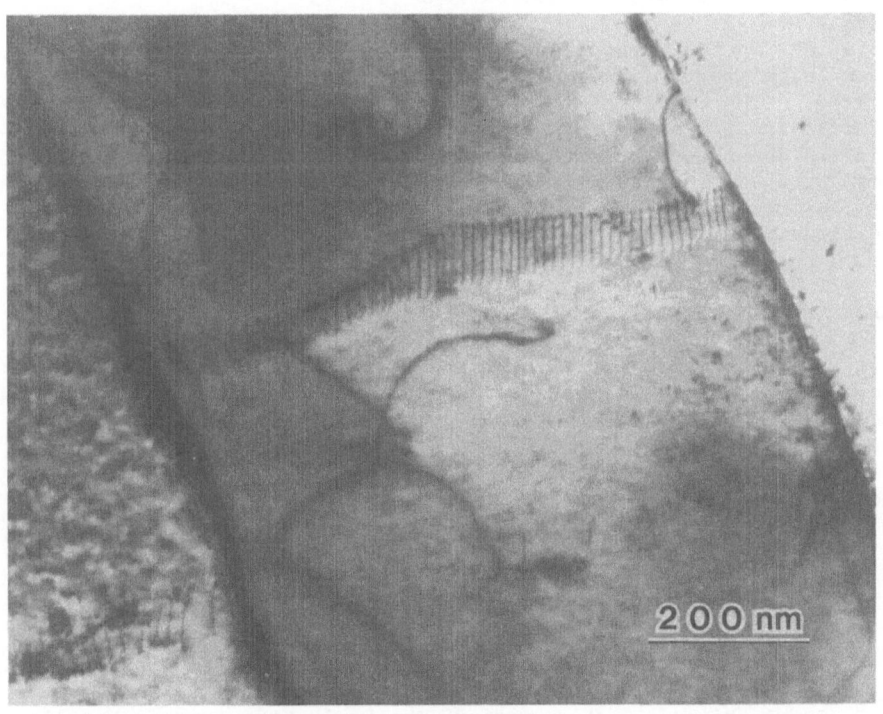

Fig. 9b. Transmission electron micrograph illustrating the dislocation structure present in one sub-boundary.

However, the relative rotations about the reference lamellae seem to be random. A similar random rotation is exhibited by the lamellae in the second example in Figure 8b. For these examples (and consistent with additional work to be presented elsewhere), the maximum variation in orientation between adjacent lamellae is about 2°.

The random variations in orientation between adjacent lamellae also lead to the formation of sub-boundaries. The sub-boundary formation is apparently an attempt by the microstructure to accommodate the misorientations which develop between the lamellae[7]. The sub-boundaries can be detected by the change in contrast along the length of an individual lamellae due to the change in orientation across the sub-boundary, and several examples can be seen in Figure 9a. The sub-boundaries form approximately normal to the lamellar interface and probably develop a cell structure with the dislocations present in the partially coherent interfaces between the NiO and CaO phases. The dislocations present in one such sub-boundary are shown in Figure 9b. This sub-boundary appears to be a pure tilt boundary, and a change in orientation of 0.5° across the sub-boundary was determined from an analysis of Kikuchi line patterns.

SUMMARY

The mismatch boundaries and lamellar terminations which have been observed within eutectic grains in metallic systems were observed in the NiO-CaO system. The presence of these faults in a ceramic eutectic is reasonable since the faults form as the microstructure attempts to achieve a lower energy lamellar habit plane orientation. It has been suggested that these faults can also provide a mechanism to change the interlamellar spacing in response to growth fluctuations[17].

Variations in the crystallographic orientation of the lamellar habit plane from being exactly parallel to $\{111\}_{NiO}$ and $\{111\}_{CaO}$ were demonstrated. The origin of this type of variation is not clear but may be related to local growth fluctuations or the existence of a shallow minima in the dependence of interfacial energy on boundary orientation. Variations were also observed in the orientation of adjacent lamellae which are accommodated by the introduction of dislocation sub-boundaries into the structure.

ACKNOWLEDGMENTS

This research was supported by the National Science Foundation under Grant No. DMR 78-05741. The optical and scanning electron microscopy was performed by M. D. Brumels.

REFERENCES

1. L. M. Hogan, R. W. Kraft, and F. D. Lemkey, Eutectic Grains, in: "Advances in Materials Research Vol. 5," H. Herman, ed., Wiley-Interscience, New York (1971).
2. R. Elliott, Eutectic Solidification, Int. Met. Rev., 22:161 (1977).
3. R. Elliott, "Eutectic Solidification Processing-Crystalline and Glassy Alloys," Butterworth & Co. Ltd., London (1983).
4. J. E. Gruzleski and W. C. Winegard, The Fault Structure in Lamellar Eutectics, J. Inst. Met., 96:301 (1968).
5. R. L. Ashbrook, Directionally Solidified Ceramic Eutectics, J. Am. Ceram. Soc., 60:428 (1977).

6. V. S. Stubican and R. C. Bradt, Eutectic Solidification in Ceramic Systems, Ann. Rev. Mater. Sci., 11:267 (1981).

7. I. G. Davies and A. Hellawell, The Structure of Directionally Frozen Al-CuAl$_2$ Eutectic Alloy, Phil. Mag., 19:1285 (1969).

8. B. Cantor and G. A. Chadwick, The Growth Crystallography of Unidirectionally Solidified Al-Al$_3$Ni and Al-Al$_2$Cu Eutectics, J. Cryst. Growth, 23:12 (1974).

9. D. E. Smith, T. Y. Tien, and L. H. Van Vlack, The System NiO-CaO, J. Am. Ceram. Soc., 52:459 (1969).

10. M. P. Pechini, Method of Preparing Lead and Alkaline Earth Titanates and Niobates and Coating Method Using the Same to Form a Capacitor, U.S. Patent 3,330,697, July 11 (1967).

11. N. G. Eror and T. M. Loehr, Precision Determination of Stoichiometry and Disorder in Multicomponent Compounds by Vibrational Spectroscopy, J. Solid State Chem., 12:319 (1975).

12. R. T. Cox, A. Revcolevschi, and R. Collongues, Growth of 0^{17} Enriched Al$_2$O$_3$ Crystal by a Floating Zone Technique, J. Crystal Growth, 15:301 (1972).

13. M. D. Brummels, "Indentation Fracture of NiO-CaO Directionally Solidified Eutectic," M.S. Thesis, Michigan Technological University (1981).

14. A. S. Yue, Microstructure of Mg-Al Eutectic, Trans. AIME, 224:1010 (1962).

15. R. H. Hopkins and R. W. Kraft, Nucleation and Growth of the Pb-Sn Eutectic, Trans. AIME, 242:1627 (1968).

16. P. Berthou and J. E. Gruzleski, The Origin and Elimination of Faults in Sn-Cd Eutectic Alloys, J. Cryst. Growth, 10:285 (1971).

17. D. D. Double, Imperfections in Lamellar Eutectic Crystals, Mat. Sci. Eng., 11:325 (1973).

18. M. Fragneau and A. Revcolevschi, Crystallography of the Directionally Solidified NiO-CaO Eutectic, J. Am. Ceram. Soc., 66:C-162 (1983).

19. N. H. Fletcher, Crystal Interfaces, J. Appl. Phys., 35:234 (1964).

20. T. A. Wall, W. W. Predebon, and B. J. Pletka, The Dependence of Yield Strength on Lamellar Termination Density in Co-CoAl Eutectic Alloys, Acta. Met., 33:287 (1985).

21. G. Garmong, C. G. Rhodes, and R. A. Spurling, Crystallography and Morphology of As-Grown and Coarsened Al-Al$_3$Ni Directionally Solidified Eutectic, Metall. Trans., 4:707 (1973).

22. G. Garmong and C. G. Rhodes, Interfacial Structure of Al-CuAl$_2$ Eutectic Composites, Acta Met., 22:1373 (1974).

23. G. Garmong and C. G. Rhodes, The Structure of Interphase Boundaries in Al-CuAl$_2$ Curved Eutectic Crystals, Metall. Trans., 5:2507 (1974).

24. G. Garmong, Structure and Crystallography of Curved Al-Al$_3$Ni and Al-CuAl$_2$ Directionally Solidified Eutectic Alloys, Metall. Trans., 6A:1335 (1975).

25. G. Garmong and C. G. Rhodes, Interfacial Ledge Structures in Ni$_3$Al-Ni$_3$Cb Eutectic Composites, Metall. Trans., 6A:2209 (1975).

26. P. Hirsch, A. Howie, R. B. Nicholson, D. W. Pashley, and M. J. Whelan, "Electron Microscopy of Thin Crystals," R. E. Krieger Publishing Company, Huntington, New York (1977).

CVD-PROCESSING OF CERAMIC-CERAMIC

COMPOSITE MATERIALS

Roger Naslain and Francis Langlais

Laboratoire de Chimie du Solide du CNRS
Université de Bordeaux-I
33405 - Talence, France

INTRODUCTION

Chemical vapor deposition (CVD), i.e. the deposition of a solid by a chemical reaction involving one or several gaseous chemical species and usually thermally activated, has been used for many years in different kinds of applications (e.g. oxidation or/and wear resistant coatings for cemented carbides, steels or alloys, preforms for drawing graded-index optical fibers, thin films for integrated circuits, coatings for nuclear fuels, etc...). In most cases, the substrates considered here have a rather simple shape and are made of non-porous materials.

The same chemical reactions have been used, more recently, to densify porous materials, the solid phase being deposited (or infiltrated) within the pores of the substrate. Thus, the so-called chemical vapor infiltration (CVI) technique, which has been initially worked out for processing ceramic-ceramic composite materials (CCCM), but which has a general character, is directly derived from the previously known CVD technology. However, CVD and CVI require very different processing conditions.

The most common composite processed according to the CVI-technique has been carbon-carbon (i.e. a material made of a carbon fiber preform infiltrated, up to almost complete densification, by a carbon matrix resulting from in-situ pyrolysis of methane) [1]. Later on, the carbon-CVI matrix has been replaced partly or totally by a variety of refractory materials : silicon carbide [2], titanium carbide [3], boron carbide [4], boron nitride [5] and even oxides such as alumina [6]. It thus appears that CVI must be considered as a promising processing technique for the synthesis of many ceramic-ceramic composites provided suitable gaseous precursors and infiltration conditions could be found.

The aim of the present contribution is to emphasize the differences existing between CVD and CVI and to show how the latter can be used for processing a variety of CCCM. The advantages and drawbacks of the CVI-technique, with respect to those of the liquid phase technique (i.e. impregnation), will be discussed, on the other hand, the properties of CCCM obtained according to the CVI-route will not be analyzed here since they have been presented elsewhere [3,7].

MAIN FEATURES OF CHEMICAL VAPOR DEPOSITION (CVD)

Principle

A refractory material (e.g. a binary compound AB) can be deposited by CVD on a given substrate, if volatile compounds of its constituting elements are known (and available) and if chemical reactions involving the precursor are activated in the vicinity of the substrate. This activation is obtained either thermally (by heating the substrate alone in a cold-wall reactor, or the substrate and the whole vapor phase in a hot-wall reactor) or electrically (by applying an electrical field between the substrate and a counter electrode, thus generating a plasma within the vapor phase). Since in the field of ceramics both the substrate and the deposit are stable at high temperatures, thermal activation, is usually preferred. Under such conditions, the gaseous precursor can be a mixture of common molecules (table 1). Nevertheless, one must underline that for some refractory materials involving alkaline-earth or rare-earth elements, such as CaO, MgO or LaB_6, no suitable gaseous precursors are known so far.

Chemical reactions that occur in the vicinity of the substrate are complex. By assuming that states close to equilibrium are reached between the gas phase, on one hand, and the substrate (and later, the deposit) on the other hand, thermodynamic approaches have been used by several authors to forecast : (i) the nature of the chemical species present at equilibrium in both the vapor phase and the deposit, (ii) the thermodynamic yields, (iii) the occurence of by-products and deposit-substrate reactions, as a function of the usual CVD-parameters (temperature, total pressure and composition of the precursor) [3, 4, 8, 9, 10]. Examples of results obtained according to such an approach are given in fig. 1, for SiC-based deposits. The deposit can be either a single phase (e.g. SiC) or a mixture of two phases (e.g. SiC + C codeposits), at a given substrate temperature and total pressure, depending on the initial composition. One of the important advantages of CVD lies in the fact that the deposit composition can be modified, in a continuous or discontinuous manner, by simply adjusting the precursor composition (one of the best examples of this flexibility is the CVD-processing of the preforms used for drawing graded-index optical fibers) [2, 10].

As shown in fig. 2, the deposition of a solid from a gaseous mixture involves several steps. The most important are : (1) the diffusion of the

Table 1 : Some refractory materials deposited by CVD and the corresponding gaseous precursors.

Refractory materials	Gaseous precursors
Carbon	CH_4 ; C_3H_8 ; C_2H_2 ; C_6H_6
Boron	BCl_3-H_2 ; BBr_3-H_2
SiC	$CH_3SiCl_3-H_2$;$(CH_3)_2SiCl_2-H_2$;$SiCl_4-CH_4-H_2$
Si_3N_4	$SiCl_4-NH_3$
B_4C	$BCl_3-CH_4-H_2$; $BBr_3-CH_4-H_2$
TiC, ZrC	$MCl_4-CH_4-H_2$ (M = Ti,Zr)
TiB_2, ZrB_2	$MCl_4-BCl_3-H_2$ (M = Ti,Zr)
BN	$BCl_3-NH_3-H_2$; BF_3-NH_3
Al_2O_3, ZrO_2	$AlCl_3-H_2-CO_2$; $ZrCl_4-H_2-CO_2$

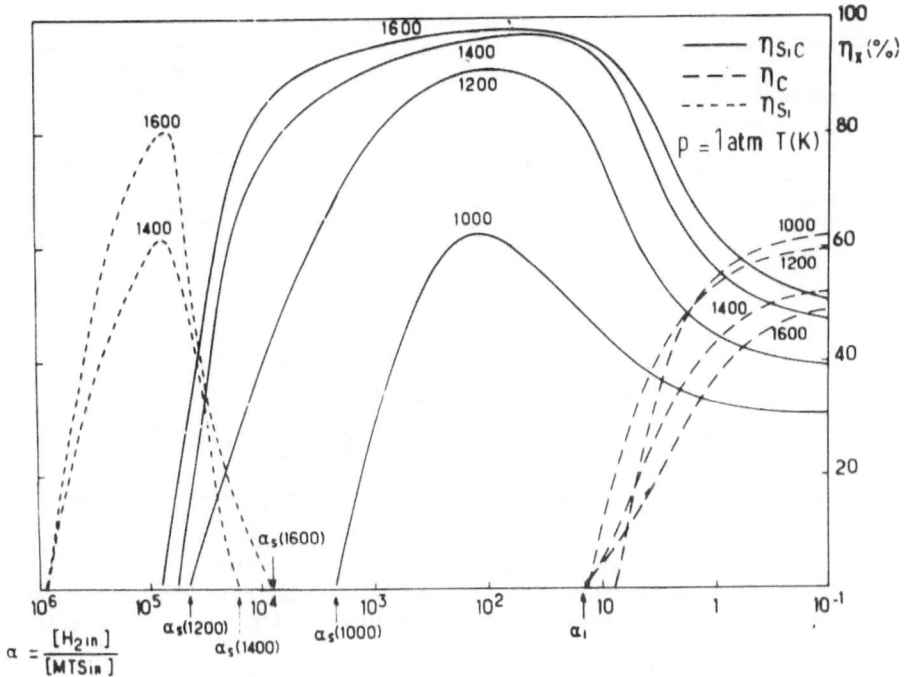

Fig. 1. CVD of SiC from CH3-Si-Cl3/H2 : variations of thermodynamic yields vs. initial composition for different temperatures, after [3].

source species from the feed gas to the substrate surface across a stagnant boundary layer δ, (2) the chemical reactions occurring between the source species adsorbed at the substrate surface, resulting in the nucleation-growth of the deposit and the formation of new gaseous species, **3** the desorption and diffusion of these new species across the boundary layer. It is generally admitted that deposition is rate-controlled by diffusion at high temperatures or/and high total pressures and by surface reaction kinec-

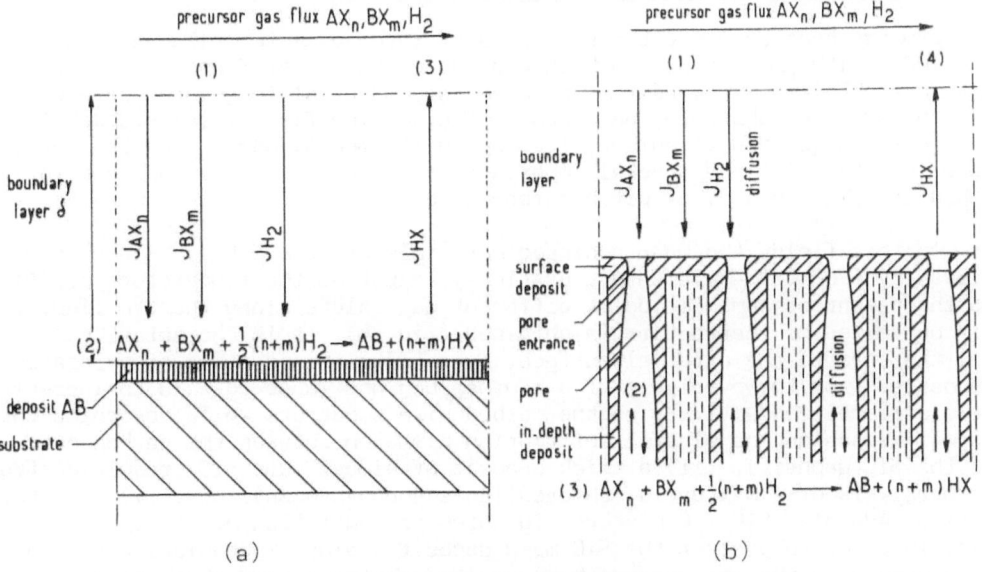

Fig. 2. Chemical vapor deposition of a solid AB from a gaseous precursor $AX_n+BX_m+H_2$: (a) on a flat substrate (CVD) and (b) within a porous substrate (CVI).

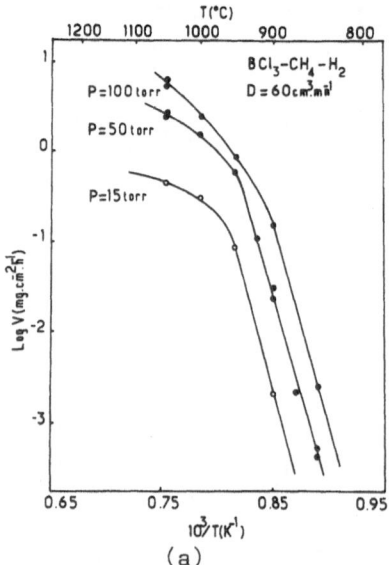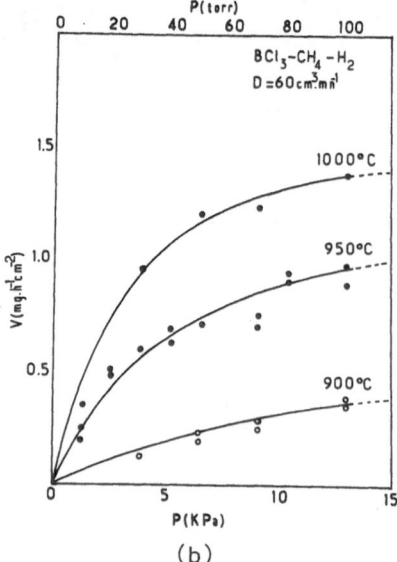

Fig. 3. Transition from surface kinetics to diffusion rate-controlled mechanisms induced by temperature or pressure in the CVD of B4C after [4].

tics in the opposite cases. Fig. 3 gives examples of transitions between these
two rate-controlling mechanisms induced by temperature or pressure for
B4C deposition [4, 9]. Within the low temperature range, the thermal variations of the deposition rate obey an Arrhenius law (characterized by the so-
called kinetics constant k and activation energy Q).

In some cases, the formation of a solid from a gaseous precursor takes
place directly within the vapor phase (homogeneous nucleation) giving rise
to soot formation, poor deposit-substrate adhesion and low yields. Such a
drawback is often overcome by lowering total pressure or temperature.

Examples of application of CVD to fibrous materials

CVD has been used for the synthesis of large diameter filaments (i.e.
from 100 to 150 μm) exhibiting both high strength and stiffness. Such materials are obtained by depositing a refractory material (e.g boron or SiC)
on a heated wire substrate, as shown in fig. 4. The first filaments developed according to this technique, i.e. boron filaments, had a tungsten core.
Since tungsten wire had several important drawbacks   it is replaced in
modern CVD filaments by ex-pitch carbon wire [12].

In this field, the most achieved materials are the SiC(C) filaments developed by AVCO (SCS-filaments). By simply adjusting the temperature profile
and the precursor composition at different gas inlets along the CVD line, a
very optimized microstructure is obtained (fig. 4). It is characterized,
from the core to the external surface, by : (i) a thin coating of pyrocarbon
(deposited from the pyrolysis of a hydrocarbon and whose role is to increase
the electrical conductivity of the carbon wire substrate while absorbing the
thermal expansion and plastic deformation mismatch between the carbon core
and the SiC-deposit), (ii) a thick deposit of almost pure SiC (resulting from
the pyrolysis of a mixture of chlorosilanes and hydrogen) and (iii) an outer
coating made of a SiC + C mixture, in which concentration profiles are optimized in order to protect the SiC main deposit against mechanical abrasion
(role played by the soft pyrocarbon) and to achieve a good chemical compatibility with inorganic matrices (role played by SiC) (fig. 5). Such materials,
that could not be obtained by another technique, have excellent mechanical
properties ($\sigma_R \sim 4$ GPa ; $E \sim 425$ GPa ; $\rho = 3.00$) [13, 16].

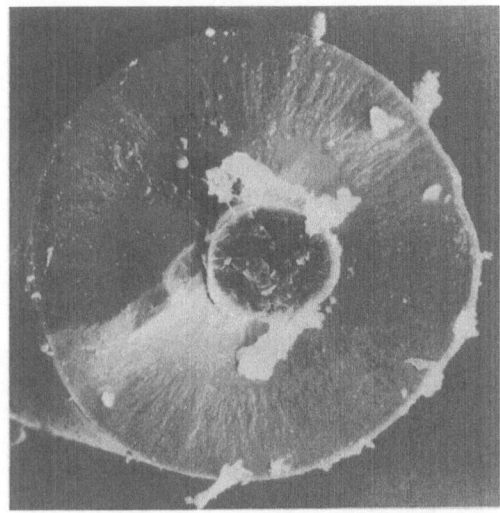

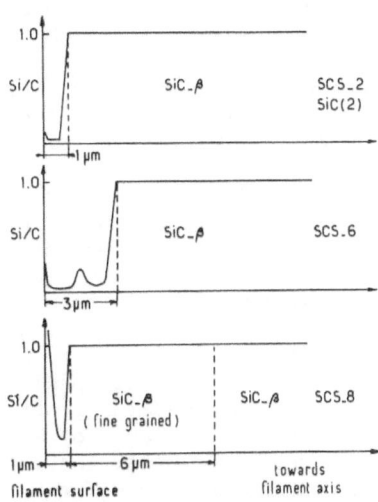

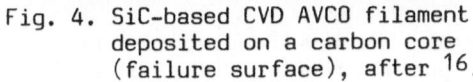

Fig. 4. SiC-based CVD AVCO filament
deposited on a carbon core
(failure surface), after [16].

Fig. 5. SiC-based CVD filaments :
composition profiles across the
outer surface coating, after [15].

CVD is also an efficient way for coating fibers with diffusion barrier or
wetting promotors. Thus, boron filaments have been coated with B4C, SiC or
BN diffusion barriers [12, 17]. In the same manner, several attempts have
been reported to coat yarn carbon fibers with various refractory materials
(particularly with titanium compounds to enhance their wetting ability by
liquid metals) [18,17].

CHEMICAL VAPOR INFILTRATION (CVI)

CVI, i.e. the densification of a porous material by a refractory material
deposited from a gaseous precursor within its pore network, is a technique
which has been more recently developed. It has been used, first, for the
processing of carbon-carbon and, then, extended to other refractory materials
such as SiC [2, 20-22], TiC [3, 7, 9], BN [5, 7, 23], B4C [4, 7] and Al2O3
[6]. Two kinds of porous materials have been densified according to the CVI-
technique : (i) porous sintered ceramics (e.g. reaction bonded SiC or Si3N4)
and (ii) porous preforms made of refractory fibers. As it is often the case
for new processing techniques, many aspects of CVI are still imperfectly understood.

Principle

As schematically shown in fig. 2, CVI is in its principle quite different from CVD although both techniques are often based on the same gaseous
precursors and chemical heterogeneous reactions. In CVD, deposition conditions are chosen in order to obtain a coating of the surface of the substrate,
as homogeneous in thickness and as dense as possible, whatever the substrate
porosity. In CVI, the aim is to fill as completely as possible the open pore
network and, thus, to favor in-depth deposition with respect to surface coating (which can never be totally avoided) by preventing the pore entrances
to become prematurely closed.

As already discussed for CVD, two steps still play a key role in CVI :
(i) surface reactions occurring between species adsorbed on the inner wall
of the pores (which represent a much higher area than the external substrate

149

surface) and (ii) mass transfers, of source species and of species resulting from chemical reactions, which are controlled by diffusion and occur both within the pores themselves and across the stagnant boundary layer surrounding the whole substrate. In CVI, the source molecules must have very important mean lifetimes and mean free paths within the heated substrate to reach the bottoms of the pores before reacting.

From the above discussion, it comes out that deposition must not be rate-controlled by diffusion but by kinetics of surface reactions (if diffusion was the rate-controlling phenomenon, deposition will preferably take place on the external surface and near the pore entrances). Furthermore, the gas phase supersaturation (particularly for the source species) at the boundary layer/substrate interface must be high enough and the deposition rate near the pore entrances as low as possible, in order to feed the pores with a sufficient flux of source species to counterbalance the depletion of the gas phase due to chemical reactions as the source species diffuse deeper in the pores. On the basis of such considerations, it comes out that CVI cannot be performed under the conditions previously established for CVD. This is particularly true for two key parameters, i.e. temperature and total pressure.

The substrate temperature must fall within the low-temperature part of the Arrhenius plot, corresponding to the surface reaction rate-controlled mechanism, previously drawn for CVD (fig. 3a). This first requirement, which has been clearly established for SiC, B$_4$C and TiC CVI, is illustrated in fig. 6 (a-c) [4, 9, 20]. When deposition is performed at high temperatures (e.g. 1050°C for TiC), it is rate-controlled by diffusion and, thus, limited to the external surface of the porous substrate. On the contrary, when deposition is performed at lower temperatures (e.g. 950° C for TiC), i.e. at the upper limit of the temperature range where surface reaction is the rate-controlling step, it also occurs within the open pore network (the observed discontinuous distribution of TiC being due to the texture of 2D-C-C preforms, as it will be discussed below).

In the same manner, total pressure must be low enough, as illustrated in fig. 6 (d-f) for TiC. At high total pressures (e.g. 0.20 atm), deposition is limited to the external surface of the substrate whereas at low pressures (e.g. 0.03 atm) it also takes place within the pores. These results are again consistent with the occurrence of a transition from a surface reaction kinetics-controlled mechanism to a diffusion-controlled mechanism, as total pressure is increased (fig. 3b) [4].

The total gas flow rate has, on the contrary, a much weaker effect. It must be rather low to induce a low deposition rate, but not to low to maintain a high enough supersaturation at pore entrances [4, 9].
Under such low temperature and total pressure conditions, the thermodynamic yields are often very limited and the amounts of unreacted species and by-products quite significant. Furthermore, since the infiltration of a porous substrate requires low deposition rates (to prevent an early sealing of pore entrances), a full densification needs long infiltration durations, at least when it is performed in a classical isothermal hot wall reactor.

Theorical approaches

Two theorical approaches of the CVI of porous substrates have been recently proposed [4, 9, 20, 22].

The Fitzer et al. model   According to E. Fitzer et al., the kinetics aspect of CVI is similar to that of heterogeneous catalysis within a porous catalyst, where the combined effect of chemical reaction and diffusion are characterized by the dimension-less numbers : $Da_{\parallel}$ (Damköler number) or $\emptyset$

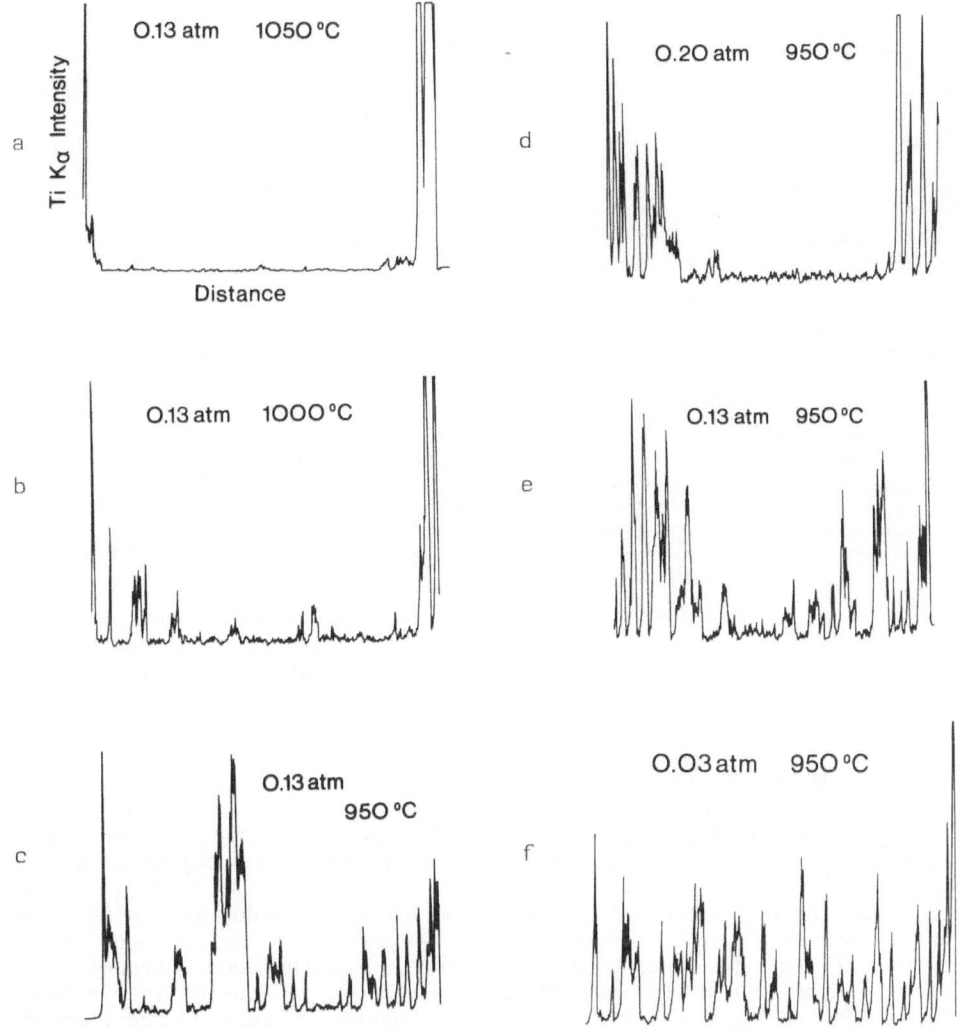

Fig. 6. Titanium (Tik_α) microprobe profiles along a cross-section of a porous 2D-C-C preform treated with TiCl4-CH4-H2 : (a-c) under a pressure of 0.13 atm and (d-f) at a temperature of 950°C, after [9].

(Thiele modulus) (fig. 7). The degree of utilization of a pore, i.e. the so-
called effectiveness factor η, is a function of $Da_{||}$ (or of $\emptyset$). A pore utili-
zation of 95-100 % is achieved when $Da_{||} < 0.4$. Thus, for a first order reac-
tion (which seems to be the case for the deposition of SiC or TiC from
CH3SiCl3 or TiCl4-CH4-H2, respectively), the impregnable pore depth L can be
calculated from the pore mean diameter R (assuming that the pores are cylindrical), the surface reaction kinetics constant k_s and the effective diffusion coefficient D_e into the pores.

The results of such a calculation are given in fig. 8 and compared
with experimental data for model materials (i.e. SiC substrates with cylindrical holes of different diameters). The substrates were densified with SiC
deposited from CH3SiCl3. It clearly appears that : (i) the agreement between
calculated and measured impregnable depths is better at low temperatures,
(ii) the impregnated depth strongly increases when the substrate temperature
is decreased and (iii) it becomes very limited (i.e. of the order of 1-2 mm)

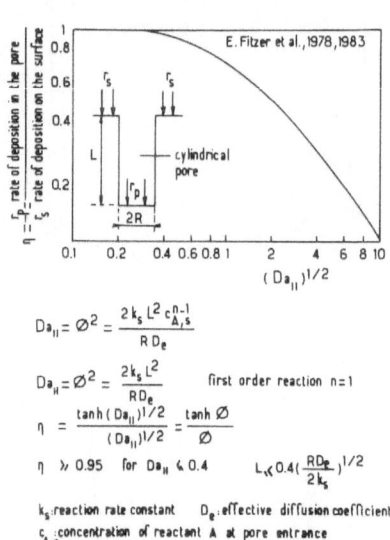

$$Da_{II} = \emptyset^2 = \frac{2k_s L^2 c_{A,s}^{n-1}}{R D_e}$$

$$Da_{II} = \emptyset^2 = \frac{2k_s L^2}{R D_e} \quad \text{first order reaction } n=1$$

$$\eta = \frac{\tanh(Da_{II})^{1/2}}{(Da_{II})^{1/2}} = \frac{\tanh \emptyset}{\emptyset}$$

$$\eta \geqslant 0.95 \quad \text{for } Da_{II} \leqslant 0.4 \qquad L \ll 0.4 \left(\frac{R D_e}{2k_s}\right)^{1/2}$$

k_s: reaction rate constant $\quad D_e$: effective diffusion coefficient
$c_{A,s}$: concentration of reactant A at pore entrance
n: reaction order

Fig. 7. Fitzer et al. theorical of CVI [22].

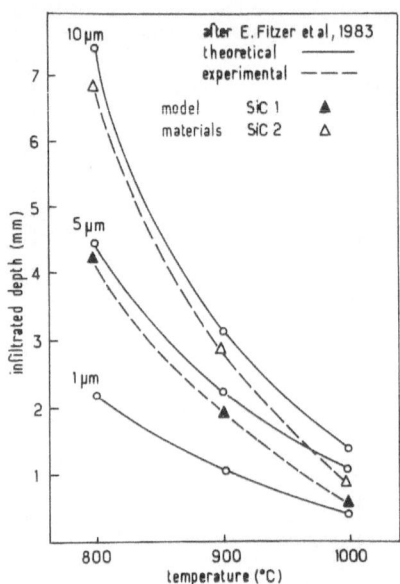

Fig. 8. Comparison between calculated and experimental depths of infiltration in cylindrical pores [22].

for the smaller pores (1 µm) [11], [20-22].

The Rossignol et al. model. Another theorical approach was developed independently by J.Y. Rossignol et al., for 2D-fibrous composite substrates, on the basis of a model previously proposed by C.H.J. Van den Breckel et al., for the deposition of a solid from a gaseous precursor, within capillary tubes (diameter of the order of 0.1-1 mm). This latter model gives the thickness G(z) of the deposit at a distance z from the tube entrance, as a function of the tube dimensions and of a coefficient K taking into account diffusion and reaction kinetics. Assuming that the deposition reaction order n is 1 with respect to the species which governs the surface reaction kinetics, K does not depend on reactant concentrations but only on the tube radius R, the diffusion coefficient D, the surface reaction kinetics constant k_0 and the activation energy E, at a given temperature T [24].

J.Y. Rossignol et al. have assumed that a given elemental volume of a 2D-porous composite preforms can be represented by an equivalent cylindrical pore of radius R and length L_e, to which the Van den Breckel et al. model can be applied provided two important modifications are made : (i) D must take into account both ordinary gaseous diffusion (D_0) and Knudsen diffusion (D_k), inasmuch as CVI is usually performed at low pressures and (ii) the decrease in size of the pore due to the deposit, particularly at pore entrance, must also be taken into consideration (fig. 9). The calculations, performed according to an iterative procedure, lead to the thickness profile of the ceramic deposit within the equivalent pore medium at any stage of densification and, particularly, at the end (i.e. when the thickness of the deposit G(0) is equal to the initial radius of the pore R, at pore entrance). They also give the volume fraction F(X) of the initial porosity which can be filled by the ceramic deposit up to pore sealing. For numerical applications, the only unknown parameter L_e is derived from a comparison between calculated and experimental f(X) values. It has been found to be of the order of 10-20 cm for the 2D-C-C preforms (∅ = 8 ; h = 15 mm) described below [4], [9].

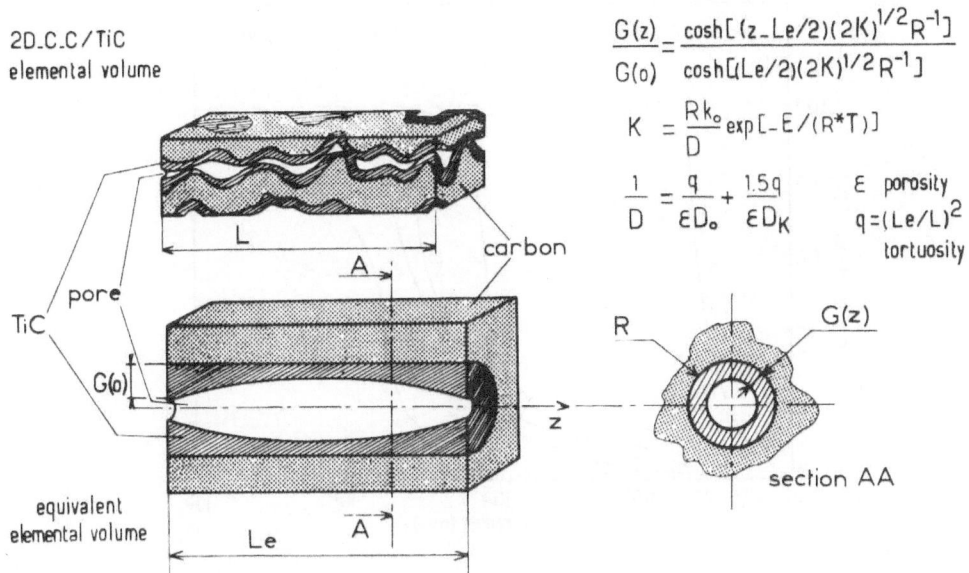

2D.C.C/TiC elemental volume

$$\frac{G(z)}{G(o)} = \frac{\cosh[(z-Le/2)(2K)^{1/2}R^{-1}]}{\cosh[(Le/2)(2K)^{1/2}R^{-1}]}$$

$$K = \frac{Rk_o}{D}\exp[-E/(R^*T)]$$

$$\frac{1}{D} = \frac{q}{\varepsilon D_o} + \frac{1.5q}{\varepsilon D_K}$$

ε porosity
$q = (Le/L)^2$ tortuosity

section AA

Fig. 9. Rossignol et al. model for the CVI of porous 2D-preform [9].

Examples of computed thickness profiles of the ceramic deposit within the equivalent porous medium are given in fig. 10 for boron carbide deposited from BCl₃-CH₄-H₂ [4]. It comes out that the ceramic distribution is rather homogeneous at low temperature and total pressure but exhibits a marked tendency to be limited to the vicinity of the external surface as temperature and total pressure are increased. Furthermore, the total gas flow rate appears to have but a weak effect on the ceramic distribution. Thus, all these com-

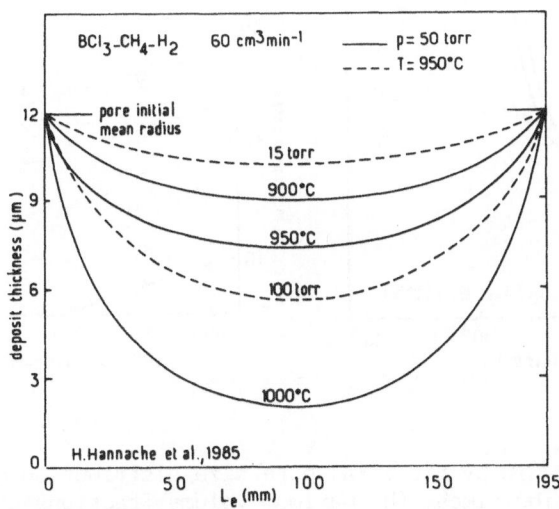

Fig. 10. Theorical B₄C distribution profiles in a model porous medium equivalent to a 2D-C-C preform [4].

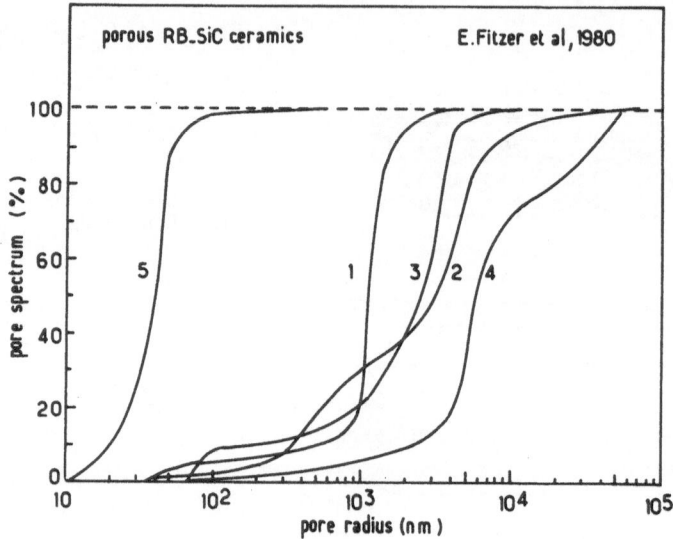

Fig. 11. Pore size distribution of RB-SiC ceramics before CVI densification, after [21].

puted results are qualitatively in good agreement with both the experimental data (see fig. 6) and the general considerations discussed above.

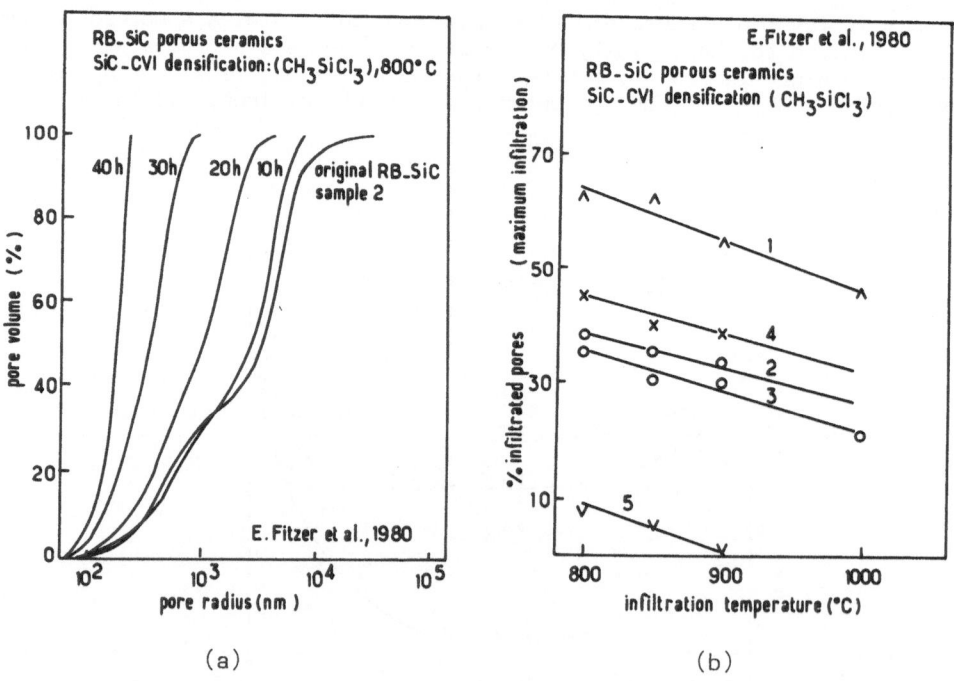

(a)   (b)

Fig. 12. CVI of RB-SiC by SiC : (a) pore size distribution after various infiltration times, (b) maximum volume fractions of infiltrated pores, after [21].

Application of CVI to the densification of porous sintered ceramics

The CVI-technique may be used for the post-sintering densification of porous ceramics provided the residual pores are interconnected and have a large enough mean diameter. It results either in a chemically homogeneous material or in a composite ceramics depending on whether the infiltrated deposit is chemically identical or not to the porous substrate.

Examples of such an application of CVI have been given by E. Fitzer et al. for porous reaction bonded SiC or Si_3N_4 ceramics characterized by different pore spectra (fig. 11) [11, 20-22]. Silicon carbide was deposited from CH_3SiCl_3 in a hot wall reactor and silicon nitride from $SiCl_4 + NH_3$ in a cold wall reactor (to avoid gas phase nucleation, as discussed below).

As shown in fig. 12 b the degree of pore filling strongly depends on both the pore spectrum and the infiltration temperature (the value of total pressure, another key parameter in CVI, was not given). The best results, i.e. 65 %, are obtained for the lowest acceptable infiltration temperature, i.e. 800°C (below this value, polymeric intermediates formed during the pyrolysis of CH_3SiCl_3 are no longer fully decomposed) and for substrates (e.g. sample 1) having at least a fraction of open pores of large radii (R > 1 μm) to permit an easy diffusion of the source species towards the inner zones. Under such conditions, all pores having radii larger than 0.1 μm are fully densified in about 40 hours (fig. 12a). Even under the best conditions, a density gradient remains within the substrate, at least when CVI is performed in an isothermal reactor, between the core, which is only partially densified, and the external surface where all the pores are filled. The CVI-densification of porous RB-SiC ceramics improves the oxidation and creep resistance as well as flexural strength [11, 20, 21].

On the other hand, the CVI-densification of porous RB-Si_3N_4 by Si_3N_4 deposited from $SiCl_4$-NH_3 seems to be more difficult to achieve due to : (i) a much higher deposition rate resulting in an early sealing of the pore entrances and (ii) a tendency to gas phase nucleation. As a consequence, only 20 % of the available open porosity has been filled in the best case. A better result is obtained when RB-Si_3-N_4 is CVI-densified by SiC with a similar improvement in the oxidation and creep resistance (but with no strength increase) [11, 20, 21].

Application of CVI to the synthesis of fibrous ceramic-ceramic composite materials

Fibrous CCCM can be obtained according to a three step process, i.e. (i) preparation of a fibrous preform, (ii) consolidation of the preform with a small amount of ceramic matrix and (iii) densification of the consolidated preform with the ceramic matrix. Steps (ii) and (iii) are performed according to either a liquid (i.e. impregnation) or a gas phase (i.e. infiltration) techniques (and eventually a combination of both techniques).

Fibrous preforms   The preforms are made from refractory fibers characterized by high strength and stiffness as well as low density. Moreover, fibers must have a high thermal and chemical stability to withstand long exposures at high temperatures with no strength alteration and no significant reaction with the matrix. Finally, fibers must have a small enough diameter, i.e. of the order of 10 μm, to be easily woven. There is but a small number of fibers that fulfill all these requirements, namely carbon, silicon carbide and α-alumina yarn fibers. Additives, which are sometimes utilized to facilitate spinning or to stabilize green fibers before firing (e.g. SiO_2 or/and B_2O_3 for alumina fibers and oxygen for carbosilane fibers), often lower the fiber thermal stability.

Depending on the degree of anisotropy which has to be achieved, the preform is made of : (i) a simple stack of unidirectional fiber monolayers or fabrics (1D or 2D-preforms), (ii) a multidirectional fiber (or fibrous rod) architecture (e.g. 3D or AD-preforms) or (iii) a quasi-isotropic felt.

Consolidation of the preform Before densification, the preform is consolidated with a small amount of ceramic matrix. This consolidation can be already performed according to CVI-technique. At this stage, the material is already composite. It is strong enough to be machined. However, its porosity is still very high (e.g. of the order of 30-50 % for a 2D-preform and even more for a felt). As already discussed, the pore spectrum is an important parameter. The best results, as far as CVI is concerned, are achieved when the pore spectrum contains a significant part of large pores which are necessary to feed properly the small pores located far from the external surface. In 2D-preforms, these large canals are located between adjacent consolidated fabric layers.

Densification of the preform by CVI Finally, the consolidated preform is densified by CVI, the ceramic being deposited within the open pores from one of the gaseous precursors listed in table I. The CVI parameters (i.e. temperature, total pressure and gas flow rate) have to be experimentally adjusted (see fig. 6) to take into account the nature of the preform (their pore spectrum and thickness), the number of preforms and the size of the CVI chamber. Densification is usually performed in hot-wall isothermal reactors. The main advantage of such a technique lies in the fact that several preforms can be simultaneously densified. On the other hand, it requires low temperature, low total pressure and low deposition rate in order to prevent an early sealing of the pore entrances. As a result, rather long infiltration times are necessary to achieve full densification [3, 9]. To overcome this drawback, thermal-gradient and pressure-gradient processes have been suggested. Such techniques lead to higher infiltration rates but require more complicated CVI-chamber and are not suitable for the simultaneous densification of a large number of preforms [25].

As shown in fig. 13 and 14 for a 2D-C-C preform densified with B_4C from BCl_3-CH_4-H_2 at 950°C and 20 torr, the pore spectrum is shifted towards the small pore diameters as the open porosity is filled with boron carbide. It is noteworthy that almost all pores having diameters larger than 1 µm are filled when densification is nearing completion and that almost 90 % of the initial pore volume is filled with the ceramic matrix. These results show that prematured sealing of pore entrances can be avoided by a proper choice of the CVI-conditions (no surface machining was made during densification) [4].

Fig. 15 shows that the overall deposition rate (i.e. corresponding to both the surface deposited mass M_S and that deposited within the pores M_p) decreases as the pores are progressively filled. The rate of surface deposition remains constant whereas that of in-depth deposition decreases with time. This feature, also reported by E. Fitzer for porous RB-SiC ceramics, could be justified by the fact that deposition is rate-controlled by surface reaction kinetics and that the total inner surface of the pores decreases with time whereas the external surface of the preform remains almost unchanged [3, 9, 20]. Furthermore, J.Y. Rossignol et al. have established, for 2D-preforms, that $X = Ln [(Mo - Mp)/Mo]$ decreases linearly with time when the CVI-parameters are properly chosen (Mo being the maximum mass of matrix which can enter the open porosity Vpo of the initial preform) [9]. It clearly appears from fig. 16 that such a law fits well the experimental data when total pressure and gas flow rate are low enough, at a given temperature. This is no longer true when densification is performed at too high total pressure and gas flow rate. In such a case, deposition preferably

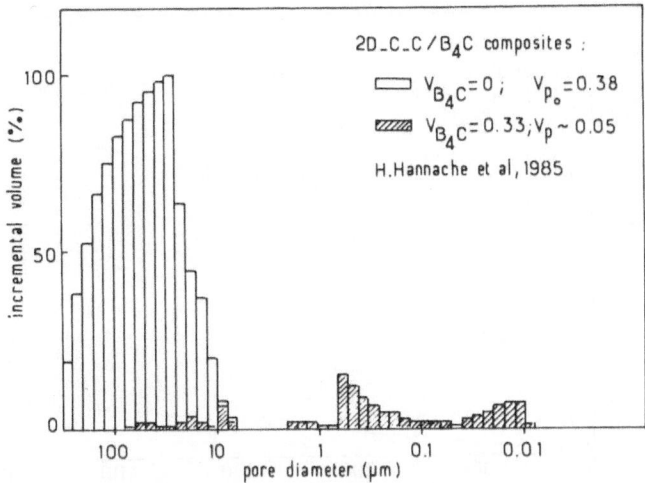

Fig. 13. Pore size distribution of 2D-C-C/B₄C composite materials before and after CVI-densification by B₄C, after [4].

occurs near the pore entrances which become sealed after a certain time. Thus a departure from the X = f(t) straight line indicates unsuitable CVI-conditions [4].

The theorical model which has been proposed by J.Y. Rossignol et al. (see above) has been used to compute the weight increase of the preform during CVI-densification. The computed and experimental curves are in good agreement, at least in their main features. However, the computed time necessary to achieve full densification is much shorter than that found experimentally (by a factor of 100). A better fit would require a more detailed knowledge of infiltration kinetics [4, 9].

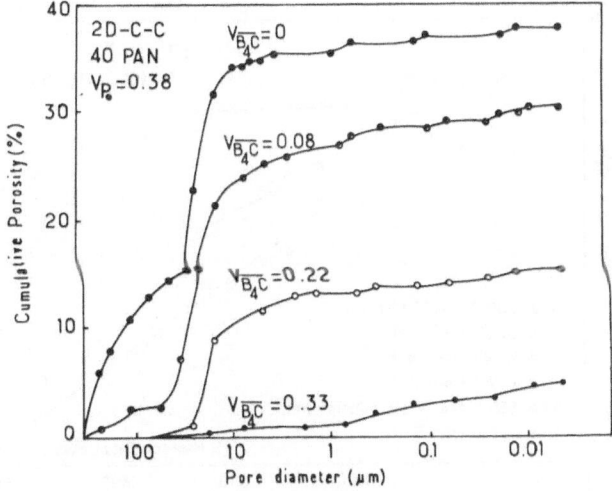

Fig. 14. Cumulative open porosity of 2D-C-C/B₄C composite materials at various stages of CVI densification by B₄C, after [4].

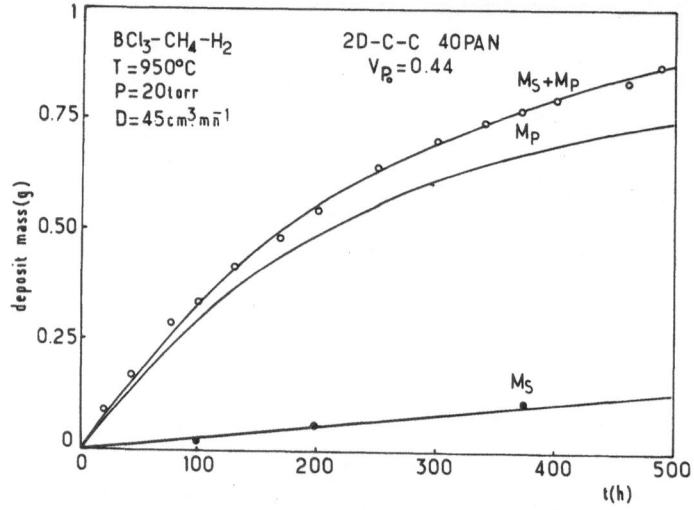

Fig. 15. Kinetics of CVI of 2D-C-C preforms by B₄C, after 4.

<u>CCCM obtained according to CVI-technique</u> CVI was used first for the synthesis of carbon-carbon, from carbon fiber preforms densified with pyrocarbon (pyrolysis of CH_4 at 1000-1200°C). Depending on CVI conditions, different· matrix microstructures are observed, i.e. smooth laminar (SL), rough laminar (RL) and isotropic (ISO) among which only RL can be graphitized by heating at high temperatures [1]. Carbon-carbon have many advantages (e.g. low density, good mechanical properties at high temperatures, biocompatibility, etc...), their main drawback being a poor oxidation resistance. Carbon-carbon were developed initially for aerospace applications at extremely high temperatures (e.g. re-entry heat shields and rocket nozzles) and used, more recently, in brake discs and prosthetic devices [26-28].

SiC-based CCCM were proposed as a way to overcome the poor oxidation

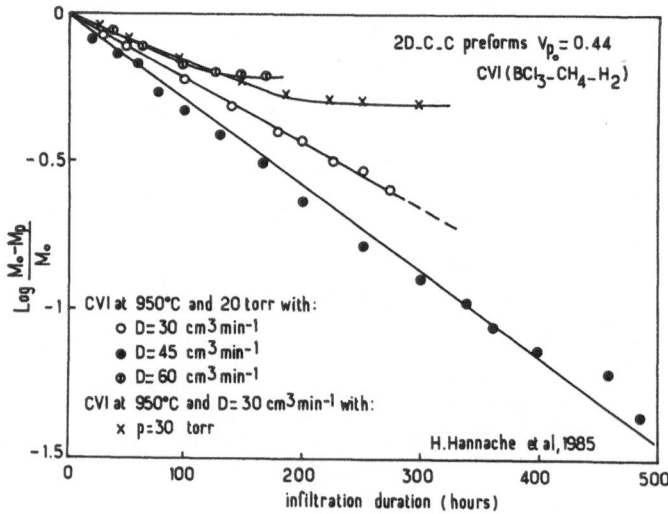

Fig. 16. Kinetics of CVI of 2D-C-C preforms by B₄C in semi-log scale, according to [4].

resistance of carbon-carbon and to improve the toughness of SiC-ceramics.
The feasibility of the CVI-technique for densifying fibrous preforms by
SiC was established in 1977 by F. Christin et al. and confirmed independently by E. Fitzer in 1978 [2, 29, 20]. It is based on the pyrolysis of a
silane (usually CH_3SiCl_3) under controlled conditions of temperature, total
pressure and gas flow rate. The deposit is made of pure SiC when CH_3SiCl_3
is mixed with a significant amount of hydrogen, and of SiC + C when hydrogen
poor CH_3SiCl_3-H_2 mixtures or CH_3SiCl_3-Ar mixtures are used [2]. A variety of
preforms, including 1D-preforms made of SiC-CVD filaments and 2D-preforms
made of carbon or SiC yarn (Nicalon) fabrics as well as quasi-isotropic
carbon felts, characterized by open porosity Vpo ranging from 90 to 20%,
have been successfully densified by SiC up to residual porosity Vp less
than 10%. The mechanical, thermal and chemical properties of 2D-C-C/SiC
composite materials (i.e. materials made of a 2D-C preform consolidated
with a small amount of pyrocarbon and densified by SiC) have been analyzed,
as a function of V_{SiC} (or Vp since for a given preform Vp + V_{SiC} = Vpo =
constant), by R. Naslain et al. [3, 7]. Replacing partly (or even totally)
carbon matrix by SiC in 2D-C-C improves mechanical properties, oxidation
resistance (provided the material has been fully densified and has not received any post-CVI surface machining) and lowers the strong anisotropy
related to the 2D-texture of the carbon preform. 2D-SiC-SiC composite materials, made from SiC yarn fibers (e.g. Nicalon), have still a better oxidation resistance. Such materials have a strength which is comparable to
that of sintered SiC-ceramics (and which is even higher for 1D-SiC-SiC
made from high strength CVD-filaments). It is noteworthy that the flexural
strength of 2D-SiC-SiC at 1500°C (in an Ar-H_2 atm) is still 50 % of their
strength at 20°C (fig. 17) [30-33]. One of the main features of these materials is the improved toughness that they exhibit when properly processed
(i.e. for optimized fiber-matrix interfaces) with respect to sintered SiC-
ceramics, as it is apparent from their ability of maintaining stress at
high strain values, their non-catastrophic failure and the occurence of
fiber pull-out (fig; 18, 19). Such unique properties make them attractive
for applications in engines las structural ceramics. One must emphasize
that SiC-SiC composite materials are already manufactured on an industrial
basis [31].

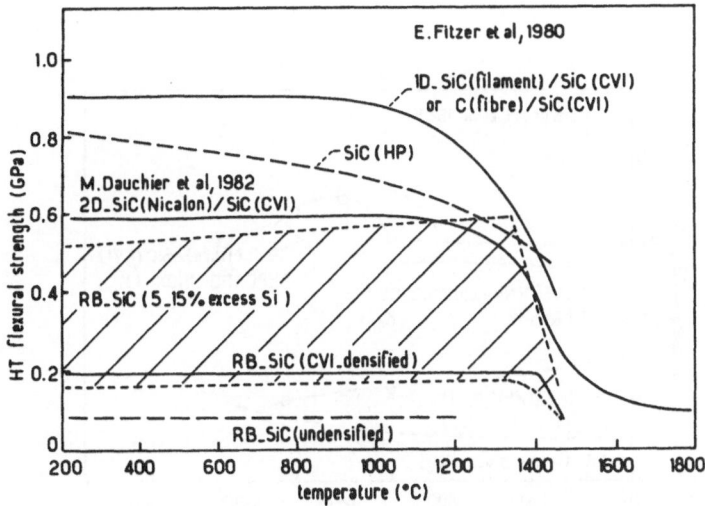

Fig. 17. High temperature flexural strength of SiC-based ceramics, after
21, 31.

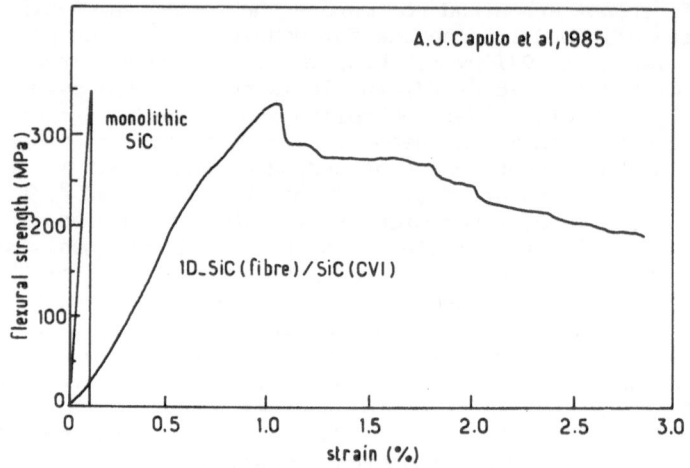

Fig. 18. Stress-strain curves of 1D-SiC-SiC composite and SiC monolithic
materials, after [25].

The feasibility of the CVI-technique has been also established for
TiC [3], [9], B$_4$C [4], BN [5], [23] matrices. A comparative analysis of the mechanical
behavior of 2D-C-C/ceramic composite materials has been recently reported
by J.Y. Rossignol et al. [7].

Finally, the CVI-technique was also used for the synthesis of alumina
matrix composite materials by R. Colmet et al. [6]. Alumina was deposited at
900-1000°C from AlCl$_3$-H$_2$-CO$_2$ within the pores of carbon, SiC (Nicalon) or
alumina (Saffil from ICI or FP from du Pont) preforms (2D, 1D or quasi-isotropic). Such CVI-conditions thus preclude the use of Al$_2$O$_3$-SiO$_2$-B$_2$O$_3$ fibers
due to their limited thermal stability. Alumina-alumina composite materials
were found to keep a significant percentage of their room temperature flexural strength in the 1000-1500°C temperature range. Due to their excellent
behavior in oxidizing atmospheres, alumina-based composite materials could

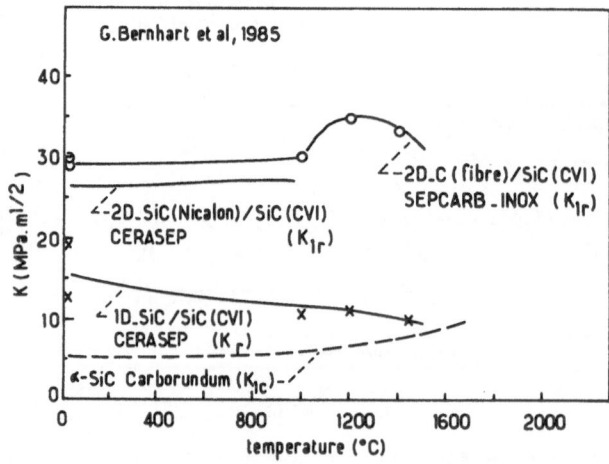

Fig. 19. High temperature toughness of SiC-based ceramics, after [33].

also be used as structural ceramics in engines.

The CVI-technique could be obviously extended to other ceramic matrices provided suitable gas precursors and kinetics conditions could be found.

Comparison of CVI and liquid impregnation techniques

With respect to liquid impregnation techniques, the CVI-processing of CCCM is characterized by advantages and drawbacks.

On the basis of the available data, CVI-processed CCCM seem to have better characteristics due to the high quality of CVI-matrices (i.e. small grain size, rather low residual porosity, excellent embedment of the fibers within the yarns). Moreover, under the conditions required by CVI (i.e. low temperature and pressure), the fibers are less damaged than during the high temperature firing step which follows liquid impregnation. Small pores of the preforms are much easily filled when utilizing gaseous precursors and wetting problems (which are common in liquid impregnation) are avoided. In CVI-processing, full densification can be achieved in one single run and apparatus (when the CVI-conditions are properly chosen), whereas it requires several impregnation/firing sequences, inasmuch as the pyrolysis of a liquid precursor results in a porous matrix. Furthermore, fiber surface treatments with low shear strength materials (e.g. pyrocarbon or hexagonal BN), which have been suggested to lower the fiber-matrix bond in order to increase toughness, can be performed, in-situ, by a simple adjustement of the feed gas composition prior to densification [25, 34].

On the other hand and as discussed above, the times necessary to achieve complete densification of rather large preforms are rather long (e.g. several weeks) with isothermal CVI reactors. However, this drawback can be partly counter-balanced by the fact that several preforms are simultaneously densified in the same deposition chamber. Moreover, the use of temperature-gradient or/and pressure-gradient processes increases infiltration rates. Furthermore, the CVI-technique is limited to refractory compounds for which gaseous precursors are readily available. Thus, refractory compounds of most alkaline-earth and rare-earth elements are practically precluded. The CVI-technique cannot be easily utilized for the deposition of complex matrices of given composition (e.g. solid solutions or ternary oxides). Finally, CVI can be in some cases a rather costly process due to : (i) the nature of the starting materials, (ii) the low thermodynamic yields and deposition rates, and (iii) the high technology which is required.

From this analysis, CVI-techniques appear to be particularly suitable for processing high performance composite ceramic parts such as those which will be necessary in advanced jet or reciprocating engines. However, more work is obviously needed on infiltration kinetics and on parameters which control failure modes, damaging mechanisms or toughness (e.g. interfaces).

ACKNOWLEDGEMENTS

Most data presented here were taken from reports of research sponsored by Société Européenne de Propulsion, the French Ministry of Defense (DRET) and the French Ministry of Research and Technology and from the theses of F. Christin, J.Y. Rossignol, H. Hannache and R. Colmet. The authors wish to acknowledge the assistance of P. Martineau and Mrs. Pinna, from Institut des Matériaux Composites, in the realization of the manuscript.

REFERENCES

1. H.O. Pierson and M.L. Lieberman, The Chemical Vapor Deposition of Carbon on Carbon Fibers, Carbon, 13 : 159 (1975)

2. F. Christin, R. Naslain and C. Bernard, A Thermodynamic and Experimental Approach of Silicon Carbide CVD. Application to the CVD-infiltration of Porous Carbon Composites, in Proc. 7th. Int. Conf. CVD (T.O. Sedwick and H. Lydin, eds.), p. 499, The Electrochem. Soc., Princeton (1979)

3. R. Naslain, J.Y. Rossignol, P. Hagenmuller, F. Christin, L. Heraud and J.J. Choury, Synthesis and Properties of New Composite Materials for High Temperature Applications Based on Carbon Fibers and C-SiC or C-TiC Hybrid Matrices, Rev. Chimie Minérale, 18 : 544 (1981)

4. H. Hannache, J.Y. Rossignol, F. Langlais, R. Naslain and P. Hagenmuller. Boron-Carbide LPCVD from BCl_3-CH_4-H_2 Gas Mixtures. Applications to the Synthesis of 2D-C-C/B_4C Composite Materials by CVI, (submitted to J. Less-common Metals)

5. H. Hannache, R. Naslain and C. Bernard, Boron Nitride Chemical Vapour Infiltration of Fibrous Materials from BCl_3-NH_3-H_2 or BF_3-NH_3 Mixtures. A Thermodynamic and Experimental Approach . J. Less-common Metals, 95 : 221 (1983)

6. R. Colmet, I. Lhermitte-Sebire and R. Naslain. Fibrous Alumina-alumina Composite Materials obtained According to a CVI-Technique (submitted to J. Amer., Ceram. Soc.)

7. J.Y. Rossignol, J.M. Quenisset, H. Hannache, C. Mallet and R. Naslain. Mechanical Behavior in Compression Loading of 2D Composite Materials Made of Carbon Fabrics and a Ceramic Matrix (submitted to J. Mat. Science)

8. R. Colmet, R. Naslain, P. Hagenmuller and C. Bernard, Thermodynamic and Experimental Analysis of Chemical Vapor Deposition of Alumina from $AlCl_3$-H_2-CO_2 Gas Phase Mixtures. J. Electrochem. Soc., 129 : 1367 (1982)

9. J.Y. Rossignol, F. Langlais and R. Naslain. A Tentative Modelization of Titanium Carbide CVI within the Pore Network of Two-dimensional Carbon-carbon Composite Preforms. Proc. 9th. Int. Conf. CVD (Mc. D. Robinson et al., eds.), p. 596, The Electrochem. Soc., Pennington (1984)

10. I. Lhermitte-Sebire, R. Colmet, R. Naslain and C. Bernard, the CVD of Alumina from $AlCl_3$-H_2-CO_2 on a Stoichiometric TiC Substrate : A Thermodynamic Approach (submitted to J. Less-common Metals)

11. E. Fitzer and D. Hegen, Chemical Vapor Deposition of Silicon Carbide and Silicon Nitride-Chemistry's Contribution to Modern Silicon Ceramics, Angew Chem. Int. Ed. Engl., 18 : 295 (1979)

12. V. Krukonis, Chemical Vapor Deposition of Boron Filament, in "Boron and Refractory Borides" (V.I. Matkovich, ed.), chap. D_I : 518, Springer-Verlag, Berlin-Heidelberg (1977)

13. R.L. Crane and V.J. Krukonis. Strength and Fracture Properties of Silicon Carbide Filament, Ceram. Bull., 54 : 184 (1975)

14. F.W. Wawner, A.Y. Teng and S.R. Nutt, Microstructural Characterization of SiC (SCS) Filaments, SAMPE Quarterly, April : 39 (1983)

15. T.F. Foltz, SiC-Fibers for Advanced Ceramic Composites, private communication 1985

16. P. Martineau, M. Lahaye, R. Pailler, R. Naslain, M. Couzi and F. Cruege, SiC-Filament/Titanium Matrix Composites Regarded as Model Composites. 1. Filament Microanalysis and Strength Characterization, J. Mater. Sc., 19 : 2731 (1984)

17. D. Morin, Boron Carbide-Coated Boron Filament as Reinforcement in Aluminium Alloy Matrices, J. Less-common Met., 47 : 207 (1976)

18. L. Aggour, E. Fitzer, E. Ignotowitz and M. Sahebkar, Chemical Vapor Deposition of Pyrocarbon SiC, TiC, TiN, Si and Ta on Different Types of Carbon Fibres, Carbon, 12 : 358 (1974)

19. W. Meyerer, D. Kizer and S. Paprocki, Versatility of Graphite-Aluminum Composites, Proc. Int. Conf. Comp. Mater. 2, Toronto, Apr. 16-20, B. Noton et al. eds., p. 141, TMS-AIME, Warrendale (Pa), (1978)

20. E. Fitzer, Chemical Vapor Deposition of SiC and Si3N4, Proc. Int. Symp. on Factors in Densification and Sintering of Oxide and Non-oxide Ceramics, Hakone, Japan, p. 40 (1978)

21. E. Fitzer, D. Hegen and H. Strohmeier, Possibility of Gas Phase Impregnation with Silicon Carbide, Rev. Int. Hautes Temper. Réfract., 17 : 23 (1980)

22. E. Fitzer, W. Fritz and R. Gadow, Possibilities for Fibre Reinforcement of Silicon Carbide, Proc. Symp. on Advanced Ceramic Materials, Tokyo Institute of Technol., Yokohama, Oct. (1983)

23. H. Hannache, J.M. Quenisset, R. Naslain and L. Heraud, Composite Materials Made from a Porous 2D-Carbon-carbon Preform Densified with Boron Nitride by Chemical Vapor Infiltration, J. Mater. Sc., 19 : 202 (1984)

24. C.H.J. Van den Breckel, R.M.M. Fonville, P.J.M. Van der Straten and G. Verspui, CVD of Ni, TiN and TiC on Complex Shapes, Proc. of 8th. Int. Conf. CVD (J.M. Blocher et al., eds.), p. 142, The Electrochem. Soc., Pennington (1981)

25. A.J. Caputo, W.J. Lackey and D.P. Stinton, Development of a New, Faster Process for the Fabrication of Ceramic Fiber-Reinforced Ceramic Composites by Chemical Vapor Infiltration, Proc. 9th. Annual Conf. Composites and Adv. Ceram. Mater., Cocoa Beach, Florida, Jan. (1985)

26. E. Fitzer, Carbon Based Composites, J. Chimie Physique, 81 : 717 (1984)

27. P.J. Lamicq, Propriétés et Utilisations des Composites carbone-carbone, J. Chimie Physique, 81 : 735 (1984)

28. J. Delmonte, Technology of Carbon and Graphite Fiber Composites, chap. 13, Van Nostrand Reinhold, New York (1981)

29. F. Christin, R. Naslain, P. Hagenmuller and J.J. Choury, Pièce poreuse carbonée densifiée in-situ par dépôt chimique en phase vapeur de matériaux réfractaires autres que le carbone et procédé de fabrication, Brevet français 77/26979, Sept. (1977)

30. L. Heraud, F. Christin, R. Naslain and P. Hagenmuller, Properties and Applications of Oxidation-Resistant Composite Materials obtained by SiC-Infiltration, Proc. of 8th. Int. Conf. CVD (J.M. Blocher et al., eds.), p. 782, The Electrochem. Soc., Pennington (1981)

31. M. Dauchier, P. Lamicq and J. Mace, Comportement thermomécanique des composites céramique-céramique, Rev. Int. Hautes Temp. Réfract., 19 : 285 (1982)

32. M. Dauchier, G. Bernhart and C. Bonnet, Properties of Silicon Carbide Based Ceramic-ceramic Composites, Proc. 30th. Nat. SAMPE, Anaheim (March 19-21, 1985), vol. 30, pp. 1519-1525, SAMPE, Covina (Cal.), 1985

33. G. Bernhart, P. Lamicq and J. Mace, Fiabilité des composites céramique-céramique, Industrie Céramique, 790 : 51 (1985)

34. D. Neuilly, J.M. Quenisset, F. Langlais, R. Naslain and L. Heraud, private communication (1985).

CVD FABRICATION OF IN-SITU COMPOSITES OF NON-OXIDE CERAMICS

Toshio Hirai and Takashi Goto

The Research Institute for Iron, Steel and Other Metals
Tohoku University, Sendai 980, Japan

INTRODUCTION

Generally, materials can be divided into three groups: polymer, metal and ceramics. Composites are fabricated by combinations of these materials for the purpose of using superior properties of each material. In many cases, separately prepared materials A and B are combined to fabricate the composite. We call the composite prepared by the process a "phase-joining composite"[1].

$$A + B \longrightarrow Composite(A+B)$$

A typical example of the phase-joining composite is FRP or FRM (fiber reinforced plastic or metal). Sintering additives are often contained in ceramic articles, therefore the resulting sintered bodies are considered to be composites. The recent trend on multi-component ceramics studies are shifting from the search of the additives to promote the sintering process to the development of a high-performance ceramic composite, as shown by the main theme of this conference.

On the other hand, composites are also prepared by a "phase-separating method"[1], in which the composite is obtained by proper treatment of a homogeneous material (x) containing both A and B. We call this composite the "in-situ composite" in this paper.

$$X_{(A+B)} \longrightarrow in\text{-}situ\ Composite(A+B)$$

The phase-separating method applied to ceramics is seen in directionally solidified ZrO_2-X eutectic[2] and partially crystallized glass ceramics[3], etc. The powders of Fe_2O_3-MnO[4] and BaO-Fe_2O_3[5] prepared by the sol-gel and hydrothermal processes, respectively, are one of the in-situ ceramic composites in a broad sense.

Recently, many kinds of in-situ ceramic composites have been prepared by the CVD process using gases as source material. As examples of the in-situ ceramic composite in the form of powder, Si_3N_4-TiN[6], Si_3N_4-C[7] and ZrO_2-Al_2O_3[8] composite-powders have been reported. TiN promotes the crystallization of amorphous Si_3N_4 in Si_3N_4-TiN composite-powder, and fully dense Sialon ceramics are obtained by the sintering of the mixture of Si_3N_4-C composite-powder and Al_2O_3 powder. For ZrO_2-Al_2O_3 composite-powder, high-toughness ceramics are prepared by the sintering of the ZrO_2-Al_2O_3 composite-powder[9]. Here it should be noted that the microstructures of these composite-powders are altered by the heat-treatments during the sintering process.

The in-situ ceramic composites in the forms of plate and film prepared by the CVD process have a unique microstructure which is almost impossible to obtain by the conventional phase-joining method.   These in-situ composites often indicate interesting properties and can be useful as new materials.   In the present paper, we review the microstructures and properties of the in-situ ceramic composite-plate and -film of non-oxide ceramics.

CVD-FABRICATION OF IN-SITU CERAMIC COMPOSITES

The CVD chambers are classified as hot wall and cold wall types. When more than two kinds of raw gases are used in the CVD process and these gases are mutually reactive , the cold wall type is preferred in laboratory scale.   In the chamber of the hot wall type, homogeneous gas phase reactions may easily take place, and it becomes difficult to obtain the deposits on the substrate because of the by-product formation in the gas phase.   Even in the cold wall type, a nozzle to prevent premature reactions may be neccessary depending on the properties of the gases used. Fig.1 shows the chamber which we are using to prepare the CVD in-situ ceramic composites.   The most important parameter in the chamber is the distance between the substrate and the nozzle[10].

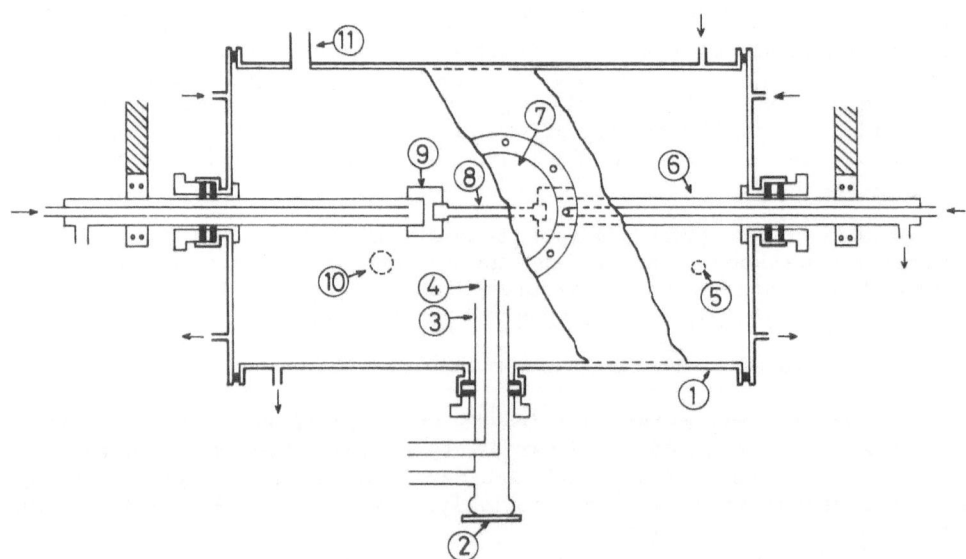

Fig.1   CVD chamber for fabricating in-situ ceramic composites[10].

(1) water-cooled vacuum chamber; (2),(7) quartz glass windows; (3),(4) gas inlets; (5) pressure gauge; (6) water-cooled copper electrode; (8) graphite heater(substrate); (9) graphite socket; (10),(11) gas outlets.

STRUCTURE AND PROPERTIES OF CVD IN-SITU CERAMIC COMPOSITES

Fig.2 shows the brief comparison of the fabrication process and structure between sintered and CVD ceramic composites. In many cases, the secondary phase is present at the grain boundary in the sintered ceramic composites, while the secondary phase disperses within the grain for the CVD in-situ ceramic composites. It is quite difficult to fabricate the ceramic composite whose matrix is amorphous by the sintering process, though relatively easy by the CVD process.

The typical structures of the CVD in-situ ceramic composites are indicated in Fig.3. These structures vary depending on the types of raw gases and the CVD conditions. Table 1 summarizes the variety, source gas and structure of the CVD in-situ ceramic composites reported in references.

C-based in-situ Ceramic Composites

The CVD carbon described in this section is called pyrolytic carbon (PyC). The structure of the PyC changes from turbostratic to hexagonal, and the interlayer spacing (C_0) varies widely depending on the deposition temperature and other CVD conditions.

The $C-B_4C$ CVD in-situ composite is prepared in the temperature range of 1100° to 2000°C[11]. When the B content is in the range of 0.5 to 1.0 wt%, B atom is substitutionally dissolved into the C matrix, and it will dissolve interstitially with increasing B content. Above 1.0 wt% B, the B atom begins to form B_4C. The C containing small amount of B has promise as a thermocouple material because of its high thermoelectromotive force[12]. However, the properties of the $C-B_4C$ composite have been not reported in literature.

The C-SiC CVD in-situ composite is obtained in the temperature range of 1440° to 2025°C[13]. The SiC (β-type about 2000 Å in dia. and about 200 Å in thickness) is dispersed in the C matrix when the composite is prepared at a relatively low temperature (1440°C). This feature is shown in Fig.4. A very small amount of α.SiC is dispersed in the deposit prepared at 2025°C. The oxidation resistance of these composites is superior to that of PyC.

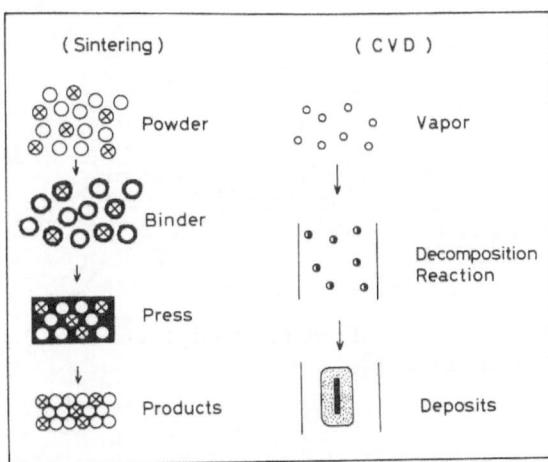

Fig.2 Brief comparison of the fabrication process and structure between sintered and CVD ceramic composites.

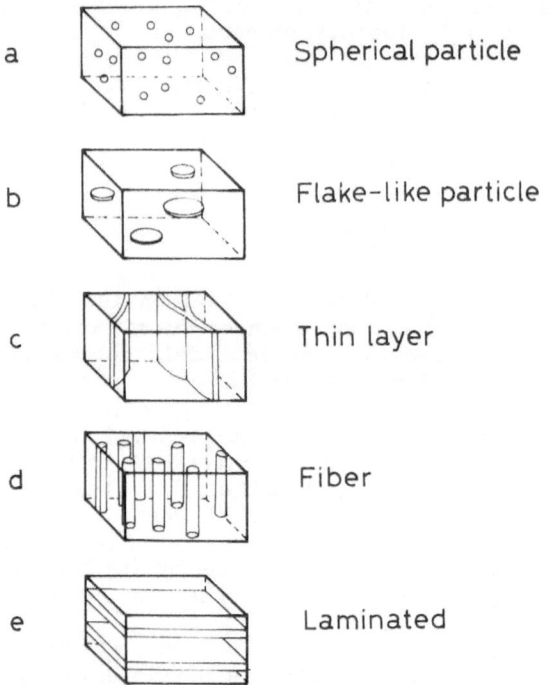

a   Spherical particle

b   Flake-like particle

c   Thin layer

d   Fiber

e   Laminated

Fig.3   Typical structures of the CVD in-situ ceramic composites.

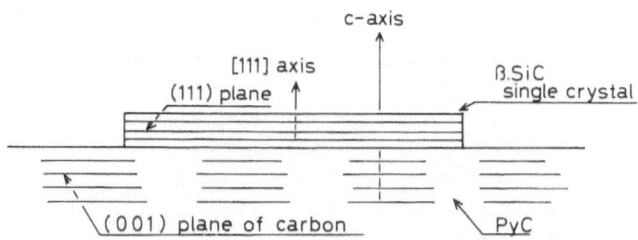

Fig.4   Mutual arrangement of carbon matrix and β.SiC in the C-SiC CVD in-situ composites13.

Table 1 Variety, source gas and structure of
CVD in-situ Ceramic Composites.

Matrix	Dispersion	Source Gases	Type[*]	Ref.
C	B_4C	C_xH_y-BCl_3	a	11
	SiC	C_3H_8-$SiCl_4$	b	13
	TiC	CH_4-$TiCl_4$	a	14
	ZrC	C_xH_y-$ZrCl_4$	a	16
	HfC	C_xH_y-$HfCl_4$	–	16
	BeO	C_xH_y-$Be(C_5H_7O_4)$	a	17
BN	C	BCl_3-NH_3-C_2H_2	–	19
	Si_3N_4	BCl_3-NH_3-$SiCl_4$	–	20
	TiN	BCl_3-NH_3-$TiCl_4$	a	20
Si_3N_4	C	$SiCl_4$-NH_3-C_3H_8	a	21
	AlN	SiH_4-NH_3-$AlCl_3$	–	24
	AlN	$SiCl_4$-NH_3-$AlCl_3$-O_2	–	25
	BN	$SiCl_4$-NH_3-B_2H_6	e	26
	BN	$SiCl_4$-NH_3-BCl_3	–	20
	TiN	$SiCl_4$-NH_3-$TiCl_4$	a,d	28
	SiO_2	SiH_4-NH_3-O_2	–	30
B_4C	C	BCl_3-C_3H_8	–	31
SiC	C	$SiCl_4$-C_3H_8	a	32
	B_4C	$SiCl_4$-BCl_3-C_3H_8	–	–
	TiC	$SiCl_4$-$TiCl_4$-C_3H_8	–	–
	Si_3N_4	$Si(CH_3)_4$-NH_3	–	35
ZrC	C	$ZrCl_4$-CH_4	–	36
Ti_3SiC_2	TiC	$TiCl_4$-$SiCl_4$-CCl_4	c	38
Ti-B-N	TiB_2	$TiCl_4$-BCl_3-N_2	a	39

[*];Refer to Fig.3.

The C-TiC CVD in-situ composite is prepared in the temperature range of 1200° to 2200°C[14]. When the Ti content is over 10 wt%, TiC particles (about 6 μm in dia.) are dispersed in the C matrix. The dimensions of the composite are almost unchanged after irradiation[15], that is, the composite is a radiation resistant material.

The C-ZrC and C-HfC CVD in-situ composites are prepared in the temperature range of 1300° to 1500°C[16]. Their dispersion type is uncertain, but may be the (a)-type in Fig.3. The bending strength of these composites is higher than that of usual PyC, hence they are effective as a thrust chamber.

The C-BeO CVD in-situ composite is obtained in the temperature range of 1600° to 2000°C[17]. BeO particles (about 1 μm in dia.) are dispersed in the C matrix.

BN-based in-situ Ceramic Composites

The structure of the CVD BN varies from turbostratic to hexagonal depending on CVD conditions in a similar manner as PyC[18]. However, the structure of the BN matrix in the BN-based in-situ ceramic composite is often turbostratic (may be expressed as amorphous).

The BN-C CVD in-situ composite is prepared at 1700°C[19]. Hexagonal BN is the matrix and the C is present in a continuous network, but the sizes or shapes of the C are uncertain. The composites containing more than 10 wt% C become electrically conductive by the continuous carbon network.

The structure of the $BN-Si_3N_4$ CVD in-situ composite obtained at 1400°C is considered to be consisted from the turbostratic BN matrix and very fine dispersed amorphous Si_3N_4 particles (less than 10 wt%) [20]. Generally, the CVD BN prepared at lower temperatures are optically transparent but those prepared at higher temperatures are opaque white. The trasparency of the usual CVD BN is unstable in the presence of moisture. On the contrary, the $BN-Si_3N_4$ CVD in-situ composite is transparent as shown in Fig.5 and the transparency of the composite is stable and does not change in the presence of moisture and also even after heat-treatment at 1600°C.

For the $BN-Si_3N_4$ in-situ composite prepared at 1800°C, the particles of $\beta.Si_3N_4$ (10 wt% and 300 to 500 Å in dia.) are dispersed in the turbostratic BN matrix. The delamination phenomenon which is a typical property of CVD BN is prevented in the composite.

The BN-TiN in-situ composite is prepared at 1400°C[20]. The particles of TiN (less than 1.9 wt% and 30 to 50 Å in dia.) are dispersed in the turbostratic BN matrix. The resulting composite is black, and the ability for heat insulation is four times higher than that of usual CVD BN.

Si_3N_4-based in-situ Ceramic Composites

The Si_3N_4(amorphous)-C CVD in-situ composite is prepared in the temperature range of 1100° to 1300°C[21]. The carbon particles (about 1000 Å in dia.) in the amorphous Si_3N_4 matrix are present in a three dimensional continuous network. This composite containing more than 0.2 wt% C is electrically conductive while the usual CVD amorphous Si_3N_4 is insulative[22]. The resistance against both HF corrosion and high temperature evaporation is improved by the addition of carbon into the amorphous Si_3N_4 matrix[23].

Fig.5 Outside appearance of the BN-Si$_3$N$_4$ CVD in-situ composite[20]. (a) as-deposited, (b) polished.

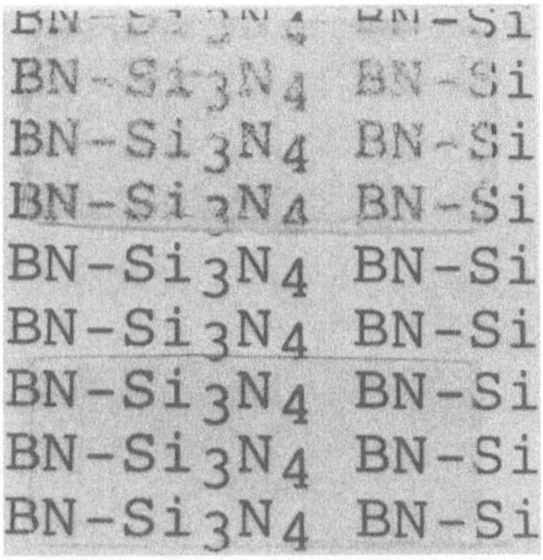

The Si$_3$N$_4$(amorphous)-AlN CVD in-situ composite is obtained in the temperature range of 600° to 1100°C[24]. The sizes of the AlN particles dispersed in the amorphous Si$_3$N$_4$ matrix vary from 150 to 500 Å depending on the deposition temperature, and the volume fraction of the AlN particles in the composite varies from 0 to 1. The electron trap density of the composite is higher than that of the amorphous Si$_3$N$_4$, and the level of trap state of the composite is lower than that of amorphous Si$_3$N$_4$. Therefore the composite is considered to be a promising material for the memory device.

The preparation of Sialon by CVD has been tried, and the composite containing both AlN and amorphous Si$_3$N$_4$ was obtained[25]. Heat-treatments in air, N$_2$ or H$_2$ atmospherers at 1400°C for < 100 h have shown no effect on the nature of the composite. The sizes or shapes of the AlN are uncertain.

The Si$_3$N$_4$(amorphous)-BN CVD in-situ composite is prepared in the temperature range of 1100° to 1300°C[26]. The amorphous Si$_3$N$_4$ and the turbostratic BN is alternately stacked in the composite as shown in Fig.6[27]. Fig.7 shows the transmittance of light in the wave length from ultra-violet to visible region for the composite whose amount of Si$_3$N$_4$ and BN is approximately equal. About 20 % of the light is not transmitted because of the surface reflection, therefore the absorption of light by the composite is very small. This composite maintains a superior transparency even after heat-treatment at 1600°C for several hours, therefore the composite may be a suitable material for a high temperature viewfinder.

The structure of the Si$_3$N$_4$-TiN CVD in-situ composite varies according to the CVD conditions[28], as shown in Fig.8. When the deposition temperature is 1050° to 1200°C, TiN particles (10 to 30 wt% and about 30 Å in dia.) are dispersed in the amorphous Si$_3$N$_4$ matrix. When the deposition temperature is 1250°C, TiN particles (5 wt% and about 100 Å in dia.) are dispersed in the α.Si$_3$N$_4$ matrix. When the deposition temperature is raised to 1350°∿1450°C, the matrix of the composite is β.Si$_3$N$_4$ and TiN becomes fibrous (about a few ten Å in dia. and about 2 μm in length)[29]. Fig.9 shows a schematic of the microstructure for the β.Si$_3$N$_4$-TiN composite. The lattice of the TiN is not coherent to that of α.Si$_3$N$_4$ but coherent to β.Si$_3$N$_4$ lattice. Table 2 indicates the crystal lattice relationships between TiN and β.Si$_3$N$_4$ in the composite. This coherence is thought to be the reason why the TiN becomes fibrous in the β.Si$_3$N$_4$ matrix.

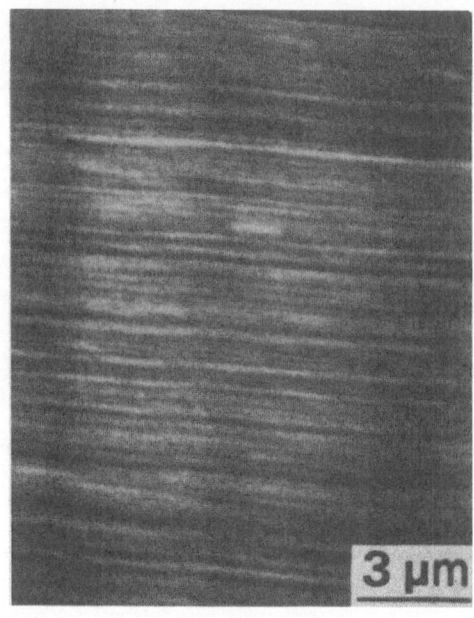

Fig.6   Cross-sectional appearance of the Si_3N_4(amorphous)-BN CVD in-situ composite[27].

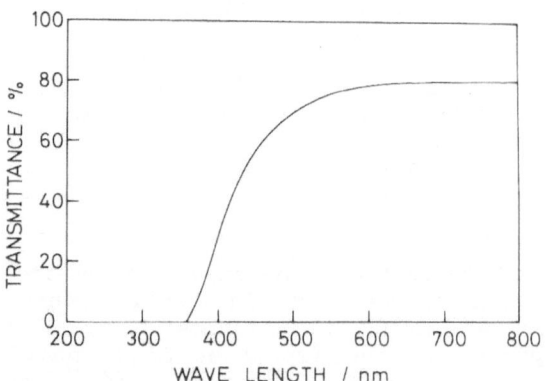

Fig.7   Transmittance of light from ultra-violet to visible region of the Si_3N_4(amorphous)-BN CVD in-situ composite.

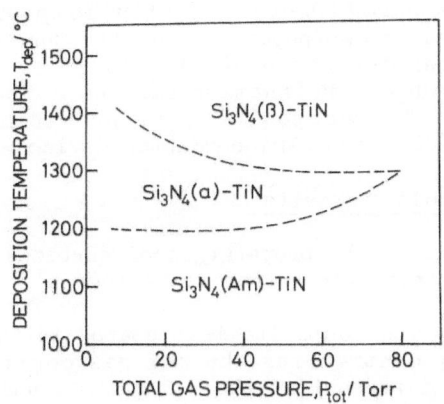

Fig.8　Effect of CVD conditions on the structure of the Si_3N_4-TiN CVD in-situ composite[28].

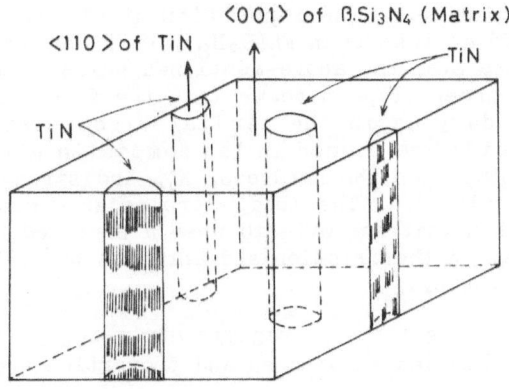

Fig.9　Schematic microstructure of the $\beta.Si_3N_4$-TiN CVD in-situ composite[29].

Table 2　Crystal Lattice Relationship between TiN and $\beta.Si_3N_4$ in the Composite[29].

$$\text{TiN fiber axis } // [001]_{Si_3N_4}$$
$$(001)_{TiN} // (001)_{Si_3N_4}$$
$$d(110)_{TiN} \approx d(001)_{Si_3N_4}$$
$$d(200)_{TiN} \approx d(300)_{Si_3N_4}$$
$$d(222)_{TiN} \approx d(330)_{Si_3N_4}$$

The Si_3N_4(amorphous)-SiO_2 CVD in-situ composite is obtained at $900^{\circ}C^{30}$, and SiO_2 is also amorphous. The structure of the composite is unknown. The composition ratio of SiO_2 to Si_3N_4 varies from 0 to 1. The density and the index of refraction for the composite depend on the composition ratio. The composite is a promising material for an insulating film of MIS or MIOS semi-conductor devices.

B_4C-based in-situ Ceramic Composites

For many occasions in the preparation of carbides by CVD, the resulting deposits often contain free carbon depending on the CVD conditions.

The B_4C-C CVD in-situ composite is prepared in the temperature range of 1400° to $1800^{\circ}C$ by controlling the raw gas composition. The free carbon is thought to be present at the B_4C grain boundary (about a few μm in dia.). The mechanical properties of the B_4C-C composite deteriorate with increase of the amount of free carbon[31].

SiC-based in-situ Ceramic Composites

The SiC-C CVD in-situ composite is obtainable by controlling the raw gas composition in the temperature range of 1500° to $1800^{\circ}C^{32}$. Fig.10 shows the effect of the propane gas flow rate, $FR(C_3H_8)$, on the carbon content of the composite. The deposition of SiC having a theoretical density is observed at less than $FR(C_3H_8)$ of 55 cm^3/min, and the SiC-C composite is obtained over the above-mentioned value. While the state of the free carbon is uncertain, it seems that the free carbon is present at the SiC grain boundary as in the similar circumstances as the B_4C-C composite. The voids are formed in the composite with increase of the amount of free carbon, then the hardness, K_{IC} and strength are lowered by the effect of the voids[33]. The load where seizure occurs for the SiC-C composite is less than that for CVD-SiC when a sintered SiC is used as the counterpart material in the friction and abrasion tests[34]. Fig.11 shows the results of the tests.

The preparation of SiC-B_4C or SiC-TiC CVD in-situ composites has been attempted, but the detailed structures and properties are unknown.

The SiC-Si_3N_4 CVD in-situ ceramic composites are prepared in the temperature range of 1300° to $1500^{\circ}C^{35}$. The deposits prepared at low temperatures are heterogeneous. SiC mainly deposits in the central part of the substrate and Si_3N_4 in the outer part. Increasing temperature results in homogeneous deposits. The sizes or shapes of the Si_3N_4 and SiC in the composite are uncertain.

Other Nonoxide-based in-situ Ceramic Composites

The ZrC-C CVD in-situ ceramic composite is prepared in the temperature range of 1550° to $2100^{\circ}C^{36}$. The free carbon segregates at the grain boundary of ZrC when the amount of the carbon is about 0.4 wt%. When the amount of the free carbon is almost 3 wt%, the particles of the free carbon (10 to 30 μm in dia.) are also present in the grains. Thermodynamic calculations on the CVD conditions for this system were performed[37], and revealed that the ZrC-C composite is obtainable at the conditions of higher CH_4 and lower $ZrCl_4$ partial pressures.

The Ti-Si-C system CVD in-situ ceramic composite is prepared by the gas-phase reaction of $TiCl_4$, $SiCl_4$ and CCl_4 mixture in the tempetature range of 1000° to $1300^{\circ}C^{38}$. The composition of the deposit changes from $Ti_5Si_3C_x$ + $TiSi_2 \longrightarrow (Ti_3SiC_2$ + $TiSi_2) \longrightarrow (Ti_3SiC_2$ + TiC)$\longrightarrow$ TiC + SiC by increasing the amount of CCl_4. The mixtures as indicated between the

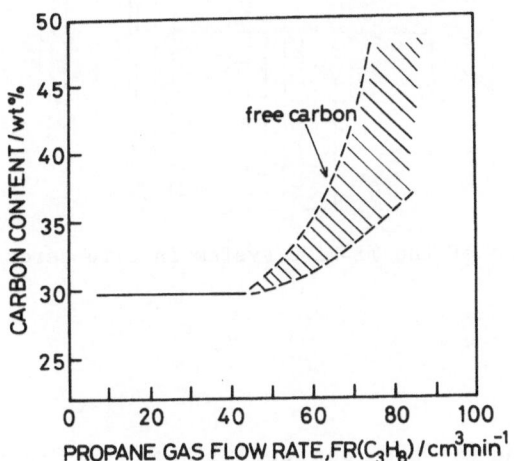

Fig.10 Effect of the propane gas flow rate on the carbon content of the SiC-C CVD in-situ composite[32].

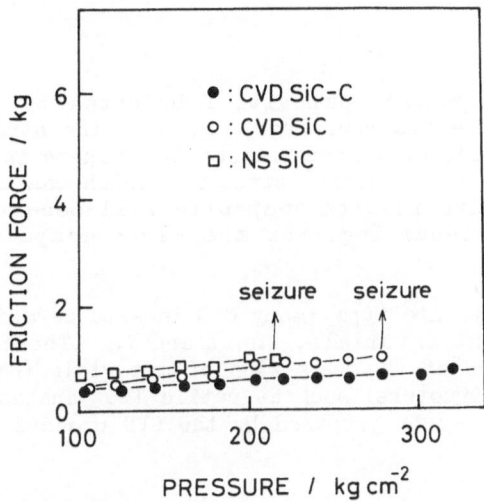

Fig.11 Results of the friction and abrasion tests[34].

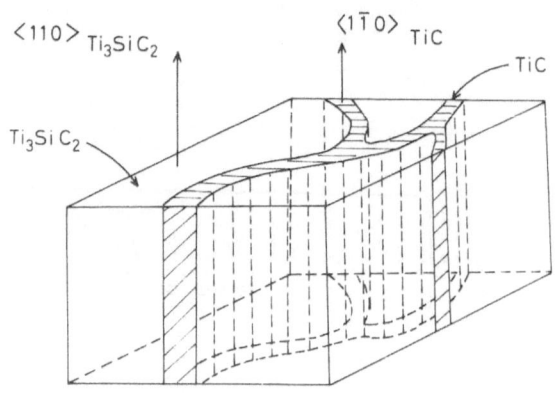

Fig.12 Structure of the Ti-Si-C system in-situ ceramic composite[38].

parentheses are the (c)-type composites shown in Fig.3. Fig.12 depicts its structure.   In this case, a peculiar coherence between the matrix and the dispersions is observed.   The properties of these composites are unknown.

Attempts to prepare ternary Ti-B-N compounds have been made by CVD in the temperature range of 1050° to 1500°C[39].   The cubic titanium boronitride and the composite of boronitride and TiB_2 are obtained. The detailed structures of the composite are unknown, however, the composites may find an application for wear resistant layers on cutting tools.

SUMMARY

The in-situ ceramic composite fabricated by the CVD of multi-component gas mixture was reviewed.   Most of the structural analyses on the composite are conducted macroscopically.   There is little information available on the micro-composite structure which can be prepared only by CVD process.   The CVD in-situ composite will become useful only when exotic properties resulting from the micro-composite structure are discovered.

Beyond the composite type, many CVD in-situ ceramic composites are a mixture of different materials, i.e. X and Y.   The composite material which is the mixture of the same kind of materials (same composition but different crystal structure) such as rutile TiO_2-anatase TiO_2, $\alpha.SiC$-$\beta.SiC$ or $\alpha.Si_3N_4$-$\beta.Si_3N_4$, may be prepared by the CVD process in the near future.

REFERENCES

1. T.Hirai,"Emergent Process Methods for High-Technology Ceramics", Materials Science Research Series, vol.17, edited by R.F.Davis, H.Palmour III and R.L. Porter, Plenum, N.Y.,(1984), p.329.
2. J.Echigoya, S.Hayashi, K.Sakai and H.Suto, J. Japan Inst. Metals, 48 (1984) 430.
3. D.G.Grossman, J. Amer. Ceram. Soc., 55 (1972) 446.
4. Y.Bando, Bull. Ceram. Soc. Japan, 19 (1984) 483.

5. M.Kiyama, Bull. Chem. Soc. Japan, 49 (1976) 1855.
6. E.J.Mehalchick and R.N.Kleiner, U.S. Patent, (4,145,224), (1979).
7. H.W.Jacobson, U.S. Patent, (4,036,653), (1977).
8. S.Hori, "Zirconia Ceramics 1", edited by S.Somiya, Uchida Rokakuho, Tokyo, (1983), p.21.
9. S.Hori, M.Yoshimura, S.Somiya, R.Kurita and H.Kaji, J. Mater. Sci. Letters, 4 (1985) 413.
10. K.Niihara and T.Hirai, J. Mater. Sci., 11 (1976) 593.
11. R.O.Grisdale, A.C.Pfister and W.van Roosbroeck, Bell System Tech. J., 30 (1951) 271.
12. C.A.Klein and M.P.Lepie, Solid-State Electron., 7 (1964) 241.
13. S.Yajima and T.Hirai, J. Mater. Sci., 4 (1969) 424.
14. A.S.Schwartz and J.C.Bokros, Carbon, 5 (1967) 325.
15. R.J.Price and J.C.Bokros, ibid., 9 (1971) 205.
16. J.S. Waugh, L.Hagen, R.Donadio and J.Pappis, Raytheon Tech. Memo., T-745 (1967).
17. J.J.Gebhardt, Decompos. Organometal. Compounds Refract. Ceram., Metals, Metal Alloys, Proc. Inst. Symp., (1967), p.319.
18. T.Matsuda, N.Uno, H.Nakae and T.Hirai, J. Mater. Sci., in press : T.Matsuda, N.Uno, Y.Matsunami, H.Nakae and T.Hirai,Proc. 5th European Conf. on Chemical Vapor Deposition, edited by J.-O.Carlsson and J.Lindstrom,(1985), p.420.
19. S.H.Chen and R.J.Diefendorf, Proc. 3rd Int. Carbon Conf., (Barden-Barden, 1980), p.44.
20. H.Nakae, Y.Matsunami, N.Uno, T.Matsuda and T.Hirai, Proc. ERATO symposium (Japan,1983) in press : Proc. 5th European Conf. on Chemical Vapor Deposition,edited by J.-O.Carlsson and J.Lindstrom,(1985), p.242.
21. T.Hirai and T.Goto, J. Mater. Sci., 16 (1981) 17.
22. T.Goto and T.Hirai, ibid., 18 (1983) 383.
23. Idem., ibid., 18 (1983) 3387.
24. S.Zirinsky and E.A.Irene, J. Electrochem. Soc., 125 (1978) 305.
25. R.L.Landingham and R.W.Taylor, "Energy and Ceramics", Materials Science Monographs, 6, edited by P.Vincenzini, Elsevier, N.Y.,(1980), p.494.
26. T.Hirai,T.Goto and T.Sakai, in Ref.1, p.347.
27. T.Hirai and T.Goto, "Amorphous Material, Physics and Technology", Editorial Committee of the Special Project Research on Amorphous Material, (1983), p.130.
28. T.Hirai and S.Hayashi, J. Mater. Sci., 17 (1982) 1320.
29. S.Hayashi, T.Hirai, K.Hiraga and M.Hirabayashi, ibid., 17 (1982) 3336.
30. T.L.Chu, J.R.Szedon and C.H.Lee, J. Electrochem. Soc., 115 (1968) 318.
31. K.Niihara, A.Nakahira and T.Hirai, J. Amer. Ceram. Soc.,67(1984) c-13.
32. T.Hirai, T.Goto and T.Kaji, Yogyo-Kyokai-Shi, 91 (1983) 502.
33. K.Niihara, A. Suda and T.Hirai, Proc. 1st Int. Symposium on Ceramic Components for Engine, edited by S.Somiya, E.Kanai and K.Ando, KTK Scientific, (1983), p.480.
34. M.Sasagawa, H.Kurosawa, Y.Hoshi, A.Ohkubo and T.Hirai, Proc. 95th meeting of Japan Inst. Metals, (Hiroshima,1984), p.149.
35. J.F.Lartique, M.Ducarroir and B.Arms, Proc. 9th Int. Conf. on Chemical Vapor Deposition, edited by McD.Robinson, C.H.J. van den Brekel, G.W.Cullen, J.M.Blocher,Jr. and P.Rai-Choudhury, Electrochem. Soc., Pennington, (1984), p.561.
36. M.P.Lepie, Trans. Brit. Ceram. Soc., 63 (1964) 431.
37. P.Salles, M.Ducarroir and C.Bernard, in Ref.35, p.615.
38. J.J.Nickl and K.K.Schweitzer and P.Luxenberg, J. Less-Common Met., 26 (1972) 335.
39. J.L.Peytavy, A.Lebugle, G.Montel and H.Pastor, High-Temp. High-Press., 10 (1978) 341.

PREPARATION OF BORON NITRIDE/BORON CARBIDE CERAMICS

BY PYROLYSIS OF BORIC ACID-GLYCERIN CONDENSATION PRODUCT

Hiroaki Wada, Kazuyuki Kuroda, and Chuzo Kato

Department of Applied Chemistry
Waseda University
Shinjuku-ku, Tokyo 160, Japan

INTRODUCTION

Pyrolytic methods of polymers or organic compounds for the synthesis of ceramics have been actively investigated in recent years.[1] Various non-oxide ceramics including silicon carbide, silicon nitride, boron nitride, and boron carbide have been prepared by this method. However, boric acid-polyol condensation products have not been considered as a precursor, though they can be easily synthesized by dehydration.

During the course of the study on the synthesis and properties of boron-containing inorganic polymers, we have found that the boric acid ester with glycerin is relatively stable to thermal treatment whereas the esters of monohydric and dihydric alcohols have lower boiling points and show higher volatility.

As to the structure of the boric acid-glycerin condensation product, the details have not been clarified yet. However, the following proposed structure is usually accepted.[2]

In this study, boron-containing ceramics were prepared by using the boric acid-glycerin condensation product as a precursor. The preliminary report has been published by us very recently.[3] In this reaction the following preferable points should be noted.

1) Ammonia gas, boron trichloride gas, and cyanide compounds which have been employed for the synthesis of boron nitride are not used in this process.

2) Metallic contamination can be avoided because only boric acid and glycerin are used and pyrolyzed under nitrogen atmosphere.

3) Both boron nitride and boron carbide were formed at the same time by this process. In other words, composite ceramics which contained both boron nitride and boron carbide could be formed by the pyrolysis of borate esters directly.

EXPERIMENTAL

The condensation product was prepared by heating an equimolar mixture of boric acid and glycerin at 150°C in an evaporator. The end-point of the dehydration was determined by the measurement of the weight loss.

The condensation product was set in an alumina boat, which was then placed in a mullite tube and heated to various temperatures (900°C-1400°C) in N_2 or Ar flow at a rate of 500 cm^3 min^{-1}. The heating rate was 7°C min^{-1} up to 600°C and 5°C min^{-1} above 600°C. The time kept at a given temperature was 2 h, then the products were cooled at 5°C min^{-1} to 600°C. The thermal treatment for 4 h was also conducted in the case of nitrogen atmosphere.

The mixture of boric oxide and carbon black was also prepared (molar ratio 1:3) and treated thermally in a same manner for comparison.

The products obtained were analyzed by XRD, XRF, and IR. ^{11}B-NMR was used for the analysis of the condensation product.

RESULTS AND DISCUSSION

Condensation Product

The condensation product was a colorless transparent glassy solid. It showed spinnability on heating. The chemical analysis of the condensation product was as follows, B:10.48%, C:35.73%, H:5.01%. These data was in good agreement with the calculated data from the proposed structure assuming the fully dehydrated product $(C_3H_5O_3B)_n$ (B:10.82%, C:36.08%, H:5.06%).

Fig.1 shows the infrared spectrum of the condensation product dissolved in chloroform. The absorption bands around 3200 cm^{-1} and 1640 cm^{-1} due to OH vibrations disappeared, indicating the complete dehydration. The bonding of B-O-C was supported by the absorption peak at 1040-1100 cm^{-1} due to C-O linkage and B-O stretching vibration at around 1300-1400 cm^{-1}.

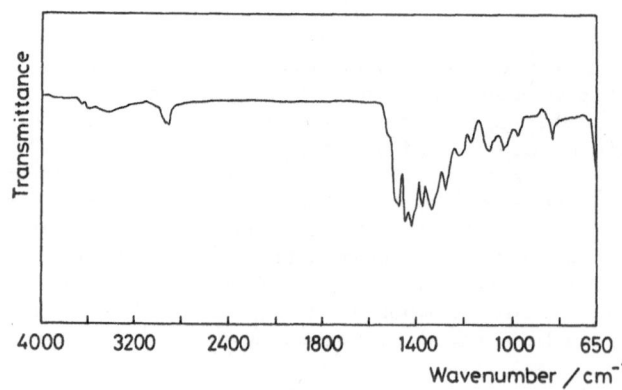

Fig. 1 Infrared spectrum of the condensation product.

^{11}B-NMR spectrum of the condensation product (Fig.2) shows two distinct signals at d=18.37 ppm and 22.86 ppm. The signal at d=18.37 ppm indicated the three coordinated boron with planar structure, suggesting the above proposed structure. The signal at d=22.86 ppm probably indicates the six-membered ring of the borate ester, though the d=18.37 ppm showed the presence of five-membered ring of the ester. The monomeric structure has been already denied and our present results also showed the oligomeric nature.

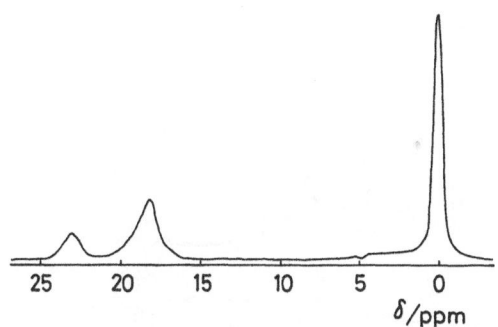

δ/ppm

Fig. 2 ^{11}B-NMR spectrum of the condensation product dissolved in CHCl$_3$ (external standard; BF$_3$O(C$_2$H$_5$)$_2$)

Pyrolyzed Products

When the condensation product was heated between 900°C and 1200°C, the product was black regardless of the kind of gas used and the heating time. XRD and IR results indicated the formation of boric oxide. Thermal analysis also showed the decomposition of the product at about 430-440°C. This result showed that the decomposition in the temperature range described above only produced the amorphous carbon and B$_2$O$_3$.

Thermal treatment of the condensation product above 1300°C in N$_2$ yielded nitrided products. XRD profiles are shown in Fig.3. The diffraction peak at 2θ=26.6° could be ascribed to d$_{002}$ of BN.   B$_4$C was also detected by the peaks at 2θ=34.9° and 37.8°. The formation of these nitride and carbide was confirmed by IR(Fig.4).

The crystallinity of the heat-treated products increased with an increase of the heating temperature. The effect of the heating time was also found in the crystallinity of BN, but B$_4$C crytals were not affected so much. Approximate crystallite size calculated by the Scherrer equation[4] was about 50Å and 100Å at 1300°C and 1400°C, respectively.

In Fig.3, the peak due to d$_{102}$ could not be detected, indicating the turbostratic nature of the obtained BN. The major peak at 2θ=26.6° was ascribed to BN, because the possibility of the presence of graphite was mostly denied by the fact that the color of the product formed at 1400°C was gray and that the IR absorption peaks at 1400cm^{-1} and 800cm^{-1} due to BN[5] became stronger than that of B$_4$C (1080cm^{-1}) with the increase of the reaction temperature.

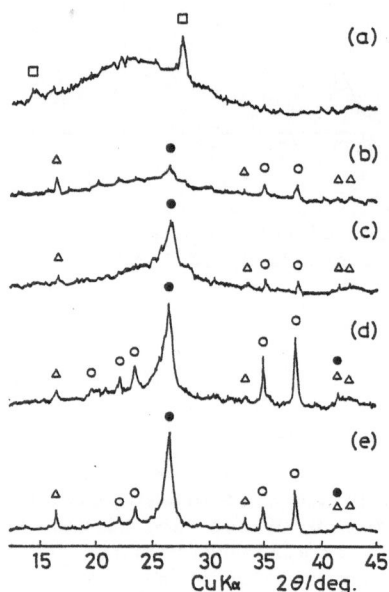

Fig.3   XRD patterns of heat-treated samples of boric acid-glycerin condensation product pyrolyzed in N_2 (a)1200°C, 2h; (b)1300°C, 2h; (c)1300°C, 4h; (d)1400°C, 2h; and (e)1400°C, 4h.

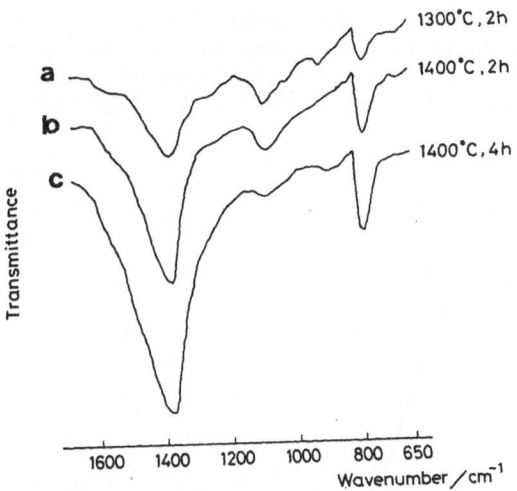

Fig.4   IR spectra of heat-treated samples of boric acid-glycerin
condensation product pyrolyzed in N_2 (a)1300°C, 2h;
(b)1400°C, 2h; and (c)1400°C, 4h.

Fig.5 shows the XRD pattern of the product heated at 1400°C for 2h in
Ar flow. Under the reaction in Ar flow, boron carbide was clearly detected.
The peak at $2\theta = 26.6°$ was attributed to the presence of graphite because
there was no nitrogen source in this case.

Both figures of Fig.3 and 5 showed other peaks, which were identified
by XRD and XRF to be the reaction products[6,7] of boric oxide and alumina
used as a reaction vessel.

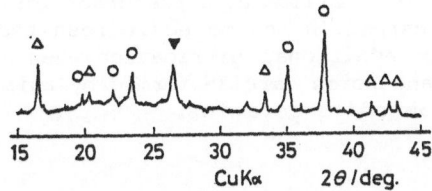

Fig.5   XRD pattern of heat-treated sample of boric acid-glycerin
condensation product pyrolyzed at 1400°C for 2h in Ar.

The physical mixture of B_2O_3 with carbon black were treated thermally at 1400°C for 2h in N_2 flow. The XRD result is shown in Fig. 6, showing the formation of B_4C. However, its intensity was weaker than that of the ester and boron nitride could not be detected.

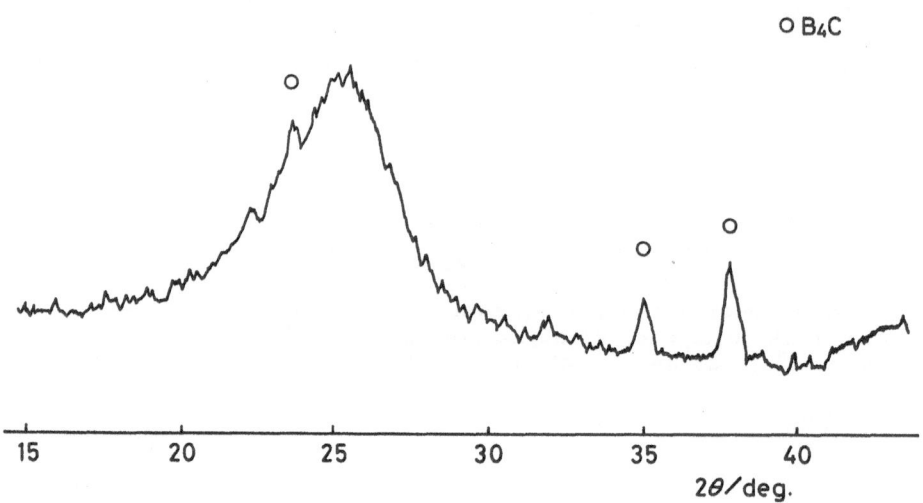

Fig. 6   XRD pattern of the heat-treated sample of boric oxide and carbon black mixture(B_2O_3:C=1:6) heated at 1400°C for 2h in N_2.

Application to Other Boric Acid-Polyol Systems

In order to apply this process to other polyol systems, triethanolamine ($N(C_2H_4OH)_3$) and diethanolamine ($NH(C_2H_4OH)_2$) were used for the synthesis of   condensation products. After the reaction with boric acid under a similar dehydration condition, the products were pyrolyzed in N_2 flow.

Fig.7 shows the XRD pattern of the heat-treated product obtained from the condensation product of boric acid and triethanolamine (molar ratio 1:1).   This figure also indicates the formation of boron nitride and boron carbide.   It is worth noting that the crystallinity is higher than that of the product from boric acid-glycerin condensation product. The pyrolyzed product from the boric acid-diethanolamine system also indicated the formation of boron nitride with a higher crystallinity. However, the formation of boron carbide was affected by the molar ratio of boric acid and diethanolamine.

Thus, we found that boric acid esters with glycerin, diethanolamine, and triethanolamine could be utilized as a precursor for B-containing ceramics for the first time. The carbon in the molecule resulted in the occurrence of carbide in Ar flow and the additional nitridation when nitrogen was used. The coexisting boron nitride and boron carbide formed by this process is considered to be a sort of unique composite ceramics.

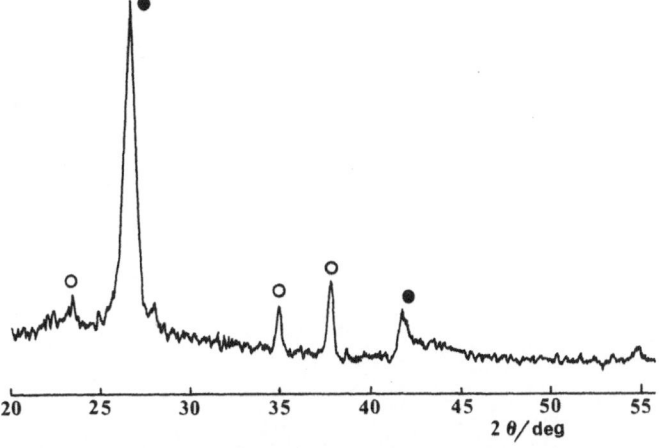

Fig.7 XRD pattern of the heat-treated sample of boric acid-triethanolamine condensation product pyrolyzed at 1400°C for 2h in N_2.

Acknowledgements

The authors wish to express their sincere thanks to Kawakami Memorial Fund for the financial support. They are indebted to the experimental assistance by Miss Hiromi Shibasaki, Mr. Kazuaki Nojima and Mr. Shuichi Ito.

REFERENCES

1. R. W. Rice, Am. Ceram. Soc. Bull., 62:889(1983).
2. H. Steinberg, "Organoboron Chemistry" Vol. 1 (Boron-oxygen and boron-sulfur compounds) Interscience,(1964).
3. H. Wada, K. Kuroda, and Chuzo Kato, Chem. Lett., 691(1985).
4. H. P. Klug and L. E. Alexander, "X-ray Diffraction Procedure", John Wiley(New York)(1954).
5. E. G. Brame, Jr., J. L. Margrave, and V. W. Meloche, J. Inorg. Nucl. Chem., 5:48(1957).
6. JCPDS files, 29-10 and 32-3.
7. Amer. Ceram. Soc. ed., "Phase Diagram for Ceramists" (1969)Supplement figs. 308 and 2339.

POROUS AND DENSE COMPOSITES FROM SOL-GEL

E. J. A. Pope and J. D. Mackenzie

Materials Science and Engineering Department
University of California
Los Angeles, CA 90024

INTRODUCTION

The sol-gel method has been widely studied recently for the preparation of bulk glasses, thin films, and porous solids.[1] One of its advantages as a process for the fabrication of ceramics is the possibility of the use of low temperatures. Another advantage is the fact that liquid solutions are mixed and, hence, good homogeneity is easily achieved. In this paper, the preparation of one type of composite based on the dispersion of solid particles in a sol-gel matrix is described. A second type of composite utilizing the connective pores of the sol-gel matrix has also been made by the impregnation of organic monomers followed by polymerization. These are thus oxide--organic polymer composites. Finally, results are presented for "triphasic" composites made by the impregnation of polymers into a porous two-phase composite of the first group. These three types of composite are described by figure 1. It should be mentioned that other composites have also been made by the exploitation of the uniqueness of the sol-gel technique. These include glass-ceramics[2], "diphasic" glass-ceramics[3], "diphasic" ceramic--metal xerogels[4], ceramic-ceramic composites[5], and moisture sensitive glasses[6].

POROUS COMPOSITES

If an inert "filler" is dispersed in a sol-gel solution, then a porous composite will be formed on gelation. Silicon carbide powders dispersed in silica gel is such an example. The general technique for the preparation of such a composite is shown in figure 2. In addition to SiC-SiO_2, we have prepared composites based upon the following systems: SiO_2-Al_2O_3, SiO_2-Si_3N_4, SiO_2-TiC, SiO_2-Al, SiO_2-SiO_2 microspheres, and SiO_2-fumed SiO_2. In figure 2, it is shown that a catalyst is required for the preparation of these porous composites. It must be mentioned that the particular catalyst used can affect the porosity very significantly.[7] In turn, the properties are then much affected.

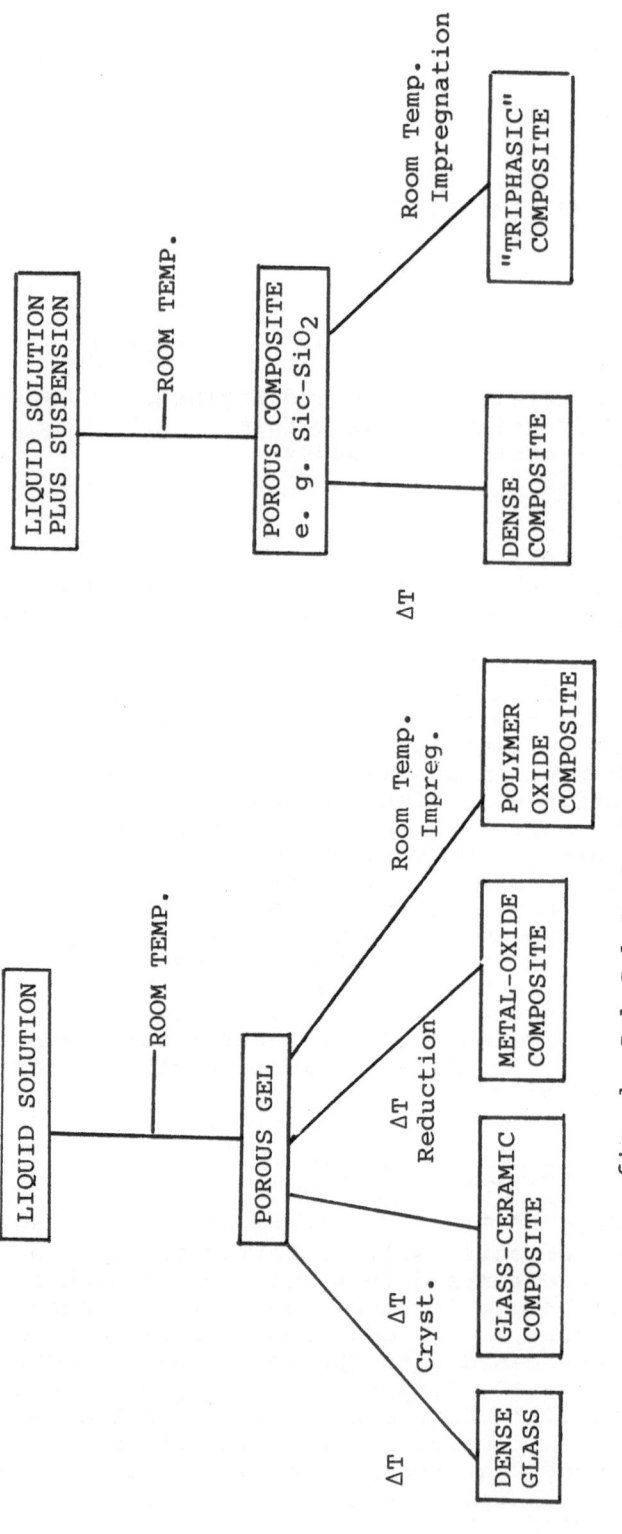

fig. 1. Sol-Gel Route to Porous and Dense Composites.

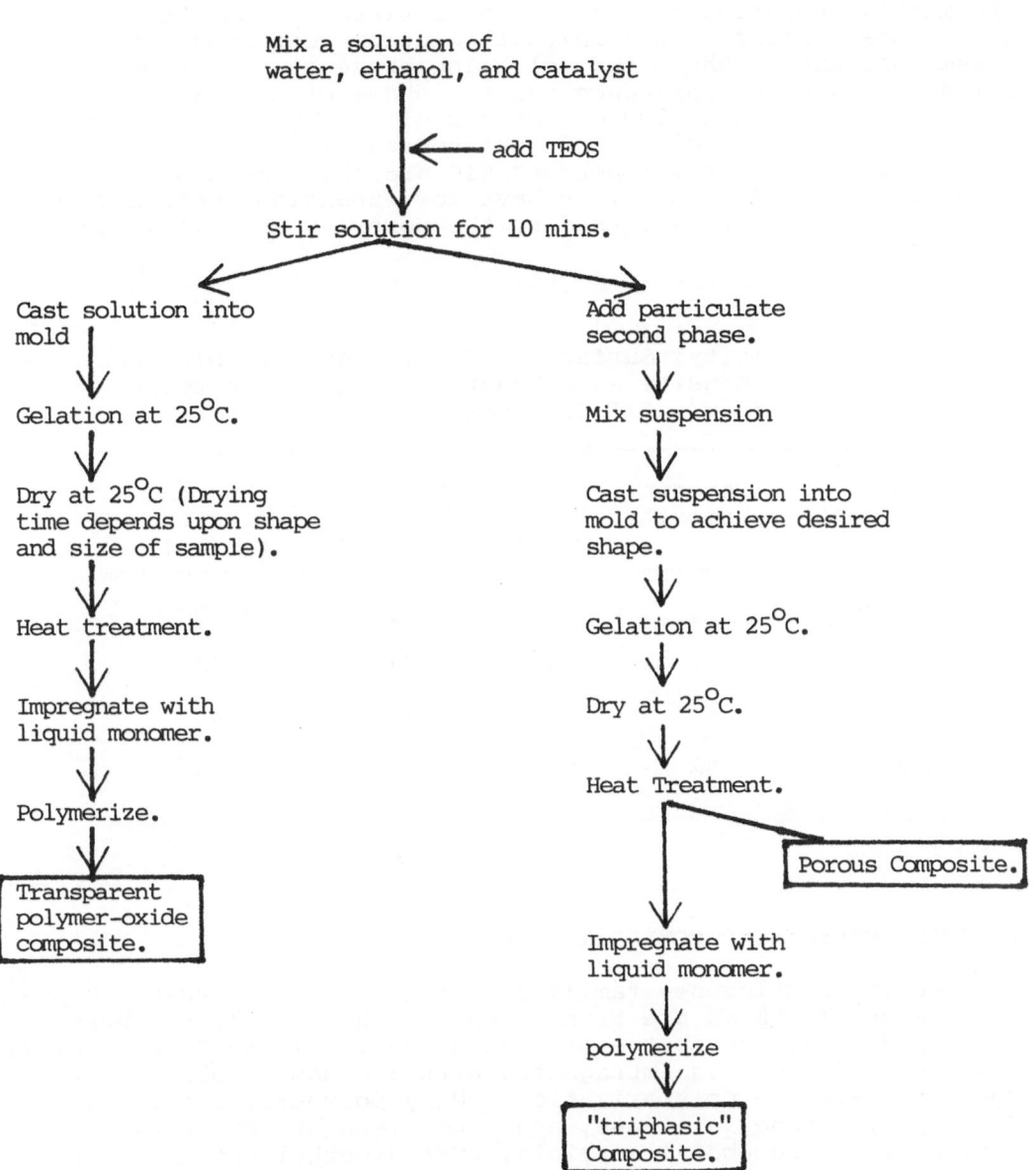

Fig. 2. Detailed schematic of preparation procedures in the processing of gel-derived composites.

During gelation and drying, large shrinkage of the gel matrix occurs. The degree of shrinkage is influenced by the filler used. This is clearly shown in figure 3. It is seen that at room temperature, the liquid mixture for the preparation of silica gel has shrunk by as much as 85% after drying. However, the addition of 10 volume percent of SiC whiskers to the initial solution can reduce the shrinkage from 85% to 30%. Ten volume percent in the initial solution corresponds to 60 volume percent in the dried gel. The presence of various fillers is seen to influence the shrinkage of silica gels even at temperatures up to 800°C. Some preliminary results obtained of the porosity, density, and surface areas of a SiC-SiO$_2$ composite with 33 weight percent SiC are shown in table 1. These light-weight composites have low expansion coefficients. Their properties are governed by the porosity as well as by the microstructure.

Table 1. Porosity, Surface Area, Average Pore Diameter, and Density as a Function of Heat Treatment of 33 w/o SiC-SiO$_2$ Composite.

Heat Treatment Temperature	Percent Porosity	Average Pore Dia.(A)	Surface Area (sq. m/gm)	Apparent Density (gm/cc)
25°C	68	92	397	2.19
600°C	72	89	418	2.40
800°C	64	96	193	2.40
1100°C	35	262	18	2.40

POLYMER-IMPREGNATED COMPOSITES

An interesting new family of inorganic oxide-organic polymer composites which are transparent in the visible has been developed according to the process shown in figure 2. Basically, a porous oxide gel is impregnated with a monomer followed by room temperature polymerization. Many polymers, including PMMA, polystyrene, silicone, and copolymers of PMMA-butyl acrylate, styrene-butyl acrylate, PMMA-dimethyl butadiene, and silicone-styrene have been impregnated into porous silica gel of 65 % porosity. Some examples of transparent composites of this type are shown in figure 4. Some properties of the composites are shown in table 2. The silica gel matrix was dried at 25°C and, hence, does not have a fully developed SiO$_2$ polymeric network. It is, therefore, difficult to calculate the relevant properties of these composites from the properties of their components. However, it can be seen that these transparent composites do have relatively high strengths and low densities. A study of the mechanical properties as a function of oxide-polymer ratio is in progress.

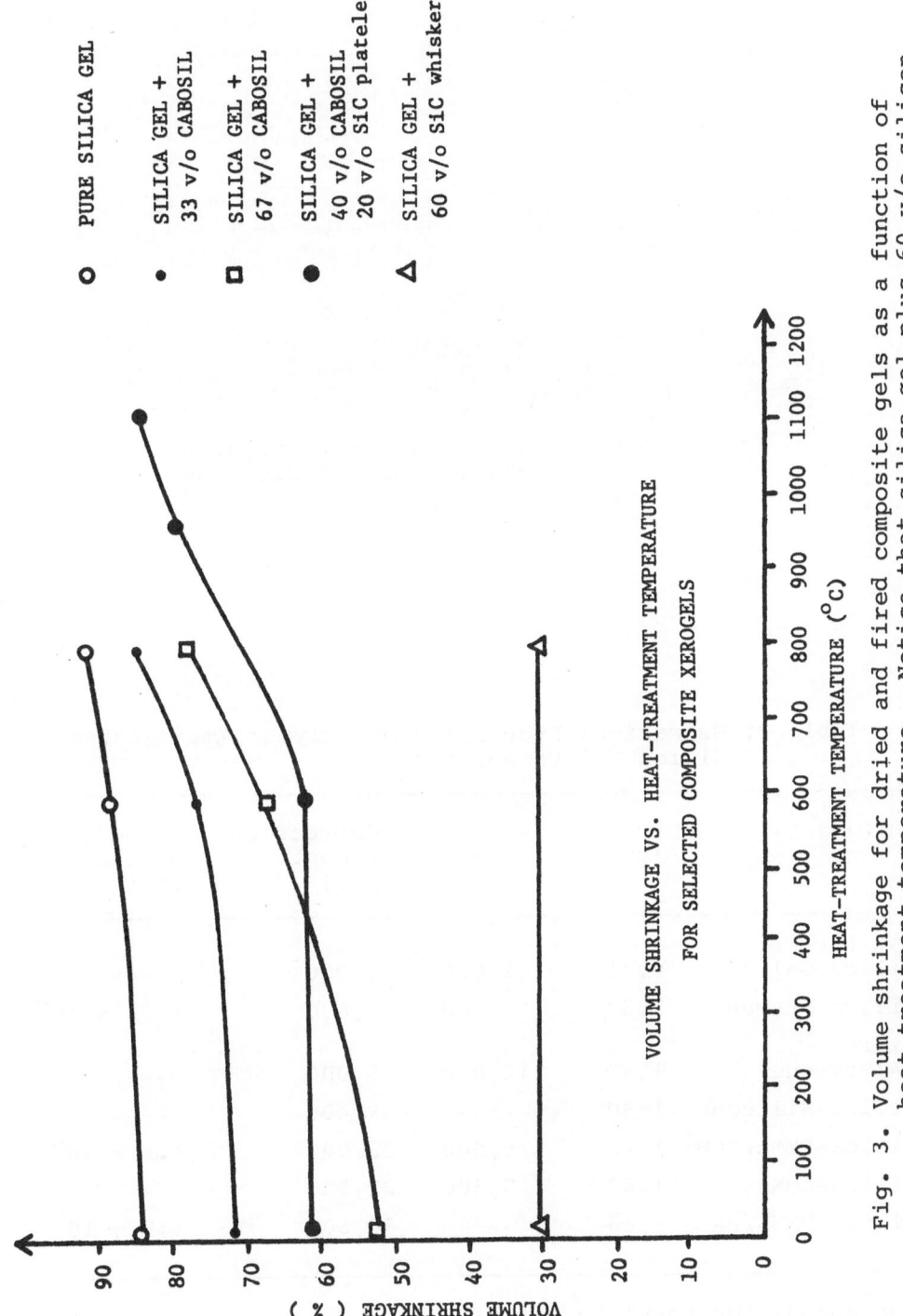

Fig. 3. Volume shrinkage for dried and fired composite gels as a function of heat treatment temperature. Notice that silica gel plus 60 v/o silicon carbide whiskers shows no further shrinkage after drying up to 800°C.

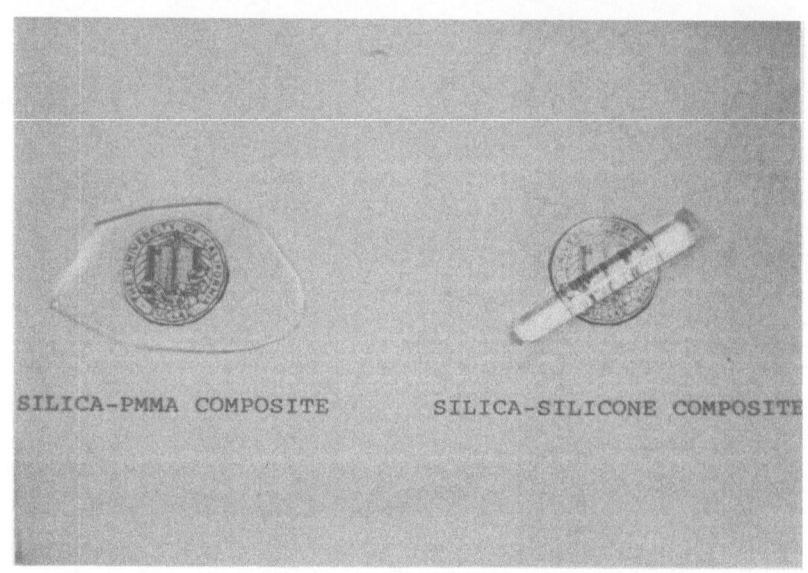

Fig. 4. Photograph of highly transparent
Silica-PMMA and Silica-Silicone
composites.

Table 2. Mechanical Properties of Polymer Impregnated
Silica Gel Composites.

Sample	Density (gm/cc)	M.O.R. (psi)	Compression Strength (psi)	Young's Modulus (psi)
65% Porous Silica Gel	0.71	2,000	5,900	-----
Silica-Styrene	1.37	5,800	26,800	1.02×10^6
Silica-BA/Styrene[a]	1.43	10,000	31,000	-----
Silica-Silicone	1.30	----	9,400	-----
Silica-PMMA/DMB[b]	1.42	9,500	33,500	1.5×10^6
Silica-PMMA	1.42	9,300	22,500	-----
Silica-PMMA/BA	1.46	----	30,500	1.5×10^6

[a]BA stands for butyl acrylate.
[b]PMMA and DMB represent polymethyl methacrylate and
dimethyl butadiene, respectively.

If the SiC-SiO$_2$ porous composites described above are now impregnated according to the process shown in figure 2, a new family of "triphasic" composites can be made. Preliminary results for a PMMA-SiC-SiO$_2$ composite containing about 70 volume percent PMMA are shown in table 3. The PMMA is seen to have increased the modulus of rupture very significantly. In the measurement of compressive strength, the compressive strain at fracture was 13.5 percent. This is somewhat surprising considering that the ceramic phase was continuous. A scanning electron micrograph of such a triphasic composite is shown in figure 5.

Table 3. The Effect of PMMA Impregnation on the Modulus of Rupture and Compression Strength of SiC-SiO$_2$ Composite.

Matrix Heat Treatment Temperature	Sample Condition	Porosity (%)	M.O.R. (psi)	Compressive Strength (psi)
Room Temp.	porous	68	314	
Room Temp.	impregnated	0	9,000	
600°C	porous	72	347	2,400
600°C	impregnated	0	10,000	28,500

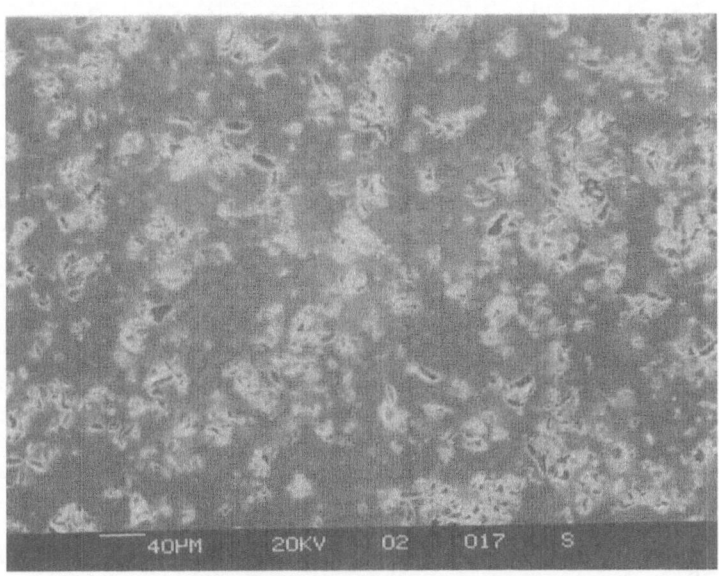

Fig. 5. Scanning electron micrograph of PMMA-SiC-SiO$_2$ triphasic composite. Large particles are silicon carbide.

CONCLUSIONS

1. A variety of highly porous, high temperature stable composites are possible via sol-gel processing.

2. The porosity of dried and fired gels can be controlled through the selection of the catalyst.

3. a) The addition of a solid, particulate second phase to the initial solution reduces the volume shrinkage of the dried and fired gel.
   b) The extent to which the volume shrinkage is reduced at different heat treatment temperatures depends upon the amount, shape, and composition of the second phase constituent.

4. Highly transparent, light-weight, polymer-oxide composites, possessing good strength have been obtained by the sol-gel route.

5. Light-weight "triphasic" composites that have high fracture ductility have also been obtained via the sol-gel route.

ACKNOWLEDGEMENTS

This work was supported by the Directorate of Chemical and Atmospheric Sciences, Air Force Office of Scientific Research. We would like to acknowledge the valuable assistance of Dr. Paul Lu and Doris Plenert.

REFERENCES

1. J. D. Mackenzie, Ultrastructure Processing of Ceramics, Glasses, and Composites, Ed. by L. L. Hench and D. R. Ulrich (J. Wiley and Sons, N.Y.,1984).

2. S. P. Mukherjee and J. Zarzycki, J. Am. Ceram. Soc., 62 (1979).

3. D. W. Hoffman, R. Roy, and S. Komarneni, J. Am. Ceram. Soc., 67 (1984) 468.

4. R. A. Roy and R. Roy, Mats. Res. Bull., 19(1984) 169.

5. J. J. Lannutti and D. E. Clark, papers B4.6 and B4.7 at 1984 Spring Meeting of the Materials Research Society, see program and abstracts.

6. G. S. Moore, N. Toghe, and J. D. Mackenzie, "Colored Glasses by the Sol-Gel Method", 36th Annual Pacific Coast Meeting, Am. Ceram. Soc.,1983.

7. E. J. A. Pope and J. D. Mackenzie, J. NonCryst. Sol., in press.

NON-EQUILIBRIUM SURFACE CONDITIONS AND MICROSTRUCTURAL CHANGES
FOLLOWING PULSED LASER IRRADIATION AND ION BEAM MIXING OF Ni
OVERLAYERS ON SINTERED ALPHA-SiC

K. L. More and R. F. Davis

Department of Materials Engineering, Box 7907
North Carolina State University
Raleigh, NC   27695-7907

ABSTRACT

   Pulsed laser irradiation and ion beam mixing of thin Ni overlayers on
sintered α-SiC have been investigated as potential surface modification
techniques for the enhancement of the mechanical properties of the SiC.
Each of these surface processing methods are non-equilibrium techniques;
materials interactions can be induced at the specimen surface which are not
possible with conventional thermal techniques. As a result of the surface
modification, the physical properties of the ceramic can be altered under
the correct processing conditions. Following laser irradiation using a
pulsed ruby or krypton fluoride (KrF) excimer laser, the fracture strength
of the SiC was increased by approximately 50% and 20%, respectively. However, ion beam mixing of Ni on SiC resulted in no change in fracture
strength. Cross-sectional transmission electron microscopy (X-TEM), scanning electron microscopy (SEM), secondary ion mass spectroscopy (SIMS), and
Rutherford backscattering techniques (RBS) have been used to characterize
the extent of mixing between the Ni and SiC as a result of the surface
modification and to determine the reason(s) for the observed changes in
fracture strength.

INTRODUCTION

   As a means of directly altering the physical properties of polycrystalline α-SiC, the modification process should either change the size
and/or position of the critical flaws in the SiC or induce a significant
residual compressive stress at and/or near the specimen surface. Under the
correct processing conditions, materials interactions induced at the surface can result in either of the above changes and, thus, a strength
increase will be observed.

   Ion beam mixing and pulsed laser irradiation are two such modification
processes. As a result of the surface processing, non-equilibrium or
metastable surface phases can be achieved. Ion beam mixing is a process by
which high energy ions are implanted into the specimen surface and are used
to mix layers of two different compositions, e.g. thin Ni overlayers on
SiC. Pulsed laser mixing ideally involves the melting of the coating and
the near surface of the substrate materials and the subsequent mixing of
the two species by liquid phase diffusion. Each of the modification techniques involves extremely rapid heating and cooling rates such that unique

surface microstructures are created and "quenched in" during cooling in rates approaching 10^9 K/s.

Both ion beam mixing and pulsed laser irradiation, although relatively new surface modification techniques, have been extensively studied and reviewed.[1-4] These techniques have most recently been applied to metal-semiconductor[5-9] and metal-insulator[10,11] systems. Investigations have also been made into the surface modification of single crystal β-SiC thin films grown in the authors' laboratory[12-15] as well as Al_2O_3 and α-SiC ceramic systems.[16,17]

The emphasis of this paper is on the microstructures formed as a result of the surface modification and the subsequent effect on the fracture strength of the SiC. Complete fracture strength results for α-SiC have been reported in previous papers.[18,19]

EXPERIMENTAL PROCEDURE

The material under investigation is an ≈98% dense α-SiC[+] produced by the pressureless sintering of a mixture of powders of submicron SiC and densification aids containing ≈0.4 wt% B and ≈0.5 wt% C at a temperature of approximately 2373K. The sintered α-SiC plates were subsequently cut into standard flexure test bars (0.64 cm X 0.32 cm X 5.08 cm) and the surfaces sequentially ground using 100, 400, and 600 grit diamond wheels. The edges of each bar were bevelled at 45° to prevent corner flaws from acting as fracture origins. The SiC surfaces were sputter cleaned in a vacuum of approximately 10^{-7} Torr followed by the evaporation of 20 nm, 50 nm, or 100 nm thick Ni overlayers. Each side of a bar was coated separately; the other sides were masked by glass slides to prevent additional deposition. The maximum specimen temperature reached during cleaning and Ni deposition did not exceed 323K.

Ion beam mixing was conducted on the 50 nm Ni coated SiC using either 350 keV Xe^+ ions at a dose of 2.5 X 10^{16} Xe^+/cm^2 or 140 keV Si^+ ions at a dose of 1 X 10^{17} Si^+/cm^2. The cross-section of the ion beam was circular having a diameter of ≈2.54 cm; thus, to mix the entire surface, there was considerable overlap of the beam spots. Laser irradiation was conducted using one of two lasers having the following parameters a pulsed ruby laser (λ = 693 nm, τ = 25 ns, nominal pulse energy density = 1.2 J/cm^2) or a KrF excimer laser (λ = 248 nm, τ = 45 ns, pulse energy density = 1.85 J/cm^2 and 3.0 J/cm^2). The surfaces of each specimen were sequentially irradiated with laser pulses until all four sides were completely annealed (the size of the laser beam is dictated by the pulse energy density used).

Each modified SiC bar was broken in 4-point flexure to determine the fracture strength and compared to the fracture strength values similarly obtained for the as-received SiC. Inner and outer span lengths of 1.27 cm and 2.54 cm, respectively, and a crosshead speed of 8.5 X 10^{-4} cm/s were used for the fracture research.

The surfaces of the laser processed and ion beam mixed SiC were examined using X-TEM samples prepared from small test bar sections by mechanical thinning to approximately 75 μm, dimpling, and subsequent low energy (5-6 kV), low angle (12') Ar^+ ion beam thinning. The X-TEM results were compared to chemical analyses derived from SIMS (CAMECA IMS-3f ion microprobe) and RBS. The SIMS profiles were obtained using an ≈50 μm diameter O^+ primary beam at a beam current of ≈1700 namp and an accelerating

----

[+] HEXOLOY®, SOHIO Chemicals and Industrial Products Co., Niagara Falls, NY

voltage of ≈14.8 kV. 1 MeV, 2 MeV, and 2.5 MeV He$^+$ beams were used for the RBS analyses. Critical flaw identification and fracture surface analysis of the as-received, laser irradiated, and ion beam mixed SiC were conducted via SEM.

RESULTS AND DISCUSSION

Starting Material

The microstructure of the α-SiC consists of fairly equiaxed α-SiC grains having an average size of approximately 8.5 μm, smaller and darker B$_4$C grains, and particles of free C, as shown in the polished surface of Figure 1(a). Fracture in sintered α-SiC occurs transgranularly, as shown in Figure 1(b). The average strength of the as-received SiC was found to be ≈310 MPa, as determined from 4-point bend experiments.

Following the 4-point bend investigations, the surfaces of the SiC were examined by SEM to determine the critical flaws which ultimately caused failure. In nearly all the SiC bars, the fracture origin was found at or near the tensile face of the specimen. These flaws resulted from processing or machining. The most common processing flaws were large SiC agglomerates or residual surface porosity (as a result of inhomogeneous binder distribution prior to sintering). Since the SiC surface was ground using 600 grit diamond as the final processing step, machining flaws were also prevalent. Common flaws found in the starting material are shown in Figure 2.

An ≈100 nm as-deposited Ni overlayer is shown in Figure 3. From the X-TEM results, the Ni-coating was found to be fairly uniform in thickness across the SiC surface, but was non-conformal around rough surface areas.

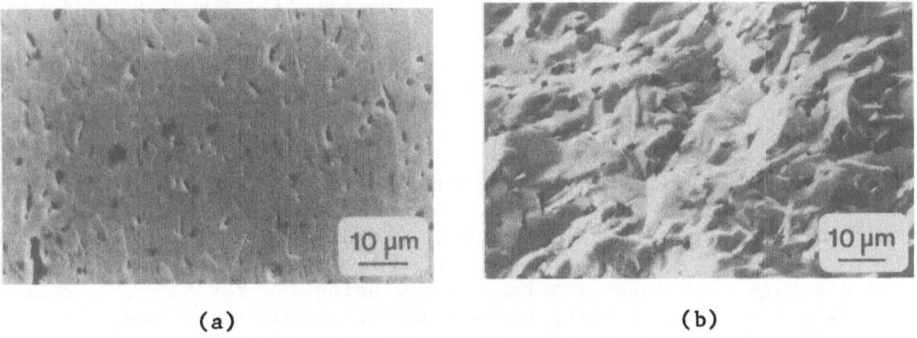

(a)   (b)

Figure 1. Microstructure of sintered α-SiC: (a) a polished surface and (b) a fracture surface.

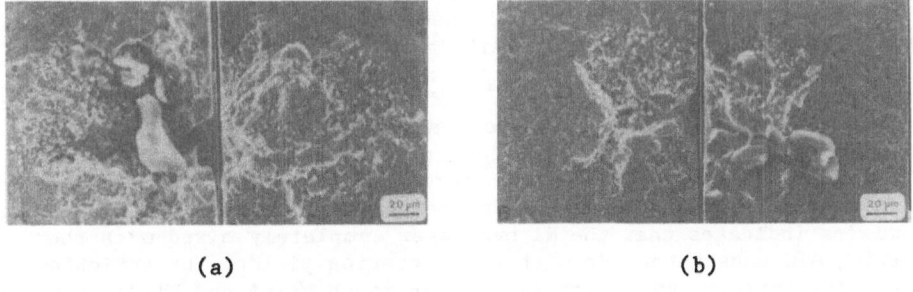

(a)   (b)

Figure 2. Common flaws found in sintered α-SiC: (a) a large SiC agglo-merate and (b) a surface crack associated with surface porosity.

197

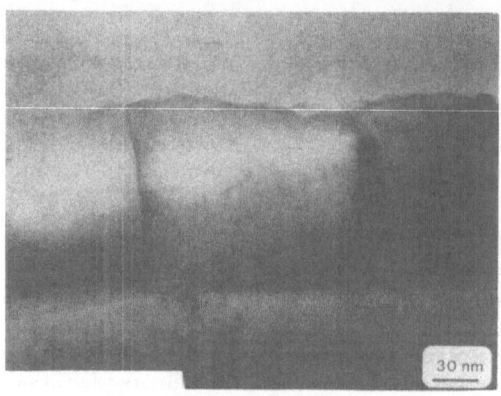

Figure 3.   100 nm as-deposited Ni overlayer on α-SiC.

Ion Beam Mixing

If a thin overlayer of material B is deposited on a substrate A and subsequently bombarded with high energy ions, unique surface microstructures, denoted in Figure 4 as A_mB_n, can result from a series of contributing effects.[3] The energy of the incident ions is adjusted such that the maximum energy is deposited at the interface. Each ion incident on the specimen surface creates a collision cascade of atoms which in turn results in the production of a huge number of mobile defects. The large number of collision cascades establish concentration gradients of defects (both vacancy and intersticial) and, thus, the atoms of the substrate and coating can mix by radiation enhanced diffusion. The atoms of the deposited film can be implanted into the substrate by direct impurity knock-on collisions (recoil implantation). Mixing of the atoms can also occur by the atomic motion of atoms within the cascades.[4]

For the initial ion beam mixing experiments, the 50 nm Ni coated SiC bars were bombarded with 350 keV Xe^+ at a dose of 2.5×10^{16} Xe^+/cm^2. The projected range of Xe^+ at 350 keV in Ni is approximately 55 nm, which coincides with the Ni-SiC interface. The resulting surface microstructure is shown in Figure 5(a). Due to the heavy dose of the heavy Xe^+ ions, there was significant bubble formation at the Ni-SiC interface (see arrows in Figure 5(a)). For this reason and for the compatibility of using a self ion, the SiC bars were later ion beam mixed using 140 keV Si^+ at a dose of 1×10^{17} Si^+/cm^2. The resulting surface microstructure was much more uniform with no bubble formation at the interface, as shown in Figure 5(b). In both ion beam mixing experiments, the Ni and SiC were completely mixed; the new surface microstructures consist of four distinct regions, as labelled on the X-TEM micrographs: (a) a residual Ni overlayer and Ni-rich amorphous SiC surface (as determined from SIMS analysis), (b) a Ni-SiC mixed region, (c) an amorphous SiC layer caused by ion implantation damage, and (d) a faulted region in the SiC substrate. The RBS spectra for the ion beam mixed SiC using Xe^+ and Si^+ are shown in Figures 6(a) and 6(b), respectively. Each spectra shows a comparison between the as-coated SiC and the ion beam mixed SiC. Before mixing, the scattering edges of the Ni and Si in SiC are very sharp corresponding to the planar nature of the unmixed interface. The spread in the scattering yields following ion beam mixing indicates that the Ni overlayer completely mixed with the underlying SiC substrate. From these scattering yields, the stoichiometric Ni/Si ratio in the mixed regions approach Ni_2Si and Ni_3Si_2 for the Xe^+ and Si^+ bombarded specimens, respectively (this contains no phase information, but is strictly a stoichiometric ratio).

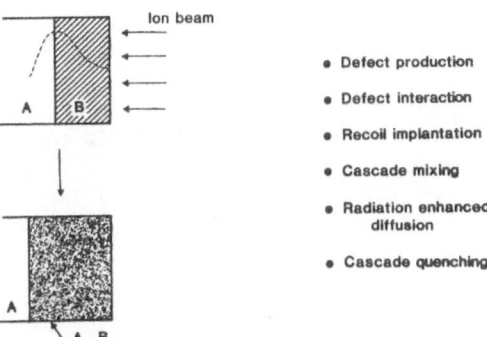

Figure 4. Schematic of interacting effects in
an ion beam mixing process.[4]

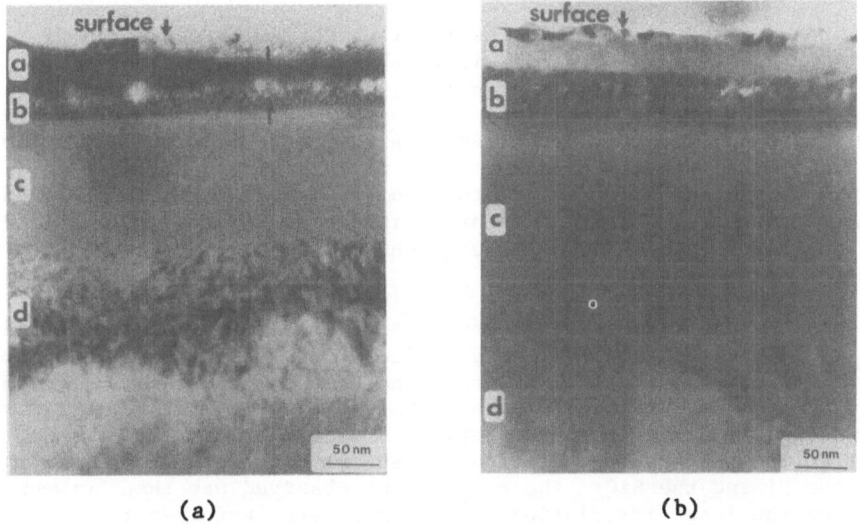

(a)   (b)

Figure 5. X-TEM micrographs of ion beam mixed 50 nm Ni-coated SiC using
(a) 350 keV Xe$^+$ at a dose of 2.5 X 10^{16} Xe$^+$/cm^2 and (b) 140 keV
Si$^+$ at a dose of 1 X 10^{17} Si$^+$/cm^2. The arrows in 5(a) point to
the bubble formation at the Ni-SiC interface.

    Following ion beam mixing, there was no observed change in fracture
strength. This surface modification process resulted in no change in the
size or position of the critical flaws, nor was there a residual compressive stress induced at the surface. The energy deposited in the collision
cascades is rapidly dissipated through atomic motion in times approaching
10^{-13} sec,[3] but even with this extremely rapid quenching effect within the
cascades themselves, the overall surface temperature does not exceed 323K.
There was no thermal contraction difference between the Ni and the SiC as a
result of ion beam mixing, thus, no effective residual stress was induced
at the surface. After processing only four specimens, this approach was
discontinued. At least within the limits created by the use of Ni coatings, the ion species, and parameters of bombardment in this work, it was
concluded that ion beam mixing is not a viable method for increasing the
fracture strength of α-SiC.

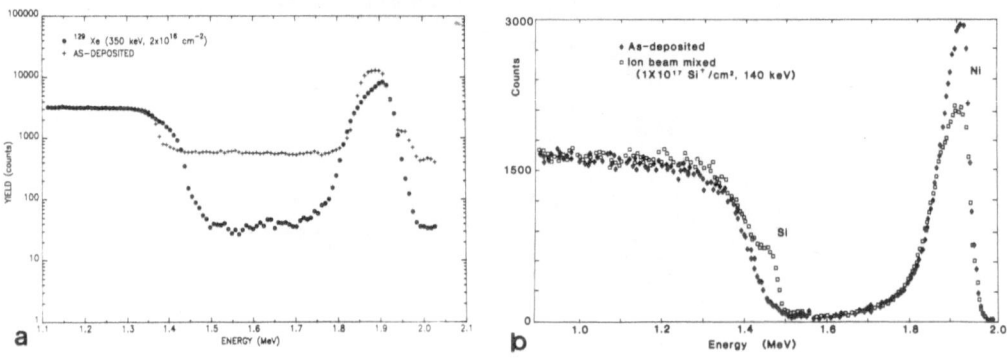

Figure 6. 2.5 MeV He$^+$ RBS spectra comparing as-deposited SiC to ion beam mixed SiC using (a) 350 keV Xe$^+$ at a dose of 2.5 X 10^{16} Xe$^+$/cm^2 and (b) 140 keV Si$^+$ at a dose of 1 X 10^{17} Si$^+$/cm^2.

Pulsed Laser Mixing

The pulsed laser mixing process is illustrated in Figure 7. If a substrate of material A is coated with a thin layer of material B and subsequently irradiated with laser pulses, depending on the properties of B and at sufficient energy densities, the energy of the laser is absorbed by the B electrons and rapidly converted to heat. As this heat is rapidly transferred to the substrate, A and B can mix via liquid diffusion. If the melting threshold is not exceeded, the two may mix as a result of solid state mass transport. Laser irradiation involves extremely rapid heating and cooling rates ($\approx 10^9$ K/s). As such, unique surface microstructures can be created and preserved.

Initial experiments were performed using pulsed ruby laser irradiation of SiC coated with 100 nm of Ni at a nominal pulse energy density of about 1.2 J/cm^2. Cross-sectional TEM micrographs of the surface microstructure following ruby laser irradiation are shown in Figure 8. As noted from the X-TEM results, no mixing occurred between the Ni overlayer and the SiC substrate. The RBS spectrum of Figure 9 shows that there was no mixing between the Ni and the SiC; the only mixing observed was that between O and C since the laser irradiation experiments were performed in air at atmospheric pressure and hydrocarbon species existed on the Ni surface prior to irradiation. Essentially, all of the energy of the laser beam was absorbed by the Ni coating and used to heat it to ever increasing temperatures. The Ni melted, flowed into the near-surface regions of the many pre-existing surface voids, and rapidly resolidified as a polycrystalline layer of non-uniform thickness across the specimen surface. Areas were also found where there was absolutely no Ni on the SiC surface. The melting threshold of the SiC was never reached for a number of reasons: (1) the initial Ni thickness, (2) the low pulse energy density, and (3) the extremely short pulse duration.

Following pulsed ruby laser irradiation, the fracture strength of the SiC was increased by $\approx 50\%$. This strength increase was found to be due mainly to a change in the size and position of the critical flaw as a result of ruby laser annealing.[18] It is reasoned that the much larger remaining volume of each of the Ni-containing voids acted as a thermal insulator which mitigated heat loss from the Ni for a time sufficient to allow for a chemical reaction between the Ni and the walls of the voids. During rapid cooling, the greater contraction of the Ni induced a compressive stress in the near-surface regions of the voids. Thus, the large

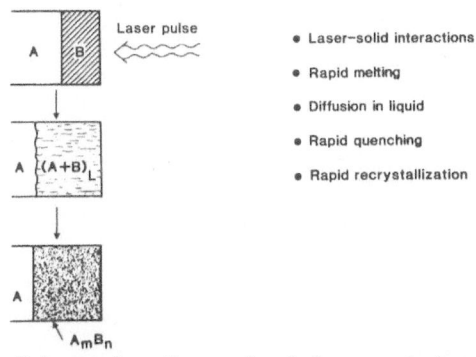

Laser pulse

• Laser-solid interactions

• Rapid melting

• Diffusion in liquid

• Rapid quenching

• Rapid recrystallization

Figure 7. Schematic of a pulsed laser mixing process.[4]

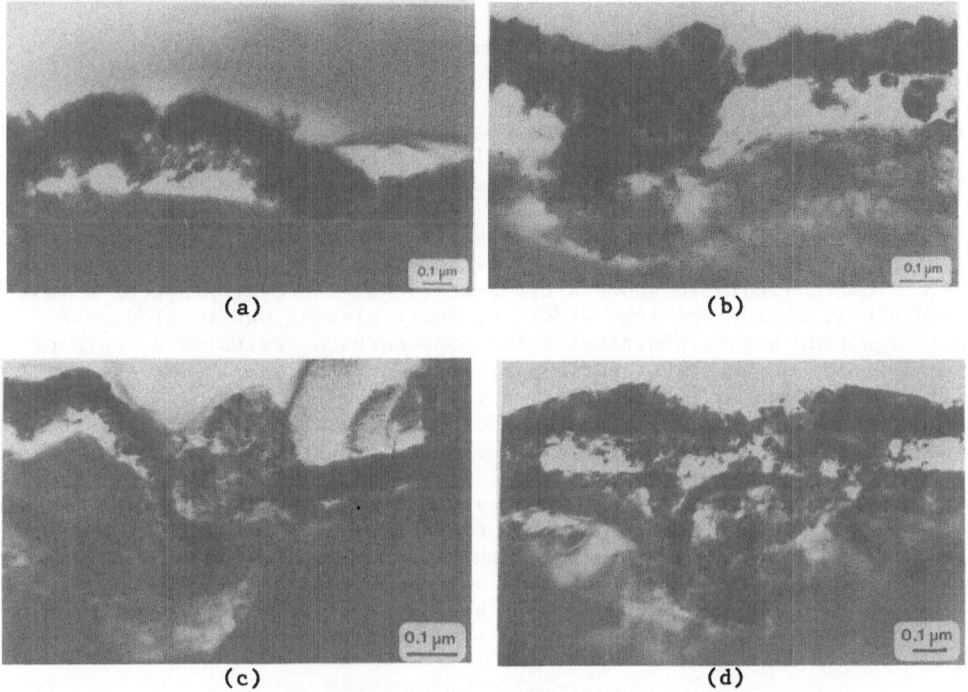

(a)   (b)

(c)   (d)

Figure 8. X-TEM micrographs of ruby laser irradiated 100 nm Ni coated SiC.

surface flaws present in the starting material were prevented from acting
as fracture origins; instead, subsurface flaws became the strength controlling flaws. The compressive stress was found to be ≈52 MPa using a
method described elsewhere[19] which accounts for 35% of the strength
increase.

In an attempt to bring about mixing between the Ni overlayer and the
α-SiC substrate, a KrF excimer laser was used for subsequent experiments.
This laser was chosen for three primary reasons: (1) the wavelength of the
excimer laser (λ = 248 nm) is much shorter than that of the ruby laser
(λ = 693 nm); thus, the photon energy of the KrF laser is significantly
greater than the band gap of the α-SiC (E_g = 3.023 eV) which can allow for
absorbtion of the laser radiation by both the Ni coating and SiC substrate,

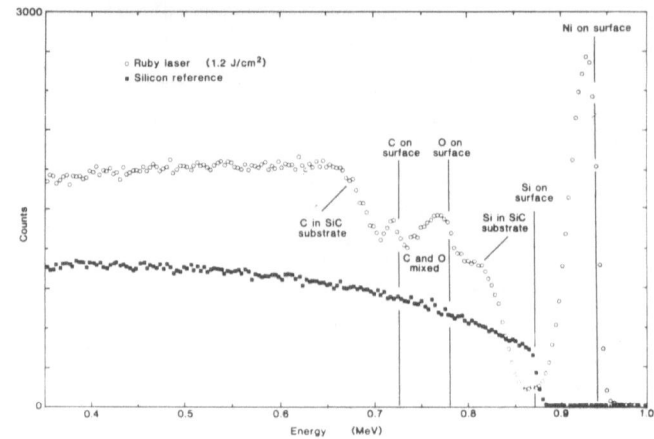

Figure 9.   1 MeV He$^+$ RBS spectrum comparing a Si
            reference standard to ruby laser irradiated
            SiC (1.2 J/cm^2).

(2) the pulse duration is twice as long for the KrF laser as for the pulsed
ruby laser allowing for a longer reaction time for mixing, and (3) the KrF
laser beam is much more homogeneous than the ruby laser beam.

    To determine the effect of KrF laser irradiation on the microstructure
of α-SiC, uncoated SiC specimens were laser "glazed" at an energy density
of 3.0 J/cm^2. Figure 10 shows a cross-sectional TEM micrograph of a laser
glazed SiC surface. The main effect of laser glazing was to produce a
highly faulted, highly plastically strained surface region to a depth of
approximately 500 nm (labelled (a) on the TEM micrograph). The energy
density of the excimer laser appeared to be high enough to melt the SiC
surface. Thermodynamically, one would expect the SiC to sublime at atmospheric pressure, but due to the extremely rapid heating and cooling rates
associated with laser irradiation, the SiC melted. Microtwins formed to
compensate for the stresses produced by the high thermal gradients and were
"quenched in" as heat was transferred to the bulk crystal. It was also
determined from SIMS depth profiling that the SiC surface was Si-rich;[18] a
small amount of C leaves the specimen surface, probably as CO_2, as a result
of the reaction between the extremely hot SiC and air.

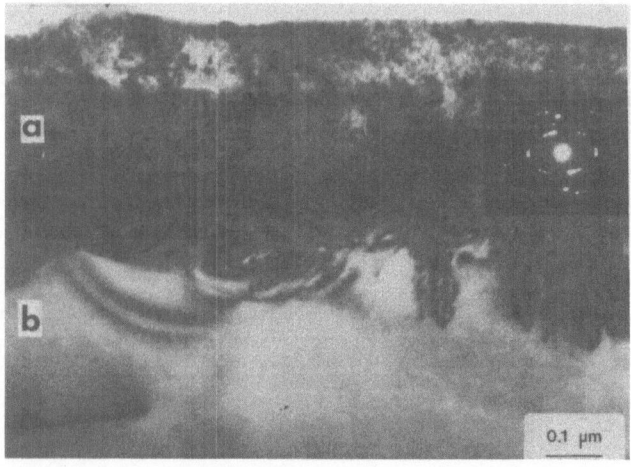

Figure 10.   X-TEM micrograph of a laser glazed
             SiC surface irradiated at 3.0 J/cm^2.

If the SiC was coated with 100 nm of Ni and subsequently irradiated with the KrF laser at a pulse energy density of 1.85 J/cm^2, the resulting surface microstructure appears as shown in Figure 11. An RBS spectra for a similarly processed specimen is shown in Figure 12. The RBS spectra shows that KrF laser irradiation of Ni-coated SiC resulted in a chemically mixed surface layer having an approximate thickness of 150 nm. Three distinct regions are observed and labelled on the surface micrograph of Figure 11: (a) a chemically mixed surface layer, (b) a distinct interface between the mixed layer and the SiC substrate, and (c) a highly faulted SiC substrate. The energy absorbed during KrF laser irradiation was sufficient to cause complete melting of the Ni overlayer. Although the SiC did not melt, the temperatures reached during irradiation were high enough to cause interdiffusion of SiC into the Ni overlayer (the solubility of SiC in molten Ni is moderately high). Therefore, the mixed layer was not caused by the melting of both Ni and SiC (mixing by diffusion between the two liquids) but by the limited interdiffusion of the SiC into the molten Ni coating during irradiation. The interdiffusion of SiC and Ni enhanced the bonding between the two materials at the interface. The mixing process also appeared to keep the thickness of the mixed layer uniform across the specimen surface. The SiC substrate is highly faulted below the mixed layer and resembles the structure in the sub-surface, microtwinned region of the laser glazed SiC.

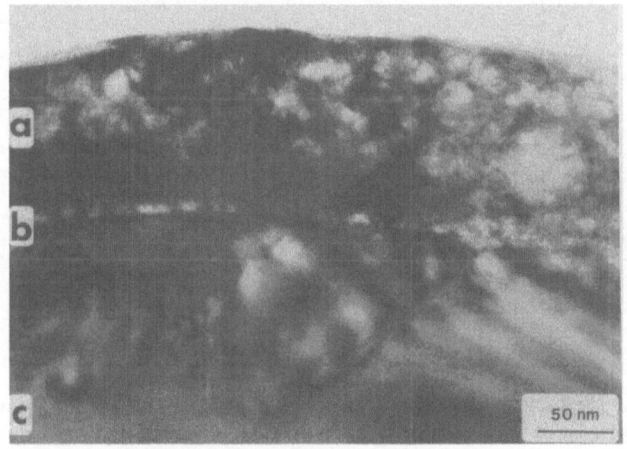

Figure 11. X-TEM micrograph of 100 nm Ni coated SiC KrF laser irradiated at 1.85 J/cm^2.

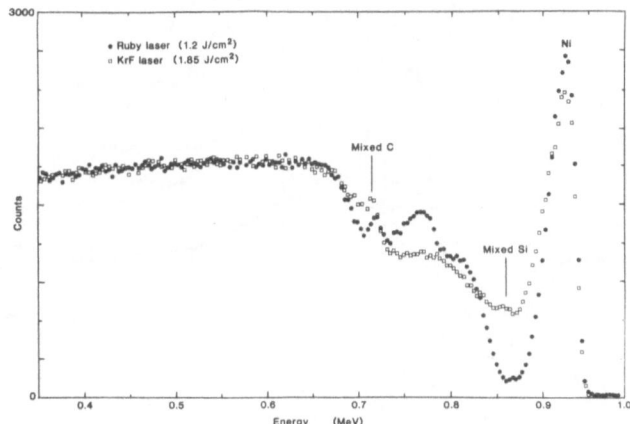

Figure 12. 1 MeV He$^+$ RBS spectrum comparing ruby laser irradiated SiC to 100 nm Ni coated SiC irradiated with a KrF laser at 1.85 J/cm^2.

If the pulse energy density of the KrF laser was increased to 3.0 J/cm^2 and the initial Ni thickness was 100 nm, the surface microstructure appears as shown in Figure 13(a). The corresponding RBS spectra for a similarly processed specimen is shown in Figure 13(b). The RBS spectra shows that for this case, the Ni and SiC were again chemically mixed, but at this significantly higher energy density, the mixed layer did not remain uniform across the specimen surface. During irradiation, the mixed layer appears to have agglomerated into large, circular "clumps" or balls across the entire surface, but the layer did not separate from the SiC substrate. The distinct interface structure was still apparent (as shown by the arrow in Figure 13(a)) indicating that the mixing occurred by chemical interdiffusion of SiC into the molten Ni during irradiation. If the initial Ni thickness was reduced to 20 nm but the energy density was kept at 3.0 J/cm^2, a similar, non-uniform surface microstructure was obtained, as shown in Figure 14(a). The corresponding RBS spectra of Figure 14(b) shows that in this case, a Ni-SiC mixed layer was also formed.

(a)   (b)

Figure 13. (a) X-TEM micrograph and (b) 2.0 MeV He$^+$ RBS spectrum for 100 nm Ni coated SiC irradiated with the KrF laser at a pulse energy density of 3.0 J/cm^2.

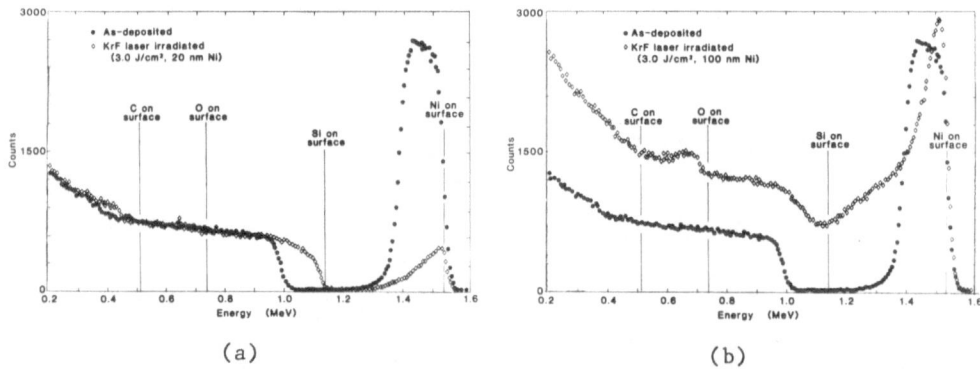

(a)   (b)

Figure 14. (a) X-TEM micrograph and (b) 2.0 MeV He$^+$ RBS spectrum for 20 nm Ni coated SiC irradiated with a KrF laser at a pulse energy density of 3.0 J/cm^2.

204

Following KrF laser irradiation, the fracture strength of the SiC was increased by about 20% no matter what set of parameters were used. An examination of the fracture surfaces of the KrF laser irradiated specimens showed that the fracture origins found for these specimens were similar in position and size to those flaws found in the as-received SiC. The significant increase in strength resulted from an induced compressive stress at the specimen surface. The compressive stress was caused by two interacting effects: (1) the enhanced bonding at the Ni-SiC interface as a result of the interdiffusion of SiC into the Ni overlayer during irradiation and (2) the difference in thermal contraction between the Ni and SiC; on cooling, the Ni contracts to a greater extent than the SiC.[18] The magnitude of the compressive stress was determined using methods described elsewhere[19] and was found to be approximately 50 MPa, which accounts for nearly the entire fracture strength increase.

CONCLUSIONS

Ion beam mixing of 50 nm Ni overlayers on α-SiC resulted in the formation of a new surface microstructure which consisted of four distinct regions: (a) a residual Ni overlayer and Ni-rich amorphous SiC surface, (b) a Ni-SiC mixed region, (c) an amorphous SiC region due to ion implantation damage, and (d) a faulted region in the SiC substrate. There was, however, no increase in the fracture strength of the SiC as a result of the surface modification. Ion beam mixing only served to implant Ni into the specimen surface; the processing did not change the nature of the intrinsic flaws, nor did it induce an effective compressive stress at or near the surface.

KrF laser irradiation of either 100 nm or 20 nm Ni overlayers on SiC at a pulse energy density of either 1.85 J/cm^2 or 3.0 J/cm^2 resulted in a mixed Ni-SiC surface region. A significant 20% increase in fracture strength occurred as a result of the surface processing, irregardless of the parameters used. The strength increase was caused by the presence of a residual compressive stress at the specimen surface. The magnitude of the compressive stress was previously found to be ≈50 MPa.

Ruby laser irradiation resulted in the greatest enhancement in strength, an approximate increase of 50%. There was no mixing between the Ni and the SiC; the strength increase was caused by the flow of Ni into the surface regions of pre-existing flaws and the subsequent reaction between the molten Ni and the walls of the voids. The flow of Ni into the surface voids prevented these flaws from acting as fracture origins and the reaction between the Ni and the SiC within the voids themselves induced a small compressive stress in the voids. The magnitude of the effective stress was found to be approximately 52 MPa which accounted for about 35% of the strength increase.

REFERENCES

1. S. D. Ferris, H. J. Leamy, and J. M. Poate, eds., *Laser Solid Interactions and Laser Processing - 1978*, American Institute of Physics, No. 50 (1979).
2. C. W. White and P. S. Peercy, eds., *Laser and Electron Beam Processing of Materials*, Academic Press, New York (1980).
3. B. R. Appleton, *J. Materials for Energy Systems*, 6(3), 1984, p. 200-11.
4. B. R. Appleton in: *Ion Implantation and Beam Processing*, J. S. Williams and J. M. Poate, eds., Academic Press, New York (1984).

5. J. M. Poate and J. W. Mayer, eds., *Laser Annealing of Semiconductors*, Academic press, New York (1984).

6. J. W. Mayer, B. Y. Tsaur, S. S. Lau, and L. S. Hung, *Nucl. Inst. and Meth.*, 182/183, 1981, p. 1-13.

7. B. Y. Tsaur, Z. L. Liau, and J. W. Mayer, *Appl. Phys. Lettr.*, 34(2), 1979, p. 168-70.

8. D. Fathy, O. W. Holland, and J. Narayan, *J. Appl. Phys.*, 58(1), 1985, p. 297-301.

9. S. S. Lau, B. Y. Tsaur, M. von Allmen, J. W. Mayer, B. Stritzker, C. W. White, and B. R. Appleton, *Nucl. Inst. and Meth.*, 182/183, 1981, p. 97-105.

10. G. C. Farlow, B. R. Appleton, L. A. Boatner, C. J. McHargue, C. W. White, G. J. Clarke, and J. E. E. Baglin, Presented at the Fall Meeting of the Materials Research Society, San Fransisco, CA (1984).

11. C. W. White, G. Farlow, J. Narayan, G. J. Clarke, and J. E. E. Baglin, *Materials Letters*, 2(5A), 1984, p. 367-72.

12. J. Narayan, D. Fathy, O. W. Holland, B. R. Appleton, R. F. Davis, and P. F. Becher, *J. Appl. Phys.*, 56(6), 1984, p. 1577-82.

13. D. Fathy, J. Narayan, O. W. Holland, B. R. Appleton, and R. F. Davis, *Materials Letters*, 2(4B), 1984, p. 324-7.

14. D. Fathy, O. W. Holland, J. Narayan, and B. R. Appleton, *Nucl. Inst. and Meth.*, B 7/8, 1985, p. 571-5.

15. J. Narayan, D. Fathy, O. W. Holland, and B. R. Appleton, *Materials Letters*, 3(7,8), 1985, p. 261-4.

16. C. J. McHargue, G. C. Farlow, C. W. White, J. M. Williams, B. R. Appleton, and H. Naramoto, *Mat. Sci. and Eng.*, 69, 1985, p. 123-7.

17. B. R. Appleton, H. Naramoto, C. W. White, O. W. Holland, C. J. McHargue, G. Farlow, J. Narayan, and J. M. Williams, *Nucl. Inst. and Meth.*, B1, 1984, p. 167-75.

18. K. L. More and R. F. Davis, Presented at The Fourth International Symposium on the Fracture Mechanics of Ceramics, Blacksburg, VA, 1985, to be published in conference proceedings.

19. K. L. More and J. J. Mecholsky, to be published.

QUANTITATIVE MICROSTRUCTURAL CHARACTERIZATION

AND DESCRIPTION OF MULTIPHASE CERAMICS

R.T.DeHoff

Department of Materials Science and Engineering
University of Florida
Gainesville, FL

INTRODUCTION

The microstructural state possessed by a material is the legacy of its history, the key to its properties, and the portent of its response to subsequent processing. Thus, a knowledge of this microstructural state plays a central role in developing an understanding of how a material behaves. In a broad sense, the concept contained in "microstructural state" may imply a hierarchy of aspects of the structure of the material (electronic, crystal, defect, phases) and a range of attributes of its structural elements (chemical. physical, mechanical, thermodynamical).

This paper focusses upon the geometrical aspects of the microstructural state, including the kinds of features that may exist, measures of their quantities, their scale, and their distribution in space. The primary tool for estimating quantitative values of the geometric properties of microstructural features is quantitative stereology (1-3).

The presentation first discusses three levels of characterization of the geometry of microstructures. embodied in the qualitative, quantitative and topographic states (4). The methodologies and fundamental relationships of stereology are then briefly reviewed, concentrating on those which provide rigorous and realistic information for real three dimensional microstructures. A brief and general discussion of potential applications of these techniques complete the presentation.

LEVELS OF CHARACTERIZATION

Three levels of characterization of the microstructure of a multiphase materials may be usefully defined:
    a. the qualitative mircostructural state;
    b. the quantitative microstructural state; and
    c. the topographic state.
Concepts involved in these levels of description are presented in this section.

The Qualitative Microstructural State

From a geometric point of view, all microstructures may be considered to be **tesselations** of three dimensional space, i.e., a subdivision of space into cells which fill it. (If porosity is present, it is considered to be one of the phases.) The geometry of individual cells may be simple or complex: e.g., spherical particles, polyhedral grains, or a multiply connected labyrinth like the pore phase in a powder stack. Together, they fill space. In general, no two cells are identical.

In a multiphase microstructure, each cell may be labelled according to the phase that occupies it. Two cells , i and j, meet to form an ij interface; three cells i, j and k, meet along ijk triple lines in space; four cells , i, j ,k and l, meet at an ijkl quadruple point. This suggests an unambiguous notation for identifying each class of microstructural feature according to the labelling of the cells whose incidence produces the feature. With this identification scheme in mind, **Table 1** lists the classes of features that may exist in one, two, three and four phase structures. It may be that in any given structure, not all of the features are present; absence of some classes of features is useful information.

The qualitative microstructural state is simply a list of the geometric features, volumes (grains), surfaces (interfaces), space curves (triple lines), and points (quadruple points) that are found by inspection to exist in the microstructure. In a multiphase material, compilation of

Table 1. List of three, two one and zero dimensional feature classes in structures with one, two three and four phases.

Dimension of Feature	One Phase	Two Phase	Three Phase	Four Phase
Three (cells)	i	i,j	i,j,k	i,j,k,l
Two (interfaces)	ii	ii,ij,jj	ii,jj,kk ij,ik,jk	ii,jj,kk,ll ij,ik,il, jk,jl,kl
One (tple lines)	iii	iii,jjj, iij,ijj,	iii,jjj,kkk, iij,iik,ijj, ikk,ijk,jjk, jkk	iii,jjj,kkk, lll,iij,iik, iil,ijj,ikk, ill,ijk,ijl, ikl,jjk,jjl, jkk,jll,jkl, kkl,kll
Zero	iiii	iiii,iiij, iijj,ijjj jjjj	15 quadruple points	35 quadruple points
Max No.	4	14	34	69

208

Table 2. Maximum number of features that may exist in each dimensional class for structures with up to ten phases.

No. of Phases	1	2	3	4	5	6	7	8	9	10
Features										
Cells	1	2	3	4	5	6	7	8	9	10
Surfaces	1	3	6	10	15	21	28	36	45	55
Lines	1	4	10	20	35	56	84	120	165	220
Points	1	5	15	35	70	126	210	330	495	715
Total	4	14	34	69	125	209	329	494	714	1000

this list is not necessarily a trivial task. Table 2 shows the (maximum) number of distinguishable features that may exist in each class for structutes that may contain up to ten phases.

The typical field observed in the microscope is a two dimensional section through the three dimensional structure, Figure 1. Geometrically speaking, features observed on the section are intersections between the plane of polish and the three dimensional space filling structure. Sections through cells in the three dimensional tesselation appear as two dimensional cells or areas, and are labelled according to the three dimensional cell sectioned. Two such cells meet at a boundary which is the trace on the plane of polish of the interface between the corresponding cells in three dimensional space. Triple lines in space appear as triple points on the section. Evidence for the existence of specific quadruple points in three dimensional space is indirect; they have no observable intersection with a sectioning plane. Thus, the experimental determination of the qualitative microstructural state may be carried out by simply inspecting a few fields under the microstructure and noting classes of cells, boundaries and triple points.

The resulting list serves two primary purposes:
a. It identifies the features that are missing in the specific structure, thus calling attention to rudimentary aspects of microstructural arrangements.
b. It provides an exhaustive list of the features that are present, permitting a systematic listing of the properties that may be measured in determining the quantitative microstructural state.

The Quantitative Microstructural State

Each of the geometric entities that may exist in the three dimensional structure, volumes, surfaces, space curves and points, may be assigned numerical values of one or more geometric properties that it possesses. These properties may be feature specific when applied to a single unit of the feature class, such as the area of interface bounding a

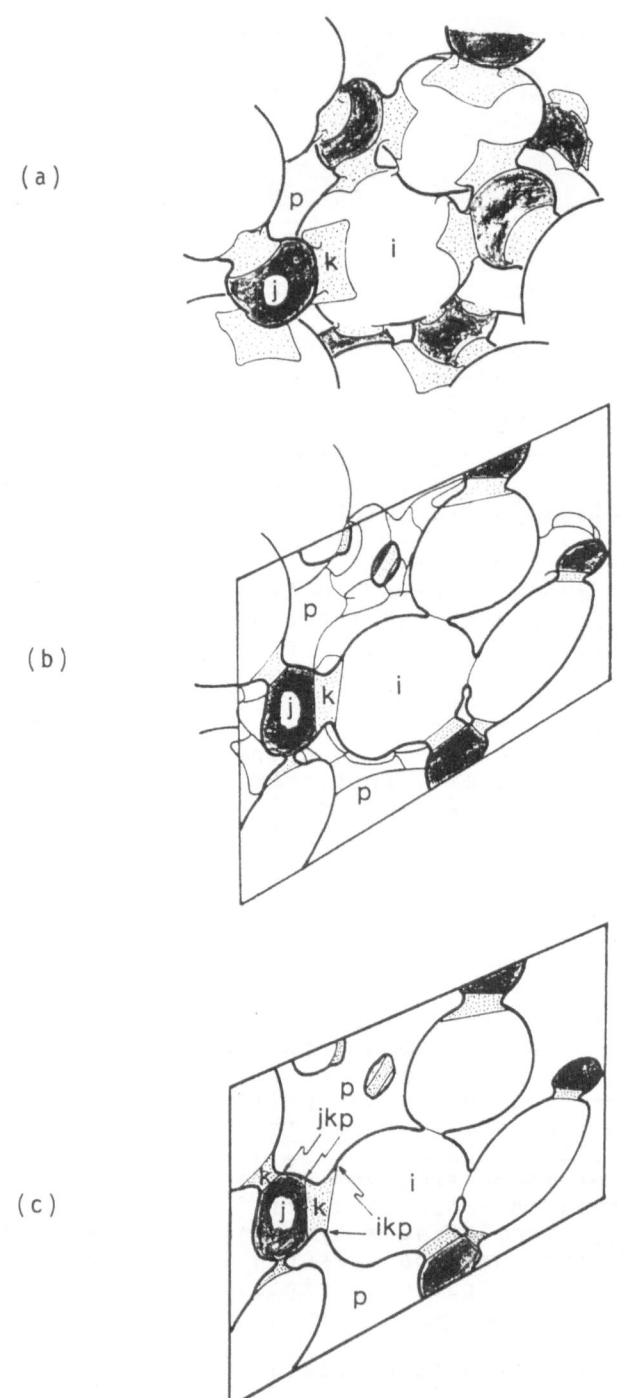

Figure 1.Phase i reacts with phase j to form k in a porous matrix (a): a section (b) produces the four phase micro-structure (c), exhibiting its variety of feature classes.

single particle of the jth phase embedded in an ith matrix. Alternatively, the property may be specified for the whole collection of features of that class in the structure, (all jk interface in the system), in which case it is called a global property. Global geometric properties of feature classes in real three dimensional microstructures are accessable to stereological estimation without simplifying assumptions; features specific properties are not. This presentation foccusses upon those properties that have unambiguous meaning, and can be unambiguously estimated.

The list of global geometric properties may be divided into two major classes: topological properties and metric properties, Table 3.

Table 3. Global geometric properties of microstructures

```
---------------------------------------------------------------
Notation                    Property
---------------------------------------------------------------
```

Topological Properties

N_V - Number of disconnected parts per unit volume
C_V - Connectivity of features per unit volume

Metric Properties: Measures of Extent

V_V - Volume of feature in unit volume of structure
S_V - Area of surface of feature in unit volume
L_V - Length of lineal feature in unit volume

Metric Properties: Curvature Measures

M_V - Integral mean curvature of surface in unit volume
k_V - Integral curvature of lineal feature
t_V - Integral torsion of lineal feature

```
---------------------------------------------------------------
```

Topological properties are independent of the shape and size of the features in the class; their value is invariant under a topological transformation. A topological transformation may be thought of as an arbitrary deformation of the feature (stretching, expanding, bending it), so long as it is not severed or joined upon itself. Topological properties of primary interest are the **number of disconnected parts** of the feature collection, and the **connectivity** of the collection, which may be thought of as the number of redundant connections existing between elements of the feature enselmble. For more specific information about topological properties, consult references 5 and 6. The experimental determination of the topological properties requires serial sectioning analysis.

The global metric properties of a feature ensemble depend explicitly upon the shape and size of features in the class. Two subcategories of these properties are:
 a. measures of extent; and
 b. measures of curvature.
Measures of extent report the total quantity of the specific feature collection: the **volume** of cells; the **area** of surfaces; the **length** of lines. These are the most accessible properties, easiest to visualize and readily determined

experimentally. The measures of curvature are integrals of local curvature values of an element of the feature class over the extent of the feature. They are not significantly more difficult to measure than are the measures of extent, but their interpretation requires more sophistication in geometric concepts.

A variety of geometric properties of the structure may be generated by combining the global properties. For example, dividing any global metric property by its corresponding number of disconnected parts yields the average value of that property, per particle. The average surface area of particles in a feature class is the total surface area divided by the number of disconnected surfaces bounding features in the class. Ratios of measures of extent report measures of the scale of the structure, like the mean lineal intercept, l_3, or the mean cross sectional area, A, of particles in a class. Certain curvature averages are also accessable from ratios of global properties, Table 4.

Table 4. Averages obtained as ratios of global properties

Property	Symbol	Relation

Metric Averages

Mean lineal intercept	$l_3 =$	$4V_V/S_V$
Mean Area Intercept	$A_3 =$	V_V/M_V
Average mean surface curvature	$H =$	M_V/S_V

Number Averages

Average particle volume	$V =$	V_V/N_V
Average particle area	$S =$	S_V/N_V
Average curvature of particles	$M =$	M_V/N_V
Average length of line	$L =$	L_V/N_V

The stereological literature abounds with relationships that require simplifying assumptions about the geometry of the microstructure (11 - 17).The most common of these relations based on simplifying shape assumptions lead to techniques for estimating particle size distributions from section measurements, (11-15). Theoretical statistical analyses of these procedures have revealed that, although they provide unbiased estimates of the number of particles in each size class, the estimates are statistically unstable (18,19), i.e., they are very sensitive to small variations in section observations. Since no two features are alike in real microstructures, use of such approximate relations should be avoided, or at least undertaken with full knowledge of the difficulties involved.

The Topographic State

Global geometric properties report a single number for the whole structure. By themselves, they contain no information about patterns of nonuniformity within the structure.

A third level of characterization, the topographical state, reports such nonuniformities.

Geometric properties of microstructures may vary with
a. position (**gradients**);
b. orientation (**anisotropies**); and
c. spatial distribution (**associations**)
within the system. Evidently, **characterization** of the topographic state contemplates a sequence of measurements as a function of position or orientation; the effort required increases proportionately.

Gradients. Any of the normalized global topological and metric properties may vary systematically with position in real microstructures. In a tapped powder stack, small particles may collect at the bottom and increase the density of the system in that region. The volume fraction of porosity, surface area and integral mean curvature per unit volume of the pore solid interface all vary systematically with height. This variation may be quantified by measuring and averaging these properties on a set of fields that are all at the same height, and then repeating this procedure at a selection of height positions. Plots of the resulting averages versus the position at which they were measured within the gradient provides a quantitative characterization of this positional nonuniformity.

Anisotropies. Not all geometric properties can exhibit anisotropic behavior. The volume fraction and topological properties in three dimensions, by their nature, cannot vary with orientation. Surface area and line length are, on the other hand, obvious candidates for the characterization of anisotropy. Anisotropy of surface area may be described by defining a distribution function which gives the fraction area of elements in the structure with normals oriented in a given direction (20). Although this concept provides a rigorous definition of anisotropy, the distribution function cannot be stereologically estimated. A variety of models for characterizing anisotropy have been proposed and invoked (21-25). All procedures rely upon a complete or partial determine of the "rose-of-the-number-of-intersections", which is an empirical description of this aspect of microstructural inhomogeniety.

Associations. Features in a microstructure may tend to associate with each other (cluster) or avoid each other (order). Alternatively, one class of feature may tend to be associated with another (particles on grain boundaries) or avoid some specific feature class. These are aspects of the spatial distribution of features in a class. Some aspects of spatial distribution may be quantified for real three dimensional structures; other important aspects are not accessable. Contiguity parameters report the tendency for the features that bound one phase to be shared with another phase. For example, in a three phase structure (i, j, k), the total surface area per unit volume bounding the jth phase is

$$S_{Vtot} = S_V(ij) + S_V(jk)$$

The fraction of this area that is shared with the kth phase is

$$c_{kj} = S_V(jk)/S_{Vtot}$$

The fraction, c_{kj}, is a measure of the tendency for the kth phase to be contiguous with the jth phase. Similar measures may be concocted for triple lines bounding a phase, or for measures involving grain boundary parameters. The meaning of parameters of this type is unambiguous; their significance in a given characterization application depends upon the context.

There are no rigorous and straightforward measures of the tendency of sets of particles to cluster, be random, or order. Interesting and sophisticated procedures have been devised for two dimensional structures, employing the Vornoi polygon construction, or its equivalent in Minkowski set algebra (3, 28). While these approaches have three dimensional analogues in concept, there exist no general stereological relations that permit the connection from two to three dimensions. Implementation of these concepts for the characterization of real three dimensional microstructures awaits the development of automated serial sectioning analysis (29).

METHODS AND PROCEDURES

Experimental estimates of the geometric properties that are necessary to define the microstructural state are based upon the fundamental relationships of stereology (1-3,7,10). The measurements required in these basic equations are best visualized as they are made "manually", i.e., without the aid of automatic equipment, and that is how they will be presented in this section. Visualization of the automation of these procedures is then relativey straightforward.

Manual Stereological Measurements

Visualize a field in the eyepiece of a microscope that contains a multiphase microstructure. Imagine a square grid of lines superimposed on the field. In practice, this can be done by inserting a retile in the eyepiece, or simply overlaying a transparency on a projection screen or photograph. The grid serves the function of probing the structure with either
    a. the points at which the grid lines cross (point probes);
    b. the lines that make up the grid (line probes); or
    c. the area outlined by the border of the grid (plane probes).
The measurements that are typical of stereology are simple counts of events that arise from the interaction of the superimposed grid with the appropriate elements of the microstructure. They are summarized in Table 5.

Note that each type of measurement may be applied independently to each appropriate class of feature that may exist in the structure. Thus, the point count may be separately applied to the i, j and k phases in a three phase structure, although in this case the third count may be obtained by difference. The line intercept count may be applied to each of the interface classes listed in Table 1

Table 5. Measurements made in stereology

Symbol	Name	Description
P_P	Point Count	Fraction of points in the grid that lie in the phase of interest
P_L	Line Intercept Count	Number of intersections between a test line and a lineal feature, per unit length of test line
P_A	Area Point Count	Number of some identifiable point feature (e.g., triple points) per unit area of field viewed
N_A	Feature Count	Number of features of interest (e.g., particle sections) per unit of field viewed
T_A	Tangent Count	Net number of tangents formed when a test line is swept across the field, per unit area viewed.

Column 3; the triple point count may be made on each class of triple point listed in the same column. The area tangent count may be applied each class of interphase interface in the structure; since grain boundaries do not separate an "inside" from an "outside" of a phase, it is not possible to assign a sign to tangents that are formed at grain boundaries. Because independent information may be obtained for each feature class, it is necessary to indicate the class subject to analysis in the notation for the quantity measured. This is somewhat cumbersome, but is essential in dealing with multiphase materials where the number of different feature classes may be large. Thus, for example, $P_P(j)$ indicates a point count for phase j; $P_A(ijj)$ indicates a triple point count for triple lines of the class ijj; etc. All of the stereological measurements listed in Table 5 are simple counts of specific events; none require measurements of dimensions of individual features in the microstructure.

Machine Assisted Measurements

Many of the manual measurements described above may be performed automatically or semiautomatically on modern image analysing computers. Semiautomatic instruments use the human operator to discriminate the features to be measured, and to enter the outlines of these features manually into a microcomputer from a bitpad or other input device. Fully automatic instruments scan the image with a television camera, digitize it into a large number of pixels, associate a gray shade with each picture point and load this information into memory. Discrimination algorithms then identify features to be measured on the basis of their absolute gray shade or other more sophisticated combinations of criteria. The primary advantages of semiautomatic instruments are their much lower cost, and the use of the human operator as a basis for discrimination. On fully automatic instruments, discrimination of the appropriate features to be measured is always a compromise, though the problem is mitigated by editing capabilities that permit acceptance, rejection or modification of individual features in the image. This automatic detection problem is compounded in multiphase microstructures in which discrimination may be based upon the juxtaposition of two or three phases.

Once the detected image is established in memory the geometric analysis is relatively straightforward. In principle, any algorithm that can be unambiguously formulated to compute geometric properties of individual features, such as areas, perimeters, specified dimensions, and the like. Set operations may be performed: (set additions and subtractions that give rise to erosions, dilatations, and like image processing operations (3)). Indeed, one of the potential problems associated with machine assisted measurements is the ease with which reams of feature specific information may be produced. Such information is of limited stereological value, because it is not unambiguously related to three dimensional geometric information. It may be used to estimate particle size distributions and number of particles, but these procedures require geometric assumptions (all the particles are perfect spheres) which are unrealistic for most microstructrures. All of the simple counting measurements described in Table 5 are available in virtually all image analysing devices.

The Fundamental Relations of Stereology

The simple counting measurements just described each provide an estimate of a geometric property of the corresponding features in the structure; these properties have been listed in Table 3. The equations that relate stereological measurments to geometric properties are elegant in their simplicity. The mathematical concepts and methods that produced these equations involve geometric probability theory, differential and integral geometry and topology (1-3,7,10).These equations have the status of expected value theorems in statistics. Thus, their validity presumes that the entire population of test probes (points, lines or planes) that may be constructed in three dimensional space is included in the evaluation of the population mean. In practice, an (isotropic, uniform, random) sample of the set of probes is included in the experiment. The sample mean is then used to estimate the population mean, with a confidence interval that is computed in the usual way in statistical analysis.

The set of mean value theorems that constitute the fundamnetal equations of stereology are summarized in Table 6. In a multiphase structure the geometric properties of individual feature classes may be estimated by simply applying the appropriate counting measurements to features in that class. Thus a line intercept count of grain boundaries in the jth phase (jj boundaries) provides an unbiased estimate of the surface area per unit volume of such grain boundaries:

$$P_L(jj) = S_V(jj)/2$$

Table 6. Fundamental Relations of Stereology.

Property	Relationship
Volume Fraction	$V_V = P_P$
Surface Area	$S_V = 2P_L$
Line Length	$L_V = 2P_A$
Integral Mean Curvature	$M_V = 2\pi N_A = \pi T_A$

Serial Sectioning Analysis

The topological properties listed in Table 3 are the most rudimentary of geometric properties, and potentially of great interest in microstructural characterization. Unfortunately, there exist no general, fundamental stereological equations that permit estimation of these topological properties. A variety of model based estimates have been proposed (1,2,11,12), but these necessarily lack generality, and are of doubtful value for characterizing real microstructures.

Topological properties can only be unambiguously estimated through serial sectioning (5,29,30). In this procedure, a series of closely spaced microsections is prepared by alternately grinding away a small layer, polishing, etching and photographing; and then repeating the procedure. The resulting · stack of photographs provides a three dimensional reconstruction of the structure in which counts of number of separate parts, connectivity and the like may be made. Evidently, the estimation of topological properties requires significantly more effort than the more ordinary stereological measures. There is great potential for advance in the automation of serial sectioning procedures as automatic sectioning equipment becomes available along with prodigious memory capacity of new generations of computer systems (29).

APPLICATION IN MUTLIPHASE CERAMICS

Realistic quantitative specification of the microstructural state of a multiphase ceramic may find application in three broad categories:
1. the evolution of microstructure in processing;
2. the development of structure-property relationships; and
3. the comparison of similar microstructures that are behaving differently.
The potentials for stereological analysis in each of these categories of research are briefly explored in this section.

Evolution of Microstructure in Processing

Processes are studied by examining a sequence of micro-structures that are generated during the process. It is essential that such a sequence must lie along a path of microstructural change, if a realistic analysis of the process is sought. This strategy requires that each microstructure in the sequence be produced from its predecessor by incrementing some process variable. Note that isothermal studies satisfy this strategic requirement, while isochronal studies do not: each structure that results from processing for a given length of time at a sequence of different temperatures has a different history. The same criticism characterizes microstructures produced by cold pressing to different green densities, followed by processing. Microstructures in such a sequence are not expected to be simply related to eachother, and are not amenable to kinematical analysis.

During the processing of multiphase ceramics it is expected that the relative amounts of each of the phases in the structure, including the pore phase, will change with time. Areas of the various interphase interfaces will also vary with time. These interphase interface areas are regions where chemical reactions and phase changes are focussed; thus, a knowledge of their extent is crucial to understanding the kinetic phenomena in progress. These areas are bounded by one or more classes of triple line; the length of these triple lines is a measure of their state of subdivision of the areas where reactions occur. All of these geometric properties are stereologically accessible for real microstructures.

Reaction rates are related to velocities of migration of these interfaces; growth or shrinkage of a phase may be described in terms of the distribution of velocities of elements of area that form its boundary. If succeeding microstructures lie along a path, then kinematic information about interface velocities may be obtained from an analysis of rates of change of the global stereological properties (31,32). In the case of multiphase ceramics these formulations are compicated by changes in the total volume of the system as porosity shrinks or expands, distorting the reference frame for defining interface velocities. A rigorous examination of the problems that arise in attempting to apply these kinematic formulations to real multiphase ceramics is feasible, but has not yet been undertaken.

Structure Property Relations

It is a central tenet in materials science that micro-structure has an important influence on properties of the material. This tenet is the basis for the control of properties by processing. The development of quantitative structure-property relations clearly presumes quantitative knowledge of the microstructure. Thus, microstructural characterization should play a central role in research in this area.

Two important factors serve to significantly impede

progress in the development of structure-property relations; these two factors operate separately on the two sides of the equation.

The first point is best illustrated in the measurement of mechanical properties. Numerical values that are reported reflect not only the state of the material, but a standardized process through which the material is being taken. Thus, in ductile materials, properties like yield strength, flow stress, ultimate stress and fracture stress are benchmarks on a stress strain curve. It is true that the process that gives rise to these values is a well defined and widely accepted tensile test. It is also well known that all of these parameters will have different values if the conditions of the test are altered. Thus, the numbers we call mechanical "properties" are determined as much by the process by which they are measured as by the microstructural state of the material.

On the structural side of the relationship it is likely that many of the properties of interest depend, not upon the average values of geometric properties of the microstructure, but upon certain extreme values. Fracture paths may seek a minimum cross sectional area in penetrating a porous structure. Initiation of fracture may depend critically upon the size of the largest preexisting flaws. Ductility may be sensitive to the largest inclusions in the microstructure. Stereological measures report representative geometric properties for the microstructure. They are averages, and if the property of interest is sensitive to some extreme value, then a correlation will exist with the stereologically estimated properties only insofar as the extreme values parallel the average in a comparison among similarly produced microstructures.

These two problems in structure property correlations are significant obstacles to the development of a quantitative science in this area. On the one hand, it seems necessary to seek to identify measures of the behavior of materials that are more truly fundamental than benchmarks on a stress strain curve. On the other, it appears essential to seek more sophisticated measures of microstructural state that yield more direct information about extreme values of interest. This is not to say that attempts to apply state of the art stereology to these problems are unprofitable. It is perhaps fair to say that stereology has not been tried in this context: there are very few examples of attempts to apply the tool in the literature, and applications to multiphase ceramics are practically nonexistent. Indeed, progress in this area can only occur by beginning to quantify microstructural state with the tools that are available, and exploring the resulting correlations.

Comparisons of Microstructures

In quality control applications, and in failure analysis, it is frequently of interest to make comparisons of microstructures that are presumed to be similar to eachother. Stereology is particularly well suited to this kind of application because its results are statistically determined. It is only necessary to obtain estimates of the parameters of interest, along with their associated confidence intervals, to test the statistical hypothesis embodied in the comparison.

Multiphase ceramic materials are particularly well suited for this kind of test, precisely because they have so many measurable geometric properties. For example, the structure depicted in Figure 1 consists of four phases: i, j, k and p. Even neglecting the presence of grain boundaries within the phases, there are six kinds of interfaces (ij, ik, ip, jk, jp, kp) and four kinds of triple lines (ijk, ijp, ikp, jkp). Thus, if only volume fraction, surface area and triple line length were measured, each structure would still yield 13 numbers in its characterization (the four phases exhibit only three independent volume fractions). Clearly, there is a bountiful store of microstructural information that would permit easy discrimination between two such similar structurtes.

SUMMARY

The microstructural state of multiphase ceramic materials provides a fertile area for the application of the tools provided by stereology. The qualitative microstructural state is complicated, because the list of feature classes is long. Even a rudimentary quantitative characterization of the microstructural state is likely to be fruitful because the structure exhibits a high density of information, and because the field is unexplored. The aspect of the topographic state that is likely to be of most immediate interest is embodied in the measures of contiguity of phases resulting from their shared surface areas. All of these measures provide the basis for following microstructural evolution during processing, for formulating structure - property relationships, and for making comparisons among microstructures.

ACKNOWLEDGEMENTS

The author is grateful to the Army Research Office for its support of a stereological study of powder processing, which fostered the development of this paper.

REFERENCES

1. R.T. DeHoff and F.N. Rhines, "Quantiative Microscopy," McGraw-Hill Book Co., Inc., New York, NY (1968).

2. E.E. Underwood, "Quantitative Stereology," Addison-Wesley Press, Reading,MA (1970).

3. J. Serra, "Image Analysis and Mathematical Morphometry", Academic Press, New York, NY (1981).

4. R.T. DeHoff, "Topography of Microstructures", Metallography, 8:71 (1971)

5. R.T.DeHoff, E.H.Aigeltinger and K.R. Craig, "Experimental determination of the topological properties of microstructures", Journal of Microscopy. 95,pt.1:69 (1972).

6. E.H.Aigeltinger and R.T. DeHoff, Quantitative determination of the topological and metric properties during sintering of copper powder", Metallurgical Transactions. 6A:1853 (1975).

7. R.T. DeHoff and S.M.Gehl, "Stereology of space curves", in "Proc. Fourth International Congress for Stereology," R. DeWit and G.M.Moore, ed., National Bureau of Standards Special Publication No. 431, U.S. Government Printing Office, Washington, DC (1975) p.29.

8. R.T. DeHoff, "Geometrical meaning of the integral mean surface curvature", in "Microstructural Science, Vol. 5". J.L. McCall, ed., Elsevier-North Holland Press, New York, NY (1977), p. 331.

9. R.T.DeHoff, "Integral mean curvature and platelet growth", Met. Trans. 10A:1948 (1979).

10. R.T.DeHoff, "Stereological meaning of the inflection point count", Journal of Microscopy, 121:13 (1981).

11. S.D.Wicksell, "The corpuscle problem II: case of ellipsoid corpuscles", Biometrika. 18:151 (1926).

12. E.Schiel and H.Wurst, "Statische Gefugeuntersuchengen II: Messung der raumlichen Kristallgrosse", Z.Metallk. 28:340 (1936).

13. W.A.Johnson, "Estimation of spatial grain size", Met. Prog.. 49:87 (1946).

14. J.W.Cahn and R.L.Fullman, "On the use of lineal analysis for obtaining particles size distribution functions in opaque samples", Trans.AIME. 206:610 (1956).

15. S.A.Saltykov, "The determination of the size distribution of particles in an opaque material from a measurement of the size distribution of their sections", in "Stereology", H. Elias, ed., Springer-Verlag, New York, NY (1967) p.167.

16. R.T.DeHoff, "The estimation of particle size distributions from simple counting measurements made on random plane sections", Trans.AIME. 233:25 (1965).

17. E.E.Underwood, "Quantitative Stereology". Addison-Wesley Pub. Co. Reading, MA (1970) Chapter 4.

18. R.S.Anderssen and A.J.Jakeman, "Abel type integral equations in stereology II: computational methods of solution and the random spheres approximation", Journal of Microscopy. 105:135 (1975).

19.   R.S.Anderssen and A.J.Jakeman, "Stable procedures for the   inversion of Abel's equation",   Jnl.Inst.Maths.Applics. 17:329 (1976).

20.   J.E.Hilliard, "Specification and measurement of micro-structural anisotropy", Trans.AIME. 224:1201 (1962).

21.   S.A.Saltykov, "Stereometric Metallography",second edit-ion, Metallurgizdat, Moscow (1958).

22.   E.E.Underwood "Quantitative Stereology", Addison-Wesley Pub. Co., Reading, MA (1970) Chapter 3.

23.   B.R.Patterson and R.T.DeHoff, "Measurement of lineal features of anisotropic microstructures: the Saltykov and tetrakiadecahedron models", in "Microstructural Science, Vol.   7", J.L.McCall ed., Elsevier North Holland, New York, NY (1979) p.445.

24.   W.J.Whitehouse, "The quantitative morphometry of aniso-tropic trabecular bone", Journal of Micorscopy. 101,pt2:153 (1974).

25.   T.P.Harrigan and R.W.Mann, "Characterization of micro-structural anisotropy in orthotropic materials using a sec-ond rank tensor", Jnl.Mats.Sci. 19:761 (1984).

26.   J. Gurland, "The measurement of grain contiguity in two phase alloys", Trans AIME. 212:452 (1958).

27. J.W.Cahn and J.E.Hilliard, The measurement of grain con-tiguity in opaque samples", Trans. AIME. 215:759 (1959).

28.   D.A.Aboav,"The arrangement of cells in a net, III", Metallography.17:383 (1985).

29.   R.T. DeHoff, "Quantitative serial sectioning anlaysis: preview", Journal of Microscopy. 131:259 (1983).

30. M.Yanuka, F.A.L.Dullien and D.E.Elrick, "Serial section-ing and digitization of porous media for two and three di-mensional analysis and reconstruction", Journal of Micro-scopy. 135:159 (1984).

31.   R.T. DeHoff, Dynamics of microstructural change", in "Treatise on Materials Science and Technology, Vol.1," H. Herman, ed., Academic Press, New York, NY (1972) p.247.

32.   R.T. DeHoff, "Stereology and metallurgy", Metals Forum. 5:4 (1982).

DISPLACIVE TRANSFORMATION MECHANISMS IN ZIRCONIA CERAMICS

AND OTHER NON-METALS

W. M. Kriven

University of Illinois at Urbana-Champaign, Materials
Research Laboratory and Department of Ceramic Engineering
Urbana, IL 61801

INTRODUCTION

Phase transformation mechanisms in a variety of non-metals can be studied from the point of view of structural aspects and nucleation, and compared with metallurgical classification schemes.[1-3] In a "reconstructive" transformation first coordination bonds or nearest-neighbor interactions are broken and remade when converting to a new structure. Such processes require a high activation energy and are usually slow and sluggish. They proceed by thermally-activated growth across an interface. "Displacive" transformations on the other hand, involve no rupture of first cordinations, but merely a distortion of the crystal lattice. The activation energy is much lower and the kinetics are fast. Displacive transformations are not necessarily martensitic. This has been a frequent source of confusion, evident in the literature. Martensitic transformations are a subset of displacive transformations.[2] A martensitic mechanism is a "lattice-distortive, virtually diffusionless structure change having a dominant deviatoric component and associated shape change, such that strain energy dominates the kinetics and morphology during the transformation."[2]

Non-metallic materials such as electronic ceramics having ferroic, ionic, optical, or magnetic properties tend to transform by shuffle-dominated[2] or dilatational mechanisms or sometimes merely by electronic transitions, and generally have negligible volume changes.[4] Displacive transformations, often accompanied by large changes in volume or coordination number, occur in inorganic and organic compounds, minerals, ceramics and some crystalline components of cement.[5,6] They are summarized in Table 1 and many exhibit martensitic characteristics.

The aim of this paper is to discuss some of the experimentally-studied examples above as they illustrate factors which play important roles in phase transformations, microstructure and hence mechanical properties of composite ceramics. Specifically, the critical resolved strain criterion for martensitic nucleation and the effect of volume change versus shape change accompanying a transformation will be examined.

Table 1. Non-metals with Lattice-distortive Transformations

Inorganic Compounds   (Structure-Type)

Alkali and ammonium halides:
    MX, NH_4X (NaCl-cubic $\rightleftharpoons$ CsCl-cubic)
Nitrates and carbonates:
    $RbNO_3$ (NaCl $\rightleftharpoons$ rhombohedral $\rightleftharpoons$ CsCl)
    KNO_3, $TlNO_3$, $AgNO_3$ (orthorhombic $\rightleftharpoons$ rhombohedral
    $CaCO_3$ (aragonite-type $\rightleftharpoons$ calcite-type)
Sulfides:
    MnS   (zinc-blende $\rightleftharpoons$ NaCl-cubic)
                                    (wurtzite $\rightleftharpoons$ NaCl-cubic
    BaS   (NaCl-cubic $\rightleftharpoons$ CsCl-cubic)
    NiS   (hexagonal $\rightleftharpoons$ rhombohedral)

Ferroic Compounds[4]

Minerals

Pyroxene layer silicates:
    Enstatite ($MgO.SiO_2$)   (orthorhombic $\rightleftharpoons$ monoclinic)
    wollastonite ($CaO.SiO_2$)   (monoclinic $\rightleftharpoons$ triclinic)
    ferrosilite ($FeO.SiO_2$)   (orthorhombic $\rightleftharpoons$ monoclinic)

Ceramics

Oxides: ZrO_2, HfO_2 (tetragonal $\rightleftharpoons$ monoclinic)
Boron nitride: BN   (wurtzite $\rightleftharpoons$ graphite)
Carbon: C   (wurtzite $\rightleftharpoons$ graphite)

Cement

Dicalcium silicate: $2CaO.SiO_2$ (trigonal $\rightleftharpoons$ orthorhombic $\rightleftharpoons$
                                    monoclinic)
                                    (monoclinic $\rightleftharpoons$ orthorhombic)

Glass

Nickel sulfide: NiS   (rhombohedral $\rightleftharpoons$ hexagonal)

Organic Compounds

Polyethylene chain polymers: (orthorhombic $\rightleftharpoons$ monoclinic)
    $(CH_2 - CH_2)_n$

CONFINED PARTICLE MECHANISMS: TRANSFORMATION TOUGHENING OF ZIRCONIA CERAMICS

The modifications of ZrO_2 at ambient pressures are:[7]

$$\text{Melt} \xrightarrow{2680°C} \underset{\substack{\text{cubic}}}{\text{I}} \xrightarrow{2200°C} \underset{\substack{\text{tetragonal}\\CN(Zr)=8}}{\text{II}} \underset{1170°C}{\overset{950°C}{\rightleftharpoons}} \underset{\substack{\text{monoclinic}\\CN(Zr)=7}}{\text{III}}$$

The tetragonal to monoclinic transformation is martensitic in the
bulk. There is a volume increase of +3% at 950°C increasing to +4.9% at
room temperature, and the transformation was notorious for causing the
material to shatter. ZrO_2 can be fully or partially stabilized by
suitable additions of CaO, MgO or Y_2O_3.[7]

224

The factors determining the metastability of confined zirconia particles in the partially stabilized ZrO_2 (PSZ) and Al_2O_3-ZrO_2 systems have been reviewed.[8,9] Particles are retained in tetragonal symmetry at room temperature by (i) matrix constraint (ii) chemical composition and (iii) nucleation barrier to transformation. After the optimum microstructure has been tailored by suitable choice of the first two factors, it is the nucleation barrier (itself a function of particle size, shape and location) which ensures that particles remain metastable until transformed by the stress field of a crack tip.[10]

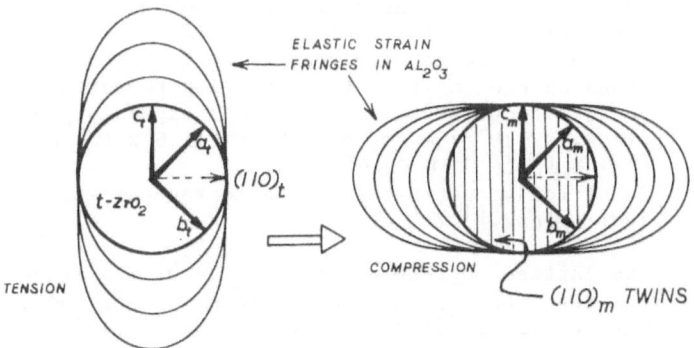

Fig. 1.   Schematic model of transformation mechanism of spherical ZrO_2 particle confined in Al_2O_3

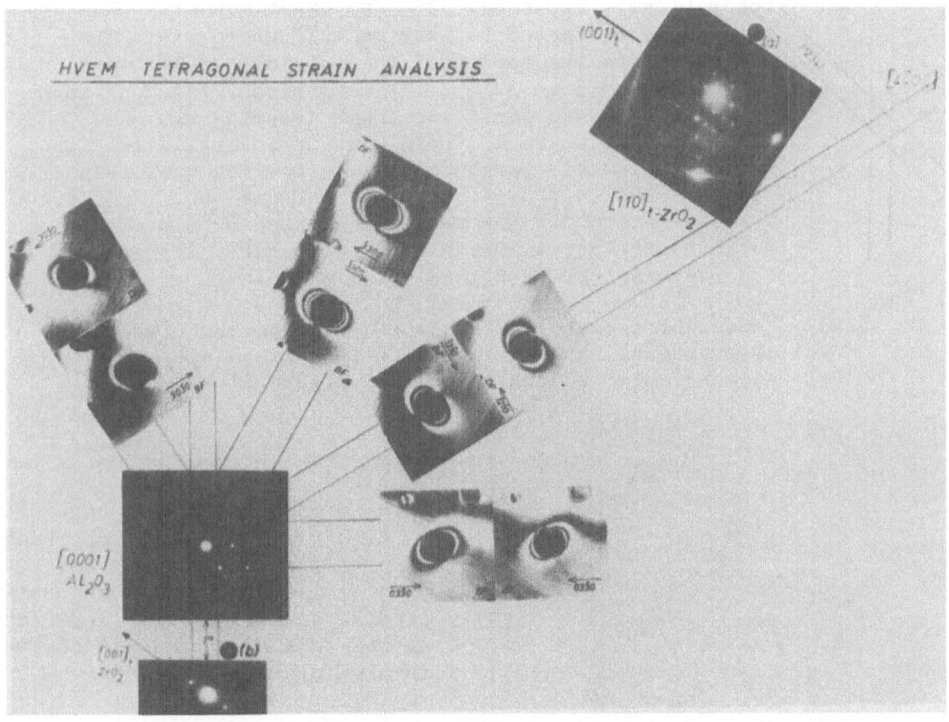

Fig. 2.   HVEM strain analysis by systematic 3-D tilting and imaging under BF and DF 2 beam dynamical conditions.

The transformation mechanism correlating crystallography and elastic strain fields before and after transformation has been elucidated.[11] In situ studies of a model system in which spherical (~0.3 μm) zirconia particles are randomly dispersed in a large-grain Al_2O_3 ceramic were made in a 1MeV HVEM, to simulate bulk conditions. The experimentally-deduced transformation mechanism is schematically summarized in Fig. 1.[11] Essentially it depicts the following features in a 0.3 μm diameter particle:

(i) Even before transformation, there was a residual tensile strain field surrounding the particle at room temperature. It was a dipole whose principle axis was parallel to the internal c_t axis of ZrO_2 (Fig. 2). This was in accordance with thermal contraction mismatch which arose on cooling from the sintering temperature of 1500°C to room temperature. The thermal expansion coefficients are 8.1×10^{-6}/°C for Al_2O_3 with 11.6×10^{-6}/°C in a_t and 16.8×10^{-6}/°C in c_t of ZrO_2. Quantitative estimates of the tetragonal strain field are in progress.[12]

(ii) The lattice correspondence adopted during transformation was such that a_t, b_t and c_t become a_m, b_m and c_m respectively (lattice correspondence C)[13] with (110) twinning in the monoclinic phase.

(iii) The particle transformed suddenly to a twinned monoclinic particle surrounded by both long range and short range strain fields (Fig. 3). Arising from the ends of oppositely shearing twins, the short-range strain field was localized to the particle-matrix interface, as theoretically expected.[14] The experimentally observed strain was estimated to vary as r^{-n} where r was the distance from the center of the particle and n had a value between 3 to 5.[15] By comparison, the long range monoclinic strain field was lower, varying as $\sim r^{-2}$. It also appeared to be a dipole whose principal direction was approximately perpendicular to the internal twin planes. Due to the +4.9% volume increase on transformation[10] the particle exerted a compressive force on the matrix which was considerably larger than the pre-existent tetragonal strain field.

(iv) Hence, by internal crystallographic and experimental observations, the tetragonal strain field was found to be perpendicular to the monoclinic strain field.

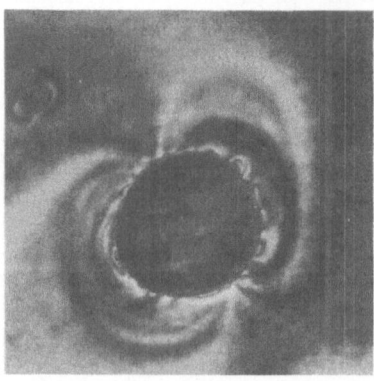

Fig. 3. Twinned monoclinic ZrO_2 particle in Al_2O_3. It has both long range compressive strain field and short-range strain fields localized at particle-matrix interface, at the ends of oppositely shearing twins.

Theoretical martensitic calculations have been made using lattice parameters corrected to the bulk, high transformation temperature of 950°C.[16] Three lattice correspondences were examined, (LC A where $a_t \rightarrow c_m$, LC B where $b_t \rightarrow c_m$ and LC C where $c_t \rightarrow c_m$)[13] in conjunction with experimentally observed twinning modes[17,18] as lattice invariant shear (LIS) systems. Martensitic solutions were found for $(110)_m$ LIS twinning with LC A; $(110)_m$, $(011)_m$ and $(00\bar{1})_m$ twinning with LC B; and $(011)_m$ twinning only with LC C. Comparison of experimental observations of LC C with $(110)_m$ was inconsistent with predictions of martensitic theory using 950°C lattice parameters. An unresolved question is whether the room temperature lattice parameters give a theoretical solution in agreement with experimental observations. Such martensitic calculations remain to be done.

It is noted that a formal confirmation of a martensitic mechanism requires agreement between theoretical predictions based on the invariant plane strain (IPS) formulation and experimental observations of habit planes, lattice invariant shear system and orientation relations between parent and product phase. Generally, as exemplified in martensitic calculations for the zirconia t→m transformation,[13,16,19] orientation relations and LIS systems are relatively insensitive indicators of a specific martensitic mechanisms. Observation of theoretically predicted habit planes is a much more sensitive indicator of a specific O.R. and LIS system. Unfortunately, however as the confined ZrO_2 particles essentially "click" into the monoclinic structure no habit plane observations could be made. However, even in the absence of habit plane data, if no martensitic predictions occur for the observed orientation relation and lattice correspondence one must conclude that more relevant martensitic calculations need to be done.

If the room temperature lattice parameters modify the crystallography only slightly, it indicates that either (i) the classical invariant plane strain (IPS) formulation of martensite theory does not apply to confined particle mechanisms. It may need to be significantly modified to take account of the elastic constraint imposed by the matrix, the particle shape and particle-matrix interface; or (ii) the mechanism is not martensitic in the "IPS" sense but merely displacive and accompanied by transformation or deformation twinning. Transformation twinning refers to twins in the product phase which originate from mirror planes in the parent phase, which is not necessarily the case in mechanical twinning.[1] At this stage, however, the zirconia confined particle mechanism appears to be consistent with the general definition of a martensitic mechanism quoted in the introduction.[2]

THE NUCLEATION BARRIER TO TRANSFORMATION

Recently, a new model for martensitic nucleation in composite ceramics has been proposed[9] and further refined.[20] It visualizes martensitic nucleation occurring at sites of anisotropic strains such as corners or edges of particles where the strain singularities scale with particle size.[21] In this way nucleation is a function of particle size rather than pre-existent embryo defects. This model is a viable solution to the stale-mate situation existing in the literature for over 30 years, due to the close and ambiguous relationship between dislocations, defects, deformation and transformation microstructures in metals. The high Pieirl's barrier to dislocation formation in ceramics removes this ambiguity.

Quantitatively, the condition for nucleation may be expressed as:[20]

$$\frac{\varepsilon_{ij}^{R'}}{\varepsilon_{ij}^{T}} > \eta = 0.2113$$

where $\varepsilon_{ij}^{R'}$ is the resolved shear strain in the particle and composed of any residual strain, ε_{ij}^{R} (eg. due to thermal expansion mismatch), plus an applied strain, ε_{ij}^{A}, such as due to a crack tip stress field;

$$\varepsilon_{ij}^{R} = \varepsilon_{ij}^{R} + \varepsilon_{ij}^{A}.$$

ε_{ij}^{T} represents the unconstrained transformational shear strain due to the change in unit cell and hence, particle shape. Preliminary estimates of the residual strain field around spherical tetragonal zirconia particles indicate that it is insufficient to overcome the nucleation barrier.[11] In this material where particles were dispersed in random orientation, no transformation zone could be identified in in situ straining experiments.

THE CRITICAL RESOLVED STRAIN CRITERION FOR MARTENSITIC NUCLEATION.

Organic polyethylene crystals consist of long chain molecules foled and packed in a regular crystallographic manner. Single crystals of polyethylene $(CH_2-CH_2)_n$ may be grown in flat, ridge-lozenge (diamond-shaped) morphology with chains parallel to the Z crystallographic axis of an orthorhombic unit cell (Fig. 4).[22] In a lozenge, the long and short morphological axes lie along the crystallographic [a] and [b] axes, respectively, while (110) planes form prism faces in which the molecules follow tight folds whose surface is perpendicular to the chain axis (Fig. 5). From the physical distinction between equivalent (110) and (1$\bar{1}$0) planes, two types of fold surfaces arise, resulting in two types of sectors per lozenge (Fig. 6). These have good fit on the morphological [a] axis and poor fit on the [b] axis. The monoclinic structure can be achieved by a stress-induced transformation.

The different types of deformation and transformation modes in crystalline polyethylene have been comprehensively studied.[22-28] Single crystals 100-200 Å thick were deposited in random orientation on support films which were drawn by different amounts of up to 40% extension, to impose a directional tensile stress. Strains of different magnitude could be resolved into components with respect to crystallographic axes within each crystal as illustrated in (Fig. 7).[24] With reference to the surface fold geometry in different sections, the dependence of slip, twinning and transformation modes on the resolved shear strain experienced by a crystal was experimentally deduced (Fig. 8).

The crystallography of the observed deformation modes in single crystals indicated that twinning, repeated twinning and slip occurred on (110) and (310) systems depending also on fold sector geometry.[22-26] This was consistent with strong, covalently bonded polymer chains shearing past each other, but in themselves remaining undistorted. This also implied that the orthorhombic to monoclinic transformation interface was restricted to containing the chain axis i.e. being of the $\{hk0\}_0$ type.

As well as the several deformation systems, two orientation relations for the transformation were observed.[22] By the general theory of transformation strains[27] the geometry of twinning and martensitic phase transformation for numerous systems was analyzed.[28]

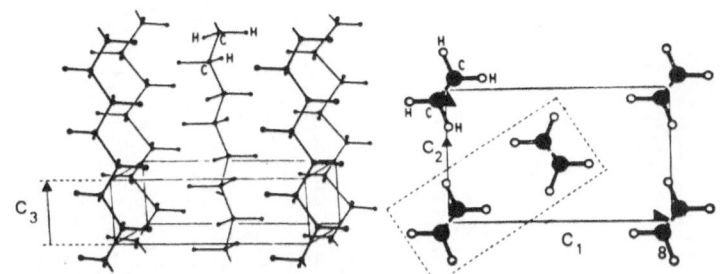

Fig. 4. Crystal structure of orthorhombic polyethylene[22]

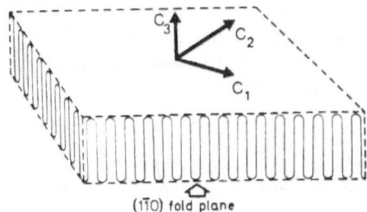

Fig. 5. Lozenge-shaped crystal morphology.[22]

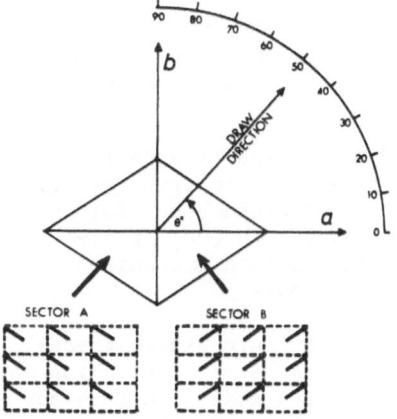

Fig. 6. Two types of {110} fold surfaces give rise to two sectors in crystal. (22)

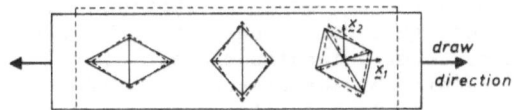

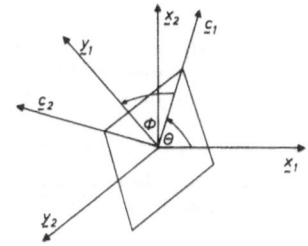

Fig. 7. Crystal lozenges in random orientation were non-hydrostatically stretched. Strains could be resolved on to crystallographic [a] and [b[axes.[22]

The martensitic geometry simplified to a two dimensional problem since lattice correspondences were chosen so that a principle strain axis ($\vec{e}_3$) was parallel to the undistorted chain axis and hence equal to zero. The other eigen vectors ($\vec{e}_2$, $\vec{e}_3$) were then less than and greater than zero respectively. The condition for an invariant plane strain (IPS) therefore were satisfied without the necessity for lattice invariant shear (LIS) slip or twinning. A martensitic mechanism could thus be formulated for any two structures exhibiting 2, 4 or 6-fold symmetry about a chain axis and related by integral unit cell repeat distances along that axis. The structure change then primarily involved perpendicular shearing or displacements of whole chains in directions having components perpendicular to the chain axis.

229

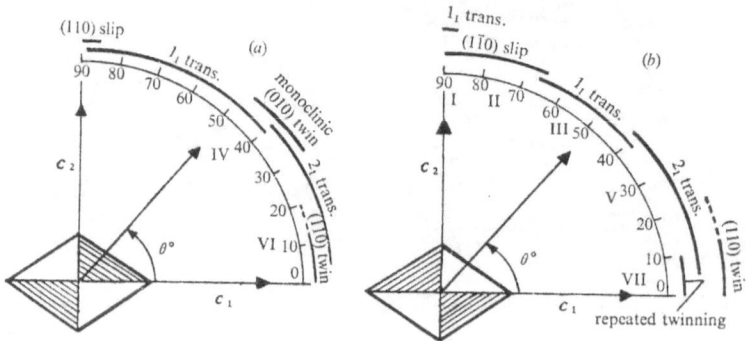

Fig. 8. Dependence of martensitic transformation or deformation
mode on crystallographic direction of resolved strain.
Summary of experimental observations.[22]

The observed orientation relations in polyethylene agreed to within
1° with martensitic transformation mechanisms denoted 1_1, and 2_2. Their
lattice correspondences, fold geometry and irrational planes of shear
(habit planes) in (001) projection were deduced. The volume change was
0.73% and in both mechanisms the folds under went minimum reorientation
and distortion.

Thus, organic single crystal studies showed that a critical
resolved shear strain criterion determines which possible deformation or
transformation mode occurs on application of uni-directional stress.
The magnitude of the shear strain must be small since the configuration
of tight fold geometry resticted the operation of large deformation
modes. The transformation or deformation mechanism favoured was the one
with the simplest shuffles or displacements of molecular chains. The
theoretical martensitic treatment in organic polymer crystals may have
general applicability for example to pyroxene layer silicates.

PYROXENE MINERALS, STEATITE CERAMICS: ENSTATITE.

The pyroxene minerals have crystal structures in which SiO_3 chains
parallel to the Z axis are arranged in sheets on (100) planes. The
layers are separated by M^{2+} cations at the centers of octahedra or distorted
octahedra formed by oxygen atoms between sheets (Fig. 9).[5] A fairly
well studied pyroxene is enstatite, (magnesium metasilicate, $MgO.SiO_2$).
Wollastonite ($CaO.SiO_2$) and ferrosilite ($FeO.SiO_2$) share similar structures
and transformational behavior as has been reviewed elsewhere. (28,5)

Enstatite is the major component of steatite ceramics[29-32] which
play an important role as insulators for high frequency application
because of their excellent mechanical properties and low power losses in
the high frequency field. They have high dielectric resistance at room
and high temperatures. The structural modifications of enstatite are
summarized in Fig. 10. Ortho- and clinoenstatite are structurally
related in that OE is essentially composed of two CE cells of β= 108.3°
twinned on (100) and translated by a b-glide. In comparison with ZrO_2
where the monoclinic βangle is ~9°, CE has double the shape change of
the unit cell. Clinoenstatite is a metastable phase formed by quenching
of PE or grinding of OE (Fig. 10). Both are displacive transformations
where PE →CE is accompanied by a −5.5% negative volume change, and OE→CE
has a negligible volume change.[5]

In the T-P phase diagram CE has a true stability field given by T-566° + (4.5° per K bar)P. Under hydrostatic conditions the OE to CE transformation is sluggish, requiring the uses of fluxes and long reaction times, as is characteristic of a reconstructive transformation. In contrast, however, it occurs with rapid kinetics on application of non-hydrostatic stresses along (100)[001] above a threshhold shear of 13.3° which arises from the lattice correspondence experimentally adopted and illustrated in Fig. 11.[33] The OE →CE phase boundary is thus raised by 200°C-300°C per Kb of non-hydrostatic shear stress. The displacive transformation involves cooperative glide of (100) partial dislocations with b = 0.83 [001] and a (100) interface. The shape change is the same as the unit cell shape change without any LIS twinning or slip, although various amounts of macroscopic kinking accompany the slip. Similar behavior was found in ferrosilite (FeSiO₃) and monoclinic to triclinic wollastonite (CaSiO₃).

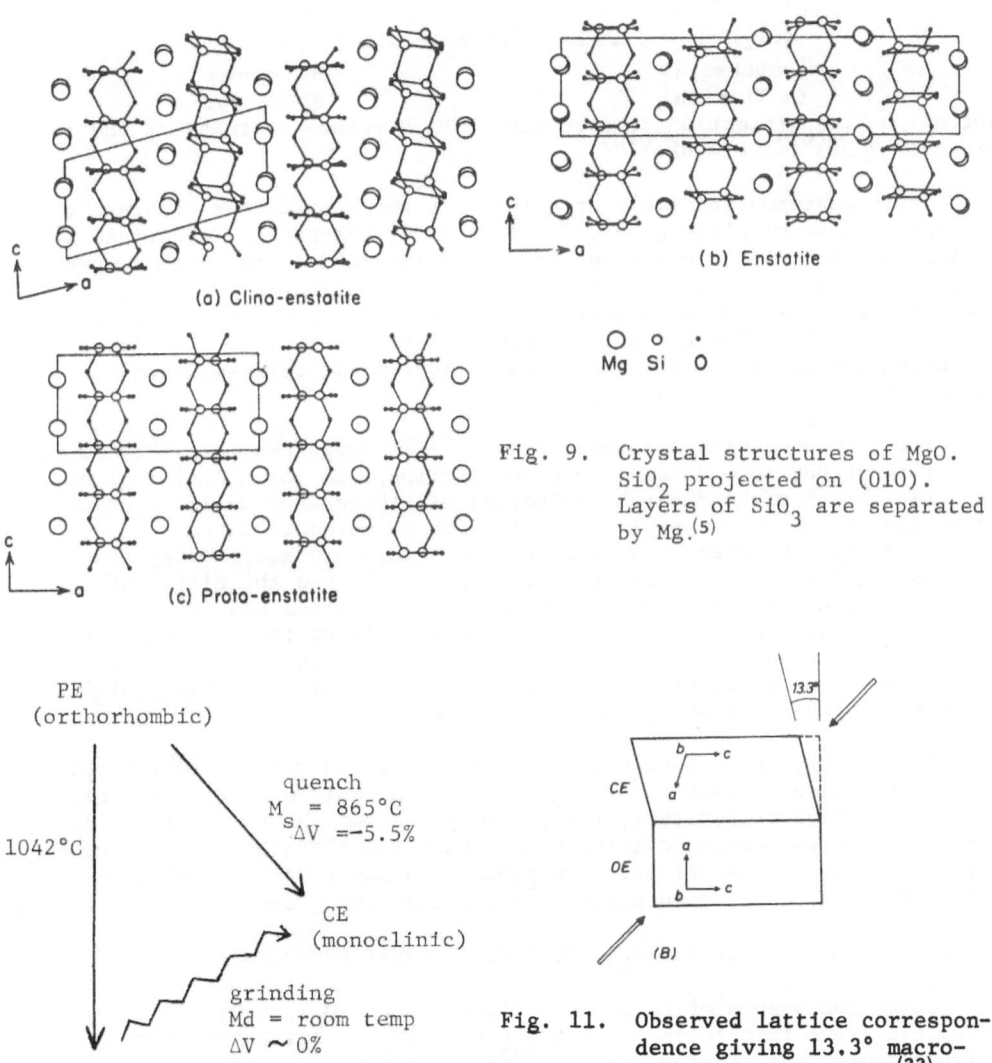

(a) Clino-enstatite

(b) Enstatite

○ ○ ·
Mg Si O

(c) Proto-enstatite

Fig. 9. Crystal structures of MgO. SiO₂ projected on (010). Layers of SiO₃ are separated by Mg.[5]

PE
(orthorhombic)

1042°C

quench
M_s = 865°C
ΔV =-5.5%

CE
(monoclinic)

grinding
Md = room temp
ΔV ∼ 0%

OE
(orthohombic)

Fig. 10. Polymorphs of enstatite.

Fig. 11. Observed lattice correspondence giving 13.3° macroscopic shape change.[33]

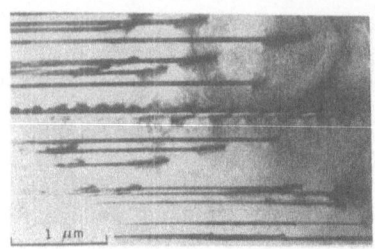

Fig. 12. TEM micrograph showing lamellae of CE growing in OE. Strain contrast at the tips is associated with "transformation partials" of 5/6 [001].

NICKEL SULPHIDE (NiS), MILLERITE, GLASS COMPONENT

The NiS phase diagram is relatively complex due to the formation of several intermediate compounds.[40] Of these, stoichiometric NiS occurs as two polymorphs:

$$\beta-NiS \xrightarrow{\quad 379^0C \quad} \alpha-NiS$$

β-NiS   α-NiS

rhombohedral   hexagonal

(millerite)   (NiAs-type)

The $\alpha \rightarrow \beta$ transformation is accompanied by a volume increase of the order of 2.3%[41] to 4.0%.[40]

The transformation was responsible for the spontaneous fracture of thermally tempered plate glass.[42-45] During processing, spherical or ellipsoidal NiS inclusions were metastably retained in the high temperature α-form at room temperature. The transformation of 80-110 μm diameter particles to the β phase caused destruction of the body, when the particles were located in the internal stress field of the glass. The accompanying volumetric expansion nucleated microcracks which propagated.

Fracture mechanics considerations[43,44] suggested that there were two critical particle sizes for transformation. In the annealed glass the critical size above which nucleation of microcracks could occur was 32 μm diameter experimentally determined as compared to 22 μm theoretically predicted. In tempered glass this was modified by the location of the inclusion in the body which reflected the different levels of stress state. The problem in tempered glass was solved by discouraging the NiS transformation by chemically doping with NiSe or NiAs.[41] Alternatively, the internal partial pressure of oxygen was increased during processing to encourage the formation of Ni_7S_6, Ni_3S_2 or even Ni, and so to avoid the NiS phase field.

In passing, it is interesting to note that the rare earth borates (eg. $LuBO_3$) also exhibit the volume increase anomaly on cooling. This occurs when high-temperature, vaterite-type polymorphs transform to low-temperature, less dense, calcite-type structures.[46] It is a matter of speculation as to whether this transformation may be beneficial when dispersed in chemically-compatible, boron-containing compounds such as glass.

DICALCIUM SILICATE ($2CaO.SiO_2$), BELITE, A CEMENT COMPONENT

Major components of cement are alite ($3CaO.SiO_2$), belite ($2CaO.SiO_2$) and tricalcium aluminate ($3CaO.SiO_2$). Belite has five polymorphs at ambient pressures:[47-54]

$$Melt \xrightarrow{2150^0C} \alpha \xrightarrow{1425^0C} \alpha'_H \xrightarrow{1177^0C} \alpha'_L \xrightarrow{675^0C} \beta \xrightarrow{525^0C} \gamma$$

hexagonal   orthorhombic   orthorhombic   monoclinic   orthorhombic

It could be said that transformation toughening was first discovered as "transformation weakening" when it was recognized that fracture and degradation of steatite ceramics resulted from the ~5.5% negative volume charge accompanying the PE→CE transformation.[32] The problem was avoided by keeping below a critical grain size of ~7 μm, so that apparently a particle size effect was in operation.[30-32] This adverse affect of a negative volume change was recently confirmed[34] in pure, synthetic protoenstatite, and PE dispersed in cordierite which shared a phase boundary in the $MgO-SiO_2-Al_2O_3$ ternary system. Chemical stabilization of PE was achieved by 1-2 mole% additions of MnO, while MgF_2 and LiF acted as mineralizers in PE formation and affected the PE → CE transformation.[35] Thermal expansion coefficients of the enstatites are of similar magnitudes to those in ZrO_2 being approximately 9.8×10^{-6} (PE), 12.0×10^{-6} (OE) and 13.5×10^{-6} (CE) up to 1000°C.[35]

In the OE → CE transformation which is achievable by grinding at room temperature,[36,37] the relative contributions of volume change versus shape change to transformation toughening may be explored. While the volume change is essentially zero, the unit cell shape change (β= 90° in OE to β= 108.3° in CE) is formally double that of ZrO_2 where (Δβ=9⁰).[13] In practice due to the lattice correspondence adopted,[33] the macroscopic shape change is 13.3° in the absence of lattice invariant shear.

The microstructure of a natural enstatite ($Mg_{0.854}$ $Fe_{0.141}$ $Ca_{0.005}$) SiO_3, i.e. bronzite, was examined by transmission electron microscopy at various stages of high pressure deformation at a constant strain rate.[38,39] The OE→CE transformation occurred by propagation of "transformation partials" originating from [001] dislocations dissociating according to:

[001]   →   5/6 [001]   +   1/6 [001]

dislocation   transformation partial   partial

Strain contrast originating from transformation partials was visible in TEM micrographs (Fig. 12).[38] The model of Fig. 13 was deduced, and it illustrates how shearing of half the OE unit cell by 5/6 is able to convert OE to CE.[38]

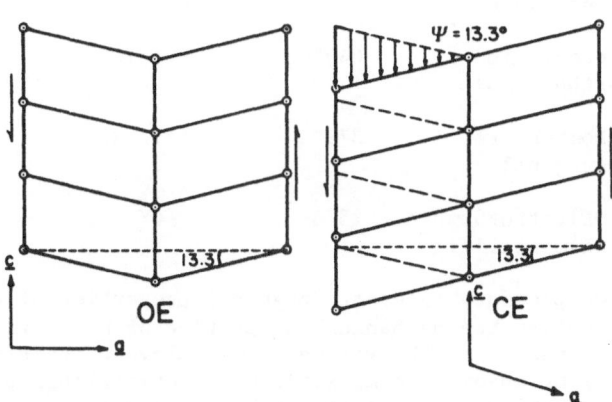

Fig. 13. Twinned relationship between OE and CE unit cells. Transformation can be achieved by dissociation of [001] dislocations into glissile 5/6 [001] partials and sessile 1/6 [001] partials.

Microstructural studies by TEM of the α→β sequence of transformation products suggested that displacive, possibly martensitic mechanisms had occurred.[52-54] Preliminary observations revealed internally twinned plates associated with $\alpha \rightarrow \alpha_H^{\prime}$ and $\alpha_L^{\prime} \rightarrow \beta$.

The dusting phenomenon of Portland cement is due to the β → γ transformation which is accompanied by an ~ 9.4% volume increase on cooling. The monoclinic unit cell (in which β= 94.5⁰)changes shape to orthorhombic where β = 90⁰. Uncertainty of composition and hence exact lattice parameters have retarded further quantitative analysis. The β → γ transformation shares the delayed transformation and fracture characteristics of steatite and tempered glass due to the enstatite and millerite transformations respectively. In each system, slow residual crack growth may provide the stresses needed to overcome the martensitic nucleation barrier as described in the first section of this paper.

Most recently, the toughening effect of the β → γ transformation was demonstrated.[55] Within the $CaO-SiO_2-ZrO_2$ ternary phase diagram ~15 vol% dispersions of β-phase particles were able to increase the toughness of a calcium zirconate $CaO.ZrO_2$ matrix by 2-2.5 times.

CONCLUSION

In summary, the current status of experimental investigation into displacive, possibly martensitic phase transformations in non-metallic systems has been reviewed. Elucidation of the transformation crystallography and three-dimensional interaction of elastic strain fields in confined zirconia particles has led to a new model of martensitic nucleation. It is proposed to occur at sites of anisotropic strain such as at corners and edges of particles. In addition, the work has questioned the applicability of bulk martensite theories to predict the crystallography of confined particle mechanisms. The experimental observations imply that matrix constraining forces also need to be considered.

Table 2. Summary of Possible Transformation Tougheners alternative to ZrO_2

Compound	Crystal Symmetries	Transformation Temperature (T_o)	Volume Change (ΔV)	Unit Cell Shape Change
ZrO_2	tetragonal → monoclinic	950°C	(+) 4.9% (R.T.)	9°
$2CaO.SiO_2$	monoclinic → orthorhombic	640°C	(+) 10%	4.6°
NiS	rhombohedral → hexagonal	379°C	(+)2.3%	--
$MgO.SiO_2$	(OE)orthorhombic → (CE)monoclinic	25°C	~0%	18.3°

Table 2 above presents relevant physical properties of other non-metals which can effect the mechanical properties of a material. It is suggested that the nature of the volume change plays a decisive role here. The variety of chemical compositions and crystallography evident opens up an area in which little research has been done, but which has great potential scientific relevance as well as technological application.

234

ACKNOWLEDGMENT

This work was supported by the Division of Materials Sciences of the U.S. Department of Energy under contract DE-AC02-76ER01198.

REFERENCES

1. J.W. Christian, Theory of Transformations in Metals and Alloys, Pergamon Press 2nd Edit. (1975).

2. M.Cohen, G.B. Olson and P.C. Clapp, pp. 1-11 Proc. Int. Conf. Martensitic Transformations, (ICOMAT), M.I.T., (1979).

3. M.Cohen and C.M. Wayman in Treatises in Metallurgy, ed. J.K. Tien and J.F. Elliott, (1981).

4. R.E.Newnham, Crystal Chemistry of Non-Metallic Materials, Springer-Verlag (1975).

5. W.M. Kriven, "Shear Transformations in Inorganic Materials" (Review), pp. 1507-32 Proc. An Int. Conf. on Solid→Solid Phase Transformations, Edited H.I. Aaronson, D.E. Laughlin, R.F. Sekerka and C.M. Wayman, Pittsburgh, (1982).

6. H.D. Megaw, pp. 86-89 Crystal Structures: A Working Approach, W.B. Saunders Publ. London (1973).

7. Proc. First Int. Conf. on Science and Technology of Zirconia, Advances in Ceramics, Vol. 3, Edited by A.H. Heuer and L.W. Hobbs. The American Ceramic Society, Columbus, OH, (1981).

8. A.H. Heuer, N. Claussen, W.M. Kriven and M. Rühle, J. Am. Ceram. Soc. 65 [12] 642-650 (1982).

9. M. Rühle and W.M. Kriven, Ber. Bunsenges. Phys. Chem. 87, 222-228 (1983).

10. A.G. Evans, pp. 193-212 in Advances in Ceramics, Vol 12, edited by N. Claussen, M. Ruhle and A.H. Heuer. The American Ceramic Soc., Columbus, OH, (1984).

11. W.M. Kriven, pp. 64-77 in Advances in Ceramics, Vol. 12, Edited by N. Claussen, M. Ruhle and A.H. Heuer. The American Ceramic Soc., Columbus, OH, (1984).

12. W. Mader, W.M. Kriven and M. Rühle, to be published.

13. W.M. Kriven, W.L. Fraser and S.W. Kennedy, pp. 82-97 in Advances in Ceramics, Vol. 3, edited by A. H. Heuer and L.W. Hobbs. The American Ceramic Society, Columbus, OH, (1981).

14. A.G. Evans, N. Burlingame, M. Drory and W.M. Kriven, Acta Metall. 29, 447-56 (1981).

15. M. Rühle and W.M. Kriven, pp. 1569-1573 in Proc. An. Int. Conf. on Solid →Solid Phase Transformations, edited by H.I. Aaronson, D.E. Laughlin, R.F. Sekerka and C.M. Wayman, AIME, Pittsburgh, (1982).

16. M.A. Choudhry and A.G. Crocker, pp. 46-53 in Advances in Ceramics, Vol. 12, edited by N. Claussen, M. Ruhle and A.H. Heuer. The American Ceramic Soc., Columbus, OH, (1984).

17. W.M. Kriven, pp. 168-83 in Advances in Ceramics, Vol. 3, edited by A.H. Heuer and L.W. Hobbs. The American Ceramic Soc., Columbus, OH, (1981).

18. E. Bischoff and M. Rühle, J. Am. Ceramic Soc. 66 [2] 123-27 (1983).

19. P.M. Kelly and C. J. Ball, J. Am. Ceram. Soc. (1985), in press.

20. M. Rühle and A.H. Heuer, pp. 14-32 in Advances in Ceramics, Vol. 12, edited by N. Claussen, M. Ruhle and A.H. Heuer. The American Ceramic Soc., Columbus, OH, (1984).

21. First proposed by A. G. Evans in Ref. 5.

22. M.J. Bevis and P.S. Allen, (Review) Sur. Defect Prop. Solids 3, 93-131, (1974).

23. F.C. Frank, A. Keller and A. O'Connor, Philos. Mag. 3, 64-73 (1958).

24. P. Allan, E.B. Crellin and M. Bevis, Philos. Mag. 27, 127-145 (1973).

25. P. Allan and M. Bevis, Proc. Roy. Soc. London A 341, 75-90 (1974).

26. P. Allan and M. Bevis, Philos. Mag. 31 [5], 1001-9 (1975).

27. A.F. Acton, M. Bevis, A.G. Crocker and N.D.H. Ross, Proc. Roy. Soc. London A 320, 101-113 (1970).

28. M. Bevis and E.B. Crellin, Polymer 12, 666-684 (1971).

29. R.S. Coe, Contrib. Mineral. and Petrol. 26 247-264 (1970).

30. H. Thurnauer and A.R. Rodriguez, J. Am. Ceram. Soc. 25, [15], 443-450 (1942).

31. M.D. Rigterink, J. Am. Ceram. Soc. 30 [7], 214-218 (1947).

32. E.C. Bloor, J. British Ceramic Soc. (1964), [1-2], 309-316.

33. R.S. Coe and W.F. Muller, Science 180, 64-66 (1973).

34. B.H. Mussler, unpublished work (1984).

35. J.F. Sarver and F. H. Hummel, J. Am. Ceram. Soc. 45, [4], 152-156 (1962).

36. R.E. Riecker and T. P. Rooney, Geol. Soc. Amer. Bulletin, 78, 1045-1054, (1967).

37. C.B. Raleigh, S. H. Kirby, N. L. Carter and H. G. Ave Lallemant, J. Geophy. Reasearch 76, [17], 4011-4022 (1971).

38. R.S. Coe and S. K. Kirby, Contrib. Mineral. Petrol 52, 29-55 (1975).

39. S.H. Kirby, in _Electron Microscopy in Mineralogy_, Edited by H. R. Wenk, Springer Verlag, 465-472 (1976).

40. G.Kullerud and R.A. Yund, _G. Petrol. 3_, 126-175 (1962).

41. L. Merker, _Glastechn. Ber. 47_, [6] 116-121 (1974).

42. E.R. Ballantyne, C.S.I.R.O. Division of Building Research, Melbourne, Australia, Report 061-5 (1961).

43. M.V. Swain, _J. Mater Sci. 16_, 151-158 (1981).

44. C.C. Hsiao, _Fracture 1977, 3_, ICF4, Waterloo, Canada, 985-989 (1977).

45. R.Wagner, _Glastechn. Ber. 50_, [11], 296-300 (1977).

46. E.M. Levin, R.S. Roth, J.B. Martin, _Amer. Mineral 46_, 1030-1055 (1961).

47. H.E.Schwiete, W. Kronert and K. Deckert, _Zement, Kalk, Gips 9_ (in German) 359-366 (1968).

48. A.Guinier and M.Regourd, Principal Paper I, _V-ISCC Tokyo_ (1968), _1_, 1-41 (1969).

49. M. Regourd and A. Guinier, Principal Paper I, Proc. _VI Int. Congr. Chemistry of Cement_, held in Moscow 1-82 (1974).

50. H. Midgley, ibid, Suppl. Paper, Sect I, 4-14 (1974).

51. S.N. Ghosh, P. Bhaskara Rav, A.K. Paul and K. Raina, _J. Mater. Sci 14_, 1554-1566 (1979).

52. J.W. Groves, _J. Mater. Sci 16_, 1063-1070 (1981).

53. J.W. Groves, _J. Mater. Sci. 18_, 1615-24 (1983).

54. J.W. Groves, _Cement and Concrete Research 12_, 619-624 (1982).

55. J.S. Moya, P. Pena and S. de Aza, _J. Am. Ceram. Soc. 68_ [9], C259-262 (1985).

ON PRECIPITATE MORPHOLOGY IN ZrO_2, α-Al_2O_3 AND $FeTiO_3$ MATRICES

K.P.D. Lagerlöf, V. Lanteri and A.H. Heuer

Department of Metallurgy and Materials Science
Case Western Reserve University
Cleveland, Ohio 44106

ABSTRACT

Precipitate morphology in ceramic matrices depends on minimization of strain and interfacial energies between precipitates and matrix. The strain energy is determined by the mismatch in both lattice parameters and elastic constants, and can readily be calculated using an elegant theory due to Khachaturyan. We have applied this theory to precipitation of tetragonal zirconia ($\underline{t}$-ZrO_2) in Mg, Ca and Y partially-stabilized ZrO_2's (PSZ's), to precipitation of TiO_2 and $Fe_{2-x}Ti_xO_3$ (x=0;1/2;1) in α-Al_2O_3 and to precipitation of Fe_2O_3 in $FeTiO_3$. Calculated precipitate morphologies are compared with experimental observations.

KEY WORDS: Strain energy, Precipitate, Morphology, PSZ, Zirconia (ZrO_2), Sapphire (α-Al_2O_3), Ilmenite ($FeTiO_3$), Hematite (Fe_2O_3), Rutile (TiO_2).

INTRODUCTION

It is now well established that a ceramic compound can be toughened by the addition of small amounts of zirconia in the form of tetragonal ($\underline{t}$-ZrO_2) particles [1-7], utilizing the so called transformation-toughening mechanism, during which the metastable $\underline{t}$-ZrO_2 particles transform to monoclinic symmetry ($\underline{m}$-ZrO_2) in the presence of a stress field [2]. The mechanism was first observed in partially stabilized zirconia (PSZ) [1-4], in which the cubic phase ($\underline{c}$-ZrO_2) is stabilized through the addition of aliovalent solutes, i.e. MgO, CaO and Y_2O_3. During aging in the two phase field, $\underline{t}$-ZrO_2 particles precipitate and grow in the cubic matrix, where they transform into monoclinic symmetry in crack tip stress fields.

Although the $\underline{t}$-ZrO_2 precipitates are coherent with the $\underline{c}$-ZrO_2 matrix in all PSZ systems, transmission electron microscopy (TEM) show quite different precipitate morphologies depending on the stabilizing solute (Fig.1). The morphology in Mg-PSZ (Fig.1a) consists of oblate sphereoids with $\{001\}_c$ habit planes. Electron diffraction studies indicate $[001]_t$ to be parallel to $\langle 001 \rangle_c$ and that $[001]_t$ is normal to the habit plane, giving rise to three possible precipitate variants. In Ca-PSZ (Fig.1b) the precipitate morphology consists of more equiaxed particles with $\{101\}_c$ habit planes, and electron diffraction again shows $[001]_t$ to be parallel to $\langle 001 \rangle_c$. A third type of morphology is observed in Y-PSZ (Fig.1c), consisting of twinned lamellae in a "colony" structure. Two tetragonal variants with the $\vec{c}$-axis parallel to $[100]_c$ and $[001]_c$ share the same $(101)_c$ habit plane.

Rühle and Heuer [8] attempted to explain the different precipitate morphologies observed in the various PSZ systems. The analysis was based on the lattice parameters of the cubic and tetragonal phases and the volume changes during precipitation. Their success was, however, limited and gave only a qualitative understanding of the differences in the precipitate morphology.

The precipitate morphology represents the minimum energy configuration taking both the strain and interfacial energies into account. During the early stages of the precipitation process, the strain energy is in general small, and the habit plane will most likely correspond to the plane with the lowest interfacial energy, usually a low index plane. As the precipitate grows, the strain energy component becomes increasingly important, and could cause a change in the habit plane. The precipitate morphology will then be such that the total energy is minimized keeping the volume constant (i.e. minimizing the interfacial area at constant volume gives rise to a spherical precipitate in an isotropic matrix, provided the interfacial energy also is isotropic).

Khachaturyan [9,10] developed an elegant method to calculate the strain energy of a coherent precipitate in a matrix, as a function of the elastic constants of the matrix, the mismatch in lattice parameters between the matrix and the precipitate, the precipitate morphology, and the precipitate distribution. The theory can easily be applied to matrices with cubic symmetry [11] and Mayo and Tsakalakos [12] worked out an analytical expression that also can be applied to materials with hexagonal, tetragonal and orthorhombic symmetries. An analytical expression for crystals with lower symmetries, such as trigonal, triclinic and monoclinic, is straight-forward, although quite tedious to obtain. However, the strain energy can easily be calculated for a matrix of any symmetry using numerical methods.

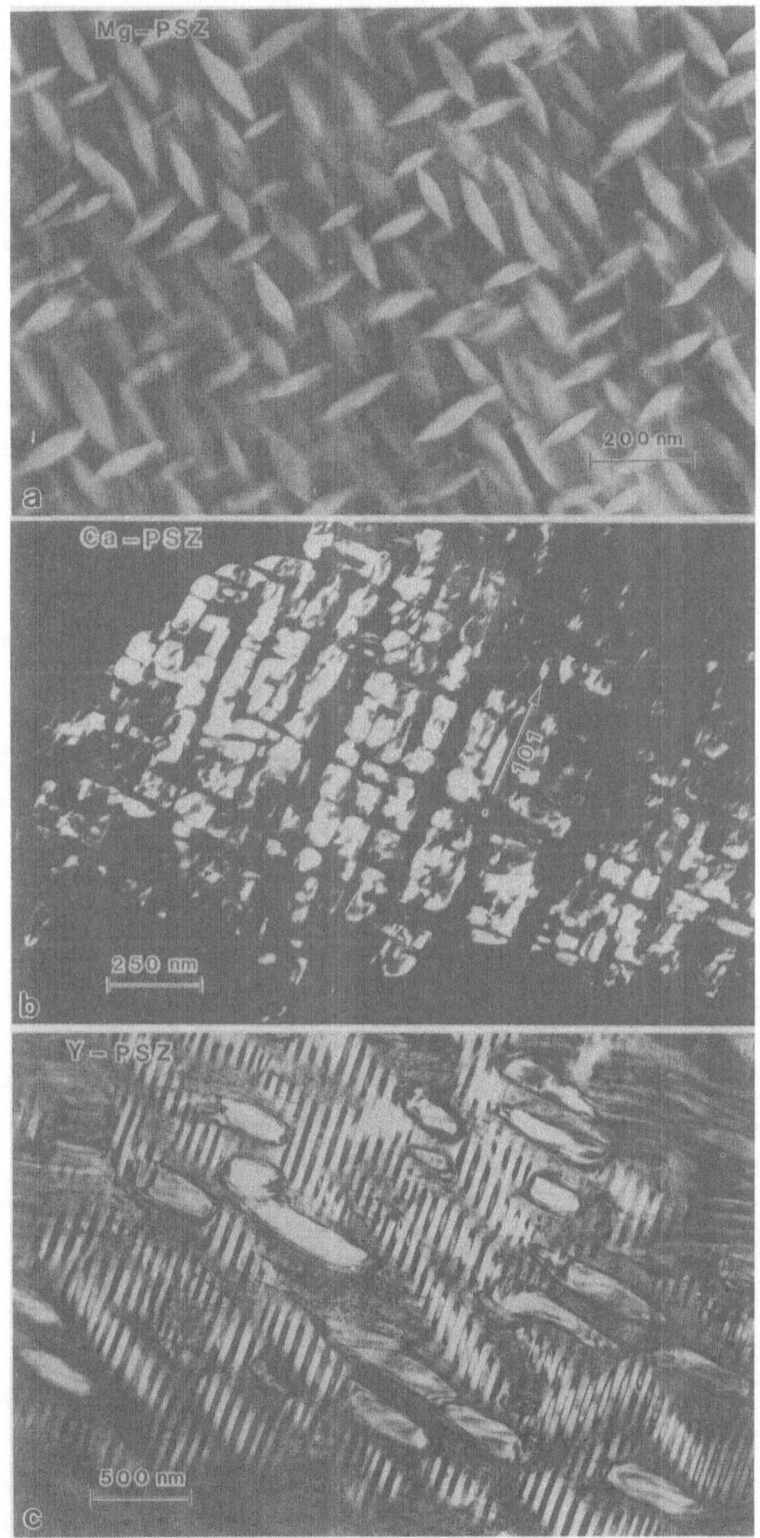

Figure 1. Precipitate microstructures in Mg-PSZ (a), Ca-PSZ (b), and
Y-PSZ (c).(from Ref.[11]).

Khachaturyan [10] further points out that degeneracy of directions corresponding to the minimum strain energy may affect the precipitate morphology. If all directions in a plane (hkl) give rise to equal minimum strain energy, the precipitate morphology would be a needle, with the needle direction perpendicular to (hkl). From a crystallographic point of view, if all the planes of a zone are potential habit planes because they give rise to the same minimum strain energy, then a needle will form along the zone axis. On the other hand, if only one direction, i.e. one habit plane normal $\vec{n}_o$, corresponds to the minimum strain energy, and the strain energy for all directions perpendicular to $\vec{n}_o$ are equal, i.e. the plane (hkl) is isotropic, then the precipitate morphology would correspond to an oblate sphereoid with the shortest dimension along $\vec{n}_o$. For intermediate situations, other morphologies may be possible, such as parallelepipeds, cuboids, etc., depending on the relative values of the strain energy for the various directions $\vec{n}$.

The theory by Khachaturyan [10] has been used in this investigation to calculate the strain energy per unit volume for a variety of habit planes in the Mg, Ca and Y-PSZ systems. In order to test the applicability of the theory to systems with semi-coherent precipitates, the theory was also applied to rutile (TiO_2) precipitates in sapphire (α-Al_2O_3)(Fig.2a), precipitates from the ilmenite-hematite series ($Fe_{2-x}Ti_xO_3$) in sapphire (Fig.2b), and hematite precipitates in ilmenite (Fig.2c).

THEORY

The theory developed by Khachaturyan [9,10] to calculate the strain energy between a coherent precipitate with an arbitrary shape in a homogenous anisotropic matrix, is based on the simple model first used by Eshelby [13,14]. Six conceptually distinct steps are involved:

1. Cut out small clusters of material from the matrix.
2. Allow these clusters to transform under stress-free conditions.
3. Apply surface stresses to the transformed clusters to recover the shape before transformation.
4. Insert the clusters into the voids of the matrix to form precipitates.
5. Weld the precipitates to the matrix and recover coherency.
6. Allow the system to relax.

The elastic strain energy arising from steps 3 and 6 can be calculated assuming that the elastic constants are equal in both the precipitate and the matrix, and is given by

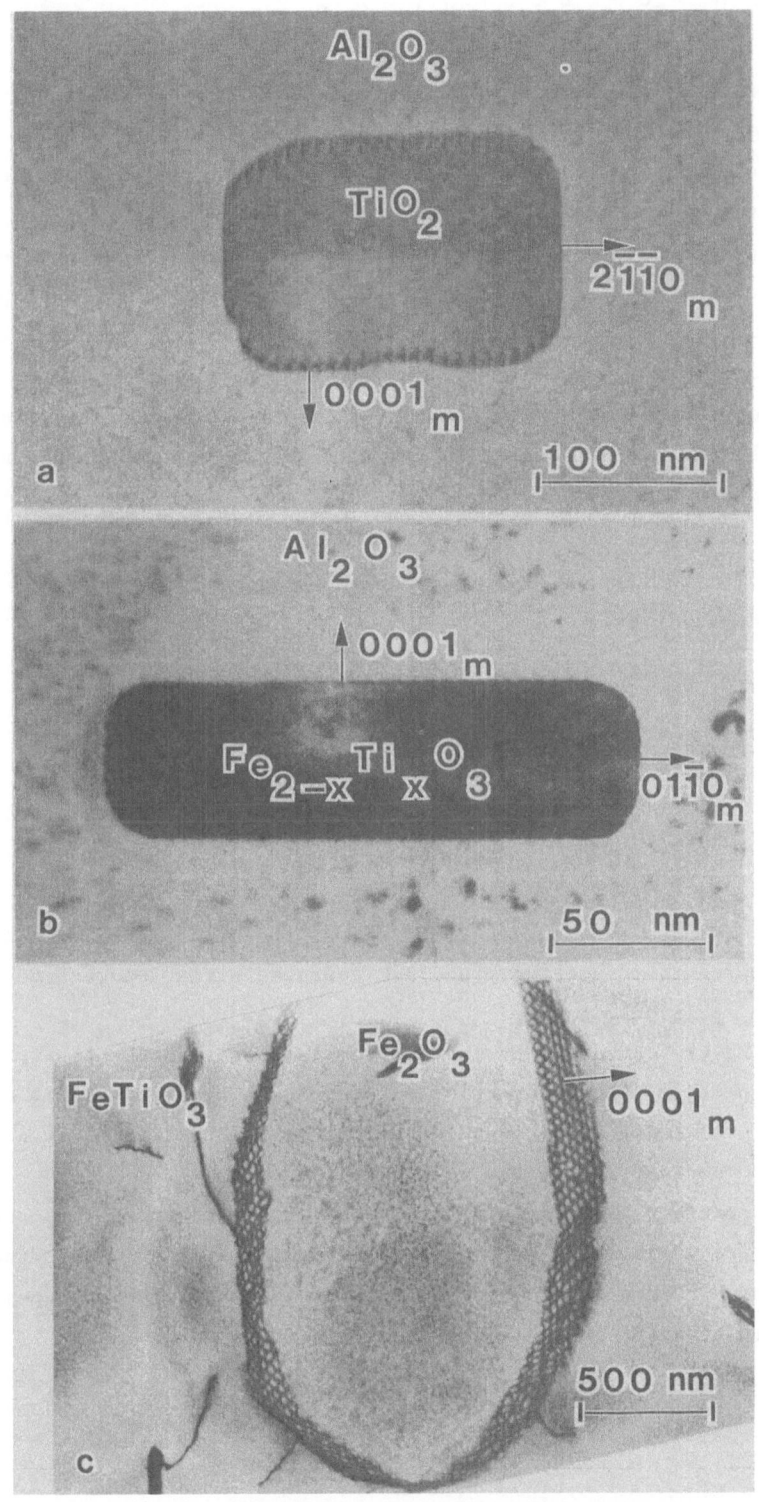

Figure 2. Precipitate microstructures of TiO_2 in α-Al_2O_3 (from Ref.[23]) (a), $Fe_{2-x}Ti_xO_3$ in α-Al_2O_3 (from Ref.[22]) (b), and Fe_2O_3 in $FeTiO_3$ (from Ref.[26]) (c).

$$E_{el} = \frac{1}{2} \sum_p V_p \lambda_{ijkl} \varepsilon_{ij}^o(p) \varepsilon_{kl}^o(p) + \int_V dv \left[-\sum_p \sigma_{ij}^o(p) \Theta_p(\vec{r}) \varepsilon_{ij} + \frac{1}{2} \lambda_{ijkl} \varepsilon_{ij} \varepsilon_{kl} \right] \qquad (1)$$

where V_p is the volume of type p precipitates, V is the total volume of transformed phase, λ_{ijkl} is the elastic constant tensor, $\Theta_p(\vec{r})$ is the shape function of type p precipitates (equal to unity inside and zero outside the precipitate), ε_{ij}^o is the stress-free strain tensor and ε_{ij} is the actual strain experienced by the precipitate after the relaxation process. The actual strains, ε_{ij}, are unknown for an arbitrary precipitate shape, and Eq.(1) can only be applied to ellipsoidal morphologies for which ε_{ij} can be calculated [13,14].

However, Khachaturyan [10] found that the actual strains could be obtained by the use of Fourier techniques, giving the following expression for the strain energy,

$$E_{el} = \frac{1}{2} \sum_{p,q} \int\!\!\int\!\!\int_{-\infty}^{+\infty} \frac{d^3\vec{k}}{(2\pi)^3} \left[\lambda_{ijkl} \varepsilon_{ij}^o(p) \varepsilon_{kl}^o(q) - n_i \sigma_{ij}^o(p) \Omega_{jk}(\vec{n}) \sigma_{kl}^o(q) n_l \right]$$
$$\times \Theta_p(\vec{k}) \Theta_q^*(\vec{k}) \qquad (2)$$

where $\vec{n} = \vec{k}/|\vec{k}|$ is a unit vector in reciprocal space, $\Omega_{ij}(\vec{n})$ is the inverse tensor to

$$\Omega_{ij}^{-1} = \lambda_{iklj} n_k n_l \qquad (3)$$

and

$$\sigma_{ij}^o = \lambda_{ijkl} \varepsilon_{kl}^o \qquad (4)$$

$\Theta(\vec{k})$ and $\Theta^*(\vec{k})$ are the Fourier transform of the shape function and its complex conjugate respectively and $d^3\vec{k}$ represents the volume integration element in reciprocal space.

Since Eq.(2) contain a summation over various precipitate types or variants, p and q, the interaction between these precipitates is accounted for and it is always possible to solve Eq.(2) if the details of a multi-phase microstructure are well established.

In the present investigation, our interest is to understand the morphology of an isolated precipitate which minimizes the total strain energy; this in turn should provide information about the prefered habit plane. Equation (2) for this case can be written as

$$E_{el} = \frac{1}{2} \int\!\!\int\!\!\int_{-\infty}^{+\infty} \frac{d^3\vec{k}}{(2\pi)^3} \, B(\vec{n}) |\Theta(\vec{k})|^2 \qquad (5)$$

where

$$B(\vec{n}) = \lambda_{ijkl} \varepsilon^o_{ij} \varepsilon^o_{kl} - n_i \sigma^o_{ij} \Omega_{jk}(\vec{n}) \sigma^o_{kl} n_l \qquad (6)$$

is the strain energy per unit volume in the reciprocal lattice direction $\vec{n}$ and

$$|\Theta(\vec{k})|^2 = |\int_V dv \exp(-i\vec{k}\vec{r})|^2 \qquad (7)$$

Now, since

$$E_{el} > \frac{1}{2} B(\vec{n}_o) V \qquad (8)$$

where $B(\vec{n}_o)=B_{min}(\vec{n})$, the equality in Eq.(8) holds for a thin platelet whose normal is $\vec{n}_o$. Thus, if the strain energy contribution to the total energy dominates over the interfacial energy, the habit plane would then be normal to the vector $\vec{n}_o$ which minimizes $B(\vec{n})$.

CALCULATIONAL PROCEDURE

It can be seen from Eq.(6) that $B(\vec{n})$ is a function of the elastic constant tensor, λ_{ijkl}, the stress-free strains, ε^o_{ij}, and the vector $\vec{n}$. As already noted, Mayo and Tsakalakos [12] developed an expression for $B(\vec{n})$ that can be used for materials with isotropic, cubic, hexagonal, tetragonal or orthorhombic symmetries. For crystals with lower symmetries, i.e. sapphire with trigonal symmetry, $B(\vec{n})$ was obtained by using numerical methods, which only requires various summations and a simple inversion of a (3x3) matrix. A computer routine was developed for a general system defined by λ_{ijkl} and ε^o_{ij}, which calculates $B(\vec{n})$ for either specific directions $\vec{n}=[u,v,w]/(u^2+v^2+w^2)^{1/2}$ or scanned directions within a plane (h,k,l) where $\vec{n}$ was defined using spherical coordinates and $hn_x+kn_y+ln_z=0$. The calculated values of $B(\vec{n})$ for a variety of directions in the systems investigated are shown in Figures 4-8, where the different directions $\vec{n}$ used for the cubic and trigonal matrices, respectively, are illustrated in the streographic projections shown in Fig.3. The elastic constants used in the calculations are given in Table 1 and the stress-free strains are shown in Table 2, which were obtained from the lattice parameters in Table 3 and the orientation relations in Table 4.

TABLE 1
Elastic constants (10^{11} Pa)

MATRIX	C_{11}	C_{12}	C_{13}	C_{14}	C_{33}	C_{44}	Ref.
$\underline{c}$-ZrO$_2$	3.940	0.910	–	–	–	0.560	(15)
α-Al$_2$O$_3$	4.968	1.636	1.109	−0.235	4.981	1.474	(16)
Fe$_2$O$_3$	2.420	0.549	0.157	−0.127	2.280	0.853	(17)

TABLE 2
Stress-free strains

MATRIX	PPT	ε_{11}^{o}	ε_{22}^{o}	ε_{33}^{o}
Mg-$\underline{c}$-ZrO$_2$	Mg-$\underline{t}$-ZrO$_2$	−0.0005906	−0.0005906	0.0202756
Ca-$\underline{c}$-ZrO$_2$	Ca-$\underline{t}$-ZrO$_2$	−0.0070175	−0.0070175	0.0097466
Y-$\underline{c}$-ZrO$_2$	Y-$\underline{t}$-ZrO$_2$	−0.0031177	−0.0031177	0.0048714
α-Al$_2$O$_3$	TiO$_2$	0.046238	−0.00546	0.060967
α-Al$_2$O$_3$	FeTiO$_3$	0.067465	0.067465	0.0873144
α-Al$_2$O$_3$	Fe$_{1.5}$Ti$_{0.5}$O$_3$	0.0627364	0.0627364	0.0725111
α-Al$_2$O$_3$	Fe$_2$O$_3$	0.058008	0.058008	0.0575429
FeTiO$_3$	Fe$_2$O$_3$	−0.00886	−0.00886	−0.0273808

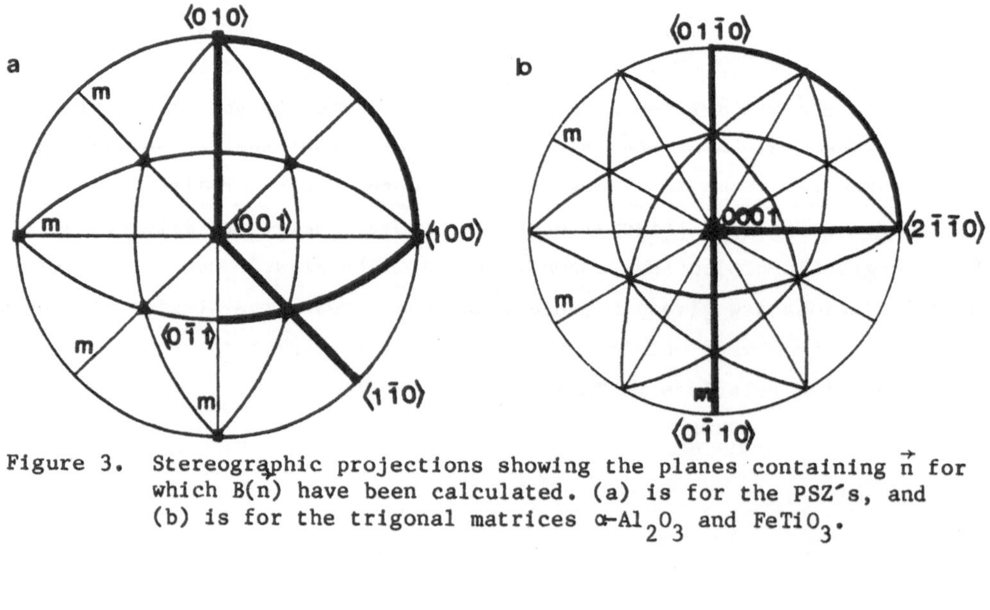

Figure 3. Stereographic projections showing the planes containing $\vec{n}$ for which $B(\vec{n})$ have been calculated. (a) is for the PSZ´s, and (b) is for the trigonal matrices α-Al$_2$O$_3$ and FeTiO$_3$.

TABLE 3
Lattice parameters (nm)

COMPOUND	cubic a_c	tetragonal a_t	c_t	trigonal* a	c	Ref.
Mg–PSZ[+]	0.5080	0.5077	0.5183	–	–	(8)
Ca–PSZ[+]	0.5130	0.5049	0.5180	–	–	(8)
Y–PSZ[+]	0.5132	0.5116	0.5157	–	–	(8)
α–Al_2O_3	–	–	–	0.4758	1.2998	(18)
Fe_2O_3	–	–	–	0.5079	1.4134	(19)
$FeTiO_3$	–	–	–	0.5034	1.3747	(19)
TiO_2	–	0.4594	0.2955	–	–	(20)

* The lattice parameters are based on a hexagonal unit cell.
+ Tetragonal lattice parameters are based on a distorted fluorite unit cell.

TABLE 4
Orientation relations

MATRIX	PPT			Ref.
$\underline{c}$–ZrO_2	$\underline{t}$–ZrO_2	$\langle 100 \rangle_m // \langle 100 \rangle_{ppt}$	$\{100\}_m // \{100\}_{ppt}$	(8)
		$\langle 010 \rangle_m // \langle 010 \rangle_{ppt}$	$\{010\}_m // \{010\}_{ppt}$	
		$\langle 001 \rangle_m // \langle 001 \rangle_{ppt}$	$\{001\}_m // \{001\}_{ppt}$	
$FeTiO_3$	Fe_2O_3	$\langle 2\bar{1}\bar{1}0 \rangle_m // \langle 2\bar{1}\bar{1}0 \rangle_{ppt}$	$\{2\bar{1}\bar{1}0\}_m // \{2\bar{1}\bar{1}0\}_{ppt}$	(21)
α–Al_2O_3	$FeTiO_3$	$\langle 01\bar{1}0 \rangle_m // \langle 01\bar{1}0 \rangle_{ppt}$	$\{01\bar{1}0\}_m // \{01\bar{1}0\}_{ppt}$	(22)
		$[0001]_m // [0001]_{ppt}$	$(0001)_m // (0001)_{ppt}$	
α–Al_2O_3	TiO_2	$(2\bar{1}\bar{1}0)_m // (011)_{ppt}$	--	(23–25)
		$[01\bar{1}0]_m // [01\bar{1}]_{ppt}$	--	
		$[0001]_m // [100]_{ppt}$	$(0001)_m // (100)_{ppt}$	

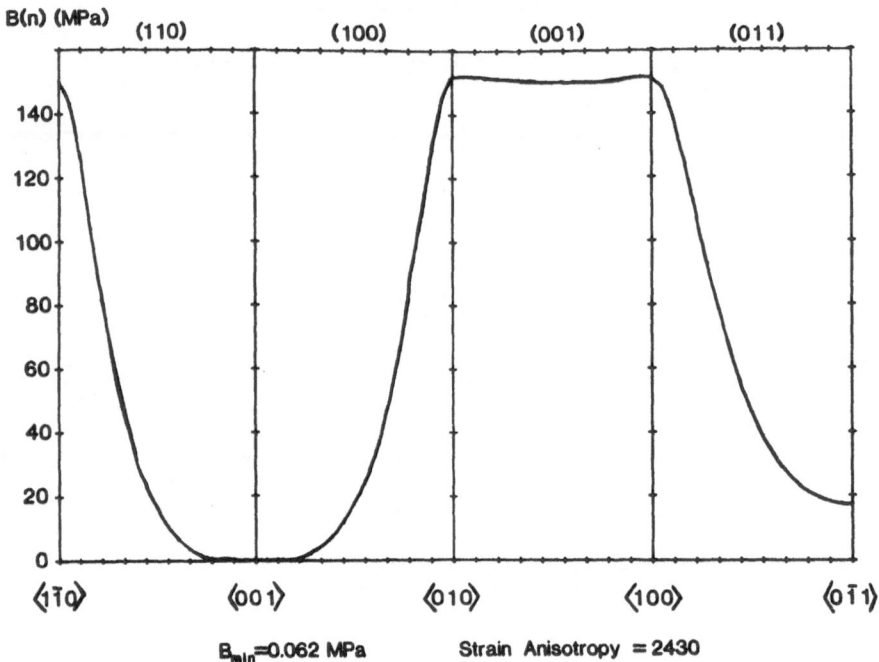

a

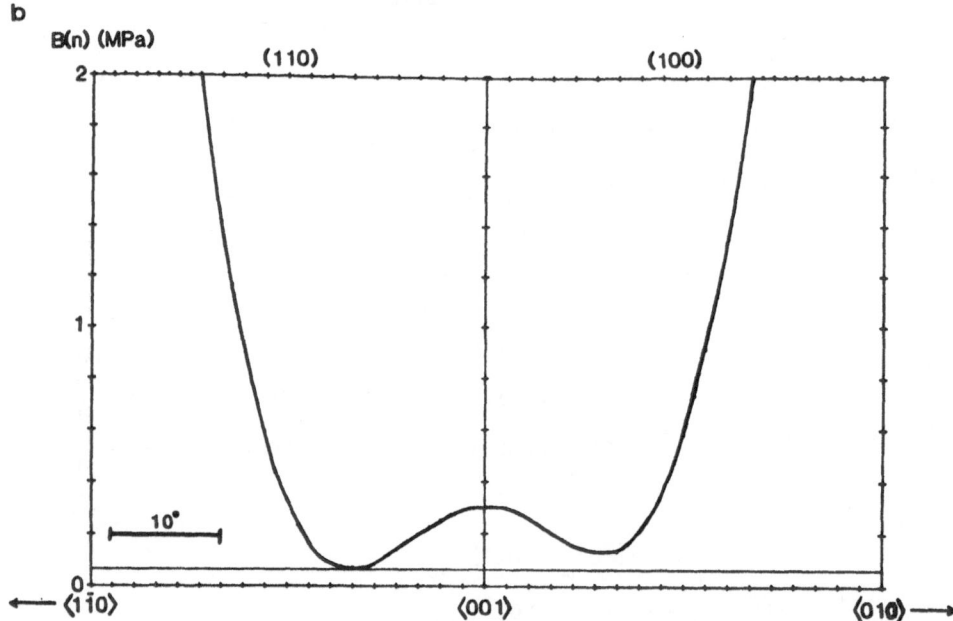

b

Figure 4.   B(n) as a function of orientation for four different planes in
Mg-PSZ. (b) shows an enlargement of the region around $\vec{n}$=<001>,
which is close to B($\vec{n}_o$).

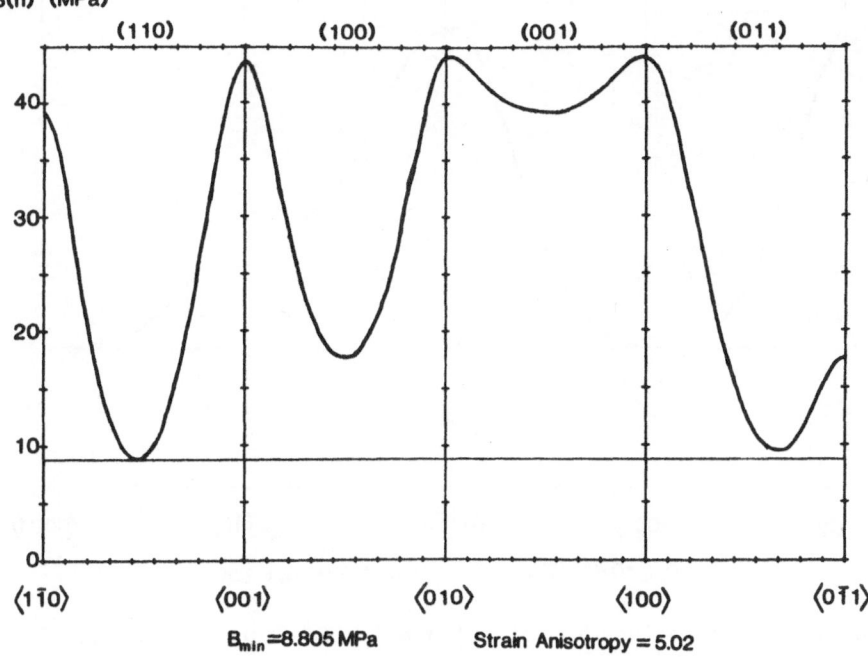

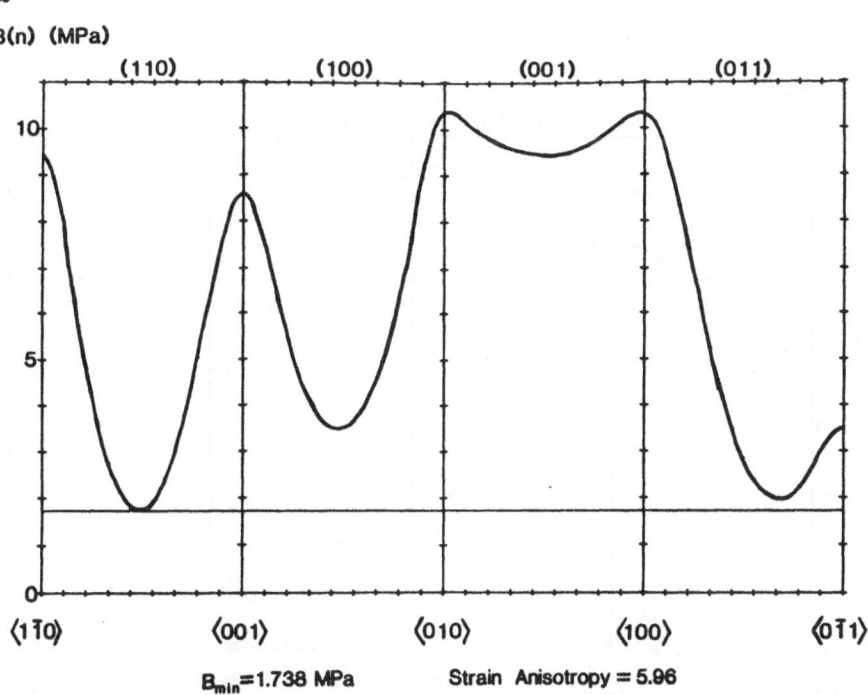

Figure 5. B($\vec{n}$) as a function of orientation for four different planes in Ca-PSZ (a), and Y-PSZ (b).

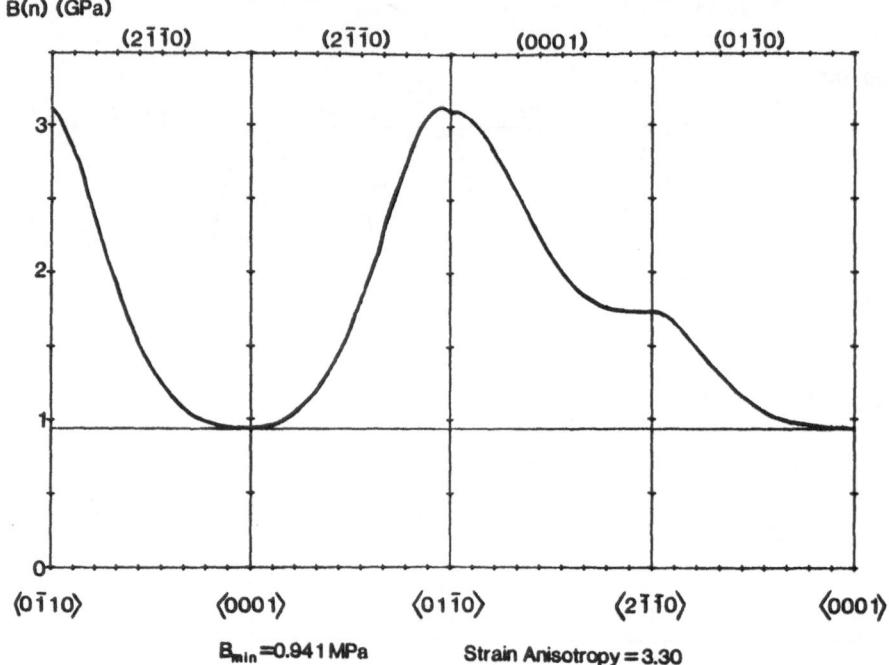

Figure 6. $B(\vec{n})$ as a function of orientation for three different planes for α-Al_2O_3 containing TiO_2 precipitates.

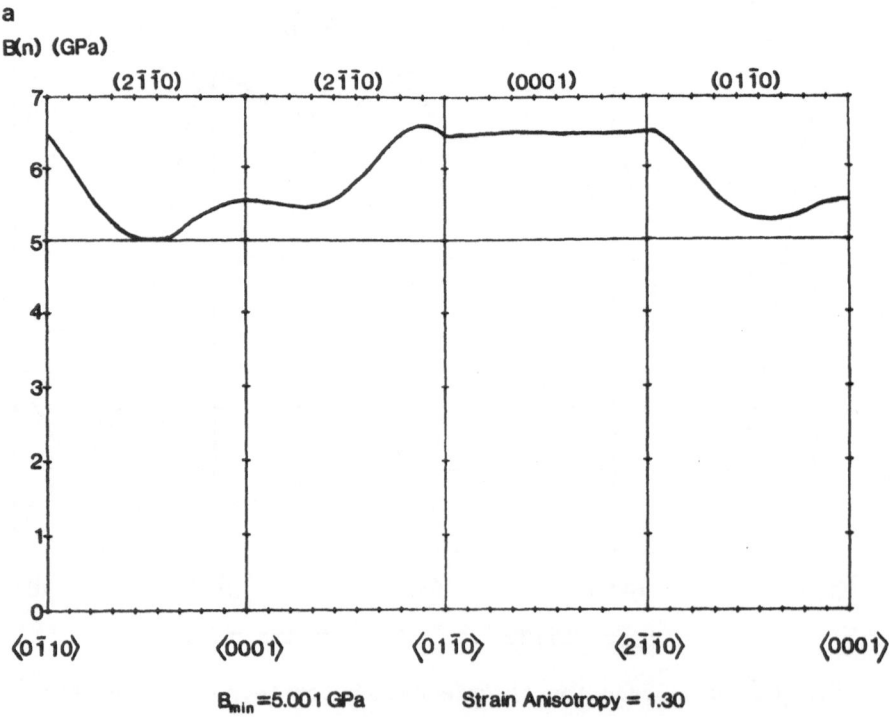

Figure 7.

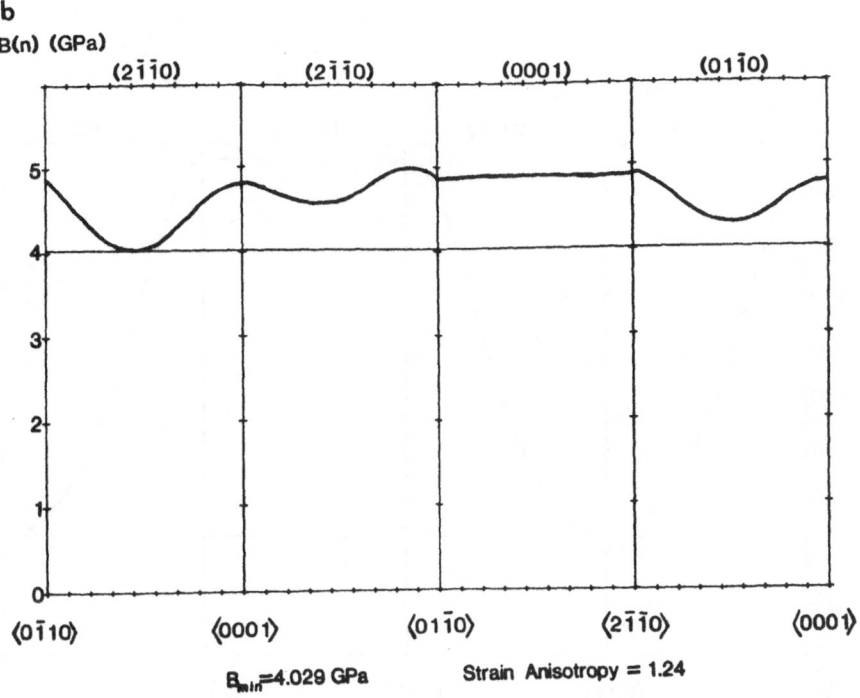

b

B(n) (GPa)

B_{min}=4.029 GPa   Strain Anisotropy = 1.24

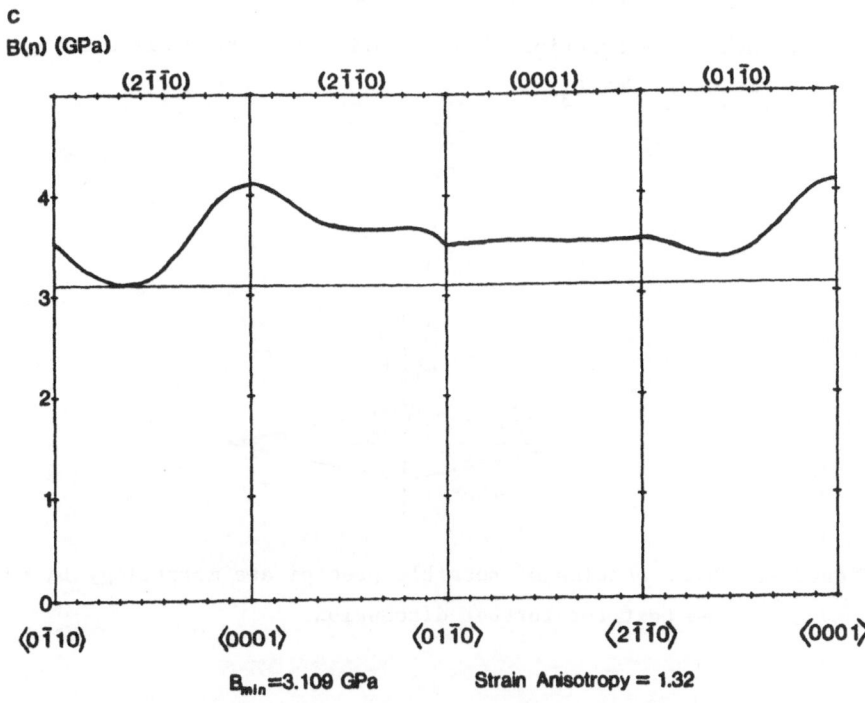

c

B(n) (GPa)

B_{min}=3.109 GPa   Strain Anisotropy = 1.32

Figure 7. B($\vec{n}$) as a function of orientation for three different planes for
α-Al$_2$O$_3$ containing FeTiO$_3$ (a), Fe$_{1.5}$Ti$_{0.5}$O$_3$ (b), and Fe$_2$O$_3$ (c).

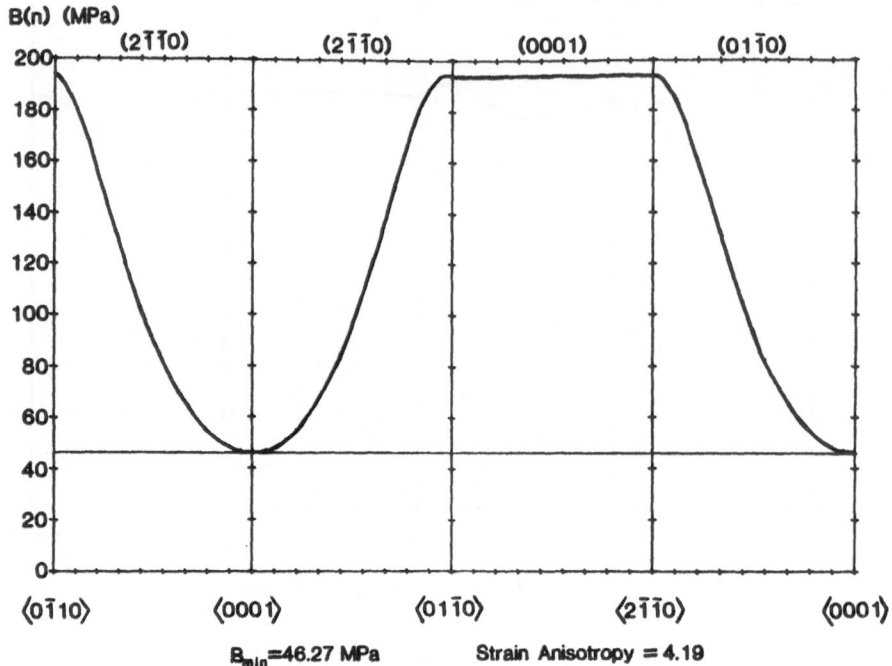

Figure 8. $B(\vec{n})$ as a function of orientation for three different planes for Fe_2O_3 in $FeTiO_3$.

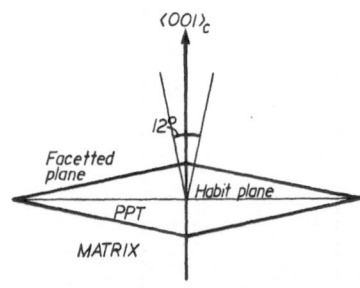

Figure 9. Cross-section of possible precipitate morphology in Mg-PSZ. See text for further discussion.

RESULTS AND DISCUSSION

Mg-PSZ

The values of $B(\vec{n})$ for directions $\vec{n}$ in the (110), (100), (001) and (011) planes of the matrix, are shown in Fig.4a. It can be seen in Fig.4a that the strain anisotropy is very large, $A=B_{max}/B_{min}=2432$, which implies that the strain energy is likely to control the precipitate morphology. Figure 4b shows an enlargement of the region around $\vec{n}=\langle 001 \rangle$, indicating that $B_{min}=B(\vec{n}_o)$ occurs for $\vec{n}_o$ perpendicular to $(2\bar{2}\,13)$, or about 12° from $\langle 001 \rangle$ in the (110) plane. Due to the large strain anisotropy, the precipitate habit plane is likely to be (001), the low index plane closest to $(2\bar{2}\,13)$. As discussed above, at the early stages of the precipitation process, the precipitate is expected to be a thin platelet on the (001) habit plane. As the platelet grows, the expected precipitate morphology would be an oblate sphereoid, with the shortest dimension along $\langle 001 \rangle$, due to the minimization of the strain and interfacial energies.

Since the minimum strain energy occurs for $(2\bar{2}\,13)$, one would expect the oblate sphereoid to be facetted. It can be seen in Fig.4b that the minimum strain energy in the (100) plane occurs for $\vec{n}$ perpendicular to (04 21), or about 11° from $\langle 001 \rangle$. By constructing a roto-symmetric precipitate with facetted planes approximately 12° from $\langle 001 \rangle$ (Fig.9), one would obtain a precipitate with an aspect ratio (diameter:thickness) close to 4.8. Both the aspect ratio and the precipiate morphology are very close to the TEM observations shown in Fig.1a.

Ca-PSZ and Y-PSZ

The strain energy per unit volume for directions in the (110), (100), (001) and (011) planes for Ca-PSZ and Y-PSZ are shown in Figures 5a and 5b, respectively. Although the magnitude of $B(\vec{n})$ differs by a factor of five (5), the relative anisotropy and shapes of the curves are very similar. The minimum values are obtained for $\vec{n}_o$ normal to (445) and (334) for Ca-PSZ and Y-PSZ respectively, both of which are close to {111}. However, during the early stages of the precipitation process, the interfacial energy is likely to control the habit plane, and TEM studies indicate that {101} habit planes are favored over {111} planes. As the precipitates grow, the strain energy becomes more important and one would expect the habit plane to change into the plane for which $B(\vec{n})$ is minimum. On the other hand, it can be seen from Fig.4 that the strain anisotropy between the minimum value, $\vec{n}_o$ and $\vec{n}=[0\bar{1}1]$ are small (about 2) for both Ca-PSZ and Y-PSZ. Thus, the anisotropy in the strain energy may be balanced or overshadowed by anisotropy in the interfacial energy. Furthermore, since each habit plane can contain two types of precipitate variants, with $[001]_t$ perpendicular to each other, interaction between the variants must be taken into account in the strain energy calculations. Khachaturyan's theory predicts for this case that the total energy can be lowered by forming layers of precipitates with one specific variant (Fig.10a). As these layers grow together, a lamellae structure can be formed, where the two variants form a twin-related "colony", as shown in Fig.10b. The observed microstructures of Ca-PSZ (Fig.1b) and Y-PSZ (Fig.1c) corresponds to the early stages and fully developed "colonies" respectively. The differences in the kinetics can partly be explained by the differences in the strain energy per unit volume, being of the order of 5 times larger for Ca-PSZ compared to Y-PSZ, resulting in more sluggish behavior for Ca-PSZ. However, the details of the precipitation interaction energies have not been investigated, and other factors, such as the diffusivity of Ca and Y in ZrO_2, may also play major roles in the differences in precipitation kinetics between the two systems.

$\alpha-Al_2O_3 - TiO_2$

The strain energy per unit volume for rutile (TiO_2) precipitates in sapphire ($\alpha-Al_2O_3$) are shown in Fig.6 for directions $\hat{n}$ in the $(2\bar{1}\bar{1}0)$, (0001) and $(01\bar{1}0)$ planes. The minimum value is obtained for $\vec{n}_o$ about 4° from [0001] towards $[0\bar{1}10]$, which is close to the normal of (0001), the expected habit plane for a thin platelet taking the interfacial energies into account. As the platelet grows, and the strain energy will become more important, one would expect the minimum energy morphology to be a parallelepiped along $\langle 01\bar{1}0 \rangle$ with a secondary habit plane corresponding to $\{2\bar{1}\bar{1}0\}$, normal to the direction with the lowest energy in the basal plane. Due to the three-fold rotation symmetry around [0001], three variants of precipitates along the three equivalent $\langle 01\bar{1}0 \rangle$ direction can form. These qualitative predictions are in full agreement with the experimental observations by Phillips et al.[23-25] (Fig.2a).

$\alpha-Al_2O_3 - Fe_{2-x}Ti_xO_3$

The strain energy per unit volume in various directions for $Fe_{2-x}Ti_xO_3$ precipitates in sapphire are shown in Fig.7a for x=1, Fig.7b for x=1/2 and Fig.7c for x=0. The minimum strain energy, $B(\vec{n}_o)$, represents planes 43°, 50° and 57° from (0001) towards $(0\bar{1}10)$ for the three cases respectively. Based on the strain energy calculations alone, one would expect the precipitate morphology to be parallelepipeds along $\langle 2\bar{1}\bar{1}0 \rangle$, where the normal

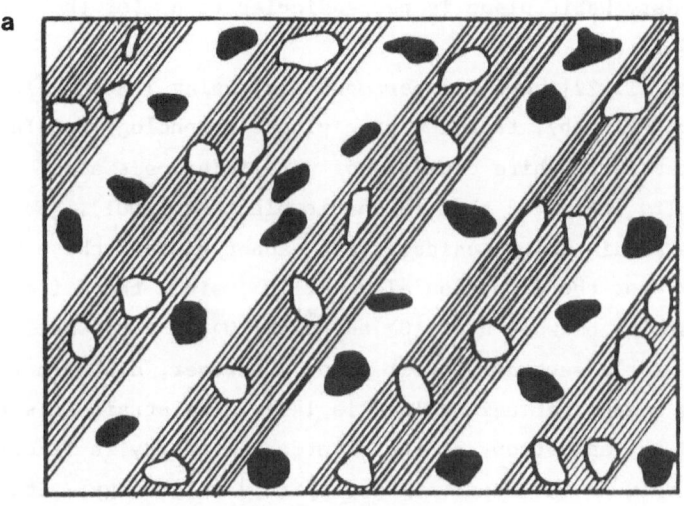

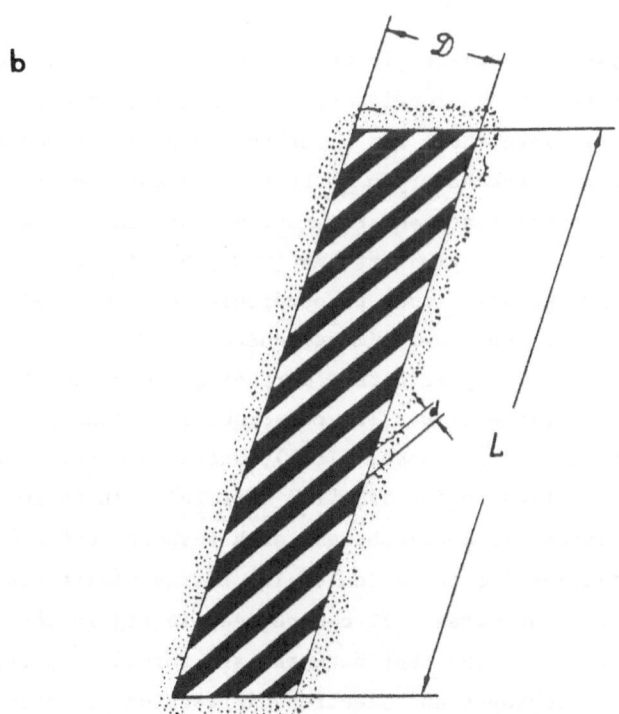

Figure 10. Schematic drawings of tetragonal precipitates in a cubic matrix showing "ordering" of precipitates due to strain-induced coarsening. (a) is a partially-ordered and (b) is a fully-ordered microstructure (from Ref.[10]).

of a secondary habit plane is perpendicular to $\vec{n}_o$ (of the type $[0h\bar{h}1]$ in the $(2\bar{1}\bar{1}0)$ plane).

Moon et al.[22] showed experimentally, using TEM and electron diffraction (Fig.2b), that the precipitate morphology for $Fe_{1.5}Ti_{0.5}O_3$ precipitates in sapphire consists of parallelepipeds along $\langle 2\bar{1}\bar{1}0 \rangle$, although their results indicated that the habit plane is (0001). For a (0001) habit plane, the strain energy calculations cannot explain the precipitate morphology, nor the direction along $\langle 2\bar{1}\bar{1}0 \rangle$, since the difference in the strain energies between $\vec{n}=\langle 01\bar{1}0 \rangle$ and $\vec{n}=\langle 2\bar{1}\bar{1}0 \rangle$ is very small, and gives no preference for a secondary habit plane. However, although the magnitude of the strain energy per unit volume is large, the strain anisotropy is very small, and the anisotropy of the interfacial energy is likely to play an important role in the resulting precipitate morphology. Thus, for precipitates from the ilmenite-hematite series in sapphire, the precipitate morphology appears to be determined by the anisotropy in the interfacial energy.

$FeTiO_3 - Fe_2O_3$

The final system considered in this investigation consists on hematite (Fe_2O_3) precipitates in an ilmenite ($FeTiO_3$) matrix. The elastic constants used for the matrix were those for hematite, which is reasonable since one of the assumptions by Khachaturyan is that the elastic constants are the same for both the precipitate and the matrix. As can be seen in Fig.8, the strain energy calculations predicts the (0001) basal plane to be the habit plane. The energy for all planes perpendicular to the basal plane are larger by a factor of four, and almost constant. Thus, the strain energy calculations predicts the precipitate morphology to be an oblate sphereoid with a (0001) habit plane (a smaller cross section along [0001] compared to a direction contained in the basal plane), which is consistent with experimental observations using TEM [21] (Fig.2c). It is important to note that the precipitates are semi-coherent with respect to the matrix, and that the possibilities for formation of low energy misfit dislocation networks has to be considered. It can be seen in Fig.2c that the precipitate is facetted, and that networks are formed on particular planes. Analysis of TEM micrographs and electron diffraction patterns indicate that networks are preferentially formed on $\{01\bar{1}2\}$ and $\{0\bar{1}14\}$ planes as well as the (0001) basal plane, which is consistent with the formation of misfit dislocations with short Burgers vectors [26].

SUMMARY

The theory developed by Khachaturyan to calculate the strain energy between a precipitate and its matrix has successfully been applied to explain the morphology of tetragonal precipitates in a cubic ZrO_2 matrix in Mg, Ca and Y-PSZ, rutile precipitates in a sapphire matrix and hematite precipitates in an ilmenite matrix. The results for precipitates from the ilmenite-hematite series in sapphire are ambiguous, and the precipitate morphology is believed to be determined by the anisotropy in the interfacial energy rather than that of the strain energy.

Although the precipitate morphology cannot be explained by strain energy calculations alone, such calculations can give useful information about probable morphologies and an added understanding about the growth of a precipitate. Further, this investigation show that the theory developed by Khachaturyan may be applied to semi-coherent as well as to fully coherent precipitates.

ACKNOWLEDGEMENTS

The authors would like to acknowledge Dr. A.R. Moon, the New South Wales Institute of Technology, Australia, for useful discussions concerning Australian star sapphire. This research was supported by the Department of Energy under grant no. DEFG0284ER45110.

REFERENCES

1. R.C. Garvie, R.H.J. Hannink, and R.T. Pascoe, Nature (London), 258, 703 (1975).

2. D.L. Porter and A.H. Heuer, J. Am. Ceram. Soc., 60, 183 (1977).

3. D.L. Porter, A.G. Evans, and A.H. Heuer, Acta Metall., 27, 1649 (1979).

4. A.G. Evans and A.H. Heuer, J. m. Ceram. Soc., 63, 241 (1980).

5. N. Claussen, J. Am. Ceram. Soc., 59, 49 (1976).

6. T.K. Gupta, F.F. Lange, and J.H. Bechtold, J. Mater. Sci., 13, 1464 (1978).

7. N. Claussen, M. Rühle and A.H. Heuer, Advances in Ceramics, Vol. 12, "Science and Technology of ZrO_2 II", Ed. M.Rühle, N.Claussen and A.H.Heuer, The American Ceramic Society, Inc., Columbus, Ohio 1984.

8. M. Rühle and A.H. Heuer, ibid. Vol.12, p. 14.

9. A.G. Khachaturyan, Sov. Phys. Solid State 8, 2163 (1967).

10. A.G. Khachaturyan, <u>Theory of Structural Transformations in Solids,</u> J. Wiley and Sons, New York, N.Y. 1983.

11. V. Lanteri, T.E. Mitchell and A.H. Heuer, J. Am. Ceram. Soc. (to be published).

12. W.E. Mayo and T. Tsakalakos, Metall. Trans. <u>11A</u>[10], 1637 (1980).

13. J.D. Eshelby, Proc. Roy. Soc. <u>A241</u>, 378 (1957).

14. J.D. Eshelby, <u>ibid.</u> <u>A252</u>, 561 (1959).

15. H.M. Kandil, J.D. Greiner and J.F. Smith, J.Am.Ceram.Soc. <u>67</u>,341(1984).

16. J.B. Wachtman, Jr., W.E. Tefft, D.G. Lam, Jr. and R.P. Stinchfield, J. Research of the N.B.S. <u>64A</u>[3], 213 (1960).

17. <u>Handbook of Physical Constants</u>, Ed. S.P. Clark, Jr., The Geological Society of America, New York, N.Y. 1966.

18. R.E. Newnham and Y.M. DeHaan, Z. Krist. <u>117</u>, 235 (1962).

19. <u>Powder Diffraction File</u>, The American Society for Testing and Materials (ASTM), Philadelphia, Pa. 1967.

20. M.E. Straumanis, T. Ejima and W.J. Jones, Acta Cryst. <u>14</u>, 493 (1961).

21. J.S. Lally, A.H. Heuer and G.L. Nord, Jr., <u>Electron Microscopy in Mineralogy</u>, Ed. H.-R. Wenk et al., Springer Verlag, Berlin 1976.

22. A.R. Moon and M.R. Phillips, Micron and Microscopica Acta <u>15</u>[3], 143 (1984).

23. D.S. Phillips, A.H. Heuer and T.E. Mitchell, Philos.Mag. <u>42</u>[3],385(1980).

24. D.S. Phillips, A.H. Heuer and T.E. Mitchell, <u>ibid.</u> <u>42</u>[3], 405 (1980).

25. D.S. Phillips, T.E. Mitchell and A.H. Heuer, <u>ibid.</u> <u>42</u>[3], 417 (1980).

26. K.P.D. Lagerlöf, A.H. Heuer and T.E. Mitchell, <u>Proceeding of the Electron Microscopy Society of America (EMSA)</u>, p. 212, Ed. G.W. Bailey, San Fransisco, Ca. 1980.

PARTICLE TOUGHENING IN PARTIALLY STABILIZED ZIRCONIA

INFLUENCE OF THERMAL HISTORY

R.H.J. Hannink and M.V. Swain

CSIRO, Division of Materials Science
Advanced Materials Laboratory
P.O. Box 4331, Melbourne, Vic. 3001, Australia

ABSTRACT

The thermomechanical properties of partially stabilized zirconia ceramics may be dramatically influenced by thermal treatments. This review describes how a variety of cooling processes and isothermal ageing treatments may be used to control the microstructure and hence make magnesia-partially stabilized zirconia (Mg-PSZ) one of the strongest and toughest sintered ceramics known. By careful selection of thermal treatment cycles the strength and fracture properties of Mg-PSZ may be tailored so as to make the material suitable for a variety of industrial and engineering applications.

INTRODUCTION

It has now been ten years since the first paper on the stress induced transformation toughening[†] of zirconia was published (1). Since that time, the concept that the constrained transformation of tetragonal (t) zirconia to the monoclinic (m) polymorph, may be used to enhance the toughness and strength of ceramics, has been widely studied and utilized (2,3). The benefits of the transformation are derived from the associated volumetric dilation ($\sim 4\%$) and deviatoric shear strains ($\sim 10\%$).

The range of different zirconia and ceramic matrices to which the toughening mechanism may be applied are numerous, and have been tabulated by Claussen (4) under the generic term of zirconia toughened ceramics (ZTC). Three main groups of transformation toughening ceramic (TTC) emerge as industrially and scientifically interesting materials.

[†]Transformation toughening (TT) is the generic term now used to describe property enhancement due to a stress induced transformation of zirconia containing materials. Toughening mechanisms, other than associated with the volume expansion and shear, eg. microcrack, crack deflection and branching also contribute to the overall toughening increment as a result of the transformation or thermal expansion mismatch strains.

These may be classified as partially stabilized zirconia (PSZ), tetragonal zirconia polycrystals (TZP) and dispersed zirconia ceramics (DZC).

In PSZ materials $t-ZrO_2$ is dispersed as precipitates within a cubic stabilized zirconia (CSZ) matrix. Calcia, magnesia or yttria are commonly used as the stabilizer (Ca-PSZ, Mg-PSZ and Y-PSZ respectively). TZP materials are generally composed entirely of $t-ZrO_2$ (although some cubic and monoclinic may also be present (5)). Such materials are fabricated from zirconia using yttria stabilizer additions (Y-TZP) or other rare earth oxide stabilizers which have a partial solubility in the t and m phases. In the DZC class of materials the matrix is of a second ceramic, eg. α- or β-alumina, silicon nitride or carbide, thoria, zinc oxide or glass, in which are dispersed zirconia particles; generally up to 30 vol%, which may or may not have yttria dissolved. In this case the critical factor for toughening enhancement is the particle sizes and location and modulus of the matrix used. The size dependance will determine whether transformation or microcrack toughening will be the predominant toughening mechanism.

Fabrication of the three groups of materials (PSZ, TZP and DZC) follows a number of routes. The process in common to all three routes is that fabrication normally follows a powder route, as opposed to melting which is reserved for single crystal production (6). In the case of PSZ materials the firing temperature used, determined by the stabilizer content, is well above that required to achieve sintering. This high temperature (~ 1700 - 1800°C) is necessary to attain a single cubic phase (c), from which precipitation occurs on cooling. Since precipitation is not a pre-requisite for the production TZP materials, firing temperatures high enough to achieve sintering and tetragonal solid solution are used ~ 1450°C (7). Fabrication of DZC materials with non-zirconia matrices will depend upon the matrix type. The most commonly toughened matrix has been α-alumina (zirconia toughened alumina, ZTA; or transformation toughened alumina, TTA) which is generally sintered as for pure alumina at ~ 1550°C. TZP and ZTA systems both benefit from a final hot isostatic pressing treatment at <1600°C. This step reduces residual porosity, with the added benefit of curtailing grain growth due to the short time at temperature required to achieve near full density.

Of the three classes of transformation toughened ceramics, Mg-PSZ has found the widest commercial application because of its facility for microstructural adaptation, making it suitable in a variety of industrial and engineering situations (8,9). In this paper we shall review the type of microstructures produced and mechanical properties attained, in Mg-PSZ, by utilizing a variety of thermal treatments which control precipitate size, distribution and stability.

PREAMBLE.

Thermodynamic considerations of the t → m transformation

The main aim in the fabrication of transformation toughened ceramics is to develop the $t-ZrO_2$ phase into a metastable state such that it will transform to the $m-ZrO_2$ form in the presence of an applied stress, at or near room temperture. The transformation occurs by a martensitic (M) reaction, ie. it is fast and diffusionless. The temperature at which it occurs is known as Ms (martensite start). Thus the aim of thermal treatment is to produce an Ms temperature for the t-phase just sufficiently below room temperature so that spontaneous transformation does not occur.

The ease with which the transformation occurs will depend upon a number of physical and structural factors. The various energetic states of the t and m phases may be depicted schematically with the use of a free energy diaagram, shown in fig 1. On the left of the diagram is shown free energy of the constrained t phase, F_t, at room temperature T_0. On the right hand side of the figure is shown the free energies for the various states of the resultant m phase, F_m, as a function of the initial t size. At the top right of the diagram is the F_m of small m particles at room temperature (ie. Ms is significantly below room temperature) and leads to a situation where the free energy of the m particle inside the constraining matrix is higher that the untransformed t particle, hence there will be no net driving force to initiate the transformation. Also in this situation the activation barrier for nucleation of the transformation, ΔF^*, will be quite large and virtually insurmountable.

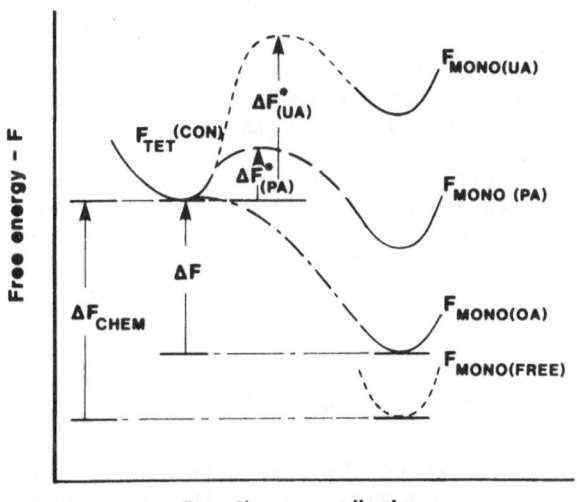

Fig. 1. Schematic representation of the various free energy forms associated with the t to m transformation normalized to the free energy of the initial t phase), of constrained (CON) t particles as a function of the initial particle size - under-aged (UA), peak-aged (PA) and over-aged (OA). For discussion see text.

If we now consider a particle whose Ms is above room temperature it will naturally transform on cooling as ΔF^*, derived from kT (k=Boltzman's constant), will be exceeded when a particular temperature is attained sufficiently below Ms, ie. sufficient undercooling. The intermediate case for the transformation occurs when Ms, for the paticles, is just below room temperature, In this situation the ΔF^* term may become very small and can be surmounted as a result of an applied stress, ie. the particle is sufficiently metastable such that a very small applied stress will be required to induce the transformation. As will be appreciated from these descriptions there exists a critical size or state below which the particles will transform (either spontaneously or with the aid of an applied stress) and above which they will not. As mentioned at the outset it is the aim of the thermal treatment to bring the particles to this critical state. The actual nucleation mechanism for the transformation in various TTC systems is still a source of debate. However, it is agreed that once the transformation is nucleated it is very rapid (martensitic). Heuer has described the thermodynamic considerations of the t to m transformation more fully (10).

Our description is only intended to give an appreciation of the thermodynamic considerations involved with the t to m transformation and the aims of optimising particle stability.

Toughening Mechanisms Resulting from the t → m Transformation

A large number of factors may be associated with the precipitates which can influence the final toughening increment, ΔK, obtained in a ceramic as a result of the confined t to m transformation. Two recent reviews have rigorously analysed the toughening contributions of the various precipitate/matrix features as they influence the passage of a propagating crack (11,12). The total microstructural contribution to the critical stress intensity factor, K_c, will be derived from:

    a)   transformation toughening, ΔK_{cT},
    b)   transformation induced microcrack toughening, ΔK_{cM}
    c)   crack deflection toughening, ΔK_{cD}.

All three mechanisms will contribute to K_c, the magnitude of their contributions will depend upon the size, composition and volume fraction of the transforming particle and the temperature at which the crack was propagating (or measurements made).

The total critical fracture stress of a material can be expressed as;

$$K_c = K_o + \Delta K_c \tag{1}$$

where K_c is the measured critical stress intensity factor, K_o is the fracture resistance of the matrix and ΔK_c the incremental contribution from the other toughening mechanisms operating inside the material at the time of crack propagation.

Analyses of the various contributions of ΔK_c lead to the following general expressions (12):

a) the toughening increment, ΔK_{ct}, from a dilatant transformation in a steady state situation is given by;

$$\Delta K_{cT} = \eta \, E_o \, V_f \, \varepsilon^T_{11} \, \sqrt{h}/(1-\nu) \tag{2}$$

where E_0 is the modulus of an unmicrocracked matrix, V_f the volume fraction of tranformed precipitates, ε^T_{11} the uniaxial transformation strain, h the width of the process zone, ν Poissons ratio and η is a factor that depends on zone shape about the crack tip. For a zone distorted by principal tensile stresses $\eta = 0.22$ whereas for shear dominated zones $\eta = 0.38$.

b) the potential toughening contribution from microcracking, K_{cM}, as a result of pre-transformed particles is given by the expression;

$$\Delta K_{cM} = \frac{0.07}{(1-\nu)} \ E_0 \ \epsilon_{ii}^T \ \sqrt{h} \ (1.2 \ f^{\frac{1}{3}} \ - \ f) \qquad (3)$$

where f is the volume fraction of precipitates contributing to the toughening. The equation assumes that the microcracks will fully extend through the matrix material between the particles. The equation indicates that transformation microcracking can make a significant contribution to the critical stress intensity factor provided (i) f is small, (ii) the modulus of the matrix is not affected by the transformation and (iii) the microcracks do not form co-planar with the propagating crack.

Fig 2 illustrates the effective contributions of the transformation crack shielding and microcrack contributions in pre-transformed material. The figure indicates that, with up to 0.2 volume fraction of contributing particles, microcrack toughening is more effective than stress induced transformation toughening. Beyond this value the stress induced transformation component becomes dominant. However a combination of both mechanisms could optimize the toughness.

c) Crack deflection toughening, ΔK_{cD}, may arise, particularly at low volume fractions of zirconia additives from (i) localized residual stress fields, resulting from the transformation, or (ii) the fracture resistance of the precipitate particles (13). Calculations to determine the actual contribution to ΔK_{cD} are complicated by the morphology, orientation and volume fraction of the particles and the nature of the moving crack front. Calculations have been performed and suggest that for Mg-PSZ the contributions to K_c, as a result of crack deflection could be as high as 1.5 (13). However this mechanism has the advantage over the previous mechanisms that it is temperature insensitive.

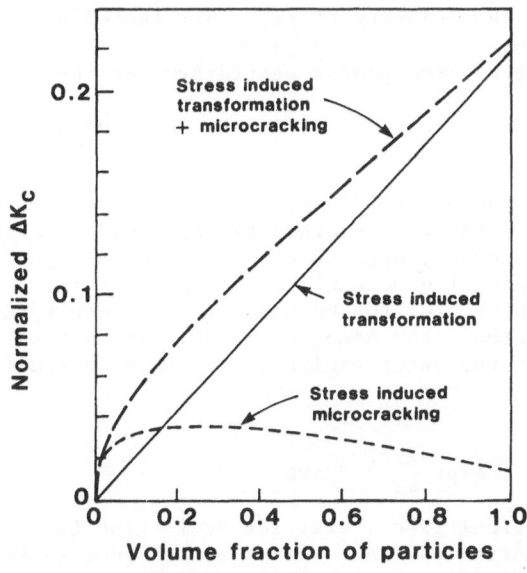

Fig. 2. Contribution to normalized K_c of the volume fraction of particles available for crack shielding by stress induced transformation, transformation plus microcracking, and stress induced microcracking in pretransformed material. (after Evans and Cannon, 12)

It is thus apparent that optimization of toughness can be achieved via a number of contributing mechanisms, which result from the transformation of metastable tetragonal precipitates in a zone around a propagating crack. We shall now examine the thermal treatment and microstructural features which can be used to achieve optimal mechanical properties.

THERMAL TREATMENT OF PSZ

Fabrication/Precipitation/Coarsening

PSZ materials are fabricated from high purity zirconia and a stabilizing oxide powder eg. CaO, MgO or Y_2O_3. The morphology of the final precipitated phase will be determined by the type of stabilizer used (14). The main detrimental impurity in the zirconia powder is usually silica. This contaminant causes destabilization of the cubic zirconia matrix by reactions with the stabilizer. At low contamination levels (<0.2%), additives may be used to purge the silica from the sintering zirconia compact (15). We shall use mainly Mg-PSZ as the example for the fabrication and thermal treatment of a typical precipitate toughening system.

Fig 3 shows the ZrO_2 rich end of the $MgO-ZrO_2$ phase diagram (16). Typical commerical Mg-PSZ compositions occur in the range 8-10 mol% MgO, indicated by the dotted region. The materials are sintered and solution treated in the cubic single phase region, just above the phase boundary. After firing, the materials may be cooled in a number of ways. The prime aim of the cooling cycle is to achieve:

a) a uniform distribution of the precipitates within the grains, and
b) retention of the majority of the precipitated phase in the tetragonal form.
c) prevention of massive pro-eutectoid ZrO_2 at the grain boundaries

The three aims are interdependent and may be achieved by a number of routes.

A uniform distribution of precipitates and prevention of pro-eutectoid ZrO_2 is most readily attained by rapid undercooling below the single phase field boundary, such that a classical homogeneous nucleation and precipitation mechanism becomes operative. If cooling is too slow large pro-eutectoid ZrO_2 boundary phases result, due to heterogeneous nucleation. For homogeneous nucleation the rate, $\dot{N}$, is strongly dependent on the under cooling, ΔT, through the free energy term, ΔF^*, such that;

$$\dot{N} \; \alpha \; \exp(\frac{-\Delta F^*}{kT}) \quad \text{and} \quad \Delta F^* \; \alpha \; (\frac{1}{\Delta T})^2 \qquad (4)$$

where ΔF^* is the critical free energy for nucleation (17). It can be appreciated from equation [4] that whilst ΔF^* is reduced with increasing ΔT, and hence $\dot{N}$ becomes greater, the lattice diffusion constant rate which controls particle growth is decreasing. If a fast cooling rate is maintained to below 1000°C very small particles will result. Therefore while the initial under cooling should be rapid this must be followed by a process which will allow coarsening of the percipitates to the room temperature metastable state.

Fig. 4 is a diagrammatic representation of three types of cooling

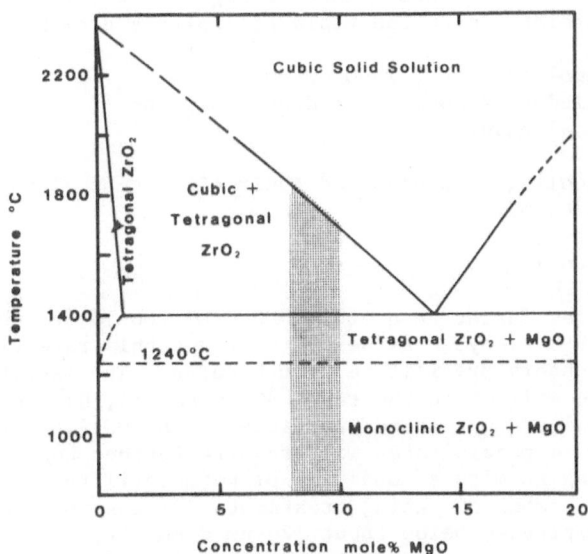

Fig. 3. Zirconia rich end of the ZrO_2-MgO phase diagram (after
Grain, 16). Shaded region shows the composition and firing
temperature most commonly used for commercial Mg-PSZ alloys.

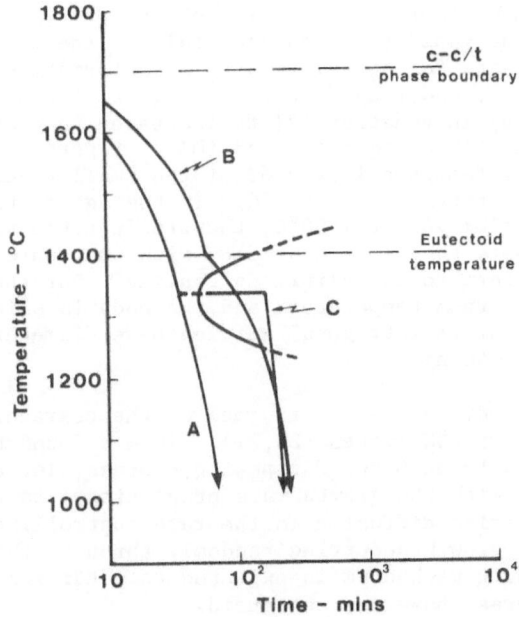

Fig. 4. Possible cooling routines for a 9.7 mol % MgO-ZrO_2 alloy.
The C-curve is the region beyond which rapid secondary
precipitate growth occurs (18). See text for a description
of the various cooling sequences.

sequences which can follow the solution treatment, superimposed on a C-curve[+] which represents a region of rapid secondary precipitate growth. The cooling sequences shown in figure 4 may be described on the basis of their relations to the rapid precipitate growth boundary as:

A) rapid cool (quench), >500°C/hour
B) controlled slow cool or sliding cool, and
C) isothermal hold.

Each of these cooling sequences and their effects on microstructure will be discussed in turn.

A) Rapid Cool

Cooling rates faster than about 500°C h^{-1} would be considered rapid for 9.7 mol% Mg-PSZ alloys. As seen from fig 4 this rate will miss the nose of the secondary precipitate growth curve. The resulting precipitate size will be in the range 30-60 nm (14,18) and occur as small ellipsoidal or lens shaped particles with (001) habit planes [fig 6(a)]. These precipitates will require further heat treatment (ageing) to bring them to a condition of metastability at room temperature. A number of ageing treatments are possible to achieve this aim. With precipitates below about 120 nm diameter, only ageing temperatures above the eutectoid temperature (1400°C, fig 2) are considered. This is because below this temperature the eutectoid decomposition reaction will dominate the ageing process (19). We shall see later how ageing below the eutectoid temperature with appropriate sized precipitates can be very beneficial for the mechanical properties.

Coarsening of the rapidly cooled specimens may be depicted schematically as shown in fig 5, in terms of the precipitate size distribution. In the rapidly cooled material all the particles will be below the critical size for stress induced transformation, Δd_{crit}. The material is said to be under-aged, UA. Ageing at 1420°C causes the particles to grow (V_f in equation [2] to increase) to a stage of peak-aged (PA), and if continued to over-aged (OA). Figure 6 (a-c) shows the coarsening sequence, depicted in fig 5, of the small precipitates following an ageing treatment at 1420°C. If the ageing treatment is too long, for this material >4h at 1400°C, the precipitates will grow beyond a critical size, where they lose coherency with the cubic matrix, and spontaneously transform to monoclinic on cooling. For these materials the Ms will be above room temperature and the body is said to be OA (fig 6(c). Optimal or PA tetragonal patcles have diameters of ~ 200 nm and a thickness of ~ 40 nm.

An extensive study has been performed on the coarsening of the precipitates in the Ca-PSZ system (20,21). It was found the driving force growth occured by an Ostwald ripening process, ie. a reduction in interfacial energy, with the growth rate proportional to $t^{1/3}$ (the L-S-W mechanism, where lattice diffusion is the rate controlling process) (17), with particle growth occurring randomly throught the grain. A similar type of growth mechanism is expected to occur for particle growth at temperatures above the eutectoid.

[+]This diagram is a modification of the metallurgical time-temperature-transformation curve (TTT) in that here it indicates the region of secondary precipitate growth (18), and not an indication of the eutectoid decomposition process.

It should be pointed out that it is only 'static' or isothermal type heat treatments which allow the expected t-ZrO$_2$ precipitate content (according to the equilbrium phase diagram) to be realized. Equilibrium contents are important for the determination of V_f and f in equations [2] and [3]. Continuous cooling will not allow the equilibrium content to be achieved.

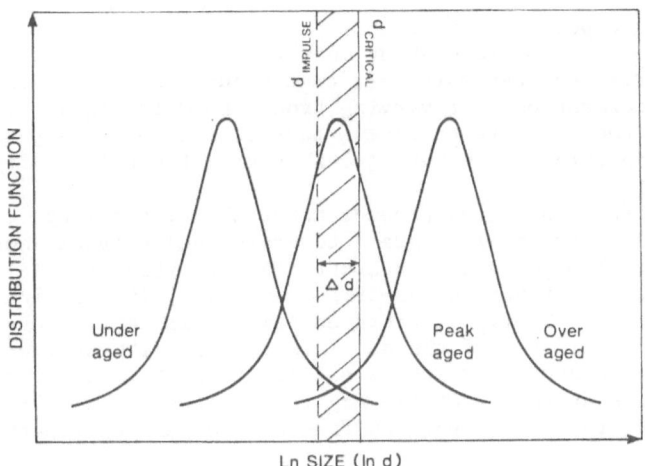

Fig. 5. Schematic illustration of the particle size distribution as a function of the particle size for the variously aged conditions. All particles to the right of d$_{CRIT}$ transform to m on cooling, Δd is the volume fraction of particles capable of being transformed from t to m by the application of a stress. When the area within Δd is maximum a material is said to be peak aged.

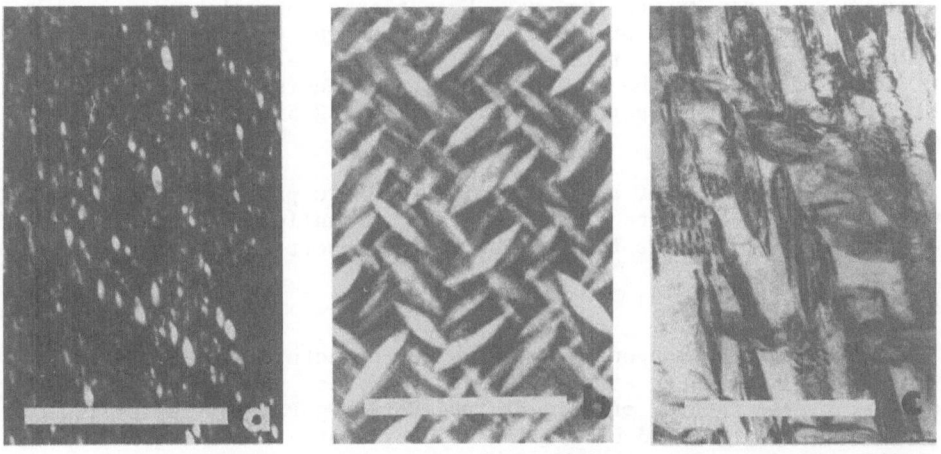

Fig. 6. Coarsening of t particles in Mg-PSZ material that had received the A-quench in figure 4. (a) As-fired (dark field TEM image, (b) aged at 1420°C for 2h and (c) for 4h, (bright field TEM images. Bar length equals 0.5 μm.

B) Controlled Cool

An economically more appealing method of cooling the sintered/solution treated bodies is by a controlled 'slow' cool. These treatments were initially preferred, not as a matter of choice but more of necessity as cooling simply involved switching off the furnace. This process lead to the early PSZ materials with very variable properties, generally over-aged because of the large thermal mass of the ceramic kiln.

Subsequent experiments have shown that PSZ alloys may be cooled at rates which produce materials at or near the peak-aged condition (22). Such cooling treatment can vary from continuous linear (sliding) to rate arrests and acceleration, ie. varying from B to C in fig 4. The main purpose of the cool will be to have a body reside for some period of time in the secondary precipitate growth region (c-region), fig 4.

These cooling conditions produce materials with a range of precipitate sizes varying from grain to grain, and probably not having attained the equilibrium volume content. Figure 7(a) shows a scanning electron micrograph of such an as-fired material. The lenticular nature of the particles is clearly illustrated (two variants are edge on whilst the third is in the plane of the sample surface). The optimum precipitate size should be a little smaller than the peak-aged materials, such that the materials may be subjected to a sub-eutectoid ageing treatment to achieve optimum fracture toughness properties.

C) Isothermal hold

In a recent study of the precipitation behaviour of Mg-PSZ, it was observed that in the temperature range 1400 to 1250°C, precipitate coarsening occured by a secondary precipitate growth process (18). This process was different from the normal Ostwald ripening in that the growth step was nucleated at the grain boundaries and moved into the grains as a growth front consuming the small precipitates. At this front 'primary' precipitate particles, probably determined by a favourable stress state, undergo an accelerated growth situation which continues until the precipitate impinges upon a growing neighbour undergoing the same growth process. When the isothermal hold time was short, two very distinct precipitate sizes occured within the same grain. Figure 7(b) shows this bi-modal precipitate distribution. Large particles occurring at the periphery of the grain (size about as in fig 6(b)), whilst the small particles (similar to fig 5(a)) are present in the central region.

Optimum hold time of about 120 min at 1340°C produced material with a very uniform precipitate size, requiring virtually no further heat treatment to optimize the fracture strength properties, see fig 9.

Sub-Eutectoid Ageing

Earlier work has shown (23) that a sub-eutectoid ageing treatment can produce the toughest sintered ceramic material yet produced. The best ageing temperature, of suitably pre-fired Mg-PSZ bodies is 1100°C (24). Suitably pre-fired materials are those which contain near optimally sized precipitates, and are most readily produced by treatments B or C in fig 3.

The microstructural influence, hence mechanical properties, of the treatment will also depend on the precipitate size of the pre-fired

materials, and may be summarized as four main features (25). Firstly
the anticipated decomposition reaction, according to the phase diagram
(fig 3, cubic ZrO_2 to m–ZrO_2 plus MgO) occurs at the grain boundaries
and around pores. Whilst this reaction is rapid in fully stabilized
zirconia (14 mol% MgO–ZrO_2) (26), it is not significant for the ageing
times used here. The reaction may also be considerably slowed down with
the addition of suitable sintering aids (15). There are three other
processes that occur within the grains:

a) the formation of an ordered anion vacancy phase, $Mg_2Zr_5O_{12}$
    (δ–phase) (27), at the precipitate matrix interface,
b) the precipitation of very small tetragonal precipitates within
    the cubic matrix of the precipitate laden grains, and
c) transformation, on cooling, of some of the original tetragonal
    precipitates.

For the ageing times used the original precipitates show no
significant increase in size. The main cause of the transformation to
the room temperature form is the additional strain induced into the
particle interface as a result of the δ–phase precipitation (25).

Depending upon the ageing time at 1100°C varying mechanical
properties may result; these are discussed later. The main difference
in the microstructural aspects between the ageing times, and hence the
different properties, is the amount of original tetragonal converted to
monoclinic on cooling. The form of this monoclinic consists of small
particles 150–200 nm in diameter and twinned on a fine scale. The
remaining tetragonal particles are exceedingly metastable such that
their nucleation free energy, ΔF^*, is vitually zero. These particls
will transform on electron microscope specimen preparation (ion beam
thinning) or as a result of the thermal stresses induced by an electron
beam. Similarly their extreme sensitivity to applied stresses results
in very large process zones (h in equation [2]) and a rising crack
resistance (R–curve) behaviour for a propagating crack (11).

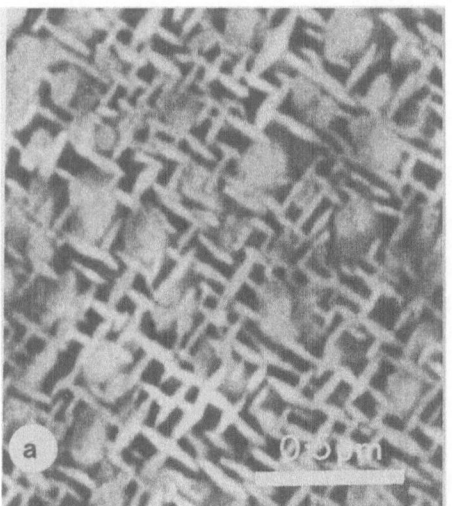 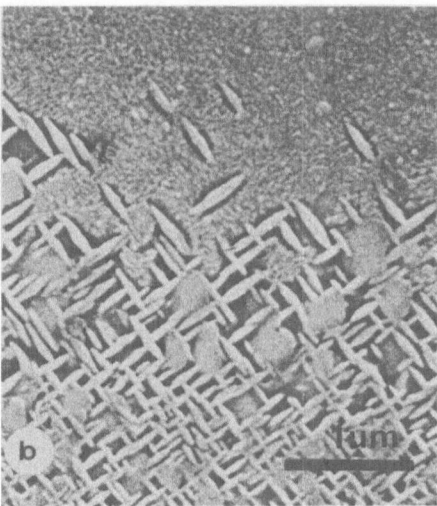

Fig. 7.   Near peak aged precipitates in Mg–PSZ, produced by two
          different cooling routes. (a) Controlled continuous cool
          through and (b) isothermal hold in the C–curve region of
          fig. 4. See text for further descriptions.

The thermal expansion of PSZ materials are a very sensitive function of monoclinic phase content. This is because of the considerable thermal expansion difference between the tetragonal ($\sim 9 \times 10^{-6}$) and monoclinic ($\sim 6 \times 10^{-6}$) phases. The difference leads to a hysteresis in the thermal expansion of materials when the transformation occurs.

Figure 8 (a) shows the thermal expansion data for a variety of ageing times at 1400°C of an Mg-PSZ alloy. It can be seen from the data that virtually no inflection (no m to t transformation) occurs until the materials approach the over-aged condition. The inflection is quite sharp, and its actual temperature will be a sensitive function of the ageing time at 1400°C.

Thermal expansion data for the 1100°C aged materials are shown in fig 8 (b). The expansion behaviour of these materials are considerably different from those of the 1400°C ageing sequence. The form of the curves will depend upon the initial precipitate size. The temperature at which the transformation commences is generally lower, as a result of the smaller monoclinic precipitate size, ie. the back transformation temperature will also be a function of the precipitates participating in the transformation, and is quite small, fig 8 (b), even after 16h at 1100°C.

After prolonged ageing at either temperature a second major inflection is observed at 1200°C. This inflection occurs as a result of the coarse monoclinic grain boundary phase transformation.

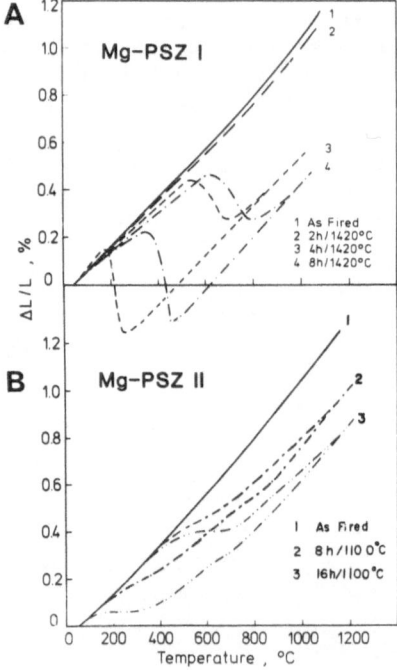

Fig. 8.   Normalized thermal expansion curves for Mg-PSZ following two different types of heat treatment. (A) An A-type cooling sequence (fig. 4) followed by a 1420°C ageing treatment and (B) a B-type cooling treatment (fig. 4) followed by a 1100°C ageing treatment.

Strength

The strength of Mg-PSZ material is strongly influenced by the
precipitate size prior to the final heat treatment (ageing). As
mentioned previously, different ageing treatments, depending on the
ageing temperature, lead to either precipitate coarsening, rapid grain
boundary decomposition or δ-phase development. Examples of the effect
of the heat treatment temperature on the room temperature bend strength,
for materials aged between 1000-1500°C for various times, as shown elsewhere . The as-fired specimens, when produced using a B-type (fig. 4)
cooling sequence, contained t-ZrO_2 precipitates of $\sim$ 150 nm in the long
dimension. Such materials are ideally suited to low temperature ageing
treatments. In these materials the optimum properties were obtained
after 8h ageing at 1100°C. The rapid deterioration of the strength in
these materials after high temperature heat treatment is an indication
that precipitate coarsening and destabilization of the t-ZrO_2
precipitates occurs on cooling from the ageing temperature.

A large family of similar curves, but with maximum properties
occurring after shorter or longer ageing times, at other temperatures
than 1100°C, may be obtained from the same material if the as-fired t-
ZrO_2 precipitate size were different. The role of initial precipitate
size on the strength developed in a material aged only at 1100°C
illustrates this critical size dependence, fig 9(a). Fig. 9(b) shows
the m-ZrO_2 contents of these same materials in the as-ground
condition. As shown in fig.1, when the precipitates are too small they
are not readily transformed by grinding or by the stress field about a
crack tip. The precipitates become more metastable when aged, this
leads to improvements in strength and toughness. However if the 1100°C
ageing treatment is too long, decomposition and coarsening of the m-ZrO_2
grain boundary phase will lead to a degradation of the properties
(28). If the t-ZrO_2 precipitates are initially too metastable then

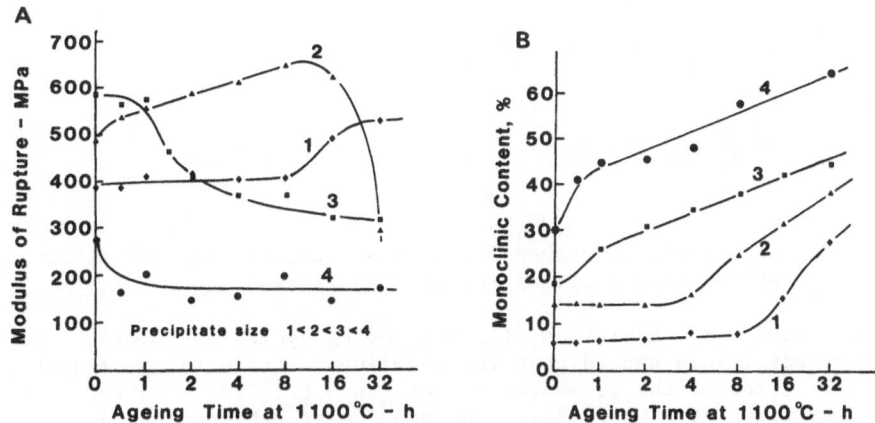

Figure 9.   Influence of initial precipitate size on the time dependence
of strength a), and monoclinic content of a ground surface
b), with sub-eutectoid heat treatment at 1100°C.

short ageing periods will lead to a destabilization and corresponding reduction in properties.

The strength of Mg-PSZ material decreases with increasing test temperature because of the increasing stability of the t-ZrO$_2$ precipitates. This corresponds to an increase in the $\Delta F*$ and a reduction in the chemical driving force, namely ΔF_{chem} in fig. 1. Typically at temperatures of only 500-600°C the strength is half the room temperature values but remains constant thereafter to 1100°C. Associated with the decrease in strength with test temperature of optimum strength material is a decrease in the Weibull modulus. The long term stability of Mg-PSZ at moderate temperatures and in humid conditions has recently been addressed (29). It was found that exposure to temperatures of 800°C and above leads to a slight degradation in strength due to the destabilization of the t-ZrO$_2$ precipitates. Also, unlike Y-TZP materials, exposure to water vapour between 200-400°C leads to a minimal change in the strength of the Mg-PSZ component.

Stress-Strain Relationships

Room temperature stress-strain curves for ceramics are usually linear to fracture. However recent studies with Mg-PSZ and TZP materials suggests that these curves can become substantially non-linear prior to fracture (30). Examples of such behaviour for optimal Mg-PSZ materials, C type in figure 4 after 120 minute hold at 1340°C, after various sub-eutectoid heat treatment times is shown in fig. 10. These results were obtained by attaching a strain gauge to the tensile surface

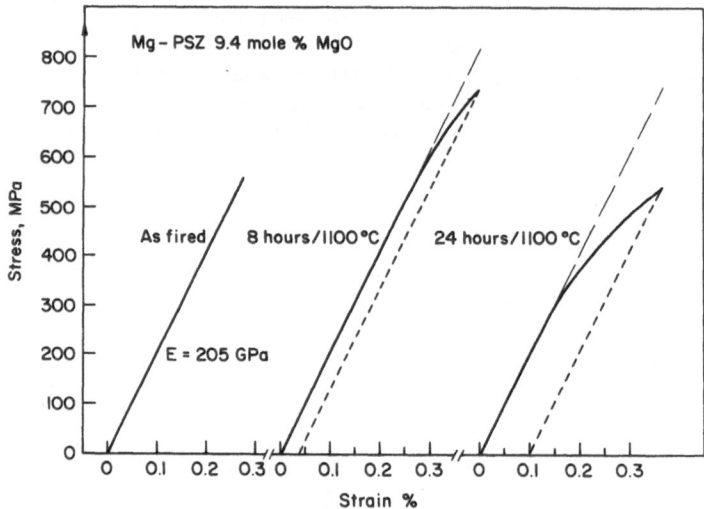

Figure 10. Stress-strain curves for type C material (fig. 4) after sub-eutectoid ageing for various times.

of a bend test bar. Additional load-unloading cycles at progressively higher stress levels showed that the non-linear strain was permanent and the total extent of the permanent offset was ~ 0.1% for materials aged for longer times at 1100°C (31). The non-linear behaviour is associated with stress induced transformation of some t-ZrO$_2$ precipitates. The transformation only occurs on the tensile surface. The initiation stress and extent being controlled by the precipitates metastability. The sample aged for 24h at 1100°C in fig. 10 displayed only an increase of 4-6% in the m-ZrO_2 content in the tensile surface, indicating that very few precipitates were able to respond to the applied stress. Marshall (32) has recently reported that 2-3% of the t-ZrO_2 precipitates transforming, reverse transform on the relaxation of the applied stress. The small increment of permanent offset in fig. 10 may be accounted for by the total volume of m-ZrO_2 generated by the applied stress.

An optical interference contrast image of the tensile surface of the 24h aged material is shown in fig. 11. The low magnification image, fig. 11a, reveals substantial surface relief of a previously featureless (flat) surface. Also present are a number of bands perpendicular to the tensile axis and many microcracks. The bands indicate that the deformation on the surface is not uniform and that isolated clumps of transformed material exist fig. 11(b). The volume dilation associated with the transformation is responsible for the surface uplift (relief). Some of the microcracks developed during flexure have linked up which will ultimately lead to specimen failure (fig. 11(c)).

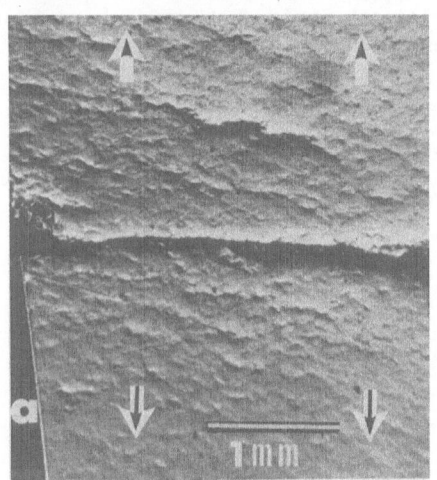
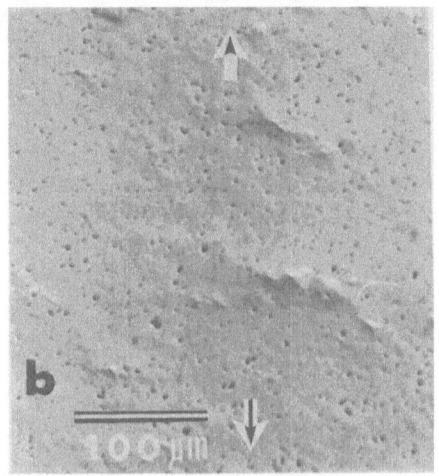
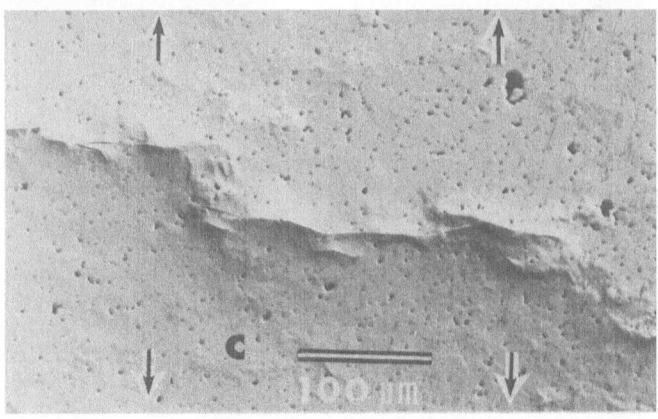

Figure 11. Optical micrographs taken using interference contrast of the deformation on the tensile surface of the 24 hour heat treated material of fig. 10. a) Low power observation of the region adjacent to the point of failure of the bar. b) Isolated clumps of transformed areas on the tensile surface. c) Micrograph of microcrack linking and associated transformation zones.

A consequence of increased precipitate metastability, due to sub-eutectoid ageing is that the critical stress to trigger the transformation is reduced. This in turn enables a much larger transformation zone to be developed around a crack and therefore a higher fracture toughness to be achieved. With increasing temperature the critical stress to trigger the t → m ZrO_2 transformation increases and the stress-strain curves remain essentially linear to fracture (30). As mentioned this arises because of the increased stability of the t-phase so that transformation will not occur prior to fracture.

Fracture Toughness

The fracture toughness of Mg-PSZ, like the strength, is a function of precipitate size and heat treatment (33). the optimum toughness is more readily and reliably achieved by sub-eutectoid rather than pro-eutectoid ageing treatments (28). The data in fig. 12 compares work of fracture following two different ageing treatments, the greatly enhanced toughness of the 1100°C aged material is apparent. This material at the peak values displays crack growth stability during notched fracture test whereas the 1400°C aged material behaves in a more conventionally brittle manner (34).

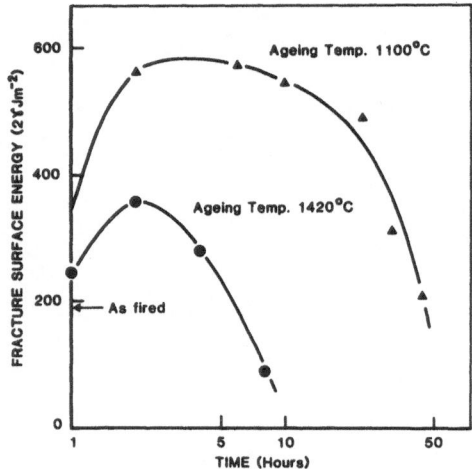

Figure 12. Comparison of the fracture surface energy with ageing time for sub-eutectoid and pro-eutectoid heat treated Mg-PSZ.

Variations in the initial precipitate size and ageing times at 1100°C enables the critical stress intensity factor, K_c, of Mg-PSZ alloys to be varied in the range 3-15 MPa m $^{1/2}$. At the lower K_c this value corresponds to the case where the precipitates are too small to be transformed by the crack tip stresses, at the other extreme the critical stress is comparable to the bend strength and the transformation zones may extend up to 50 μm from the crack tip (35). Such data may be used to test the viability of equation [2], equating the toughness increment to the volume fraction of transforming precipitates, zone size, etc. Assuming the modulus, volume fraction of precipitates and dilational strain to remain constant, the results shown in fig. 13 confirm the toughness increment dependence as the square root of the zone size. The value of η from these results in 0.43, which is close to that predicted for a shear zone dominated crack tip (12). Optical interference images

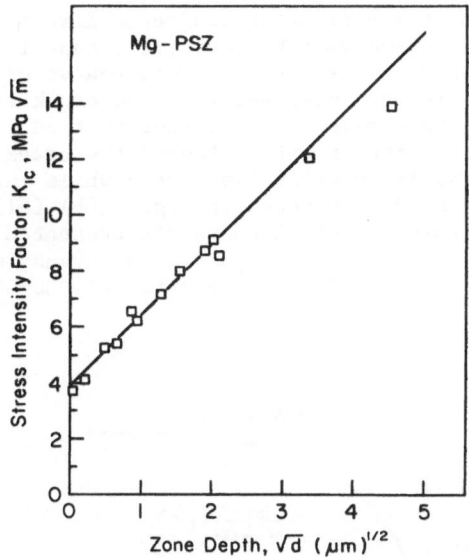

Figure 13. Relationship between transformed zone depth and critical stress intensity factor for Mg-PSZ materials containing 9.4 mole % MgO.

indicate the presence of large shear zones (bands) inclined between 45-60° to the forward crack direction.

The fracture toughness, like the strength, decreases with increasing test temperature. This feature is shown in fig. 14 for the same Mg-PSZ material having received five different ageing treatments. The 8h/1400°C material contained predominantly m-ZrO$_2$ precipitatess whereas the remainder were essentially tetragonal. The 8h/1400°C material also showed the least temperature sensitivity. As mentioned previously, the transformation zone size decreases rapidly with temperature thereby accounting for the decrease in toughness (33).

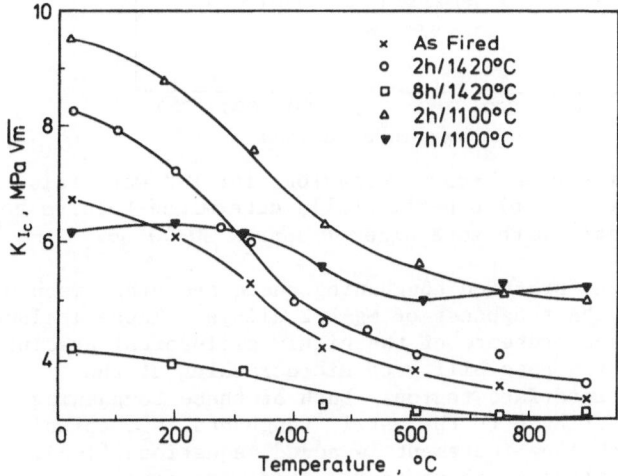

Figure 14. Temperature dependence of Mg-PSZ specimens given various prior heat treatments.

Another feature of transformation toughened materials is that they are predicted to display R-curve behaviour (12), that is increasing fracture toughness with crack extension. This behaviour occurs because steady state toughness is only achieved when the crack tip is completely shielded from the applied stress by the transformation zone; that is a wake of transformed precipitates exists behind the crack tip. The form of the predicted R-curve is shown in Fig. 15(a) while the experimentally determined R-curve for Mg-PSZ is shown in fig. 15(b) (31). More details of the methods of determining this R-curve are presented elsewhere (35). The transformed zone size surrounding the crack, used to establish the data in fig 15 (b), was approximately 30–40 µm

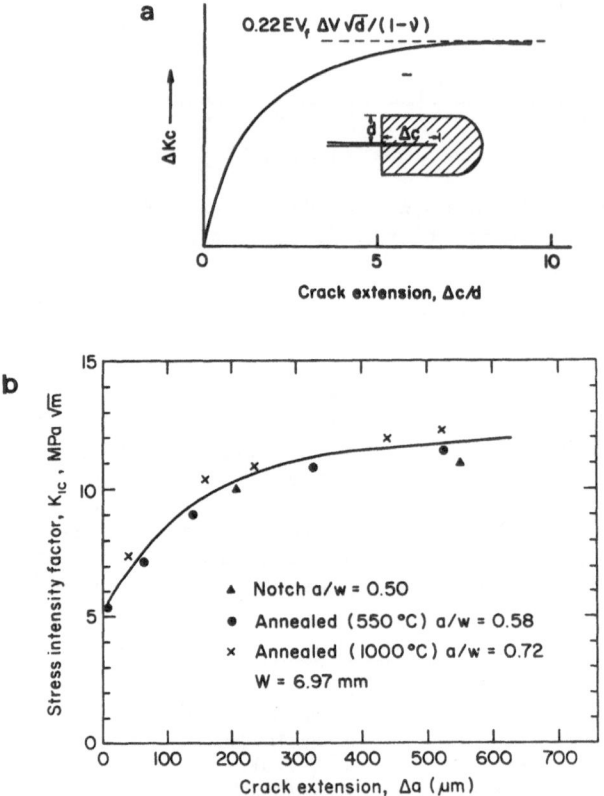

Figure 15.   a) Predicted R-curve behaviour for PSZ materials with η = 0.22. b) Experimentally determined R-curve for Mg-PSZ specimen with zone size of approx 30–40 µm.

Apart from transformation toughening there are other mechanisms that contribute to the toughness of Mg-PSZ alloys. These include; crack deflection due to the presence of the highly ellipsoidal preciptates as shown figure 6 and the possibility of microcracking at the matrix/precipitate interface region. Both of these toughening mechanisms will contribute to the matrix toughness, K_0, to which the transformation toughening increment is added, equations (1–3). A superb example of crack deflection about coarse precipitates in pro-eutectoid over aged Mg-PSZ is shown in fig. 16. This extreme tortuosity coupled with crack branching and precipitate bridging behind the crack tip would be anticipated to contribute to R-curve behaviour in this

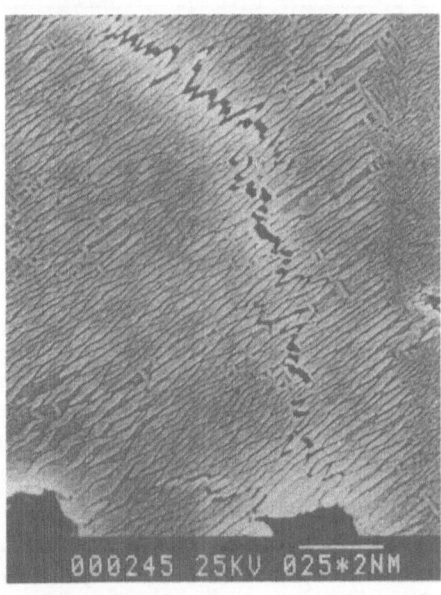

Figure 16. The tortuous crack path in overaged pro-eutectoid heat treated Mg-PSZ.

material. Similarly microcracking in the m-ZrO_2 grain boundary of sub-eutectoid aged material also contributes to the R-curve behaviour (26).

Strength-Toughness Relationships

An alternative approach to plotting strength and/or toughness as a function of ageing temperature and time is to plot the strength versus toughness (30). The usual relationship between these two terms is:

$$\sigma = K_c / Y\sqrt{c} \tag{5}$$

where c is the critical flaw size and Y a geometric term ~ 1.2 for surface cracks. Room temperature bend strength and K_c results for a series of Mg-PSZ alloy compositions and thermal treatments are shown in fig. 17.

The data in fig. 17 indicates that the strength passes through a maximum at a K_c value of 10-11 MPa m$^{1/2}$ and thereafter decreases. The curve may be conveniently divided into two regimes; to the left of the maximum the results fit the anticipated linear relationship of equation [5] with a critical flaw size of 100 μm. At higher toughness the strength decreases and closely follows the calculated value for the critical stress to trigger the transformation (30). That is, the strength shows a transition from being flaw size limited to transformation initiation limited. Similar curves are also found for the strength-toughness relationships for TZP materials although in these materials the flaw size is much smaller (25 μm), due in part to the smaller grain size (36).

An alternative, although more involved interpretation of these strength-toughness observations is based on the R-curve dependence of the toughness (31).

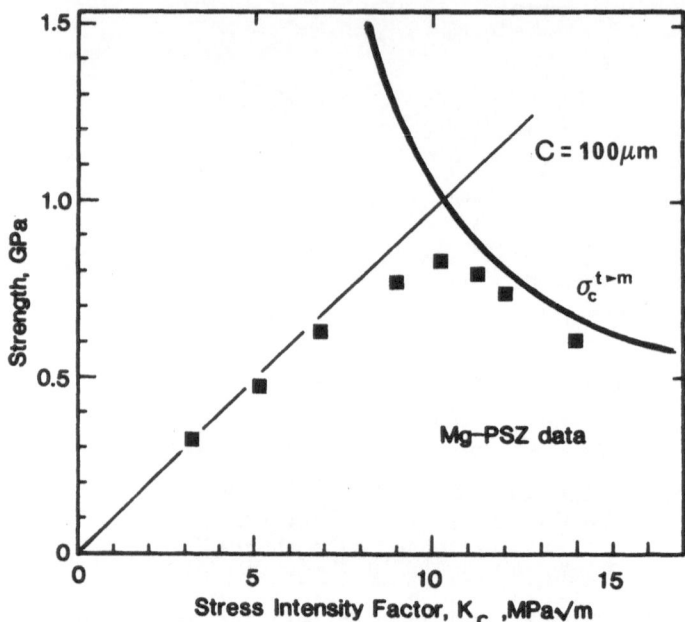

Figure 17. Variation in bend strength of Mg-PSZ versus K_{1c} for materials of various heat treatments but containing mainly t-ZrO$_2$ precipitates.

CONCLUSIONS

The strength and toughness of partially stabilised zirconia materials are strongly dictated by the size of the initial t-ZrO$_2$ precipitates and subsequent thermal history. It has been shown that optimally sized precipitates in Mg-PSZ alloys may be developed by various cooling procedures that cut the nose of the secondary precipitation curve. Fine tuning for the development of peak values of strength and toughness may be achieved by subsequent sub-eutectoid or pro-eutectoid ageing. The best values of toughness in Mg-PSZ materials have been obtained by sub-eutectoid ageing. With this treatment strengths in excess of 800 MPa and K_{1c} values of 12-14 MPa √m at room temperature may be had in Mg-PSZ alloys containing 9.7 mole % MgO. These values decrease with increasing temperature due to the decrease in chemical driving force for the t to m transformation. Future work in these alloys is likely to concentrate on the effect of various additives on the secondary precipitation curves and optimisation of microstructure to develop materials with a suitable compromise between strength-toughness and thermal stability.

REFERENCES

1. R.C. Garvie, R.H.J. Hannink and R.T. Pascoe, 1975, Nature, 258, 703.
2. A.H. Heuer and L.W. Hobbs eds., "Science and Technology", Advances in Ceramics Vol 3.
3. N. Claussen, M. Ruhle and A.H. Heuer eds. "Science and Technology of Zirconia II", Advance in Ceramics Vol 12.
4. N. Claussen, in [3] p 325.
5. M. Ruhle, N. Claussen and A.H. Heuer, in [3] p 352.
6. R.P. Ingel, D. Lewis, B.A. Bender and R.W. Rice, in [3] p408.

7. H.G. Scott, 1975, J. Mat. Sci., 10, 1527.
8. R.H.J. Hannink, M.J. Murray and M. Marmach, in "Wear of Materials 1983", K.C. Ludema ed, ASME, p181.
9. P.A. Janeway, 1984, Ceramic Industry, 122, 40.
10. A.H. Heuer, in [2] p 98.
11. M.V. Swain and L.R.F. Rose, "Advances in Fracture Research", 1984, ICF6, S.R. Valluri, D.M.R. Taplin, P. Rama Roa, J.F. Knott and R. Dubey eds, p 473.
12. A.G. Evans and R.M. Cannon, Acta Met., to be published.
13. K.T. Faber and A.G. Evans, 1983, Acta Met., 31, 565.
14. R.H.J. Hannink, 1978, J. Mat. Sci., 13, 2487.
15. R.H.J. Hannink and J. Drennan, this volume.
16. C.T. Grain, 1967, J. Am. Ceram. Soc., 50, 288.
17. J.W. Martin and R.D. Doherty, "Stability of Microstructure in Metallic Systems, Cambridge University Press, Cambridge, 1976.
18. R.R. Hughan and R.H.J. Hannink, 1985, submitted to J. Am. Ceram. Soc.
19. D.L. Porter and A.H. Heuer, 1979, J. Am. Ceram. Soc., 62, 298.
20. R.H.J. Hannink, K.A. Johnston, R.T. Pascoe and R.C. Garvie, in [2], p 116.
21. J.M. Marder, T.E. Mitchell and A.H. Heuer, 1983, Acta Met. 31, 387.
22. R.H.J.Hannink and R.C. Garvie, 1982, J. Mat. Sci., 17, 2637.
23. M.V. Swain and R.H.J. Hannink, in [3], p 225.
24. U.S. Patent 4,279,655 (1981).
25. R.H.J. Hannink, 1983, J. Mat. Sci., 18, 457.
26. M.V. Swain, R.C. Garvie and R.H.J. Hannink, 1983, J. Am. Ceram. Soc., 66, 358.
27. H.J. Rossell and R.H.J. Hannink, in [3], p 139.
28. R.H.J. Hannink and M.V. Swain, 1983, J. Aust. Ceram. Soc. 18, 53.
29. M. Marmach and M.V. Swain, 1985, J. Aust. Ceram. Soc. 20.32
30. M.V. Swain, 1985, Acta Metallurgica in press.
31. M.V. Swain and L.R.F. Rose, 1985, submitted to J. Am. Ceram. Soc.
32. D.B. Marshall 1985 submitted to J. Am. Ceram. Soc.
33. M.V. Swain, R.H.J. Hannink and R.C. Garvie, 1983, Fracture Mechanics of Ceramics. 6 p 339
34. M.V. Swain, 1983, ibid. p 355
35. M.V. Swain and R.H.J. Hannink in [3], p 225
36. M.V. Swain, 1985, Fracture Mechanics of Ceramics, Vols 7 and 8 to be published Plenum Press.

FABRICATION AND PROPERTIES OF TRANSFORMATION-TOUGHENED

SODIUM BETA"-ALUMINA

David J. Green

Department of Materials Science and Engineering
The Pennsylvania State University
University Park, PA   16802

INTRODUCTION

Sodium β''-Al_2O_3 is a fast ion conductor for sodium and is useful for a variety of applications[1-3]. In particular, it is the prime candidate as the solid state electrolyte in the Na-S battery. The technical feasibility of this battery has been demonstrated but its lifetime is often limited by electrolytic degradation. Two failure modes have been identified[4] and the analysis of one of these modes [5-7] indicated that the critical current density for the initiation of failure should depend on the fracture toughness of the electrolyte. Thus, techniques to increase the fracture toughness of the ceramic would be beneficial in terms of the lifetime of the battery. Moreover, if the increase in toughness can be translated into improved strength, this would lead to improvements in other aspects of the mechanical reliability.

Transformation toughening is one approach that has been particularly successful in improving the strength and toughness of ceramics. This phenomenon is well established both theoretically and experimentally[8-10] and involves the inducement by stress of a phase transformation in the vicinity of a crack tip. The transformation that is most utilized and studied is that of tetragonal (t-) to monoclinic (m-)ZrO_2. The tetragonal phase is usually only stable at high temperatures but provided the ZrO_2 is kept below a critical size, it can be supercooled to room temperature. The subsequent stress-induced phase transformation involves a large volume increase and shear and the ensuing residual stress distribution reduces the crack extension force[8-10]. It is a reasonable approach, therefore, to determine whether this mechanism can be utilized in Na β''-Al_2O_3.

In preliminary studies[11-12], it was shown that ZrO_2 was compatible with β''-Al_2O_3, that it could be retained in the tetragonal phase and that increases in both fracture toughness and strength were feasible. The aim of this paper is to review previous work on the transformation toughening of β''-Al_2O_3 and to discuss the influence that the ZrO_2 additions have on the electrical and mechanical properties.

INFLUENCE OF ZIRCONIA ADDITIONS ON PROPERTIES

Mechanical Properties

Theoretical analyses have been performed to show that the stress-induced ZrO_2 phase transformation decreases the stress intensity factor (K_I) at a crack tip[8,9,13-15]. This reduction in the crack driving force means that a higher stress needs to be applied to a body before a crack of given size will propagate. The critical value of K_I is denoted as K_{IC} and this parameter is commonly referred to as the fracture toughness of the material. These theoretical calculations have shown that the increase in fracture toughness produced by a stress-induced phase transformation is approximately $0.21V_v Ee^T(h)^{1/2}/(1-\nu)$, where E is the Youngs' modulus of the composite, V_v is the volume fraction of material that transforms, e^T is dilational transformation strain, h is the width of the transformation zone and ν is Poisson's ratio. As the elastic properties of β''-Al_2O_3 and ZrO_2 are similar [16,17] the increment in fracture toughness is simply proportional to $V_v(h)^{1/2}$. There is no data available for h in β''-Al_2O_3/ZrO_2 composites but if we consider h to be independent of volume fraction, we can estimate the increment in K_{IC} from typical K_{IC} values for the end members of the composite system. Figure 1 shows the results of such a calculation using K_{IC} values of 3 and 9 $MPa.m^{1/2}$ for single phase β''-Al_2O_3 and ZrO_2 respectively. Thus, for example, the addition of 20 vol% ZrO_2 would give rise to 40% increase in fracture toughness.

In most ceramics, the strength (σ_f) is determined by the critical flaw size (c) and fracture toughness of the material through the relation $K_{IC} = Y\sigma_f c^{1/2}$, where Y is a constant that depends on the geometry of the loading and the crack. The implication is that increases in fracture toughness can only be translated into increased strength if the critical flaw is not significantly changed. This is important in the β''-Al_2O_3/ZrO_2 system as it will be shown later, that additions of ZrO_2 can actually reduce the critical flaw size. Thus, for this system, increases in strength are often greater than the increase in toughness.

Electrical Properties

Although addition of t-ZrO_2 should increase fracture toughness, it is clear that it will act as an inert phase in terms of sodium ion conduction and will therefore increase the ionic resistivity. There are a variety of approaches to calculate the effect of the addition of a second phase on conductivity. One of the more successful approaches is that of effective-medium percolation theory. In its original form, this theory considers the overall conductivity of a random mixture of particles of two different conductivities under the assumption that the inhomogeneous surroundings of a particle can be replaced by an effective medium[18-19]. The effective resistance ρ_m for the random mixture of particles with resistivities ρ_1 and ρ_2 is given by

$$\rho_m = 4\rho_1\rho_2/\{\rho_1(3V_v-1) + \rho_2(2-3V_v) + \{[\rho_1(3V_v-1) +$$
$$\rho_2(2-3V_v)]^2 + 8\ \rho_1\rho_2\}\}^{1/2} \qquad (1)$$

where V_v is the volume fraction of the phase with resistivity ρ_2. For the β''-Al_2O_3/ZrO_2 system we can consider the ionic resistivity of the ZrO_2 (ρ_2) to be extremely high and thus equation 1 reduces to $\rho_m = 2\rho_1/(2-3V_v)$. Figure 2 shows a plot of this latter function and it is seen that there is a gradual rise in resistivity until $V_v \cong 0.6$, after which the resistivity increases dramatically. This transition is generally referred to as the percolation threshold. For the addition of 20 vol % ZrO_2 the ionic resistivity is expected to increase by 43%. It is interesting to note when

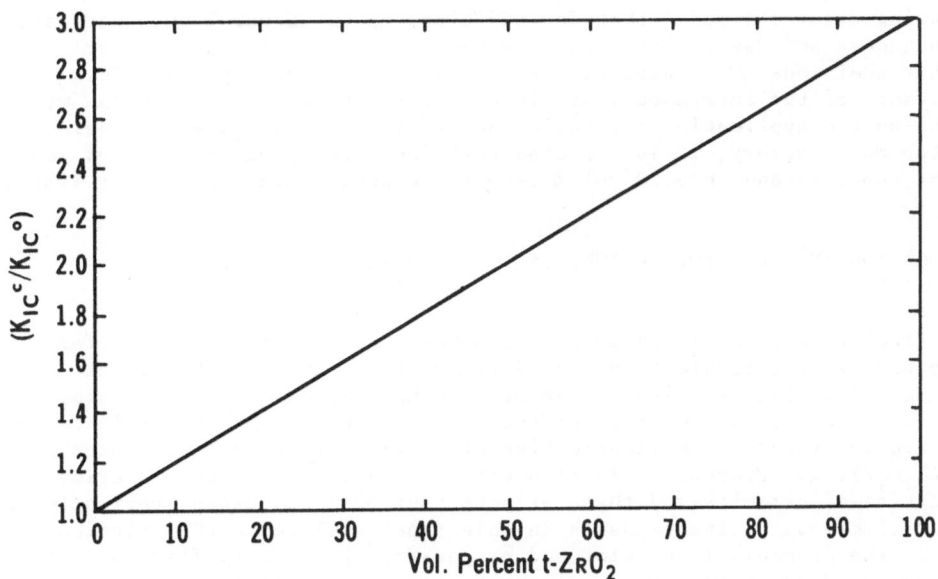

Figure 1. Estimated fracture toughness for $\beta''-Al_2O_3/ZrO_2$ composites $(K_{IC}{}^c)$ as a function of the fracture toughness of $\beta''-Al_2O_3$ $(K_{IC}{}^o)$ and the vol.% ZrO_2.

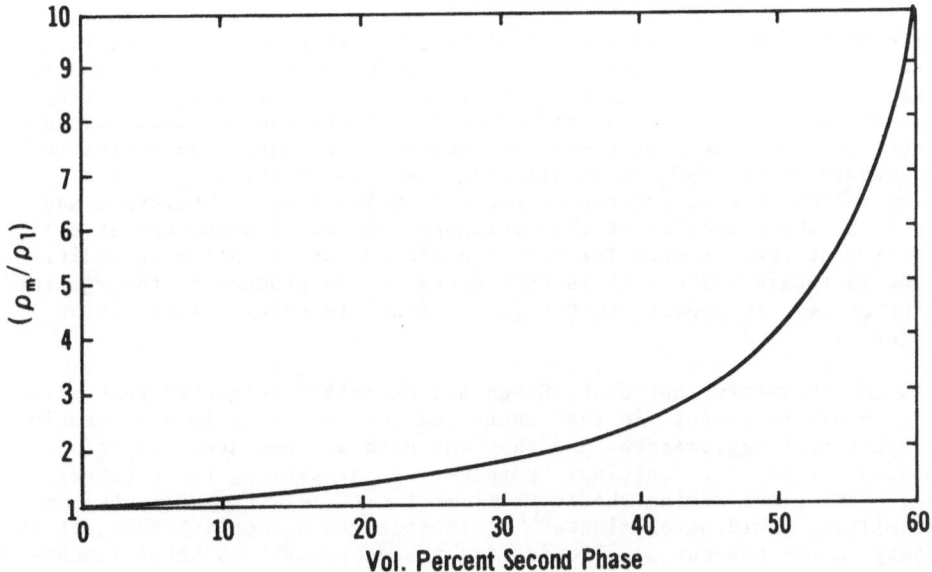

Figure 2. The influence of the addition of a non-conducting phase on the resistivity of a two-phase composite.

comparing Figures 1 and 2 that for additions up to 20 vol % the increases
in toughness and resistivity are expected to be very similar but for
further additions the resistivity should increase more rapidly. The
importance of the increased resistivity caused by the ZrO_2 additions will
depend on the application for which the electrolyte is being considered.
For the Na-S battery, it is expected that increases greater than 40% would
be unacceptable and thus 20 vol % ZrO_2 represents a practical upper limit.

FABRICATION AND FLAW POPULATIONS IN $\beta''-Al_2O_3/ZrO_2$ PARTICULATE COMPOSITES

Fracture origins in ceramics are often related to inhomogeneities
introduced in the fabrication procedure and thus, the reliability of
ceramics is intimately tied to the processing. It is worth discussing,
therefore, some of the sources of these failure origins, especially those
that are important in the fabrication of particulate composites, such as
$\beta''-Al_2O_3/ZrO_2$. Processing is also critical in terms of the electrical
properties of ceramics and these effects have been discussed previously for
the β-aluminas.[20] The emphasis in this paper will be on the relationship
between the mechanical behavior and processing, especially flaw populations
related to agglomerates, voids, impurity inclusions and large grains.

Most ceramic powders contain agglomerates, which may be classified as
hard or soft. In hard agglomerates the crystals are sintered or bonded
together and this is usually a result of a drying or calcination process.
Soft agglomerates, however, are usually a result of electrostatic forces
and these can be broken apart by suitable surfactants. The major problem
caused by these agglomerates is that they can have different densification
kinetics than the surrounding powder. Figure 3 shows the presence of a
hard ZrO_2 agglomerate in $\beta''-Al_2O_3/ZrO_2$ and these have been identified as
failure sources.[21] It is clear that the ZrO_2 agglomerate underwent more
densification than the surrounding material, leaving a deleterious flaw.
For particulate composites, agglomerates can also result from poor mixing
of the two components. Agglomerates can be deleterious in other ways, in
that they can lead to large grains and can act as sources of microcracks
when aided by residual stresses. These effects will be discussed later.

The initial studies on the $\beta''-Al_2O_3/ZrO_2$ system used dry pressing of
powders to form specimens with wet milling being used to mix the powders
and presumably, reduce the agglomerate size[11,12]. It is worth pointing out
that there were some important differences in these two studies. Lange et
al[11] attempted a variety of powder preparation techniques and indicated it
was necessary to use Y_2O_3 as an alloying addition to the ZrO_2. This
addition reduces the unconstrained phase transformation temperature and
thus aids in the retention of the tetragonal phase. Viswanathan et al[12]
used $NaZrO_3$ as their source for ZrO_2 and did not use an alloying additive.
In order to retain $t-ZrO_2$, it is then necessary to produce a finer-grained
microstructure. It appears that this was feasible using a fast firing
technique[12].

As an alternative approach, Green and Metcalf[21] suggested that slip
casting should be useful, in that under the correct conditions it should
break apart soft agglomerates and that the hard agglomerates can be
sedimented out prior to casting. Moreover, by dispersing their powders in
a liquid, excellent mixing should be ensured and any silica impurity produced
by wet milling could be eliminated[11]. In order to accomplish this, it is
necessary to use non-aqueous liquids for the dispersant as water leaches
out sodium from the Na $\beta''-Al_2O_3$ structure and there is a dramatic increase
in pH. It was found that several alcohols could be used as the dispersant
without the need for any additional surfactant.[21] For the two commercial

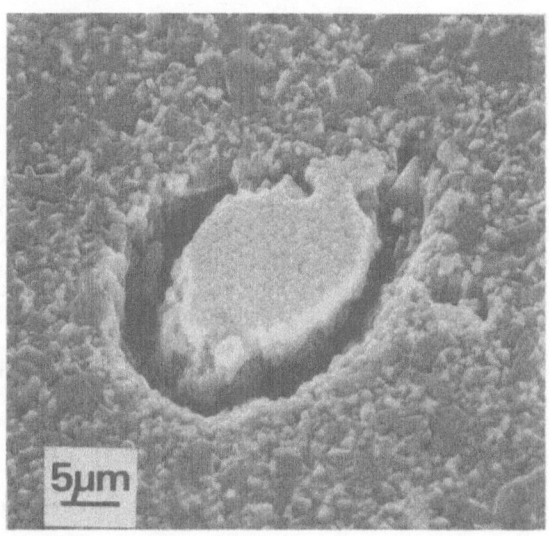

Figure 3. The presence of ZrO_2 aggregates in the composites led to 'void-like' microstructural defects (light phase is ZrO_2).

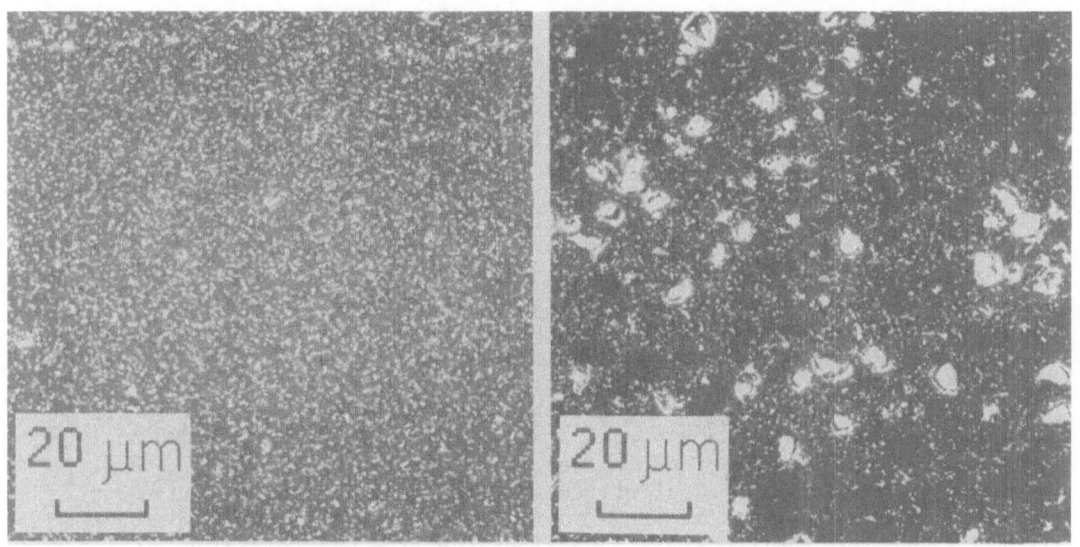

Figure 4. Polished sections of a β''-Al_2O_3/10 vol.% ZrO_2 specimen showing the top (left) and bottom (right) surfaces of the casting. The latter surface shows the presence of the sedimented aggregates and their associated defects.

powders used in this study it was found that although the majority of particles in the powders were less than 1 μm, there were some hard agglomerates that were significantly larger. The presence of the agglomerates was particularly noticable in the fired slip-cast body as the aggregates settle to the bottom of the casting, as shown in Figure 4. In subsequent work, the large ZrO_2 agglomerates were removed by sedimentation prior to the casting. Although a similar procedure was attempted for the $\beta''-Al_2O_3$, it was found that sedimentation led to increases in ionic resistivity[22]. This effect could be a result of the smaller $\beta''-Al_2O_3$ grain size produced by the sedimentation, but as the vol % of agglomerates is relatively small, it was more likely that the sedimentation was preferentially removing some unreacted phases from the powder and hence changing the final composition. As shown in Figure 5, the $\beta''-Al_2O_3$ agglomerates appear to densify without causing any noticable flaw. It is however, important to note that if the $\beta''-Al_2O_3$ grains become too large they may become the strength-controlling flaw population as the fracture toughness of single crystal of $\beta''-Al_2O_3$, especially along its cleavage plane, is substantially lower than that of the surrounding matrix. The importance of eliminating large grains from these materials will be discussed later.

Voids are also common failure origins and are often considered to be a result of incomplete sintering. In some cases, however, the voids that act as failure origins are substantially larger than the voids that remain as a result of the powder packing. In some cases, it is clear that the voids are a result of the burn-out of organic matter from the powder compact. The source of the organic matter may be lint, hair, plastic, etc. Figure 6 shows a void that caused failure in a $\beta''-Al_2O_3/ZrO_2$ composite. It was suspected that it may be a chip from the plastic stirrer that was used in the slip preparation. The use of clean room conditions is thus important for the removal of such flaws.

Large inclusions can also give rise to failure and although these may be impurities, it may be one of the major phases for particulate composites. As the thermal expansion of the inclusion may differ from the surrounding material residual stresses will arise during cooling from the sintering temperature. These stresses may lead to spontaneous or stress-induced microcracking. It is now established that such effects can be avoided provided the size of the inclusion is less than a critical size.[23] It is expected that there is a difference in the thermal expansion of $\beta''-Al_2O_3$ and ZrO_2 and so this is an important consideration for these composites. Anisotropic thermal expansion or phase transformations can give rise to similar effects. Inclusions also give rise to stress concentrations due to differences in elastic properties or the inclusion may possess a lower fracture toughness than the surrounding material. This latter effect is particularly important in $\beta''-Al_2O_3/ZrO_2$ composites, as $\beta''-Al_2O_3$ is known to cleave easily along its basal plane. This effect is compounded by the ease of exaggerated grain growth that occurs in single phase $\beta''-Al_2O_3$[24], such that in the single phase material failure origins are often associated with large grains. In this respect, the addition of ZrO_2 to $\beta''-Al_2O_3$ can be more useful than increasing K_{IC}, in that it can be used to completely eliminate exaggerated grain growth. In order to achieve this control, the ZrO_2 must be located at the grain boundaries of the $\beta''-Al_2O_3$ and its effectiveness will depend on the distribution and volume fraction of the ZrO_2, but in both $\beta''-Al_2O_3/ZrO_2$ and Al_2O_3/ZrO_2 composites, it was found that additions > 7.5 vol % of the ZrO_2 was sufficient[22,25]. Figure 7 compares the microstructures of single phase $\beta''-Al_2O_3$ with a composite in which the exaggerated grain growth has been eliminated.

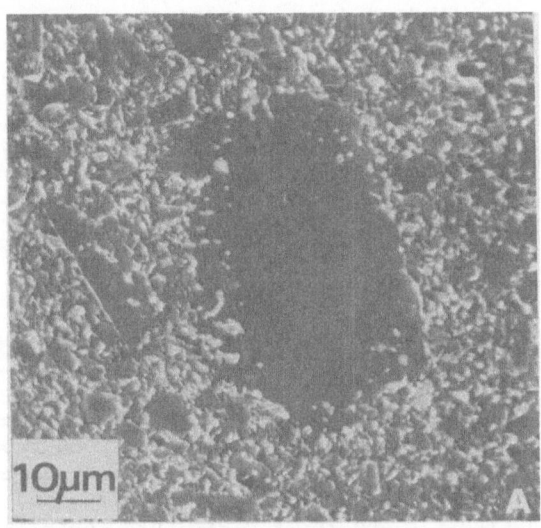

Figure 5. The presence of β''-Al_2O_3 aggregates led to larger grains in the final microstructure but no densification defects.

Figure 6. Failure origin due to the presence of a void.

IONIC RESISTIVITY OF β''-Al$_2$O$_3$/ZrO$_2$ COMPOSITES

As indicated earlier, the addition of ZrO$_2$ to β''-Al$_2$O$_3$ is expected to increase the ionic resistivity (ρ). For β''-Al$_2$O$_3$, ionic resistivities are generally determined at 300°C. Lange et al[11] studied only a composite containing 15 vol % ZrO$_2$ and found ρ = 7.7 Ω .cm. Compared to their measurements on a commercial single phase β''-Al$_2$O$_3$ (3.7 Ω .cm), this value represented more than a 100% increase in resistivity. Their composite contained a large fraction of β-Al$_2$O$_3$, which is known to have a higher resistivity than the β'' phase.[26] It is important to note that Lange et al.,[11] did not compare the resistivity of the composite to that of β''-Al$_2$O$_3$ processed in the same way. Viswanathan et al[12], in their study, reported that additions of up to 8.6 vol % ZrO$_2$ usually did not increase resistivity at all, but their value for β''-Al$_2$O$_3$ was significantly higher than that reported by Lange et al[11] (6.4 Ω .cm). This difference, however, could simply be a result of differences in the processing procedure. For the higher volume fractions, Viswanathan et al[12] did not report any data but indicated that the resistivity was very high. More extensive data has been measured by Green[22] on composites containing t-ZrO$_2$ and cubic ZrO$_2$. These data are shown in Figure 8 and it is clear that there is a relatively gradual increase in resistivity. Compared with Figure 2, however, the increase is greater than that predicted by percolation theory. For example, at 20 vol % the resistivity is increased by 70%. A feasible explanation for the additional increase may be the reduction in grain size caused by the ZrO$_2$ addition. For example, Youngblood et al[26], have shown that changing the β''-Al$_2$O$_3$ grain size from 100 to 2 μm, increased resistivity by a factor of ~1.6. The increase could also be a result of impurities introduced from the ZrO$_2$ powder. Figure 9 shows ionic resistivity values for some of the composites over a limited temperature range. It appears that the addition of ZrO$_2$ does not significantly influence the conduction process in terms of the activation energy.

MECHANICAL BEHAVIOR OF β''-Al$_2$O$_3$/ZrO$_2$ COMPOSITES

Previous studies have shown that the addition of t-ZrO$_2$ to β''-Al$_2$O$_3$ increases both the fracture toughness and the strength. Lange et al[11] reported values of K_{IC} = 4.5 MPa.m$^{1/2}$ and an average strength of 350 MPa for a 15 vol % ZrO$_2$ composite and compared this with literature data to indicate that these values are higher than β''-Al$_2$O$_3$. Viswanathan et al[12] concentrated on a composite containing 8.7 vol % ZrO$_2$ and compared their data with single phase β''-Al$_2$O$_3$ processed in the same way. They concluded that the fracture toughness increased by more than a factor of 2 but that these values were high because of the residual surface stresses that are induced by grinding. The stresses associated with grinding promotes the formation of m-ZrO$_2$ preferentially at the surface and the compressive stresses associated with the transformation leads to an apparent increase in toughness. For the same composite, they found the strength was increased by a factor of 1.5 to 1.8 and for a 15 vol % composite by a factor of 2.2 to 2.5. Viswanathan et al.[12] concluded that the strength increase was probably a result of the surface compression. Increases in strengths were measured by Green and Metcalf[21]. Figure 11 shows the type of strength increases that were observed for composites containing various volume fractions of ZrO$_2$[27]. For the 15 vol % composite, the values are similar to both of the previous studies[11,12]. Green and Metcalf[21], however, concluded that as the increase in strength is larger than the increase in fracture toughness, there must be a reduction in the critical flaw size when one adds the ZrO$_2$ and that this is probably related to the elimination of exaggerated grain growth. In other words, the control of the grain size eliminates failure associated with the large β''-Al$_2$O$_3$ grains. In the composites containing Y$_2$O$_3$, the ZrO$_2$ transformation is

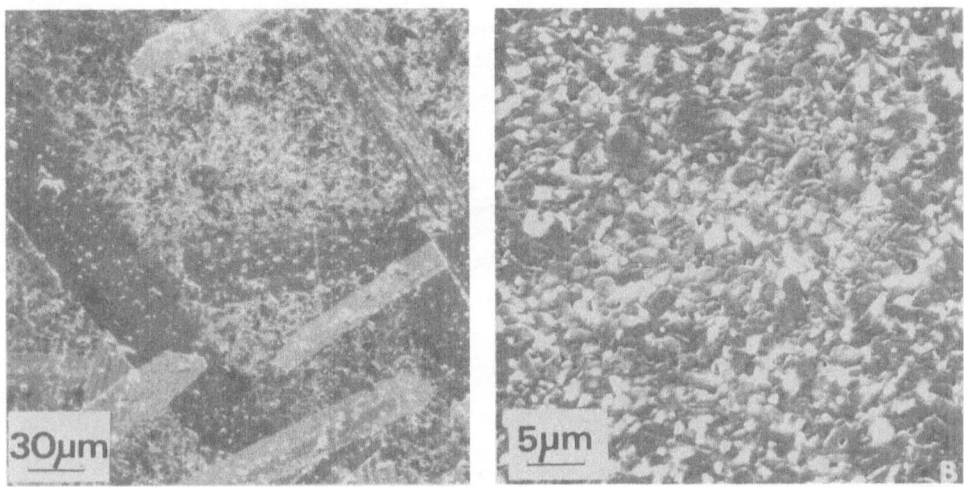

Figure 7. A comparison of single phase β''-Al$_2$O$_3$ (left) with a composite containing 15 vol% ZrO$_2$ (right) shows the elimination of exaggerated grain growth by the second phase addition.

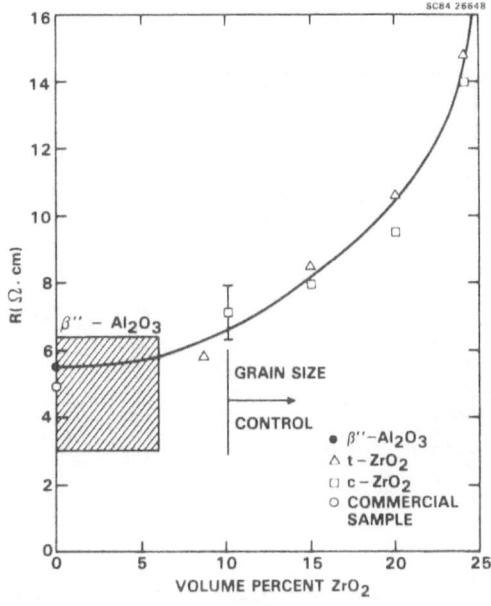

Figure 8. The change in ionic resistivity (300°C) of β''-Al$_2$O$_3$ with additions of cubic or tetragonal ZrO$_2$.

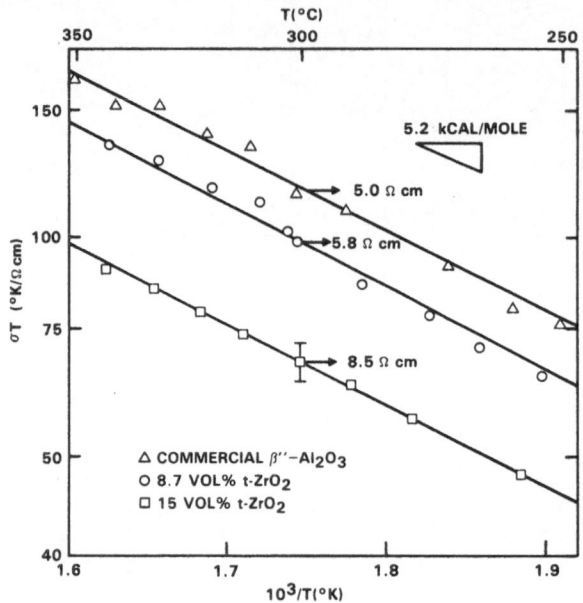

Figure 9. Ionic resistivity (300°C) of commercial β''-Al_2O_3 and composites containing tetragonal ZrO_2.

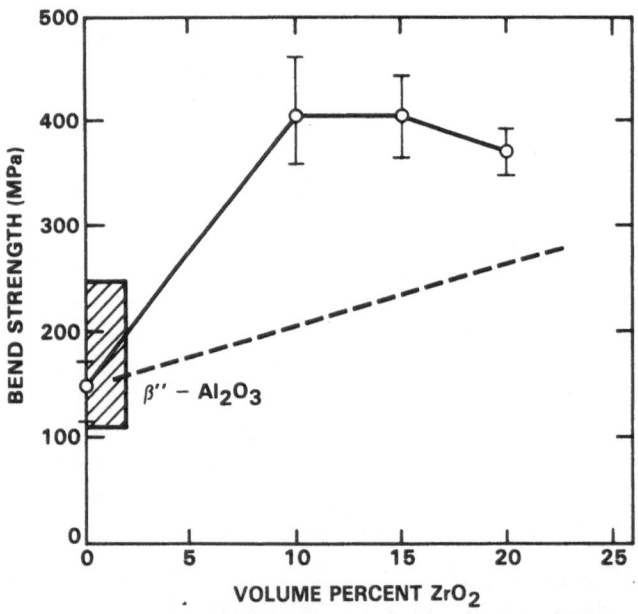

Figure 10. The strength of β''-Al_2O_3/ZrO composites as a function of the amount of t-ZrO_2. The dotted line indicates the strengthening predicted from the fracture toughness data assuming no change in critical flaw size.

expected to be more difficult than for the unstabilized ZrO_2 and Green and Metcalf observed evidence of a stress-induced phase transformation on fracture surface but not on polished surfaces[27]. It seems unlikely, therefore, that surface compression was increasing their K_{IC} values which were measured on polished surfaces using indentations. It was also doubtful that surface compression was influencing their strength values as all the failures occurred at volume rather than surface defects.

In order to determine whether the increase in strength in $\beta"-Al_2O_3$ /ZrO_2 composites was a result of the increase in fracture toughness due to transformation toughening, changes in the critical flaw size or surface compression, Green[22] studied composites containing $t-ZrO_2$ or cubic ZrO_2. If the reduction of the critical flaw size is important then one would expect strength increases to occur in both systems but the strengthening would be somewhat larger for the $t-ZrO_2$ system as a result of the transformation toughening. Table 1 compares the K_{IC} values for the two systems and shows that the $t-ZrO_2$ is more effective in increasing toughness. The increase in fracture toughness is similar to that predicted in Figure 1. The K_{IC} data was also used to predict in the increase in strength that one would expect if the critical flaw size was not influenced by the ZrO_2 additions. It was found that for both the cubic and $t-ZrO_2$ additions that the strength was increased compared to a similarly-processed single phase $\beta"-Al_2O_3$ and that the measured strength was considerably in excess of that predicted assuming no change in flaw size. The conclusion, therefore, was that it was a reduction in critical flaw size that was the major strengthening mechanism for both systems and that this was probably a result of the elimination of the exaggerated grain growth associated with the $\beta"-Al_2O_3$.

Table 1. The Influence of Additions of Tetragonal and Cubic ZrO_2 on Fracture Toughness

Volume Percent ZrO_2	$K_{IC}(MPa \cdot m^{1/2})$	
	Cubic ZrO_2	Tetragonal ZrO_2
0	3.0	3.0
10	3.5	3.3
15	3.2	4.1
20	3.3	4.1
25	3.9	4.5

In order to study the effect of surface compression on the K_{IC} data, Green[28] used a heat treatment to remove Y_2O_3 from the surface. In this work the effect of the residual surface stress was studied using the parameter $P/c^{3/2}$, where P is the load applied to a Vicker's indentor and c is the radius of the radial cracks formed at the indentation corners. For a stress-free surface, this parameter is expected to be proportional to K_{IC}[29] and hence is a constant, but it will change with crack length for a residually-stressed surface[30]. The results of the study are shown in Figure 11 and it was found that the surface compression increased the load needed to form the indentation cracks and that although $P/c^{3/2}$ was constant as expected for the stress-free surface, the toughness could be apparently increased by more than a factor of 2 by the surface compression. It is interesting to note that the trend observed in Figure 11 is opposite to that seen in the work of Viswanathan et al., and this has been discussed in recent publications [31,32]. The introduction of surface compression in β''-Al_2O_3 may be a useful strengthening mechanism and an additional technique to improve the resistance to electrolytic degradation.

CONCLUSIONS

It has been conclusively demonstrated that additions of t-ZrO_2 can be used to increase the fracture toughness and strength of β''-Al_2O_3. The increased toughness is associated with transformation toughening but the strength depends on the flaws introduced by processing. It is critical, therefore, to identify and remove the sources of these flaws. Indeed, further increases in strength over those reported here are feasible, if further improvements in processing are made. It has been found that the addition of an inert phase, such as ZrO_2, can be useful in eliminating exaggerated grain growth in β''-Al_2O_3 and that in some cases, the ensuing removal of failure origins associated with large grains can be a major strengthening mechanism. The ionic resistivity of β''-Al_2O_3/ZrO_2 composites depends primarily on the volume fraction of ZrO_2 added and is such that the resistivity invariably increases. The use of the toughening and strengthening mechanisms associated with the ZrO_2 will therefore depend on the resistivity requirements for a particular application. Future studies are needed to determine the influence of the ZrO_2 additions on the lifetime of the electrolyte.

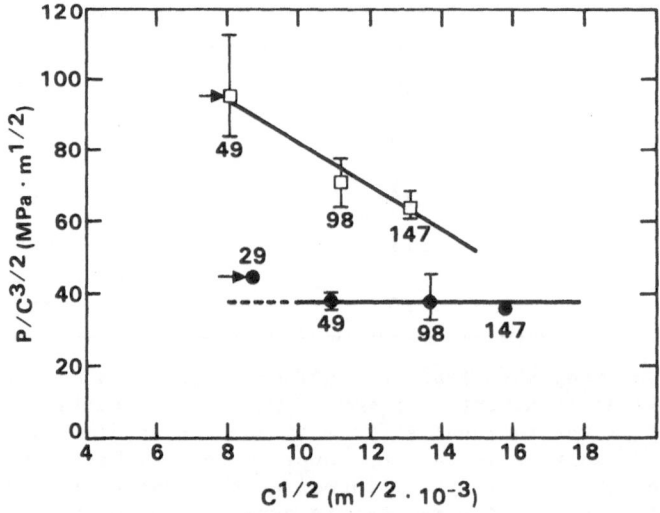

Figure 11. Comparison of indentation behavior for a compressively-stressed surface (open symbols) and a stress-free surface (closed symbols) in β''-Al_2O_3/ZrO_2 composites. (The numbers indicate the indentation load (N) and the arrows, the load at which radial cracks were first observed.)

ACKNOWLEDGMENTS

The author would like to acknowledge the critical discussions with F. F. Lange. The preparation of the manuscript was supported by a grant from the Faculty Research Fund, The Pennsylvania State University.

REFERENCES

1. G. Holzapfel and H. Rickert, Solid State Chemistry - New Possibilities for Research and Industry, Naturwiss 64:53 (1977).
2. B. Steele, Ions in the Solid State, New Scientist, 78:705 (1978).
3. T. A. Wheat, Exploring Ionic Conductors, Clay and Ceramics 52:17 (1978).
4. L. C. DeJonghe, L. A. Feldman and A. Buechele, Slow Degradation and Electron Conduction in Sodium Beta-Aluminas, J. Am. Ceram. Soc. 16:780 (1981).
5. R. H. Richman and G. J. Tennenhouse, A Model for Degradation of Ceramic Electrolytes in the Na-S Batteries, J. Am. Ceram. Soc. 58:63 (1975).
6. D. K. Shetty, A. V. Virkar and R. S. Gordon, Electrolytic Degradation of Lithia-Stabilized β"-Alumina, in: "Fracture Mechanics of Ceramics, Vol. 4," R. C. Bradt et al., Plenum Press, New York (1978).
7. L. A. Feldman and C. DeJonghe, Initiation of Mode 1 Degradation in Sodium Beta-Alumina Electrolytes, J. Mater. Sci. 17:517 (1982).
8. A. G. Evans and A. H. Heuer, Transformation Toughening in Ceramics: Martensitic Transformations in Crack Tip Stress Fields, J. Am. Ceram. Soc. 63:241 (1980).
9. F. F. Lange, Transformation Strengthening: Thermodynamic Approach to Phase Retention and Toughening, in: "Fracture Mechanics of Ceramics, Vol. 6," R. C. Bradt et al., eds., Plenum Press, New York (1983).
10. N. Claussen, M. Ruhle and A. H. Heuer (eds), "Science and Technology of Zirconia II, Advances in Ceramics, Vol. 12," American Ceramic Society, Columbus, OH (1984).
11. F. F. Lange, B. I. Davis and D. O. Raleigh, Transformation Strengthening of β"-Al_2O_3 with Tetragonal ZrO_2, J. Am. Ceram. Soc. 66:C-50 (1983).
12. L. Viswanathan, Y. Ikuma and A. V. Virkar, Transformation Toughening of β"-Alumina by Incorporation of Zirconia, J. Mater. Sci. 18:109 (1983).
13. R. McMeeking and A. G. Evans, Mechanics of Transformation-Toughening in Brittle Materials, J. Am. Ceram. Soc. 65:242 (1982).
14. B. Budiansky, J. Hutchison and J. Lambroupolos, Continuum Theory of Dilatant Transformation Toughening in Ceramics, Int. J. Solids Struct. 19:337 (1983).
15. D. B. Marshall, M. D. Drory and A. G. Evans, Transformation Toughening in Ceramics, in: "Fracture Mechanics of Ceramics, Vol. 6," R. C. Bradt et al., eds., Plenum Press, New York (1983).
16. A. V. Virkar and R. S. Gordon, Fracture Properties of Polycrystalline Lithia-Stabilized β"-Al_2O_3, J. Am. Ceram. Soc. 60:58 (1977).
17. F. F. Lange, Transformation Toughening: IV, Fabrication, Fracture Toughness and Strength of Al_2O_3/ZrO_2 Composites, J. Mater. Sci. 17:247 (1982).
18. R. Landauer, The Electrical Resistance of Binary Metallic Mixtures, J. Appl. Phys. 23:779 (1952).
19. D. G. Ast, Evidence for Percolation-Controlled Conductivity in Amorhous As_xTe_{1-x} Films, Phys. Rev. Letters 33:1042 (1974).

20. R. W. Powers and S. P. Mitoff, The Influence of Crystal Structure and of Microstructure on Some Properties of Polycrystalline Beta-Alumina, in: "Solid Electrolytes", P. Hagenmuller and W. Van Gool (eds.), Academic Press, New York (1977).

21. D. J. Green and M. G. Metcalf, Properties of Slip-Cast Transformation-Toughened β''-Al_2O_3/ZrO_2 Composites, Am. Ceram. Soc. Bull. 63:803 (1984).

22. D. J. Green, Transformation Toughening and Grain Size Control in β''-Al_2O_3/ZrO_2 Composites, J. Mater. Sci. 20:2639 (1985).

23. D. J. Green, Microcracking Mechanisms in Ceramics, in: "Fracture Mechanics of Ceramics, Vol. 5," R. C. Bradt et al., eds., Plenum Press, New York (1983).

24. G. E. Youngblood, A. V. Virkar, W. R. Cannon and R. S. Gordon, Sintering Processes and Heat Treatment Schedules for Conductive, Lithia-Stabilized β''-Al_2O_3, Am. Ceram. Soc. Bull. 56:206 (1977).

25. D. J. Green, Critical Microstructures for Microcracking in Al_2O_3/ZrO_2 Composites, J. Am. Ceram. Soc. 65:610 (1982).

26. G. E. Youngblood, G. R. Miller and R. S. Gordon, Relative Effects of Phase Conversion and Grain Size on Sodium Ion Conduction in Polycrystalline, Lithia-Stabilized β''-Al_2O_3, J. Am. Ceram. Soc. 61:86 (1978).

27. D. J. Green and M. G. Metcalf, Unpublished Data.

28. D. J. Green, Improved Beta"-Aluminum Oxide Electrolytes Through Transformation Toughening, Final Report, Subcontract No. 4523010 (LBL), May 1984.

29. D. B. Marshall, B. R. Lawn and A. G. Evans, Elastic/Plastic Indentation Damage in Ceramics: The Median/Radial Crack System, J. Am. Ceram. Soc. 63:574 (1980).

30. D. B. Marshall and B. R. Lawn, An Indentation Technique for Measuring Residual Stresses in Tempered Glass Surfaces, J. Am. Ceram. Soc. 60:86 (1977).

31. D. J. Green, Discussion of "Crack Size Dependence of Fracture Toughness in Transformation-Toughened Ceramics, J. Mater. Sci. To be published, (1985).

32. Y. Ikuma and A. V. Virkar, Response to Discussion of "Crack Size Dependence of Fracture Toughness in Transformation-Toughened Ceramics, J. Mater. Sci. To be published, (1985).

PHASE TRANSFORMATION AND TOUGHENING IN MgO DISPERSED WITH ZrO_2

Yasuro Ikuma, Atsushi Yoshimura, Kuniaki Ishida and
Wazo Komatsu

Ikutoku Technical University
1030 Shimoogino, Atsugi, Kanagawa 243-02, Japan

INTRODUCTION

Transformation of tetragonal ZrO_2 to monoclinic ZrO_2 has widely been used to increase the toughness of ceramic materials. The materials being used as a matrix include cubic ZrO_2,[1,2,3] α-alumina,[4] β"-alumina,[5] ZnO,[6] Si_3N_4,[7] etc. For the toughening of alumina one can use either monoclinic ZrO_2 or tetragonal ZrO_2. The results of most of the works indicated that the enhancement of toughness was greater when the tetragonal ZrO_2 was dispersed in the matrix.[8] This is, of course, due to the fact that stress induced transformation is responsible for the toughening.

Extending the idea, we expect that MgO can be used as both matrix and stabilizer, i.e., we could enhance the toughness of MgO by incorporating ZrO_2. We have shown[9] that this could be done by sintering MgO+9.8mol% ZrO_2 at 1300-1400°C. The fracture toughness estimated by indentation technique was higher for MgO+9.8mol% ZrO_2 than for MgO.

In this work we studied the phase changes of ZrO_2 in the composites of MgO+5-20mol% ZrO_2 and related the polymorphic forms of ZrO_2 in the composites to the toughening of the material.

EXPERIMENTAL

ZrO_2* at varying fractions(0, 5, 10, 15, and 20mol%) was mixed with MgO* by using ball milling technique. Cylindrical alumina was used as milling medium. Disks of 1cm in diameter and 0.5cm thick were uniaxially pressed at 147MPa and subsequently treated by isostatic pressing at 127MPa. Sintering was performed at 1400°C, 1500°C and 1600°C for 1hr in air.

Linear shrinkage and density of the sintered specimen were measured. Density measurement was performed by immersing the specimen in butyl alcohol. X-ray diffraction analysis was performed on the sintered specimen. The height of X-ray diffraction peaks at the highest intensity (peaks of (111) for cubic ZrO_2, (101) for tetragonal ZrO_2 and (11$\bar{1}$) for monoclinic ZrO_2) was used to represent the amount of each phase in the specimen. The peak height may not be absolutely proportional to the amount of each

---

* Kanto Chemical Co., Tokyo, Japan.

phase.[10] However, it is good enough to estimate the relative change of each phase.

Fracture toughness was measured using indentation technique on the polished surface of specimen. The polishing process might alter the K_{IC} value of the specimen[11] if the toughening was caused by phase transformation of ZrO_2. This could be avoided by annealing the polished specimen. However, we did not anneal the polished specimen at high temperatures, because it would change the composition. Although there is a question as to the validity of the absolute value of K_{IC} measured in this way, it will not pose any problem if we discuss the relative change of K_{IC}.

The compositions studied in this work lie in MgO rich portion of the phase diagram shown in Fig. 1. If the system reached the equilibrium at sintering temperatures, ZrO_2 should be present as cubic ZrO_2 or tetragonal ZrO_2.

Some of the specimens of MgO+10mol% ZrO_2 sintered at 1600°C for 1hr were annealed at temperatures between 1000 and 1400°C for 1 to 24hrs and then the measurements similar to those done on the sintered specimens were performed.

RESULTS

Density

The relative densities of MgO-ZrO_2 ceramics sintered at 1400-1600°C are shown in Fig. 2. The densities of MgO and ZrO_2 utilized for the calculation of theoretical densities of composite materials are listed in

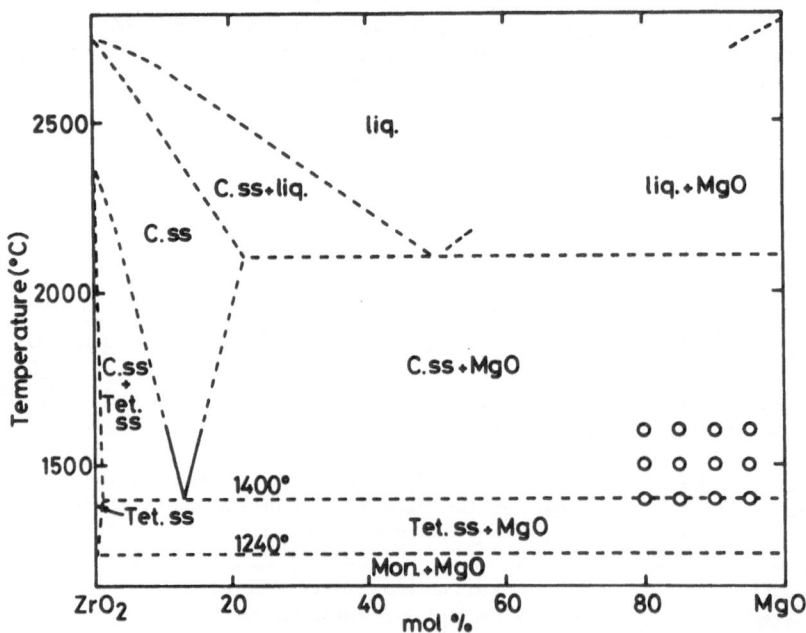

Fig. 1. Phase diagram of MgO-ZrO_2 system showing compositions and temperatures investigated in the present study.

Table 1. Theoretical densities

Composition	Density(g/cm^3)
MgO	3.58
m-ZrO_2(0%MgO)[12]	5.83
t-ZrO_2(1mol% MgO)[13]	5.82
c-ZrO_2(15mol% MgO)[14]	5.69
MgO+5% ZrO_2 {1400°C	3.78
1500, 1600°C	3.78
MgO+10% ZrO_2 {1400°C	3.97
1500, 1600°C	3.97
MgO+15% ZrO_2 {1400°C	4.14
1500, 1600°C	4.15
MgO+20% ZrO_2 {1400°C	4.30
1500, 1600°C	4.31

Table 1. The results of X-ray diffraction analysis showed that ZrO_2 in the specimen was cubic, tetragonal or monoclinic depending on the sintering temperature. We have calculated the theoretical density of composite materials assuming that polymorphic form of ZrO_2 with the highest concentration

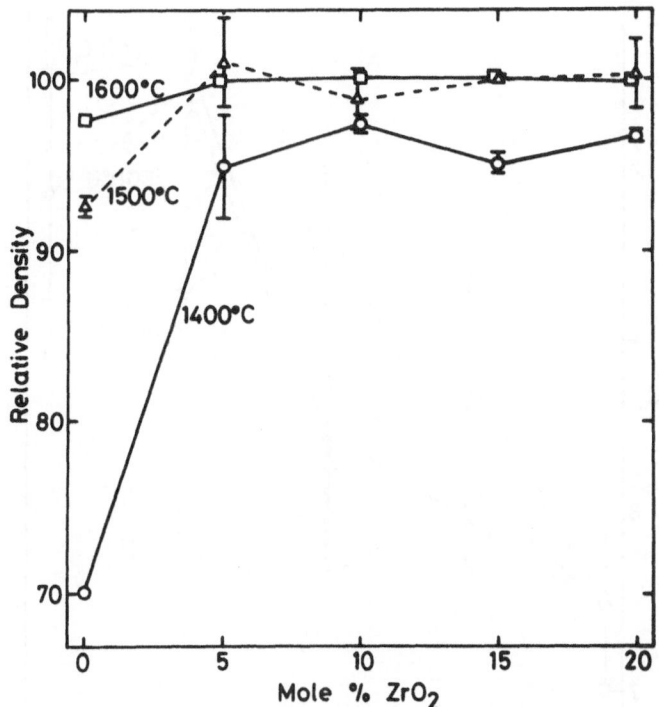

Fig. 2. Relative density of MgO+ZrO_2 composite after sintering at various temperatures(1400, 1500, and 1600°C). Error bars on the data points represent the standard deviation. The standard deviation of some points is too small to be shown.

was the only phase of ZrO_2. This can be one of the reasons why the relative densities over 100% were obtained(Fig. 2).

At all temperatures studied, the density of the sintered body increased as MgO was mixed with ZrO_2. The effect is pronounced especially at 1400 and 1500°C. The examination of polished surfaces of these specimens revealed that there were few pores in those sintered at 1500 and 1600°C with addition of ZrO_2, supporting the results shown in Fig. 2. However, the specimens sintered at 1400°C had some porosity especially in those mixed with 5% ZrO_2.

As-sintered composites

Fracture toughnesses estimated by indentation technique for as-sintered composites are shown in Fig. 3. For the composite materials sintered at 1500 and 1600°C the values of K_{IC} increased approximately with % ZrO_2. At 20% ZrO_2, K_{IC} was nearly equal to K_{IC} at 15%. On the other hand, the values of K_{IC} for material sintered at 1400°C showed a different trend. They had a maximum at 10% ZrO_2.

The results of K_{IC} correspond nicely to the amount of phase in these

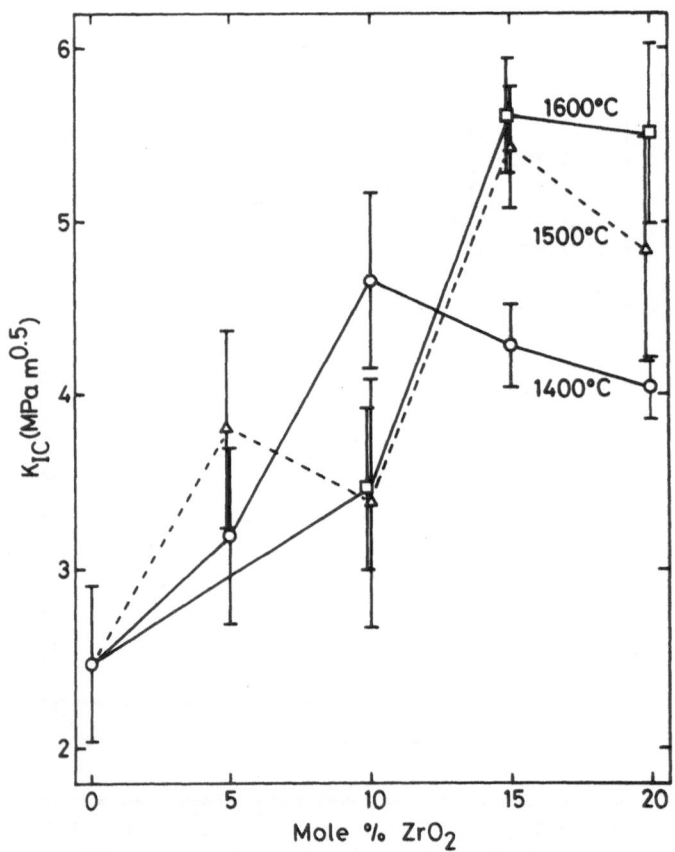

Fig. 3. Fracture toughness of $MgO+ZrO_2$ composite after sintering at various temperatures(1400, 1500, and 1600°C). Error bar is standard deviation.

materials determined from the X-ray diffraction analysis(Fig. 4). For the composite material sintered at 1500 and 1600°C, the only polymorphic form of ZrO_2 was cubic. This is in accord with the prediction of the phase diagram(Fig. 1). The concentration of cubic ZrO_2 increased as ZrO_2 was added to MgO and the K_{IC} value of the material increased at the same time although there was a little fluctuation. Therefore the toughening of MgO+ cubic ZrO_2 must be related to the presence of cubic ZrO_2.

For the composite material sintered at 1400°C, ZrO_2 was present as monoclinic and tetragonal phases(Fig. 4). The phase diagram predicts the presence of only tetragonal ZrO_2. We suspect that some tetragonal ZrO_2 has transformed into monoclinic ZrO_2 during the cooling process. The concentration of tetragonal ZrO_2 was maximum at 10% ZrO_2 where the maximum K_{IC} was observed. For the composite material at higher concentration of ZrO_2 the amount of tetragonal phase decreased and so did the K_{IC} value.

Annealed composites

Fracture toughness of MgO+10% ZrO_2 annealed at various temperatures ranging from 1000 to 1400°C after sintering at 1600°C are shown in Fig. 5. At all the temperatures investigated, the K_{IC} values had a maximum at annealing time of 1-2hrs. When the composite was annealed for more than 1-2hrs, the K_{IC} value gradually decreased.

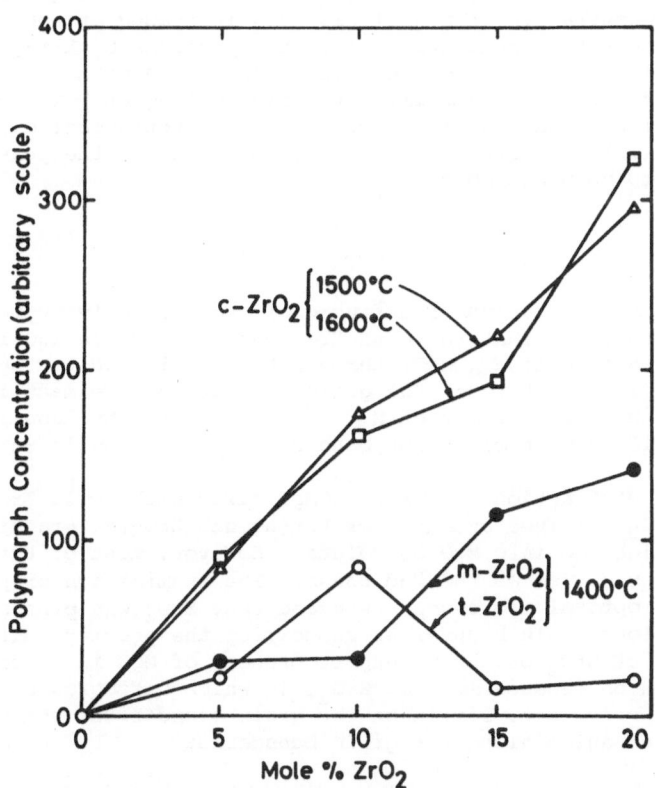

Fig. 4. Polymorph concentration in MgO+ZrO_2 composite after sintering at various temperatures(1400, 1500, and 1600°C).

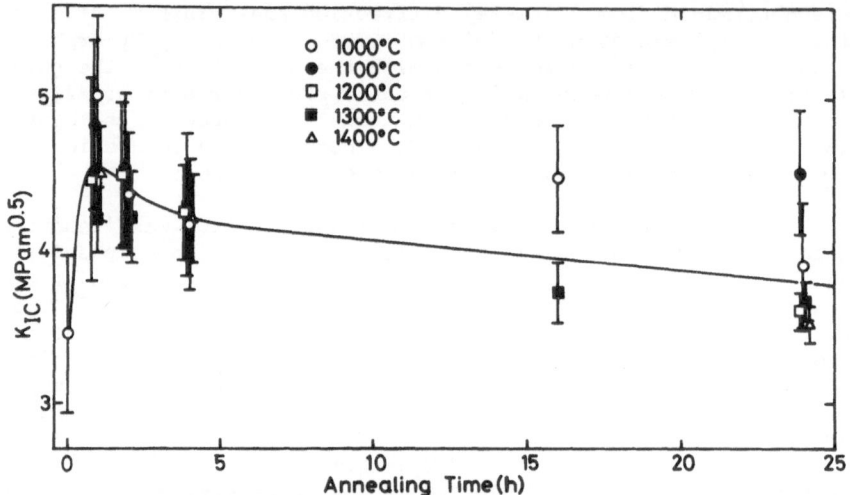

Fig. 5. Fracture toughness of MgO-ZrO$_2$ composite after sintering at 1600°C and annealing at various temperatures (1000-1400°C). Error bar is standard deviation.

The phase change of ZrO$_2$ examined by X-ray diffraction analysis during
the annealing process is shown in Fig. 6. As the annealing process continued, the concentration of monoclinic ZrO$_2$ increased at the expense of
cubic ZrO$_2$. A careful examination of these specimens by X-ray diffraction
analysis revealed the traces of tetragonal ZrO$_2$. We noticed that the peak
from (111) of cubic ZrO$_2$ gradually developed tailing at lower 2θ indicating
that tetragonal ZrO$_2$ was forming. Probably some tetragonal did exist but
the peak from (101) of tetragonal ZrO$_2$ is too close to the peak from (111)
of cubic ZrO$_2$ to be measurable.

DISCUSSIONS

After extensive studies of diffusional creep of polycrystalline MgO,
Gordon[15] has concluded that the mass transport process in MgO is controlled
by the diffusion of cation (Mg^{2+}) through the lattice while the diffusion
of anion (O^{2-}) is fast through the grain boundary of the material. We can
explain the high sinterability of MgO mixed with ZrO$_2$ by assuming the enhancement of diffusion process for both cation and anion in MgO.

The X-ray diffraction analysis of MgO mixed with 1mol% ZrO$_2$ has revealed that ZrO$_2$ remained as cubic or tetragonal ZrO$_2$.* Fractions of ZrO$_2$
mixed in MgO might go into MgO as solute. However, most of them remained
on the grain boundaries as the 2nd phase. The examination of polished surface under the optical microscope revealed that ZrO$_2$ was present as a
phase mainly along grain boundaries supporting the previous discussion. One
of the effects of ZrO$_2$ on the transport process of MgO is to enhance the
diffusion of anion because most of ZrO$_2$, in which diffusion of oxygen is
expected to be very fast, remain on the grain boundaries, hence enhancing
the transport of anion along the grain boundaries.

According to Eq. (1) we see that the concentration of V''_{Mg} increases

* The X-ray peak of ZrO$_2$ at 1% level was too weak to distinguish cubic ZrO$_2$
from tetragonal.

300

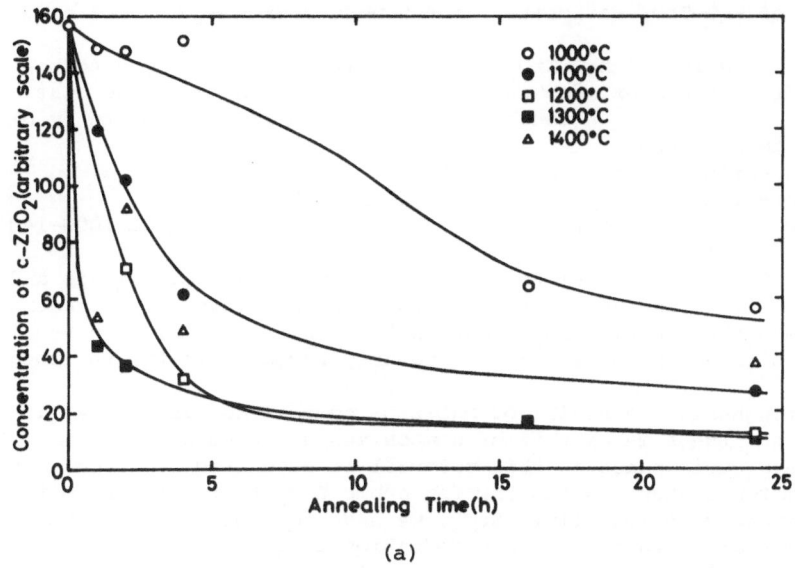

(a)

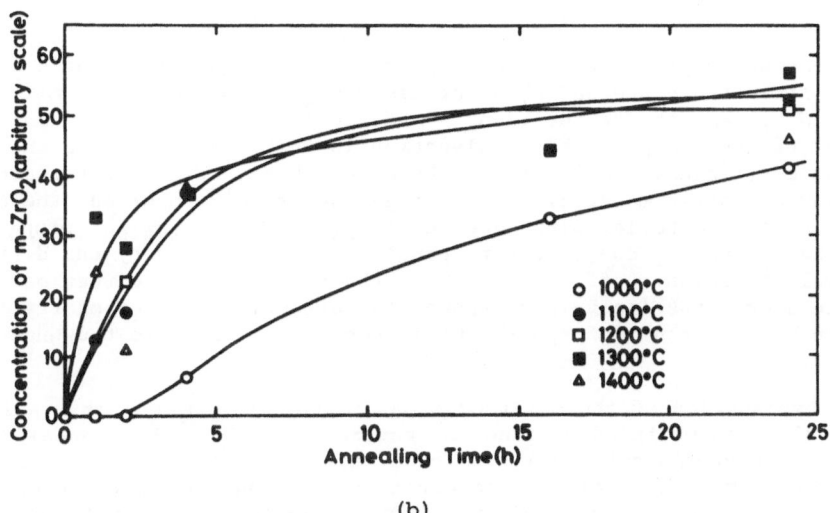

(b)

Fig. 6. Polymorph concentration in MgO-ZrO$_2$ composite after sintering at 1600°C and annealing at various temperatures(1000-1400°C). (a) cubic ZrO$_2$, and (b) monoclinic ZrO$_2$.

$$ZrO_2 \xrightarrow{\text{MgO}} Zr_{Mg}^{\bullet\bullet} + 2O_O^x + V_{Mg}'' \tag{1}$$

as the amount of ZrO$_2$ increases leading to the enhancement* of lattice diffusion of Mg^{2+} in the composite material. Although we do not know whether the overall transport process in MgO dispersed with ZrO$_2$ is controlled by anion grain boundary diffusion or cation lattice diffusion, both of them are enhanced and consequently it is very natural to find that

_____
* Schottky type defect is assumed to be dominant in MgO.[14,15]

sintering of MgO-ZrO$_2$ composite was enhanced over that of pure MgO.

The toughening of MgO-ZrO$_2$ composite is complicated due to the differences in ZrO$_2$ phases existed in the material depending on the heating history. We can discuss the toughening of the MgO-ZrO$_2$ composite by dividing them into three groups:
1) As-sintered composites at 1400°C.
2) As-sintered composites at 1500 or 1600°C.
3) Composites sintered at 1600°C and then annealed at 1000-1400°C.

MgO-ZrO$_2$ composites sintered at 1400°C contained monoclinic and tetragonal ZrO$_2$. Since the composite with the largest amount of tetragonal ZrO$_2$ showed the highest K$_{IC}$, transformation toughening such as observed in partially stabilized zirconia[1,2,3] is probably responsible for the toughening.

The toughening mechanism of MgO-ZrO$_2$ composites sintered at 1500 or 1600°C is not the same as that of composites sintered at 1400°C. ZrO$_2$ was present as a cubic phase in this material as predicted from the phase diagram. Since room temperature is too low for cubic ZrO$_2$ to transform to tetragonal and subsequently to monoclinic ZrO$_2$, we have to look for the cause of toughening in the area other than transformation toughening. In fact, when the concentration of monoclinic ZrO$_2$ was examined in these specimens before and after a ball milling lasting about 24hrs, we did not see any appreciable changes.

We can not explain the toughening by the difference in Young's moduli between matrix and dispersed phase as discussed by Swearengen et al.[16] for ThO$_2$-glass system because the Young's modulus[17] of cubic ZrO$_2$ is smaller than that of MgO. A plausible explanation is that the difference in thermal expansion coefficients (α)* between these two phases creates the thermal stress in the matrix material. As Davidge and Green[18] observed, the crack in the composite material with $\Delta\alpha > 0$ ($\Delta\alpha = \alpha_{matrix} - \alpha_{inclusion}$) propagates toward the inclusions due to the tangential stress created by the difference in thermal expansion coefficients. Then the crack changes direction to propagate around the inclusions because the effective surface energy (γ) of ZrO$_2$[19] is higher than γ of MgO.[20] Consequently the fracture toughness increases.

The explanation of the behavior of annealed MgO-ZrO$_2$ composites is not simple. Although we tried to find the tetragonal ZrO$_2$ in the annealed specimen(virgin and ground surfaces), we found only traces of them. We have to conclude that tetragonal ZrO$_2$ is not stable in the composite annealed at temperatures ranging from 1000 to 1400°C after sintering at 1600°C. Let us calculate the volume change of the eutectoid decomposition during the annealing. When the composite of 80mol% MgO+20mol% ZrO$_2$ is sintered at 1600°C, we have 76.5mol% MgO(MgO contains virtually no ZrO$_2$) and 23.5mol% cubic ZrO$_2$. The cubic ZrO$_2$ contains 15mol% MgO to stabilize the cubic phase. Let us take 1cm^3 of cubic ZrO$_2$. When the specimen is annealed at 1300°C, cubic ZrO$_2$ decomposes to yield MgO and tetragonal ZrO$_2$ which contains only 1% of MgO. Using the values listed in Table 1, we can calculate the change in volume during the decomposition. Cubic ZrO$_2$(1cm^3) transforms to tetragonal ZrO$_2$ of 0.875cm^3 and MgO of 0.081cm^3 (total = 0.956cm^3). The decomposition involves the negative volume change (-4.4%) within the matrix of MgO. Therefore, the tetragonal ZrO$_2$ is not constrained in the matrix. Unconstrained tetragonal ZrO$_2$ might transform to monoclinic ZrO$_2$ when the temperature is lowered to the region where monoclinic ZrO$_2$ is stable. However, as we have shown previously, when MgO+10% ZrO$_2$ composite was sintered at 1400°C, we found the monoclinic ZrO$_2$ as well as tetragonal. This indicates that if we fabricated the specimen at temperatures where the tetragonal ZrO$_2$ was stable, it could be retained at room temperature.

---

* α_{MgO}=13.5x10^{-6}/°C, and α_{c-ZrO_2}=10.0x10^{-6}/°C.[17]

The annealing time of 1-2hrs was not long enough to transform all the particles of ZrO_2 to tetragonal ZrO_2. The composite annealed at 1000°C for 1-2hrs had no monoclinic but showed the maximum K_{IC}. Therefore we have to conclude that there was some tetragonal ZrO_2 which was hard to detect and tetragonal to monoclinic transformation was probably the cause of enhanced K_{IC} in this short time annealing. At long time anneal, most of ZrO_2 was in the monoclinic form but the K_{IC} value was still higher than in the as-sintered composite. This implies that the monoclinic phase plays an important role in the toughening of annealed composite but we do not have any definite evidence for it.

REFERENCES

1. R. C. Garvie, R. R. Hughan, and R. T. Pascoe, Strengthening of lime-stabilized zirconia by post sintering heat treatments, in:"Materials Science Research, vol.11", H. Palmour III, R. F. Davis, and T. M. Hare, eds., Plenum Press, N.Y. (1978).
2. T. K. Gupta, F. F. Lange, and J. H. Bechtold, Effect of stress-induced phase transformation on the properties of polycrystalline zirconia containing metastable tetragonal phase, J. Mater. Sci., 13:1464 (1978)
3. D. L. Porter and A. H. Heuer, Microstructural development in MgO-partially stabilized zirconia(Mg-PSZ), J. Am. Ceram. Soc., 62:298 (1979).
4. N. Claussen, Fracture toughness of Al_2O_3 with an unstabilized ZrO_2 dispersed phase, J. Am. Ceram. Soc., 59:49 (1976).
5. L. Viswanathan, Y. Ikuma, and A. V. Virkar, Transformation toughening of β"-alumina by incorporation of zirconia, J. Mater. Sci., 18:109 (1983).
6. H. Ruf and A. G. Evans, Toughening by monoclinic zirconia, J. Am. Ceram. Soc., 66:328 (1983).
7. N. Claussen and J. Jahn, Mechanical properties of sintered and hot-pressed Si_3N_4-ZrO_2 composites, J. Am. Ceram. Soc., 61:94 (1978).
8. F. F. Lange, Transformation toughening, part 4 fabrication, fracture toughness and strength of Al_2O_3-ZrO_2 composites, J. Mater. Sci., 17:247 (1982).
9. Y. Ikuma, W. Komatsu, and S. Yaegashi, ZrO_2-toughened MgO and critical factors in toughening ceramic materials by incorporating zirconia, J. Mater. Sci. Lett., 4:63 (1985).
10. R. C. Garvie and P. S. Nicholson, Phase analysis in zirconia systems, J. Am. Ceram. Soc., 55:303 (1972).
11. Y. Ikuma and A. V. Virkar, Crack-size dependence of fracture toughness in transformation-toughened ceramics, J. Mater. Sci., 19:2233 (1984).
12. JCPDS,"Powder diffraction file, inorganic volume", No. 13-307.
13. JCPDS,"Powder diffraction file, inorganic volume", No. 24-1164.
14. K. Ando, Y. Oishi, H. Koizumi, and Y. Sakka, Lattice defect and oxygen self-diffusion in MgO-stabilized ZrO_2, J. Mater. Sci. Lett., 4:176 (1985).
15. R. S. Gordon, Understanding defect structure and mass transport in polycrystalline Al_2O_3 and MgO via the study of diffusional creep, in: "Advances in Ceramics, vol. 10", W. D. Kingery, ed., American Ceramic Society, Columbus (1984).
16. J. C. Swearengen, E. K. Beauchamp, and R. J. Eagan, Fracture toughness of reinforced glasses, in:"Fracture Mechnics of Ceramics, vol. 4", R. C. Bradt, D. P. H. Hasselman, and F. F. Lange, eds., Plenum Press, N.Y. (1978).
17. W. D. Kingery, H. K. Bowen, and D. R. Uhlmann, "Introduction to Ceramics", John Wiley Sons, N. Y. (1976).
18. R. W. Davidge and D. J. Green, The strength of two-phase ceramic/glass materials, J. Mater. Sci., 3:629 (1968).
19. M. V. Swain, R curve behaviour of magnesia partially stabilized zirconia and its significance to thermal shock, in:"Fracture Mechanics of

Ceramics, vol. 6", R. C. Bradt, A. G. Evans, D. P. H. Hasselman, and F. F. Lange, eds., Plenum Press, N. Y. (1983).

20. A. G. Evans and R. W. Davidge, The strength and fracture of fully dense polycrystalline magnesium oxide, Phil. Mag., 20:373 (1969).

EFFECT OF IMPURITIES ON MICROSTRUCTURE AND MECHANICAL PROPERTIES OF Si$_3$N$_4$-TiC COMPOSITES

S.T. Buljan and G. Zilberstein

GTE Labs., Inc., Waltham, Mass

INTRODUCTION

Attainment of desired properties in dispersoid-modified composites strongly depends on the type and degree of chemical interaction between the matrix and the dispersoid. The Si$_3$N$_4$-TiC (SNT) composites which have been developed for cutting tool and wear applications combine the high fracture toughness and thermal stability of the Si$_3$N$_4$ matrix with the hardness of the TiC dispersoid. The resultant material exhibits outstanding resistance to abrasive wear. Properties relevant to performance in the severe environment of high speed metal removal--strength, hardness, and fracture toughness--may be altered by the chemical reaction between the matrix and the dispersoid. Densification of the composite proceeds through liquid phase sintering. Formation of the liquid, consisting of SiO$_2$ (which is present in Si$_3$N$_4$ powder), added sintering aids (Y$_2$O$_3$, Al$_2$O$_3$), and TiO$_2$ (which originates from the TiC surface) marks the beginning of densification and initiates a number of chemical reactions. Impurities, mainly glass modifiers, concentrate in the liquid phase. The type of impurity varies, depending on the powder production process. Common, however, are transition metals, especially iron and Group II (Ca, Mg) elements. In addition, powders such as those prepared by carbothermal reduction of SiO$_2$ may contain free carbon.

The objective of the present study was to define and evaluate the character and the extent of influence of precursor powder purity on the chemical interaction of matrix and dispersoid and to determine its effects on the mechanical properties of Si$_3$N$_4$-TiC composites.

RESULTS AND DISCUSSION

Chemical Interactions

Precursor Si_3N_4, which was selected for this study, represents fairly well a general case of commercially available high and low purity powders. Chemical analyses (Table I) indicate major differences in iron content and free carbon presence in Si_3N_4-II (low purity) powder. Chemical analysis of TiC powder used for preparation of both composite materials is also included in Table I.

Table I

Chemical Analysis of Precursor Si_3N_4 and TiC Powders

Precursor Powder	IMPURITIES, Wt. %											
	METALS									NON METALS		
	Al	Mg	Ca	Fe	Cr	Ni	Mn	Mo	Free Si XRD*	Free C CTC*	O_2 NAA*	Cl XRF*
	Emission Spectroscopy											
SN-I	<0.01	ND	<0.0005	<0.05	ND	ND	ND	<0.005	<1.0	ND	2.0	<0.05
SN-II	0.2	0.008	0.05	0.7	0.03	0.01	0.03	ND	<<1.0	0.34	2.0	ND
TiC	0.05	ND	0.01	0.05	0.01	ND	0.001	ND	ND	0.1	0.27	ND

*XRD--X-ray diffraction phase analysis

CTC--Combustion-thermal conductivity method; Leco

NAA--Neutron activation analysis

XRF--X-ray fluorescence analysis

Si_3N_4 composites with 30 volume percent (v/o) TiC containing Y_2O_3 and Al_2O_3 as sintering aids were hot pressed in argon at 1725°C to full density. Since SiO_2 and TiO_2 concentrations in starting powders affect the composition and properties (melting point, viscosity) of the glass, which occupies 6%-8% of the densified compact volume, powders with equivalent oxygen content were selected. Thermodynamic evaluation of the reaction

$$Si_3N_4 + 3\ TiC = 3\ SiC + 3\ TiN + 1/2\ N_2 \tag{1}$$

proceeding under applied hot pressing conditions has indicated high probability of SiC and TiN formation as products. It was shown experimentally[1] that, in fact, under applied processing conditions the second reaction product in Equation (1) is not TiN, but $TiC_{1-x}N_x$. The x-ray

306

diffraction phase analysis of a number of hot pressed Si_3N_4-TiC specimens has consistently indicated the presence of $TiC_{1-x}N_x$ with lattice
parameter a_o = 0.427-0.428 nm, which approximately corresponds to
$TiC_{.5}N_{.5}$ composition, as shown in Figure 1.[2]

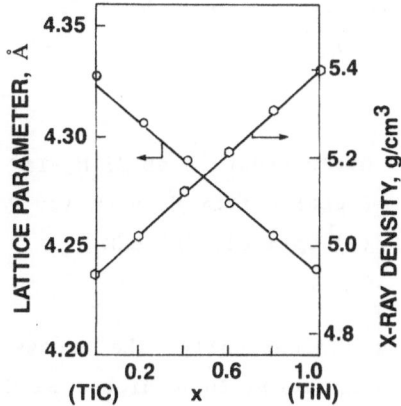

Figure 1: Lattice Parameter and X-ray Density of
$TiC_{1-x}N_x$ (0≤x≤1.0)

Based on these data, the reaction between Si_3N_4 and TiC could then be
expressed by the equation

$$Si_3N_4 + 6\ TiC = 3\ SiC + 6\ TiC_{0.5}N_{0.5} + 1/2\ N_2 \qquad (2)$$

The reactive impurities capable of changing the viscosity of the
glass phase or the carbon potential in the system are expected to have a
strong influence on the reaction kinetics and resultant microstructure.
A microstructure of the composite prepared from high purity SN-I powder
is shown in Figure 2a. As can be seen from this photomicrograph, the
reaction products which should appear as uniform zones around TiC grains
(light phase) are not resolved in optical microscopy at this magnification. Examination of this material by Scanning Electron Microscropy
(SEM) (Figure 2b) reveals the presence of reaction products which appear
as diffuse zones approximately 0.1 μm wide around TiC grains.

The microstructure of the SNT composite prepared using low purity
Si_3N_4 powder (SN-II), which contained ≈0.7% Fe and 0.34% free-carbon impurities, is shown in Figure 3. The reaction zones around TiC grains
are now well resolved in both optical (Figure 3a) and scanning electron
(Figure 3b) microscopes. Since both materials (Figures 2 and 3) were

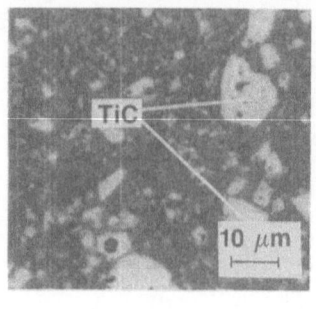
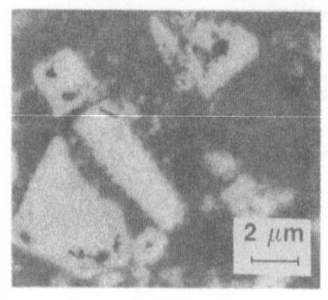

(a)   (b)

Figure 2:   Microstructure of Si_3N_4-TiC Composite
Prepared from High Purity Powder (SN-I)
(a) Optical, (b) SEM

formulated and processed identically, the excessive formation of reaction products can be attributed to a higher Si_3N_4 + TiC reaction rate due to the combined effect of impurities and, in a major part, to free carbon and iron. To determine a degree of microstructure alteration, a quantitative estimate of the Si_3N_4 + TiC reaction rate was made based on the model of solid state reaction of fine powders.[3]

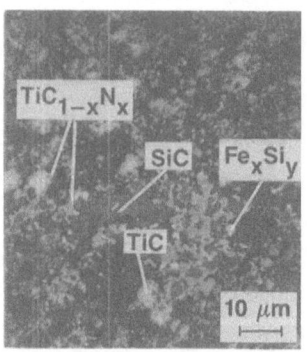
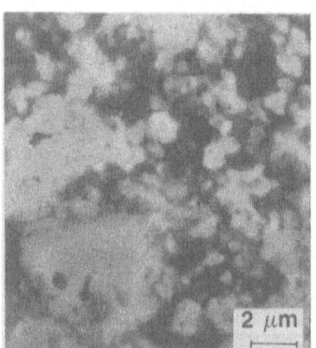

Figure 3:   Microstructure of Si_3N_4-TiC Composite
Prepared from Low Purity Si_3N_4 Powder (SN-II)
(a) Optical, (b) SEM

According to this model, the thickness of the reaction product layer (Y) forming on a grain of the radius (R) is related to the fraction of material that has already reacted (x) according to the equation:

$$Y = R(1-\sqrt[3]{1-x}) \tag{3}$$

The reaction rate constant, KD, is calculated then from the relationship:

$$(1-\sqrt[3]{1-x})^2 = (KD/R^2)t \tag{4}$$

where t is reaction time.

The TiC grain size (radius) distribution was determined by quantitative metallographic analysis of 20 photomicrographs at 1000X from randomly selected areas on polished cross sections of the SNT composite made with high-purity Si_3N_4 powder. A digital image processor was used to obtain the data and construct a histogram of the TiC grain area distribution for each photomicrograph. Examination of these histograms indicated a close similarity of TiC grain size distribution in all 20 photomicrographs. For example, 80% of the grains in each photograph fell into the 0-1 μm^2 size range, 85% into the 0-2 μm^2 size range, 95% into the 0-5 μm^2 range and so on. Hence, it was concluded that the 20 histograms of TiC areas distribution were generated from one population and that they adequately described TiC grain size distribution in a SNT composite. Applying additional data processing based on an assumption of a spherical shape for TiC grains, a mean histogram of an equivalent TiC grain radius versus its volume fraction in the SNT composite was constructed and is given in Figure 4.

The thickness of the reaction product layer (Y) on the TiC surface was measured directly from SEM photomicrographs at 5000X magnification and was ≈ 0.1 μm for high purity material (SN-I) and ≈ 0.6 μm for low purity (SN-II), independent of TiC grain size. According to Equation (2), 1 mole of TiC will produce 1 mole of $TiC_{.5}N_{.5}$ and 0.5 mole of SiC. For simplicity of calculations, it was assumed that SiC forms only as β-cubic. The lattice parameters of TiC, $TiC_{.5}N_{.5}$, and β-SiC, are very similar. (TiC, $a_o=0.4328$ nm; $TiC_{.5}N_{.5}$, $a_o=0.4274$ nm; β-SiC, $a_o=0.436$ nm). Therefore, it is assumed that upon reaction a unit volume of TiC will produce 1.5 volume unit of reaction products. To determine the volume fraction (x) of dispersoid (TiC) consumed by reaction in each material, Equation (3) was used in the form

$$x = 1-(1-Y/R)^3 \tag{5}$$

The resulting relationship between TiC grain size and its fraction consumed by Si$_3$N$_4$ + TiC reaction in high-purity (Curve I) and low purity (Curve II) composites (which were formulated and processed identically) is shown in Figure 5.

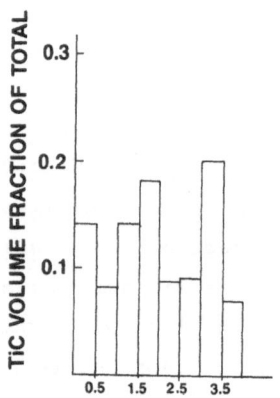

TiC GRAIN RADIUS R, μm

Fig. 4: Histogram of TiC Grain
Size (Radii) Distribution
in Si$_3$N$_4$-TiC Hot
Pressed Composites

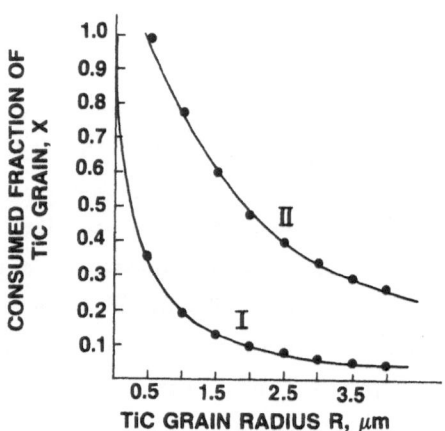

TiC GRAIN RADIUS R, μm

Fig. 5: Relationship Between
TiC Grain Size and Its
Fraction Consumed by
Si$_3$N$_4$ + TiC Reaction
Under Standard Processing
Conditions. (I) High-Purity
Material (SN-I), (II) Low
Purity Material (SN-II)

Figure 5 shows that in the low purity material the degree of TiC consumption by the matrix-dispersoid reaction is substantially higher than in the low purity material. Under identical processing conditions, the TiC grains of, for instance, 0.5 μm radius will be fully consumed by reaction in the low purity material and only $\approx$1/3 of that in the high purity material. Using Equation (4) and the data from Figure 5, the reaction rate constants KD_I and KD_{II} were calculated.

$$KD_I \approx 5.4 \times 10^{-5} \ \mu m^2/min$$

$$KD_{II} \approx 1.8 \times 10^{-3} \ \mu m^2/min$$

Comparison between KD_I and KD_{II} shows that the increased amount of impurities present in the Si$_3$N$_4$-TiC composite SN-II is capable of increasing the rate of TiC consumption by ≈30 times.

Our earlier study[1] on reaction products' formation in the Si_3N_4-TiC system has indicated that iron and carbon impurities, in addition to having a catalytic effect on Si_3N_4 + TiC reaction, are also capable of changing the mechanism of reaction product formation. Due to limited reaction, disposition of reaction products in the composites prepared from high purity powder is difficult to observe. Large reaction zones can be generated by prolonged heating at sintering temperature. The microstructure of reaction zones around TiC grains in the high-purity material for which the time at the densification temperature was extended to 4 hours is shown in Figure 6. As can be seen, the reaction products, $TiC_{1-x}N_x$ and SiC, appear as a mixture of crystals, which suggests that the SiC is formed by combining Si with C which diffused from the TiC surface.

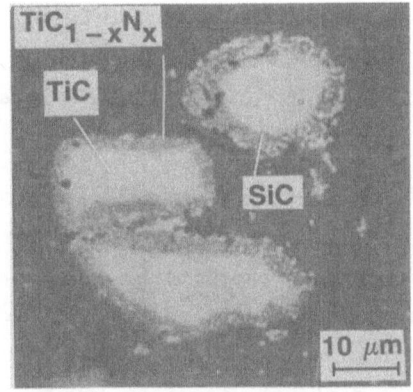

Figure 6: Reaction Products in High-Purity Si_3N_4-TiC
Composite. Extended Sintering Time (360 min)

Analytical Electron Microscopy (AEM) examination of high purity material has confirmed (Figure 7) that the Si_3N_4 + TiC reaction products are indeed concentrated at the surface of TiC grains. $TiC_{.5}N_{.5}$ crystals in many cases were found to be crystallographically coherent with the parent TiC grain, as determined by the convergent beam electron diffraction technique (see inserts in Figure 7), while SiC crystals were usually set in glassy intergranular material, near TiC.

Low initial concentration in the matrix, as well as absence of free carbon as a reaction product, results in an unimpeded carbon diffusion

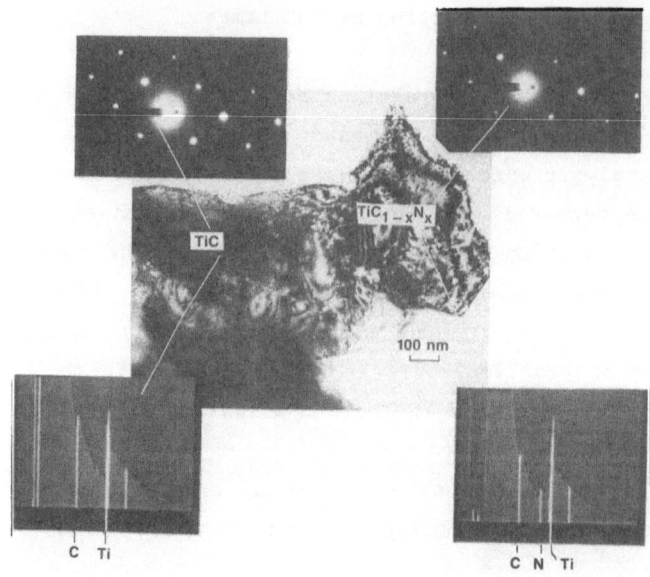

Figure 7: AEM Examination of Reacted TiC Grain in High-
Purity Si_3N_4-TiC Composite (Center) Dark
Field Image; (Left) Electron Energy Loss Spectrum
and Electron Diffraction Pattern of the TiC Grain;
(Right) The Same of the $TiC_{1-x}N_x$ Grain

toward and away from the TiC-matrix interface. Presence of carbon in
the matrix surrounding TiC is expected to reduce the rate of carbon diffusion from TiC. In a separate experiment, carbon was added to the mixture of pure powders to intensify the effect. As can be seen from a microstructure of carbon-contaminated composite (Figure 8a), the reaction
products are arranged in distinct, separate layers, the SiC forming next
to the remaining portion of unconsumed TiC grain core (and sometimes
even occupying its entire core) and the $TiC_{1-x}N_x$ building a peripheral
layer. This type of reaction products' disposition indicates that the
SiC was predominantly formed by Ti and Si interdiffusion in the TiC
grain interior. Free carbon present in the system apparently acted as a
barrier for carbon diffusion from TiC. Such disposition of reaction
products is observed in low purity SN-II powder based materials.

In Si_3N_4-TiC composites, iron as a finely dispersed powder can dissolve in the intergranular glass, lowering its melting point and viscosity. Diffusion of reacting species through such a liquid can therefore
start at a lower temperature and proceed at a higher rate. The mechanism of reaction products' formation would not be expected to change in

this case. That is, the SiC would still apparently form at the expense of carbon diffusion from a TiC surface. However, in the case of inhomogeneous distribution, iron segregations in local volumes of the composite are often encountered (Figure 3a, center of micrograph). In such areas, an excess of iron undissolved in the intergranular glass can react directly with Si_3N_4 and TiC particles. Iron and TiC interaction can produce a melt around 1410°C when an eutectic composition is reached.[4] To separate the effect of iron and carbon, iron in a form of coarse powder was intentionally added to the pure composite powder mixture. The microstructure of an area around iron inclusion is shown in Figure 8b. Examination of iron-contaminated areas of the Si_3N_4-TiC composite (Figure 3a, center, and 8b) reveals presence of iron silicides and/or iron aluminum silicides and SiC enveloped by $TiC_{1-x}N_x$. An observation that the SiC crystals are, as a rule, located in the center of such an aggregate and $TiC_{1-x}N_x$ around the SiC perimeter, together with the large size and occasionally proper geometrical shape of the crystals, suggests that both were formed by fractional crystallization from the liquid (melt). Hence, iron segregations in Si_3N_4-TiC composites appear to act as sources of substantial microstructural inhomogeneities. Iron silicides and/or iron-aluminum silicides, which usually form in such areas and which are characterized by extreme brittleness and relatively low melting points, can make a composite unsuitable for high-temperature, high-stress applications.

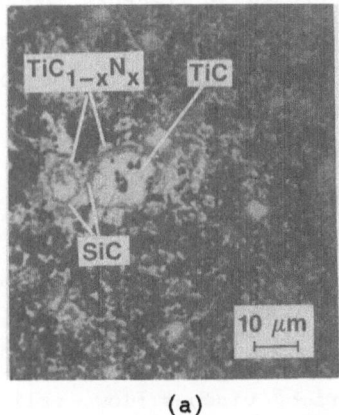

(a)

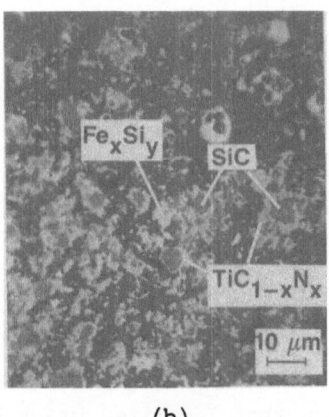

(b)

Figure 8: Microstructure of Si_3N_4-TiC Composite
Contaminated with Carbon (a) and Iron (b)

Thus far, it has been shown that free carbon and iron impurities lead to substantial alterations of Si_3N_4-TiC composite microstructures. Hardness, fracture toughness, and room and elevated temperature strength of the high-purity and low-purity Si_3N_4-TiC composites, which are summarized in Table II, appear to indicate close microstructure-property interdependence.

Table II

Mechanical Properties of Si_3N_4-TiC Composite Materials

Si_3N_4-TiC Composites	Microhardness Knoop GN/m^2	Indentation Fracture Toughness K_{IC} (MN/m$^{3/2}$)	Modulus of Rupture (MPa)		
			RT	1000°C	1200°C
SN-I	15.08 (±0.35)	4.45 (±0.22)	551.6 (±4.06)	827.4 (±83.4)	441.3 (±2.0)
SN-II	13.95 (±0.15)	3.88 (±0.14)	558.5 (±16.2)	530.9 (±5.82)	241.3 (±0.37)

As can be seen from this table, the Si_3N_4-TiC composites containing carbon and iron impurities (SN-II) are characterized by the properties consistently lower than the material prepared from high-purity powder (SN-I). The loss of hardness can most certainly be related to the greater amount of reaction products, SiC and $TiC_{1-x}N_x$, whose hardness is lower than that of TiC. For example, the Si_3N_4-based hot pressed composites containing the same volume fraction of SiC and TiN dispersoid as the materials under study exhibited hardness of 13.65 GN/m^2 and 12.99 GN/m^2, respectively. The change in microstructure associated with the matrix-dispersoid reaction also causes a decrease in fracture toughness with a very slight increase in the room temperature fracture strength. Observed change in K_{IC} and MOR translates into an apparent reduction of critical flaw size by 25%.

The elevated temperature strength is largely controlled by the properties of the intergranular glass, namely its composition and viscosity. The strength of the composite prepared from high purity Si_3N_4 powder increases at 1000°C and is somewhat reduced at 1200°C. It has been postulated[5] that at the maximum strength the viscosity of the glass is such that localized crack blunting occurs by viscous flow, relieving stresses at the crack tip. Above this temperature, the viscosity of the glass is reduced, and the strength of the material decreases rapidly. Position of the strength apex with respect to temperature relates to the volume fraction and the character of the glass. In a composite containing glass of a lower softening point, the expected increase in load bearing ability of the material due to localized crack blunting would occur at lower temperatures. Concurrently, strengths at a given temperature above maximum will be lower than those observed on compacts with more refractory glass. The softening point and viscosity of iron containing SiO_2-Y_2O_5-Al_2O_3 glass is expected to decrease with increasing impurity concentration. In addition to that, a higher rate of the matrix-dispersoid reaction in the presence of iron and free carbon may also result in an increased concentration of titanium in the glass, further decreasing its viscosity. Consequently, the strength of a composite prepared from low purity Si_3N_4 powder at 1000°C shows a small reduction and a considerable decrease at 1200°C.

SUMMARY AND CONCLUSIONS

The details of the matrix-dispersoid reaction taking place during Si_3N_4-TiC composite densification (in the presence of sintering aids and carbon and iron impurities, under standard processing conditions) have been studied.

It was shown that the optimum properties are achieved in Si_3N_4-TiC composites through the use of starting materials free from iron and carbon impurities. Loss of hardness, fracture toughness, and elevated temperature strength in Si_3N_4-TiC composites was related to a higher rate of the matrix-dispersoid reaction and changes in the mechanism of reaction products' formation caused by iron and carbon impurities during the composite's densification. Because of the detrimental effect of iron and free carbon on Si_3N_4-TiC composite microstructure, phase composition, and mechanical properties, these impurities should be regarded as particularly undesirable contaminants.

ACKNOWLEDGMENTS

The authors would like to thank Mr. E. Geary for experimental assistance in materials' preparation; Mr. T.A. Emma for AEM, EELS, and electron diffraction study; and Mrs. M. Anza for her help in preparation of the manuscript.

REFERENCES

1. G. Zilberstein and S.T. Buljan, Characterization of Matrix-Dispersoid Reactions in Si_3N_4-TiC Composites, in Second Conference on Advances in Materials Characterization, Alfred Univ., Alfred, NY
   (1984), in print.

2. M. Shimada, T. Suzuki and M. Koizumi, Fabrication and Characteriza-
   tion of $TiC_{1-x}N_x$ ($0 \leq x \leq 1$) and $Mo_2B_{2-x}C_x$ ($x=0$, 1, 2) by High Pressure
   Hot Pressing, Materials Letters, Vol. 1, No. 5, 6, pp 175-177
   (1983).

3. W. Jander, Z. Anorg, Allgem. Chem., 163, 1 (1927).

4. Lars Ramquist, Wetting of Metallic Carbides by Liquid Copper,
   Nickel, Cobalt and Iron, Int. Journal of Powder Met., 1, 4, pp 2-21
   (1965).

5. R.W. Davidge and A.G. Evans, Strength of Ceramics, Mater. Sci. Eng.,
   6, 5, pp 281-298 (1970).

ON THE MICROSTRUCTURE AND HARDNESS CHARACTERISTICS OF

COMPOSITE CERAMICS FOR TOOL APPLICATIONS

J.A. Yeomans and T.F. Page

Department of Metallurgy and Materials Science
Pembroke Street
Cambridge CB2 3QZ. U.K.

ABSTRACT

Microstructure-property relationships in three different types of ceramic tool materials (Al_2O_3-4wt% ZrO_2 , Al_2O_3-30wt% TiC/N and a β'-sialon (z=0.2)) have been investigated as part of a programme exploring the factors controlling tool life. Microstructures have been characterised using a wide range of techniques. Indentation testing has been employed to derive a range of near-surface deformation parameters including hardness as a function of both indentation scale and temperature, indentation fracture toughness and the microstructural control of fracture paths. Whilst the alumina-titanium carbonitride material generally showed the best room temperature properties, the sialon displayed better high temperature behaviour. Further, two compositions of α'/β' sialon materials have been investigated, the higher α' content material being found to possess even more superior properties.

INTRODUCTION

For several years there have been incentives to find new tool materials to supplement and replace the existing range of high speed steels and cemented carbides. Despite the unreliable nature of ceramic tools initially explored in the 1940s (eg[1]), recent advances in the development of engineering ceramics, together with the successful use of various ceramic coatings on existing tools (eg[2]), has led to renewed interest in the use of monolithic ceramics for tool applications.The general properties of high hardness and apparent chemical inertness are primarily attractive, though brittleness is an obvious disadvantage.

The advent of two new categories of composite ceramics, namely transformation toughened materials (eg[3]), and alloyed silicon nitride (the sialon family[4],) has resulted in there now being many more possibilities for manipulating ceramic microstructures to optimise required properties.

The study described in this paper is part of a larger project aimed at identifying those properties of tool materials relevant to wear, the overall objective being to establish well-founded materials selection criteria for specific applications.

It is generally accepted (eg[5]) that a tool material should exhibit:

1. resistance to gross plastic deformation
2. resistance to fracture
3. resistance to wear

with (3) embracing a multitude of poorly defined parameters such as "adequate" hardness at the working temperature, sufficient microtoughness to withstand edge chipping and abrasion, chemical stability and low adhesion to the workpiece (eg[6]).

The present study has initially concentrated on systematically exploring those mechanical properties controlling near surface deformation and conveniently accessible by indentation hardness testing i.e.:

a) hardness as a function of load and thus indentation size
   i.e. penetration resistance relevant to contact scale.

b) hardness as a function of temperature (20-1000°C)
   i.e. penetration resistance at elevated contact temperatures.

c) the microtoughness parameter K_C relevant to abrasion
   and edge retention.

d) the microstructural control of indentation fracture paths
   and hence toughness.

e) the influence of surface finish and environment on (a)-(d).

Further aspects of the project are concerned with the chemical compatibility of tool/substrate combinations and will be reported later.

The five materials which have been investigated to date have all been provided by Sandvik Hard Materials U.K. Of these, three are broadly different types of tool materials, namely a Al_2O_3-4wt% ZrO_2 zirconia-toughened alumina (ZTA), a "thermal shock resistant" alumina-30wt% TiC/N mixture, and a β'-sialon, (8wt% Al_2O_3 6wt% Y_2O_3), while the other two are further (α'/β') sialons of different α'/β' ratios.

EXPERIMENTAL TECHNIQUES

All microstructural and indentation investigations involved surfaces carefully prepared on specimens cut from unworn tool tips using a Capco Q35 annular diamond saw. Surfaces were prepared by grinding on silicon carbide papers from 180 to 1200 grit size and polishing on cloths loaded with diamond paste from 6 to 1/4 μm. X-ray diffractometry and indentation testing were performed on these surfaces without any further preparation. After indentation, the alumina-based materials were gold-coated to improve contrast in reflected light microscopy and facilitate easier measurement. All specimens were either gold or carbon coated prior to examination in the scanning electron microscope.

Microstructural Characterisation

Samples were examined using a combination of reflected light microscopy (RLM), scanning electron microscopy (SEM) and X-ray diffractometry (XRD). XRD was performed with a standard Philips vertical diffractometer. Phase distributions and grain sizes were determined using either a Quantimet 720B image analyser, or simple lineal analysis, on photomicrographs. Several SEM imaging techniques were used on both CamScan

S4 and Cambridge Instruments SII microscopes, including secondary electron imaging (SEI) and backscattered electron imaging (BEI) (using a solid state detector in the S4) . Since cathodoluminescense (CL) imaging is known to be useful in studying the microstructures of some engineering ceramics[7], the SII was modified for CL mode imaging by replacing the scintilator assembly with a simple perspex light guide. Prior to microscopy, some of the ZTA samples were etched either in boiling orthophosphoric acid for 5 mins, or thermally in air for 30 mins at 1550°C.

Indentation Testing

Load-variant micro-indentation hardness testing was performed using a standard Leitz Miniload instrument fitted with a Vickers profile indenter and using loads in the range 100-1000g. Results from these tests were anaylsed using a computer program due to Sargent[8] and described by Sargent and Page[9] in which hardness load data is fitted to an expression of the form:

$$P=ad^n$$

where P is the load, d is the indentation size, a is a constant (in fact a hardness calculated at a constant unit indentation size) and n is the indentation size effect (ISE) index. If n=2, then hardness is invariant with load while values of n<2 imply increasing hardness with decreasing indentation size. The parameters "a" and "n" not only allow characterisation of individual materials but also enable hardness to be predicted for any contact size.

For indentation toughness testing, and to use loads sufficiently high to ensure that radial cracks (of half-length c) had progated well outside any plastic zone (i.e. 2c>d), standard macro-Vickers hardness testing equipment was used in the load range 2.5-15kg. All measurements were made using the superior optics of the Leitz instrument. Data was analysed using the equation due to Anstis et al.[10] viz.

$$K_c=\beta(E/H)^{1/2}(P/c^{3/2}) \text{ where } \beta=0.016\pm0.004$$

Vickers microhardness tests in the temperature range 20-1000°C were performed in vacuum (10^{-5}-10^{-6} torr) in a Wilberforce Scientific Instruments hardness tester as described by Naylor and Page[11].

RESULTS AND DISCUSSION : THE THREE INITIAL MATERIALS

Materials Characterisation

The ZTA material consists of an equiaxed α-alumina matrix of grain size 1-5μm, ~4 vol% zirconia particles and laths of β-alumina. The zirconia particles occur mainly at triple points and are in the size range 0.5-1μm i.e.covering the critical size for transformation[3]. XRD showed the zirconia to be present as both the tetragonal and monoclinic polymorphs. In both SEI (Fig.1a) and BEI (Fig.1c) SEM images the zirconia particles appear bright, due to atomic number contrast. Fig.1b shows a RLM micrograph of the material. The β-alumina laths have been revealed by etching in orthophosphoric acid. The presence of this third phase in ZTA has recently been noted by Drennan[12] who has identified it as β'''-alumina. SEM CL imaging allows ready observation of all three phases (Fig.1d) with the zirconia particles appearing brighter than the matrix (indicating an enhanced luminescense in the spectral range of the detector (blue-green)) and the β-alumina laths appearing darker (indicating suppressed luminescense with respect to the parent α-alumina). Comparision of Figs.1c and 1d shows that not all the zirconia particles detected by atomic number contrast are apparent in the CL

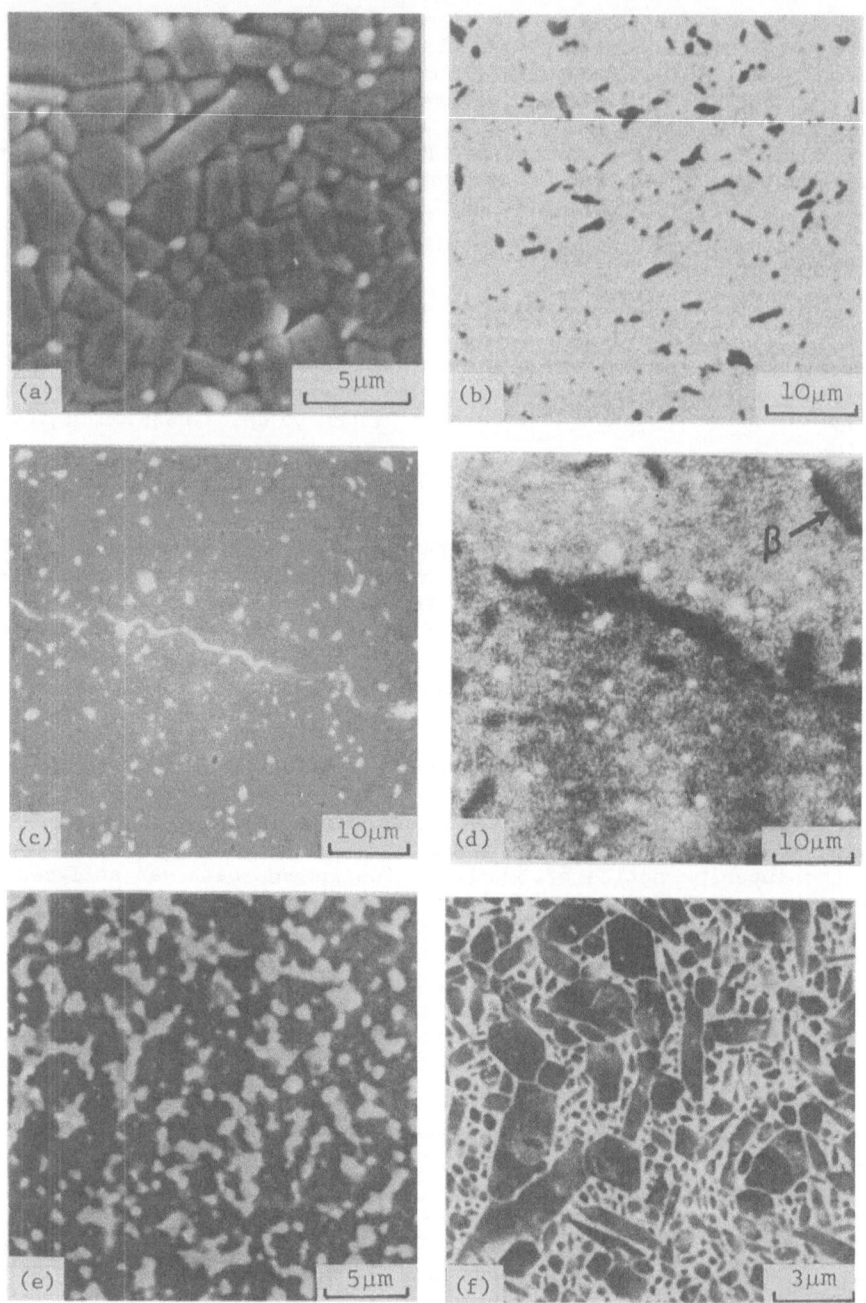

Fig.1   Microstructures of three types of tool material. (a) SEI (SEM) of
thermally etched ZTA showing bright zirconia particles (mainly at
alumina grain boundaries).(b) RLM of ZTA etched in orthophosporic
acid, the β- alumina laths appearing dark.(c) and (d) BEI/CL (SEM) of
the same area. Not all the zirconia particles detected by atomic
number contrast appear in the CL image, supposedly because they are
not monoclinic. Note that CL also detects the β-alumina laths
undetected in the BEI image.(e) BEI (SEM) showing atomic number
contrast in Al_2O_3-TiC/N. The alumina is the darker phase.(f) BEI
(SEM) of the β'-sialon. The sialon grains appear dark with respect to
the higher atomic number grain boundary glass.

image. However, since only monoclinic zirconia is strongly luminescent[7], this suggests that not all of the near-surface zirconia particles have transformed.

The Al_2O_3-TiC/N specimen shows similar constrast in RLM and SEI and BEI SEM modes. A SEI micrograph is shown in Fig.1e. The alumina appears dark with respect to the ~25 vol% titanium carbonitride. The grain sizes of both phases are comparable and in the 1-5μm range. The SEM contrast arises from differences in atomic number, alumina being lower than TiC/N; while in RLM, alumina has the lower reflectivity. XRD shows the C:N ratio to be 1:3 in the titanium carbonitride, calculated on the assumption that the (111) d-spacing varies linearly with composition.

The β'-sialon microstructure is shown in the BEI of Fig.1f. The composition of the sialon is thought to be close to $Si_{5.8}Al_{0.2}O_{0.2}N_{7.8}$. All the yttrium is assumed to be in the glass which totals ~15 vol% and images bright with respect to the β' grains suggesting that it has a higher average atomic number.

Indentation Size Effect at Room Temperature

Fig.2 shows the hardness:load behaviour of the three materials. As with most ceramics, all three show an increase in hardness with decreasing load, indicative of ISE indices <2. The ISE characteristics "n" and a hardness value calculated at a constant contact size of 10μm ($H_{10\mu m}$) are shown in Table 1.

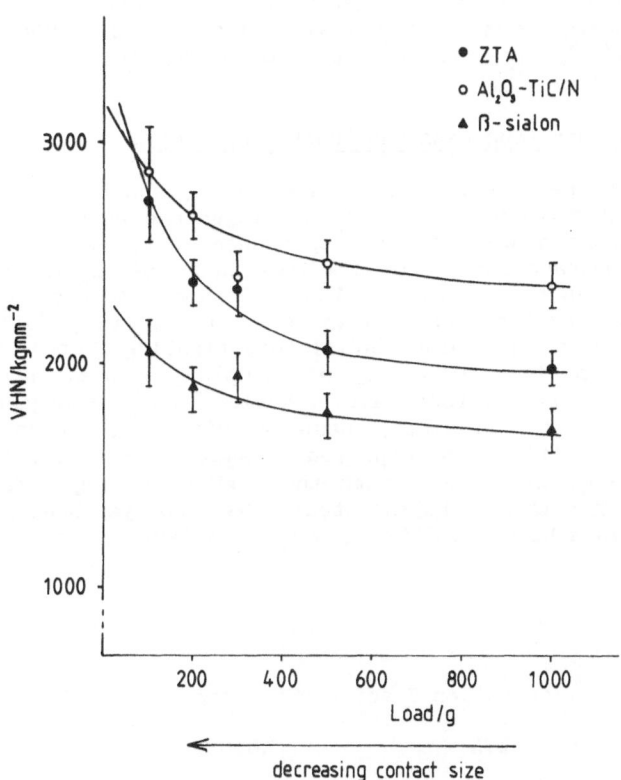

Fig.2   The hardness:load (i.e. contact scale) behaviour of the three types of material initially studied.

Table 1: Room Temperature Values of ISE indices and $H_{10\mu m}$

	ISE	$H_{10\mu m}/kgmm^{-2}$
ZTA	1.71±0.06	2670
Al_2O_3-TiC/N	1.80±0.08	2826
β'-sialon	1.79±0.06	2145

Over the majority of loads (and thus contact sizes) the Al_2O_3-TiC/N is the hardest material and β'-sialon the softest. However, at low loads, the curves cross and the ZTA becomes the hardest material. Clearly, the ZTA has a lower ISE index than the other two materials, indicating a higher degree of sensitivity of hardness to scale. It would appear that an addition of 30wt% TiC/N decreases this sensitivity but does not eliminate it.

The crossover in the curves of Fig.2 exemplifies the need to examine relative hardness values over a range of contact sizes and not to simply assume that a single hardness value-often measured at scales large compared to the scale of the contact size of the application-fully characterise a materials' behaviour. Various combinations of "n" and bulk hardness will achieve high hardness at the requisite contact scale. Thus materials of low bulk hardness can be compensated by low "n" values, while higher bulk hardness materials only need values of "n" close to 2 to achieve the same response. While bulk hardness is known to be a fairly intrinsic materials property and "n" is known to be very sensitive to both microstructural scale and environment[8,13], the exact interplay of these parameters still requires further exploration.

Indentation Fracture Toughness and Crack Morphology

Indentation fracture toughness values are given in Table 2. The results were obtained on surfaces which looked damage-free when examined with RLM. However, subsequent studies on the dependence of measured toughness values on the surface preparation of the ZTA have shown that K_c varies from 4.8, on a surface which has been ground on 1200 SiC paper and then polished for 30 mins on 14µm diamond paste only, down to 2.5 for a fully polished specimen annealed at 500°C for 1hr. We are still investigating whether this effect is due to transformation-toughening or simple residual stress left by polishing (as noticed by, for example, Naylor[14] on non-toughened ceramics). Since, when used as tool inserts, these materials are prepared with ground faces, ZTA would initially be expected to show superior fracture toughness values to those given above, though the high operating temperature might compensate for this later. Similar studies have not yet been carried out for the other materials but this effect may be important.

Table 2: Indentation Fracture Toughness Values, $K_c/MPam^{1/2}$

ZTA	3.8±0.2
Al_2O_3-TiC/N	3.9±0.2
β'-sialon	4.6±0.2

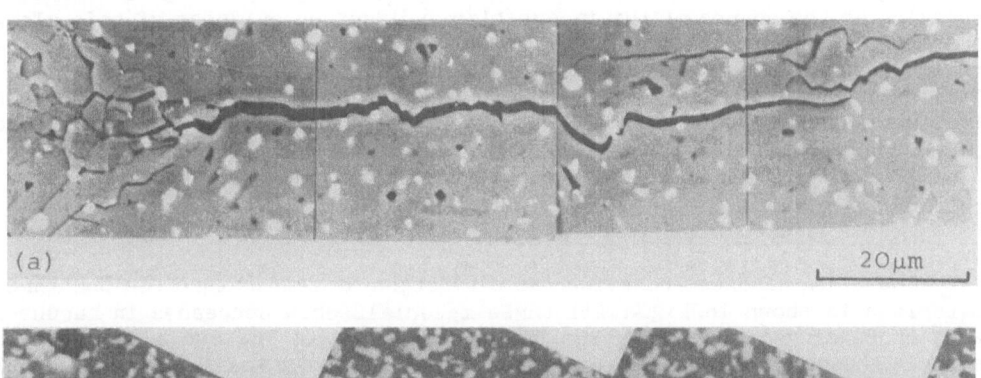

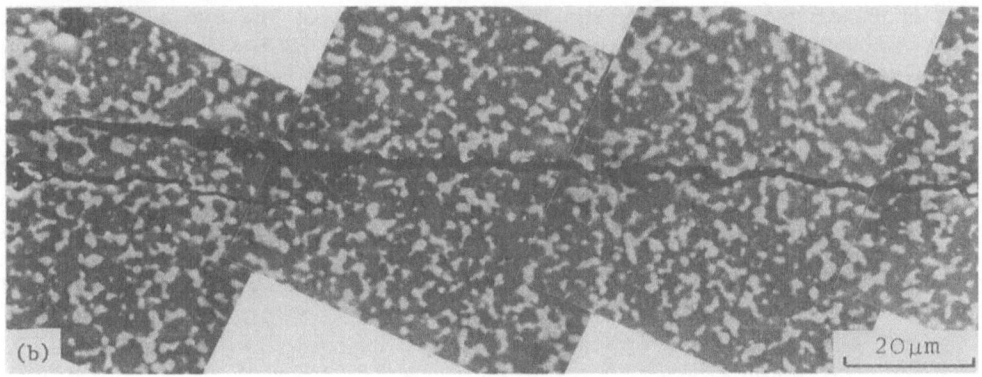

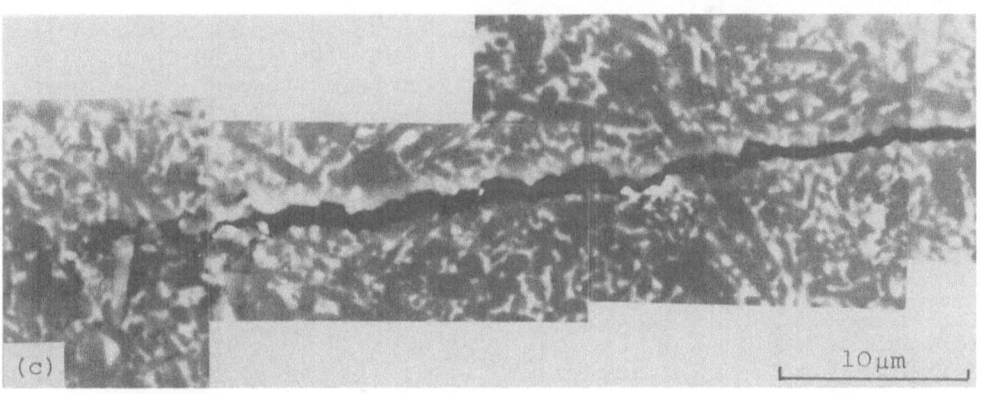

Fig.3   BEI (SEM) of the radial cracks in the three commercial materials (a) ZTA, showing severe grain boundary cracking at the corner of the hardness indentation and also crack branching along the radial.(b) Al$_2$O$_3$-TiC/N also showing the grain boundaries to be the preferred crack path.(c) β'-sialon exhibits microstructural control of the fracture path, but of a more complicated nature.

Fig.3 shows BEI SEM images of radial cracks in all three materials. In all cases the cracks are not straight and show microstructural path control. The formula due to Anstis et al. presupposes that the radials are straight. Clearly, any deviations from this will result in calculated values higher than the true toughness.

Fig.3a shows a radial crack in ZTA. There is significant grain boundary cracking at the corner of the indentation and further evidence of this along the length of the crack. In Al₂O₃-TiC/N (Fig.3b) the indications are that the grain boundaries are weak and crack-controlling in this material also. Similarly, the β'-sialon (Fig.3c) shows clear microstructural control of the fracture path although the preferred crack route seems to vary from place to place. Further studies on the control of crack morphology should prove most useful in optimising toughness through microstructure.

Hardness as a Function of Temperature

The 1000g hardness behaviour as a function of temperature for the three materials is shown in Fig.4. All three materials show decreases in hardness with increasing temperature, but at different rates. The two alumina-based materials show similar behaviour (i.e. a fairly rapid loss), while the loss of hardness for the β'-sialon is less rapid. Critically, the order of hardness of the three materials changes with temperature. At 500°C the β'-sialon curve crosses that of the ZTA and that of the Al₂O₃-TiC/N at 800°C. Hence, at the temperatures that the tools are likely to experience during service (around 1000°C), the β'-sialon is the hardest of the three studied and this may reflect the difficulty of slip through this complex crystal structure even at high temperatures. The changes in the order of hardness with temperature again emphasise the need not to rely on simple single hardness values relecting relative hardness behaviour over a wide range of conditions.

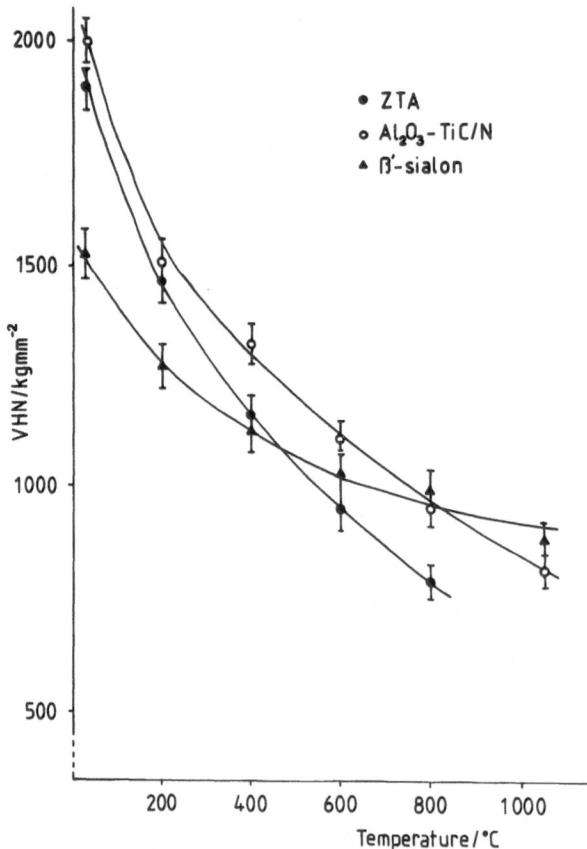

Fig.4   The hardness:temperature behaviour of the three initial materials. The β'-sialon curve crosses the other two and it is the hardest of the three materials at temperatures > 800°C.

Summary

    Room temperature work shows that the Al_2O_3-TiC/N is the hardest
material over most loads (>100g) and that it has a fracture toughness
comparable with the other two materials. On the basis of these mechanical
properties alone, it would be considered as the most suitable tool material.
However, since the tool operating temperature, at high metal cutting speeds,
is around 1000°C, the superior high temperature behaviour of the β'-sialon
suggests that it may outperform the alumina-based materials at least on this
criterion. Thus, it was decided to study some other sialon materials in order
to investigate the possiblity of further improving behaviour with α'/β'
phase mixtures.

RESULTS AND DISCUSSION : THE TWO α'/β' SIALONS

Materials Characterisation

    The microstructures of the two α'/β'-sialons are shown in the BEI
micrographs, Figs.5a and 5b. Assuming that the dark grey grains are β'-sialon,
the mid-grey grains are α'-sialon (containing yttrium and therefore having a
higher average atomic number) and the white areas correspond to a grain
boundary glass (of even higher atomic number and also containing yttrium),
then image analysis of many fields of view gives the phase proportions as:

    (a) 52% α' 38% β' 10% glass

    (b) 37% α' 51% β' 12% glass

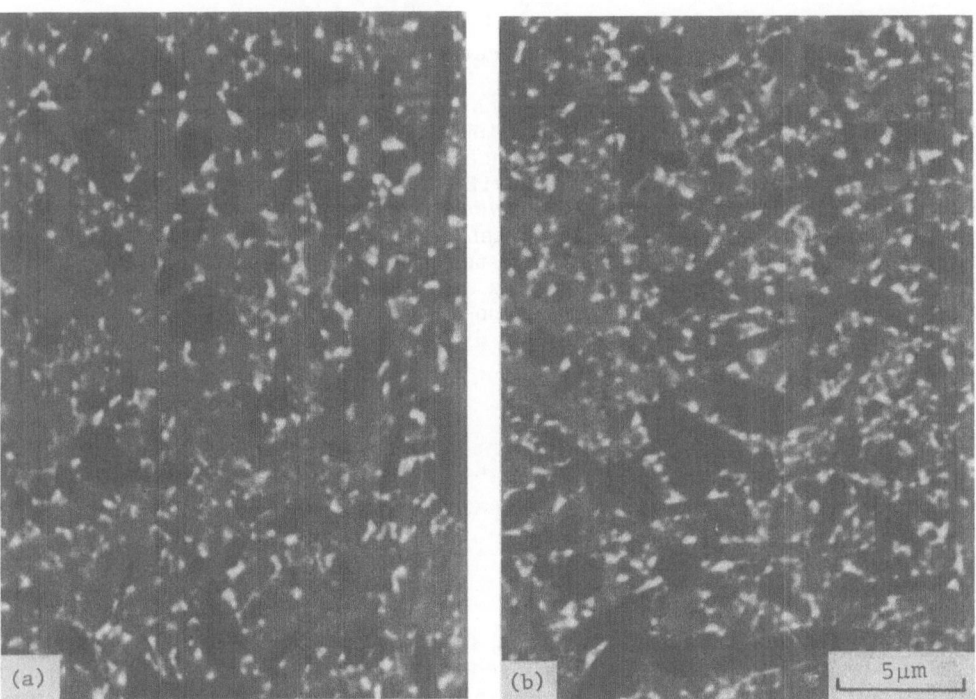

Fig.5   BEI (SEM) of the two α'/β'-sialons.(a) is the higher α' content
         material and has the higher proportion of mid-grey areas.

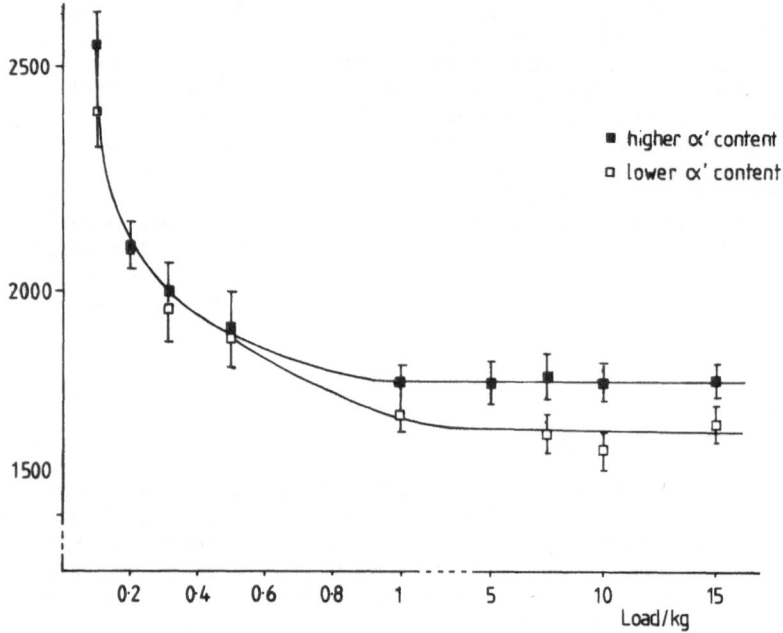

Fig.6   The hardness:load (i.e. contact scale) behaviour of the two α'/β'-sialons. The curves are indistinguisable at low loads but then diverge, the higher α' content material having the higher "bulk" hardness.

Indentation Size Effect at Room Temperature

The room temperature hardness behaviour of the two composite sialons is shown in Fig.6. The two are indistinguishable at low loads, and both have ISE indices of 1.75±0.05. Hence the presence of more alumina than in the simpler β'-sialon has increased the sensitivity of hardness to contact scale. At loads greater than 1000g, the two curves diverge and it can be seen that the "bulk" hardness of the higher α' content sialon is higher than that of the higher β' content one. This may be due to the difference in crystal structure of the α' and β', the more complex α' having interstices between the structural units rather than the channels in the more ordered β'. Thus slip (and other deformation mechanisms) is expected to be more difficult in α'.

Indentation Fracture Toughness

The α'/β' sialons show a slight decrease in indentation fracture toughness compared with the straight β'-sialon. Both have values of 4.3MPam$^{1/2}$. Reasons for this and the microstructural control of fracture paths are still being investigated.

Hardness as a Function of Temperature

Fig.7 shows 1000g hardness as a function of temperature for the two α'/β'-sialons. The β'-sialon is included for comparison. It can be seen that the higher α' content sialon is consistently the hardest and that it also retains its hardness to high temperatures. The lower α' is still always harder than the β'-sialon but does show a roll-off of hardness around 800°C.

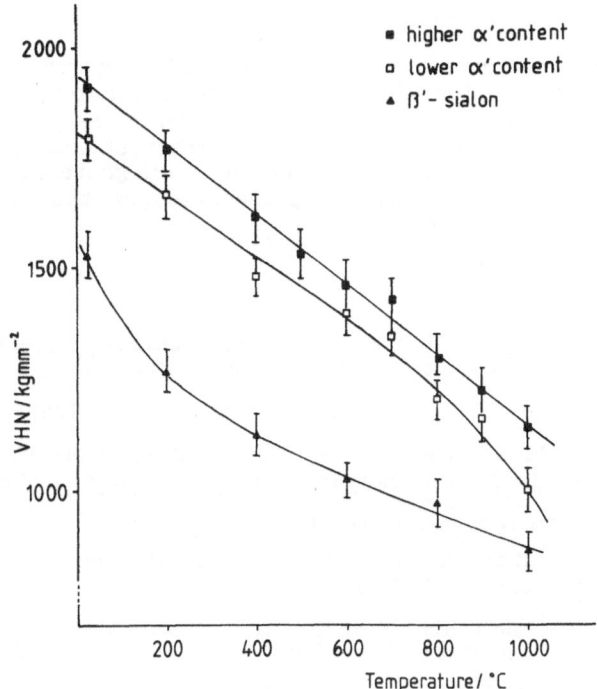

Fig.7   The hardness:temperature behaviour of the three sialon materials.
The higher α' content sialon has the superior hardness over all
temperatures tested.

Summary

The α'/β'-sialons are harder than the β'-sialon at all temperatures
tested and again in terms of hardness, the higher α' content sialon is
superior to that of the lower α' content . The fracture toughness for the
composite sialons is slightly lower than that for the β'- sialon but this is
not thought to be significant. Hence, the higher α' content material looks
very promising in terms of mechanical properties alone.

CONCLUSIONS

Our studies to date have demonstrated the value of characterising
surface deformation and fracture behaviour by indentation hardness testing
over a range of conditions. It has been clearly established that single
hardness values do not give reliable guidance as to the order of merit of
materials, the effects of contact size and temperature at least having to be
considered. Further work is progressing which will integrate the mechanical
properties data here with studies of the chemical compatiblity of the
various materials with a range of workpieces. Also the effects of surface
finish on microtoughness and the mechanisms of microstructural control of
fracture are being further investigated.

ACKNOWLEDGEMENTS

The authors are grateful to Prof. D. Hull, and formerly Prof. R.W.K. Honeycombe, F.R.S. F.Eng, for the provision of laboratory facilities. JAY acknowledges the tenure of a SERC CASE award with Sandvik Hard Materials UK. Dr David Jack (of Sandvik) is thanked for interest and support. Helpful discussions with the Cambridge Tribology Group are gratefully acknowledged. Brian Barber kindly helped with the preparation of the figures for this paper.

REFERENCES

1. E. Dow Whitney, "Modern Ceramic Cutting Tool Materials," Metals/Materials Tech. Series No. 8201-083, ASM (1982).

2. E.M. Trent, "Metal Cutting," Butterworths, London (1984).

3. A.H. Heuer, N. Claussen, W.M. Kriven and M.Ruhle, Stability of Tetragonal ZrO_2 Particles in Ceramic Matrices, J. Am. Ceram. Soc. 65:642-650 (1982).

4. K.H. Jack, Sialon Tool Materials, Metals Tech. 9:297-301 (1982).

5. N.P. Suh, New Theories of Wear and Their Implications for Tool Materials, Wear 62:1-20 (1980).

6. R. Komanduri and J.D. Desai, Tool Materials, in: "Kirk-Othmer: Encyclopedia of Chemical Technology, Vol 23," Wiley, New York (1983).

7. J.T. Czernuszka and T.F. Page, Cathodoluminescence: a Microstructural Technique for Exploring Phase Distributions and Deformation Structures in Zirconia Ceramics, to be publ. in Comm. Am. Ceram. Soc. (1985).

8. P.M. Sargent, Factors Affecting the Microhardness of Solids, Ph.D. Thesis, University of Cambridge, (1979)

9. P.M. Sargent and T.F. Page, The Influence of Microstructure on the Microhardness of Ceramic Materials, Proc. Brit. Ceram. Soc. 26:209-224 (1978).

10. G.R. Anstis, P. Chantikul, B.R. Lawn and D.B. Marshall, A Critical Evaluation of Indentation Techniques for Measuring Fracture Toughness: I, Direct Crack Measurements, J. Am. Ceram. Soc. 64:533-538 (1981).

11. M.G.S. Naylor and T.F. Page, The Effect of Temperature and Load on the Indentation Hardness of Silicon Carbide Engineering Ceramics, in Proc. of the Fifth Int. Conf. on Erosion by Solid and Liquid Impact (ELSI V), Publ. by Cavendish Laboratory, Cambridge (1979).

12. J. Drennan, The Observation of β-alumina Phases in Zirconia- Toughened Alumina (ZTA), J. Mat. Sci. Lett. 4:725-727 (1985).

13. J.T. Czernuszka and T.F. Page, A Problem in Assessing the Wear Behaviour of Ceramics: Load, Temperature and Environmental Sensitivity of Indentation Hardness, Proc. Brit. Ceram. Soc. 34:145-156 (1984).

14. M.G.S. Naylor, Microhardness, Friction and Wear of SiC and Si_3N_4 Materials as a Function of Load, Temperature and Environment, Ph.D. Thesis, University of Cambridge, (1982).

MECHANICAL PROPERTIES AND WEAR RESISTANCE OF SILICON NITRIDE-TITANIUM
CARBIDE COMPOSITES

J.G. Baldoni, M.L. Huckabee and S.T. Buljan

GTE Laboratories, Incorporated

Waltham, Massachusetts

INTRODUCTION

Ceramic matrix composites that contain particulate or whisker dispersoids represent one of the more prominent areas for the future development of materials for use in severe environments.   In addition to
thermal stability and corrosion resistance, these applications often simultaneously require high strength, fracture toughness and wear resistance. Successful composite design and property prediction is contin-
gent upon the understanding of the composite's fracture mechanics and
the resolution of the intricate mechanical interactions of its components. The lack of sufficient information on systems which utilize polyphase, polycrystalline matrices and mechanically anisotropic components severely restricts the ability to predict their properties. The
fracture toughness, for example, of such ternary or quaternary composites often does not follow predictions based upon consideration of single phase, isotropic matrix based composites. A silicon nitride-titanium carbide composite is an example of a brittle ternary composite.
The matrix consists of two phases, polycrystalline silicon nitride and
an intergranular glass phase based on silica and sintering aids,
throughout which titanium carbide particulate dispersoids are evenly
distributed. The aim of this present work was to evaluate room and elevated temperature properties and the abrasive wear resistance of this
composite and to examine some of the microstructural parameters affecting the composite's fracture resistance.

Fully dense, hot pressed billets of Si_3N_4 and Si_3N_4 containing TiC were used in this study. The dispersoid was added at concentrations of 10, 20 and 30 volume percent. The billets were fabricated using high purity powders with 6 weight percent (w/o) Y_2O_3 as a densification aid. The Al_2O_3 content of these compositions was approximately 2 w/o.

Modulus of rupture samples were rectangular bars, 0.127 x 0.259 x 2.5 cm , machined from the billets with the tensile face perpendicular to the hot pressing direction. The machine lay was parallel to the length of the bars and the edges were chamfered. The tensile face of each sample was wet-polished to a 1 μm diamond paste finish.

For fracture toughness measurements, randomly selected groups of test bars of the monolith and the 30 v/o TiC composite were precracked with a single Knoop indentation. A 2-3 kg load was used, with the long axis of the indentation aligned perpendicular to the tensile stress direction of the bars. The precracked samples were annealed in vacuum at 1200°C and slowly cooled (60°C/hour) to room temperature in order to relieve the residual stresses produced by the indentation process.[1] Test bars of the materials, both precracked and non-precracked, were broken in a four point loading fixture placed in a tungsten mesh furnace. Inner and outer loading points were 1.016 and 2.286 cm, respectively. Room temperature testing was performed in air and high temperature testing in flowing argon. A crosshead speed of 0.05 cm/min was used for all tests. The fracture surfaces of each broken precracked test bar were examined by optical microscopy and the individual induced surface flaw sizes were measured from photomicrographs. Fracture toughness, K_{IC}, was calculated using the relationship:[1]

$$K_I = \sigma M(\pi a/Q)^{1/2} \tag{1}$$

where σ is the maximum outer fiber tensile stress, M is a numerical factor related to flaw and specimen geometry, a is the flaw depth, and

$$Q = \Phi^2 - 0.212(\sigma/\sigma_y)^2 \tag{2}$$

where σ_y is the tensile yield stress and Φ is the elliptic integral which is tabulated in standard mathematical tables.

The value of M for a small, semicircular flaw is 1.03, and this value was used for all calculations.[1] The plastic zone correction factor, $0.212 (\sigma/\sigma_y)^2$, is considered negligible for all calculations except for those temperatures where nonlinear force-time response in MOR testing is observed.

Samples for microstructural characterization and fracture toughness determination by indentation[2,3] [indentation fracture toughness, (IFT)] were polished sections cut from the same hot pressed billets used for test bars. The polished surface was perpendicular to the hot pressing direction. Microhardness values were obtained using a Knoop indenter. The microstructure of each material was analyzed quantitatively from optical photomicrographs.

Wear tests were conducted with a basic pin-on-disc type tester in which the disc was a removable metal bonded 45 μm diamond surface. The pin test surface (1.25 x 1.25 cm square) was swept and rotated (20 rpm) against the spinning (40 rpm) abrasive disc. All tests were performed under dry, abrasive conditions in a flowing argon atmosphere. The test specimens were abraded for one minute intervals, ultrasonically cleaned, hot air dried and subsequently weighed to 0.0001 grams. The abrasive disc was cleaned with a detergent solution and dried between test intervals. All specimens were abraded for a total time of five minutes. The density of each material was determined by the standard immersion technique, and volume loss was calculated from the measured weight loss and density.

RESULTS AND DISCUSSION

Mechanical Properties

Analysis by x-ray diffraction revealed β-Si_3N_4 to be the major phase of the monolithic material and the matrix phase of the composites. The diffraction patterns of the composite indicated limited reaction between the particulate phase and the matrix during hot pressing, which resulted in the formation of trace amounts of $TiC_{1-x}N_x$ and SiC. Dispersoid-matrix interactions have been reported to be related to processing parameters and composition.[4,5]

Optical microstructural analysis of the composites showed the TiC grains to be randomly distributed in the Si_3N_4 matrix (Figure 1). Quantitative image analysis showed a distribution of dispersoid particle sizes from submicron to 8 μm with an average size of approximately 1.5 μm. The majority of the observed TiC particles appeared to be single grains. However, some particulate agglomerates were observed.

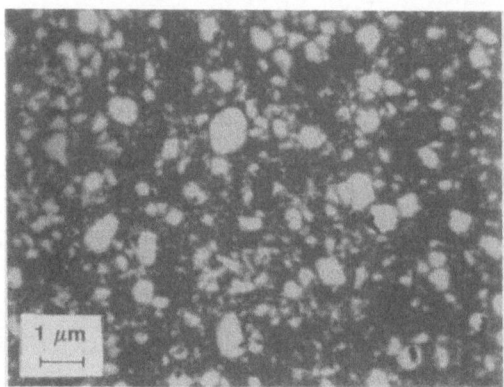

Figure 1: Microstructure of a Si_3N_4 (2 w/o Al_2O_3+6 w/o Y_2O_3) Matrix - 30 v/o TiC Composite

The room temperature microhardness of the investigated materials was found to increase linearly with increasing TiC content (Figure 2a). However, the observed hardness increases were less than expected from rule of mixture behavior based on stoichiometric TiC (KHN≈24 GPa).[6] The lower hardness values are attributed to loss of carbon from and partial nitridation of the carbide phase during hot pressing, since both sub-stoichiometric titanium carbide, TiC_{1-x}, and TiN exhibit lower values of microhardness compared to TiC.[6,7]

As opposed to theoretical predictions based on crack-particle in-teraction[8-10], the room temperature fracture toughness, as measured by indentation, was found to be statistically invariant with increasing TiC content (Figure 2b) and equivalent to that of monolithic silicon ni-tride.

The room temperature moduli of rupture of the composites are com-pared to monolithic Si_3N_4 in Figure 3, along with critical flaw size calculated from strength and fracture toughness data using Equation (1) and assuming a semicircular flaw shape. The observed decrease in strength and the apparent increase in critical flaw size with increasing dispersoid content can be attributed to the increased probability of TiC agglomerate formation which has been suggested[11] as the strength con-trolling flaw for these types of composites.

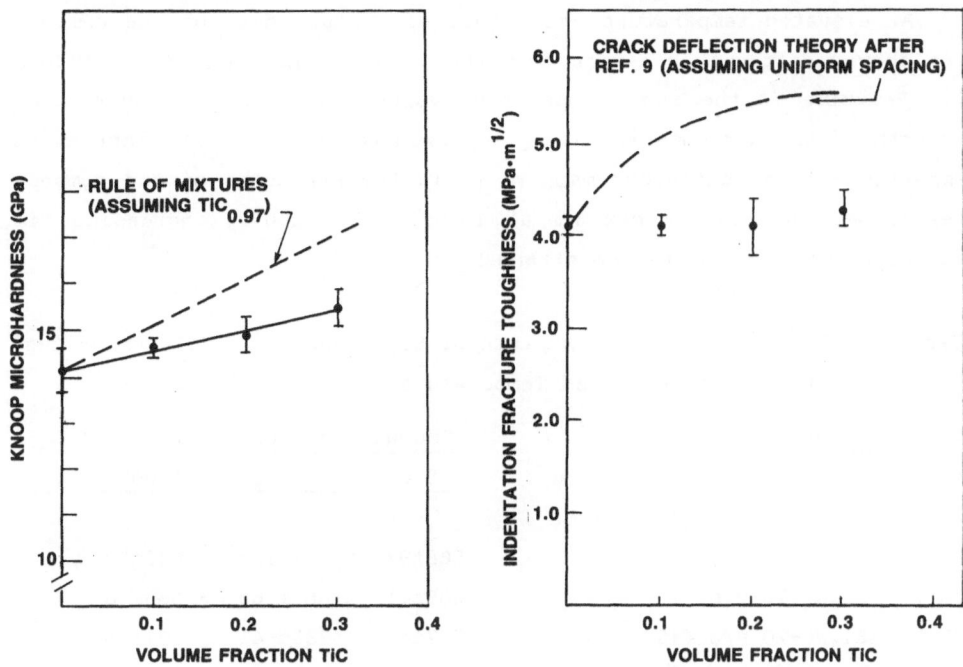

Figure 2: Effect of Dispersoid Content on the Measured Room Temperature Knoop Microhardness (A) and Indentation Fracture Toughness (B) of Si_3N_4 Based Composites

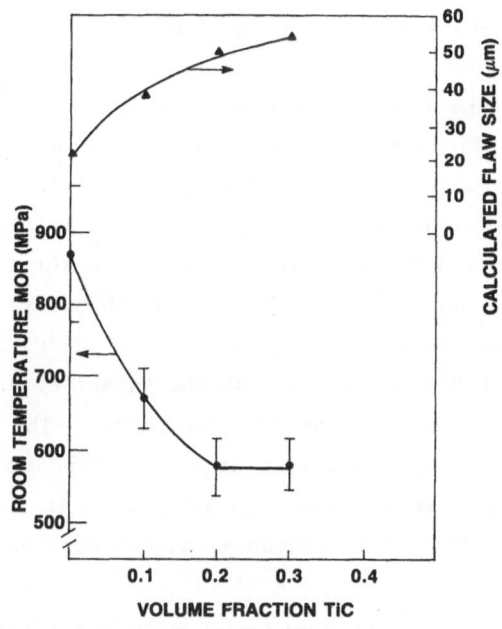

Figure 3: Effect of Si_3N_4 Based Composite Dispersoid Content on Strength and Calculated Flaw Size

At elevated temperature the moduli of rupture data of the composites showed different behavior compared to the monolithic Si_3N_4 (Table I). Relative to the room temperature value, at 1000°C and above the strength of the noncomposite Si_3N_4 appeared to decrease with increasing temperature in a continuous manner, while the strengths of the composites showed an apparent maximum at 1000°C. At 1200°C, increasing TiC content resulted in decreased strength.

Table I. Measured Moduli of Rupture of Si_3N_4 and Si_3N_4-TiC Composites at Room and Elevated Temperatures

Material	Modulus of Rupture (MPa)		
	25°C	1000°C	1200°C
Si_3N_4	868±97 >	717±118 >	636±32
Si_3N_4+10 v/o TiC	669±41 ≤	685± 63 >	585± 8
Si_3N_4+20 v/o TiC	577±42 <	735± 41 >	521± 5
Si_3N_4+30 v/o TiC	582±35 <	698± 55 >	516±21

In order to more comprehensively evaluate mechanical property behavior at elevated temperature, additional testing was performed using the monolithic Si_3N_4 and specimens prepared from a second billet of the Si_3N_4 + 30 v/o TiC composite. Modulus of rupture and fracture toughness, using the controlled surface flaw technique, were measured at room temperature and in the range of 800 to 1200°C.

The values of the room temperature fracture toughness of the monolithic and 30 v/o TiC composite materials obtained by the controlled surface flaw technique were found to be equivalent (Figure 4). At temperatures above 900°C, the measured K_{IC} of both materials was observed to increase with increasing temperature. However, the relative increase was higher for the Si_3N_4 matrix-30 v/o TiC composite compared to the Si_3N_4 monolithic material. Subcritical crack growth was observed in both materials at 1100°C and above, but the extent of crack growth prior to fast fracture was higher for the composite. The amount of crack growth at 1200°C for the monolithic Si_3N_4 was approximately equivalent to that of the composite at 1100°C (Figure 5). At the crosshead rate employed, nonlinear force-time response was not observed.

The values of modulus of rupture of both materials measured up to 1250°C are shown in Figure 6. Relative to room temperature, the strength of the Si_3N_4 + 30 v/o TiC composite appeared to decrease

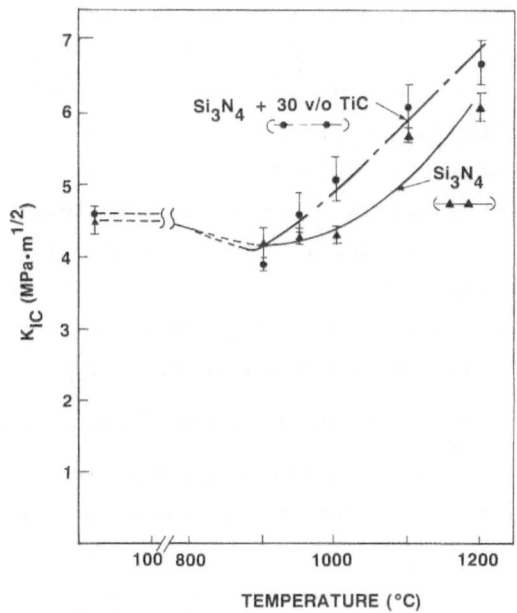

Figure 4: Fracture Toughness of Si₃N₄ and Si₃N₄-30 v/o TiC Composite Measured by the Controlled Surface Flaw Techniques at Room and Elevated Temperature

Figure 5: Observed Subcritical Crack Growth During Controlled Surface Flaw Tests of (A) Si₃N₄ at 1200°C (B) Si₃N₄+30 v/o TiC at 1100°C

slightly at 800°C and then increase to a maximum at 1000°C. At temperatures above 1000°C, a rapid loss in strength was observed. Above 900°C, the modulus of rupture of the monolithic Si₃N₄ was found to decrease with increasing temperature up to 1150°C and subsequently a strength increase at 1200°C was indicated.

The strength increase near 1000°C observed for the Si₃N₄ + 30 v/o

TiC composite is similar to that observed for Al_2O_3 based ceramics containing appreciable amounts of SiO_2, which forms an intergranular glass phase. The increase in strength at elevated temperature has been associated with plasticity of the glass phase where it is postulated that at the maximum in strength the viscosity of the glass is such that localized crack blunting occurs by viscous flow, relieving stresses at the crack tip. Above this temperature the viscosity of the glass phase is reduced and the strength of the material decreases rapidly.[13] The observed dependence of the modulus of rupture on temperature for the monolithic Si_3N_4 is similar to that reported for hot pressed Si_3N_4 containing 10% by weight CeO_2 as a sintering aid.[14], and Si_3N_4 containing MgO sintering aids.[15,16] Crack blunting mechanisms appear operative in monolithic Si_3N_4 systems in the temperature range where a strength increase is observed.

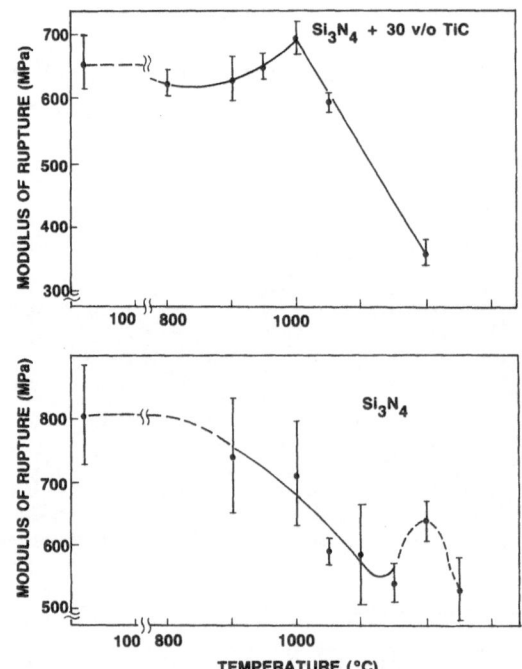

Figure 6:   Measured Four Point Modulus of Rupture of Si_3N_4 and Si_3N_4 + 30 v/o TiC Composite at Room and Elevated Temperature

As indicated by the increased subcritical crack growth above 1100°C for the Si_3N_4 composite, the intergranular phase appears to be substantially altered by the introduction of the dispersoid. ESCA analysis of as received powders has shown the presence of TiO_2 on the surface of the TiC particles. STEM microscopy coupled with energy dispersive x-ray analysis of the glass phase at triple points in Si_3N_4-TiC composite microstructures has shown the presence of titanium in the glass.[4] Apparently the glass viscosity at a given temperature is appreciably lowered by the addition of TiO_2 or TiC, which allows subcritical crack growth to occur at a more rapid rate above 1100°C and produces a maximum in strength for the composite at 1000°C.

Since the room temperature fracture toughness of the Si_3N_4 monolithic material (as measured by two techniques) was found to be unaffected by the addition of TiC, the cracks produced with the Vickers indenter for IFT determinations were examined for evidence of crack deflection. Figure 7 shows the observed crack paths in (a) the monolithic material and (b) a 20 v/o $TiC-Si_3N_4$ matrix composite. For the composite, cracks were found to propagate with little apparent deflection. Limited deflection was observed within a number of TiC crystals, grain A in Figure 7b, for example. The predominant crack-dispersoid interaction was crack penetration of the larger (≈ 2 μm) TiC crystals.

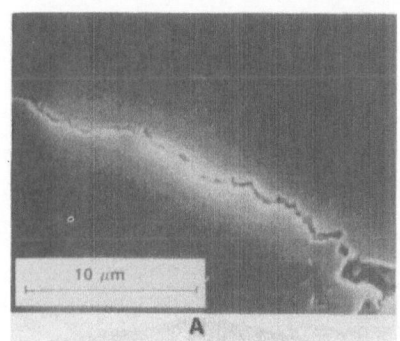 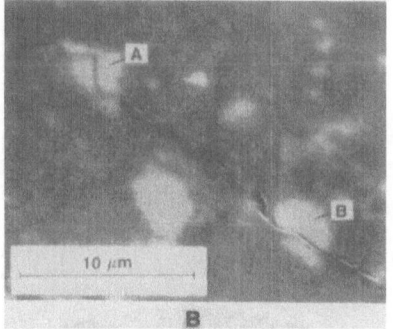

Figure 7: Observed Crack Trajectories Produced at Room Temperature by Indentation with a Vickers Indenter in (A) Monolithic Si_3N_4 and (B) Si_3N_4 + 20 v/o TiC Composite (Crack Propagation Direction ↖)

Figure 8 shows the cumulative frequency distribution of the measured crack propagation angles for a limited number (approximately 200 crack segments per sample) of indentation cracks. The crack angle at 0.5 cumulative frequency for both materials is approximately twenty-five degrees, indicating no increase in crack deflection by the addition of TiC to Si_3N_4. While toughening by second phase crack deflection has been observed in glass matrix composites[10,17] and polycrystalline matrix composites with little or no intergranular phase[18,19], published data for particulate additions to Si_3N_4 often conflicts with crack-particle interaction toughening theories.[8-10] The fracture energy of Si_3N_4 has, in general, been shown to decrease with the addition of SiC particulates with the relative value being lowered by increasing the dispersoid con-

tent.[20] Other investigations have shown a maximum in room temperature
fracture toughness at 15-20 V/O TiC addition to Si_3N_4[11,21] These cited
data do not follow predictions based on crack interaction mechanisms,
which predict increased fracture toughness with increased dispersoid
content.

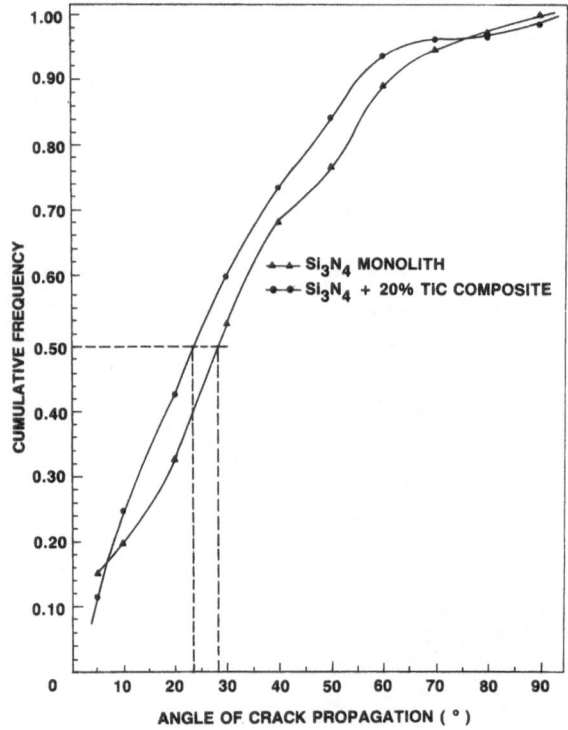

Figure 8: Measured Cumulative Distribution of Crack Propagation Angle
of Si_3N_4 and Si_3N_4 + 20 v/o TiC Composite

Microstructural analysis of Si_3N_4 monolithic materials containing
different sintering aids has indicated that the intergranular glass
phase in these materials is continuous[25] The room and elevated temperature modulus of rupture of Si_3N_4 containing a Y_2O_3 sintering aid has
been shown to be affected by the amount and composition of the glass
phase. It has been suggested that the properties of the continuous
glass phase should strongly influence the mechanical properties of a
dense Si_3N_4 body.[26] The additions of TiC to Si_3N_4 were found to modify
the composition, and possibly the amount, of the intergranular glass
phase, which resulted in decreased resistance to subcritical crack
growth above 1000°C. This modified glass phase could possibly lower the
inherent fracture toughness of the matrix.

Cracks were observed to propagate through the matrix phase of the
composite along a less tortuous path compared to cracks in the mono-

lithic Si$_3$N$_4$ (Figure 7). This may also be a consequence of the modification of the intergranular glass phase which may affect liquid phase sintering and reduce the grain size or aspect ratio of the acicular β-Si$_3$N$_4$ grains, a phenomenon reported to reduce the fracture toughness of silicon nitride.[27] These changes in the matrix phase of the composite matrix could offset any gains in fracture toughness produced by limited crack deflection around or within TiC grains or by microcracking of TiC grains.

Additionally, it may be speculated that the unique properties of TiC, as well as other groups, IVA and VA transition metal carbides, may dictate the mechanical interaction within the composite and thus affect the predicted behavior. At room temperature, TiC, due to a large Peierls stress, has a very high critical resolved shear stress (CRSS) and behaves in a brittle manner.[6,7] Single crystals of TiC cleave readily on {100} planes. The three point bend strength of as-cleaved single crystals have been measured to be 300-670 MPa.[22] The room temperature four point bend strength of polycrystalline TiC, in which the fracture mode was identified as being predominantly transgranular cleavage, has been reported to be approximately 500 MPa.[23] The CRSS of TiC is strongly sensitive to temperature, and the material experiences a brittle to ductile transition in the range 800-1000°C. Above the transition temperature, the CRSS decreases rapidly with increasing temperature, and appreciable ductility is observed with the principal operative slip system being {111} <110>.[7] Figure 9 shows the calculated radial tensile stresses due to the thermal expansion mismatch for a TiC particle in a Si$_3$N$_4$ matrix[1] upon cooling from 1800°C, 1200°C, and 1000°C compared to the CRSS of TiC. Potential residual stresses can be relieved by plastic deformation of the TiC dispersed phase from the hot pressing temperature (≈1800°C) down to approximately 1000°C. Observation of polished sections of these composites has revealed pull-outs due to cleavage from the centers of the larger TiC crystals indicating a highly strained, work hardened lattice.

The calculated tensile radial stress of a TiC particle in a Si$_3$N$_4$ matrix with a ΔT of 1000°C is 1,300 MPa. This is appreciably higher

---

[1] Calculated after reference 24 using values of Young's Modulus(E) of 427 GPa and 306 GPa, Poisson's ratios (ν) of 0.191 and 0.26, and thermal expansion coefficients of 7.4 x 10^{-6}•C^{-1} and 3.0 x 10^{-6}°C, respectively for TiC and Si$_3$N$_4$.

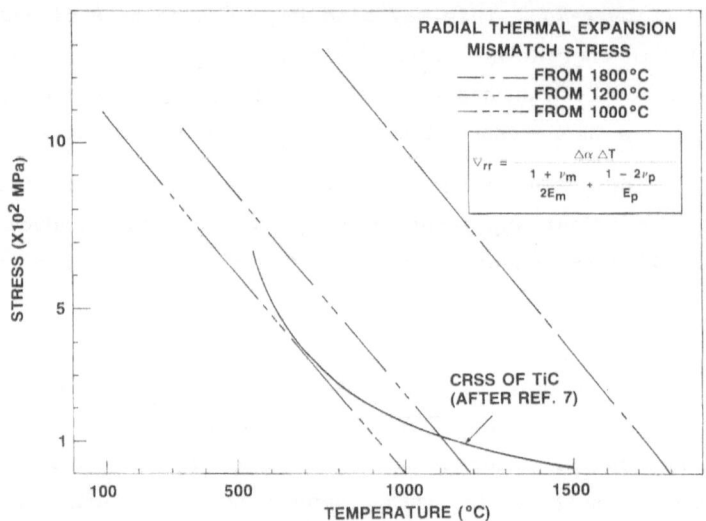

Figure 9: Calculated Thermal Expansion Mismatch Stresses Compared To the Critical Resolved Shear Stress (CRSS) of Single Crystal TiC

than the reported strength values for TiC single crystals or polycrystalline materials. Therefore, the lack of observable crack deflection may also be related to cleavage of TiC grains by the buildup of thermal stress. This cleavage could relieve the mismatch stress and reduce the driving force for crack deflection. If the dispersoid particles are not cleaved in this manner upon cooling from processing temperature, the addition of the tensile stress field of the advancing crack to the residual stress may produce TiC cleavage and allow the crack front to propagate through the dispersed phase as observed. Due to the anistropy of TiC, the orientation of the advancing crack front relative to the cleavage plane of the dispersoid should affect the direction of crack propagation through the TiC crystal, e.g., ideally a crack approaching a {100} face should pass essentially straight through a particle, while a crack approaching a {110} face should propagate through the crystal at a forty-five degree angle to the face, although jogs may be observed in either case. Indications of this phenomenon are seen in grains A and B (Figure 7b).

The addition of particulate TiC to SiC matricies has been reported to increase fracture toughness by an apparent crack deflection mechanism with no reported dispersoid penetration by propagating cracks.[18] The thermal expansion coefficient and Young's Modulus of SiC are higher than those of Si_3N_4, $4.8 \times 10^{-6} \, ^\circ C^{-1}$ and 440 GPa for SiC and $3.0 \times 10^{6} \, ^\circ C^{-1}$ and

306 GPa for Si_3N_4. This results in a calculated residual tensile radial mismatch stress, using a ΔT of 1000°C, which is approximately 300 MPa lower than that calculated for the Si_3N_4-TiC system. Apparently this reduction is sufficient to negate TiC crystal cleavage.

An additional consideration related to the unique properties of TiC may contribute to the observed increased fracture toughness of the composite relative to the Si_3N_4 monolith at elevated temperatures. Since TiC undergoes a brittle-ductile transition at 800-1000°C, it is anticipated that the dispersoid-matrix interface is essentially stress free at and above the transition temperature. Thus crack deflection is less likely. However, additional toughening mechanisms, crack blunting or line tension, related to the plastic dispersoid may become operative.

Wear Resistance

The room temperature fracture toughness and hardness of Si_3N_4 based ceramics are compared to Al_2O_3 based materials in Table II.

Table II. Mechanical Properties of Wear Resistant Materials

Material	Hardness (GPa)	Fracture Toughness (MPa•m$^{1/2}$)
Si_3N_4	14.1±0.5	4.1±0.1
Si_3N_4 + 30 v/o TiC	15.5±0.4	4.3±0.2
Al_2O_3	15.3±0.4	2.5±0.1
Al_2O_3 + 30 v/o TiC	17.0±0.5	3.2±0.1

The two body abrasive wear resistance of metals has been the subject of many investigations and it has been shown that the wear resistance is determined by the metal's hardness.[28] The abrasive wear resistance of the tested ceramics was found to be related not only to the material's hardness but also to its fracture toughness.

The abrasive wear resistance (inverse wear rate normalized to that of Al_2O_3) of the Si_3N_4 based monolith and composite are compared to Al_2O_3 based materials in Figure 10. The harder alumina based materials, with lower values of fracture toughness compared to silicon nitride based ceramics, were found to have lower abrasive wear resistance. For

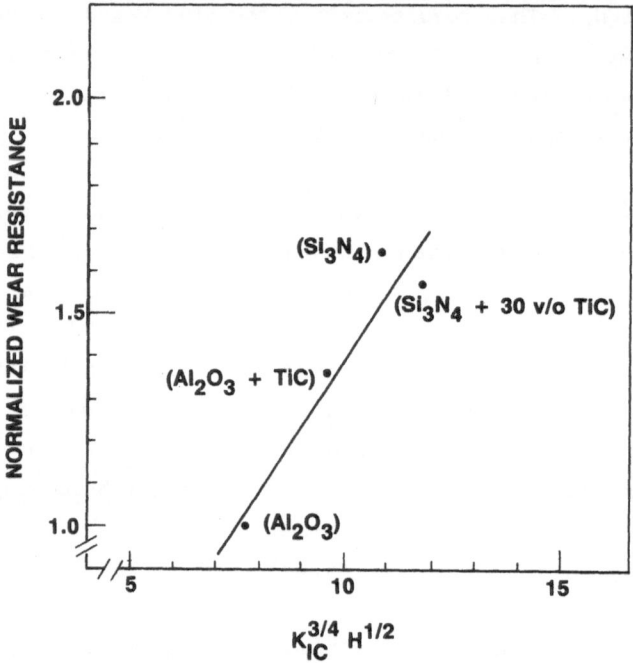

Figure 10: Abrasive Wear Resistance of Tested Ceramics

brittle substances, such as ceramics and ceramic composites, material removal by fracture that occurs in abrasion can be assumed to take place when lateral cracks of adjacent indentations caused by penetration of sharp surface protrusions (or abrasive particles as in pin on disc tests) of the opposing surface intersect.[29] The removed volume (V) is then:

$$V_i = r_i h_i \ell_i \tag{3}$$

where r_i is critical indentation separation, h_i is the depth of the indentation, and ℓ_i is the sliding distance.

Considering the dependence of the size of the indentation and the length of the cracks emanating from such angular indentations on the hardness (H) and fracture toughness (K_{IC}), respectively, the following expression for maximum volume removed by the system of indenters in a grinding operation was derived:[29]

$$V \propto \frac{1}{K_{IC}^{3/4} H^{1/2}} \sum_{i=1}^{i=N} P_i^{5/4} \ell_i \equiv \frac{1}{K_{IC}^{3/4} H^{1/2}} N\bar{P}^{5/4} \bar{\ell} \tag{4}$$

where N is the number of abrasive particles and $\bar{P}$ is a vertical force on the particle and $\bar{\ell}$ is the sliding distance. From the experimental wear data obtained, the abrasion resistance for the Al_2O_3 and Si_3N_4-based ceramic materials was found to be directly proportional to $K_{IC}^{3/4}H^{1/2}$ (Figure 10). This abrasive wear resistance parameter provides, to a first approximation, a relative ranking of brittle materials and shows that increases in either hardness or fracture toughness provide increases in abrasive wear resistance. The increased fracture toughness of silicon nitride based materials produces an improved abrasive wear resistance compared to harder alumina based ceramics.

CONCLUSIONS

The values of the room and elevated temperature fracture toughness of monolithic silicon nitride and Si_3N_4-TiC composites were determined. At room temperature, the fracture toughness of the composites was found to be statistically invariant with increasing TiC content and equivalent to its dispersoid free counterpart. The absence of toughening by crack-particle interactive mechanisms was attributed to antagonistic effects produced by chemical interaction of matrix and dispersoid during densification leading to modification of the intergranular phase. Additionally, it could be speculated that the easy cleavage and anisotropy of the dispersoid could contribute to the observed behavior. From these results it is apparent that for systems which are far removed from model binary systems, where an isotropic dispersoid is considered in a homogeneous single phase matrix, the predicted toughening may not be observed.

Modification of the intergranular glass phase produced by TiC additions results in decreased subcritical crack growth resistance at elevated temperature which affects the measured elevated temperature strength and fracture toughness.

ACKNOWLEDGMENTS

The authors would like to thank Mr. Glenn McCloud and Mr. Peter Ness for experimental assistance in materials' preparation and testing, and Mrs. M.Anza for her help in preparation of the manuscript.

REFERENCES

1. J.J. Petrovic, et al., Controlled Surface Flaws in Hot Pressed Si_3N_4, J. Am. Ceram. Soc., 58 [3-4]: 113 (1975).

2. A.G. Evans and E.A. Charles, Fracture Toughness Determinations by Indentation, J. Am. Ceram. Soc., 59 [7-8]: 371 (1976).

3. J.G. Baldoni, et al., Particulate Titanium Carbide-Ceramic Matrix Composites, Proc. Second International Conf. on the Science of Hard Materials, Rhodes, Greece, 1984 (in print).

4. G. Zilberstein and S.T. Buljan, Characterization of Matrix-Dispersoid Interactions in Si_3N_4-TiC Composites, Second Conference on Characterization, Alfred Univ., Alfred, NY, 1983 (in print).

5. S.T. Buljan and G. Zilberstein, Effect of Impurities on the Microstructure and Mechanical Properties of Si_3N_4-TiC Composites (these Proceedings).

6. L.E. Toth, Transition Metal Carbides and Nitrides, Academic Press, New York (1971).

7. W.S. Williams, Transition-Metal Carbides in "Progress in Solid State Chemistry Vol. 6", H. Reiss and J.O. McCaldin, ed., Pergamon Press, Oxford and New York (1971).

8. F.F. Lange, The Interaction of a Crack Front with a Second Phase Dispersion, Phil. Mag., 22: 983 (1970).

9. K.T. Faber and A.G. Evans, Crack Deflection Processes - I. Theory, Acta Metall., 31 [4]: 565 (1983).

10. K.T. Faber and A.G. Evans, Crack Deflection Processes - II, Experiment, Acta. Metall., 31 [4]: 577 (1983).

11. T. Mah, et al., Fracture Toughness and Strength of Si_3N_4-TiC Composites, Bull. Am. Ceram. Soc., 60 [11]: 1229 (1981).

12. G.D. Quinn and J.B. Quinn, Slow Crack Growth in Hot Pressed Silicon Nitride "Fracture Mechanics of Ceramics, Vol. 6," R.C. Bradt, et al., ed. Plenum Press, New York (1983).

13. R.W. Davidge and A.G. Evans, The Strength of Ceramics, Mater. Sci. Eng. 6 [5]: 281 (1970).

14. D.C. Larsen and G.C. Walther, Property Screening and Evaluation of Ceramic Turbine Engine Materials, Interim Technical Report No. 5, Contract F33615-75-C-5196, IITRI (1978).

15. R.K. Govila, Indentation-Precracking and Double-Torsion Methods for Measuring Fracture Mechanics Parameters in Hot Pressed Si_3N_4, J. Am. Ceram. Soc.

16. S.H. Knickerbocker, et al., Displacement Rate and Temperature Effects in Fracture of a Hot Pressed Silicon Nitride at 1100°C to 1325°C., J. Am. Ceram. Soc., 67 [5]: 365 (1984).

17. J.C. Swearangen, et al., Fracture Toughness of Reinforced Glasses, "Fracture Mechanics of Ceramics, Vol. 4," R.C. Bradt, et al., ed. Plenum Press, New York (1978).

18. G.C. Wei and P.F. Becher, Improvements in Mechanical Properties in SiC by the Addition of TiC Particles, J. Am. Ceram. Soc., 67 [8]: 571 (1984).

19. K.T. Faber and A.G. Evans, Intergranular Crack Deflection Toughening in Silicon Carbide, Comm. Am. Ceram. Soc., C-94 (1983).

20. F.F. Lange, Effect of Microstructure on Strength of Si_3N_4-SiC Composite System, J. Am. Ceram. Soc., 56 [9]: 445 (1973).

21. O.N. Grigorev, et al., Optimization of the Properties of a Tool Material Based on Silicon Nitride, Poroshkovaya Metalluryiya, No. 7, no. 7 (223): 73 (1981).

22. W.S. Williams and R.D. Schaal, Elastic Deformation, Plastic Flow and Dislocations in Single Crystals of Titanium Carbide, J. Appl. Phys. 33 [3]: 955 (1962).

23. A.P. Katz, et al., Mechanical Behavior of Polycrystalline TiC, J. Mat. Sci., 18: 1983 (1983).

24. J. Selsing, Internal Stresses in Ceramics, J. Am. Ceram. Soc., 44 [8]: 419 (1961).

25. L.K.V. Lou, et al., Discussion of Grain Boundary Phases in a Hot Pressed MgO Fluxed Silicon Nitride, J. Am. Ceram. Soc., 61 ([9-10]: 462 (1978).

26. J.T. Smith and C.L. Quackenbush, Phase Effects of Si_3N_4 Containing Y_2O_3 or CeO_2: 1, Strength, Bull. Am. Ceram. Soc., 59 [5]: 529 (1980).

27. F.F. Lange, Fracture Toughness of Si_3N_4 as a Function of the Initial α-Phase Content, J. Am. Ceram. Soc., 62 [7-8]: 428 (1979).

28. M.M. Kruschov, Principals of Abrasive Wear, 28: 69 (1984).

29. A.G. Evans and T.R. Wilshaw, Quasi-Static Solid Particle Damage in Brittle Solids--I. Observations, Analysis and Implications, Acta Met. 24: 939 (1976).

THE STRUCTURE AND PROPERTIES OF INTERFACES

IN REACTION-BONDED SILICON CARBIDES

J.N. Ness and T.F. Page

Department Of Metallurgy And Materials Science
Pembroke Street
Cambridge CB2 3QZ U.K.

ABSTRACT

Results are presented of a detailed microstructural study of the various interfaces occurring in reaction-bonded silicon carbides. All the interfaces are found to be remarkably clean due to the nature of the reaction-bonding process; although impurity-filled inclusions are found at both epitaxial interfaces and grain boundaries and can affect the high temperature strength. Grain boundaries contain a thin (~10Å) amorphous silicon carbide film and are strong, as are the epitaxial interfaces. By comparison, the SiC:Si interfaces appear partially epitaxial and possibly contain an amorphous silicon carbide film, but are weak in nature.

INTRODUCTION

Reaction-bonded silicon carbides (RBSC) are composite ceramics formed by infiltrating silicon into a porous preformed compact of a fine (≤1μm) graphite dispersion and silicon carbide (~10μm) powder. The silicon reacts with the graphite to form 'new' silicon carbide which bonds the material together.[1] The resultant microstructure typically consists of a matrix of composite silicon carbide grains with ~10% silicon filling the residual pore space necessary for complete penetration of the silicon during the reaction.

In brief, the microstructural evolution entails the progressive solution of the graphite in the liquid silicon, followed by the epitaxial nucleation, growth and coalescence of β-SiC* on the original α-SiC grains.[3,4] These β-SiC coatings then transform to the same α polytypes as the underlying substrates to produce composite grains of uniform crystal structure.[3,5] This transformation is driven by the brief (~few minutes) high temperature rise (to ~2000°C) due to the exothermic nature of the Si+C reaction.[6] A typical reflected light micrograph (RLM) of a polished section is shown in fig.1(a), whilst fig.1(b) illustrates the microstructure

---

* β-SiC refers to the cubic SiC polytype, whilst α-SiC is used collectively for all the hexagonal and rhombohedral polytypes. The Ramsdell[2] notation describes a polytype by a number representing its c-axis layer repeat and a letter for its Bravais lattice type (e.g. 3C,6H,15R). This notation will be used throughout this paper.

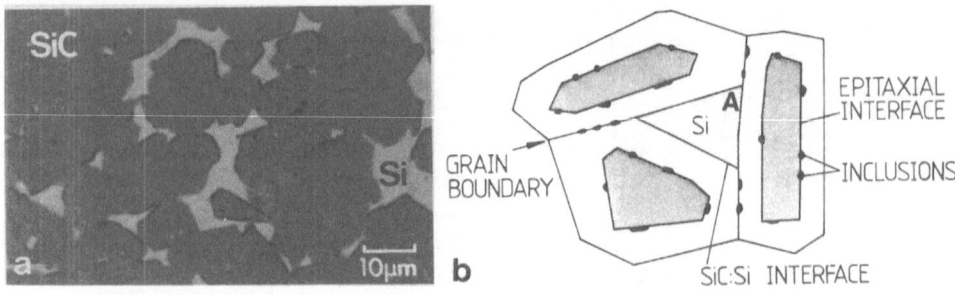

Fig. 1. Microstructure of reaction-bonded silicon carbide. (a) Reflected
light micrograph of a polished, uncoated, section displaying
polishing damage at the SiC:Si interfaces; (b) schematic showing the
types of interface and composite SiC grain structure.

schematically and shows the three main types of interface[*] produced during
fabrication viz. (in order of formation):

1) SiC:SiC epitaxial interfaces between the 'old' and 'new' SiC;

2) SiC:SiC grain boundaries between the newly-formed SiC deposits of
   different grains

3) SiC:Si interfaces between the newly-formed SiC and the residual Si.

These interfaces are important because they have a controlling effect
on the mechanical properties of the material (e.g. creep, bend strength and
fracture). In addition, their structure and properties is also important in
helping to understand the properties of other SiC/Si composites e.g. SiC
fibre composites.

This paper presents the results of a detailed microstructural study of
these interfaces, using a variety of microscopical techniques. In general,
each type of interface has required investigation at a variety of levels,
down to near-atomic resolution. Besides the intrinsic structure of each
interface, the distribution of impurities at the various interfaces can be
important in determining mechanical properties, and these have also been
investigated. Of considerable importance in the reaction-bonding process is
the way in which the infiltrating liquid silicon transports most elemental
impurities up the compact.[3] Thus, such impurities are usually left
concentrated either in the residual silicon or as inclusions in the various
interfaces. Further, since the silica coating on the original SiC particles
is removed by reaction with graphite, there are few impurities available to
form impurity bonding films at interfaces. This is in direct contrast to
most hot-pressed or conventionally sintered silicon carbides, where the
impurity-based liquids, added to promote liquid-phase sintering, solidify to
form glassy grain boundary films with commensurate mechanical property
deterioration.[7] Thus, the final part of this report is concerned with the
effect of the various interfaces and inclusions on fracture strength and
behaviour.

---

*Also present are $\beta \rightarrow \alpha$ polytype transformation interfaces left in areas of
incomplete transformation, but they will not be dealt with here
further [eg 5,6]

Epitaxial Interfaces

As mentioned above, the new SiC formed during the process deposits epitaxially as β-SiC on the original α-SiC grains, eventually producing a fairly uniform coating from the coalescence of a number of growth nuclei.[3] Subsequently, the coating undergoes a solid-state β→α transformation to produce a grain of uniform crystal structure; the transformation usually being seeded by the underlying crystal structure of the original grain.[3,4] Thus, the epitaxial interfaces act as nucleation sites for both the original deposition and the β→α transformation, and are the regions at which the cohesion of the 'old' and 'new' SiC occurs. Hence, their structure is important in understanding both the formation and final properties of the material.

The evidence for epitaxy derives principally from electron-optical techniques,[4] though transmitted polarised light techniques can also be used. Backscattered electron micrographs reveal uniform channelling contrast, and hence crystallographic orientation, across each grain.[4,6] This has been confirmed by TEM lattice imaging of the interfaces, the detailed structure of the interface depending on the polytype of the original grain. Most of the grains are 6H, and an epitaxial interface occurring in one such grain is shown in fig.2(a). This centred dark-field TEM image reveals a fully coherent interface with no misfit or strain, but decorated with occasional inclusions. These inclusions provide the only way of detecting such coherent interfaces in TEM specimens (and BEI images of polished samples), and contain remanant impurity material (see later) from the bonding process. For comparison, fig.2(b) shows part of an epitaxial interface from a grain of a polytype based on 4H (in fact 111R). As well as showing similar inclusions to the 6H interface, some localised strain is apparent at the interface. Even in those regions where good fringe matching is still apparent, occasional

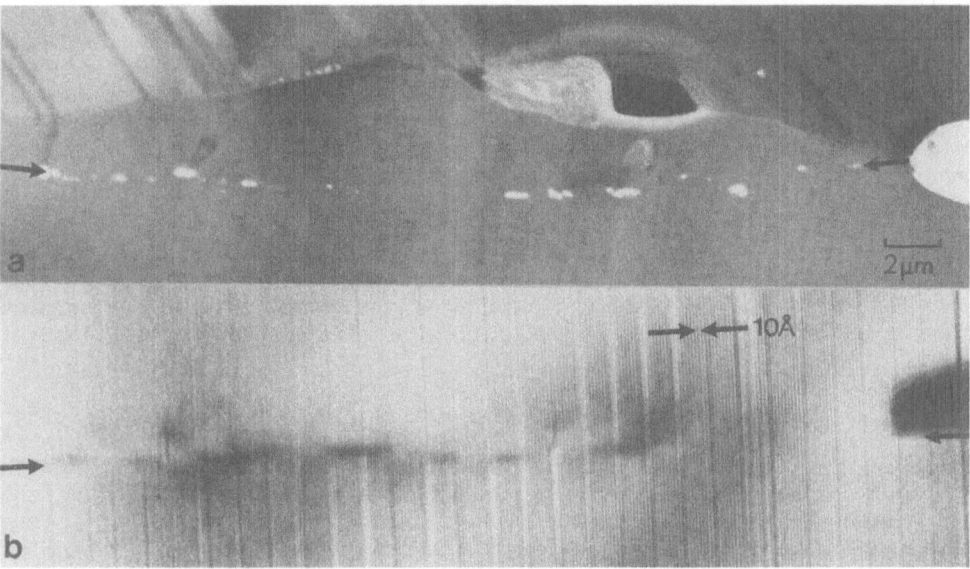

Fig. 2.   Epitaxial interfaces. (a) Centred dark-field TEM image showing 'lit-up' inclusions along the interface (arrowed) in a 6H grain. Note the continuity of stacking faults across the interface; (b) bright field lattice image showing local strain at an interface in a 111R grain.

misfit dislocations are present. This type of appearance is common in grains based on 15R and 4H, the next two most common polytypes after 6H. The difference in appearance can be explained by considering the relative molar volumes of the various polytypes.[8] 6H and 3C have very similar molar volumes, whilst 15R and 4H are both ~1% more dense. Bearing in mind that the new material initially deposited as 3C, the difference in interface appearance seems to be explained by the good initial molar volume matching between 6H seeds and 3C overgrowths, and the poorer matching for 4H(15R) and 3C. This difference is also apparent as strain at $\beta \rightarrow \alpha$ transformation interfaces.[9,10]

Despite these small structural differences and the presence of inclusions, the behaviour of all epitaxial interfaces is very similar; that is, they are strong and never appear as preferential fracture paths at any temperature.

Grain Boundaries

As the new epitaxial SiC deposits grow they eventually coalesce and form grain boundaries. These boundaries are expected to provide the principal limit on structural cohesion of the compact, and hence their detailed structure is vitally important in understanding the properties of the composite. A variety of TEM techniques have been utilised in their study and we have found that these boundaries contain a thin (~10.5Å) wetting film of pure amorphous SiC. The evidence for this is as follows:

1) By orienting the boundary 'edge-on', and plotting Fresnel fringe spacing versus defocus (fig.3(a)), we have estimated the boundary thickness - the minimum in the plot is taken to correspond to this.[12] The change in intensity of these fringes can also be used to estimate the mean inner potential change in the boundary relative to the grains. This method reveals a drop in the potential ($\Delta V/V$) of 2-4%,[9] which corresponds well to what would be expected if amorphous SiC were present in the boundary film.

2) The boundaries can be imaged in dark field by placing a small objective aperture on part of the amorphous haloes diffracted by the film e.g. fig.3(b). These images show maximum boundary intensities at reciprocal space positions corresponding to the reciprocal interatomic spacings expected in amorphous SiC. In addition, the ion-beam thinning process used to produce the specimens tends to create a thin layer of amorphous SiC on the specimen surfaces. This can be used as a calibration in that it is always found to 'light-up' in dark-field at the same time as the boundary films (e.g. fig.3(b)).

3) EDS and EELS analyses in the TEM and Auger analysis of in-situ fractured boundaries have failed to reveal any detectable impurities, with a detection limit of ~0.5%. This indicates that the boundary films are no more impure than the surrounding grains, ruling out segregation to the boundary films.

The morphology of the boundaries is determined by the cubic morphology of the growing epitaxial material. In general, the boundaries are planar reflecting the {111}, {110} and {100} cubic facets developed during growth of the epitaxial material.[3] However, in some cases, a serrated growth front will produce a stepped or serrated grain boundary e.g. fig.3(c), the steps again reflecting the original cubic crystallography.

Most of the grain boundary area consists of these thin film boundaries. However , various inclusions are also usually present which are similar in nature to those found along the epitaxial interfaces. In particular they contain:

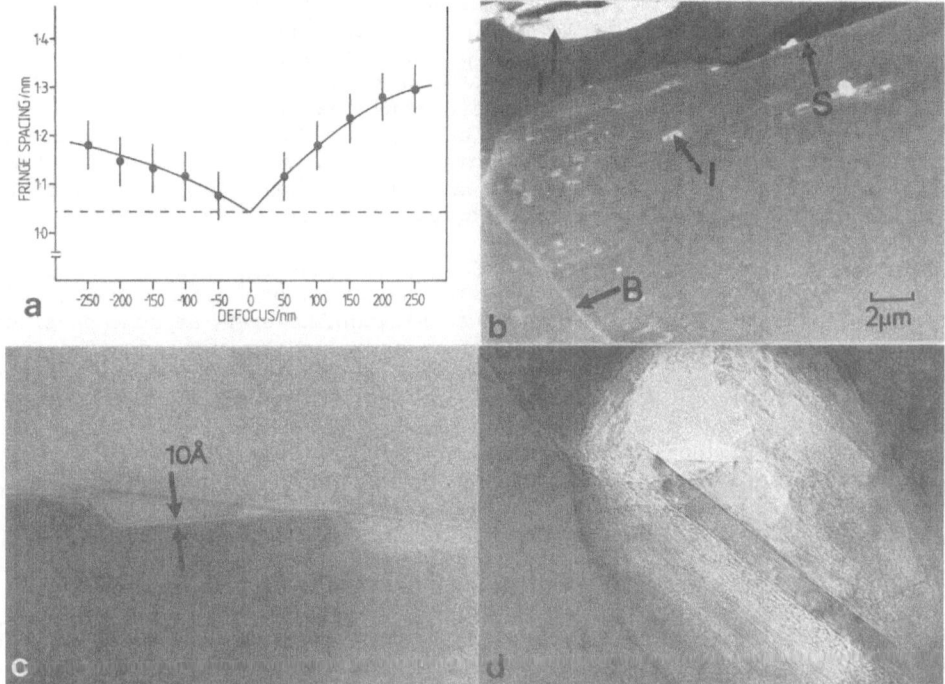

Fig. 3. Grain boundaries. (a) Plot of boundary Fresnel fringe spacing versus
defocus; (b) centred dark field at reciprocal space position
corresponding to the Si-Si spacing in SiC (~3Å) and showing the
SiC:SiC(B) and SiC:Si(S) boundary films and graphite (d(0002)=3.3Å)
inclusions (I); (c) bright field image of a stepped boundary; (d)
graphite inclusion.

1) residual silicon or graphite left over from the reaction-bonding
process, e.g., fig. 3(d), and

2) various low melting point silicides e.g. calcium, iron and titanium
silicides.[3] These are produced because of the very low solubility of most
elements in solid silicon.

The presence of these inclusions seems to have little effect on the
room temperature strength of the boundaries - they always appear to be
fracture resistant and are not preferential fracture paths. However, as is
explained later, they can have a detrimental effect on strength at higher
temperatures.

SiC:Si Interfaces

As the residual silicon solidifies, SiC:Si interfaces are formed. These
interfaces occur in all SiC:Si composites, but despite their common
occurrence, they seem to have received little attention as to their
structure. In terms of RBSC, they are found to be the weakest interface,
being preferential fracture paths and crack initiating sites. They are also
the first to undergo polishing damage as can be seen in fig.1(a). Thus,
whilst the SiC:SiC interfaces provide the mechanical cohesion, the SiC:Si
interfaces appear to provide the strength limit. Further, dark-field TEM has
clearly established that the silicon regions rarely protrude for any
distance along the grain boundaries impinging onto them.[9]

Fig.4 shows an HREM image of part of one such interface. The SiC is only partially transformed to α, some β being still present. Inspection of the interface with the β-SiC reveals good planar matching between the inclined $(111)_{Si}$ and $(111)_{β-SiC}$ plane stacks, with only occasional misfit. This relationship would be expected from their respective crystal structures, i.e., $(222)_{Si}$ // $(220)_{β-SiC}$ and $d(220_{Si}) \sim d(222_{SiC}) \sim 1.54A$. For the 6H-SiC:Si part of the interface this good matching is still seen to occur for that half of the twin-related 6H unit cell which is in a parallel orientation to the original cubic stack, poorer matching occurring for the other half.

This good matching suggests that the silicon may tend to nucleate preferentially on any areas of untransformed β-SiC, producing the observed polycrystalline structure of the residual silicon. However, at some parts of the interface, image contrast with defocus suggests the presence of a thin amorphous film of possibly amorphous SiC. This is supported by dark-field imaging (e.g. fig.3(b)). However, it is not yet clear whether this film is genuinely present or is simply an artefact of specimen preparation (e.g. preferential etching during ion-beam thinning). If such a film were present, it might help to explain the surprisingly weak nature of these 'epitaxial' interfaces (see next section).

THE EFFECT OF INTERFACES ON MECHANICAL PROPERTIES

As outlined in the previous section, SiC:SiC interfaces seem to provide cohesion, whilst SiC:Si interfaces limit strength. This is illustrated in fig.5, which shows a RLM of a polished section of RBSC containing an indentation fracture crack path. In many cases the crack passes through SiC grains, but never preferentially through grain boundaries or epitaxial interfaces (this is confirmed by comparison with secondary electron images of the same crack in which the epitaxial interfaces are more readily visible by impurity-controlled contrast). Conversely, the SiC:Si interface is often fractured showing its inherent relative weakness. For comparison, fig.6 shows two parts of the same fracture surface, one part containing normal microstructure, the other a very large area of residual silicon. In the first case, mainly transgranular cleavage is apparent but, in the second case, fracture has occurred exclusively at the SiC:Si interfaces producing a surface 'replica' of the SiC grain structure.

This fracture behaviour can change with elevated temperature. Fig.7 shows a plot of modulus of rupture (MOR) versus temperature for RBSC made

Fig. 4.   HREM image of part of a SiC:Si interface showing the epitaxial nature (i.e. plane-matched) of the silicon on the β-SiC. The behaviour , with varying focus, of the white line at the interface suggests possible evidence of an amorphous film (see also fig. 3(b)).

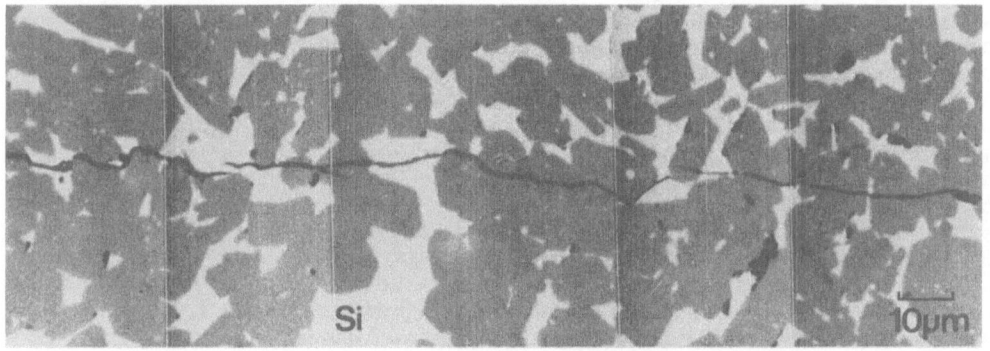

Fig. 5.   Reflected light micrograph of an indentation fracture crack, from a 20kg Vickers indentation, in a polished section of RBSC.

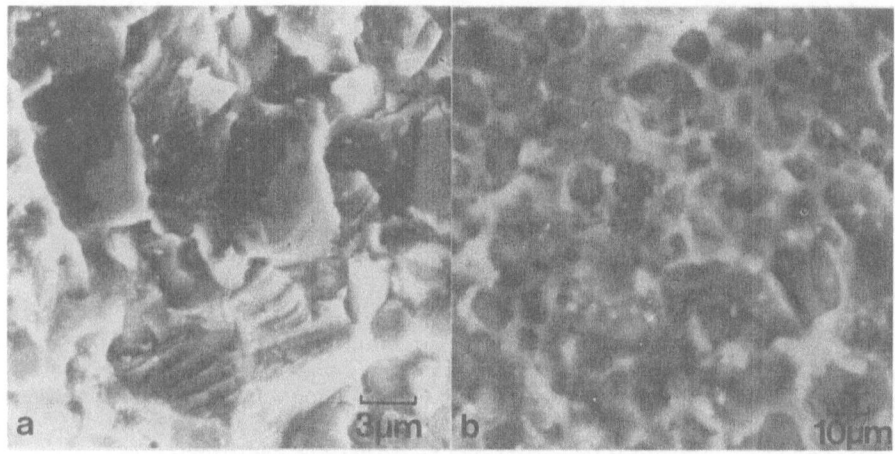

Fig. 6.   Room temperature fracture surface of RBSC. (a) Mainly transgranular failure; (b) failure at the SiC:Si interface at a large region of residual silicon.

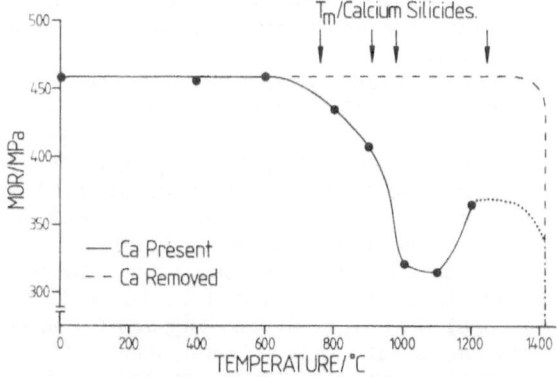

Fig. 7.   MOR versus temperature for RBSC made from both calcium containing[11] and pure starting material. The standard error in MOR is ±20MPa. The melting temperatures (T_m) of the various calcium silicides are arrowed.

with both pure and impure (calcium containing) starting materials.[11] It can be seen that whilst the pure material retains its room temperature strength up to the silicon melting temperature, the impure material shows progressive strength losses beyond ~700°C. These strength losses occur at temperatures which correspond well to the melting points of various calcium silicides. Inspection of higher temperature fracture surfaces of these materials also shows an increasing amount of intergranular failure, with silicides present on the facets. At first sight, it would seem that these results could be explained by the melting of the silicides contained in the grain boundaries. However, this might imply that the epitaxial interfaces should also be weaker, and there is no evidence for this. A better explanation is produced by considering the effect of a low melting point silicide in the residual silicon pools, at the grain boundary edges and nodes (e.g. 'A' in fig.1(b)). The melting of the silicide here would leave a wedge geometry having a considerable stress raising effect on the adjoining grain boundary, making it more likely to fail - this being <u>enhanced</u> by the presence of inclusions weakening the boundaries. This is confirmed by the observation that almost total intergranular failure is observed above the silicon melting temperature.[9] No such analagous effect can occur because of the inclusions in the epitaxial interfaces since they are completely enclosed within the grains and this stress raising geometry cannot occur. The subsequent rise in strength with temperature may correspond to formation of the higher melting point silicide. This loss of strength can be completely eliminated by reducing the level of calcium in the starting materials - other silicides melting at sufficiently high temperatures not to affect the strength.

DISCUSSION

The results presented in the previous sections demonstrate the noteworthy 'cleaning' action of the reaction-bonding process, that is, the interfacial structures - and hence mechanical properties - are dominated by structural matching and not by the impurity films that are found in other (e.g. hot-pressed, conventionally sintered) silicon carbides.

For instance, the relatively strong epitaxial interfaces never show any signs of the silica coating present on the original grains before reaction. In fact, these interfaces are usually fully coherent, being only detectable in TEM images by lines of pores. Interestingly, where strain is formed by epitaxial growth of β-SiC onto grains with polytypes of different molar volumes (e.g. 4H,15R), this strain still persists after the β-SiC has transformed to the same polytype of the underlying crystal. This would imply that the β→α transformation does not involve enough long-range diffusion (if any) to allow the near-interface strains to be relieved, supporting the belief that this transformation involves only short-range diffusional interchange and not large scale volume diffusion.[10]

The grain boundaries between new epitaxial deposits also show no sign of any impurity segregation, containing only a relatively pure, thin (~10Å) film of amorphous silicon carbide, which seems to provide good cohesion, making the boundaries relatively strong. By comparison, there is also some evidence for a similar amorphous film at the SiC:Si interfaces, which are, however, relatively weak. This film may possibly be an attempt to provide a better degree of structural matching between the α-SiC and the silicon which is not necessary between β-SiC and silicon where full matching seems possible. This is supported by the image contrast of fig.4 where the amorphous film seems to be less thick (or not present) at the β-SiC:Si interface. The exact cause of the weak nature of this interface is still uncertain, as is the exact locus of fracture paths at the interfaces (i.e. exactly at the boundary, in the silicon or in the silicon carbide).

This study has demonstrated how interfacial structures between ceramic phases can be controlled by matching of the adjacent crystal structures where thin impurity films are absent. Further, the mechanical properties of the interfaces we have studied vary from strong (SiC:SiC epitaxial interfaces), fairly strong (SiC:SiC grain boundaries) to weak (SiC:Si).

SUMMARY AND CONCLUSIONS

A variety of microstructural techniques have been used to characterise the various interfaces in RBSC. Intrinsically most interfaces are clean, due to the silicon carrying most soluble metallic impurities up the compact. Reduction of silica (from the surface of the SiC grits) is also important in producing boundary cleanliness. Epitaxial interfaces are found to be largely coherent with occasional inclusions of graphite/silicon/silicides lying along them. In cases where the original SiC particles consist of polytypes other than 6H (e.g.4H,15R) misfit dislocations and strain are present and due to the molar volume difference between 15R/4H and 3C polytypes during growth of the epitaxial layer. That this strain persists after transformation of the 3C overlayer to the same polytype as the substrate suggests that little or no long range diffusion is involved in the transformation. Grain boundaries are bounded by β-SiC facets and always contain a thin (~10Å) wetting film of pure amorphous SiC, there being no evidence of any preferential segregation of impurities to the film. Similar inclusions to those found along epitaxial interfaces are also present. The grain boundaries are mechanically strong but can be weakened by the presence of low melting point silicides both in the boundaries and in the silicon adjacent to the boundaries. However, this effect can be removed by eliminating calcium (which forms very low melting point silicides) from the starting materials. SiC:Si interfaces seem to be epitaxial, the remanant β-SiC acting as nucleation sites for silicon solidification. There is some evidence for a thin film of amorphous SiC at the interfaces, but this is not conclusive. These interfaces are weak and act as fracture-initiation sites and preferential fracture paths - they are the strength limiting interface.

ACKNOWLEDGEMENTS

The authors would like to thank Prof.D.Hull and previously Prof.R.W.K.Honeycombe FRS,FEng for provision of laboratory facilities. JNN acknowledges the finance of an SERC CASE award with UKAEA Springfields. Mr.P.Kennedy, Mr.J.O.Ware and members of the Cambridge Tribology Group are all thanked for useful discussions and suggestions.

REFERENCES

1. C. W. Forrest, P. Kennedy, and J. V. Shennan, 1972, The Fabrication And Properties Of Reaction-Bonded Silicon Carbides, in: "Special Ceramics 5", P. Popper, ed., BCRA, Stoke-On-Trent.

2. L. S. Ramsdell, 1947, Studies In Silicon Carbide, Am. Min., 32:64.

3. J. N. Ness and T. F. Page, Microstructural Evolution In Reaction-Bonded Silicon Carbide, J. Mater. Sci., (submitted, 1985).

4. G. R. Sawyer and T. F. Page, 1978, Microstructural Characterisation Of 'REFEL' (Reaction-Bonded) Silicon Carbides, J. Mater. Sci., 13:885.

5. T. F. Page and G. R. Sawyer, 1980, Discussion of 'Microstructural Characterisation Of 'REFEL' (Reaction-Bonded) Silicon Carbides':Authors' Reply, J. Mater. Sci. (Letters), 15:1850.

6.   J. N. Ness and T. F. Page, Polytype Formation And Transformation In Reaction-Bonded Silicon Carbide, to be published, <u>Bulletin Mineralogie</u>, 1985.

7.   Yo. Tajima and W. D. Kingery, 1982, Grain-Boundary Segregation In Aluminium-Doped Silicon Carbide, <u>J. Mater. Sci.</u>, 17:2289.

8.   W. von Münch, 1982, <u>in:</u>'Landolt-Börnstein, Numerical Data And Functional Relationships In Science And Technology, vol.III/17a', K.-H. Hellwege, ed., Springer-Verlag, Berlin.

9.   J. N. Ness, unpublished work.

10.   N. W. Jepps and T. F. Page, 1983, Polytype Transformations In Silicon Carbide, <u>J. Cryst. Growth And Charact.</u>, 7:259.

11.   P. Kennedy, UKAEA Springfields, private communication.

12.   N. W. Jepps, T. F. Page and W. M. Stubbs, 1981, Method for the TEM Characterization of Grain Boundary Films in Ceramics, <u>Proc. Materials</u> Research Society, 5:45, (Eds. H. J. Leamy et al.) North Holland.

SOME FACTORS AFFECTING MECHANICAL AND MICROSTRUCTURAL

ANISOTROPY IN REACTION-BONDED SILICON CARBIDES

J.N. Ness and T.F. Page

Department of Metallurgy and Materials Science
Pembroke Street
Cambridge CB2 3QZ U.K.

ABSTRACT

This paper explores the microstructural anisotropy, and the resultant anisotropy in mechanical properties, in reaction-bonded silicon carbides (RBSC). The anisotropy in both extruded and pressed α-RBSC is due to alignment of particles based on $\{10\bar{1}0\}_\alpha$ and $\{11\bar{2}0\}_\alpha$ cleavage during preforming. This alignment produces a change of effective grain size with direction (and a concurrent increase in weak SiC:Si interface length), which can be used to explain the change in fracture strength, indentation fracture toughness and crack paths with specimen orientation. β-RBSC shows strong $\{111\}_\beta$ texture, but no anisotropy in mechanical properties because of its equiaxed grains.

INTRODUCTION

Anisotropy can occur in composite ceramics for a variety of reasons e.g. texturing by plastic deformation, preferential grain growth, purposely aligned fibres and mechanical particle alignment during fabrication, see Rice.[1] It is the last of these that provides the source of anisotropy in reaction-bonded silicon carbide (RBSC). The alignment is a consequence of the non-equiaxed nature (due to cleavage) of the α-SiC[*] grit and occurs during the shaping and compaction stages of the production process (see below). The purpose of this paper is to report a study of this alignment by quantitative microscopy and x-ray texture techniques and to quantify its effect on two specific mechanical properties i.e. modulus of rupture and indentation fracture toughness. The aim is then to explain any observed anisotropy in these properties in terms of the microstructure.

RBSC is produced by infiltrating molten silicon into a porous preformed compact of graphite and α-SiC (10-15μm grain size). The graphite reacts with the silicon to produce 'new' silicon carbide which bonds the material

---

[*] β-SiC refers to the cubic SiC polytype, whilst α-SiC is used collectively for all the hexagonal and rhombohedral polytypes. The Ramsdell[2] notation describes a polytype by a number representing its c-axis layer repeat and a letter for its Bravais lattice type (e.g. 3C,6H,15R). This notation will be used throughout this paper.

together, the final composite microstructure consists of a matrix of silicon carbide with about 10% silicon occupying the residual pore space necessary for complete silicon infiltration during fabrication.[3,4] The microstructural evolution consists of the continuous dissolution of graphite into the liquid silicon and the subsequent precipitation of β-SiC onto the original SiC grits to form a continuous ,fairly uniform, epitaxial coating.[5] This coating then undergoes a solid-state β→α transformation to produce grains of uniform crystal structure.[5,6]

The original green compact is formed by a variety of standard processes[3] (with the aid of a polymeric binder), during which the anisotropy is introduced into the material by alignment of the original particles before any new SiC is formed. Two forming routes, extrusion and pressing, have been investigated in this study are presented in the following section.

RESULTS

Mechanical Properties

Most mechanical properties might be expected to show anisotropic behaviour as a consequence of preferential particle alignment in a material. Two interrelated room-temperature properties have been measured in this work - modulus of rupture (MOR) in three point bending and indentation fracture toughness (K_c), together with the parallel anisotropy in indentation fracture paths. Table I shows the MOR for an extruded compact,[7] with specimens cut both parallel (E) and radially (R) to the extrusion direction. Alongside this are average crack length and K_c measurements for cracks formed along the same directions (i.e. the extrusion and radial directions) in a single polished section using 1kg Vickers indentations. Also shown are the results for pressed material, using the pressing normal and a direction in the pressing plane as principal directions.

Examination of these results shows anisotropy in both MOR and K_c, with a slightly greater effect for the toughness measurements. For the extruded sample, the MOR value in the extrusion direction is higher than that in the radial direction. However, it must be borne in mind that the controlling flaw direction for the MOR will be the one at right angles to the specimen length. Thus, the radial MOR value is controlled by flaws in the extrusion direction, and vice-versa. Therefore, it can be deduced that the flaws in the radial direction are shorter than those in the extrusion direction. This is reflected in the crack length and K_c results from the indentation tests, which show that the extrusion direction is the weaker locus for crack propagation. Indentation results for the pressed sample similarly establish that the pressing plane is the preferred crack path.

Fig.1 shows a reflected light micrograph of part of an indentation in extruded material used to determine the K_c values, illustrating the fracture paths of the indentation-induced cracks. It is shown in a companion paper,[8] and can be seen here, that fracture occurs preferentially along the SiC:Si interfaces, most other fracture loci being non-preferential and transgranular across the SiC grains. Thus it might be expected that the MOR and K_c values would be controlled by the distribution of SiC:Si interfaces. This, in turn, should be controlled by the SiC grain distribution. Therefore an explanation of the property anisotropy requires a quantitative knowledge of the microstructure and this is explored in the following section.

358

TABLE I: MOR and K_c Variation in Extruded and Pressed RBSC Samples

MATERIAL	MOR_1/MPa (FLAWS $11\bar{1}2$)	MOR_2/MPa (FLAWS $11\bar{1}1$)	K_c^1/MPam$^{\frac{1}{2}}$	c^1/μm	K_c^2/MPam$^{\frac{1}{2}}$	c^2/μm
EXTRUDED	431±10	321±15	2.6±0.3	48±2	4.2±0.2	36±1
PRESSED	422±14	*	2.9±0.2	44±1	4.3±0.3	35±2

*MOR Specimens unavailable Errors are one Standard Error in the mean

For Extruded 1 = Extrusion Direction   2 = Radial Direction

Pressed 1 = Pressing Plane   2 = Pressing Normal

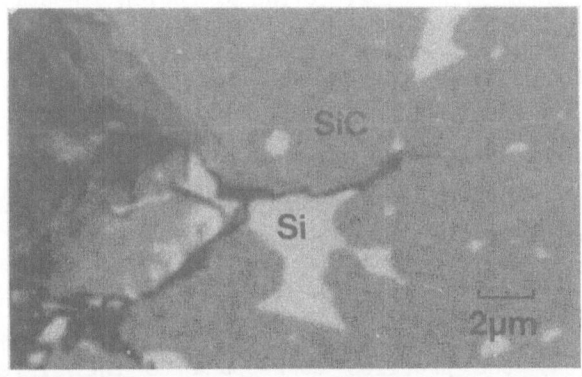

Fig. 1. Reflected light micrograph of crack from a 1kg Vickers indentation, showing preferential failure of SiC:Si interfaces.

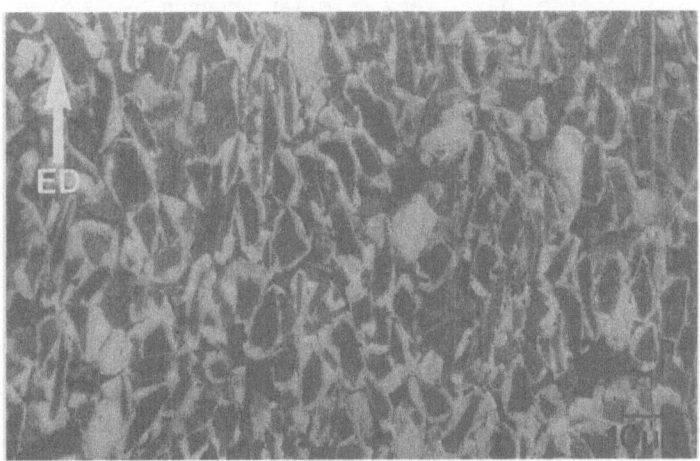

Fig. 2. Secondary electron image of a polished section of an extruded specimen, showing alignment of grains in the extrusion direction. The composite grains usually display dark angular centres (the original grits) with lighter epitaxial overlayers.

Microstructural Anisotropy

In order to study the grain size and shape distribution in RBSC, several secondary electron scanning electron micrographs were taken on each section. Fig.2 shows an example of one such micrograph, taken on a section of an extruded sample containing the extrusion direction, showing obvious grain alignment in that direction. The silicon appears as the darkest irregular regions in the image, the different grey levels in the silicon carbide grains being controlled by different impurity levels.[4,5] The mainly dark angular core in each composite grain corresponds to the original SiC grit, the lighter imaging material being the SiC formed in the reaction-bonding process.[4] Tracings were taken off these micrographs showing both the cores and the final grains, and then processed on a Quantimet 720B image analyser. Measurements of linear grain intercepts in various directions were made and in this way polar plots of mean linear intercept (MLI) versus direction were produced.

Fig.3(a) shows an example of such a plot for a section of an extruded specimen containing the extrusion direction (the section used to generate the K_c data in table I). A marked anisotropy in effective grain size is apparent, the maximum lying close to the extrusion direction. Examination of several such sections suggests that the anisotropy can be explained by simple alignment of plate-like particles along the extrusion direction. Similar results are found for pressed material with the particles being flattened into the pressing plane. The quantitative relation of MOR, K_c and this effective grain size distribution will be explored in later discussion.

As mentioned previously, the grain distribution would be expected to control the SiC:Si interface distribution, which, in turn, would be expected to control the fracture behaviour. The Quantimet results suggest that the effective SiC:Si interface length should be greater in, say, the extrusion direction of an extruded specimen. Since this is directly related to the size of the residual silicon regions, a simple way to investigate this hypothesis is to take surface roughness profiles on polished sections. The features shown by the profiles correspond to the difference in height between the silicon and the silicon carbide, the silicon polishing preferentially. Fig.3(b) shows two surface roughness traces for the principal directions of fig.3(a), their appearance being remarkably different. Both traces have very similar roughness values, but quite different average feature spacings, the radial trace having about half the spacing of that parallel to the extrusion direction. This is very similar to the ratio of effective grain sizes in the principal directions of fig.3(a), confirming that the SiC:Si interface length is indeed controlled by the effective SiC grain size.

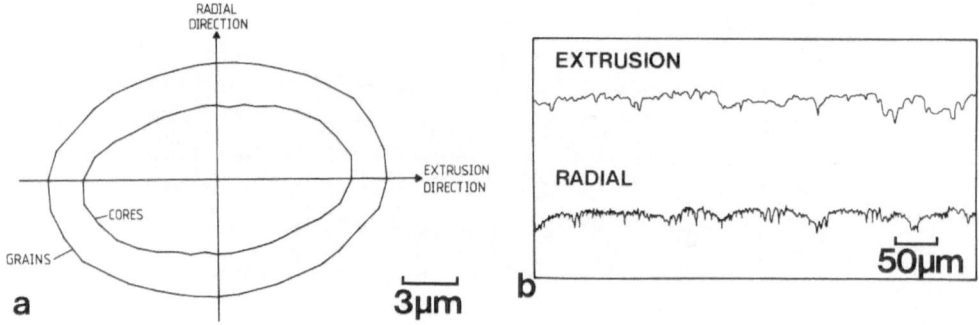

Fig. 3. Extruded material. (a) Plot of mean linear intercept versus direction for a section containing the extrusion direction; (b) surface roughness traces for the two principal directions in (a).

X-Ray Texture Pole Figures

The previous section has illustrated the presence of physical anisotropy in RBSC, but has not covered the possibility that this is directly related to the crystallography of the material. We have investigated this by means of x-ray texture pole-figures.

The most dominant polytype present in the material is 6H and so reflections from this polytype were used. Ideally, we would like to have used the $\{0006\}_{6H}$ reflection (the first non-zero intensity c-axis reflection), but the stacking sequence of tetrahedral layers in the 6H structure makes this reflection coincident with $\{10\bar{1}2\}_{6H}$ and thus renders it useless for texture determination. Of the next most obvious reflections, $\{10\bar{1}0\}_{6H}$ is systematically absent and $\{11\bar{2}0\}_{6H}$ is again coincident with other reflections. Thus, the first simple available non-zero reflection is $\{10\bar{1}1\}_{6H}$ (multiplicity of six - point group $\bar{6}m2$), which was the one used.

Fig.4 shows x-ray texture pole-figures for extruded and pressed specimens. The extruded figure is taken from the section used for fig.3, whilst the pressed figure is from an analogous one containing the pressing direction and pressing plane. Both figures show very similar and pronounced anisotropy, which would be expected from the previous section. However, the figures do not correspond to the anisotropy expected by the alignment of $\{0001\}_{\alpha}$ platelike particles, as we originally thought. It was therefore necessary to investigate the crystallography more thoroughly, as will now be shown.

Growth Of New SiC

As mentioned in the introduction, the new SiC formed in the process initially deposits as β-SiC onto the original α grains, before subsequently transforming to the underlying polytype. It thus grows with morphologies characteristic of β-SiC i.e. $\{111\}_{\beta}$, $\{110\}_{\beta}$ and $\{100\}_{\beta}$.[5] These cubic planes have a simple relation to the underlying α crystallography and thus the growth habit can be used to examine the crystallography of the original grains. The first stage in this process was to incorporate large single crystals of known habit into green compacts and then to examine the appearance of the new material deposited onto them e.g. fig.5(a). The morphologies so produced could then be compared to new material formed on normal sized (10-15μm) grit e.g. fig.5(b). Such studies, coupled with serial

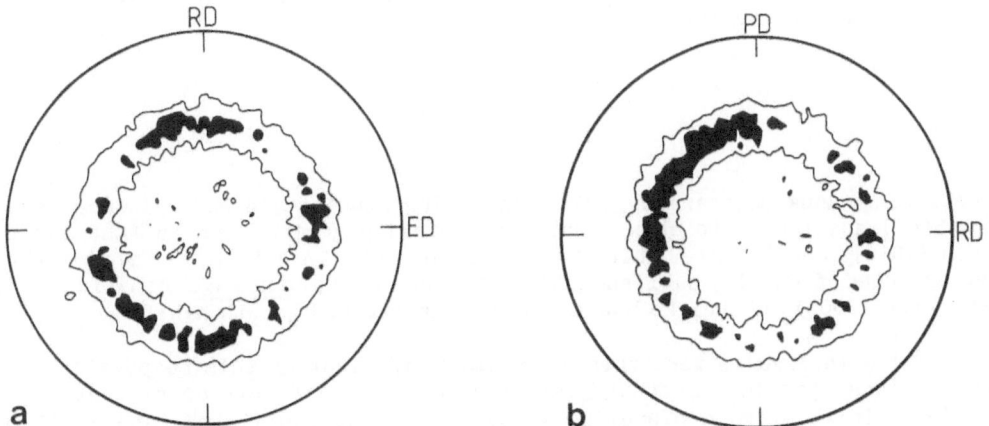

a   b

Fig. 4. $\{10\bar{1}1\}_{6H}$ x-ray texture pole-figures (CoK_{α} radiation) for (a) section used for fig.3 and (b) an analogous section in pressed material containing the pressing plane and pressing direction (Black = highest intensity).

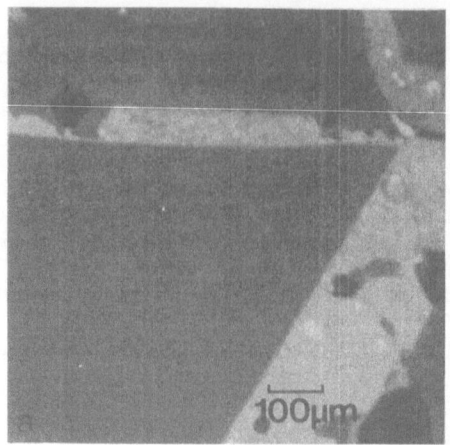 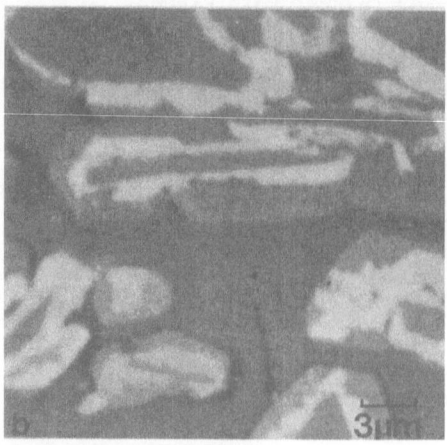

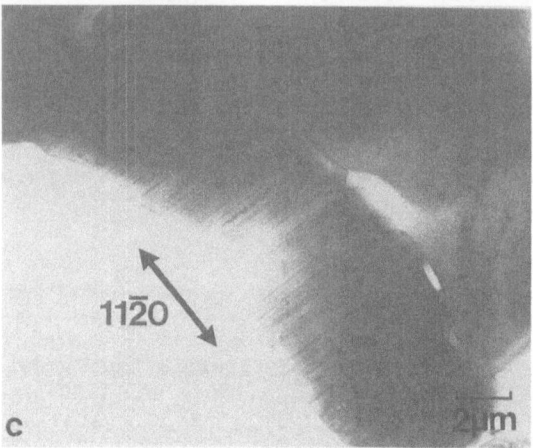

Fig. 5. Morphology of SiC grains. (a) Secondary electron image (SEI) of a
large single crystal, with a light-imaging epitaxial layer; (b) SEI
of standard material showing similar morphology to (a). The growth
facets of known crystallography in (a) can also be identified in (b)
and can be used to identify the particle habit; (c) TEM bright field
of platelike grain.

sectioning, Laue photography, SEM channelling studies and TEM evidence (e.g.
fig.5(c)) enabled us to show[5] that the platelike grains were in fact based
upon $\{10\bar{1}0\}_\alpha$ and $\{11\bar{2}0\}_\alpha$ habits and not on $\{0001\}_\alpha$. This is indeed what might
be expected from the published information on the <u>cleavage</u> behaviour of
SiC,[9] the $(0001)_\alpha$ platelike habit being the preferred <u>growth</u> habit.

The pole-figures were then reexamined in light of this reappraisal of
the original particle morphologies and found to be fairly consistent with
preferentially aligned platelike particles of the two $\{10\bar{1}0\}_\alpha$ and $\{11\bar{2}0\}_\alpha$
habits. However, the fit was by no means perfect, because of a mixture of the
$\{11\bar{2}0\}$, $\{10\bar{1}0\}$ and $\{0001\}$ habits occurring. For extruded material, the
alignment is along the extrusion direction; for pressed material the
alignment is such that the particles lie in the pressing plane.

Normally, reaction-bonded material is made using α-SiC starting material, but some material has been made using fine-grained (~1μm) β-SiC grits. This is made in the usual way but, in this case, the new material deposits as β-SiC and does not transform e.g. fig.6(a). We have been unable to find any shape anisotropy in any section of the material using quantitative microscopy, nor any significant variation in the indentation fracture toughness (K_c = 4.28(±0.17)MPm$^{1/2}$), which was a good confirmation of the unbiased nature of the indentation tester used. However, {111}$_\beta$ pole figures show marked crystallographic alignment. Fig.6(b) shows a {111}$_\beta$ pole-figure for a pressed specimen, the section being taken in the plane of pressing, showing a pronounced alignment around the pressing direction. Since it would be expected that cubic material would form preferentially larger {111} faces (growth normal to these faces being favoured[5]), this may simply reflect the grains' tendency to rest on their larger faces.

Thus, this material provides a marked example of one which shows pronounced crystallographic texture, but no anisotropy in shape or mechanical properties.

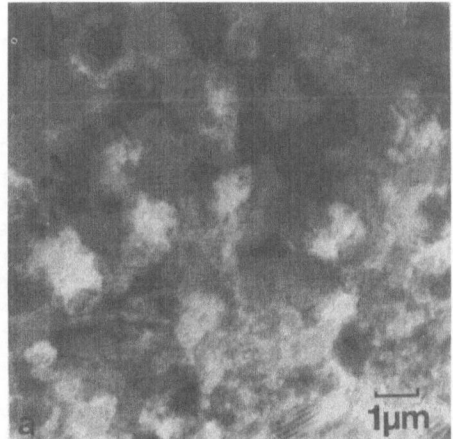 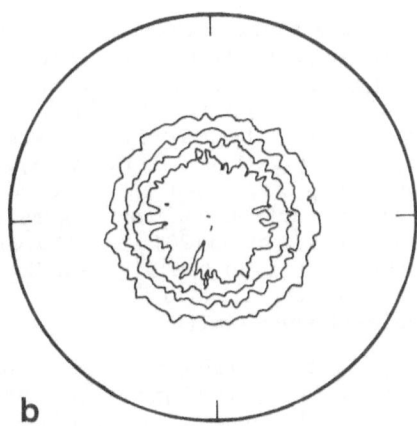

Fig. 6. β-RBSC. (a) TEM bright field showing equiaxed grains of uniform size; (b) {111}$_\beta$ x-ray texture pole-figure of pressed specimen cut parallel to the pressing plane.

DISCUSSION

The results presented in the previous section can be used to give a quantitative explanation of the anisotropy in mechanical properties. Consider first the relationship between MOR and effective grain size. It would be expected that the MOR for a specimen cut in a particular direction would be related to the grain size (and hence flaw size) in the perpendicular direction (i.e. $MOR_E \propto (MLI_R)^{-1/2}$ and vice-versa). Table II shows that this is indeed the case, if the ratio of the two values is considered. The ratio of the two perpendicular MORs is very similar to the square root of the ratio of the mean linear grain intercepts in the transverse directions. This is also reflected in the surface feature spacing measurements of fig.3(b). Also shown in the table is the ratio of the average indentation crack lengths. This is also in good agreement with the MLI ratio. Thus, in both cases, the anisotropy in mechanical properties corresponds well to the microstructural anisotropy of the grain size and hence defect size.

TABLE II: Comparison of Mechanical and Microstructural Anisotropy

MOR_E/MOR_R	C_E/C_R	$(\frac{MLI_R}{MLI_E})^{\frac{1}{2}}$	FEATURE SPACING RATIO$^{\frac{1}{2}}$
1.34	1.35	1.32	1.41

As mentioned in the previous section, these results are best explained by considering the distribution of the SiC:Si interfaces, which are known to be the weakest part of the microstructure. The alignment of grains in the extrusion direction (with analogous alignment for pressed material) produces an effective increase in SiC:Si interface length, as supported by surface roughness measurements. Thus, the quantitative microscopy has enabled us to explain the mechanical property anisotropy in terms of simple particle alignment. However, our attempts to quantify the crystallography proved more difficult.

The basic problem we encountered in trying to produce meaningful x-ray texture pole-figures was that the most useful reflection(s) viz. $\{0001\}_\alpha$ was coincident with $\{10\bar{1}2\}_{6H}$. Thus, a less geometrically convenient reflection had to be used i.e. $\{10\bar{1}1\}_{6H}$. However, careful interpretation of this data showed that the pole-figures were incompatible with the alignment of $\{0001\}_\alpha$ plates, as we had initially expected. Reexamination of the crystallography of the grits, based on evidence from the SiC nucleation and growth process during reaction-bonding, suggested that platelike particles were more likely to be based on $\{11\bar{2}0\}_\alpha$ and $\{10\bar{1}0\}_\alpha$ cleavage, and this showed a better, if not perfect, fit to the figures. However, it is clear that the crystallographic habit of the aligned particles is sufficiently convoluted to make the use of pole-figures limited.

By comparison, the cubic RBSC showed an extremely simple anisotropy in its $\{111\}_\beta$ pole-figure. The ease of interpretation lay largely in the equiaxed nature of the grains and hence full equivalence of the $\{111\}$ planes. However, this crystallographic anisotropy had no effect on mechanical properties which were again controlled by the shape of the grains.

CONCLUSIONS AND SUMMARY

We have investigated both the microstructural and mechanical property anisotropy in α and β-RBSC. α-RBSC showed an alignment of platelike particles in both extruded and pressed materials, in the extrusion direction and in the pressing plane respectively. This alignment of particles produces an anisotropy in the residual silicon regions and hence, critically, in the weak SiC:Si interface distribution. The grain/interface anisotropy corresponded well to the anisotropy in MOR and indentation fracture toughness, the MOR being controlled by the flaws in transverse directions to the specimen length. Attempts to quantify the crystallography proved difficult both because simple 6H-SiC reflections could not be used and because the resultant $\{10\bar{1}1\}_{6H}$ pole-figures did not correspond to the expected alignment of $\{0001\}_\alpha$ plates. However, reassessment of the crystallography of the grits, by comparison with the crystallography of the newly-formed SiC, showed that platelike grains were, in fact, based on $\{10\bar{1}0\}_\alpha$ and $\{11\bar{2}0\}_\alpha$ and not the $\{0001\}_\alpha$ growth habit. This provided a better fit to the pole-figures, but the technique was felt to be limited by the mixture of habits present in the material. By comparison, the β−RBSC showed strong {111}$_\beta$ anisotropy, simply interpretable as equiaxed grains preferring to rest on their larger faces. However, this anisotropy did not affect mechanical properties, which were found to be isotropic in line with the equiaxed grain structure.

ACKNOWLEDGEMENTS

The authors would like to thank Prof.D.Hull and previously Prof.R.W.K.Honeycombe FRS,FEng for provision of laboratory facilities. JNN acknowledges the finance of an SERC CASE award with UKAEA Springfields. Mr.P.Kennedy, Mr.J.O.Ware and members of the Cambridge Tribology Group are all thanked for useful discussions and suggestions.

REFERENCES

1. R. W. Rice, 1977, Microstructure Dependence Of Mechanical Behaviour Of Ceramics, in: "Treatise On Materials Science And Engineering, vol. II", Academic Press, New York.

2. L. S. Ramsdell, 1947, Studies In Silicon Carbide, Am. Min., 32:64.

3. C. W. Forrest, P. Kennedy, and J. V. Shennan, 1972, The Fabrication And Properties Of Reaction-Bonded Silicon Carbides, in: "Special Ceramics 5", P. Popper, ed., BCRA, Stoke-On-Trent.

4. G. R. Sawyer and T. F. Page, 1978, Microstructural Characterisation Of 'REFEL' (Reaction-Bonded) Silicon Carbides, J. Mater. Sci., 13:885.

5. J. N. Ness and T. F. Page, Microstructural Evolution In Reaction-Bonded Silicon Carbides, J. Mater. Sci., (submitted, 1985).

6. T. F. Page and G. R. Sawyer, 1980, Discussion of 'Microstructural Characterisation Of 'REFEL' (Reaction-Bonded) Silicon Carbides':Authors' Reply, J. Mater. Sci. (Letters), 15:1850.

7. J. O. Ware, UKAEA Springfields, Private Communication.

8. J. N. Ness and T. F. Page, The Structure And Properties Of Interfaces In Reaction-Bonded Silicon Carbides, this volume.

9. R. W. Keyes, 1960, Morphology Of Commercial Silicon Carbide Crystals, in: "Silicon Carbide 1959", Pergamon, London.

MICROSTRUCTURE-MECHANICAL PROPERTY RELATIONSHIPS IN 94% ALUMINA CERAMICS

John R. Hellmann* and John Matsko**

*Ceramics Development Division 1845
**Ceramics Processing Section 7476-1
Sandia National Laboratories
Albuquerque, New Mexico

Stephen W. Freiman and T.L. Baker

Inorganic Materials Division
National Bureau of Standards
Gaithersburg, Maryland

INTRODUCTION

Alumina ceramics are experiencing an increased application over conventionally used glasses in the manufacture of high voltage electrical components due to their higher strength, fracture toughness, resistance to dielectric breakdown, and lower gas permeability. Fabrication of most electrical components requires sealing the ceramic to metallic portions of the assembly. In general, the seals must be strong and hermetic to permit proper operation of the component. The most common ceramic-to-metal seal processes involve "metallizing" the ceramic surface at elevated temperature (nominally 1500°C, wet hydrogen atmosphere, for 45 minutes) using a paint comprised of refractory metal particles, then "brazing" the metallized ceramic to the metal component at a lower temperature (typically 800-1000°C).

The use of 94% aluminas, where the balance is comprised of a MgO-CaO-Al_2O_3-SiO_2 glass phase, permits the hermetic sealing of the components at a markedly lower temperature than possible if no glass phase was present. The role of liquid phase reactions between the ceramic and metallized layer during joining has been recognized for many years.[1,2] Previous studies illustrated the importance of proper selection of metallization slurry particle size distribution, metallization sintering atmosphere, and glass phase composition in the ceramic on the integrity of ceramic-to-metal seals.[2,3]

More recent studies[4,5] have led to an improved understanding of the effect of joining operations on the physical properties of the ceramic near the ceramic-to-metal interface. Previous observations of grain growth[6] and glass phase crystallization[7,8] in the ceramic illustrated that microstructural evolution occurs in alumina during thermal excursions typical of component fabrication. However, few studies have addressed the effect of microstructural evolution on the bulk physical properties of the ceramic and how those properties relate to the reliability of components fabricated from the ceramic.

The purpose of this investigation is two-fold:

1. Evaluate the effect of processing conditions on microstructural evolution in two 94% alumina ceramics prepared from commercial powders,

2. Quantify the effect of microstructure on the strength, toughness, and fatigue resistance of those ceramics.

EXPERIMENTAL PROCEDURE

Two 94% alumina powders which are routinely used in electron tube manufacture were selected for this study. Emission spectrographic analysis,[*] scanning electron microscopy,[&] X-ray sedigraphic techniques,[$] nitrogen absorption isotherms,[#] X-ray powder diffraction,[@] and gas pycnometry were employed to characterize the powders after they were calcined at 500°C for 2 hours in air to remove the organic binders.

It was readily recognized that a large number of samples would be required to adequately quantify the effect of processing conditions (cold consolidation pressure, sintering temperature, and soak time) on the evolution of microstructure. For this reason, a factorial experimental design was employed to insure sample economy while maintaining a high level of significance in data. A full description of the statistical methods used are beyond the scope of this paper and are described in detail elsewhere.[9-11] In general, experiments were conducted at the corners and centerpoint of a cube design in which the cube axes represent isostatic compaction pressure, sintering temperature, and sintering time (Figure 1). Ranges for the experimental variables were selected to bracket known processing conditions for the materials. The number of replicates required at each experimental condition was calculated from previous data of experimental variability in similar studies.

---

* National Spectrographic Laboratories, Cleveland, OH

& JEOL (USA), Peabody, MA, Model 35C

$ Micromentics Instrument Corp., Norcross, GA, Model 5000ET

Quantachrome Corp., Syosset, NY, Quantasorb

@ Sieman-Allis Inc., Cherry Hills, NJ, Model D-500

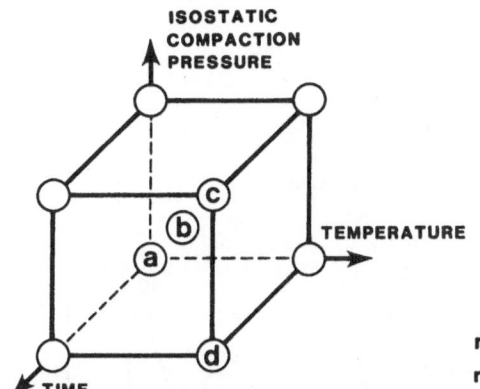

PROCESSING VARIABLES

18 ksi ≤ P ≤ 30 ksi

1 hr ≤ t ≤ 5 hr

1600 ≤ T ≤ 1705°C

EXPERIMENTAL RESPONSE

Bulk density, grain size and
shape distribution

REPLICATES

r = 2 at corners

r = 3 at centerpoint } 19 experiments

Figure 1. Factorial experimental design employed in the sintering study.

Specimens were isostatically compacted at room temperature into bars
nominally 2.54 cm square x 25 cm long for the sintering study and 5 cm
square x 25 cm long for subsequent fabrication of mechanical test specimens.
All specimens were bisque fired to simulate routine component processing,
machined into appropriate configurations for testing, and high fired in a
$MoSi_2$[##] element furnace in air (Figure 2).

Specific gravity was determined using ASTM C329-56 in which K-10
kerosine was substituted for water as the immersion medium. The kerosine
specific gravity was determined according to ASTM D891-59, method B.

Specimens were polished to <1 micron finish on diamond paste, then
thermally etched for reflected light microstructural analysis. Quantimet
image analysis[+] was employed to quantify average grain size, largest grain
size, and grain morphology. Microprobe elemental analysis[**] was performed on
polished sections to determine grain boundary glass phase compositions.
Scanning transmission electron microscopy (STEM)[$$] and selected area
electron diffraction[&&] were employed to characterize grain boundary
crystallization.

All mechanical properties[1] evaluations were determined on 2.5 cm
diameter x 0.1 cm thick discs. A Vickers indentor was used to introduce a
flaw at the center of the as-fired surfaces using loads up to 300 N.
Indentations were made in air and allowed to sit for 30 minutes before
testing. In the case of inert strength measurements, a drop of silicone oil

---

Deltech, Inc., Denver, CO

+ Cambridge Instruments, Monsey, NY, Quantimet Model 900

** Cameca Instr. Inc., Stamford, CT, Model MBX

$$ JEOL (USA), Peabody, MA, Model 100C

&& JEOL (USA), Peabody, MA, Model 200CX

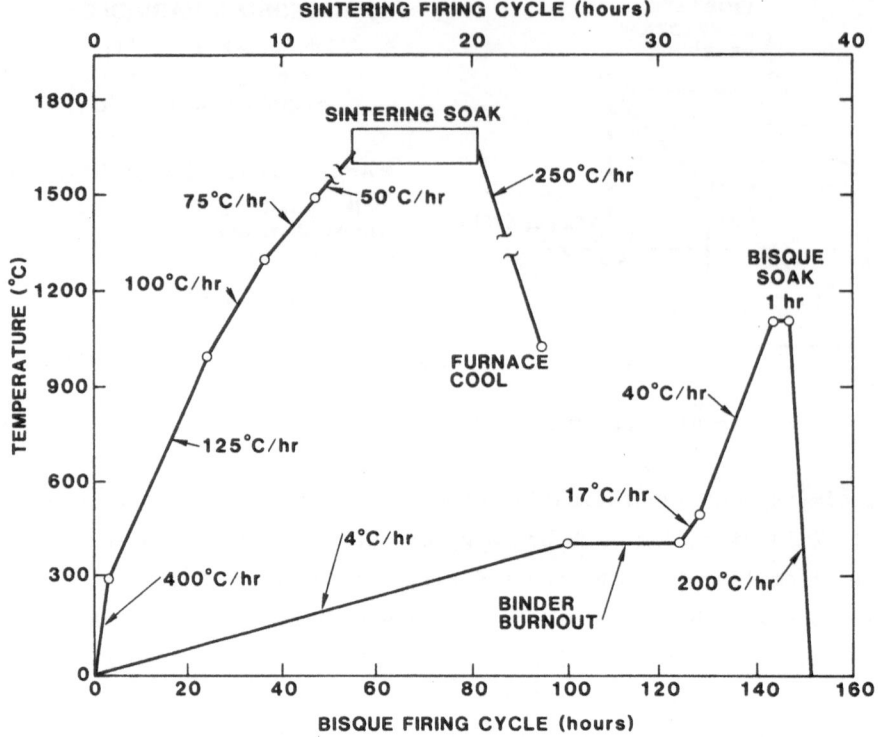

Figure 2. Bisque and high firing cycles employed in the sintering study.

was placed at the point of indentation before loading to minimize environmentally enhanced crack growth. The discs were broken in biaxial tension (flat on a 3 point support) in which the inner support diameter was 4 mm and the outer support diameter was 20 mm. The indentation was oriented on the tension side of the specimen and strengths were calculated from previously developed flexural formulas[12]. Since it was necessary to keep breaking times below 10 s to avoid significant slow crack growth during inert strength measurements, failure loads were measured using a piezoelectric load cell. All specimens were examined after failure to determine whether the failure had initiated from the indentation site.

Dynamic fatigue tests were performed in biaxial tension on discs which had been indented at a load of 10N. The specimens were loaded in water at stress rates ranging from 10^{-3} to 10^{5} MPa/s. Stress corrosion exponents were calculated from plots of Log(failure stress) vs Log(stressing rate) according to previously developed models.[14]

RESULTS AND DISCUSSION

Bulk powder characteristics, after calcining to remove organic binders, are summarized in Table 1. X-ray diffraction analysis revealed no change in crystalline phase assemblages after calcining. Significant differences in the relative concentrations of several glass modifier cations (Ca^{+2}, Mg^{+2},

Table 1. Bulk powder characteristics after calcining

Emission Spectrographic Analysis

Element (wt. %)	Alumina 1	Alumina 2
Al	Major	
Si	0.5/5.0	
Ca	0.5/5.0	0.24
Cr	0.5/5.0	—
B	0.021	0.006
Cu	< 0.001	
Fe	0.057	0.034
Mg	0.44	1.00
Mn	< 0.001	
Sr	—	—
Ti	0.076	0.01
Zr	0.01	0.042
Ba	0.003	—
Li	0.008	0.002
K	0.055	0.015
Na	0.085	0.045

Not detected: Sb, As, Pb, W, Ge, Bi, Cb, Be, Mo, Sn, V, Cd, Ag, Cu, Zn, Co.

	Alumina 1	Alumina 2
Particle Size (microns)	4.4	3.1
Surface Area (m^2/g)	2.2	3.1
Specific Gravity	3.96	3.80

B^{+3}, Ba^{+2}) and glass network intermediate cations (Zr^{+4}, Ti^{+4}) were observed between the two aluminas. The importance of these cations to glass phase viscosity, nucleation, and growth of crystalline phases during cooling the ceramic from processing temperatures is well known[7,15-16] and will be discussed in more detail later.

Density determinations were performed on samples which had been processed at the centerpoint of the experimental design (24 ksi isostatic compaction, sintered at 1653°C for 3 hours), then crushed in a steel mortar and ground to -200 mesh in a tungsten carbide lined mill. The density values in Table 1 represent an average of four gas pycnometric density measurements. Standard deviations ranged from 0.002 to 0.007 g/cc.

The results of the sintering experiments were analyzed in terms of a quadratic model which accounts for interactions between the processing variables (P,T, t).[9] The empirical models permit prediction of densification and microstructural evolution as a function of processing conditions for subsequent experiments and are conveniently represented in the graphical form shown in Figure 3. The results indicate that optimum densification occurred at the low temperature, long time region of processing space with a negligible effect of consolidation pressure. Consolidation pressure did not effect densification since the materials exhibit liquid phase sintering above 1500°C. However, pressure did affect grain size in both ceramics, possibly due to alterations in the particle size distribution due to fracture of aggregates in the powders during pressing at the pressures employed in this study[17]. Results for alumina 2 yield curves very similar to those shown in Figure 3.

Grain growth in 94% aluminas as a function of glass phase composition has been discussed briefly by Reed[7]. He pointed out the importance of the effect of substituting various glass modifier cations on the size and location of the primary crystallization field for $\alpha-Al_2O_3$ and how it affects crystal growth during cooling the ceramic from sintering temperatures. For example, substitution of MgO for CaO in the glass phase

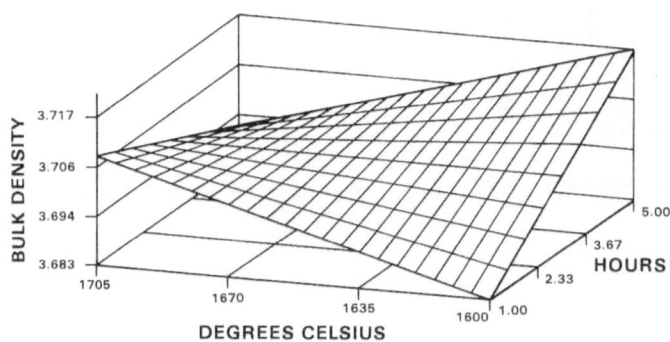

$$\rho_{bulk} = 3.10 + 0.00036\ (T) + 0.2005\ (t) - 0.00012\ (T)(t)$$

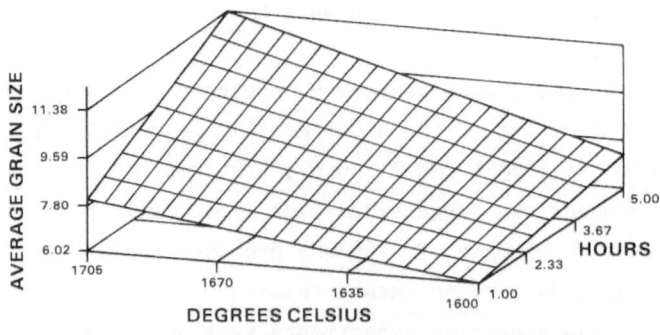

$$A.G.S. = 4.64 - 0.00036\ (T) - 6.9913\ (t) - 0.9120\ (P)$$
$$+ 0.00493\ (T)(t) + 0.00061\ (T)(P) - 0.0244\ (t)(P)$$

Figure 3. Graphical representations of the responses of bulk density and average grain size of alumina 1 to processing conditions.

yields a negligible change in viscosity[15] but substantially decreases the size of the primary crystallization field for alumina. As a result, grain growth in aluminas possessing lower CaO:MgO ratios should be less severe than in ceramics where the glass phase is modified by CaO alone. Microprobe analysis of the glass phase compositions (Table 2) revealed that alumina 1 possesses a substantially higher CaO to MgO ratio (5:1) than alumina 2 (0.3:1). The evolution of microstructure in alumina 1 differs substantially from the alumina 2 in that coarsening occurred in a markedly more acicular manner (Figure 4). Calculation of the activation enthalpy for grain growth (from the slope of Log(average grain size) vs. 1/T plots) yielded no differences between the two aluminas, suggesting that differences in microstructural evolution are due mainly to modification of the primary crystallization field for alumina and not to differences in glass phase viscosity.

Glass phase composition also affects microstructural evolution at temperatures below where grain growth is expected to be active (<1450°C). Figure 5 shows that grain boundary crystallization occurs during typical joining processes. STEM and electron diffraction analysis revealed the crystalline phase is anorthite ($CaAl_2Si_2O_8$, JCPDS #12-301), in agreement with published phase equilibrium diagrams.[18]

MECHANICAL PROPERTIES EVALUATIONS

Microstructural coarsening and grain boundary crystallization can markedly alter the mechanical properties of ceramics.[19-20] The effects of grain size, shape, and grain boundary crystallization on the fracture toughness, strength, and fatigue resistance were evaluated on aluminas 1 and 2 that had been processed at conditions known to produce a 50% increase in grain size (24ksi, 1653°C, 3 hrs. and 30ksi, 1705°C, 5 hrs., Figure 4.).

Table 2. Microprobe elemental analysis of glass phases

Mole % oxide	Alumina 1	Alumina 2
SiO_2	40.5	43.6
Al_2O_3	29.4	34.2
CaO	22.4	4.4
MgO	4.7	12.4
Na_2O	1.1	0.5
K_2O	1.4	0.1
B_2O_3	0.9	4.1
BaO	0.04	0.01
TiO_2	0.4	0.1
ZrO_2	0.1	0.3
FeO	0.2	0.1
Cr_2O_3	0.7	0.01
P_2O_5	0.7	0.1

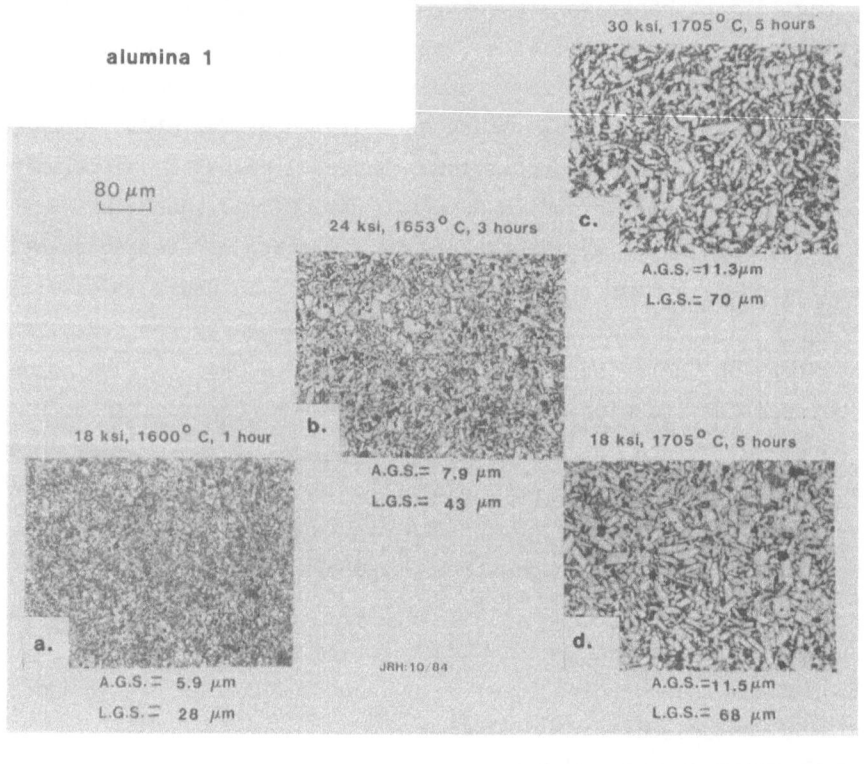

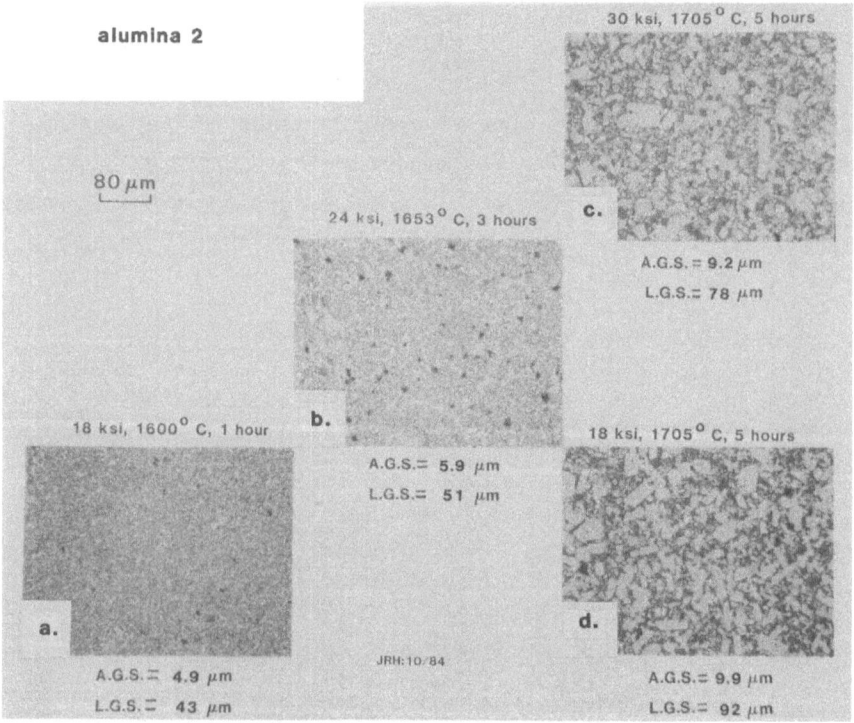

Figure 4. Microstructural evolution as a function of processing conditions.
Photos a-d represent processing conditions shown in figure 1.

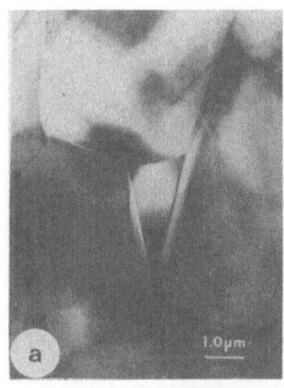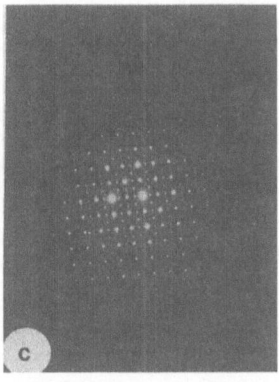

Figure 5. a. Transmission electron micrograph of glass phase at a grain
boundary triple point in "as-sintered" alumina 1.
   b. Heavily twinned crystalline phase forms during annealing at
      1000°C for 1 hour.
   c. Selected area electron diffraction analysis confirmed the
      presence of anorthite ($CaAl_2Si_2O_8$).

Figure 6 shows a comparison of the inert strength data obtained for
aluminas 1 and 2 sintered at 1653°C for 3 hours plotted as log strength
versus log indentation load. The dashed line (slope=-1/3) represents
strengths expected on the basis of an indentation-fracture model[21]. The
fracture toughness of the material, K_{1c}, calculated from the position of the
dashed line along the indentation load axis[21] are shown in Table 3. The
larger K_{1c} for alumina 1 may be attributed to its more acicular
microstructure, in agreement with data previously obtained on silicon
nitride ceramics[20].

It has been previously shown that departures from the indentation model
at small indentation loads as a function of grain size in polycrystalline
materials[22,23] are indicative of a direct microstructural influence on the
stress intensity factor at a flaw, such as due to localized stresses arising
from thermal expansion anisotropy. At indentation loads of 3N or less many
specimens failed from "natural" flaws in the specimen surface rather than
from the indentation induced crack. These data are shown as the cross-
hatched boxes in Figure 6 and represent the upper limit to the strength of
these aluminas in the "as-fired" condition. One can characterize the regime
over which failure is dominated by microstructurally related flaws by
determination of the quantity P^*. This quantity represents the intersection
of the strength level of the natural flaws with the dashed line describing

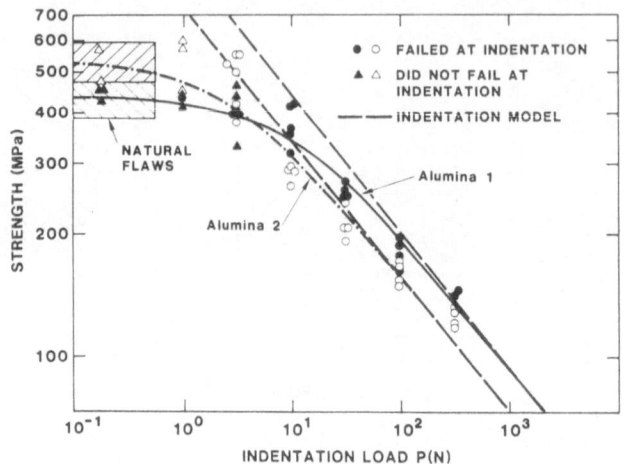

Figure 6. Typical strength vs. indentation load results for aluminas 1 and 2 processed at 24 ksi, 1653°C, 3 hours.

the indentation model[23]. Alumina 1 exhibits a higher P^* than alumina 2 for both sintering conditions (Table 3), indicating that microstructure plays a more important role in the fracture of alumina 1.

The slope of the curve obtained for the dynamic fatigue measurements gives an indication of the relative fatigue resistance of the two materials. The difference in stress corrosion susceptibilty between the two materials can be quantified by calculating "n", the exponent in the crack growth equation: $V=V_o(K_1/K_{1c})^n$, where V is the crack velocity, V_o is an empirical constant, and K_1 is stress intensity. For the case of indentation induced cracks, $n=4/3[(1/s)-1]-2/3$, where s is the slope of the dynamic fatigue

Table 3. Mechanical properties summary

	σ PLATEAU (MPa)	P^* (N)	K_{IC} (MPa m$^{1/2}$)	A.G.S.' (microns)
ALUMINA 1				
24 ksi, 1653°C, 3 hours	431 ± 44	11.4	5.0	7.9
30 ksi, 1705°C, 5 hours	418 ± 21	10.4	4.7	11.3
ALUMINA 2				
24 ksi, 1653°C, 3 hours	537 ± 64	5.9	4.0	5.9
30 ksi, 1705°C, 5 hours	465 ± 79	5.5	4.4	9.2

'AVERAGE GRAIN SIZE

curve (Figure 7). The values for the stress corrosion exponent "n" differ markedly between the two aluminas; no significant difference in "n" was observed as a function of processing conditions (Figure 7 inset). The differences in the stress corrosion susceptibility between the two materials may be attributed to differences in grain boundary glass composition (Table 2). We are currently evaluating the effect of varying the CaO/MgO ratio on the slow crack growth behavior in bulk glasses of compositions similar to the grain boundary phases measured in this study.

Strength and fatigue resistance were also affected by grain boundary crystallization. In general, strengths measured in the low indentation load regime decreased on the order of 20% after crystallization of the grain boundaries. The decrease in strength may be due to residual stresses produced by differential thermal contraction of the crystalline phases in the grain boundaries during cooling. The stress corrosion exponents for the two aluminas converged to a value of n=41 after crystallization, suggesting that environmental effects on fracture may be largely determined by the stress corrosion resistance of the crystalline grain boundary phase(s).

SUMMARY

A statistically designed experiment was implemented to quantify the effect of processing conditions on microstructural evolution in two 94% alumina ceramics. Models developed from experimental data were used to select processing conditions for the synthesis of microstructures of interest (acicular vs equiaxed grains) for mechanical properties evaluations. Differences in microstructural evolution are attributed to

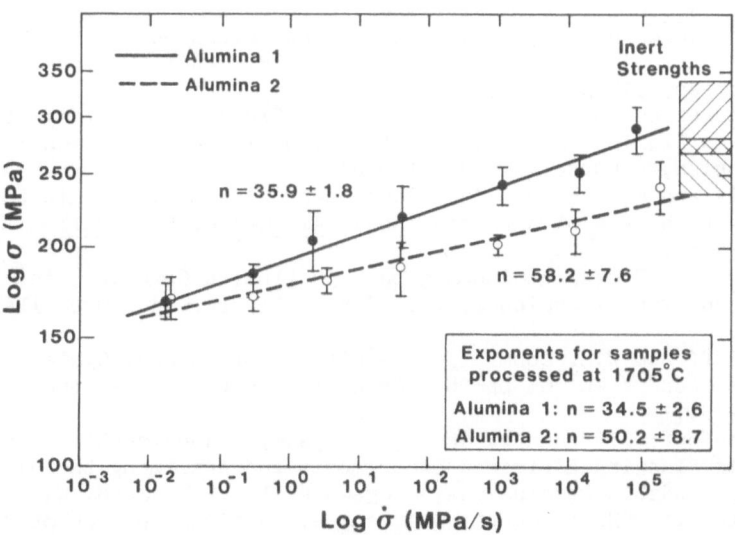

Figure 7. Dynamic fatigue results for aluminas 1 and 2 processed at 24 ksi, 1653°C, 3 hours.

glass phase compositional effects on crystallization of $\alpha-Al_2O_3$ from the glass phase during cooling from sintering temperatures.

Acicular microstructures exhibited slightly higher fracture toughnesses than equiaxed microstructures. Both materials showed direct effects of localized microstructural stresses on strength for small flaw sizes. Dynamic fatigue results indicate a significant difference in stress corrosion susceptibility between the two materials which is attributed to compositional differences in the glassy grain boundaries.

Thermal treatments typical of ceramic-to-metal sealing processes yielded crystallization of the grain boundaries. Crystallization decreased the strength of the materials, possibly due to stresses produced during cooling from crystallization temperatures. The stress corrosion exponent for both materials converged to a single value after crystallization suggesting that environmental effects on fracture may be dominated by the susceptibility of the crystalline grain boundary phase to stress corrosion.

ACKNOWLEDGMENTS

We gratefully acknowledge the assistance of B. Botsford, J. Bumgarner, W. Coblenz, P.F. Hlava, and W.R. Sorenson in performing much of the starting material and microstructural characterization required in this study.

REFERENCES

1. S.S. Cole, Jr. and H.W. Larisch, Theory of Adherance in Ceramic-to-Metal Seals, in: "Advances in Electron Tube Techniques," D. Slater, ed., Pergamon Press, NY (1961).
2. M.E. Twentyman, High Temperature Metallizing, Part 1. The mechanism of glass migration in the production of ceramic-to-metal seals, J. Mater. Sci. 10:765(1975).
3. M.E. Twentyman and P. Popper, High Temperature Metallizing, Part 2. The effect of experimental varibles on the structure of seals to debased aluminas, J. Mater. Sci. 10:777(1975).
4. J.C. Swearengen and W.F. Chambers, Characterization of Metallized Alumina: Structure, SAND-76-0539, Sandia National Laboratories, Albuquerque, NM, (1976).
5. J.C. Swearengen, O.L. Burchett, and J.H. Gieske, Characterization of Metallized Alumina: Properties, SAND-76-0591, Sandia National Laboratories, Albuquerque, NM, (1976).
6. S.S. Cole and F.J. Hynes, Some Parameters Affecting Ceramic-to-Metal Seal Strength of a High Alumina Body, Am. Ceram. Soc. Bull. 37:135(1958).
7. L. Reed, Heat Treatment Studies on High Alumina Ceramics, in: "Advances in Electron Tube Techniques, Vol. 2," D. Slater, ed., MacMillan Co., NY(1963).
8. H.P. LaBuff and W.G. Schmidt, Some Effects of Thermal Cycles on Seal Strength, Report No. GE pp-186, General Electric Co. Neutron Devices Div.,(1975).
9. D.H. Doehlert, P.D. Bantz, and C.P. Criste, "Experimental Strategies, A Course in Planning Efficient Experiments and Analyzing the Data to Obtain Dependable Conclusions," Edgework, Inc., Redmond, WA.
10. M.G. Natrella, "Experimental Statistics," National Bureau of Standards Handbook 91, U.S. Gov't. Printing Office, Wash., D.C.(1963).
11. W.G. Cochran and G.M. Cox, "Experimental Design," 2nd. edition, John Wiley and Sons, NY(1957).

12. D.B. Marshall, An Improved Biaxial Flexure Test for Ceramics, Am.Ceram. Soc. Bull. 59:551(1980).
13. D.B. Marshall and B.R. Lawn, Flaw Characteristics in Dynamic Fatigue: The Influence of Residual Contact Stresses, J. Am. Ceram. Soc. 63:532(1980).
14. E.R. Fuller, B.R. Lawn, and R.F. Cook, Theory of Fatigue for Brittle Flaws Originating from Residual Stress Concentrations, J. Am. Ceram. Soc. 66:314(1983).
15. E.K. Turkdogan and P.M. Bills, A Critical Review of Viscosity of CaO-MgO-Al$_2$O$_3$-SiO$_2$ Melts, Am. Ceram. Soc. Bull. 39:682(1960).
16. R.C. DeVries and E.F.Osbourn, Phase Equilibria in High-Alumina Part of the System CaO-MgO-Al$_2$O$_3$-SiO$_2$, J. Am. Ceram. Soc. 40:6(1957).
17. G.L. Messing, C.J. Markhoff, and L.G. McCoy, Characterization of Ceramic Powder Compaction, Am. Ceram. Soc. Bull. 61:857(1982).
18. E.F. Osborn, R.C. DeVries, K.H. Gee, and H.M. Kraner, Optimum Composition of Blast Furnace Slag As Deduced From Liquidus Data for The Quaternary System CaO-MgO-Al$_2$O$_3$-SiO$_2$, Trans. AIME 200:33(1954).
19. R.W. Rice, Elastic Anisotropy and the Grain Size Dependence of Ceramic Fracture Energies, J. Mater. Sci. 19:1267(1984).
20. F.F. Lange, Fabrication and Properties of Dense Polyphase Silicon Nitride, Am. Ceram. Soc. Bull. 62:1369(1983).
21. P. Chantikul, G.R. Anstis, B.R. Lawn, and D.B. Marshall, A Critical Evaluation of Indentation Techniques for Measuring Fracture Toughness: II, Strength Method, J. Am. Ceram. Soc. 64:539(1981).
22. B.R. Lawn, S.W. Freiman, T.L. Baker, D.B. Cobb, and A.C. Gonzalez, Study of Microstructural Effects on the Strength of Alumina Using Controlled Flaws, J. Am. Ceram. Soc. 67:C-67(1984).
23. R.F. Cook, B.R. Lawn, and C.D. Fairbanks, Microstructure-Strength Properties in Ceramics: I. Effect of Crack Size on Toughness, submitted to J. Am. Ceram. Soc.

NEW LOW EXPANSION MAGNETIC MATERIALS-A COMPOSITE APPROACH

Dinesh K. Agrawal and Rustum Roy

Materials Research Laboratory
The Pennsylvania State University
University Park, PA 16802

INTRODUCTION

A composite ceramic is fabricated by mixing and processing two or more phases which are supposed to be thermodynamically compatible with each other and thus controlling or tailoring a desired property in the composite by varying the amount of each component or by modifying the processing technique. The di/multiphasic concept permits the selection of both materials – both in terms of any special property each phase might have and quantity of each phase. While this concept opens the door wide for tailoring a special property, there are several problems that one may encounter – the intimacy of mixing, the strength of the sintered body, the reactivity of the constituents and the interplay of the elastic forces all combine to determine the effectiveness of the composite material.

In earlier work, the authors have first produced a new family of negative and 'zero' expansion ceramics and then successfully adopted this diphasic approach[1,2] to produce near zero thermal expansion composite ceramics. The strategy was to mix two thermodynamically compatible ceramic phases – one with negative coefficient of thermal expansion (α) and the second one with positive α – and process the mixture by a standard ceramic technique. The negative α phase was chosen from a new structural family of materials now known as [NZP] or [CTP] – discovered recently for low α applications[3,4]. The positive α phase can be selected from several compositions such as $GdPO_4$, Nb_2O_5, $ZrSiO_4$, $Mg_3(PO_4)_2$ and $Zn_3(PO_4)_2$ etc. Figures 1 and 2 demonstrate some of the thermal expansion data of some compositions belonging to $CaZr_4P_6O_{24}$ + Nb_2O_5, $ZrSiO_4$ and MgO systems. It is evident from these results that several compositions exhibit near zero expansion profiles over wide temperature ranges.

In order to fabricate new low thermal expansion magnetic materials the same diphasic strategy was adopted in which Yttrium Iron Garnet (YIG-$Y_3Fe_5O_{12}$) was chosen as a magnetic material with positive α and $Na_4Zr_2Si_3O_{12}$ (another member of NZP-family) as primary phase to control the thermal expansion of the composite ceramic produced. Here in this paper, we present the results of the thermal expansion behavior of various compositions in $Na_4Zr_2Si_3O_{12}$ + YIG system.

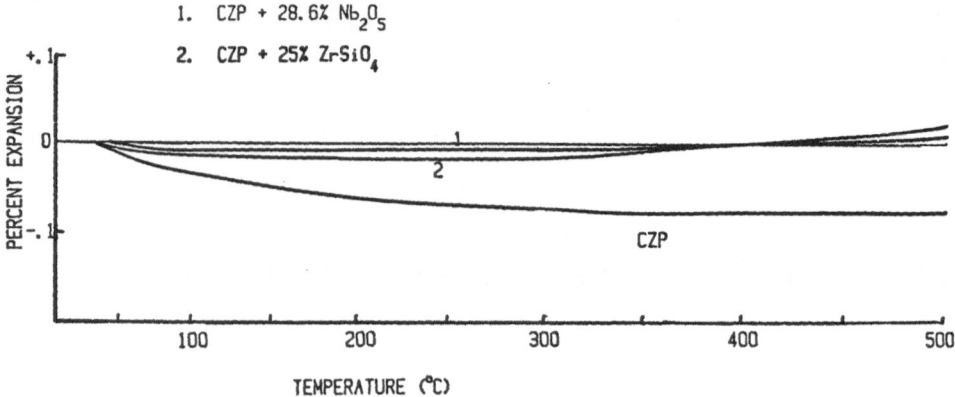

Fig. 1. Thermal Expansion of CZP + Nb_2O_5 and $ZrSiO_4$ systems.

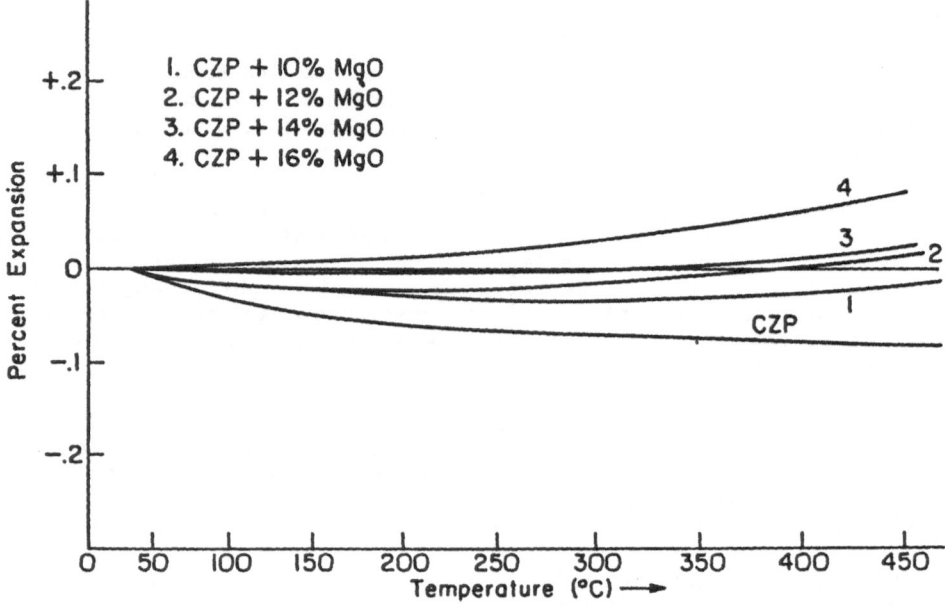

Fig. 2. Thermal Expansions of CZP + MgO System.

EXPERIMENTAL

Single phases of $NaZr_2P_3O_{12}$, $CaZr_4P_6O_{24}$ and $Na_4Zr_2Si_3O_{12}$ (called NZP, CZP and NZS respectively henceforth) were synthesized by solid state reaction using oxide powders as precursors. The stoichiometric amounts of reagent grade $CaCO_3$, Na_2CO_3 (both Fisher Scientific Company), ZrO_2 (Alfa Products) SiO_2 powder (Fisher Scientific Co.) and $NH_4H_2PO_4$ (J.T. Baker Chemical Company) were mixed and homogenized in acetone by hand mixing or ball milling and then air dried. The dry powder was calcined at 200°, 600° and finally at 900°C to drive off the volatiles. The calcined powder was consolidated into 1 inch pellets by cold pressing at a pressure of 20,000 psi. The pellets were then heat treated and sintered at 1000°C/16 hrs and 1200°C/2 days to form single phases of NZP, CZP and NZS. The phase identification of the sintered material was carried out on a GE x-ray diffractometer using a graphite monochromator and CuKα radiation. These sintered pellets were ground to a fine powder (−60 mesh). The diphasic compositions were prepared by adding preformed $Y_3Fe_5O_{12}$ (YIG) powder to NZP, CZP or NZS, in various proportions. The mixture was homogenized in acetone and pelletized again and heat treated at different temperatures. Dilatometric measurements were made on small rectangular bars (~ ·5 x 1 x 1 cm^3) cut out from the sintered samples in a Harrop Dilatometric Analyzer. The dilatometer was first calibrated using a fused silica standard and adjusting the heating rate to about 1°/minute in the programmer.

RESULTS AND DISCUSSION

The various compositions which were examined in this investigation are listed in the following Table; the sintering temperature and time are also given. The phase composition of the sintered materials was determined by x-rays and is presented in the last column of the Table.

It was observed that only $Na_4Zr_2Si_3O_{12}$ is thermodynamically compatible with $Y_3Fe_5O_{12}$ to form a composite material; NZP and CZP had reacted with YIG to produce Yttrium and Iron phosphates. The composition #3 when sintered at 1200°C produced a partially melted material in which YIG had reacted with NZS but when the same composition was sintered at lower temperature (1040°C) there was no reaction.

The thermal expansion curves of the NZS + YIG system are presented in Figure 3. ΔL/L (or percent expansion) was measured from room

TABLE 1 Compositions Studied, Firing Temperature, Time and Phases Present after Sintering

Composition	Temperature (°C)	Time (hrs.)	Phases Present
1. $NaZr_2P_3O_{12}$ + $Y_3Fe_5O_{12}$ (2:1)	1200	18	NZP + YPO_4 + $FePO_4$
2. $CaZr_4P_6O_{24}$ + $Y_3Fe_5O_{12}$ (2:1)	1200	18	CZP + YPO_4 + $FePO_4$
3. $Na_4Zr_2Si_3O_{12}$ + $Y_3Fe_5O_{12}$ (3:1)	1200	16	NZS + YPO_4 + glass (partially melted)
4. $Na_4Zr_2Si_3O_{12}$ + $Y_3Fe_5O_{12}$ (3:1)	1040	15	NZS + YIG
5. $Na_4Zr_2Si_3O_{12}$ + $Y_3Fe_5O_{12}$ (3:2)	1000	15	NZS + YIG
6. $Na_4Zr_2Si_3O_{12}$ + $Y_3Fe_5O_{12}$ (0.85:0.15)	1000	15	NZS + YIG

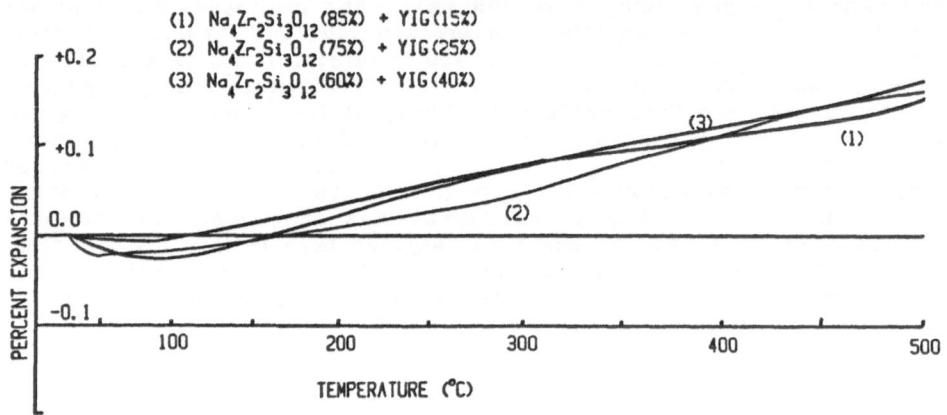

Fig. 3. Thermal Expansion of NZS + $Y_3Fe_5O_{12}$ System.

temperature to 500°C at a heating rate of 1°/min. All the three compositions show near 'zero' average thermal expansion between 30° and ~150°C, and up to 500°C α is approximately 3.4×10^{-6}°C^{-1}. It is interesting to note that up to 100°C these compositions demonstrate slightly negative expansion.

It is concluded that by adopting a diphasic approach one can control and tailor a special property in the composite ceramic by varying the amounts of individual components and modifying the processing parameters. In the system $Na_4Zr_2Si_3O_{12} + Y_3Fe_5O_{12}$, some compositions were found to exhibit low thermal expansion behaviour up to 150°C.

ACKNOWLEDGEMENT

This research work was supported by the U.S. Air Force under the contract number AFOSR-83-0291.

REFERENCES

1. R. Roy and D.K. Agrawal, 1985, Successful Design of New Very Low Thermal Expansion Ceramics, in: Mat. Res. Soc. Symp. Proc., Vol. 40, 1985, E.A. Giess, King-Ning Tu and E.A. Uhlman, Eds., Mat. Res. Soc. Publication, Pittsburgh.
2. D.K. Agrawal and R. Roy, 1985, Composite Route to 'Zero' Expansion Ceramics, J. Mat. Sci. (in press).
3. J. Alamo and R. Roy, 1984, Ultra Low Expansion Ceramics in the System $Na_2O-ZrO_2-P_2O_5-SiO_2$, J. Am. Cer. Soc. 67:C78.
4. R. Roy, D.K. Agrawal, J. Alamo, and R.A. Roy, 1984, [CTP]: A New Structural Family of Near Zero Expansion Ceramics, Mat. Res. Bull., 19:471.

ELECTROCERAMIC COMPOSITES

R. E. Newnham

Materials Research Laboratory
The Pennsylvania State University
University Park, Pennsylvania 16802

INTRODUCTION

Composite materials have found a number of structural applications but their use in the electronics industry has been relatively limited. As the advantages and disadvantages of electroceramic composites are better understood we can expect this picture to change.

In this paper we review some of the basic ideas underlying composite electroceramics: sum and product properties, connectivity patterns leading to field and force concentration, the importance of periodicity and scale in resonant structures, the symmetry of composite materials and its influence on physical properties, polychromatic percolation and coupled conduction paths in composites, varistor action and other interfacial effects, coupled phase transformation phenomena in composites, and the important role that porosity and inner surface area play in many composites.

PROPERTIES OF COMPOSITE MATERIALS

For convenience in understanding, the physical and chemical properties of composites can be classified as sum properties, combination properties, and product properties. The basic ideas underlying sum and product properties were introduced by Van Suchtelen.[1] For a sum property, the composite property coefficient depends on the corresponding coefficients of its constituent phases. Thus the stiffness of a composite is governed by the elastic stiffnesses of its component phases and the mixing rule appropriate to its geometry. In general the property coefficient of the composite will be between those of its constituent phases.

This is not true for combination properties which involve two or more different coefficients. Poisson's ratio is a good example of a combination property since it is equal to the ratio of two compliance coefficients. As is well known, some composite materials have extremely small values of Poisson's ratio, smaller than those of the materials used to make the composite. Wood is a good example from nature.[2]

Product properties are more complex and more interesting. The product properties of a composite involve different properties in its constituent phases with the interactions between the phases often causing unexpected results.

Sum Properties

Dielectric constant will be used to illustrate a simple sum property. Series and parallel mixing rules represent bounding conditions for the dielectric constant $\bar{K}$ of a diphasic composite:

$$\bar{K}^n = V_1 K_1^n + V_2 K_2^n + \ldots$$

where K_1 and K_2, are the dielectric constants of the constituent phases, and V_1 and V_2, are their volume fractions. The exponent n is +1 for parallel mixing, and n = -1 for series mixing. For many composites, the geometric arrangement is partly series and partly parallel. In this case, $\bar{K}$ can often be described by a logarithmic mixing rule for which the exponent n $\approx$ 0.

There are, of course, many other mixing rules in addition to the series and parallel models. These represent only the limiting conditions. A more complete discussion of the dielectric properties of heterogeneous materials is given in the classic article by Van Beek.[3]

As examples of composite ferroelectrics, consider the temperature of capacitors. Depressors are materials that are added to a high K capacitor formulation to depress the peak at the Curie point, resulting in a flatter temperature dependence. Bismuth stannate and magnesium zirconate are often used as depressors for barium titanate multilayer ceramics. These additives form a second phase in the grain boundary regions of the ceramic. The grain boundary phase has a much lower dielectric constant than $BaTiO_3$ and depresses the dielectric constant of the ceramic, largely through the series mixing rule. Low K boundary phases in series with the high K grains have a much greater effect on the permittivity than do those in parallel. The brick wall model gives a good description of diphasic ceramic dielectrics.[4]

Combination Properties

For simple mixing rules, the properties of the composite lie between those of its constituent phases, but combination properties involve two or more coefficients which may average in a different way.

An example of interest is the acoustic wave velocity which determines the resonant frequency of piezoelectric devices. For a long, thin rod, the velocity of waves propagating along the length of the rod is $v = (E/\rho)^{1/2}$ where E is Young's modulus and ρ is the density. Fiber-reinforced composites often have very anisotropic wave velocities. Consider a complaint matrix material reinforced with parallel fibers. Long, thin rods fashioned from the composite have different properties when the fibers are oriented parallel or perpendicular to the length of the rod. Much faster wave velocities are measured for longitudinal orientation (v_L) than for transverse orientation (v_t) of the fibers.

Experimental data[5] for composites made from steel filaments embedded in epoxy conform closely to the equations for v_L and v_t. It is interesting to note that v_t, the wave velocity for waves traveling transverse to the fibers, is less than the velocity of both epoxy and steel, the two phases which make up the composite. The slowness of this wave is caused by the fact that density and stiffness depend differently on volume fraction. This difference in mixing rules for E and ρ cause the combination property v_t to lie outside the range of the end members. The longitudinal wave, v_L, behaves more normally. In this case E and ρ follow the same mixing rule and the values for v_L lie between those of the end members.

Product Properties

A product property utilizes different properties in the two phases of a composite to produce yet a third property through the interaction of the two phases. By combining different properties of two or more constituents, surprisingly large properties are sometimes obtained with a composite. Indeed, in a few cases, product properties are found in composites which are entirely absent in the phases making up the composite.

A magnetoelectric composite made from ferroelectric $BaTiO_3$ and ferrimagnetic cobalt titanium ferrite has been studied by scientists from Philips Laboratory.[6,7] A dense diphasic ceramic of the perovskite and spinel-structure phase was obtained by directional solidification, and then electrically poled to make the $BaTiO_3$ phase piezoelectric. When a magnetic field is applied to the composite, the ferrite grains change shape because of magnetostriction. The strain is transmitted to adjacent piezoelectric grains resulting in an electrical polarization. Magnetoelectric effects a hundred times larger than those in Cr_2O_3 are obtained this way. Subsequent research[8] has led to the development of a broadband magnetic field probe with an exceptionally flat frequency response to 650 kHz.

CONNECTIVITY

Connectivity[9] is a key feature in property development in multiphase solids since physical properties can change by many orders of magnitude depending on the manner in which connections are made.

Each phase in a composite may be self-connected in zero, one, two or three dimensions. It is natural to confine attention to three perpendicular axes because all property tensors are referred to orthogonal systems. If we limit the discussion to diphasic composites, there are ten connectivities: 0-0, 1-0, 2-0, 3-0, 1-1, 2-1, 3-1, 2-2, 2-3, and 3-3. Connectivity patterns for more than two phases are similar to the diphasic patterns, but far more numerous. There are 20 three-phase patterns and 35 four-phase patterns. For n phases the number of connectivity patterns is $(n + 3)!/3!n!$.

During the past few years we have been developing processing techniques for making piezoelectric composites with different connectivities.[10] Extrusion, tape-casting, injection molding, fugitive phase and methods have been especially successful. The 3-1 connectivity pattern, for instance, is ideally suited to extrusion processing. A ceramic slip is extruded through a die giving a three-dimensionally connected pattern with one-dimensional holes, which can later be filled with a second phase. Another type of connectivity well-suited to processing is the 2-2 pattern made up of alternating layers of the two phases. The tape-casting of multilayer capacitors with alternating layers of metal and ceramic is a way of producing 2-2 connectivity. In this arrangement both phases are self-connected in the lateral directions but not self-connected perpendicular to the layer.

Stress Concentration

The importance of stress concentration in composite materials is well known from structural studies but its relevance to electroceramics is not so obvious. Stress concentration is a key feature of many of the piezoelectric composites made from polymers and ferroelectric ceramics.[10] By focusing the stress on the piezoelectric phase, some of the piezoelectric coefficients can be enhanced by others reduced.

As an example, consider the hydrostatic voltage coefficient $g_h = (d_{31} + d_{32} + d_{33})/\varepsilon_{33}$ in a 1-2-3-0 composite. This composite is made up of

PZT fibers in the poling direction (X_3) and glass fibers in the X_1 and X_2 directions. The fibers are embedded in a foamed polymer matrix. In terms of the 1-2-3-0 symbol, the PZT is self-connected in one-dimension, the glass fibers in two-directions, the polymer in three dimensions, and the voids in none.

Hydrostatic stress waves are converted to uniaxial stresses inside the composite. Stress components in the X_1 and X_2 directions are carried by the glass fibers while stresses along X_3 act upon the PZT. Because of its greater compliance, the polymer matrix transfers stress to the fibers. Foaming the polymer reduces the Poisson ratio of the composite, preventing transfer of stress between the X_3 (poling) direction and the orthogonal X_1 and X_2 directions. As a result d_{33} is kept large while reducing d_{31} and d_{32}. This improves g_h because normally d_{31} and d_{32} are opposite in sign from d_{33}. As an added benefit, the dielectric permittivity ε_{33} is reduced by eliminating much of the ferroelectric PZT from the transducer. Improvements in g_h of two orders of magnitude have been demonstrated.[11]

Advantageous internal stress transfer can also be utilized in pyroelectric coefficients. If the two phases have different thermal expansion coefficients, there is stress transfer between the phases which generates the electrical polarization through the piezoelectric effect. In this way it is possible to make a composite pyroelectric which is not piezoelectric.[12]

Electric Field Concentration

The multilayer design used for ceramic capacitors is an effective configuration for concentrating electric fields. By interleaving metal electrodes and ceramic dielectrics in a 2-2 connectivity pattern, relatively modest voltages are capable of producing high electric fields.

Multilayer piezoelectric transducers are made in the same way as multilayer capacitors.[13] The oxide powder is mixed with an organic binder and tape-cast using a doctor blade configuration. After drying, the tape is stripped from the substrate and electrodes are applied with a screen printer and electrode ink. A number of pieces of tape are then stacked, pressed, and fired to produce a ceramic with internal electrodes. After attaching leads, the multilayer transducer is packaged and poled. When compared to a simple piezoelectric transducer, the multilayer transducer offers a number of advantages: (1) the internal electrodes make it possible to generate larger fields for smaller voltages, eliminating the need for transformers for high-power transmitters. Ten volts across a tape-cast layer 100 microns thick produces an electric field of 10^5V/m, not far from the depoling field of PZT. (2) The higher capacitance inherent in a multilayer design often improves acoustic impedance matching. (3) Many different electrode designs can be incorporated in the transducer to shape poling patterns, which in turn control the mode of vibration and the ultrasonic beam pattern. (4) Additional design flexibility can be achieved by interleaving layers of different composition. One can alternate ferroelectric and antiferroelectric layers, for instance, thereby increasing the depoling field. (5) Grain-oriented piezoelectric ceramics can also be tape-cast into multilayer transducers. Enhanced piezoelectric properties are obtained by aligning the crystallites parallel to the internal electrodes.[14] (6) Another advantage of the thin dielectric layers in a multilayer transducer is in improved electric breakdown strength. Gerson amd Marshal[15] measured the breakdown strength of PZT as a function of specimen thickness. The D.C. breakdown field for ceramics 1 cm thick were less than half that of 1 mm thick samples. It is likely that the trend will continue to even thinner specimens, leading to improved poling and more reliable transducers.

A good deal has been written about the importance of scale in magnetic, optical, and semiconductor materials, and many of the same effects occur in ferroelectrics: critical domain sizes, resonance phenomena, electron tunneling, and non-linear effects.

Intrinsic Size Effects

In ferromagnetic materials, there are three kinds of magnetic structures for small particles.[16] Multidomain structures are common for particles larger than a critical size; magnetization in large particles takes place through domain wall motion. Below this critical size, single domain particles are observed, and switching takes place by rotation rather than wall movement, thereby raising the coercive field. Very small particles exhibit a superparamagnetic effect in which the spins rotate in unison under thermal excitation. Only modest magnetic fields are required to align the spins of adjacent particles.

Analogous behavior in ferroelectric particles has yet to be fully established, but a variety of interesting experimental results are accumulating.[17] In $BaTiO_3$ ceramics, single domain behavior is observed in grains less than $\sim 1\ \mu m$ in size,[18] while dielectric phenomena resembling superparamagnetism are found in relaxor ferroelectrics. The fluctuating microdomains in this superparaelectric state are about 20 nm across.[19]

Composite materials made up of single domain and superparaelectric particles have yet to be investigated in a systematic way with proper control of the connectivity and surrounding environment. The controlled synthesis of submicron ferroelectric grains will do much to stimulate research in this area.

Surface treatment of the ferroelectric phase allows control of the mechanical boundary conditions. Titanyl coupling agents are effective in bonding PZT to epoxy.[20] Mechanical pull tests have been used to demonstrate the strength of the ceramic-polymer bond. Improved stress transfer and large piezoelectric coefficients in 1-3 and 0-3 piezoelectric composites are obtained as a result of better bonding.

Polymers are about a hundred times more compliant than ceramics. If a ceramic grain is surrounded by polymer the mechanical constraints are relatively small. This means that more complete poling is possible, as demonstrated in 0-3 ferroelectric composites.[21]

Electrical boundary conditions can also be controlled by adjusting the dielectric constant and conductivity of the surrounding phase.

Resonant Structures

Periodicity and scale are important factors when composites are to be used at high frequencies where resonance and interference effects occur. When the wavelengths are on the same scale as the component dimension, the composite no longer behaves like a uniform solid.

An interesting example of unusual wave behavior occurs in composite transducers made from poled ferroelectric fibers embedded in an epoxy matrix.[22] When driven in thickness resonance, the regularly-spaced fibers excite resonance modes in the polymer matrix causing the matrix to vibrate with much larger amplitude than the piezoelectric fibers. The difference in compliance coefficients causes the non-piezoelectric phase to respond far more than the stiff ceramic piezoelectric. Composite materials are therefore capable of mechanical amplification. Multiply-poled piezoelectric transducers (MUPPETS) have been fabricated from pre-poled PZT fibers mounted in a polymer matrix.[23]

Domain-divided transducers operate on a similar principle.[24] Multi-domain crystals and ceramics have been used as acoustic phase plates and high frequency transducers. Optical analogs occur in the twinned acentric crystals used to phase-match fundamental and second harmonic wavelengths.[25]

Nonlinear Phenomena

Second harmonic generation and other nonlinear optical effects are well known, but the corresponding low-frequency phenomena have not been thoroughly investigated. The recent upsurge of interest in actuators[26] is changing this situation. Electrostriction is a second order electromechanical coupling between strain and electronic field. For small fields electrostrictive strains are small compared to piezoelectric strain, but this is not true for the high fields generated in composite transducers.

Multilayer electrostrictive transducers[27] made from relaxor ferro-electrics such as lead magnesium niobate (PMN) are capable of generating strains larger than PZT. Since there are no macrodomains in PMN there are no "walk-off" effects in electrostrictive micropositioners. Moreover, poling is not required and there are no aging effects. The concentration of electric fields in composite transducers make non-linear effects increasingly important.

SYMMETRY OF COMPOSITE MATERIALS

A wide variety of symmetries are found in composite materials. Examples of crystallographic groups, Curie groups, black-and-white groups, and color groups will be given, and the resulting effect on physical properties discussed.

In describing the symmetry of composite materials, the basic idea is Curie's principle of symmetry superposition: <u>A composite material will exhibit only those symmetry elements that are common to its constituent phases and their geometrical arrangement</u>.

The practical importance of Curie's principle rests upon the resulting influence on physical properties. Generalizing Neumann's law from crystal physics:[28] <u>The symmetry elements of any physical property of a composite must include the symmetry elements of the point group of the composite</u>.

Crystallographic Groups

Laminated composites are good illustrations of composite materials conforming to crystallographic symmetry. In a unidirectional laminate the glass fibers are aligned parallel to one another, such that a laminate has orthorhombic symmetry (crystallographic point group mmm). Mirror planes are oriented perpendicular to the laminate normal, and perpendicular to an axis formed by the intersection of the other two mirrors. The physical properties of a unidirectional laminate must therefore include the symmetry elements of point group mmm. If the laminate is heated, it will change shape because of thermal expansion. Less expansion will take place parallel to the fiber axis because glass has a lower thermal expansion and greater stiffness than that of polymer. The laminate will therefore expand anisotropically but it will not change symmetry. The heated laminate continues to conform to point group mmm.

A cross-ply laminate is made up of two unidirectional laminates bonded together with the fiber axes at 90°. Such a laminate belongs to tetragonal point group $\bar{4}2m$, as indicated in Table 1. Laminated composites with $\pm\Theta$ angle-ply alignment exhibit orthorhombic symmetry consistent with point group 222.

Table 1. Symmetry groups of representative Compositives

Unidirectional laminate	mmm
Cross-ply laminate	$\bar{4}2m$
Angle-ply laminate	222
Tetragonal honeycomb extrusion	
Unpoled	4/mmm
Longitudinally poled	4mm
Transversely poled	mm2
Glass-ceramic	$\infty\infty$m
Polar glass-ceramic	∞m
Ferroelectric-Ferrimagnetic Composite	
Unpoled, unmagnetized	$\infty\infty$m
Poled, unmagnetized	∞m
Unpoled, magnetized	∞/mm'
Parallel poled and magnetized	∞m'
Transverse poled and magnetized	2'mm'

Other types of symmetry elements can also be introduced during processing. The extruded honeycomb ceramic used as catalytic substrates are an interesting example.[29] By suitably altering the die used in extruding the ceramic slip, a large number of different symmetries can be incorporated into the composite body when the extruded form is filled with a second phase.

Lead zirconate titanate (PZT) honeycomb ceramics have been transformed into piezoelectric transducers by electroding and poling. The symmetry of the honeycomb transducers depend on the symmetry of the extruded honeycomb and also on the poling direction. For a square honeycomb pattern, the symmetry of the unpoled ceramic is tetragonal (4/mmm) with four-fold axis parallel to the extrusion direction. When poled parallel to the same direction,[30] the symmetry changes to 4mm. Transversely poled composites filled with epoxy are especially sensitive to hydrostatic pressure waves,[31] and in this case the symmetry belongs to orthorhombic point group mm2.

Curie Groups and Magnetic Symmetry

The piezoelectric properties and symmetry of natural composites such as wood and bone are found to conform to texture symmetries.[32] Some texture symmetry groups belong to the 32 crystallographic point groups, but others do not. Composite bodies with texture may also belong to one of the Curie groups: $\infty\infty$m, $\infty\infty$, ∞/mm, ∞m, ∞/m, ∞2, and ∞. Polar glass-ceramics with conical symmetry illustrate the idea.[33] A glass is crystallized under a strong temperature gradient with polar crystals growing like icicles into the interior from the surface. Certain glass-ceramic systems such as $Ba_2TiSi_2O_8$ and $Li_2Si_2O_5$ show sizable pyroelectric and piezoelectric effects when prepared in this manner. Polar glass-ceramics belong to the Curie point group ∞m, the point group of a polar vector. As the glass is crystallized in a temperature gradient, it changes symmetry from spherical ($\infty\infty$m) to conical (∞m), the same as that of a poled ferroelectric ceramic.

To describe the magnetic fields and properties it is necessary to introduce the black-and-white Curie groups. Magnetic fields are represented by axial vectors with symmetry ∞/mm'. The symbol m' indicates that the mirror planes parallel to the magnetic field are accompanied by time reversal.

The magnetoelectric composite described previously is an excellent illustration of the importance of symmetry in composite materials. In combining a magnetized ceramic (symmetry group ∞/mm') with a poled ferroelectric ceramic (symmetry group ∞m), the symmetry of the composite is obtained by retaining the symmetry elements common to both groups: $\infty m'$.

An interesting feature of this symmetry description is its effect on physical properties. According to Neumann's law, the symmetry of a physical property of a material must include the symmetry elements of the point group. The symmetry of a magnetized ceramic and a poled ferroelectric both forbid the occurrence of magnetoelectricity, but their combined symmetry ($\infty m'$) allows it. By incorporating materials of suitable symmetry in a composite, new and interesting product properties can be expected to occur.

TRANSPORT PROPERTIES OF COMPOSITES

Conductor-filled composites are discussed in this section, emphasizing the importance of percolation in random and segregated mixes. Differential thermal expansion between matrix and filler sometimes leads to remarkable variations in resistance with temperature. Composite PTC and NTC thermistors and chemical sensors based on these ideas have been fabricated.

Percolation and Segregated Mixing

Most composite conductors are made up of conducting metal particles suspended in an insulating polymer matrix. Particle contact and percolation require a larger volume fraction when the metal and polymer grains are comparable in size. When the conducting particles are small, they are forced into interstitial regions between the insulating particles; this forces the conducting particles in contact with one another, resulting in a low percolation limit.

These ideas are borne out by experiments on copper particles embedded in a matrix of polyvinylchloride.[34] The critical volume fraction decreases markedly when the Cu particles are far smaller than the polymer particles. When the size ratio is 35:1, the critical volume percent is only 4% Cu. This highly segregated mixing establishes contact between conducting copper particles at a very low ratio of conductor to insulator.

Composite Thermistors

A second interesting effect is the dependence of electrical resistance on temperature. PTC thermistors are characterized by a positive temperature coefficient of electrical resistance. Doped barium titanate ($BaTiO_3$) has a useful PTC effect in which the resistance undergoes a sudden increase of four orders of magnitude just above the ferroelectric Curie temperature (130°C). The PTC effect is caused by insulating Schottky barriers created by oxidizing the grain boundary regions between conducting grains of rare earth-doped $BaTiO_3$.

Similar PTC effects are observed when polymers are loaded near the percolation limit with a conducting filler. The Polyswitch overload protector[35] is made from high density polyethylene with carbon filler.

At room temperature the carbon particles are in contact giving resistivities of only 1 Ω-cm, but on heating the polymer expands more rapidly than carbon, pulling the carbon grains apart and raising the resistivity. Polyethylene expands very rapidly near 130°C, resulting in a pronounced PTC effect comparable to that of $BaTiO_3$. A rapid increase in resistivity of six orders of magnitude occurs over a 30° temperature rise.

Combined NTC-PTC composites have also been constructed.[36] Vanadium sesquioxide (V_2O_3) has a metal-semiconductor transition near 160°K with a large increase in conductivity on heating. This material can be incorporated in a composite by mixing V_2O_3 powder in an epoxy matrix. The filler particles are in contact at low temperatures and exhibit an NTC resistance change similar to that observed in V_2O_3 crystals and single phase ceramics. On heating above room temperature, the polymer matrix expands rapidly, pulling the V_2O_3 grains apart and raising the resistance by many orders of magnitude. This produces a PTC effect similar to the Polyswitch composite. The net result is an NTC-PTC thermistor with a conduction "window" in the range -100°C to +100°C. This is a good example of the use of coupled phase transformations in composites.

REFERENCES

1. J. van Suchtelen, Product Properties: A New Application of Composite Materials, Philips Res. Repts. 27:28 (1972).
2. R.F.S. Hearmon, "Applied Anisotropic Elasticity," Oxford University Press, London (1961) pp. 40-41.
3. L.K.H. van Beek, Dielectric Behaviour of Heterogeneous Systems, Prog. in Diel. 7:69 (1965).
4. D. A. Payne, Role of Internal Boundaries Upon the Dielectric Properties of Polycrystalline Ferroelectric Materials, Ph.D. Thesis, Pennsylvania State University (1973).
5. C. A. Ross and R. L. Sierakowski, Elastic Waves in Fiber Reinforced Materials, Shock and Vibration Digest 7:1 (1975).
6. J. van den Boomgaard, D. R. Terrell, R.A.J. Born, and H.F.J. I. Giller, An in Situ Grown Eutectic Magnetoelectric Composite Material, Part 1, J. Mat. Sci. 9:1705 (1974).
7. A.M.J.G. van Run, D.R. Terrell, and J.H. Scholing, An in Situ Grown Eutectic Magnetoelectric Composite Material, Part 2, J. Mat. Sci. 9:1710 (1974).
8. L.P.M. Bracke and R.G. van Vliet, A Broadband Magnetoelectric Transducer Using a Composite Material, Int. J. Elec. 51:255 (1981).
9. R. E. Newnham, D. P. Skinner, and L. E. Cross, Connectivity and Piezoelectric-Pyroelectric Composites, Mat. Res. Bull. 13:525 (1978).
10. R. E. Newnham, L. J. Bowen, K. A. Klicker, and L. E. Cross, Composite Piezoelectric Transducers, Mat. in Eng. II:93
11. M. J. Haun, Transverse Reinforcement of 1-3 and 1-3-0 PZT-Polymer Piezoelectric Composites with Glass Fibers, M.S. Thesis, Pennsylvania State University (1983).
12. A. Halliyal, A. S. Bhalla, and R. E. Newnham, Polar Glass Ceramics— A New Family of Electroceramic Materials: Tailoring the Piezoelectric and Pyroelectric Properties, Mat. Res. Bull. 18:1007 (1983).
13. G. O. Dayton, W. A. Schulze, T. R. Shrout, S. Swartz, and J. V. Biggers, Fabrication of Electromechanical Transducer Materials by Tape Casting, Adv. in Ceramics 9:115 (1984).
14. M. Granahan, M. Holmes, W. A. Schulze, and R. E. Newnham, Grain-Oriented $PbNb_2O_6$ Ceramics, J. Amer. Ceram. Soc. 64:C68 (1981).

15. R. Gerson and T. C. Marshall, Dielectric Breakdown of Porous Ceramics, J. Appl. Phys. 30:1650 (1959).
16. I. S. Jacobs and C. P. Bean, Fine Particles, Thin Films, and Exchange Anisotropy, in: "Magnetism, Vol. III," G. T. Rado and H. Suhl, ed., Academic Press, N.Y. (1963).
17. M. Multani, The Finite Solid State Lattice, in: "Preparation and Characterization of Materials," J. M. Honig and C.N.R. Rao, eds., Academic Press, N.Y. (1981).
18. Y. Ozaki, Ultrafine Electroceramic Powder Preparation from Metal Alkoxides, Ferroelectrics 49:285 (1983).
19. V. A. Bokov and I. E. Myl'nikova, Electrical and Optical Properties of Single Crystals of Ferroelectrics with a Diffused Phase Transition, Sov. Phys.-Solid State 3:613 (1961).
20. E. Galgoci, Ceramic-Polymer Bonding in Piezoelectric Composites, Ph.D. Thesis, Pennsylvania State University (1985).
21. J. Giniewicz, (Pb,Bi)(Ti,Fe)O_3/Polymer 0-3 Composite Materials for Hydrophone Applications, M.S. Thesis, Pennsylvania State University (1985).
22. T. R. Gururaja, Piezoelectric Composite Materials for Ultrasonic Transducer Applications, Ph.D. Thesis, Pennsylvania State University (1984).
23. T. R. Gururaja, D. Christopher, R. E. Newnham and W. A. Schulze, Continuous Poling of PZT Fibers and Ribbons and Its Application to New Devices, Ferroelectrics 47:193 (1983).
24. R. E. Newnham, C. S. Miller, L. E. Cross, and T. W. Cline, Tailored Domain Patterns in Piezoelectric Crystals, Phys. Stat. Sol. 32:69 (1975).
25. D. Feng, N. Ming, J. Hong, Y. Yang, J. Zhu, Z. Yang, and Y. Wang, Enhancement of Second Harmonic Generation in $LiNbO_3$ crystals with Periodic Laminar Ferroelectric Domains, Appl. Phys. Lett. 37:607 (1980).
26. K. Uchino, S. Nomura, L. E. Cross, R. E. Newnham and S. J. Jang, Electrostrictive Effects in Perovskites and ITS Transducer Applications, J. Mat. Sci. 16:569 (1981).
27. L. E. Cross, S. J. Jang, R. E. Newnham, S. Nomura, and K. Uchino, Large Electrostriction Effects in Relaxor Ferroelectrics, Ferroelectrics 23:187 (1980).
28. J. F. Nye, "Physical Properties of Crystals," Oxford University Press, London (1957).
29. I. M. Lachman and R. N. McNally, High Temperature Monolithic Supports for Automobile Exhaust Catalysis, Ceram. Eng. Sci. Proc. 2:337 (1981).
30. T. R. Shrout, L. J. Bowen, and W. A. Schulze Extruded PZT/Polymer Composites for Electromechanical Transducer Applications, Mat. Res. Bull. 15:1371 (1980).
31. A. Safari, A. Halliyal, R. E. Newnham, and I. Lachman, Transverse Honeycomb Composite Transducers, Mat. Res. Bull. 17:301 (1982).
32. I. S. Zheludev, Piezoelectricity in Textured Media, Solid State Physics 29:315 (1974).
33. G. J. Gardopee, R. E. Newnham, and A. S. Bhalla, Pyroelectric $Li_2Si_2O_5$ Glass-Ceramics. Ferroelectrics 33:155 (1981).
34. S. K. Bhattacharya and A.C.D. Chaklader, Review on Metal-Filled Plastics. Part 1. Electrical Conductivity, Polym.-Plast. Tech. Eng. 19:21 (1982).
35. R. D. Sherman, L. M. Middleman, and S. M. Jacobs, Electron Transport Processes in Conductor-Filled Polymers, Polymer Eng. and Sci. 23:36 (1983).
36. K. A. Hu, R. E. Newnham, J. P. Runt, and A. Safari (in preparation).

FINITE ELEMENT/DIFFERENCE MODELING OF ELECTROCERAMICS

S. DaVanzo, W. Carlson, R.E. Newnham and A. Safari

Materials Research Laboratory
The Pennsylvania State University
University Park, PA 16802

ABSTRACT

The hydrostatic piezoelectric response of PZT composites and the electric field distribution around flaws in capacitor materials have been modeled using the Finite Element and Finite Different methods respectively. This paper reviews the work done at The Pennsylvania State University on these two models.

INTRODUCTION

Lead zirconate titanate (PZT) and barium titanate ($BaTiO_3$) are two of the most widely used materials in the electroceramics industry. PZT is used mainly for piezoelectric applications, while $BaTiO_3$ is used primarily as a capacitor dielectric.

This paper reviews the ongoing research efforts at the Materials Research Laboratory of The Pennsylvania State University on the modeling of electroceramic materials. It is divided into two parts, the first describes the Finite Element Method (FEM) modeling of PZT composites for hydrophone materials, while the second outlines the work on modeling the effects of flaws on the electric field distributions within capacitor materials using the Method of Finite Differences (FD).

HYDROPHONE MATERIALS

Numerous PZT composites have been fabricated for use as hydrophone materials. Although PZT has a strong uniaxial piezoelectric response, the hydrostatic piezoelectric response, d_h ($d_h = d_{33} + d_{31} + d_{32}$), is only approximately 10% of the uniaxial response due to the fact that $d_{31} = d_{32} = -(1/2)d_{33}$. A number of workers[1-5] have incorporated PZT into composites in order to improve upon the limitations of single phase PZT. Composites have been characterized by their connectivity patterns as described by Newnham[5]. The connectivity pattern refers to the number of orthogonal directions each phase within a composite is continuous. A 3-1 composite is one in which PZT is continuous in three directions and the the second phase (polymer in this case) is continuous in one.

Two types of composites, perforated 3-1 composites and 3-0 macrovoid composites, will be discussed in this review. The 3-1 perforated composites were fabricated by Safari[3] by drilling holes perpendicular to the poling direction in prepoled PZT blocks. 3-0 macrovoid composites were fabricated by Kahn[6] at the Naval Research Laboratory. In macrovoid composites ordered arrays of voids are introduced into PZT by screening a fugitive ion between layers of green PZT tapes in the desired void pattern. The parts were then constructed by stacking the ink tape combination to form a block. When the fugitive ink was burned out of the structure and the part fired, the result was an ordered array of voids.

All of the composites fabricated for hydrophone applications were designed such that when a hydrostatic load is applied to the material the distribution of stress within the composite results in a larger component of the stress in the direction of poling than in the two orthogonal directions. The effect of this stress distribution is to reduce the effect of the hydrostatic response of the negative d_{31} and d_{32} piezoelectric coefficients thus enhancing the hydrostatic piezoelectric charge coefficient d_h. In order to understand the enhancement of the hydrostatic response, as well as guide future research efforts, the FEM was employed to calculate the electric field distribution within the composite as part of a model for the hydrostatic piezoelectric large coefficient d_h.

Finite Element Calculations

The FEM has been used by DaVanzo[7,8] to model the hydrostatic response of 3-1 perforated and 3-0 macrovoid composites. The modeling of the composites follows the same procedure with one exception; modeling of the 3-0 composites requires calculations of the electric field distribution during poling. 3-1 composites are fabricated from prepoled PZT blocks whereas 3-0 composites are poled after fabrication. In the 3-1 composites poling of the ceramic phase is uniform in both direction and magnitude while in the 3-0 composites the voids distort the direction and magnitude of the electric field leading to non-uniform poling of the ceramic phase. The non-uniformity in the poling of the ceramic must be accounted for in the model. In this paper the method used for the modeling of the 3-1 composites will be outlined, the reader is referred to the references for a complete description of the model which includes poling effects[8].

The FEM is a method for solving boundary value problems in which the differential equation governing the behavior of the region of interest is expanded on a set of simple basic functions called 'shape' functions. The result of this expansion is a set of algebraic equations in which the generalized displacements (displacement, rotation, potential, etc.) of a finite number of points within the region, called 'nodes', become the independent variables. In practice the region is broken into a large number of small subregions called 'elements'. Figure 1 shows a typical FEM grid for a 3-1 perforated composite. The grid consists of cubic elements 0.25mm on a side. The above mentioned

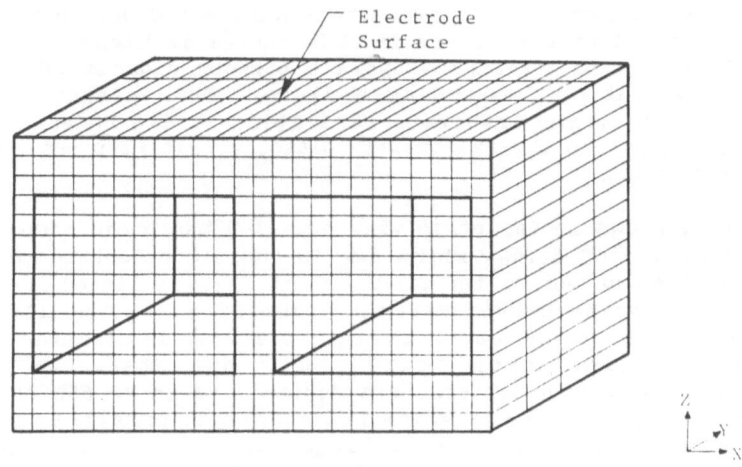

Fig. 1.   A typical Finite Element grid for a 3-1 composite.

nodes are, in this case, the corners of the cubes. In order to use the
FEM the model must be described to the FEM routine. The description
included a number for each node, the location of each node, the node
numbers which comprise each element, the material constants for each
element and any loads on the nodes or elements. The first step in the
model is to set up the FEM grid and supply the model information to the
routine. For the 3-1 perforated composites the load was a hydrostatic
load of 0.7MPa. The FEM routine calculates the displacement of each
node and the six components of the stress at the centroid of each
element. The next step in the model was to calculate a stress induced
polarization for each element. The polarization was calculated for each
element using the stress at the center of the element and the relation
$P_i = d_{ijk}\sigma_{jk}$, where d_{ijk} is the piezoelectric tensor for PZT and σ_{jk} is
the stress at the element centroid. These stress induced element
polarizations were then summed using a series parallel model to obtain
an induced polarization for the composite. The series parallel model
treats each element as a parallel plate capacitor with a charge defined
by $Q = P_3 A$ where P_3 is the component of the polarization in the direct of
poling and A is the area of the element face normal to the poling
direction. The element charges were then added using the procedure
prescribed for summing capacitor networks. From the induced
polarization for the composite the hydrostatic piezoelectric charge
coefficient was then calculated using the relation $P_3 - d_h p$ where p is the
hydrostatic load applied to the composite.

RESULT AND DISCUSSION

Figure 2 shows a plot of d_h versus the hole center to center
distance (X) for 3-1 perforated composites with a constant hole size of
2.5cm. The plot shows good agreement between experimental and FEM
results. Figures 3-5 show contour plots of the σ_{11}, σ_{22}, and σ_{22}
components of the stress on a plane, perpendicular to the y axis,
0.125mm into the composite. The relative intensity of the three

components of the stress at points between the holes should be noted. The σ_{33} component of the stress in this region is higher than the two orthogonal directions. Figure 6 shows a contour diagram of the stress induced polarization for the same plane. From either the stress diagrams or the stress induced polarization it can be seen that the major contribution to the hydrostatic piezoelectric response comes from the pillar regions.

Figure 7 shows an isometric view of a 3-0 macrovoid composite. In Table 1 the ratio of d_h/d_{33} where d_{33} is the piezoelectric coefficient in the direction of poling for single phase PZT, is listed for four 3-0 macrovoid composite configurations. The first two configurations contain voids which are approximately 6×10^{-4} cm in diameter while the second two are for voids 2.5×10^{-4} cm in diameter. In all four configurations the void patterns are designed to cover 50% of the void plane. For each diameter of voids two spacings, parallel to the void planes, were fabricated. The spacings were 3.8×10^{-3} and 7.6×10^{-3} cm respectively with the narrower spacing listed first. The Table shows good agreement between experimental and FEM results for the ratios.

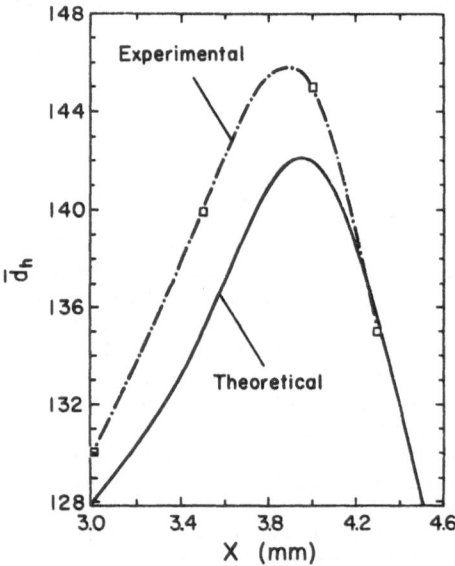

Fig. 2. Variation of the hydrostatic charge coefficient d_h as a function of X (the center to center distance for adjacent holes).

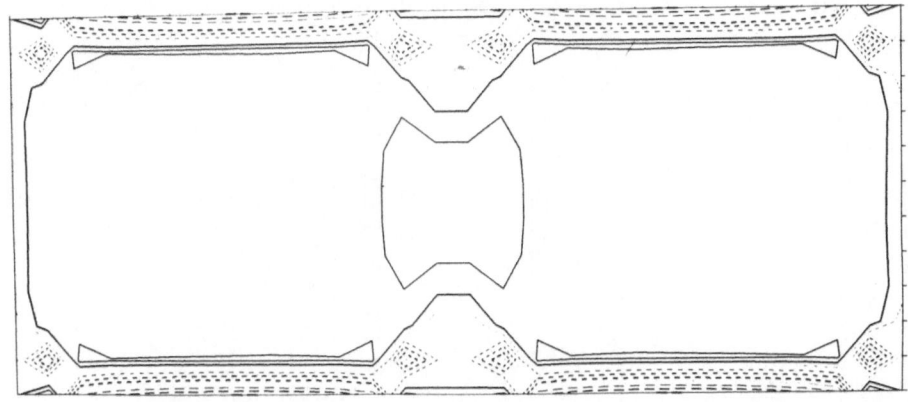

Fig. 3. A contour map of the 11 component of the stress tensor in a 3-1 composite.

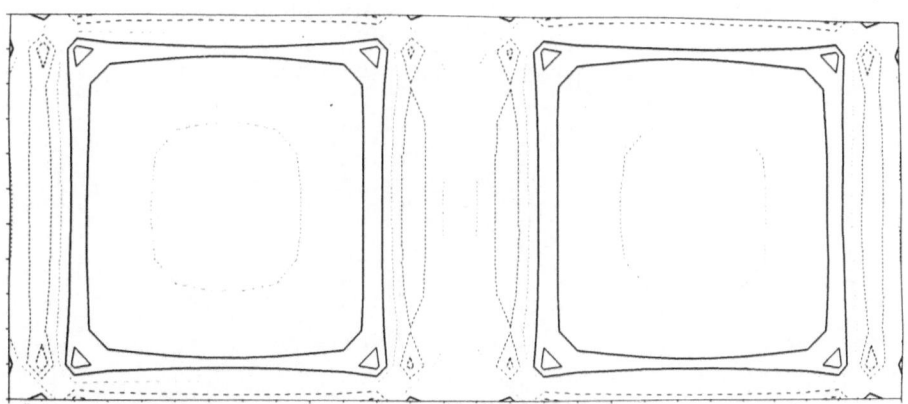

Fig. 4. A contour map of the 22 component of the stress tensor in a 3-1 composite.

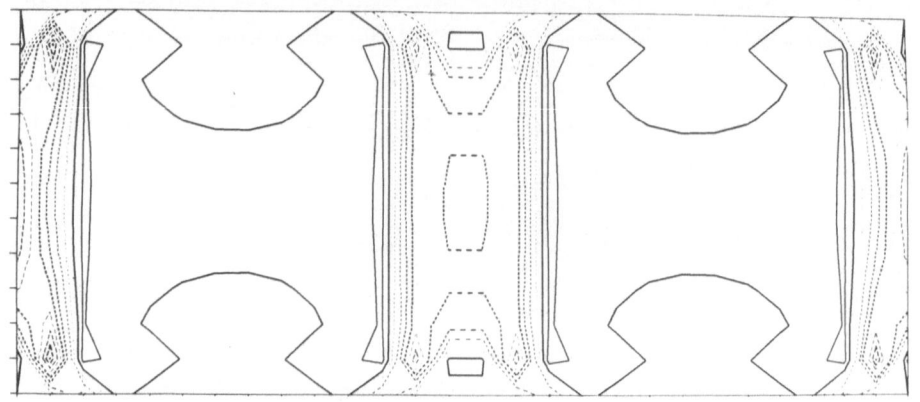

Fig. 5. A contour map of the 33 component of the stress tensor in a 3-1 composite.

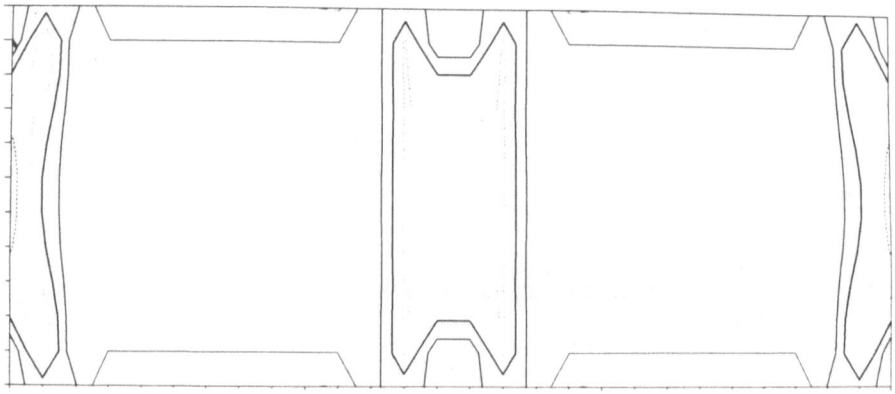

Fig. 6. A contour map of the 33 component of the stress induced polarization P_3 in a 3-1 composite.

Table 1. d_h/d_{33} Ratios for Macrovoid Composites with Circular Voids

Configuration	d_h/d_{33}	
	Exp.	FEM
Large Circles 1 layer	47%	48%
Large Circles 3 layers	42%	45%
Small Circles 1 layer	44%	49%
Small Circles 3 layers	40%	48%

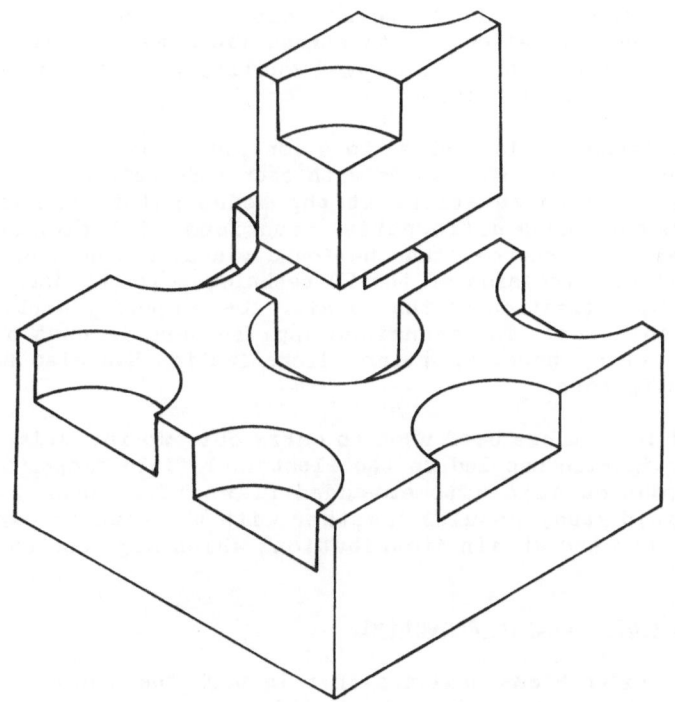

Fig. 7. An isometric view of a 3-0 macrovoid composite.

The FEM has been employed in a model for the hydrostatic response of PZT composites with very good reliability. The calculated stress fields within the composites make it possible to identify the parameters for each composite configuration which have the greatest influence on the hydrostatic response. The hydrostatic response for a number of new configurations of 3-0 macrovoid composites has been predicted and will be presented elsewhere[8].

CAPACITOR MATERIALS

Because of its high dielectric constant $BaTiO_3$ is used widely in the capacitor industry, but the use of any dielectric material is limited by the dielectric breakdown of the material. A ceramic, although traditionally considered to be a single phase homogeneous material, can be considered to be a multiphase material where manufacturing flaws (pores, delamination of electrodes, etc.), as a second phase severely limit the breakdown voltage of the material.

A FD (Finite Difference) computer model has been used to analyze non-conductive parallel plate capacitors. This model applies to which can be represented in two dimensions with an embedded second phase of either voids or conductive material of different geometries. Results of the analysis are presented as local potential or electrical field concentrations within the capacitor. Electrostrictive and piezoelectric strains can then be calculated knowing the material properties. These studies are intended to delineate the field-induced internal stress leading to mechanical fracture and electrical breakdown in $BaTiO_3$ materials. Identification of the stress field may result in alternate manufacturing techniques and improved quality control, by identifying and eliminating harmful flaws.

The FD technique is applied to a continuum that may be divided into a finite number of discrete nodes with each node representing an unknown value. Discretized equations at the nodes points are then used to replace the continuum differential equations. Solution of the set of these linear equations may then be found via direct or iterative linear procedures. The precision of the FD technique will be dependent on the manner of discretization of the domain, the computing machine, and the method of solution. The technique applied here depends on a regular rectilinear mesh, however, polar discretization has also been used by other investigators.

The FD method has been used to carry out two investigations. The first investigation has led to the electrical field distribution within a homogeneous ceramic with embedded flaws. The second investigation uses the field study results together with the elastic parameters to predict stress and strain distributions which may lead to mechanical breakdown.

Calculations of Electric Potential

For field problems analyzed in this work the Laplace equation was treated by the method of central differences. This is a 'divided difference' scheme and is commonly used to analyze field problems. The simultaneous solution of all nodes Laplacians as in a direct method will yield the potential at all points in the field. Boundary conditions are specified for both the perimeter of the field and the inclusions.

Symmetry conditions are applied where feasible but other conditions may occur where the potentials must be assumed. One such condition is the potential at a conductor interface. In this analysis the conductors are treated as constant potential surfaces. Another boundary condition is the far field condition which is defined as the undisturbed field far from an included object. The accuracy of this assumption is dependent on both the distance from the inclusion and the boundary conditions at the inclusion. In some cases the boundary condition at the inclusion may be approximated if the ratio of the permitivities between the two phases is very large. In addition to assumptions in the numerical calculations, there are further restrictions on the geometrical resolution. The continuum under study is assumed homogeneous in each phase, but the grain size limitations will alter the potential distribution and thereby place a lower limit on the validity of the numerical results. The node size is therefore limited by the physical nature of the material.

This code has been applied to flaws in uniform and non-uniform fields where the permitivity of the flaw is much lower than that of the surrounding ferroelectric. The code was used to determine local fields around spherical and elliptical flaws. Further checks have been made with closed-form solutions of spherically flawed geometries. The closed-form solutions have been derived for flaw geometries in uniform D.C. fields only. Geometries studied include spherical voids and spherical conductors about which the Laplacian differential equations have been solved electrostatic potential external to the void. Some elliptical solutions have also been derived. The local electric fields around and within flaws are easily determined from the potential derivative in the appropriate coordinate system and the particular flaw geometry. The coordinate geometry used in both cases depend on the shape of the inclusion: for the case of spherical flaws polar coordinates were used. The chief purpose of the closed form solutions are to check the field concentration around flaws in the capacitors by assuming that the far field is uniform. The check may also be useful for numerical results where acute geometries are encountered in the vicinity of cracks and delaminations.

Manufacturers have used numerical methods to predict electric field levels in various multilayer geometries, both with and without embedded flaws (delaminations, large voids, etc.). The FD model developed here is for the study of local potential concentrations around singular inclusions and is also applicable to multiple flawed ceramics. Principal assumptions of the analysis include a constant D.C. voltage, two-dimensional geometries, and the use of a square FD solution technique with Laplacian differential equations. Other features include a semi-automatic node generator requiring some user inputs, and a graphical output of electric potential values. The code is in operation on an IBM mainframe computer and requires the International Mathematics and Statistics Library software package. A summary of the case studies is shown in Table 2.

Table 2. Summary of Maximum Local Electrical Fields

Flaw Type	Electric Field inside of flaw	Electric Field exterior of flaw
Spherical Voids		
• closed-form	1.5	1.5
• roughened	2.3	2.2
• roughened w/protruding		
voids	5.1	2.8
Diamond void	3.2	2.9
Square void	2.7	1.9
Delamination 6% thickness	13	3.6
End of Conductors	–	1.2
Delamination on conductor 10% delam thickness	9.2	2.8
Misplaced Conductor	–	2.24
w/Protruding Conductor	–	3.51

Elastic Analysis

The second area of research requires both the elastic material parameters and the electrical field solutions. The goal is to describe stress and strain configurations in the ceramic under steady state and fatigue limitations of the material. In particular, interesting phenomena in the void formation, crack propagation, and fatigue from electrical loading are being investigated. Electrical field patterns are used to calculate the polarization and dielectric constants as well as to analyze local strains. The isotropic forms of electric polarization $P_i = \varepsilon_0 X_{ij} E_i$ and electrostrictive strain $\varepsilon_{ij} = Q_{ijkl} P_k P_l$ are utilized to evaluate the particular strains. Mechanical breakdown in the material for various flaw geometries is being treated with fracture mechanics theory to check stress intensity factors and to predict ultimate fracture strength. Principal assumptions are that permitivity, susceptibility, and electrostiction are constant for a given field strength and polarization. If only the maximum tensile and compressive stresses need to be determined, the electric potential data may be studied for the maximum gradients and these gradiants used to calculate strain. Electrical field data resulting in a given strain may be compared to mechanical fracture strength via stress intensity factors and assumed flaw size for the particular material. Data from recent calculations is presented in Figure 8 showing fracture strength under various D.C. fields. This data show an approximate upper bound on the mechanical strength for various crack lengths in the ceramic.

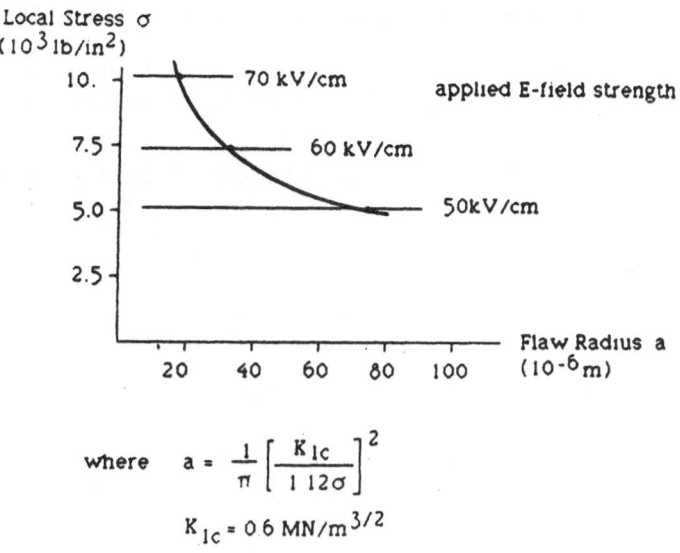

where $a = \dfrac{1}{\pi}\left[\dfrac{K_{1c}}{1\,12\sigma}\right]^2$

$K_{1c} = 0\,6\ MN/m^{3/2}$

Fig. 8. Local stress vs. flaw radius for a local field of 1.5 times the applied field.

Conclusions

Some planar numerical studies have been conducted to determine local fields within BaTiO$_3$ ceramics with embedded second phase flaws. It has been determined through these studies that local electrostriction strains may approach and exceed fracture limitation criteria as defined by fracture mechanics theory. Further study will be undertaken to determine if local eletrostriction strain and electrical fields are detrimental to the dielectric behavior of layered capacitors.

REFERENCES

1. K. Rittenmyer, T. Shrout, W.A. Schulze and R.E. Newnham. Piezoelectric 3-3 Composites. Ferroelectrics 41, 189–195 (1982).

2. D.P. Skinner, R.E. Newnham and L.E. Cross. Flexible Composite Transducers. Mat. Res. Bull. 13, 599 (1978).

3. A. Safari, R.E. Newnham, L.E. Cross and W.A. Schulze. Perforated PZT-polymer Composites for Piezoelectric Transducer Applications. Ferroelectrics 41, 197 (1982).

4. R.E. Newnham, L.J. Bowen, K.A. Klicker and L.E. Cross. Composite Piezoelectric Transducers. Mat. in Engr. 112, 93 (1980).

5. R.E. Newnham, D.P. Skinner and L.E. Cross. Connectivity and Piezoelectric-Pyroelectric Composites. Mat. Res. Bull. 13, 525 (1978).

6. M. Kahn. Ordered Macrovoid Composites. (to be published).

7. S. DaVanzo, A. Safari, R.E. Newnham. Finite Element Modeling of Perforated PZT-Polymer Composites. <u>Ferroelectric Letters</u> 3, 109-121 (1985).

8. S. DaVanzo and M. Kahn. Finite Element Modeling of Macrovoid Composites. (to be published).

MULTIPHASE INTERACTION FOR SEEKING EXOTIC PHENOMENA

Hiroaki Yanagida and Yoshinobu Nakamura

Faculty of Engineering, University of Tokyo

Tokyo, 113 Japan

INTRODUCTION

New and innovative materials have been keys to new technologies or civilization so far in history of human being. We are now entering a new era so called "information age". Various kinds of new materials are expected to overcome difficulties in finding proper materials to construct and design new devices. Scientific approaches to finding the proper materials are called "Materials Design". There are three categories in the concept of materials design[1] depending upon the stage of R&D of the materials. The first is the one to seeking candidates of new materials. This is important especially for electronic ceramics. The second is for making the candidates practically useful. This may correspond to R&D of structural ceramics. The last is for easy accesses to and from materials data bases. The design of new alloys (metals and/or dielectric ceramics) have been effective by consulting data bases. As for functional ceramics especially for atmosphere sensitive materials we need new materials which have not been known yet, since we do not have enough candidates.

Table I.   Three Ways for Seeking New Functional Ceramics

1. Crystal Chemistry Basis
   NASICON (Na ion conductor)
   $Bi_2O_3(+Y_2O_3)$ - $ZrO_2(+Y_2O_3)$(Solid Solution, oxide ion conductor)
   AlN (Heat conductor)
   Y_2O_3:Eu (Cathode luminescence)
   $ZnO-Al_2O_3-Li_2O$ (Gas Sensor)

2. Serendipity Basis
   $BaTiO_3(+La_2O_3)$(PTC effect)
   $BaTiO_3(+La_2O_3,+Bi_2O_3)$ (2-step PTC)
   SiC (IC substrate)

3. Multiphase Interaction Basis
   $TiO_2-MgCr_2O_4$ (Acid-Base, Humidity Sensor)
   $ZnO-Li_2O-M_2O_3$ (Host-Guest, Humidity Sensor)
   CuO-ZnO (p-n, Humidity and Gas Sensor)
   SiC (+BeO) (precipitation, IC substrate)
   AlN (+Y_2O_3) (Precipitation, IC substrate)
   $ZnO(+Bi_2O_3)$ (grain boundary oxidation, varistor)

There are three ways for seeking new functional materials as shown in Table I, where examples are taken from ceramic materials. The first two are discussed elsewhere in detail.[1] The present paper will be focused upon the last approach. Here consider the concepts of "homogeneity" and "heterogeneity". Multiphase mixtures sometime give rise to heterogeneity which is responsible for exotic properties. Reliability of materials may arise from homogeneity. Of course the concepts must be defined dimension by dimension. Exotic properties usually arise from heterogeneous structure of the dimension of boundaries or interface. In order for the materials with exotic properties to have reliability for practical use homogeneity is required in the stereological dimension such as deviation of grain or pore size. Homogeneity can also be defined for chemical purity, crystal symmetry, mass ratio of constituent atoms or ions of the crystal, degree of crystallinity, etc.[2] In the present paper the effectiveness of the idea to pursue heterogeneity for exotic properties and homogeneity for reliability will be discussed.

MULTIPHASE INTERACTIONS

Properties of homogeneous mixture of multiphases are given by a linear combination of that of each phase. Refractive index, density or some other properties are the cases. Design of the properties is very easily achieved in these cases. It is very natural that we consider there takes place some interaction among the phases in most of the cases, however. In composite materials we expect bests of the properties from each constituent although unfortunately and usually we find worsts. Therefore, this is not the concept concerned in the present paper. The interesting cases are those where strong interaction takes place among the phases heterogeneously dispersed or arrayed. Strong interaction between two phases is expected when the two phases have opposite properties. Of course the mixture must still be of heterogeneous structure in order to prevent neutralization of the properties. The examples of combination of opposite properties are, acid-base, host-guest, p-n junction or contact, two phase chemical reaction to give rise to precipitation, grain-grain boundary, etc.

ACID-BASE BASIS INTERACTION

Porous titania is an excellent candidate material for humidity sensors. Titania is a very stable n-type semiconductor to heat or corrosion. Pore size of the material is controlled in the range 1000~3000Å suitable for adsorption of water vapor. In order to improve its sensor characteristics various additives have been tried. Conduction mechanism is due to proton migration. It may be considered that the porous titania surface works as solid acid points. From the refractory point of view titania is an acid, too. Among the additives so far tried $MgCr_2O_4$ has been the most promising one.[3] This is a base also from the refractory classification. Both materials, TiO_2 and $MgCr_2O_4$, are stable to so called "cleaning treatment" ($500°C$, 1 min) in order to remove organic contamination since they are refractories. The heterogeneous mixture of TiO_2 and $MgCr_2O_4$ is much more sensitive to humidity. Active points are analyzed to lie nearby the interface at the two phases. It may be considered that the surface of $MgCr_2O_4$ works as solid base points while that of TiO_2 as solid acid points. Both phases must be solid to keep coexistences of acid and base points. Otherwise they should be neutralized.

HOST-GUEST BASIS HETEROGENEITY

Host-guest materials such as intercalation compounds are also excellent candidates for exotic functional materials. Zinc oxide lattice can incorporate lithium ions as guests as shown in Fig. 1.[4] Interstitial Li^+

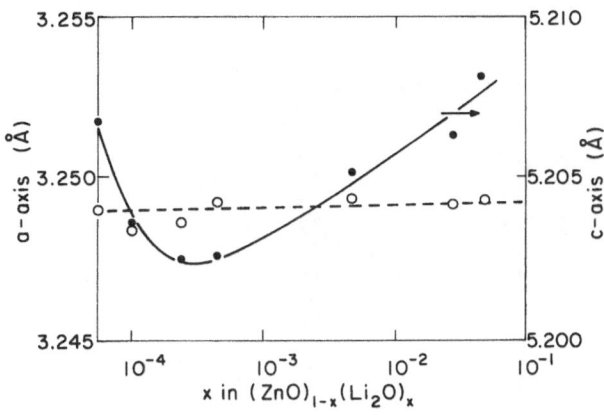

Fig. 1. Lattice Parameters of ZnO with Incorporated Li_2O. Open circle: a-axis. Closed circle: c-axis.

intercalated between Zn-O layers are sensitive to humidty while the Li^+ ions substituting lattice Zn^{2+} ions are inert to humidity at all. The Zn-O stabilized by doping Al_2O_3, Cr_3O_3 or V_2O_3.

P-N HETERO-CONTACT BASIS PHENOMENA

Hetero-contact between a p-type semiconductor oxide NiO and a n-type one ZnO shows more non-linear V-I characteristics when humid than when dry.[5] This was more enhanced when CuO was used instead of NiO and the contact became more junction like under humid conditions, where the thresholed barrier was analyzed to be around 1.0 V.[6] This hetero-contact can be used to detect dew point or liquid water. V-I characteristics changes with the kind of solvents as shown in Fig. 2.[7] We may be able to identify the solvent from the characteristics. The mechanism of enhancement of electric current under forward bias is considered to be due to ionic current through adsorbed or intercalated water between two different type semiconductors. From the p-type semiconductor electron holes are supplied into the water to give rise to protons. Electric charge of protons is liberated at the surface of the n-type semiconductor. This is an electrolysis of water enhanced by the p-n contact. Gap between the contact is of 10μm order. Bubbling of gases from the gap was observed to give an evidence of electrolysis. Sometimes precipitation of copper metal upon the surface of zinc oxide was observed to show ionic current through the liquid. The contribution, transference number, of copper ion migration to the total current was negligible small, however. Most of the current is, therefore, considered due to protonic migration. The hetero-contact between CuO and ZnO is not only sensitive to humidity or solvent but also to gases. It has a very good selectivity for carbon monoxide as shown in Fig. 3. Furthermore, the selectivity and sensitivity are tunable by changing bias as shown in Fig. 4. Since this contact is not sensitive to H_2 gas at all, it means that the working mechanism is due to adsorption of CO probably on the surface of CuO to give rise to some adsorption state but not the oxidation of the gas with the adsorbed oxygen which is observed for flammable gas sensors made of porous SnO_2 or ZnO.[8,9,10] It is very difficult to improve gas selectivity for the sensors where the working mechanism is due to oxidation. We can disperse catalysts such as Pt or Pd to improve the selectivity, of course. We have not succeeded in improving the selectivity satisfactorily yet, however. Using adsorption mechanism we can change semi-

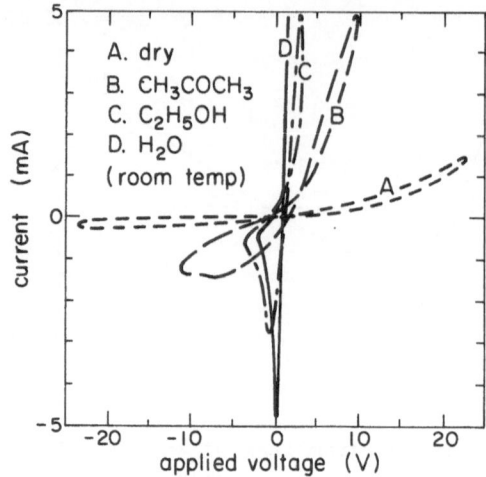

Fig. 2. Voltage-current Characteristic of CuO-
NiO Hetero contact in Various Solvents.

conductor etc. The authors have tried some combinations. So far it is
observed that the change in p-type semiconductor or the equivalent noble
metal is more effective than the change in n-type semiconductor. Furthermore the selectivity can be improved by changing bias voltage if we use
such diodal contacts.

INTERACTION GIVING RISE TO SOLID PHASE PRECIPITATION

Sintered body of silicon carbide with small amount of berylia additives
shows an excellent thermal conductivity while it is an insulator. Aluminum nitride sintered with yttria also shows a good thermal conductivity.
It is proposed that those additives extract impurities from the impure powders through forming grain boundary or intergranular precipitates. Thus
silicon carbide or aluminum nitride is much more homogenized from the viewpoint of chemical purity. These are the cases where introduction of heterogeneity can give rise to homogeneity.

CONCLUSION

It seems very promising for seeking new and exotic materials to expect
from the multiphase interaction. Among them hetero-contacts have been discussed more in detail in the present paper. We can try various combinations of materials. Furthermore by applying various biases we can control
or tune the characteristics very easily. The p-n contact between CuO and
ZnO has been shown very effective for humidity, solvent and CO gas sensors.

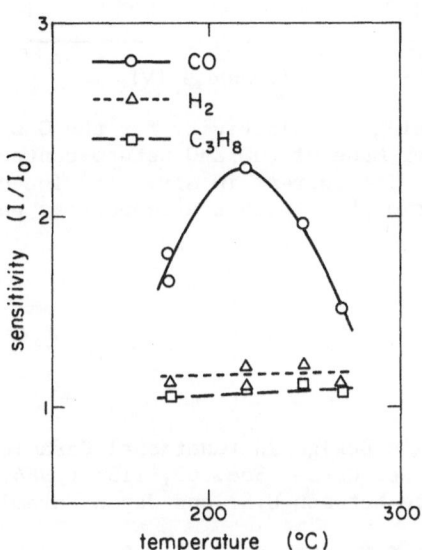

Fig. 3. Sensitivity and Selectivity
to various Gases of the CuO/ZnO Hetero-
contact. I_o: The Current in Air, I:
The current where the respective gas
was introduced by the concentration of
4000 ppm.

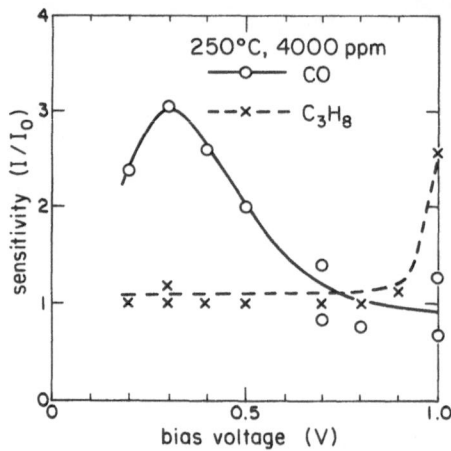

Fig. 4. Tuning of Selectivity for the Gas Species of the Sensor Made of CuO/ZnO Hetero-contact at 250°C. I_o: The current in Air. I: The current under the atmosphere with the respective gas by 4000 ppm.

REFERENCES

1. H. Yanagida, Materials Design in Functional Ceramics, Seramikkusu 19: 417 (1984), Bull. Amer. Ceram. Soc. 63, 1135 (1984), also presented at the "Small Conference between U.S. and Japan Ceramic Societies", Tokyo, May 17, 1984.
2. H. Yanagida, Design of New Ceramics, Kagaku to Jikken: 71 (1981).
3. T. Nitta and J. Terada, Ceramic Humidity Sensor - Humiceram National Technical Report 22: 885 (1976).
4. M. Miyayama, Y. Takuma and H. Yanagida, Lattice Parameters and Electrical Conductivity in Li_2O Doped ZnO, Chemistry Letters:1735 (1982).
5. K. Kawakami and H. Yanagida, Effects of Water Vapor on the Electrical Conductivity of the Interface of Semiconductor Ceramic-Ceramic Contacts, Yogyo-Kyokai-Shi 87, 112 (1979).
6. Y. Toyoshima, M. Miyayama, H. Yanagida and K. Koumoto, Effect of Relative Humidity on Current-Voltage Characteristics of Li-Doped CuO/ZnO Junction, Jpn. J. Appl. Phys. 22:1933 (1983).
7. H. Yanagida and Y. Nakamura, Humidity Sensors made of CuO-ZnO Hetero-contact, presented at the 87th Annual Meeting of the American Ceramic Society, 4-E-85, May 6, Cincinnatti (1985).
8. T. Nagira and H. Yanagida, Can We Make a Propane Gas Sensor?, Hyomen 19:252 (1981).
9. H. Yanagida, N. Miyayama, S. Saito and T. Nagira, Gas Sensing Mechanism of Porous Zinc Oxide with or without Pt Activator, presented at the 86th Annual Meeting of the American Ceramic Society, 9-JVIII-84, May 3, Pittsburgh (1984).
10. S. Saito, M. Miyayama, K. Koumoto and H. Yanagida, Gas Sensing Characteristics of Porous ZnO and Pt/ZnO Ceramics, J. Amer. Ceram. Soc. 68:40 (1985).

PROCESSING OF HETEROGENEOUS CERAMICS FOR DIELECTRIC APPLICATIONS

D. A. Payne

Department of Ceramic Engineering and Materials Research
Laboratory
University of Illinois at Urbana-Champaign
Urbana, IL 61801

INTRODUCTION

The properties of polycrystalline ceramics are greatly influenced by the boundary conditions that exist within ceramic microstructures. This is especially so for dielectric materials, which by definition are insulators; across which the lines of electric flux must pass, and be distributed, without the transport of charge. Consequently, dielectric ceramics must be processed to high quality, with a minimum of physical defects, and be of sufficiently high density that no connected porosity channels exist for parallel leakage.

On application of an electric field (E) to an ideal dielectric, charge separation is induced without conduction, which gives rise to dielectric displacement (D) and polarization (P) phenomena. The ability to store charge is given by the dielectric permittivity (ε),

$$D = \varepsilon E = \varepsilon_0 E + P \qquad , \qquad (1)$$

which, when compared with the permittivity of free space (ε_0), is an indicator of the relative permittivity (ε_r) or dielectric constant (K) of the material alone,

$$K = \varepsilon/\varepsilon_0 = 1 + \chi \qquad . \qquad (2)$$

Ceramic materials respond differently to equivalent applied fields, and the extent of polarization induced is dependent on the dielectric susceptibility (χ),

$$\chi = P/\varepsilon_0 E \qquad , \qquad (3)$$

i.e., how susceptible the material is to being polarized by the electric field.

Dielectric properties also depend critically on the measurement conditions or boundary conditions. For example,

$$K = f(E, T, \sigma, f, t...) \qquad , \qquad (4)$$

Table 1. Values of Dielectric Constant for Single Phase Ceramic Materials

	$\bar{K}(0.1$ Vac, 25°C, 1MHz$)$		
SiO_2	4.5	Bi_2O_3	30
Al_2O_3	8.6		
$3Al_2O_3 \cdot 2SiO_2$	6.5	TiO_2	120
MgO	8.2	$SrTiO_3$	310
$MgO \cdot SiO_2$	5.6		
$2MgO \cdot 2Al_2O_3 \cdot 5SiO_2$	5.0	$BaTiO_3$	2200

where the dielectric "constant" is a function of the applied field strength (e.g., weak or strong field for hysteretic ferroelectrics); temperature (T), especially if phase transitions occur; mechanical stress (σ), whether residual or applied for piezoelectric materials; cyclic frequency (f) or static time (t), for dispersion or aging phenomena in multidomain dielectrics, etc. Thus the term dielectric "constant", is used in the context of a figure of merit, compared with free-space, for the charge storage capability of the material under equivalent measurement conditions. Table 1 lists representative values of dielectric constant, for single phase ceramic materials, measured at weak voltage and at room temperature. This paper is principally concerned with high dielectric constant ceramics based upon titanate materials, e.g., $BaTiO_3$.

For heterogeneous ceramics containing mixtures of insulating phases, the electric lines of flux will tend to distribute themselves according to the relative susceptibilities of the constituent phases. This should lead to property enhancement or dilution in multiphase mixtures, depending on the design of the ceramic microstructures. In fact, properties could be optimized or "tailor-made" by the microstructural engineering of controlled chemical heterogeneities within dielectric mixtures or composite materials. For example, one type of mixing, consisting of a continuously connected interphase (or even interface) normal to flux passage - of a different chemical and dielectric nature than the major phase - would lead to internal electric field splitting, according to,

$$K_1 E_1 = K_2 E_2 \qquad , \qquad (5)$$

where there was continuity of the normal components of the dielectric displacements ($D^N_1 = D^N_2$) across the boundary. This would lead to a decrease in the saturation behavior ($K_1 = f(E_1)$) of high voltage ceramic capacitors based upon non-linear ferroelectric dielectrics like $BaTiO_3$. That is, there should be a reduction in the commerically important voltage coefficient of capacitance (C = f(V)). Similarly, the series model would be expected to stabilize the thermal characteristics associated with the Curie-Weiss transition ($K_1 = f(T)$). There would be a corresponding decrease in the temperature coefficient of capacitance (C = f(T)). Thus, opportunities exist for the tailoring of multiphase dielectrics, with improved properties-for more stingent applications-through the design and control of ceramic microstructures.

Controlled distributions of chemical heterogeneities, by appropriate ceramic processing methods, would lead to interesting combinations of dielectric properties, which could never be attained in single crystals alone. These novel properties would strictly be a function of polycrystallinity, and are developed by the rational application of ceramic engineering practices. This is especially so for new internal boundary layer devices, which are fabricated from semiconducting-insulating mixtures. Electric field spitting would now be controlled by resistivity (ρ) distributions, and it is the ability to manipulate the space-charge layer processes, that give rise to enhanced apparent dielectric constants in the range 10,000-100,000. The role of microstructural engineering of internal boundary layer devices will be outlined later in the paper. But, first, a review is given of the effects of the processing cycle, on the purposeful fabrication of heterogeneous ceramics, from insulating-insulating dielectric mixtures.

PROCESSING CYCLE

The approach taken in this paper is to consider the effects of the overall processing cycle on the dielectric properties of heterogeneous ceramics. A schematic of the processing cycle is shown in Fig. 1, in which interactive dependencies are illustrated. Since the resultant dielectric properties are very much dependent on phase assemblage (i.e., how the phases are distributed, and their relative concentrations), the microstructure-property relations are inturn dependent on the initial composition and thermal processing conditions used. High dielectric constant (K > 1000) materials based upon $BaTiO_3$ are very sensitive to impurity dilution and grain size enhancement. The precursors are usually of high purity and fine particle size, and care must be taken in the mixing of the initial formulation (not milling) so as to avoid contamination from low dielectric constant, deleterious, impurities (e.g., Al_2O_3, SiO_2). Otherwise, it would be extremely difficult to consistently

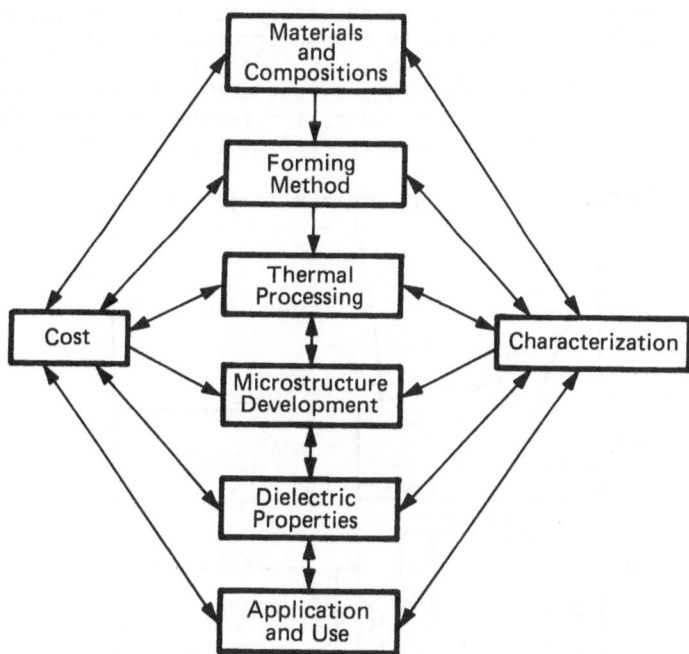

Fig. 1. Processing Cycle for the Fabrication of Dielectric Ceramics, Illustrating Interactive Dependencies.

manufacture a reliable product, with reproducible properties, for final applications and use. Similarly, the dielectric properties are very much dependent on grain size.

It is well known that the room-temperature dielectric constant of fine grain BaTiO$_3$ is much greater than coarse grain material. For single component, homogeneous, polycrystalline BaTiO$_3$, dielectric constant values of 5000-3500 have been reported for 1-2 μm grain size material[1], whereas, K values of 1500-2000 are usually measured for specimens with grain sizes greater than 10 μm. Thus, it is extremely important to control the microstructure development, so that dielectric devices can be manufactured within tolerances. Such is the case for a parallel plate capacitor, of specified electrode area (A) and dielectric thickness (d), where the capacitance (C) can vary significantly for a set geometry,

$$C = \varepsilon_0 KA/d \qquad , \qquad (6)$$

due to the aforementioned dependence of dielectric "constant" on grain size. This has been attributed to residual transformational stress (σ), and mechanical boundary conditions, in fine grain ceramics lacking 90° domains.[2]

An important aspect of the processing cycle is the feed-back obtained from characterization studies. Analytical electron microscopy could be used to obtain chemical and physical information which relate to phase composition and grain size development. Dielectric measurements, themselves, are a powerful tool, for the determination of chemical heterogeneities in polyphasic ceramics. Based upon equivalent circuits and physical models, the information carried in the lines of flux tells much about phase distributions. In fact, computer controlled firing cycles should be used, to ensure the consistent reproduction of overall thermal processing conditions, so necessary for the repetitive manufacture of reliable products. A guide to thermal processing, and phase mixing, would be the use of phase diagrams. For example, in the BaTiO$_3$-TiO$_2$ system (Fig. 2), titania rich mixtures would be processed completely by solid-state sintering at 1400°C, whereas titanate rich mixtures would be sintered initially in the presence of a liquid phase. An approach will be

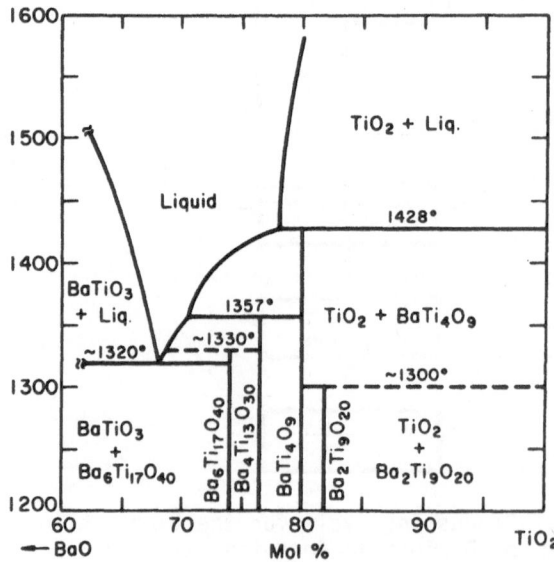

Fig. 2. Phase Diagram for the BaTiO$_3$-TiO$_2$ System.[3]

demonstrated, in this paper, for the application and sensitivity of dielectric analysis, to the characterization of dielectric mixtures and impedance boundaries (or barriers), which are designed and developed on purpose for property optimization, in heterogeneous ceramic composites.

Another important aspect of the processing cycle is the overall manufacturing cost for the final product. Often, engineering decisions have to be made, to maintain a competitive margin, without sacrificing quality or dielectric characteristics. This may influence the choice of starting materials and additives, or ceramic forming methods used. For example, an aqueous-based tape-casting system may be less expensive than an organic-based system, but not immediately adaptable to an existing doctor-blade process, due to variations in plasticity and surface chemistry. These changes could affect breakage/yield values; and chemisorption processes would alter the available surface free energy, the reduction of which is the driving force for densification in sintering. Thus, temporary processing aids (e.g., binders, plasticizers, solvents, dispersants, etc.,) may not be all that "temporary", in that they could affect microstructure development, and thereby dielectric properties, for equivalent firing cycles.

Another example of interactive dependencies, would be the use of compositional additives to reduce firing temperatures. Sintering aids could be used to reduce costs associated with high temperature furnace design, fuel and refractories; and also to reduce costs associated with co-fireable electrode systems (i.e., to allow the use of less expensive alloy system e.g., silver rich palladium alloys, rather than noble metals). By judicious choice of sintering aids, it should also be possible to inhibit grain growth, and optimize the dielectric constant-grain size relationships. In addition, if the sintering aid was an additive dielectric minor phase, which had a lower melting point than the major phase, it may also be possible to control the phase distribution and microstructure, by liquid-phase sintering techniques. For example, non-reactive liquid-phase sintering would facilitate the formation of interphase boundaries, with grain-to-grain separation; whereas, reactive liquid-phase sintering would eventually form interfacial barriers, with grain-to-grain contact, and with modified diffusion zones extending into the grain interior. There could be many variations between these two extremes, but the latter would be more sensitive to thermal processing conditions, due to the kinetics of diffusion, and the disappearance of the liquid phase. Dielectric properties would vary accordingly, and be controlled by the types of chemical heterogeneities introduced within the ceramic microstructure.

Interrelationships that exist between chemical composition and thermal processing conditions on microstructure development and dielectric properties are discussed in this paper. The first examples to be considered, are dielectrics in the system $BaTiO_3$-TiO_2.

BARIUM TITANATE-TITANIA

Following the synthesis of $BaTiO_3$, which was disclosed after World War II, a significant amount of compositional processing was carried out in the system BaO-TiO_2. This eventually led to the establishment of two distinctly different types, or classes, of dielectric fomulations, depending upon application.

Class I dielectrics, as they later became to be known, were principally developed for temperature compensating applications, with low loss at high frequencies (e.g., tan δ < 10^{-3} at 1MHz). They were based upon dielectric mixtures in the TiO_2-rich position of the phase diagram

417

(Fig. 2). The dielectric constants were low (K < 100) and of a non-ferroelectric character. The temperature coefficients of dielectric constant (ν_K) were in the parts per million, to parts per thousand, per °C, range ($10^{-6} - 10^{-3}$ °C^{-1}), where,

$$\nu_K = K^{-1} (dK/dT)_{E,\sigma} \qquad . \qquad (7)$$

Class II dielectrics, on the other hand, were of high dielectric constant (K > 1000), and were based upon compositions in the BaTiO$_3$-rich portion of the phase diagram. For general purpose use and miniaturization, high dielectric constants in the range 5000-10,000 were developed by making use of the dielectric anomaly at the ferroelectric transition temperature, i.e., by shifting (e.g., with SrSnO$_3$) and/or broadening (e.g., with CaZrO$_3$) the Curie-Weiss transition to coincide with the application temperature. Consequently, the temperature coefficients (ν_K) were of the order of parts per hundred per °C (~10^{-2}°C^{-1}). However, with time and experience, it was determined that temperature stabilized high dielectric constant materials (e.g., K ≈ 2000, ν_K ≈ 10^{-3} °C^{-1}) could be developed, by processing methods which limited grain growth and introduced chemical heterogeneities within the microstructure (i.e., as distinct form the aforementioned "shifted" homogeneous crystalline solutions, e.g., (Ba,Sr)(Ti,Sn)O$_3$). The manner by which the dielectric anomaly could be diluted, without shifting the Curie temperature, will be reviewed after the discussion on temperature compensating dielectrics.

Temperature Compensating Dielectrics: Class I

As outlined previously, temperature compensating dielectrics were first developed from compositions in the TiO$_2$-rich portion of the phase diagram. Figure 3 illustrates the dielectric properties for compositions

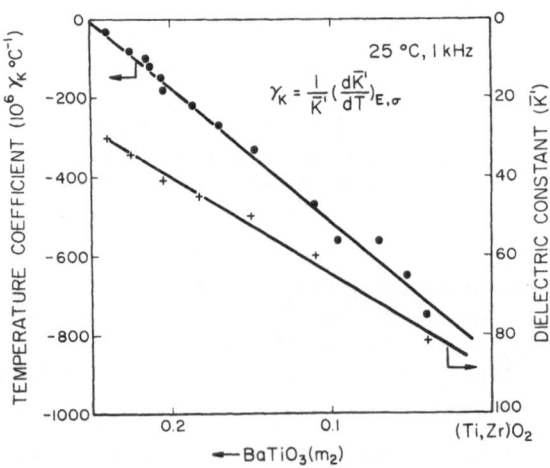

Fig. 3. Dielectric Properties of Heterogeneous Ceramics Fabricated from Titania Rich Mixtures.

fired at 1400°C in the TiO_2 + $BaTi_4O_9$ phase field (Fig. 2). The firing temperature was the minimum temperature necessary for the densification of high quality dielectrics (Q = 1/tan δ > 10^3). The dielectric properties and phase composition were somewhat dependent on furnace cooling rate and source materials used. Information summarized in Fig. 3 are for a TiO_2 source, which had been modified with a minor amount (~10^{-2}) of ZrO_2. Dielectric constant, and temperature coefficient, are illustrated as a function of $BaTiO_3$ additions to $(Ti,Zr)O_2$. The fired ceramics were heterophasic, as determined by x-ray diffraction and electron microscopy. A random distribution of phases was observed within the microstructure, for the dielectric composites prepared by solid-state sintering methods.

Figure 3 illustrates the linear mixing dependence of K and ν_K as a function of batching composition. For example, $(Ti,Zr)O_2$ corresponds to $K \approx 90$ with $\nu_K \approx$ - 850.10^{-6} °C^{-1}; whereas, 0.25 $BaTiO_3$ + 0.75 $(Ti,Zr)O_2$ corresponds to $K \approx 30$ and with $\nu_K = 0$, within experimental error of $\pm$ 10.10^{-6} °C^{-1}. The latter composition has a $BaO:TiO_2$ ratio of 1:4, which is the stoichiometry for $BaTi_4O_9$. Linear mixing of dielectric properties (K and ν_K) was determined for dielectric mixtures across the $BaTi_4O_9$-TiO_2 phase field. Thus, dielectric constants and/or temperature coefficients could be engineered, or tailor-made, by compositional and phase control of dielectric mixtures.

Dielectric mixing rules have been reviewed elsewhere[4], but linear mixing can sometimes be modeled for dielectric mixtures containing a random dispersion of insulating phases, depending upon the relative susceptibilites. For example, increasing volume fraction of TiO_2 would lead to an enhancement of dielectric constant and temperature coefficient; and vice versa, for property dilution.

Compositions based around barium tetratitanate were the mainstain of temperature stable high Q dielectrics for over 20 years. In 1968, data were reported that substantial additions of lanthanide oxides would double the dielectric constant.[5] For example, in the Nd_2O_3-BaO-TiO_2 system, K values of 65 were obtained with zero temperature coefficient, and were attributed to complex mixing of multiphase mixtures, including the important pyrochlore phase, $Nd_2Ti_2O_7$.

Today, through systemmatic variations of ionic size, valence, polarizability and structure, higher K values have been developed (e.g., $K \approx 80$-90) with temperature stability, through careful control of multiphase dielectric mixtures. The role of phase distributions on properties for higher dielectric constant ceramics, is discussed in the following section for Class II dielectrics.

High Dielectric Constant Ceramics: Class II

High dielectric constant ceramics in the BaO-TiO_2 system (Fig. 2) can be fabricated from $BaTiO_3$-rich mixtures. The phase diagram indicates that additions of TiO_2 to $BaTiO_3$ cause a progressive lowering of the liquidus temperature, with a eutectic point at approximately 1320°C. Consequently, mixtures fired above this temperature, would be sintered in the presence of a liquid phase. If the liquid phase wet the surface of the $BaTiO_3$ grains, and if there were a sufficient volume fraction, it should be possible to form a continuously connected interphase. The presence of the liquid phase, and associated surface tension forces, should enhance the rearrangement of particles and aid in densification processes.

Early work on dielectric formulations in the $BaTiO_3$-TiO_2 system, reported that TiO_2 additions also inhibited grain growth.[6] For example, 4 mol.% additions of TiO_2 (i.e., m_2 = 0.04) restricted the grain size to approximately 1 μm for ceramics fired at 1340°C; whereas, single component BaTiO₃, with no additions, had a grain size in excess 10 μm, for ceramics fired at 1300°C.[6]

A series of experiments, were carried out in this laboratory, for the processing of heterogeneous dielectrics in the system BaTiO₃-TiO₂. Fast-firing techniques were used, with the time at temperature (t_s) of 1 hour. Figure 4 illustrates the resultant dielectric constant-temperature characteristics for a series of mixtures which were densified at an equivalent sintering temperature (T_s) of 1316°C. This was the minimum temperature found necessary for the fabrication of low loss (tan δ < 10⁻²) insulators. Pure BaTiO₃ had a grain size of approximately 15 μm , whereas mixtures containing TiO₂ had a grain size of 1-2 μm.

Figure 4 illustrates the dielectric anomaly at the Curie temperature (T_c) was diluted for increasing additions of TiO₂, with the Curie temperature remaining constant ($T_c \approx 125°C$). On the other hand, the dielectric constant at room temperature was found to increase, initially, with TiO₂ additions; and this enhancement was attributed to a fine grain size effect.[1,2] However, after a critical concentration for inhibition of grain growth[6,7], the dielectric constant decreased with further additions of TiO₂. This has been attributed to a series dilution, as the volume fraction of lower dielectric constant interphase material increased. Series-mixing and Curie-Weiss behavior are discussed in the following section.

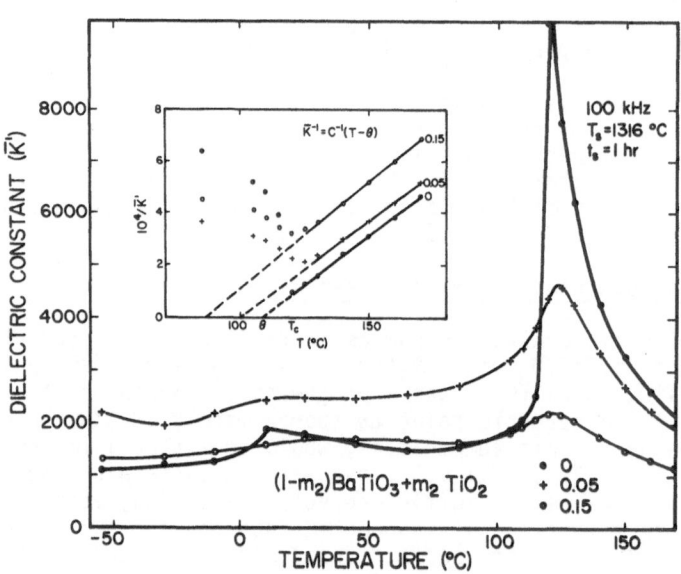

Fig. 4. Dielectric Constant - Temperature Behavior, for Heterogeneous Ceramics, Fabricated from Barium Titanate-Rich Mixtures; and Curie-Weiss Analysis.

Series-Mixing and Curie-Weiss Analysis. For a two-phase model, containing a continuously connected interphase (or even interface), and in which the second phase is of relatively low dielectric constant, compared with the major phase ($K_2 \approx 10^{-2} K_1$), the lines of flux have to cross the boundaries normal to flux passage, which leads to an equivalent circuit based upon series layers.[8] The effective dielectric constant ($\bar{K}$), for the diphasic composite, in terms of series elements, is given by

$$K^{-1} = (1 - v_2)K_1^{-1} + gv_2 K_2^{-1} \quad , \quad\quad\quad (8)$$

where, g, a shape factor, specifies the volume fraction of the second phase (v_2), connected in series. That is, there is a hyperbolic dilution of composite $\bar{K}$, with increasing volume fraction of low dielectric constant second phase material, distributed in series($\bar{K} \propto v_2^{-1}$).

The room temperature dielectric constant values given in Fig. 4, can now be explained in terms of this model, after taking into account an initial fine grain size enhancement. Above the ferroelectric transition ($T > T_c$), in the cubic paraelectric state, where no ferroelectric domains exist and there is no crystalline anisotropy, the mechanical boundary conditions should relax, and there should not be a grain size effect on dielectric constant.

For temperatures above the Curie temperature (T_c), $BaTiO_3$ grains will follow the Curie-Weiss dependence,

$$K_1 = C(T - \theta)^{-1} \quad , \quad\quad\quad T > T_c \quad\quad\quad (9)$$

where C, the Curie constant, and θ, the extrapolated Curie-Weiss temperature, are material parameters. Consequently, for heterogeneous mixtures, with a temperature dependent $BaTiO_3$ major phase, and a temperature stabilized high titanate boundary phase ($K_2 \neq f(T)$), in series connectivity,

$$K^{-1} = (1 - v_2)C^{-1}(T- \theta) + gv_2 K_2^{-1} \quad . \quad\quad\quad (10)$$

This is illustrated in figure 4, where the constancy of T_c and C are indicated.

The contribution of $BaTiO_3$, to the composite impedance, extrapolates to zero at the Curie-Weiss temperature ($K_1^{-1} = 0$ at $T = \theta$), and analysis of the data, at this temperature, gives the impedance contribution from the boundary phase,

$$K^{-1} = gv_2 K_2^{-1} \quad . \quad\quad\quad (11)$$

The efficacy of Curie-Weiss analysis, is that it is a method of dielectric characterization, which probes the impedance characteristics of the interphase/interface regions. By engineering the boundary conditions, heterophasic composites can be produced, with high dielectric constants and reduced temperature coefficients. For example, Fig. 4 illustrates that mixtures containing TiO_2 ($m_2 = 0.15$), had the same dielectric constant as pure $BaTiO_3$ at room temperature (i.e. $K_1 = \bar{K} = 1750$), but with a greatly reduced temperature coefficient, which has been attributed to series dilution.

Another example, of series mixing, for heterophasic dielectrics, is given in the following section, for compositions in the barium titanate-bismuth stannate system.

Bismuth compounds are often used as lower melting point sintering aids, which have insulating properties and useful dielectric properties at room temperature. The economic advantages which accrue from reduced temperature firing (T_s < 1300°C) were discussed in a previous section on the Processing Cycle. Also, the feasibility of stabilizing dielectric properties, by the purposeful distribution of controlled chemical heterogeneities, within the boundary regions by liquid phase sintering methods, was earlier demonstrated for Class II dielectrics. Bismuth stannate is one of many additives, or combinations of additives, which effect both cost reductions, and improved stability, for high K ceramics.

Fig. 5 illustrates the dielectric constant-temperature

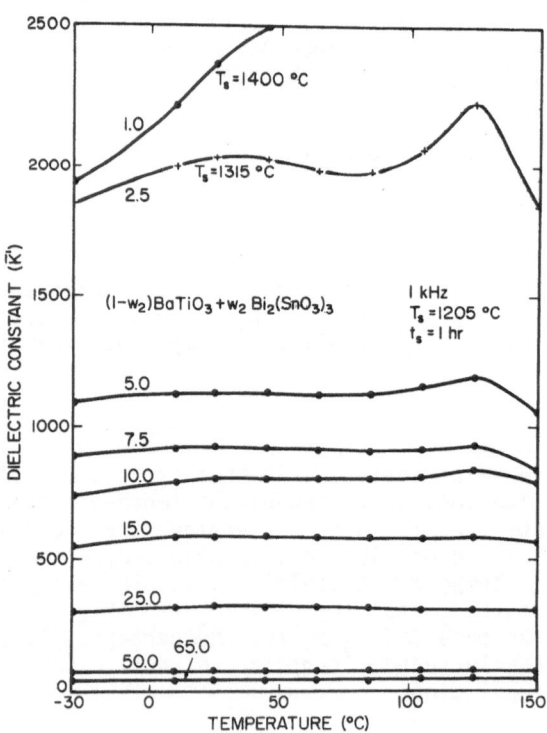

Fig. 5.   Dielectric Constant - Temperature Behavior for Heterogeneous Ceramics, Fabricated from Barium Titanate-Bismuth Stannate Mixtures

Table 2. Electric Field Dependence of Dielectric Constant
for Barium Titanate-Bismuth Stannate Mixtures

ϕ_K (0.1 Vac, 25°C, 1KHz, E dc)

Composition		Field Dependence at Bias Strengths(E)			
$BaTiO_3$ (w_1)	$Bi_2(SnO_3)_3$ (w_2)	0.5 (MV/m)	1.0 (MV/m)	1.5 (MV/m)	2.0 (MV/m)
1.00	0	0.97	0.90	0.78	0.20
0.99	0.01	0.98	0.95	0.88	0.78
0.95	0.05	0.99	0.97	0.94	0.90
0.90	0.10	1.00	0.99	0.98	0.96

characteristics for a series of experiments carried out in this
laboratory on dielectric mixtures in the system $BaTiO_3$-$Bi_2(SnO_3)_3$. The
data given are for illustrative purposes only, since the actual values of
dielectric constant were found to be sensitive to the morphology and
"activity" of the various $BaTiO_3$ powders used. As before, fast-firing
techniques were carried out to preserve chemical heterogeneity. The
temperature (T_s) indicated was the minimum temperature for the
densification of high quality insulators ($Q > 10^3$). $Bi_2(SnO_3)_3$
dissociates on heating to form the pyrochlore, $Bi_2Sn_2O_7$, and free
SnO_2.[9] X-ray diffraction determined the fired mixtures to be polyphasic,
containing the pyrochlore, tin oxide, and a slightly modified barium
titanate, with reduced tetragonality.[10] The grain size was 1-2 μm for
the fired ceramics.

Dielectric characteristics illustrated in Fig. 5, indicate a
dilution of the dielectric anomaly at T_c, and a dilution of the
dielectric constant at room temperature, for increasing additions of
"$Bi_2(SnO_3)_3$". For example, a mixture containing 5 wt. % bismuth stannate
(corresponding to "m_2" ≃0.01), densified at a reduced temperature of
1200°C, and with a room temperature K = 1100 and $\nu_K < 10^{-3}°C^{-1}$. Curie-
Weiss analysis of dielectric mixtures ($w_2 < 0.10$) determined the
transition temperature of $BaTiO_3$ grain cores to be not shifted (T_c =
125°C), with constancy of the Curie-constant; but the extrapolated Curie-
Weiss temperature (θ) was progressively moved to lower temperatures with
increasing additive content (i.e., $\theta = f(w_2)$), indicative of the
formation of impedance barriers.

The results have been analyzed in terms of the series model, in
which the thickness of the interfacial region increased with increasing
bismuth stannate content (i.e., $d_2 = f(w_2)$). Such a model accounts for
the stabilization of the temperature coefficients of dielectric constant
(ν_K), and should also give rise to reduced saturation behavior by
splitting of the electric field away from the ferroelectric major
phase. Equation 5 indicates, that for $K_2 = 10^{-2}K_1$, the effective field
across the $BaTiO_3$ grains would be reduced a hundred-fold. That is, only
1% of the applied field would be across the ferroelectric component, thus
delaying saturation behavior to higher field strength.

Table 2 lists values of the electric field coefficient of dielectric
constant (ϕ_K), measured at various bias strengths, where

$$\phi_K = K^{-1}(dK/dE)_{T,\sigma} \qquad . \qquad (12)$$

Note, for single component $BaTiO_3$ ($m_2 = 0$), the electric field coefficient of dielectric "constant" at E = 2 MV/m is ϕ_K = 0.20. That is, the corresponding capacitance has been decreased by 80% over the initial weak field value. For mixtures containing bismuth stannate, the field dependence at equivalent bias levels was much less. Table II indicates, that for $w_2 = 0.01$, $\phi_K = 0.78$; and for $w_2 = 0.10$, $\phi_K = 0.96$. That is, the voltage coefficient of capacitance (as well as the temperature coefficient of capacitance) was greatly reduced by the processing of heterogeneous dielectrics with controlled series connectivity in the boundary regions.

The applicability of the series model, is again clearly illustrated in Fig. 6, for the dependence of the room temperature dielectric constant

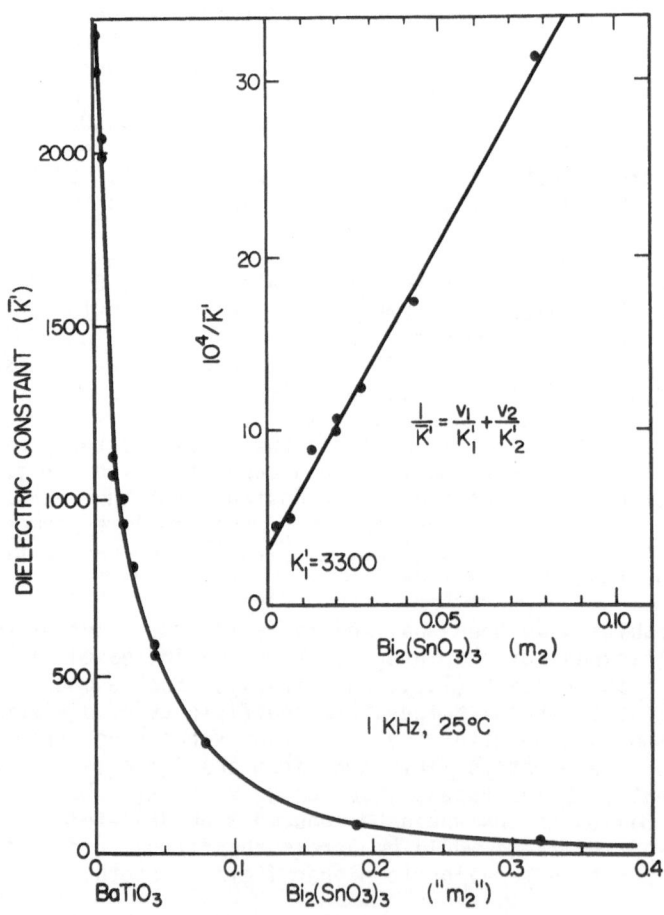

Fig. 6 Series Dilution of Dielectric Constant, for Barium Titanate - Bismuth Stannate Mixtures.

on bismuth stannate content. The hyperbolic form illustrates the sensitivity to series dilution, and the need to control precisely the additive content, for the reproducible processing of heterophasic dielectrics. Data in Fig. 6 are expressed in terms of mol. %, where m_2 = 0.01 corresponds to 3.7 wt.%, and w_2 = 0.01 is equivalent to 0.25 mol. %. Analysis of the data, according to the two-layer series model, illustrated in the inset in Fig. 6, gives an intercept value of K_1 = 3300 for pure (m_2 = 0) fine grain barium titanate, and a gradient value of K_2 = 26 for bismuth stannate. Both values are in good agreement with data reported in the literature.[1,11]

Examination of the overall properties indicates that series dilution in composite $\bar{K}$ is more than off-set by improved temperature and voltage stability. For example, $\bar{K}$ = 1500 for m_2 = 0.01 with $\nu_K < 10^{-3} {}^\circ C^{-1}$ between -50 and 150°C, and ϕ_K = 0.88 at 2Mv/M. The optimization of properties, compared with coarse grain, homogeneous $BaTiO_3$, is principally attributable to microstructure control (K_1 = 3300) with minor phase connectivity. Another example, of purposefully engineered heterophasic microstructures, is given in the following section for $BaTiO_3$-$NaNbO_3$ dielectric mixtures.

BARIUM TITANATE-SODIUM NIOBATE

Fig. 7 illustrates the microstructure of an heterogeneous dielectric fabricated in the system $BaTiO_3$-$NaNbO_3$.[12] While $NaNbO_3$ is not a low melting point sintering aid (T_m = 1390°C) it has useful dielectric properties (K_2 = 220), particularly well suited for higher dielectric constant composite applications. The dielectric illustrated in Fig. 7 had properties of $\bar{K}$ = 2000, tan δ = 0.005 and $\bar{\rho}$ = 10^{13} Ω-cm. The composition contained m_2 = 0.05, and was sintered at T_s = 1450°C for t_s = 1 hr. Again, fast-firing techniques were used to conserve chemical heterogeneity, since phase-diagram data indicated that crystalline solubility would eventually occur with extended time at temperature.

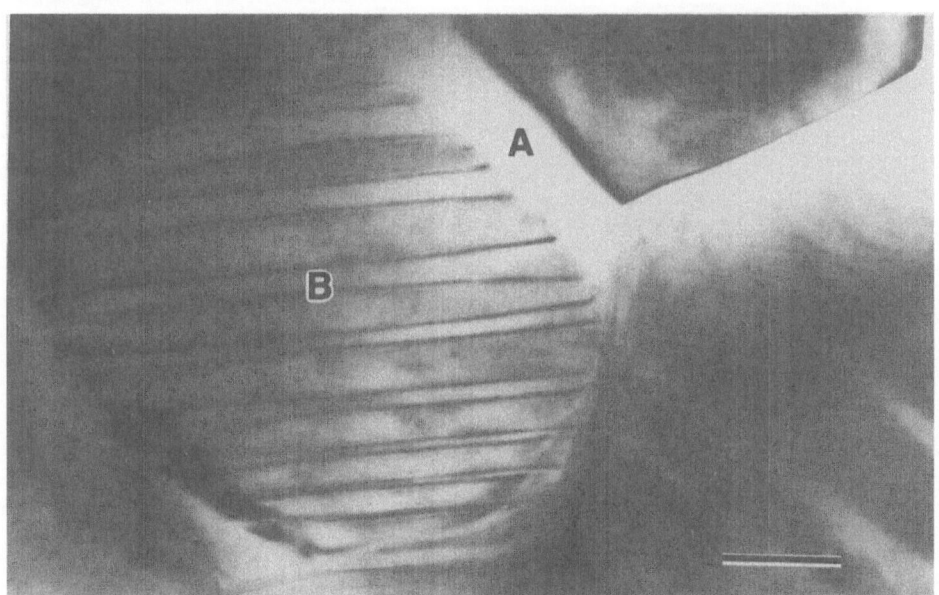

Fig. 7.   TEM Photomicrograph of the Microstructure of Fast-Fired 0.95 $BaTiO_3$ + 0.05 $NaNbO_3$ Ceramic, Illustrating Niobate Modified Boundary Regions (A) and Ferroelectric Grain Cores (B). Bar = 0.25 μm.

Similar to the previous discussion for heterogeneous mixtures with titania and bismuth stannate, the Curie temperature (T_c) for $BaTiO_3$-$NaNbO_3$ dielectrics remained constant, indicating a substantial volume fraction of unmodified $BaTiO_3$. From series mixing[8] and Curie-Weiss analysis, the presence of continuously connected impedance boundaries was inferred. Both the temperature and voltage characteristics of dielectric constant were stabilized, and sodium niobate was found to inhibit grain growth of barium titanate to 0.5-2 µm.

The TEM photomicrograph (Fig. 7) illustrates a zone of modification (A) around $BaTiO_3$ grain cores (B). Ferroelectric domains do not extend all the way to the grain boundary, but terminate at a chemically modified zone which was found to be rich in niobate. The thickness of the interfacial region depended on m_2 and t_s. Grain-to-grain separation was also identified by Fesnel imaging, at higher resolution, in the under-focus, over-focus and diffuse dark field modes.[13] The thickness of the interphase region was 2-3 nm for the sample shown.

Compositional profiles were also determined on a dedicated STEM with a 1 nm probe size. Fig. 8 illustrates the chemical heterogeneity for fast-fired mixtures (t_s = 0.5 hr), and niobate diffusion with extended time at temperature (t_s = 19 hr.). The dielectric properties were quite different between the two heat treatments, as to be expected. The former (t_s = 0.5 hr) had all the characteristics of an heterogeneous dielectric, with electric-field splitting; whereas, the latter, had all the characteristics of a substituted dielectric, with a shifted Curie-temperature and dielectric anomaly at room temperature, and pronounced dielectric saturation with applied voltage. Thus, for the same initial chemical formulation, the thermal processing conditions were critical for the engineering of controlled heterogeneities in the boundary regions. The importance of this cannot be over-emphasized for the fabrication of stabilized dielectrics with novel properties.

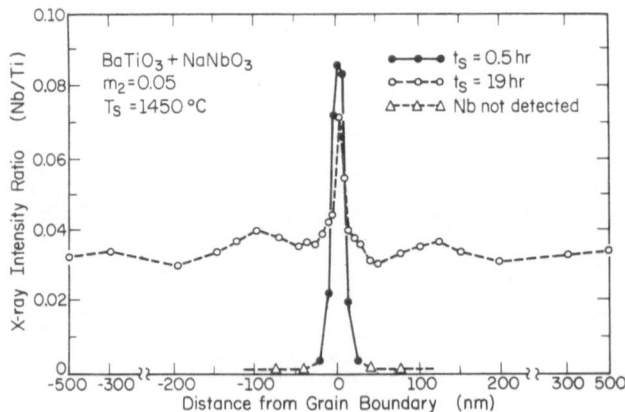

Fig. 8. STEM Analysis of Nb Distribution in the Boundary Regions, Comparing Heterogeneous Fast-Fired Mixtures with Equilibrated Extended Time Firing.

Another example of a novel property, is the enhanced apparent dielectric constant ($\bar{K}'$) obtained from boundary layer phenomena in semiconducting-insulating polycrystalline mixtures. The ceramic processing of controlled resistivity distributions, in heterogeneous ceramics, is discussed in the following section for boundary layer capacitors.

BOUNDARY LAYER CAPACITORS

Shortly after the discovery of titanate capacitors, and the development of Class I and Class II dielectrics, a third class was developed based upon reduced titanates (RT). By gaseous reduction and/or chemical doping, the titanate compositions were reduced to a conductive state (n-type), across which charge could be transported on application of an electric field. That is, the reduced titanates acted as an effective electrode material.

External Barrier Layer: Class III

Techniques were developed by which insulating barriers could be formed at the interface with applied contact electrodes.[14,15] For example, in a silver-firing operation, where a silver paste was chemically bonded to the ceramic, by the softening and diffusion of a lead bismuth borate frit which contained acceptor ions (e.g., Cu). The electric field was now supported by the surface impedance barrier(s) ($d_2 \simeq 10$ μm) which were much thinner than the bulk ceramic thickness ($d \simeq 1$ mm.). Correspondingly, the low-frequency capacitance (C_o) for a RT parallel plate capacitor (equation 6) was now enhanced by a factor d_1/d_2, of the order of 1000, where the internal conductive thickness $d_1 \simeq d$, was much greater than d_2.

From measured values of capacitance and external dimensions (d, A), the calculated value of effective dielectric constant ($\bar{K}'$) appears to be greatly enhanced, by

$$\bar{K}' = K \ (d/d_2) \qquad , \qquad (13)$$

where the electric field was only supported by resistivity distributions at the surface barriers. Thus, the external barrier layer capacitors (as they later became to be known) were an effective way of obtaining a large capacitance per unit area (~ 1 μF/cm^2), especially suited for miniaturized devices. However, they were, and remain, an intrinsically low voltage component (~ 10 V dc), since the applied field-strengths across the thin layer(s) approach a reliability limit (~ 1 MV/m). A more reliable way of distributing the applied field would be to have a series of insulating layers, effectively in cascade, within the microstructure; and this approach has given rise to the ceramic processing of internal boundary layer capacitors.

Internal Boundary Layers: Class III

Early attempts to reinsulate the boundary regions within reduced titanate ceramics, were reported for gaseous reoxidation[16] and acceptor diffusion[17]; and higher voltage ratings were obtained (25-50 V dc). Also, an important development, was the use of liquid-phase infiltration methods for the introduction of chemical heterogeneities within the boundary regions.[18-20] The latter gave rise to both interphase and interfacial regions, the relative thicknesses of which were dependent on the kinetics of reactivity with the molten phase. Again, the thermal processing conditions were extremely important for the reproducible manufacture of a consistent product, especially in the post-sintering infiltration stage. By modulating the effective thicknesses of the insulating layers, it was possible to optimize the dielectric properties for internal boundary layer capacitors, with voltage ratings up to 200 V dc.

Figure 9 illustrates the dielectric constant temperature characteristics for an internal boundary layer (IBL) capacitor, fabricated from a barium titanate-based composition. The formulation was essentially a Class II, K=6000, $(Ba,Sr)(Ti,Sn)O_3$ composition, doped with antimony and counter-doped with copper. The Curie-point was adjusted to below room temperature, at 10°C. The ceramics were insulating (10^{13} - 10^{14} ohms) with tangents δ = 0.05-0.10. Examination of the microstructure indicated a grain size (20-50 μm) sufficiently coarse that microscale electrical measurements were possible. The grains were found to be semiconducting, and the boundaries insulating. Details are reported elsewhere.[21] It is important to note, the enhanced apparent dielectric constant value of $\bar{K}'$ = 19,000, measured at weak-field, was very temperature dependent ($\nu_K \approx 10^{-2}$ $^0C^{-1}$); indicative of the Curie-Weiss behavior of the compensated interfacial region. However, stabilized internal boundary layer capacitors can now be fabricated from non-ferroelectric based compositions.

For example, $SrTiO_3$, which has a Curie point at 110°K, is very stable at room temperature. Figures 10 and 11, on the following page, illustrate the dielectric constant characteristics of an IBL capacitor, fabricated from semiconducting $SrTiO_3$, fired in a reducing atmosphere, with additional processing of liquid phase infiltration. The ceramics were insulating (10^{13}-10^{14} ohms), with tangents δ = 0.01-0.03. At room temperature (Fig. 10) the apparent dielectric constant was enhanced to $\bar{K}' \approx$ 23,000, with excellent temperature stability ($\nu_K \approx 10^{-4}$ °C^{-1}). Figure 11 illustrates the electric field dependence of dielectric constant was ϕ_K = 0.80 at 0.5 MV/m. The capacitors could withstand 200 V dc, with tangents δ < 0.05.

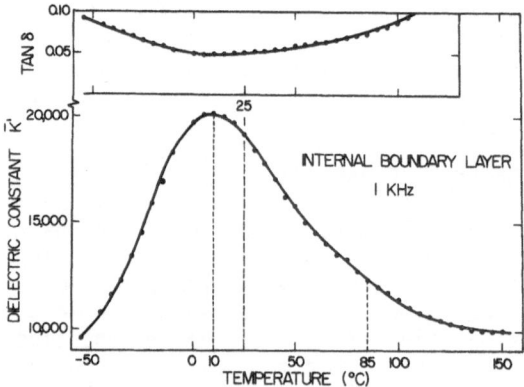

Fig. 9. Dielectric Constant-Temperature Behavior for an Internal Boundary Layer Capacitor, Fabricated from Substituted Barium Titanate.

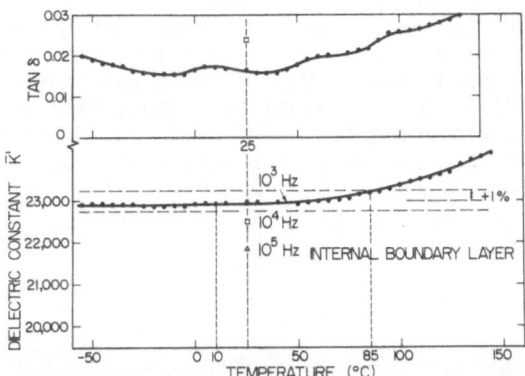

Fig. 10. Dielectric Constant-Temperature Behavior for an Internal Boundary Layer Capacitor Fabricated from Strontium Titanate.

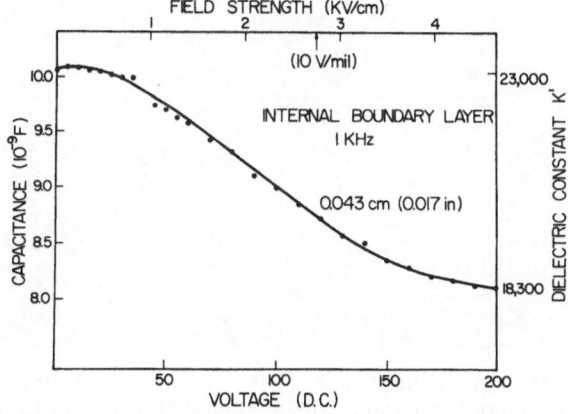

Fig. 11. Electric Field Dependence of Dielectric Constant for an Internal Boundary Layer Capacitor Fabricated from Strontium Titanate.

These remarkable properties were achieved by the purposeful microstructural engineering of electric field distributions within polycrystalline microstructures. Figure 12 illustrates the chemical heterogeneity observed in a back-scattered electron image mode, on a SEM, of high atomic number elements (i.e., Pb, Bi), which were preferentially located within the boundary regions. Note, the continuous connectivity of the chemically different boundary regions. The grain size was coarse ($d_1 \simeq 50$ μm), and the combined thickness (d_2) of the interphase and reaction zone, was approximately 0.5 μm. This was independently confirmed by STEM, EPMA and diffusion calculations. Thus, similar to equation 13, a microstructural engineering factor of d_1/d_2 would enhance the intrinsic dielectric constant (K = 200-300) by a factor of 100. This is in good agreement with values of calculated apparent dielectric constant of $\bar{K}' = 23,000$.

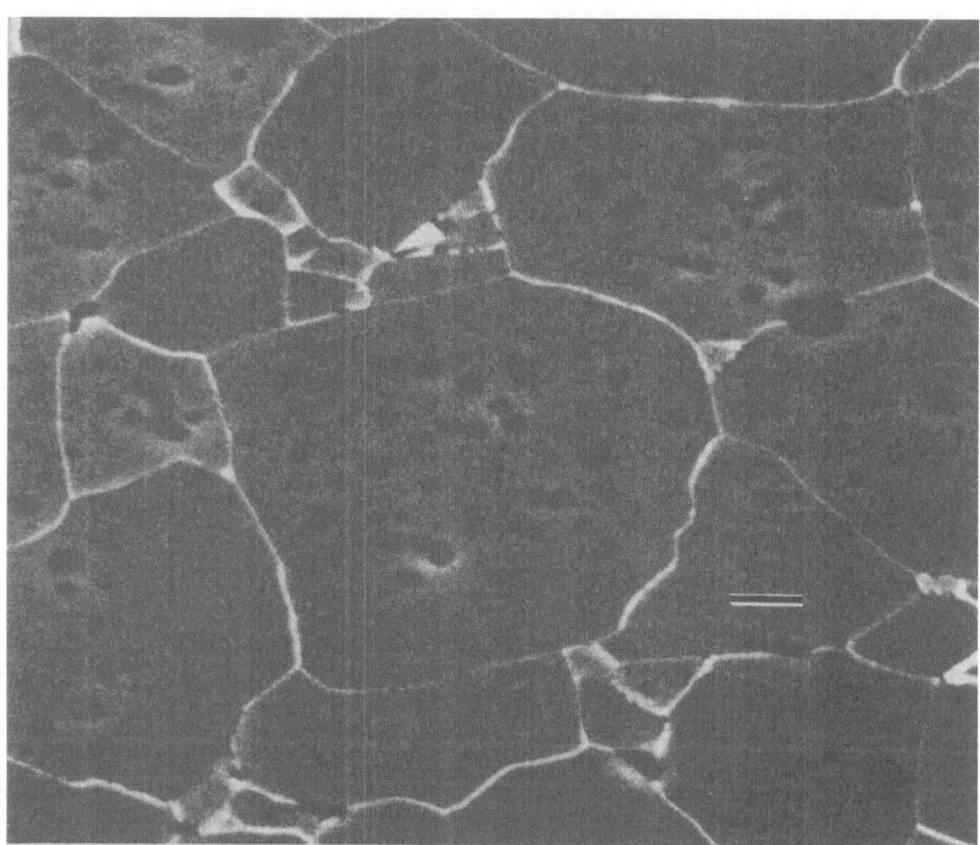

Fig. 12. SEM Compositional Contrast Image of Chemical Heterogeneities in the Boundary Regions for a Strontium Titanate Based Internal Boundary Layer Capacitor. Bar = 10 μm.

ACKNOWLEDGEMENTS

This work was supported by the U.S. Department of Energy, Division of Materials Sciences, under contract DE-AC023-76ER01198. The author gratefully acknowledges collaboration with L. E. Cross, H. D. Park, S. M. Park, Z. Xu, A. K. Mehrotra and O. C. Jahnke. The Transelco Division of Ferro Corporation supplied materials and initial support.

REFERENCES

1. K. Kinoshita and A. Jamaji, Grain Size Effects on Dielectric Properties in $BaTiO_3$, J.Appl.Phys. 47:371 (1976).

2. W. R. Buessem, L. E. Cross and A. K. Goswami, Phenomenological Theory of High Permittivity in Fine-Grained Barium Titanate, J.Am.Cer.Soc. 49:33 (1966).

3. T. Negas, R. S. Roth, H. S. Parker and D. Minor, Subsolidus Phase Relations in the $BaTiO_3$-TiO_2 System, J.Sol.Stat. Chem. 9:297 (1974).

4. D. A. Payne and L. E. Cross, Microstructure-Property Relations for Dielectric Ceramics:I. Mixing of Isotropic Homogeneous Linear Dielectrics,: "Ceramic Microstructure '76", 584, R. M. Fulrath and J. A. Pask, eds., Westview Press, Boulder Colorado (1977).

5. R. L. Bolton, Temperature Compensating Ceramic Capacitors in the System Baria-Rare Earth Oxide-Titania, Ph.D. Thesis, The University of Illinois at Urbana-Champaign (1968).

6. H. Kniepkamp and W. Heywang, Depolarization Effects in Polycrystalline Barium Titanate, Z.ang.Phys. 6:385 (1954).

7. G. H. Jonker and W. Noorlander, Grain Size of Sintered Barium Titanate, in: Science of Ceramics 1:255, G. M. Stewart, ed., Academic Press (1962).

8. D. A. Payne and L. E. Cross, Microstructure-Property Relations for Dielectric Ceramics:II. The Brick-Wall Model of Polycrystalline Microstructure, in: "Microstructure and Properties of Ceramic Materials," 380, T. S. Yen and J. A. Pask, eds., Science Press, Beijing (1984).

9. R. S. Roth, Pyrochlore-Type Compounds Containing Double Oxides of Trivalent and Tetravelent Ions, J.Res.Nat.Bur.Std. 56: (1956).

10. H. D. Park, Preparation and Properties of Dielectric Mixtures in the System $BaTiO_3$-$Bi_2(SnO_3)_3$, M. S. Thesis, The University of Illinois at Urbana-Champaign (1978).

11. K. Prohaska, The Properties and Method of Preparation of Bismuth Stannate for Use in Electrical Ceramics, Szklo i Ceramika 24:209 (1973).

12. D. A. Payne, The Role of Internal Boundaries Upon the Dielectric Properties of Polycrystalline Ferroelectric Materials, Ph.D. Thesis, The Pennsylvania State Unversity, (1973).

13. Z. Xu and D. A. Payne, Interphases and Interfaces in $BaTiO_3$-$NaNbO_3$ Ceramics, submitted to, Elect.Micro.Soc.Am. (1986).

14. R. R. Roup, A Review of Development and Current Trends in Ceramic Dielectrics Used for Capacitor Applications, J.Am. Cer.Soc. 41:499 (1958).

15. L. Maxwell, D. Freifielder and P. Franklin, Layerized High Dielectric Constant Capacitors, Elec.Des. 7:73 (1959).

16. R. M. Glaister, Barrier-Layer Dielectrics, J.Inst.Elec.Eng. 109B(22): 282 (1961).

17. H. Brauer, Grain Boundary Barrier Layers in $BaTiO_3$ Ceramic with a High Effective Dielectric Constant, Z.ang.Phys. 29:282 (1970).

18. S. Waku, Studies on the Boundary-Layer Ceramic Capacitor, Rev. El.Comm.Lab.N.T.T.Japan 15: 689 (1967).

19. S. Waku, M. Uchidate and K. Kuichi, Studies on the $(Ba,Sr)TiO_3$ Boundary Layer Ceramic Dielectrics, ibid. 18:681 (1970).

20. T. Edahiro and F. Yoshimura, Experiments on Boundary layer Dielectrics with Low Temperature Coefficients, ibid, 21: 843 (1973).

21. H. D. Park and D. A. Payne, Characterization of Internal Boundary Layer Capacitors, Adv.in Ceramics 1:242, (1981).

DIELECTRIC AND ELECTRICAL PROPERTIES OF BaTiO$_3$ COMPOSITES

W. Huebner, F.C. Jang, and H.U. Anderson

Department of Ceramic Engineering
University of Missouri - Rolla
Rolla, MO   65401

ABSTRACT

In this investigation composites of unconsolidated BaTiO$_3$ powder ($\geq$ 99.9% purity, $\leq$ 0.1 μm crystallite size) or partially-sintered BaTiO$_3$ with either air or polymer were studied. The purpose of this study was to measure the dielectric and electrical properties of the composites, and to determine how well these properties fit existing theories concerning fine-grained permittivity and dielectric mixing rules.

INTRODUCTION

Recent investigations[1,2] on high-purity BaTiO$_3$ have shown that partially sintered specimens with sub-micron grain size and high porosity showed good resistance to dry atmosphere electrical degradation. However, porous specimens exhibit low dielectric constants, low breakdown strengths, and high water permeability. It is expected that filling the porosity with a polymer would improve the breakdown strength and water impermeability. The purpose of this investigation was to measure the dielectric and electrical properties of composites made from BaTiO$_3$ and either polymer or air, and to determine how well these properties fit existing theories concerning fine-grained permittivity and dielectric mixing rules.

Numerous rules appear in the literature which predict the dielectric constant of mixtures depending upon the relative volumes and permittivities of the constituents, as well as their shape and continuity. Articles by Reynolds and Hough[3], Meredith and Tobias[4], and Van Beek[5] review the application and validity of most of these rules.

Niesel-Bruggeman[6] found composites to obey:

$$K_c = \tfrac{1}{4} \left((2E_p - E_p') + ((E_p' - 2E_p)^2 + 8K_1K_2)^{\frac{1}{2}} \right) \qquad (1)$$

where: K_c = composite dielectric constant
K_1 = dielectric constant of phase 1
K_2 = dielectric constant of phase 2
V_1 = volume fraction of phase 1
V_2 = volume fraction of phase 2

$E_p = V_1K_1 + V_2K_2$

$E_p' = V_1K_2 + V_2K_1$

Bottcher's[7] equation has been found to apply to non-dilute systems and is given by:

$$\frac{K_c - K_1}{3K_c} = V_2 \frac{K_2 - K_1}{2K_c - K_2}$$

(2)

These particular mixing rules are of interest due to their applicability to the current experimental data.

Partially-sintered $BaTiO_3$ and unsintered $BaTiO_3$ powder exhibit unusual dielectric properties believed to be due to a surface layer effect. Anliker et al.[8], while studying depolarization effects in very fine particle size $BaTiO_3$, observed a broad Curie transition which they attributed to a 100Å thick tetragonal surface layer which persisted well above the Curie temperature. Both Chynoweth's[9] observation of assymetric pyroelectric effects and Triebwasser's[10] observation of birefringence of $BaTiO_3$ single crystal surfaces have been explained by space charge layers. Numerous studies[11-17] of the switching time and dielectric constant of single crystals as a function of thickness indicate a nonferroelectric surface exists. English[18], using electron-mirror microscopy, found the surface of $BaTiO_3$ to be ferroelectric. Goswami[19] explained the absence of ferroelectric behavior in unsintered $BaTiO_3$ by a nonferroelectric surface layer. Thus there appears to be agreement concerning the existence of a surface layer, but its exact nature is not clearly understood.

Based solely on the existence of a low dielectric constant surface layer one would expect decreasing the grain size of polycrystalline $BaTiO_3$ would decrease the overall dielectric constant. However it is well known that a high dielectric constant can be obtained for sintered, dense, approximately 1μm grain size $BaTiO_3$. Numerous studies[20-23] have shown room temperature permittivities can range from approximately 3500 - 6000. For single crystal $BaTiO_3$ the room temperature permittivities are 4000 and 170 along the a and c axes respectively. Buessem et al.[24] have proposed that the high permittivity in fine-grained $BaTiO_3$ arises from the absence of $90°$ twinning which gives rise to high internal stresses. This pertains only to sintered, polycrystalline specimens in which grains are constrained by the surrounding matrix. Goswami[25] has shown that unsintered powder of comparable density to sintered specimens does not show the anomalously high permittivity. Goswami[19,25] also observed that progressive heat treatment of $BaTiO_3$ results in a gradual increase in permittivity and appearance of ferroelectricity. He ascribed this to the annealing out of lattice defects which removed the influence of a low dielectric constant, nonferroelectric surface.

It is not clear from the literature if sub-micron grain size $BaTiO_3$ can exhibit a similiar high permittivity at room temperature. Graham et al.[26] observed hot-pressed, sub-micron grain size $BaTiO_3$ exhibited a dielectric constant of 3000, but were unsuccessful in sintering specimens with sub-micron grain sizes. One of the goals of the present work is to partially-sinter high purity (> 99.9%), fine-grained (< 0.1μm) $BaTiO_3$ to study the dielectric properties of sub-micron $BaTiO_3$.

EXPERIMENTAL PROCEDURE

$BaTiO_3$ powders utilized in this study were prepared by an organometallic technique first described by Pechini[27]. Resulting powders are chemically homogeneous, uniformly-sized, and approximately 0.1μm in diameter. X-ray powder diffraction patterns showed line-broadening effects but revealed the powder to have tetragonal symmetry.

Specimens for measurement on unsintered $BaTiO_3$ compacts were pressed in a ½ inch diameter stainless steel die at various pressures up to 75000 psi without the addition of a binder. Those specimens used for the sintering study were pressed at 50000 psi with the addition of 8 weight % binder. Green densities were approximately 62% theoretical.

Porous specimens were prepared by partial-sintering at temperatures from 500 - 1000°C, for times of 1 - 4 hours in a SiC muffle tube furnace. Typical density and shrinkage curves are contained in Figures 1 and 2. Figures 3a-b contain the corresponding SEM micrographs. The process of preparing polymer composites from these specimens is as follows: 1) Disks are initially dried for 24 hours under vacuum at 200°C to minimize water vapor. 2) After cooling, disks are then immersed in a styrene monomer-initiator (0.1 weight % AIBN) solution under vacuum. 3) Initial polymerization is then accomplished by slowly raising the temperature from 30-60°C over a period of 120 hours. 4) The polymerization is completed by annealing the disks at 60°C for 48 hours under atmospheric pressure.

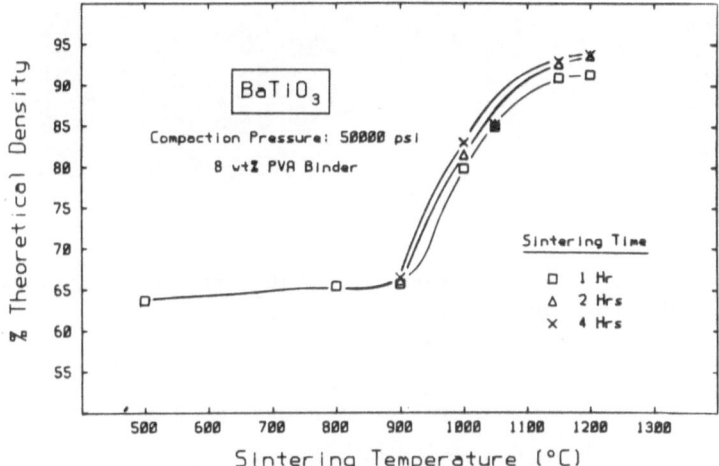

Fig. 1. Theoretical density of specimens sintered
at various temperatures and times.

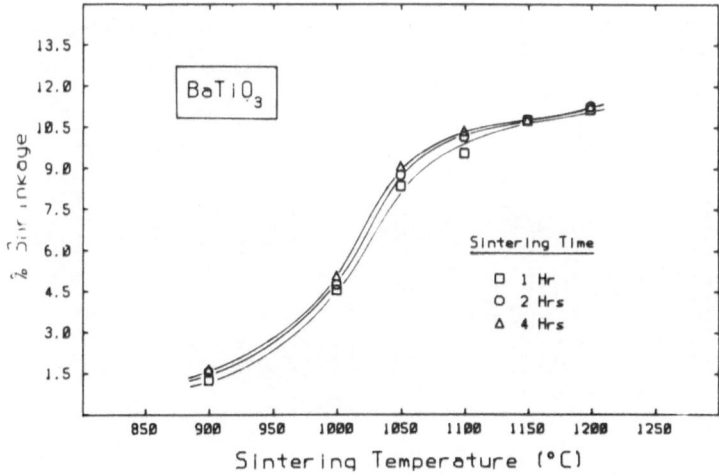

Fig. 2. Shrinkage of specimens sintered at various
temperatures and times.

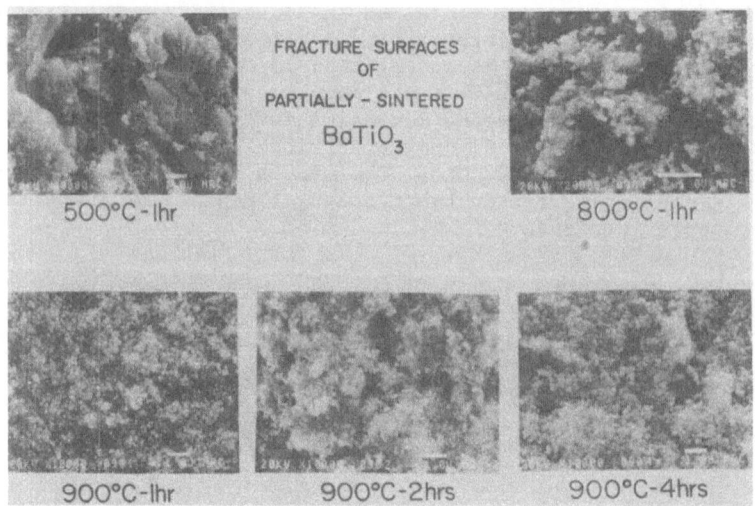

Fig. 3a. Scanning electron micrographs of BaTiO$_3$ specimens partially-sintered at 500, 800, and 900°C.

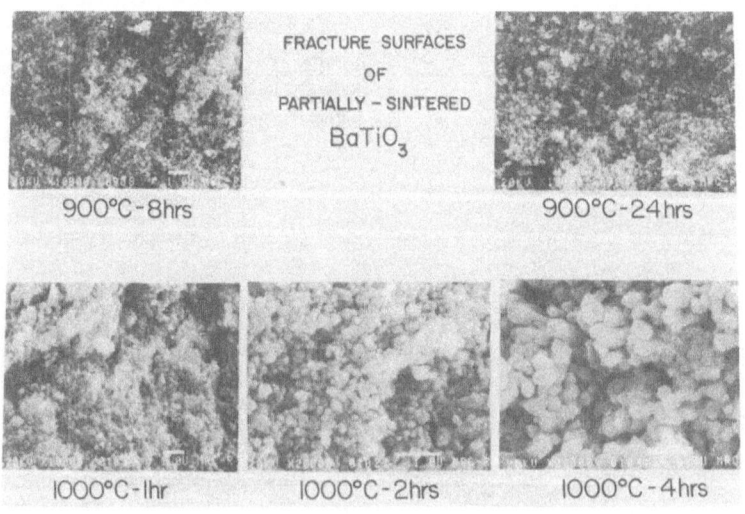

Fig. 3b. Scanning electron micrographs of BaTiO$_3$ specimens partially-sintered at 900 and 1000°C.

The filling efficiency of the polymer into the open porosity was determined by scanning electron microscopy and density measurements using:

$$\rho_c = \rho_a V_a + \rho_b V_b \qquad (3)$$

where: ρ_c = composite density

ρ_a, ρ_b = BaTiO$_3$, polymer density

V_a, V_b = BaTiO$_3$, polymer volume fraction

For all of the specimens studied the polymer filled greater than 95 percent of the total open porosity.

Thin films of polymer with up to 32 volume percent BaTiO$_3$ filler were prepared by dispersing unsintered BaTiO$_3$ powder into a liquid formed by dissolving polystyrene in dioxane. The mixture was cast onto a glass plate, the solvent allowed to partially evaporate from 20 - 60°C, and then the film was heat-cured at 60°C under vacuum. Films from this simple process are approximately 0.010 inches thick.

Air-drying silver paint was used as the electrode material for all the specimens. Capacitance and dissipation factor were measured up to 100 kHz as a function of temperature using a computer-controlled General Radio 1689 RLC bridge. Resistivity measurements were made using a Hewlett Packard 4140 pA/DC voltage source. These measurments were made in a dry atmosphere. DC breakdown strengths were measured in silicon oil at room temperature.

RESULTS AND DISCUSSION

The 1 kHz dielectric constant of a 170°C, vacuum-annealed, unsintered BaTiO$_3$ compact (≃ 60% dense) is shown in Figure 4 as a function of temperature. The dissipation factor is not shown but is less than 2% over the temperature range measured. These results agree with those of Goswami[19] as far as the magnitude of the dielectric constant and nonferroelectric behavior are concerned. The low dielectric constant can be explained by the existence of a nonferroelectric surface layer[11-17,19]. Chynoweth[9] categorized the proposed surface layers into two main groups: 1) Space charge or exhaustion layers in the range of 0.1 μm thick, which are generally ferroelectric, and 2) Chemically or mechanically disturbed layers composed of a lossy, low dielectric constant, nonferroelectric material in the range of 10 Å thick.

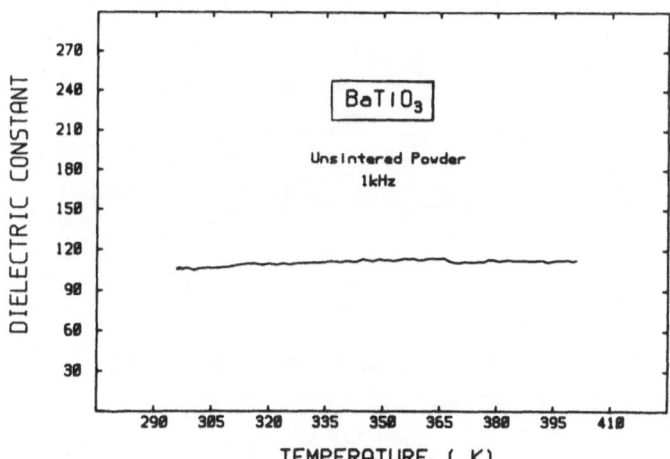

Fig. 4. Dielectric constant vs. temperature for
an unsintered BaTiO$_3$ specimen.

The effect of increasing heat treatment on the dielectric constant is shown in Figures 5-7. The increase in dielectric constant is related to the degree of grain growth and sintering. Grain size vs. sintering temperature is shown in Figure 8; little grain growth occurs for temperatures less than 900°C. These results show that 0.1 µm grains do display broad transitions at the Curie point of 120°C and as the grain size approaches 1 µm the peak becomes sharper. As can be seen in Figure 9, in the region of constant grain size the dielectric constant increases with the percent shrinkage. Thus it appears that fine-grained behavior is strictly a function of the sintering conditions and the degree of shrinkage.

In a previous report of Buessem et al.[24] it was proposed that the high dielectric constant of 1-3 µm $BaTiO_3$ is due to the absence of 90° twinning which gives rise to internal stress below the Curie temperature. Figure 9 is nearly identical in form to the permittivity vs. stress curve derived for fine-grained $BaTiO_3$ by Buessem. Due to the striking similarity this suggests that internal stress in a compact is proportional to the shrinkage (ie: the degree of intergranular contact), and that the enhancement of the dielectric constant with increasing shrinkage may be understood in these terms. It appears that internal stress increases as shrinkage increases from 1 - 1.7%.

When applying mixing rule theories to $BaTiO_3$ composites with either air or polymer it is important to distinguish between unsintered and sintered results. Figure 10 is a plot of log K vs. volume fraction air/polymer for the $BaTiO_3$ specimens studied. Application of either the Niesel-Bruggeman or Bottcher mixing rules, both of which fit the data, results in zero porosity dielectric constants of 500 and 5000 for unsintered and sintered $BaTiO_3$ respectively. These results indicate that microstructures with less than 1 µm grains can produce enhanced dielectric constants.

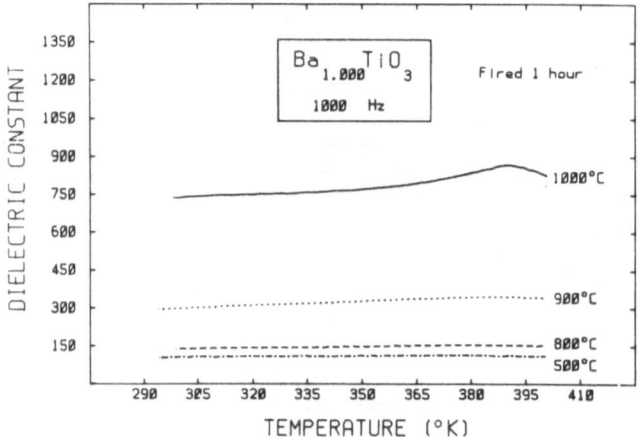

Fig 5. Dielectric constant vs. temperature for $BaTiO_3$ specimens sintered for 1 hour at various temperatures.

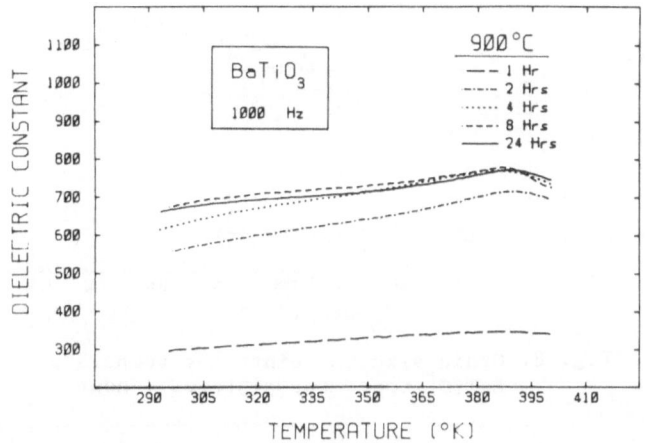

Fig. 6. Dielectric constant vs. temperature for BaTiO$_3$ specimens sintered at 900°C for various times.

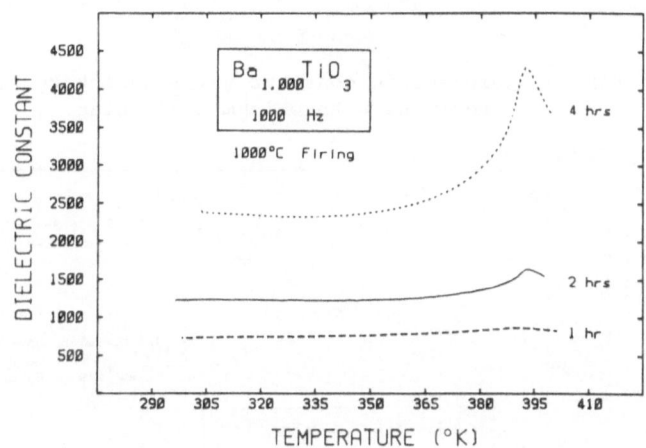

Fig. 7. Dielectric constant vs. temperature for BaTiO$_3$ specimens sintered at 1000°C for various times.

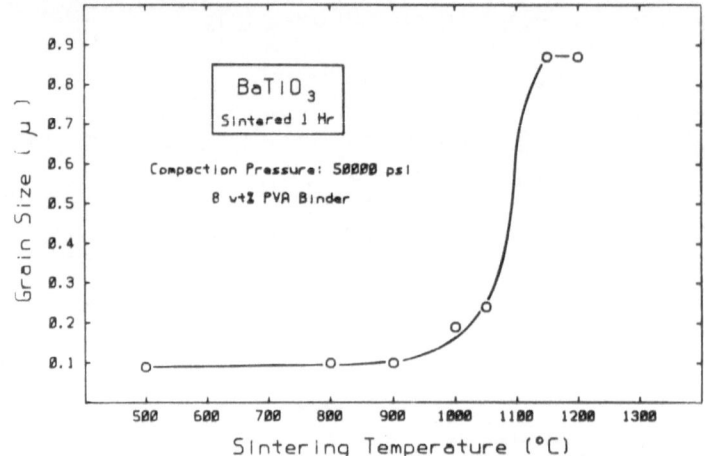

Fig. 8. Grain size vs. sintering temperature of the
BaTiO₃ specimens sintered 1 hour.

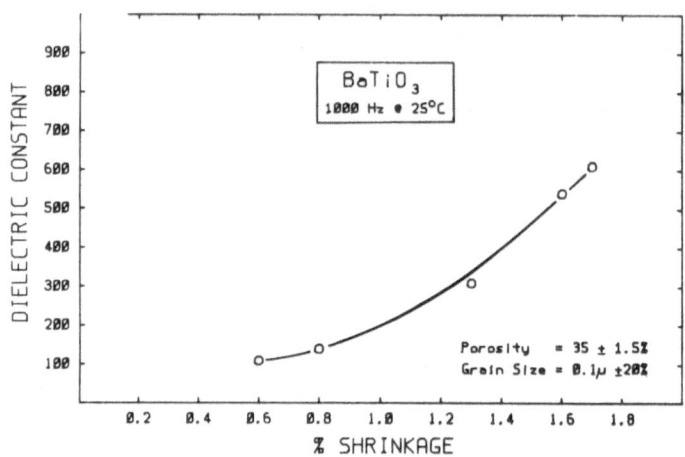

Fig. 9. Dielectric constant vs. % shrinkage for BaTiO₃
specimens with 0.1 μm grain size.

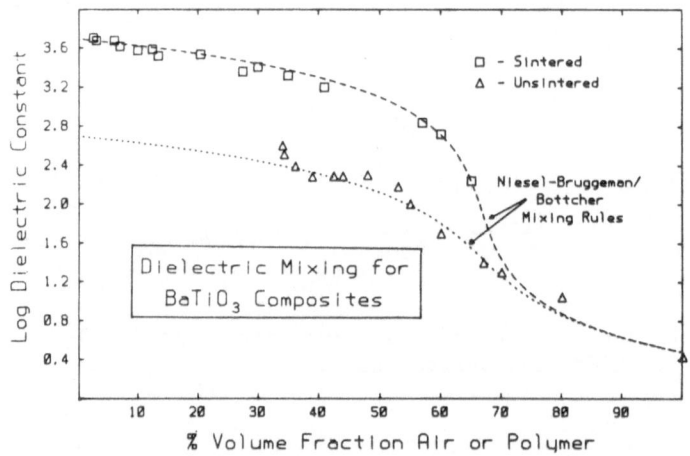

Fig. 10. Application of mixing rules to the dielectric
properties of BaTiO₃ composites.

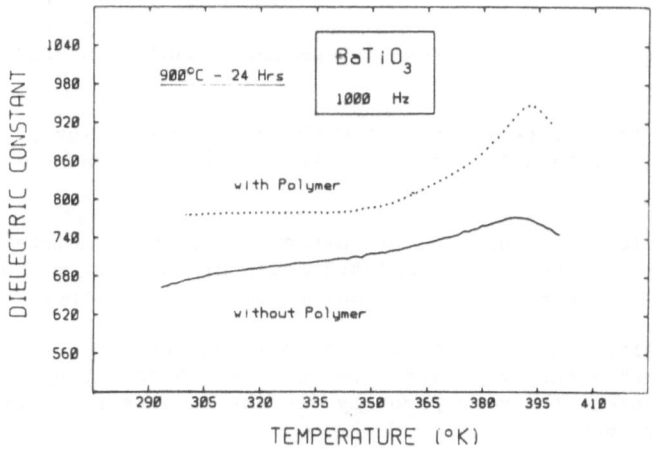

Fig. 11. Dielectric constant vs. temperature for BaTiO₃ specimens sintered at 900°C for 24 hours, with and without polymer.

As can be seen in Figure 11, addition of a polymer to porous $BaTiO_3$ enhances the dielectric constant and further develops the peak at the Curie point. It appears the polymer phase increases the internal stress by inter-granular coupling. The restoration of ferroelectricity due to the polymer phase implies the previously-proposed nonferroelectric surface layer is not due to a high concentration of lattice defects.

Breakdown strength measurements were made on specimens which were vacuum-dried at 125°C. DC breakdown strengths for $BaTiO_3$-air composites averaged approximately 90 kV/cm, and $BaTiO_3$-polymer composites averaged 175 kV/cm. This effect can be explained by the high resistivity of the polymer phase, and the elimination of $BaTiO_3$ - air interfaces. In low density $BaTiO_3$ the "weak spots" are intergranular pores, which are assumed to be the origin of the dielectric breakdown. These potential breakdown sources are eliminated by the polymer phase, resulting in an improved breakdown strength. The high-resistivity polymer phase also inhibits electron avalanche.

The resistivity measurements made on the composite specimens showed the addition of a polymer did not effect the resistivity. Room temperature resistivities were approximately 10^{14} ohm-cm, and 10^{13} ohm-cm at 85°C.

Not shown figuratively, but the present study has also found the addition of a polymer to porous $BaTiO_3$ eliminates the detrimental effect of water vapor on the dielectric properties. Porous specimens exhibit a non-linear dielectric constant and dissipation factor in the presence of a humid atmosphere, while polymer-impregnated specimens do not.

SUMMARY

1) BaTiO$_3$ with grain size less than 1.0 µm can produce enhanced dielectric constants.

2) The increase in dielectric onstant with increasing shrinkage is due to the increase in internal stress associated with increasing intergranular contact.

3) Composite dielectric constants of compacts made from unsintered powder agree well with either the Niesel-Bruggeman or Bottcher mixing rules; the zero-porosity dielectric constant extrapolates to approximately 500.

4) Composite dielectric constants of partially-sintered (shrinkage $\leq$ 5%, no grain growth) compacts also agree with the Niesel-Bruggeman and Bottcher mixing rules; the zero-porosity dielectric constant extrapolates to approximately 5000.

5) Polymer-BaTiO$_3$ composites exhibit higher dielectric onstants, lower dissipation factors, higher breakdown strengths, and impermeability to water as compared to porous BaTiO$_3$.

ACKNOWLEDGMENT

The authors would like to thank Dr. H. Yasuda for his advice concerning the polymer impregnation. This work was partially sponsored by the Office of Naval Research and Hughes Aircraft Company.

REFERENCES

1. D.A. Anderson, and W. Huebner, "Electrical Degradation of High-Purity BaTiO$_3$," Presented at the 1985 American Ceramic Society National Convention in Cincinnati, OH, May 26-30.
2. W. Huebner, H.U. Anderson, and D.E. Day, "Reliability Studies of Ceramic Capacitors," ONR Progress Report, October 1984, Contract # N00014-82-K-0294.
3. J.A. Reynolds, and J.M. Hough, "Formulae for Dielectric Constant of Mixtures," Proc. Phys. Soc. Lond., B70:769 (1957).
4. R.E. Meredith and C.W. Tobias, "Resistance to Potential Flow through a Cubical Array of Spheres," J. Appl. Phys., 31:1270 (1960).
5. L.K.H. van Beek, "Dielectric Behavior of Heterogeneous Systems," Progress in Dielectrics, 7:69-114, ed. J.B. Birks, London: Heywood Books (1967).
6. W. Niesel, "Die Dielektrizitätskonstanten heterogener Mischkörper aus isotropen und anisotropen Substanzen," Ann. d Phys., 6:336 (1952).
7. C.J.F. Böttcher, "The Dielectric Constant of Crystalline Powders," Rec. Trav. Chim. Pays-Bas, 64:47 (1945).
8. M. Anliker, H.R. Brugger, and W. Känzig, "Behavior of Colloidal Seignettoelectrics: III," Helv. Phys. Acta., 27:99 (1954).
9. A.G. Chynoweth, "Surface Space-Charge Layers in Barium Titanate," Phys. Rev., 102[3]:705 (1956).
10. S. Triebwasser, "Space Charge Fields in BaTiO$_3$," Phys. Rev., 118[1]:100 (1960).
11. W.J. Merz, "Switching Time in BaTiO$_3$ and Its Dependence on Crystal Thickness," J. Appl. Phys., 27:938 (1956).
12. M.E. Drougard and R. Landauer, "On the Dependence of the Switching Time of BaTiO$_3$ Crystals on Their Thickness," J. Appl. Phys., 30:1663 (1959).
13. E. Fatuzzo and W.J. Merz, "Surface Layer in BaTiO$_3$ Single Crystals," J. Appl. Phys., 32[9]:1685 (1961).

14. H. Schlosser and M.E. Drougard, "Surface Layers on Barium Titanate Single Crystal Above the Curie Point," J. Appl. Phys., 32[7]:1227 (1961).

15. P. Coufova and H. Arend, Czech. J. Phys., B12:308 (1962).

16. A.V. Turik, "The Problem of the Surface Layer in BaTiO$_3$ Single Crystals," Sov. Phys. Sol. State, 5:1748 (1964).

17. D.R. Callaby, "Surface Layer of BaTiO$_3$," J. Appl. Phys., 37:2295 (1966).

18. F.L. English, "Electron-Mirror Microscopy Study of BaTiO$_3$ Surface Layers," J. Appl. Phys., 39[7]: 3231 (1968).

19. A.K. Goswami, "Dielectric Properties of Unsintered Barium Titanate," J. Appl. Phys., 40[2]:619 (1969).

20. H. Kniepkamp and W. Heywang, "Depolarization Effects in Polycrystalline BaTiO$_3$," Z. Angrew. Phys., 6[9]:385 (1954).

21. G.H. Jonker and W. Noorlander, pp 255-64 in Science of Ceramics, Vol. 1 ed. G.H. Stewart, Academic Press, New York, 1962.

22. L. Egerton and S.E. Koonce, "Effect of Firing Cycle on Structure and Some Dielectric and Piezoelectric Properties of BaTiO$_3$ Ceramics," J. Amer. Cer. Soc., 38[11]:412 (1955).

23. A.A. Anan'eva, B.V. Strizkov, and M.A. Ugryumor, "Some Anomalous Properties of Chemically-Pure Barium Titanate," Bull. Acad. Sci., USSR, Phys. Ser. 24:1395 (1960).

24. W.R. Buessem, L.E. Cross, and A.K. Goswami, "Phenomenological Theory of High Permittivity in Fine-Grained Barium Titanate," J. Amer. Cer. Soc., 49:33 (1966).

25. A.K. Goswami, "Dielectric Properties of Explosively Compacted BaTiO$_3$," J. Amer. Cer. Soc., 56:100 (1973).

26. H.C. Graham, N.M. Tallan, and K.S. Mazdiyasni, "Electrical Properties of High-Purity Polycrystalline Barium Titanate," J. Amer. Cer. Soc., 54:548 (1971).

27. M.P. Pechini, U.S. Patent # 3,330,697 July 11, 1967.

COMPOSITE PIEZOELECTRIC SENSORS

A. Safari, G. Sa-gong, J. Giniewicz and R.E. Newnham

Materials Research Laboratory
The Pennsylvania State University
University Park, PA 16802

INTRODUCTION

A hydrophone is an underwater microphone or transducer used to detect underwater sound. The sensitivity of a hydrophone is determined by the voltage that is produced by a hydrostatic pressure wave. The hydrostatic voltage coefficient, g_h, relates the electric field appearing across a transducer to the applied hydrostatic stress, and is therefore a useful parameter for evaluating piezoelectric materials for use in hydrophones. Another piezoelectric coefficient frequently used is the hydrostatic strain coefficient, d_h, which describes the polarization resulting from a change in hydrostatic stress. The g_h coefficient is related to the d_h coefficient by the relative permittivity (K): $g_h = d_h/\varepsilon_o K$, where ε_o is the permittivity of free space.

A useful 'figure of merit' for hydrophone materials is the product of hydrostatic strain coefficient d_h and hydrostatic voltage coefficient g_h. The product $d_h g_h$ has the units of $m^2 N^{-1}$. Other desirable properties for a hydrophone transducer include (i) low density for better acoustical matching with water, (ii) little or no variation of the g_h and d_h coefficients with pressure, temperature and frequency, and (iii) high compliance and flexibility so that the transducer can conform to any surface and withstand mechanical shock. Compliance also leads to large dampening coefficients which prevent 'ringing' in a passive transducer.

Lead zirconate titanate (PZT) is widely used as a transducer material because of its high piezoelectric coefficients. However, for hydrophones, PZT is a poor choice for several reasons. PZT has a large piezoelectric d_{33} coefficient, but its hydrostatic strain coefficient d_h (=d_{33} + $2d_{31}$) is small because d_{33} and $2d_{31}$ are opposite in sign, and almost cancel one another. Moreover, the high permittivity of PZT (K$\cong$1800) lowers the voltage coefficient g_h to miniscule values. In addition, the density of PZT (7.9 g/cm^3) makes it difficult to obtain good impedance matching with water. PZT is also a brittle ceramic and for some applications a more compliant material with better shock resistance is desirable.

Other materials used for hydrophone applications are lead metaniobate PbNb$_2$O$_6$[1] and PbTiO$_3$[2]. Their d_h values are slightly higher than that of PZT (Table 1) and the g_h values are an order of magnitude better because of their modest dielectric constants. Unfortunately, PbNb$_2$O$_6$ and PbTiO$_3$ are also

Table 1. Dielectric and Piezoelectric Properties of Single Phase and Selected Composite Materials

	K_{33}	d_{33} pC/N	g_h $(10^{-3}Vm/N)$	d_h pC/N	$g_h d_h$ $10^{-15}m^2/N$
PZT (501)	1800	450	2.5	40	100
PbTiO$_3$	230	53	23	47	1080
PbNb$_2$O$_6$	225	85	33	67	2200
PVF$_2$=(CH$_2$-CF$_2$)$_n$	12	35	100	10	1000
0-3 PZT-Polyurethane	26	10	8	2	10
0-3 PZT-Silicone (large particle)	100	340	32	28	900
0-3 PbTiO$_3$-Chloroprene	40	60	100	35	3500
0-3 0.5 PbTiO$_3$-0.5 BiFeO$_3$	40	45	65	25	1625
0-3 0.5 PbTiO$_3$-0.5 Bi[Fe$_{0.98}$Mn$_{0.02}$]O$_3$	40	55	90	30	2700
0-0-3 PZT-Carbon-Polymer Composite	120	50	30	30	900
1-3 PZT-Epoxy	54	150	56	27	1536
1-3 Glass-Ceramics	10	10	100	10	1000
1-3 Diced Copper Encapsulated Composite	400	400	75	265	20000
1-3-0 PZT-Epoxy + glass sphere	78	180	60	41	2460
1-3-0 Foamed Polyurethane	41	180	210	73	14600
3-1 Perforated PZT-Epoxy	650	410	30	170	5000
3-2 Perforated PZT-Epoxy	375	350	60	200	12000
3-3 PZT-Silicon Rubber	45	200	45	180	8100
3-3 PZT-Epoxy	–	–	50	90	4500

dense, brittle ceramics, which undergo a large volume change at the Curie temperature, often causing fracture during preparation.

Polyvinylidene fluoride [PVF$_2$ = (CH$_2$-CF$_2$)$_n$] offers several advantages over PZT and other piezoelectric ceramics[3]. It has low density, high flexibility, and although PVF$_2$ has low d_{33} and d_h, the piezoelectric voltage coefficient g_h is large because of its low relative permittivity.

There are, however, problems associated with the use of PVF$_2$. The major problem is the difficulty in poling PVF$_2$. A very high field is necessary to pole PVF$_2$ (1.2 MV/cm), and this limits the thickness that can be poled. Pyroelectric phenomena in PVF$_2$ also produce undesirable polarization fluctuations with temperature.

It is clear that none of the single-phase materials are ideal for hydrophones and there is need for better piezoelectric materials.

One approach to the problem is to develop composite materials in which the desired properties can be incorporated through use of a combination of materials with different properties. In designing composite materials for hydrophone applications, a logical choice would be a piezoelectric ceramic and a compliant polymer. In such a composite, the ceramic produces a large piezoelectric effect, while the polymer phase lowers the density and permittivity and increases the elastic compliance.

In a composite the electric flux pattern and the mechanical stress distribution, and hence the resulting physical and electromechanical properties, depend strongly on the manner in which the individual phases are interconnected. In this regard the connectivity of a composite, defined as the number of dimensions in which each component phase is continuous[4], is of crucial importance. When referred to in an orthogonal axis system, each phase in a composite may be self-connected in zero, one, two, or three directions. For diphasic composites, there are ten connectivity patterns designated as 0-0, 0-1, 0-2, 0-3, 1-1, 1-2, 1-3, 2-2, 2-3, and 3-3. In the notation used here, the piezoelectric phase appears first.

During the past few years, a number of investigators have examined piezoelectric ceramic-polymer composites with different connectivity patterns. The method of preparation of these composites covers a wide spectrum of ceramic fabrication processes, and the piezoelectric properties of the composites depend to a large extent, on the connectivity pattern. In this paper, a brief summary of the piezoelectric properties of composite transducers with different connectivity is presented. A more extensive description of the work on other PZT-polymer composites can be found in recent review papers[5,6]. A schematic diagram of various types of composites with different connectivity is shown in Fig. 1.

COMPOSITES WITH 0-3 CONNECTIVITY

The simplest type of piezoelectric composite consists of a polymer matrix loaded with ceramic powder. In a composite with 0-3 connectivity, the ceramic particles are not in contact with each other while the polymer phase is self-connected in all three dimensions. In many ways the 0-3 composites is similar to polyvinylidene fluoride (PVF_2). Both consists of a crystalline phase embedded in an amorphous matrix, and both are reasonably flexible.

Early attempts to fabricate flexible composites of piezoelectric ceramic particles and polymers were made by Kitayama[7], Pauer[8] and Harrison[9]. The d_{33} coefficient of these composites were comparable with PVF_2, but the d_h value was lower than those of solid PZT and PVF_2 polymer (Table 1). To improve the properties of these composites Harrison[9] fabricated a composite with much larger PZT particles up to 2.4 mm. Here the particle size approaches the thickness of the composite, and since the PZT particles extend from electrode to electrode, near saturation poling can be achieved. The large rigid PZT particles can also transmit applied stress extremely well, leading to high d_{33} values when measurements are taken across the particles. Permittivity in this composite is lower than that of homogeneous PZT, resulting in an improved voltage coefficient.

An improved version of the 0-3 composite was synthesized by Banno[10,11]. Rather than using PZT as the ceramic filler, pure or modified lead titanate was employed because of its greater piezoelectric anisotropy. The lead titanate filler is produced by water-quenching the ceramic, thereby exploiting the high strain present in the material in order to produce fine powders. The average particle size was about 5μm. To fabricate composite bodies, the piezoelectric powders and chloroprene rubber were mixed and rolled into 0.5mm thick sheets at 40°C using a hot roller, and then heated at 190°C for 20 minutes under a pressure of 13 kg/cm^2. The composites were poled in a field of 100-150 kV/cm field for 30 minutes.

As shown in Table 1, the hydrostatic voltage coefficient g_h of pure $PbTiO_3$ composites is comparable to that of PVF_2 polymer. The d_h value of 35 pC/N was independent of pressure and g_h values were reduced only about 2% when pressure was increased to 40MPa[12].

Recently we have fabricated flexible composites with a more active piezoelectric material[13]. The piezoelectric ceramic used in these composites are $Pb_{1-x}Bi_xTi_{1-x}Fe_xO_3$ (PT-BF) and $0.5\ PbTiO_3-0.5\ Bi[Fe_yMn_{1-y}]O_3$ (Mn doped PT-BF) which has a very large spontaneous strain.

447

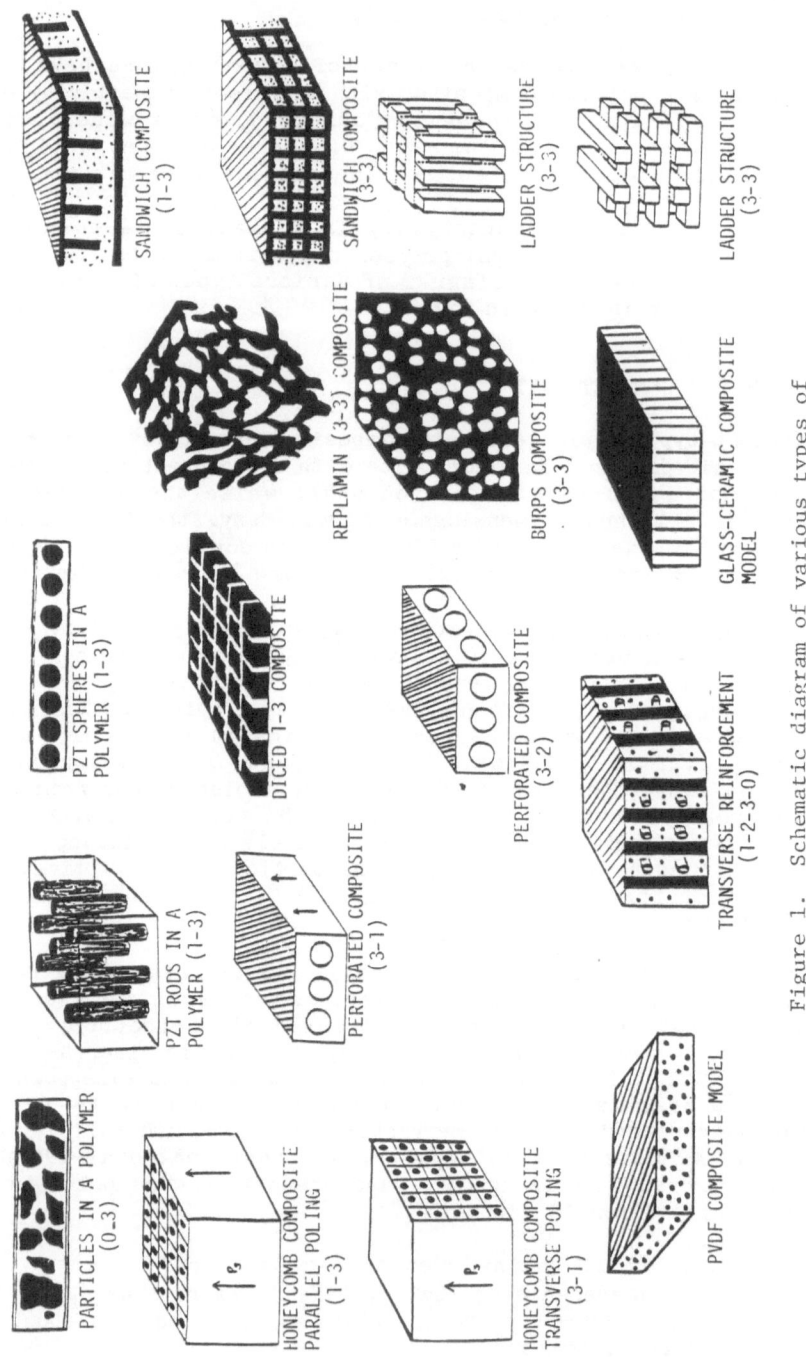

SANDWICH COMPOSITE (1-3)

SANDWICH COMPOSITE (3-3)

LADDER STRUCTURE (3-3)

LADDER STRUCTURE (3-3)

REPLAMIN (3-3) COMPOSITE

BURPS COMPOSITE (3-3)

GLASS-CERAMIC COMPOSITE MODEL

PZT SPHERES IN A POLYMER (1-3)

DICED 1-3 COMPOSITE

PERFORATED COMPOSITE (3-2)

TRANSVERSE REINFORCEMENT (1-2-3-0)

PZT RODS IN A POLYMER (1-3)

PERFORATED COMPOSITE (3-1)

PARTICLES IN A POLYMER (0-3)

HONEYCOMB COMPOSITE PARALLEL POLING (1-3)

HONEYCOMB COMPOSITE TRANSVERSE POLING (3-1)

PVDF COMPOSITE MODEL

Figure 1. Schematic diagram of various types of composites with different connectivity.

The spontaneous strain in $PbTiO_3$ is about 6%. In PZT compositions near the morphotropic boundary, it is about 2%. And since in BF-PT, the spontaneous strain is as large as 18%, we were not surprised to find a substantial increase in the hydrostatic voltage coefficients of the composites.

To fabricate the composites, the filler powder is synthesized from the system $Pb_{1-x}Bi_xTi_{1-x}Fe_xO_3$ for which there is a continuous solid solution across the entire composition range. The composition of the powders synthesized in this study lie in the range x = 0.5-0.7 and y = 0.005-.1. To prepare the filler powder, PbO, TiO_2, Bi_2O_3, Fe_2O_3 and MnO_2 were mixed and ball-milled with zirconia media. The oxides were subjected to a low temperature (700°C-800°C) primary calcination for 1.5 hours, followed by a second high temperature firing (950°C-1050°C). Water quenching produces an average particle size of 5μm. To fabricate the composites, piezoelectric ceramic powders and eccogel polymer[14] were mixed and calendered at 40°C. The calendered material is then cured at 80° under slight pressure. Composites were poled in an 80°C silicone oil bath by applying a field 100-120 kV/cm for 20 minutes. The poled composites exhibit outstanding hydrostatic sensitivity attaining values of g_h and d_hg_h well in excess of the values reported for pure $PbTiO_3$ composites (Table 1). The g_h and d_h values of these composites[12] remain virtually constant over a broad pressure range (Fig. 2).

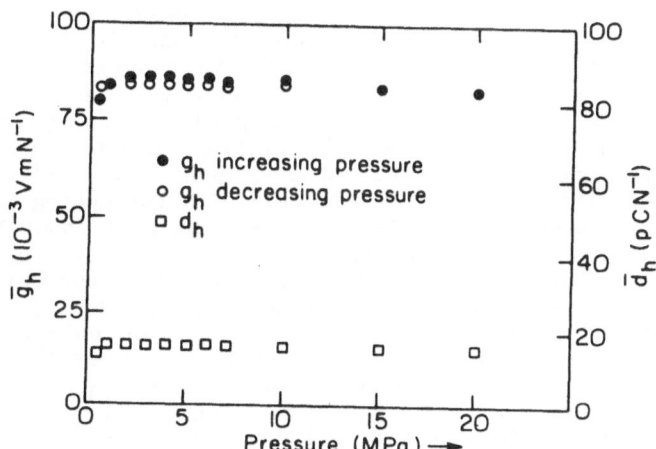

Fig. 2. Variation of g_h and d_h with pressure for $Pb_{0.5}Bi_{0.5}Ti_{0.5}Fe_{0.5}O_3$ -polymer composite.

Flexible 0-3 composite have also been developed at Bell Laboratories by Zipfel[15]. In this composite a polyurethane matrix material is mixed with 30-40 volume percent of a non-ferroelectric filler such as tartaric acid or lithium sulfate monohydrate. The liquid mixture is injected into an evacuated 3 inch diameter steel mold having two broad faces approximately 1.27mm apart. While the polymer is undergoing polymerization, the composite is polarized by applying hydrostatic pressure and an electric field. In this way the piezoelectric crystals suspended in the liquid copolymer rotate until they are electrically aligned. After polymerization, the resulting rubbery matrix holds the oriented particles in polar alignment for maximum sensitivity. The dielectric constant of this cable hydrophone is typically 4.2 and the piezoelectric sensitivity is stated to be 'comparable with most of the other piezoelectric materials.'

<u>Poling Method of 0-3 Composites</u>

As mentioned earlier, 0-3 composites prepared from PZT, PbTiO$_3$ and (Pb,Bi)(Ti,Fe)O$_3$ powders are poled at very large field strength (100-150 kV/cm) in order to achieve sufficient poling. The reason for the necessity of such large fields will be clear from the following discussion.

For a 0-3 composite consisting of spherical grains embedded in a matrix, the electric field E$_1$, acting on an isolated spherical grain is given by:

$$E_1 = \frac{3K_2}{K_1 + 2K_2} \, E_0.$$

In this equation, K_1 and K_2 are the dielectric constants of the spherical piezoelectric grains and the polymer matrix, respectively, and E$_0$ is an externally applied electric field. For a 0-3 composite of PZT powder and polymer, K_1 is about 2000 and K_2 about 5. In such a composite with an external field of 100 kV/cm, the electric field acting on the piezoelectric particles is only about 1 kV/cm which is insufficient to pole the composite. According to the above equation $E_1 \sim E_0$ only when the dielectric constant of the piezoelectric phase approaches that of the polymer phase. Most of the ferroelectric materials have very high dielectric constants and hence the above condition cannot be satisfied.

A different way to control the poling of 0-3 composites is to create a continuous electric flux path between the PZT particles. To do this, we have added a small volume fraction of a conductive third phase such as carbon, germanium, silver and silicon to the PZT-polymer composite. In preparing these composites, 68.5 volume percent PZT 501 and 1.5 volume percent carbon were mixed and dry ball-milled. After ball milling, the fillers were mixed with eccogel polymer and placed in a mold under pressure. It is found that the PZT-polymer composite with a small addition of a conducting phase can be poled in about five minutes under a field of 35-40 kV/cm at about 100°C. Fig. 3 shows the effect of the poling voltage on the d$_{33}$ values of the composite. It is observed that a duration of 5 minutes is sufficient for full poling of the composites. Similar composites were made using a powder of pure PbTiO$_3$ as a filler. It is found that in PbTiO$_3$ composites, the poling results of composites with germanium additives were better than carbon additives. Dielectric and piezoelectric properties of PZT and PbTiO$_3$ composites with conductive additives are shown in Table 1. The g$_h$ d$_h$ values of these composites are comparable with those of PVF$_2$, further details of the poling method will be reported elsewhere[16].

COMPOSITES BASED ON 1-3 CONNECTIVITY

Composites in which the piezoelectric ceramic is self connected one dimensionally and the polymer phase is self-connected three dimensionally were developed by Klicker[17]. In a 1-3 composite, PZT rods are embedded in a continuous polymer matrix. Under the idealized saturation in which the polymer phase is far more compliant than PZT, the stress on the polymer will be transferred to the PZT rods. The stress amplification on the PZT phase along with the reduced permittivity greatly enhances the piezoelectric voltage coefficient. To provide a better understanding of the composites, the piezoelectric properties were studied as a function of volume fraction PZT, rod diameter, and sample thickness.

The magnitude of the d$_h$g$_h$ product of 1-3 composites with PZT rods in polymer matrix is large, but far less than the theoretical value. Part of the reason is that the Poisson ratio of the polymer used is fairly high, thus an

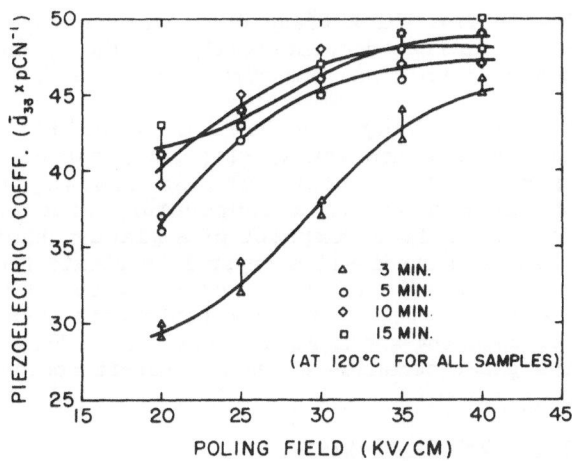

Fig. 3. Variation of piezoelectric d_{33} coefficient of PZT-carbon-eccogel composite with poling field.

internal stress exists which opposes the applied stress, and the stress amplification of a 1-3 connectivity model is greatly reduced. One way to reduce the Poisson ratio of a polymer is to introduce porosity into it. The fabrication of 1-3-0 composites with porous polyurethanes was developed by Klicker[18] who showed that porosity in the polyurethane matrix has a dramatic effect on d_h and g_h. Other types of 1-3-0 composites were studied by Lynn[19]. Porosity was introduced into different types of polymer matrices either by adding a foaming agent or by mixing commercial hollow glass spheres with the polymer. Using a foamed REN epoxy, composites with 4 volume % of 280μm diameter rods showed a three-fold increase in d_h above the unfoamed REN Epoxy. When hollow glass spheres of an average diameter of 80μm and 2μm wall thickness were mixed in REN and Spurrs epoxy, d_h increased about twice that of the ordinary epoxy composites. For all the composites with glass spheres, no pressure dependences were found, while in foamed REN epoxy composites, the pressure dependence remains a problem.

Transverse reinforcement is another technique used to enhance the hydrostatic piezoelectric coefficient[20,21]. These composites were made by mounting two types of fibers in the polymer matrix: PZT rods parallel to the poling direction and stiff glass fibers in the two transverse directions. The glass fibers carry most of the transverse stress thereby decreasing d_{31} without appreciable reduction in d_{33}. Transversely reinforced composites have 1-2-3 connectivity, or 1-2-3-0 connectivity if the polymer matrix is foamed. In many of these polymer ceramic composites a coupling agent is required to achieve good stress transfer.

Savakus[22] developed a simplified preparation technique for making composites with 1-3 connectivity. Composite piezoelectric transducers have been constructed by partially dicing PZT ceramics and back-filling with epoxy. Composites containing 10 to 70 volume percent PZT were prepared with several different rod diameters. Measured dielectric constants ranged from 200 to 1000, longitudinal piezoelectric coefficients d_{33} from 200 to 350 pC/N, and hydrostatic piezoelectric coefficients d_h from 40 to 80 pC/N. When diced

PZT ceramic capped with 2mm thick brass or stainless steel and encapsulated with alumina, d_h and g_h increased dramatically[23]. The g_h and d_h of this type of composite are 75 x 10^{-3} vm/N and 265 pC/N respectively.

Recently extensive study has been made of the dielectric and piezoelectric properties of a new family of polar glass ceramics[24,25]. Grain oriented glass-ceramics of $Li_2Si_2O_5$ and $Ba_2TiSi_2O_8$ are prepared by crystallization of the glasses in a temperature gradient. They may be regarded as diphasic composites composed of a glassy phase and one or more crystalline phases, wherein needle-like crystals nucleate from the surface and grow into the bulk of the sample in the direction of the temperature gradient. g_h and g_hd_h of these tested sample are comparable with those of PVF_2 polymer. The usual problems of depoling and aging encountered in PVF_2 and most ceramics are avoided with the glass-ceramics which are non-ferroelectric and hence do not require poling.

Composites with 3-1 and 3-2 Connectivity

Composites of PZT and polymer with 3-1 and 3-2 connectivity patterns have been fabricated[26] by drilling holes in sintered PZT blocks and filling the holes with epoxy. The influence of hole size and volume fraction PZT on the hydrostatic properties of the composite was evaluated. By decoupling the piezoelectric d_{33} and d_{31} coefficients in the composite, the hydrostatic coefficients are greatly enhanced. On samples optimized for hydrophone performance, the dielectric constants of 3-1 and 3-2 composites are 600 and 300 respectively. For two typical composites, the piezoelectric coefficients d_h, g_h, and g_hd_h for 3-1 composites are 170 (pC/N), 30 (X10^{-3} Vm/N), and 5000 (10^{-15} m^2/N) respectively, and the corresponding values for 3-2 composites are 200 (pC/N), 60 (10^{-3} Vm/N), and 12000 (10^{-15} m^2/N).

The composites are extremely rugged and show no pressure dependence up to 7MPa (1000 PSI). Recently perforated PZT blocks were fabricated by injection molding of the ceramic[27] and backfilled with polymer. g_h and g_hd_h of these composites fifteen and fifty times larger than those of PZT ceramics respectively. Similar composites can be made[28] by extruding the ceramic rather than drilling. Composites with 3-1 connectivity were fabricated by impregnating an extruded, sintered honeycomb configuration of PZT with epoxy. The composites had lower density (=3000 Kg/m^3) and lower dielectric constant (~400) than that of solid PZT. The maximum piezoelectric d_{33} coefficient of the composites was 350 pC/N. g_h and d_hg_h values of the composites were an order of magnitude higher than those of solid PZT.

The study of the 3-1 composites was carried one step further by developing a small hydrophone containing a number of perforated 3-1 elements[29]. Sixteen cubic elements were prepared with each measuring 0.4 x 0.4 x 0.4cm with a 0.2mm diameter hole filled with epoxy, the elements were encapsulated in a flexane polyurethane which was modified by foaming with nitrogen gas and mixing with 55% glass microballoons. The specimen exhibited a g_h value of 34 x 10^{-3} Vm/N which is comparable with that of single element composite.

COMPOSITES WITH 3-3 CONNECTIVITY

In a 3-3 composite each of the constituent phases is continuously self-connected in three dimensions to give two interlocking skeletons in intimate contact with one another. This type of structure is exhibited by certain polymer foams, by some phase-separated metals and glasses, by three-dimensional waves, and by natural substances such as wood and coral. The piezoelectric and pyroelectric properties of 3-3 composites have been investigated with some rather remarkable results. For certain coefficients,

452

dramatic improvements can be made over the best single-phase piezoelectrics.

Piezoelectric ceramic-polymer composites with 3-3 connectivity were first made by Skinner[30] using a lost-wax method with coral as a starting material. Among the advantages of these composites are high hydrostatic sensitivity, low dielectric constant, low density for improved acoustic impedance matching with water, high compliance to provide damping, and the mechanical flexibility needed to develop conformable transducers. Shrout[31] developed a simpler method for fabricating a three-dimensionally interconnected lead zirconate-titanate (PZT) and polymer composite with properties similar to the coral-based composites. The simplified preparation method involves mixing plastic spheres and PZT powder in an organic binder. When carefully sintered, a porous PZT skeleton is formed, and later back-filled with polymer to form a 3-3 composite. This technique is commonly referred to as the BURPS process, an acronym for burned-out plastic spheres. Since the process involves the generation and emission of gaseous hydrocarbons, the name BURPS is highly appropriate. Scientists at Mitsubishi Mining and Cement have developed several technqiues for introducing connected porosity in PZT ceramics: reactive sintering, foaming agents, organic additives, and careful control of particle size and firing conditions[32]. The porosity of the samples tested for hydrophone performance was about 50%. The average pore/size was about 100μm. g_h and $d_h g_h$ coefficient of this sample was about 50×10^{-3} Vm/N and 90 xpC/N respectively[33]. The g_h showed very slight decrease when the applied pressure was increased up to 70MPa and upon releasing the pressure, the g_h values were recovered.

SUMMARY

The dielectric and piezoelectric properties of several different types of composites are described and their figures of merit for hydrophone applications ($d_h g_h$) are discussed. Hydrophones are used at low frequencies where the acoustic signal has a wavelength much larger than the scale of the macrostructure of the composite. It is shown that the hydrostatic voltage coefficient g_h and figure of merit $g_h d_h$ of ceramic-polymer composites are an order of magnitude higher than those of single phase materials.

REFERENCES

1. G. Goodman, 'Ferroelectric Properties of Lead Metaniobate,' J. Am. Ceram. Soc. 36:368 (1953).
2. T.-Y. Tien and W.G. Carlson, 'Effect of Additives on Properties of Lead Titanate,' J. Am. Ceram. Soc. 45:567 (1962).
3. Y. Wada and R. Hayakawa, 'Piezoelectricity and Pyroelectricity of Polymers,' Japan J. Appl. Phys. 15:2041 (1976).
4. R.E. Newnham, D.P. Skinner and L.E. Cross, 'Connectivity and Piezoelectric-Pyroelectric Composites,' Mat. Res. Bull. 13:525 (1978).
5. R.E. Newnham, A. Safari, J. Giniewicz and B.H. Fox, 'Piezoelectric Sensors,' Ferroelectrics 60:15 (1984).
6. R.E. Newnham, A. Safari, G. Sa-gong and J. Giniewicz, 'Flexible Composite Piezoelectric Sensors,' IEEE Ultrasonic Symposium Proceedings, 501 (1984).
7. T. Kitayama and Sugawara, 'Flexible Piezoelectric Materials,' Rep. Proc. Gr. Inst. Elec. Comm. Eng. Japan, CPM27-17 (1972).
8. L.A. Pauer, 'Flexible Piezoelectric Materials,' IEEE Int. Conf. Res., p. 1-5 (1973).
9. W.B. Harrison, 'Flexible Piezoelectric Organic Composites,' Proc. of the Workshop on Sonar Transducer Materials, Naval Research Labs., (Feb. 1976).
10. H. Banno and S. Saito, 'Piezoelectric and Dielectric Properties of

Composites of Synthetic Rubbers and $PbTiO_3$ and PZT,' Japan. J. Appl. Phys. 22: supp. 22-2, 67 (1983).

11. H. Banno, 'Recent Developments of Piezoelectric Ceramic Products and Composites of Synthetic Rubber and Piezoelectric Ceramic Particles,' Ferroelectrics, 5 (1983).

12. R.Y. Ting, 'Evaluation of New Piezoelectric Composite Materials for Hydrophone Applications,' Ferroelectrics (to be published).

13. J. Giniewicz, '(Pb,Bi)(ti,Fe)O_3-Polymer Comspoite Materials for Hydrophone Applications,' M.S. Thesis, The Pennsylvania State University (1985).

14. Eccogel 1365 Series (Emerson and Cumming, Densey and Almy Chemical Division, W.R. Grace and Co.).

15. G.G. Zipfel, '0-3 Piezocomposite,' Bell Labs. Record., April. 1983, p. 11-13.

16. G. So-gong, A. Safari, R.E. Newnham, 'Easily poled 0-3 composites,' Ferroelectrics (to be published).

17. K.A. Klicker, 'Piezoelectric Composite with 3-1 Connectivity for Transducer Applications,' Ph.D. Thesis, The Pennsylvania State University (1980).

18. K.A. Klicker, J.V. Biggers and R.E. Newnham, 'Composites of PZT and Epoxy for Hydrostatic Transducer Applications,' J. Am. Ceram. Soc. 64:5 (1982).

19. S.Y. Lynn, 'Polymer-Piezoelectric Ceramic Composites with 3-1-0 Connectivity for Hydrophone Applications,' M.S. Thesis, The Pennsylvania State University (1982).

20. M. Haun, 'Transverse Reinforcement of 1-3 and 1-3-0 PZT-Polymer Piezoelectric Composites with Glass Fibers,' M.S. Thesis, The Pennsylvania State University (1983).

21. M. Haun, P. Moses, T.R. Gururaja and W.A. Schulze, 'Transversely Reinforced 1-3 and 1-3-0 Piezoelectric Composites,' Ferroelectrics 49:259 (1983).

22. H.P. Savakus, K.A. Klicker and R.E. Newnham, 'PZT-Epoxy Piezoelectric Transducers: A Simplified Fabrication Procedure,' Mat. Res. Bull. 16:677 (1981).

23. A. Safari, R.E. Newnham and L.E. Cross, 'Diced, Capped and Encapsulated PZT Composite,' (applied for patent).

24. A. Halliyal, A. Safari, A.S. Bhalla, R.E. Nenwham and L.E. Cross, 'Grain-Oriented Glass-Ceramic for Piezoelectric Devices,' J. Am. Ceram. Soc. 67:331 (1984).

25. R.Y. Ting, A. Halliyal and A.S. Bhalla, 'Polar Glass Ceramic For Sonar Transducers,' J. Appl. Phys. Lett. 44:9 (1984).

26. A. Safari, Perforated PZT-Polymer Composites with 3-1 and 3-2 Connectivity for Hydrophone Applications,' Ph.D. Thesis, The Pennsylvania State University (1983).

27. I. Kalnin and R. Hughes, 'Preparation of Perforated PZT by Injection Molding,' (applied for patent).

28. A. Safari, A. Halliyal, R.E. Newnham and I.M. Lachman, 'Transverse Honeycomb Composite Transducers,' Mat. Res. Bull. 17:301 (1982).

29. R.Y. Ting, 'Evaluation of New Piezoelectric Composite Materials for Hydrophone Applications,' Ferroelectrics (to be published).

30. D.P. Skinner, R.E. Newnham and L.E. Cross, 'Flexible Composite Transducers,' Mat. Res. Bull. 13:599 (1978).

31. K.R. Rittenmyer, T.R. Shrout and R.E. Newnham, 'Piezoelectric 3-3 Composites,' Ferroelectronics 41:189 (1982).

32. R.Y. Ting, 'Evaluation of New Piezoelectric Composite Materials for Hydrophone Applications,' Ferroelectrics (to be published).

WAVE ABSORPTION IN PIEZOCERAMIC-POLYMER COMPOSITES

A.E. Semple, S.M. Pilgrim, W. Thompson, Jr., and R.E. Newnham

The Pennsylvania State University
University Park, PA   16802

INTRODUCTION

As new materials have become more difficult and expensive to develop, composites have received increased interest because they provide fewer development problems and often display better properties than single phased materials. Proper design of a composite involves several considerations and principles: properties of the constituents, connectivity of the individual phases, general phase connectivity, and possible cross-coupled properties. In the past, these principles have been profitably applied to optimizing electrical flux patterns and mechanical stress distributions to improve the hydrostatic strain and voltage coefficients of piezoelectric composites.

The most easily fabricated composites are those of the 0-3 and 0-0-3 varieties which consist of discrete particles (one or two types) within a three-dimensionally connected matrix. These composites show possible application as passive or active absorbers. In order to assess the pertinent parameters of these materials when used as passive absorbers, improvements to a measurement technique known as the Transfer Function Method were developed.

The Transfer Function Method of determining the complex dynamic modulus, i.e., storage modulus plus loss factor, is particularly amenable for testing polymeric materials. It consists of exciting a mass-loaded rod into longitudinal vibrations at a given frequency. The complex acceleration ratio between the two ends is related to the complex modulus by a pair of coupled, transcendental equations derived by applying appropriate boundary conditions to the solution of the longitudinal wave equation. The method does not require a long sample, and unlike methods employing the well-known -3 dB bandwidth technique, it is particularly suited to testing high-damping materials.

However, the Transfer Function Method, in the form used by many practitioners, does have some drawbacks. First, dynamic modulus values can only be computed at the resonance frequencies of the sample. This usually means only two to five data points can be obtained in the audio frequency range. Second, the solution of these transcendental equations becomes very difficult at higher frequencies. Recent improvements made to the Transfer Function Method have substantially eliminated these two problems. Loss factor and storage modulus results can now be obtained at any frequency, provided the phase angle of the acceleration ratio does not equal 0°.

Materials selection for the preparation of test samples included a variety of polymer matrices and ceramic powders. In order to determine the utility of piezoelectric powders in enhancing the damping properties in the audio range (20 Hz –20 kHz) some understanding of the possible loss mechanisms in 0-3 and 0-0-3 composities is essential. An attempt to separate and clarify the different contributions by proper material selection was made. Composites containing 10 to 60 volume-percent of ceramic powder, incorporated in various polymeric matrices, have been produced and their loss factors have been measured.

LOSS MECHANISMS

The number of possible loss mechanisms in 0-3 and 0-0-3 composities is very large; however, many are of minor importance in the frequency range investigated here. The possible mechanisms may be categorized into three groups: particle effects (intra and inter), particle-matrix effects, and matrix effects. In most commercial materials the matrix effects are the most important. Consequently, polymeric materials that show a glass-rubber transition in the frequency and temperature range of interest (e.g., Eccogel and polychloroprene) were selected for use as matrix materials.

Many particle effects are possible within the 0-3 composites. Losses due to coupling to the piezoelectric soft modes and interparticle cavity resonances are probably negligible at the frequencies and wavelengths employed. More important effects include deagglomeration (which can be reduced by milling and sieving), particle dewetting, and particle-particle friction. Additionally, loss mechanisms based on interactions of the matrix and ceramic are possible. These include: particle-matrix stress effects (from normal thermal expansion mismatch or from residual thermal stresses), matrix conductivity losses (for piezoelectric particles), and thermal stress induced polarization. All of these mechanisms are likely and possibly of important magnitude. Losses through mechanisms associated with pores are also possible; however, proper sample preparation can reduce porosity to less than 2%.

With the large number of available loss mechanisms, it should be possible to overlap the frequency and/or temperature regions of these loss mechanisms to produce a broadband absorber. Such a broadband absorber should possess a good characteristic impedance match to the surrounding medium, and have overlapping peaks in the frequency response of the dielectric constant, the piezoelectric charge coefficient and the compliance. Of the matrices chosen, two (polyethylene and polypropylene) are thermoplastics with glass transition temperatures below room temperature and two (Eccogel and Spurrs epoxy) are thermosets with transition temperatures near or above room temperature. The ceramic powders used were TiO_2 (anatase) from WCD and PZT501A from Ultrasonic Powders. These were chosen to provide a nonpiezoelectric and a piezoelectric ceramic, both with well characterized properties.

SAMPLE PREPARATION

Composite preparation followed two basic routes, one for thermoplastic matrices, and one for thermoset matrices. Only 0-3 composites were prepared. The ceramic powders TiO_2 and PZT501A were sieved to sub 325 mesh prior to incorporation into the matrix. For thermoplastic matrices, the components were first dry mixed by hand and then mixed in a Brabender mixer for 20 minutes at 40 rpm. Mixing of polyethylene samples was done at 155°C and polypropylene samples, at 195°C. After removal from the mixer, the raw composites were hot pressed in a rectangular mold (1cm x 1cm x 8cm) at corresponding temperatures with pressures of 25 to 35 MPa for 45 to 60 minutes. Composites were then slowly cooled under pressure (25 MPa) to room temperature.

Preparation of composite samples with thermoset matrices followed two basic schemes. A few samples were simply hand mixed by sieving the powder into the polymer precursor, with subsequent hand mixing with a spatula. The majority of these samples, however, were then mixed in the Brabender mixer for 20 minutes at 40 rpm. After mixing, the composites were poured into Teflon molds with minimal enfolding of air. Temperature and cure times were kept constant at 80°C for 8 hours.

Note that none of the composites containing PZT have, to-date, been effectively polarized. The ability to accomplish this remains an active area of research. Hence, in all of the composite samples tested during this study, the PZT merely functions as a dense filler similar to lead or ferrite particles in other high-damping materials.

TRANSFER FUNCTION METHOD

Many investigators including Capps [1], Pritz [2] and Madigosky and Lee [3] have used the Transfer Function Method to determine the complex dynamic modulus of materials. Norris and Young [4] analyzed the case when a rod-like specimen was loaded with an end mass. This is of practical importance since it is common to mount an accelerometer with non-negligible mass on the undriven end to monitor the acceleration level there.

The sample is modeled as a slender rod of length L, density ρ, and constant cross-sectional area, A. The axial displacement at any position x is u(x). The longitudinal wave equation, Eq. (1), relates the displacement to the complex elastic modulus and the density, ρ, of the specimen.

$$(E' + iE'')(\partial^2 u/\partial x^2) = \rho(\partial^2 u/\partial t^2) \tag{1}$$

The ratio of the loss modulus, E'', to the storage modulus, E', defines the loss factor η. Applying the boundary conditions that exist at the ends of the rod to the solution of Eq. (1), one can solve for the displacement ratio of the driven and mass-loaded ends. The resulting expression can be separated into real and imaginary parts (T_R and T_I) as shown in Eqs. (2) and (3).

$$T_R = \cosh[\xi\Omega] \ (\cos\xi - R\xi \sin\xi) + R\xi\Omega \cos\xi \ \sinh[\xi\Omega] \tag{2}$$

$$T_I = \sinh[\xi\Omega] \ (\sin\xi + R\xi \cos\xi) + R\xi\Omega \sin\xi \ \cosh[\xi\Omega] \tag{3}$$

where,

R = the mass ratio, $M/\rho AL$

M = the end mass

ξ = the frequency parameter, $\omega L/c$

$\Omega = \tan(\delta/2)$

δ = the angle by which E'' lags E'

c = the complex phase velocity, $(E^*/\rho)^{1/2}\sec(\delta/2)$

$E^* = E'(1 + i\eta)$

To obtain the loss factor and storage modulus, the solutions for ξ and Ω (or, equivalently, δ) are inserted into Eqs. (4) and (5).

$$\eta = \tan(\delta) \tag{4}$$

$$E' = \rho(\omega L/\xi)^2 \cos^2(\delta/2)\cos(\delta) \tag{5}$$

Knowing T_R, T_I, and R, the solution of Eqs. (2) and (3) for ξ and Ω is usually obtained by an iterative procedure such as the Newton-Raphson technique. Difficulties in using the Transfer Function Method are linked to problems associated with obtaining the solution for ξ and Ω.

EXPERIMENTAL PROCEDURE

The test setup is shown in Fig. 1. The bar-like sample is harmonically vibrated along its axis, at any frequency of interest, by an Electrodyne shaker. Two BBN 501 Accelerometers, mounted on either end of the sample, supply voltage signals proportional to the end accelerations or, equivalently, the displacements. Each signal is amplified and fed into a HP 3570A Network Analyzer. The Network Analyzer computes the amplitude and phase of the ratio of the two inputs. A HP 9825B computer controls the equipment and also performs the necessary calculations. The computer is linked to both a display screen and pen plotter to facilitate outputting the results in graphic form. A thermal printer outputs the results in tabular form during the tests.

The shaker-specimen fixture, shown in Fig. 2, is a tripod structure with neoprene gaskets at the connecting bolts to reduce lateral structure-borne vibrations. The test stand is small enough to fit inside a Tenney Temperture Chamber should tests at various temperatures be desired. The sample is connected to the shaker via a small rectangular stirrup which threads into the shaker. One accelerometer is mounted inside this stirrup while the other accelerometer is attached to a metal end mass.

IMPROVEMENTS TO THE TRANSFER FUNCTION METHOD

As previously stated, several problems occur when using the Transfer Function Method. In theory, Eqs. (2) and (3) can be solved at any frequency. In practice, convergence problems, in the iteration procedure, reduce the valid solutions to just those obtained at the longitudinal resonance frequencies. An additional problem is that the solutions obtained at higher

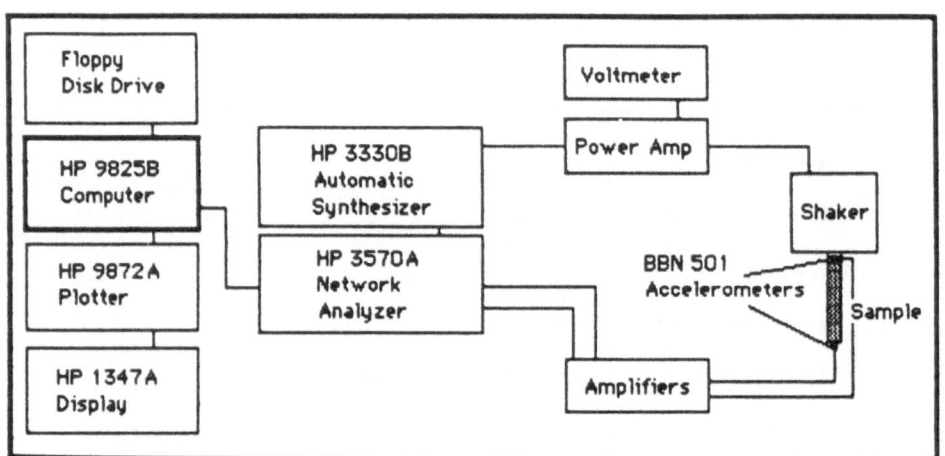

Fig. 1. Equipment setup for the Transfer Function Method tests.

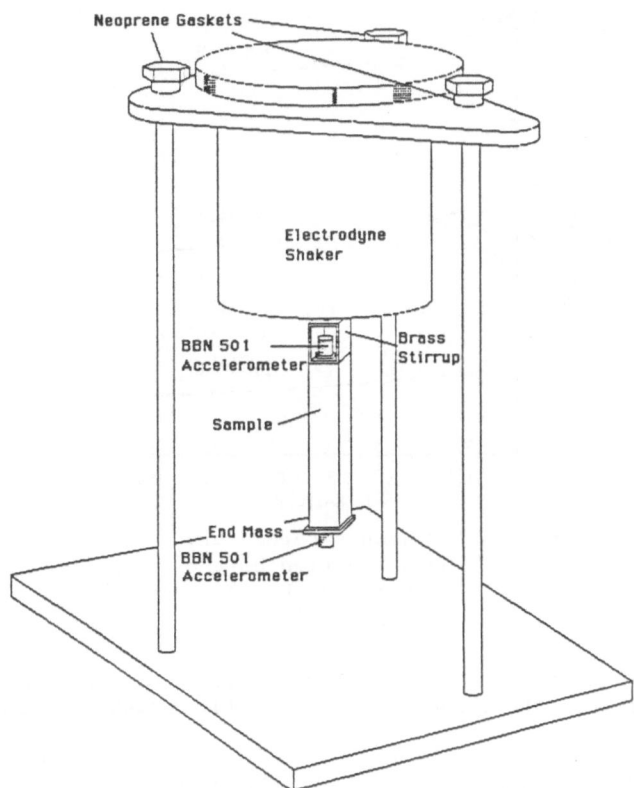

Fig. 2. Tripod shaker-specimen test stand.

mode resonances are often incorrect. In its original form the Transfer
Function Method can not provide the desired accuracy or repeatability.

Improvements made to the method during this study enable the equations to
be solved at any frequency except for those where the phase of the acceleration ratio is zero. The methodology behind these improvements was obtained by
investigating and understanding the behavior of Eqs. (2) and (3).

As the frequency increases the frequency parameter ξ also increases. The
real and imaginary components of the acceleration ratio alternately rise and
fall in magnitude with increasing frequency. The frequencies where the real
component, T_R, becomes equal to zero correspond to longitudinal resonances of
the bar. Likewise, the imaginary component, T_I, will nearly obtain a local
maximum or minimum at these frequencies. By correlating the behavior of the
two equations to the physical response of the sample, it should be possible to
determine the proper region of convergence for the iterative solution
procedure.

An improved "seed" algorithm has been developed for the starting point of
the iteration procedure, which eliminates problems with false roots and also
allows the equations to be solved at any frequency between adjacent mode
resonances. Two empirical equations comprise this improved "seed" algorithm.
The first approximates a value for ξ at the first mode resonance. It turns
out that T_R and T_I are nearly independent of the damping parameter, Ω, at the
resonances. However, the equations are still affected by changes in the value
of R. The dependency of the frequency parameter, ξ, on the mass ratio, R, is
shown in Fig. 3. Since the value for R is known and the effect of Ω can be
neglected, as far as locating the convergence region, then ξ is the control-

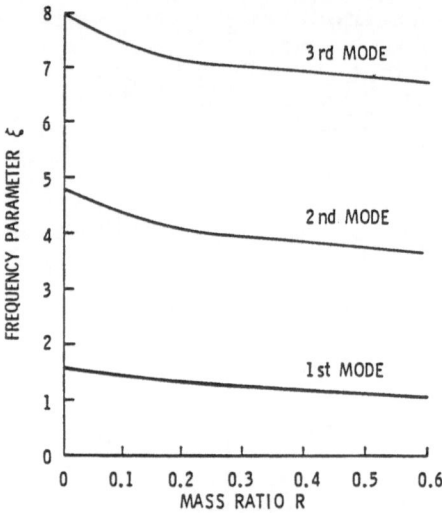

Fig. 3. Frequency Parameter, ξ,
as a function of the
mass ratio, R, for
$\tan(\delta/2) = 0.1$.

ling parameter when seeding the iteration procedure. The following empirical
equation approximates the value of ξ for the first mode resonance very
closely:

$$\xi_1 = 0.002R^2 - 0.132R + 1.254 \tag{6}$$

Once the values for Ω and ξ at the first mode resonance are determined, a
second equation facilitates solving for the ξ values at higher mode resonances. While at the first mode resonance the components of the acceleration
ratio are strong functions of ξ and R, at higher mode resonances they become
strong functions of ξ and n (mode number). The value of ξ for successive mode
resonance are very nearly separated by a value of π. The slope in Fig. 4 at

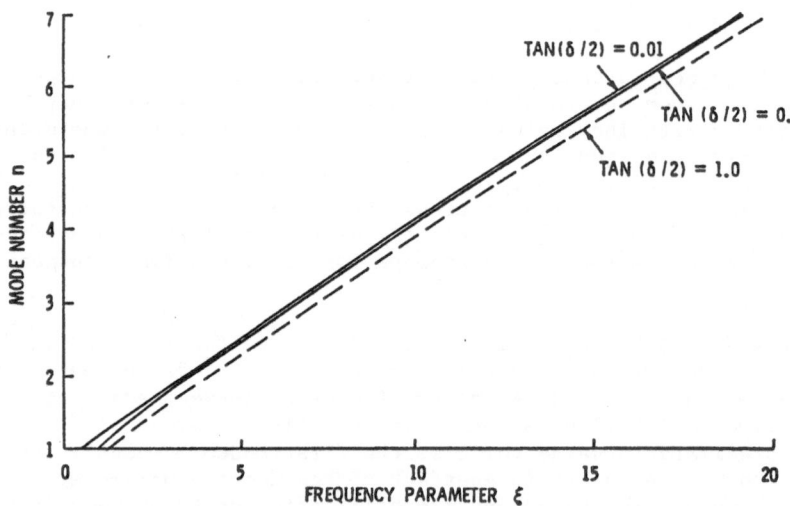

Fig. 4. Mode number, n, versus the frequency parameter,
ξ, at different damping levels for mass ratio,
R = 1.0.

higher frequencies (larger values of ξ) is close to π. The empirical equation used to seed the iterative procedure in the neighborhood of the nth mode resonance is:

$$\xi_n = (n - 1)\pi + \xi_1/n \qquad\qquad (7)$$

This approximation yields ξ values which are usually within 3% of the correct solution. For very high order modes, the error is usually within 0.5%, i.e., the error decreases with increasing mode order. It is interesting to note that the improvements made to the Transfer Function Method are most significant at high frequencies which is where the original form of the method was the most inaccurate.

With reliable results at the resonances, it is not difficult to interpolate the seed values for intermediate frequencies. This allows results to be obtained across the entire frequency spectrum except for frequencies where the phase angle of the acceleration ratio approaches zero. The solution for ξ and Ω becomes complicated when T_I equals zero (phase = 0°) because an additional solution of Eqs. (2) and (3) is always Ω equal to zero. The practice invoked here has been to disregard results where the phase angle is within 2° of zero or 180°.

RESULTS

Results are presented here for Eccogel and polychloroprene samples. Eccogel possesses high-damping properties. A plot of the storage modulus and loss factor for Eccogel without the addition of PZT is shown in Fig. 5. Notice the characteristic peak showing the frequency location of the rubber-glass transition. The addition of 10% unpolarized PZT by volume (Fig. 6) increases the storage modulus and decreases the loss factor. The addition of dense particulates into a polymer matrix will always have this effect.

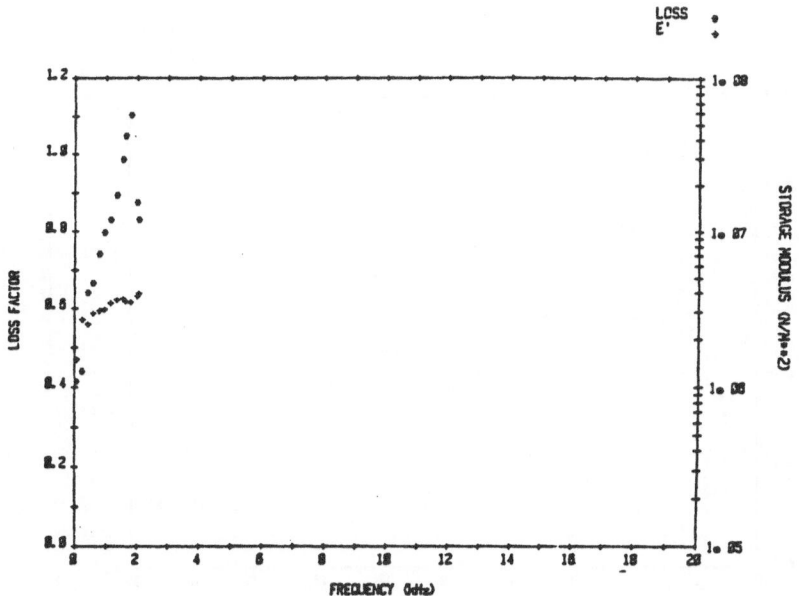

Fig. 5. Loss factor and storage modulus versus
frequency for an Eccogel sample.

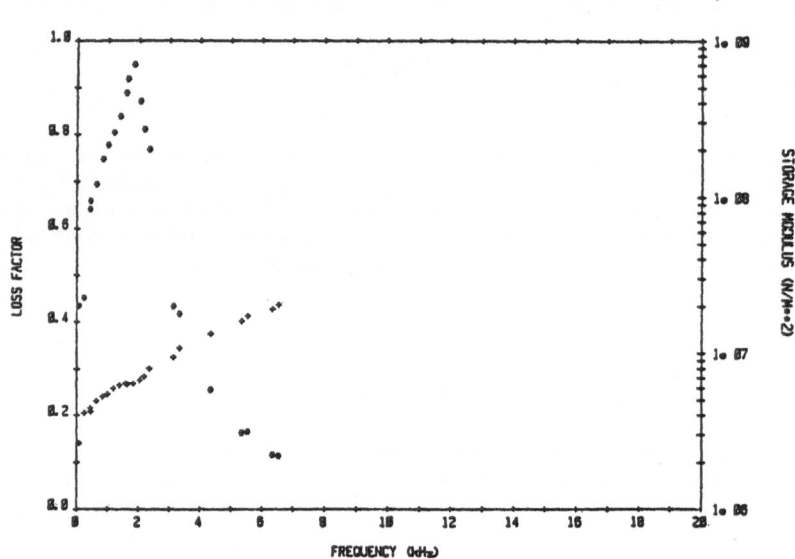

Fig. 6. Loss factor and storage modulus versus
frequency for an Eccogel/10% PZT sample.

Fig. 7 shows the components of the complex modulus for a pure polychloroprene sample. The damping factor values are close to those of the
Eccogel sample. This is to be expected. The transition peak occurs near
6 kHz. No results are shown above 7 kHz because the mass-loaded end acceleration levels were so low that they approached the noise floor of the
electronic equipment. The test was stopped when this noise floor was reached.

The plot of the results for polychloroprene with 25% unpolarized PZT
(Fig. 8) provides an interesting comparision with those for the pure

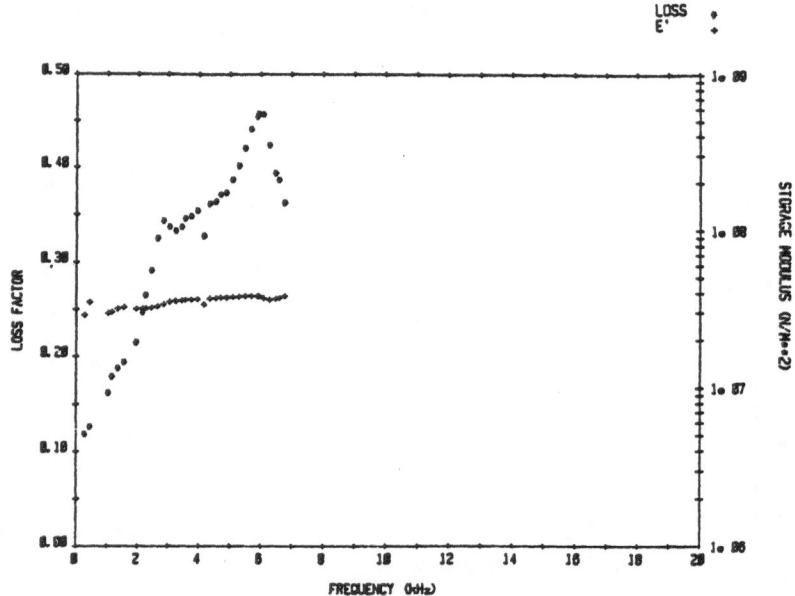

Fig. 7. Loss factor and storage modulus versus
frequency for a polychloroprene sample.

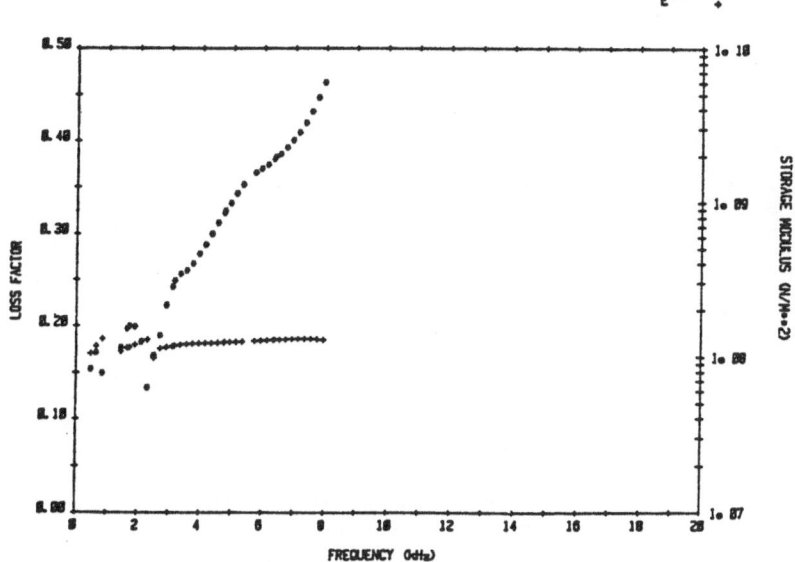

Fig. 8. Loss factor and storage modulus versus
frequency for a polychloroprene/25% PZT
sample

polychloroprene sample. The loss factor levels are approximately the same for
the two samples. This suggests that increased damping due to the particle-
matrix interaction offsets the decrease in damping which occurs due to the
addition of dense, low-damping particulates.

CONCLUSIONS

The improvements to the Transfer Function Method greatly increase the
accuracy and frequency range of the method. Tests can be run very quickly
once the sample is mounted to the shaker assembly. A full characterization of
the damping properties of most any material, in the acoustic frequency range,
can be obtained through the use of this test procedure.

The damping values for the composites can be grouped into two general
categories, high-damping materials ($\eta > 0.1$), and lower-damping materials ($\eta < 0.1$). All Eccogel and polychloroprene samples, including those with the
filler added, were high-damping. The polypropylene, polyethylene, and Spurrs
epoxy samples had loss factors between 0.01 and 0.1.

REFERENCES

1. R. N. Capps, "Dynamic Young's Moduli of Some Commercially Available
   Polyurethanes," J. Acoust. Soc. Am., 73(6), 2000-2005, (1983).

2. T. Pritz, "Transfer Function Method for Investigating the Complex Modulus
   of Acoustic Materials: Rod-Like Specimen," J. Sound Vib., 81(3), 359-376,
   (1982).

3. W. M. Madigosky and G. F. Lee, "Improved Resonance Technique for Materials
   Characterization," J. Acoust. Soc. Am., 73(4), 1374-1377, (1983).

4. D. M. Norris, Jr. and W. C. Young, "Complex Modulus Measurement by
   Longitudinal Vibration Testing," Exp. Mech., 10, 93-96, (1970).

STRUCTURED MACROVOIDS IN CERAMIC PZT

Manfred Kahn and Beatrice Kovel

Naval Research Laboratory
Washington, DC 20375-5000

Introduction

The hydrostatic response of PZT is only about 10% of its uniaxial response, $d_h = d_{33} - d_{31} - d_{32}$

$$= d_{33} \left(1 - \frac{d_{31} + d_{32}}{d_{33}}\right)$$

where $\frac{d_{31} + d_{32}}{2\, d_{33}}$ is about 0.4.

A reduction of this ratio will then raise d_h.

The piezoelectric charge output of PZT is fundamentally a response to strain. The phenomological Gibbs function for dense, tetragonal PZT[1] under hydrostatic pressure as well as experimental data from various sources have been used previously to show that[2]

$$\frac{d_{31}}{d_{33}} = \frac{S_{12}}{S_{11}} = \gamma$$

where S are compliances and γ is the Poisson Ratio (also ∼ 0.4 for dense PZT).

Composites containing PZT have higher d_h values. This has been the object of considerable attention[3,4]. Its success can be attributed to the fact that a composite, if considered a black box, can have a Poisson Ratio that is significantly less than that of dense PZT.

The minimization of the Poisson Ratio has been implemented in various configurations. The present work is an extension of the use of disordered anisotropic porosity[5] to the implementation of ordered anisotropic pores or voids.

A model of such a structure is shown in Fig. 1. It is more compressible at 90° to the longest pore dimension than parallel to it. Under hydrostatic conditions the strain ε_3 then becomes larger than ε_1. The hydrostatic strain can then be expressed as

$$\varepsilon_h = \varepsilon_3 - \varepsilon_{31} - \varepsilon_{32}.$$

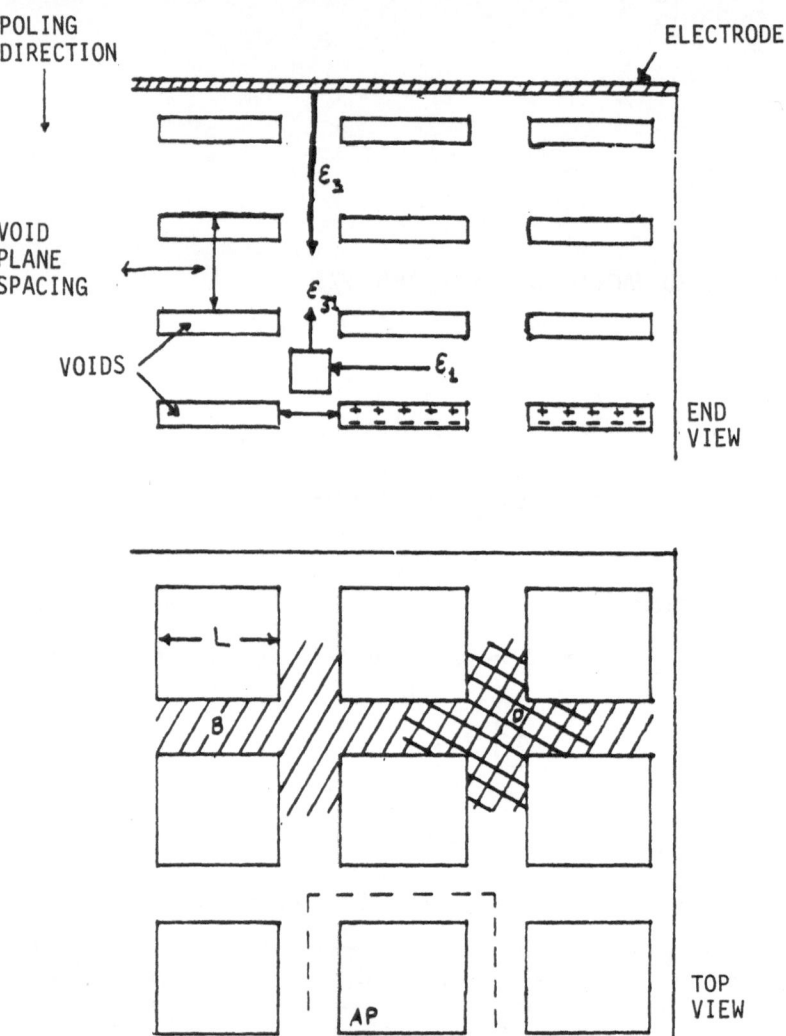

POLING DIRECTION

ELECTRODE

VOID PLANE SPACING

VOIDS

END VIEW

TOP VIEW

Figure 1. Rectangular pore model

When $\varepsilon_{31} = \varepsilon_{32}$

$$\varepsilon_h = \varepsilon_3 \left(1 - 2\frac{\varepsilon_{31}}{\varepsilon_3}\right)$$

$$= \varepsilon_3 \left(1 - 2\frac{\varepsilon_{31}}{\varepsilon_1} \cdot \frac{\varepsilon_1}{\varepsilon_3}\right)$$

$$= \varepsilon_3 \left(1 - 2\gamma \cdot a\right)$$

where $a = \frac{\varepsilon_1}{\varepsilon_3}$ and $\gamma = \frac{\varepsilon_{31}}{\varepsilon_1}$ is the Poisson Ratio of the dense material in between the voids. Since $a < 1$, ε_h and therefore d_h of this structure are higher than if it had isotropic pores.

466

The implementation of the model shown in Fig. 1 with dense external PZT walls provides voids that are sealed against ambient pressure. The empty voids exhibit a relatively high compliance since there is no need to fill them with a polymer or other second phase material. The latter could be a source of pressure dependent elasticity, thermal coefficient of expansion mismatches or, in the case of ceramic-ceramic composites, diffusion and sintering control problems.

Optimum composite design implies capability to shape and locate the second phase in the composite at will. This is done by the application of computer graphics that are used to prepare photolitographic deposition patterns. These are then used to deposit shapes of an organic (fugitive) ink onto green ceramic tapes, which are then stacked in multiples and laminated. Burnout and sintering then leave appropriately shaped voids in the ceramic. The fugitive ink technique was adapted from the methods used for the preparation of multilayer capacitors with impregnated electrodes, as originally proposed by T. Rutt.[6] As noted, the void pattern can be designed so that all voids are fully sealed inside the ceramic.

Processing Parameters

A PZT 5A (Channel 5502) slip that contained 34% binder solution was used to cast 2.0 mil thick tapes.

Fugitive ink was screened on these. It contained 22% resin, 29% fine carbon and 15% of the PZT powder in an organic solvent system, dispersed on a three roll mill.

The screen used was made with patterns for nine individual ceramic test parts. These patterns were designed to cover ~ 50% of the total area of each part.

There was a void free strip (3-10 mil wide) along each edge of the patterns, so as to provide a dense seal around the ordered voids in each individual sintered ceramic chip.

A total of 60 of the 2 mil sheets were stacked. Samples were made with ink patterns on each sheet resulting in 1.58 mil void plane spacings in the fired ceramic. In an alternate design, two unscreened sheets were inserted between the screened sheets, resulting in 3.8 mil void plane spacings. The stack was then laminated and diced, so that from each pad nine, green 0.5"x0.5", test parts were obtained. Control blanks were prepared by laminating 60 unscreened sheets.

The samples were burned out at a rate of 10°C/hour to 700°C and sintering took place at 1200°C for 30 minutes.

The resulting ceramics were silver electroded and poled at 130°C with 3KV/mm.

Test Procedures

Dielectric properties were measured on a 4270A Hewlett Packard Bridge. The uniaxial piezoelectric stress constant (d_{33}) was measured on a Berlincourt Piezo d_{33} meter (Model CPDT 3300) with hemispherical probes. A good approximation of d_{31} was derived by using the instrument probes with a thin insulator to apply stress onto 2 surfaces orthogonal to the electrodes, and touching 2 wire probes connected to the instrument to the silver electrodes. The resultant reading is multiplied by the ratio of the area touched by the insulated instrument probes to the area of the silver electrodes.

Densities were determined by weighing parts in air and immersed in water. For void shape analysis, ground and polished surfaces were gold coated and photographed on a SEM.

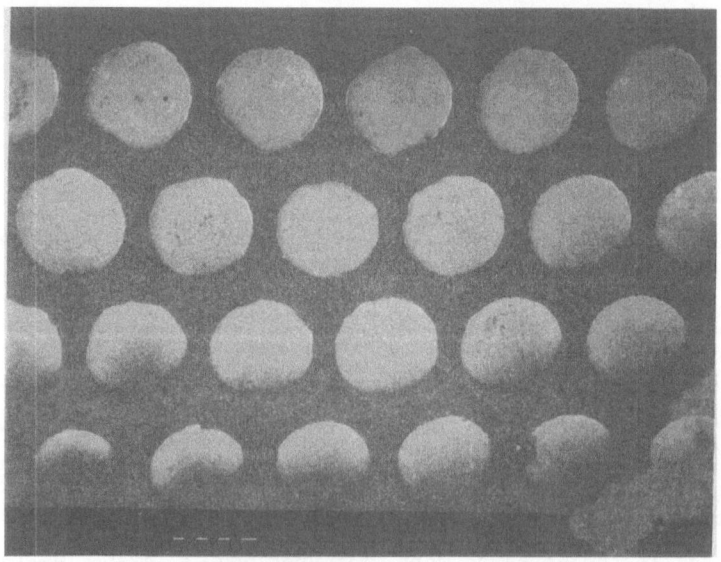

Figure 2. Ordered void structures
Top views

RESULTS AND DISCUSSION

Void Shapes and Porosities

Figure 2 shows top views of ceramic parts with their top surfaces ground. They show void patterns consisting of corner connected squares and of round voids. Figure 3 shows a cross section of such parts. The individual voids were about 0.55 mils thick.

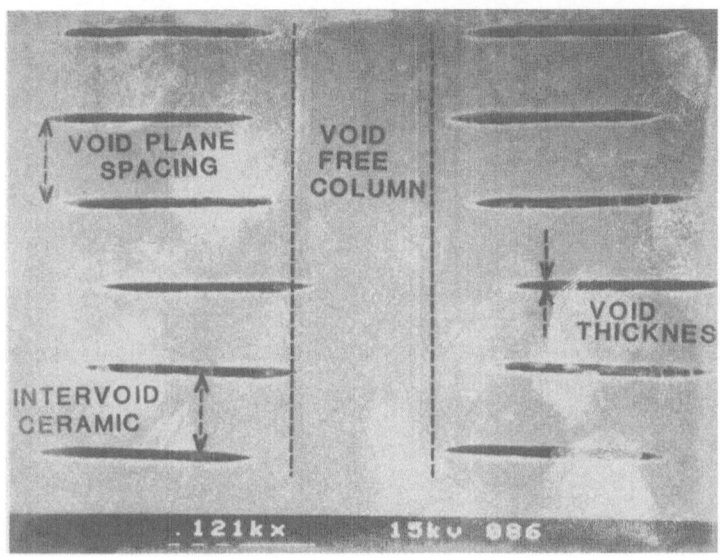

Figure 3.
Cross section
of ordered
void struc-
tures

Most of the blanks and the samples with .024" diameter round voids had less than 1% open porosity. Some samples of other designs did show higher percentages of open porosity. These were attributed to extrinsic effects, since all patterns called for a dense margin around the internal voids.

The internal porosity of blanks was 3.5%, see Table 1. The Calculated Anisotropic Porosity from void dimensions of samples with four different void patterns are also shown in this table. The Derived Anisotropic Porosities in the last column were determined by subtracting from the Immmersion Measured Porosity the applicable isotropic porosity (3.5% for samples with the wider void plane spacing). There is acceptable agreement between the Derived and the Calculated Anisotropic Porosities.

TABLE 1: VOID VOLUMES AND POROSITY

Shape and Dimensions of Ordered Voids	Calculated Void Plane Spacing Mils	Anisotropic Porosity from Void Dimensions (%)	Immersion Measured Porosity (%)	Isotropic Porosity (%)	Derived Anisotropic (1) (%)
None (Blanks)		-	3.5	3.5	
Corner connected squares 0.012"x0.012"	3.7	6.5	10.9	3.5	7.4
Discs 0.024" dia.	3.92	4.8	8.5	3.5	5
Unconnected squares 0.026"x0.026"	1.62	17	15.5	-	15.5
Orthogonal .012" bars (.033" spaced)	1.34	17	16.3	-	16.3

(1)   Immersion measured porosity minus isotropic porosity

Electrical Properties

Table 2 shows the dielectric constants of two of the more interesting void patterns, measured with a field parallel to the laminating direction, i.e., at 90° to the void plane. They are seen to range from 326 to 526, with $\frac{K_{poled}}{K_{unpoled}}$ ratios up to 1.31. It appears that higher dielectric constants are seen with the wider void plane spacing.

TABLE 2

Dielectric Constants (K) of 2 Anisotropic Void Patterns
(Measuring field at 90° to the major void plane)

Void Pattern	3.8 mil		Void Plane Spacings 1.58 mil	
	K unpoled	$\frac{K\ poled}{K\ unpoled}$	K unpoled	$\frac{K\ poled}{K\ unpoled}$
Large Discs 0.024" diam	526	1.14	415	1.26
Parallel Bars	441	1.31	326	1.3

TABLE 3

Average Unpoled Dielectric Constants (K)

with Different Sample Orientations

Orientation of measuring field relative to direction of laminating pressure	Blanks	Void Plane Spacings	
		3.8 mil	1.58 mil
Parallel	908	492	421
Perpendicular		960	979
$\dfrac{\text{K parallel}}{\text{K perpendicular}}$		0.51	0.43

Some of the samples gave dielectric constants as low as 42% of that of blanks, see Table 3. With 50% of void area (relative to total electode area) larger dielectric constants would be expected. The lower K must then be attributed to void misalignment caused during the lamination process as was observed in some cross-sectioned samples. Table 3 also compares the dielectric costants measured parallel and perpendicular to the laminating direction. The anisotropy ranged from 0.51 to 0.43.

TABLE 4

$$d_{33} \text{ Values and } \frac{d_{31} + d_{32}}{2\,d_{33}} \text{ Ratios}$$

	d_{33}	$\dfrac{d_{31} + d_{32}}{2\,d_{33}}$
	PC/N	PC/N
Pressed Pellets (PVA Binder)	400	0.41
Blank Laminates	230	0.43
Laminates with anisotropic Voids[1]	170	0.23

[1] Averages of 9 patterns, poled at 90° to the major void plane

Table 4 shows that the d_{33} values of poled blank laminates were about 58% of those of pellets pressed with PVA binder[5]. The lower d_{33} is attributed to the chemistry of the organic binder system used for the laminates, in which chlorine is thought to detrimentally interact with the PZT. The binder composition did not change the $\frac{d_{31} + d_{32}}{2d_{33}}$ ratio of the blanks that is here geometry dependent. In the laminates with voids this ratio was reduced to 0.23 or less. Also the d_{33} of samples with voids was lower, 74% of that of the blanks. The d parameters of two void patterns are shown in Table 5. The parallel bar pattern is anisotropic in the X-Y plane, and $\frac{d_{31}}{d_{33}}$ ratios of .08 to .19 (with as much as twice as high $\frac{d_{32}}{d_{33}}$ ratios) were observed, giving the average values reported in Table 5.

TABLE 5

$$d_{33} \text{ and } \frac{d_{31} + d_{32}}{2 \, d_{33}} \text{ Ratios}$$

of 4 Anisotropic Void Patterns
Poled at 90° to the Major Void Planes

Void Pattern	Void Plane Spacings			
	3.8 mil		1.58 mil	
	d_{33}	$\frac{d_{31} + d_{32}}{2 \, d_{33}}$	d_{33}	$\frac{d_{31} + d_{32}}{2 \, d_{33}}$
	pC/N	pC/N	pC/N	pC/N
Large Discs (0.024" dia)	191	0.23	194	0.21
Parallel bars	202	0.23	162	0.21

TABLE 6

Calculated Hydrostatic Responses of 2 Anisotropic Void Patterns
Poled at 90° to the Major Void Planes

Void Pattern	Void Plane Spacings				
	3.8 mil		1.58 mil		
	d_h	g_h	d_h	g_h	$d_h \times g_h$
	(pC/N)	(10^{-3}Vm/N)	(pC/N)	(10^{-3}Vm/N)	
Large Discs (.024" Dia)	103	19.4	113	24.3	2745
Parallel Bars	106	20.8	93.9	25	2350

NOTE: d_{33} of blanks of this material was 230 pC/N

The hydrostatic responses of these two void patterns are shown in Table 6 to give d_h values to 113 pC/N and g_h to 25 milliVm/N. The narrower void plane spacing, gives the highest $d_h g_h$ values as shown in the last column in this table.

SUMMARY

Ceramics with ordered void structures were prepared and the following has been established:

1) Porosity measurements and void size calculations imply that the 3.5% of isotropic porosity found in dense ceramic diffuses out in samples with void plane spacings of < 1.58 mils.

2) The dielectric constant normal to the major void plane is reduced approximately proportionally to the fractional area of the voids. Void column offset reduces it more. Fringing effects appear minimal with 1.58 mil void plane spacings.

3) The d_{33} of the ordered void structures is 3/4 of that of blanks. d_{31}/d_{33} ratios are reduced by 50% or more.

4) When comparing different structures made of material with the same d_{33}, one finds in ordered void structures more than 3 times the d_h and more than 10 times the g_h values of dense ceramincs. This occurs in a hermetically sealed and fracture resistant ceramic structure.

5) Some void configurations give a gain in flexure strength over that of void free ceramics, in particular when a surface obtained from saw cutting through voids, at 90° to their major plane, is under tension.

References

1. A.A.H. Amin, "Phenomological and Structural Studies of Pb_2rO_3-$PbTiO_3$ Piezoceramics," PhD Thesis, Pennsylvania State University, November 1979, pps. 26,33.
2. M. Kahn, "Acoustic and Elastic Properties of PZT Ceramics with Anisotropic Pores," Naval Research Laboratory, presented at the 87th Annual Meeting of the America Ceramic Society, Cincinnati, OH, May 7, 1985.
3. R.E. Newnham, D.P. Skinner, L.E. Cross, "Connectivity and Piezoelectric-Pyroelectric Composites," Mat. Res. Bull. 13, 525-536, 1978.
4. A. Safari, "Perforated PZT Polymer Composites with 3-1 and 3-2 Connectivity for Hydrophone Applications," PhD Thesis in Solid State Science, The Pennsylvania State University, 1983.
5. M.Kahn, R.W. Rice and D. Shadwell, "Preparation and Piezoelectric Response of PZT Ceramics with Anisotropic Pores," Naval Research Laboratory, presented to the 87th Annual Meeting of the American Ceramic Society, Cincinnati, OH, May 7, 1985.
6. U.S. Patent 3 829 356 T.C. Rutt 8/13/84.

COMPOSITE THERMISTORS

K.A. Hu, J. Runt, A. Safari and R.E. Newnham

Materials Research Laboratory
The Pennsylvania State University
University Park, PA 16802

INTRODUCTION

A thermistor is a temperature-dependent resistor, made from materials having a large change in electrical resistance with temperature. For critical temperature thermistors, the electrical behavior is governed by a phase transformation which alters the scattering mechanism or the concentration of charge carriers. Thermistors are used as temperature sensors, as protection against current or voltage surges, as flow-meters, and in automatic gain elements.

Doped $BaTiO_3$ ceramics exhibiting a positive temperature coefficient (PTC) effect have been known for quite some time and have been widely used as thermistors[1]. However, $BaTiO_3$ PTC materials are limited by their relatively high resistivity at room temperature, and high materials and fabrication costs. Considerable effort has been devoted to overcoming these difficulties. Probably the most successful alternative materials have been based on composites of carbon black with a semi-crystalline polymer (often polyethylene) or a paraffin[2-8].

PTC effects similar to $BaTiO_3$ ceramics are observed when crystalline polymers are loaded near the percolation threshold with carbon black. At room temperature the carbon black particles are in contact giving resistivities of approximately 1 ohm-cm. Upon heating, a large volume change occurs at the polymer melting point: the polymer expands much more rapidly than the carbon black in this temperature range, separating the carbon grains and raising the resistivity. A rapid increase in resistivity of between typically 1.5 and 8 orders of magnitude around the polymer or paraffin melting point has been reported[5-8]. Commercial carbon black-loaded polyethylene thermistors exhibit a resistivity increase of approximately 4-6 orders of magnitude[9]. It is interesting to note that only very small PTC effects have been reported for conductive composites in which amorphous polymers are used as matrices[5].

In this paper we review our research on a new class of thermistor materials which exhibit low room temperature resistivities and very large PTC anomalies[10,11]. These materials are based on conductive ceramics and either amorphous or crystalline polymers. Several of the transition metal oxide fillers used in the study undergo a negative temperature coefficient

(NTC) transformation changing from semiconductor to metal. In combination with the appropriate polymer, interesting NTC-PTC composites are produced.

EXPERIMENTAL

Three polymers were used in our studies: high density polyethylene and a flexible and rigid epoxy. Polyethylene (PE) is a semicrystalline polymer with a melting point of about 130°C. The particular PE used in our research was obtained from Phillips Petroleum Co. (Mailex 6001). The rigid epoxy ('Spurr' epoxy) was obtained from Polysciences, Inc., and has a glass transition temperature (T_g), when cured at 70°C, of approximately 65°C. The flexible epoxy (Eccogel 1365-80) was obtained from Emerson Cuming. The conductive fillers VO_2, V_2O_3, VO, Ti_2O_3, NbO_2 and TiO were obtained from Alfa Products.

Resistivity versus temperature curves (Fig. 1) for most of the fillers exhibit a rather sharp transition to a more metallic state[12]. For instance in V_2O_3 there is a transition at approximately -100°C which results in a large change in resistance. Above -100°C it exhibits metallic behavior and below -100°C it shows semiconductor behavior.

Composites with epoxy matrices were fabricated by mixing the appropriate epoxy precursors and ceramic fillers in the desired proportions and curing at temperatures ranging from 70°C to 120°C. Curing was most frequently conducted at 70°C for 8 hours. Composites with thermoplastic PE patrices were prepared by mixing the PE and ceramic filler in a Brabender mixer at 150°C for 10 minutes. All concentrations stated in this paper refer to volume percent ceramic. The d.c. resistivity of the composites was measured as a function of temperature with a Keithley digital electrometer Model 616.

RESULTS AND DISCUSSION

The relationship between resistivity and temperature for several

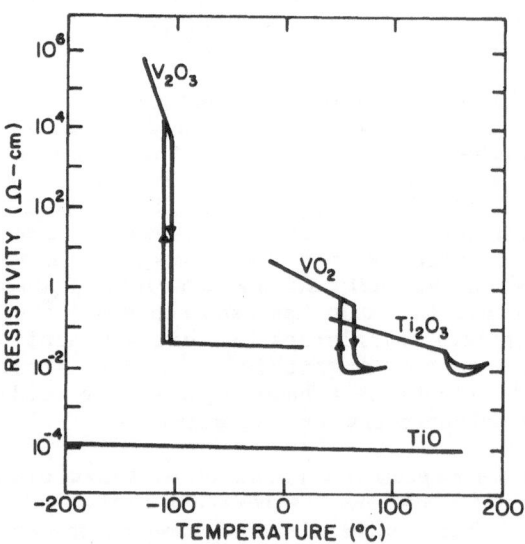

Fig. 1. Resistivity-temperature behavior in V_2O_3, VO_2, VO, Ti_2O_3 and TiO single crystals.

476

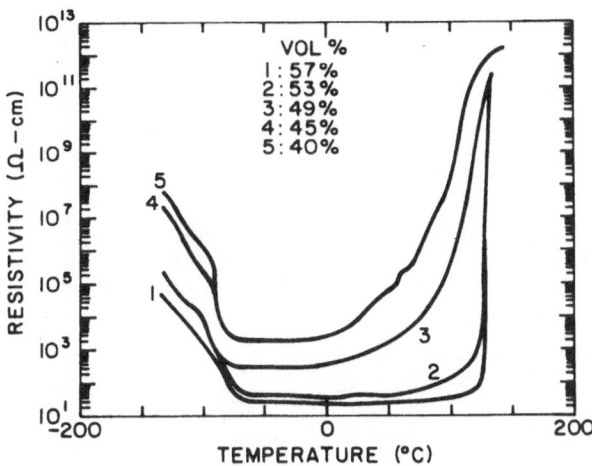

Fig. 2. Temperature dependence of the electrical resistivity of V_2O_3-polyethylene composites prepared with different volume percentages of V_2O_3.

V_2O_3-polyethylene composites is shown in Fig. 2. Two remarkable changes in resistance are observed, giving the resistivity-temperature curve a 'square-well' appearance. As evidenced by the relatively low values for the room temperature resistivity, the compositions shown in this figure are above the percolation threshold. For temperatures below the polyethylene melting point and for compositions above the percolation limit, the V_2O_3 particles are in contact and control the resistivity. On heating the composites from low temperatures, the V_2O_3 particles are in their semiconducting state with resistivities near 10^6 ohm-cm, far less than the resistivity of the polymer matrix. Near −100°C the filler particles undergo a semiconductor-metal transition accompanied by a large change in resistivity. A drop of five orders of magnitude has been reported for single crystals[12] (Fig. 1). The V_2O_3-polyethylene composites show a sizeable but somewhat smaller drop of two to three orders of magnitude.

Above the transition, the V_2O_3 particles convert to a metallic state with a resistivity of 0.1 ohm-cm. Similar behavior is observed in the composite but the resistivity is about a hundred times larger. The resistance is independent of temperature from −75°C to 100°C in both V_2O_3 single crystals (Fig. 1) and composites with V_2O_3 volume fractions greater than about 50%. As the temperature is increased further, a dramatic increase in resistivity of up to ten orders of magnitude is observed at the polyethylene melting point (~130°C).

Similar behavior has been observed for composites made with Spurrs epoxy and VO_2, V_2O_3, VO, Ti_2O_3 and TiO as conductive fillers[10,11]. Resistivity versus temperature curves for Spurrs epoxy composites with several different fillers are illustrated in Fig. 3. The volume fraction filler in these composites is above the percolation threshold and so they have quite low room temperature resistivities, although somewhat larger than the conductive fillers alone. TiO Spurrs composites (not shown) exhibit the lowest room temperature values (on the order of 1-5 ohm-cm)[11]. Composites prepared from NbO_2 (a semiconductor over the temperature range of interest) also exhibit a resistivity anomaly in the same temperature

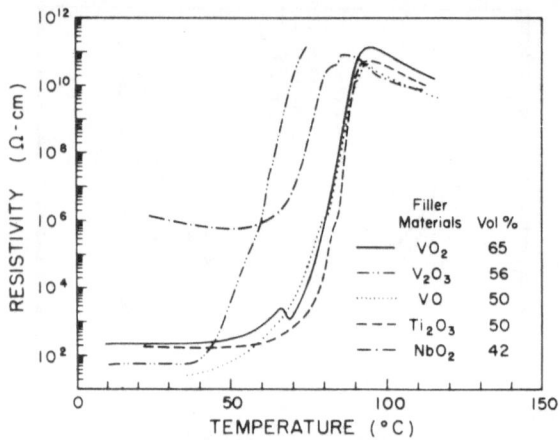

Fig. 3. Temperature dependence of the electrical resistivity of Spurrs epoxy composites cured at 70°C for 8 hours with different filler materials.

range as the other composites (Fig. 3). Since the ultimate room temperature resistivity of NbO_2 is considerably larger than that of the metallic fillers, the magnitude of the resistivity increase is much smaller (approximately five orders of magnitude).

As discussed in the introduction, the PTC behavior observed in the semi-crystalline polyethylene composite can be associated with the large volume change at the polymer melting point. However, very sizable and reproducible PTC effects are not usually anticipated in metal-insulator composites with amorphous polymer matrices. One might speculate that in the case of Spurrs epoxy composites that the differential thermal expansion is such that the oxide particles are forced far enough apart to cause the observed transition. However, we do not observe a sizable volume expansion of the Spurrs epoxy at temperatures near the PTC transition. The T_g of the Spurrs epoxy is about 65°C as measured by differential scanning calorimetry and thermal mechanical analysis and it is tempting to correlate the temperature of the PTC phenomenon with T_g. However, composites prepared with the elastometric Eccogel epoxy (T_g less than room temperature) exhibit a large PTC effect at approximately the same temperature (although the transition is not as sharp) as the Spurrs-based composites[10]. Further experimental work is in progress to understand more clearly the nature of the PTC transition in amorphous polymer-conductor composites.

SUMMARY

The semiconductor-to-metal phase transitions in metal oxides of vanadium and titanium were used to prepare composite thermistors which exhibit both NTC and PTC resistance effects. Composites were prepared by mixing V_2O_3 and other oxides in various proportions in polymers such as polyethylene and rigid and flexible epoxies. These composites showed unusually large PTC resistivity increases of up to 10 orders of magnitude. Accompanying NTC effects as large as 5 orders of magnitude were observed at the semiconductor-to-metal transitions in the filler materials. The resistance vs. temperature curve of NTC-PTC composites resembles a square well with a conducting range with three temperature regimes: semiconductor to metal to insulator.

REFERENCES

1. J.M. Hebert, 'Ferroelectric Transducers and Sensors,' Gordon and Breach, New York (1982).
2. K. Ohe and Y. Naito, _Jap. J. Appl. Phys._ 10:99 (1971).
3. F. Beuche, _J. Appl. Phys._ 44:532 (1973).
4. F. Beuche, _J. Polym. Sci. Polym. Phys. Ed._ 11:1319 (1973).
5. J. Meyer, _Polym. Eng. Sci._ 13:42 (1973).
6. J. Meyer, _Polym. Eng. Sci._ 14:706 (1974).
7. M. Narkis, A. Ram and F. Flashner, _Polymer Eng. Sci._ 18:459 (1978).
8. M. Narkis, A. Ram and Z. Stein, _J. Appl. Polym. Sci._ 15:1515 (1980).
9. F.A. Doljack, _IEEE Trans. Comp., Hybrids, Manufct. Tech._ 4:372 (1981).
10. K.A. Hu, J. Runt, A. Safari and R.E. Newnham, _Ferroelectrics_, submitted for publication.
11. K.A. Hu, J. Runt, A. Safari and R.E. Newnham, _Phase Trans._, submitted for publication.
12. F.J. Morin, _Phys. Rev. Lett._ 3:34 (1959).

GRAIN RESISTIVITY AND CONDUCTION IN METAL OXIDE VARISTORS

Herbert R. Philipp

General Electric Corporate Research and Development
P.O. Box 8
Schenectady, NY 12301

INTRODUCTION

Metal oxide varistors are ceramic devices with highly nonlinear current-voltage characteristics similar to back-to-back Zener diodes.[1-4] Typical I-V characteristics for a metal oxide varistor at 77 K and for a small range of temperatures near 300 K are shown in Fig. 1. They are produced by sintering ZnO powder together with small amounts (1 to 10 mole %) of other oxide additives. The resultant structure, which can be idealized by the "block model" shown in Fig. 2, is comprised of semiconducting n-type ZnO grains of dimension 10 to 20 μm surrounded by insulating barriers at the ZnO grain boundaries. These varistors have proved useful in a variety of applications particularly as high-quality voltage surge suppressors.[5,6]

Electrically, the varistor can be represented by the simple equivalent circuit shown in Fig. 3. Here C_p and R_p are the capacitance and resistance of the intergranular layer, respectively, and r_g is the ZnO grain resistance. The "dc response" of varistors concerns itself with the voltage-dependent behavior of R_g and the influence of r_g on this behavior. The "ac response" of varistors concerns itself mainly with the frequency-dependent dielectric behavior of the grain boundary barriers[7-9] and with the time-dependent

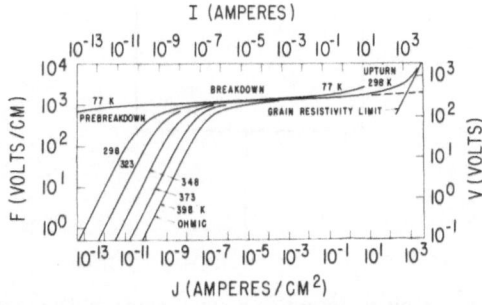

Fig. 1.   Current-voltage characteristics of a metal oxide varistor at 77 K and for a small range of temperatures near 300 K.

response of R_p, particularly to short rise-time, high current pulses.[10,11] In this paper we shall mainly limit our discussion to the dc properties. Emphasis will be placed in highlighting areas in which our understanding is not complete and where further experimental and theoretical work is needed. First, however, several aspects of the pulse response of varistors will be briefly mentioned because of their bearing on the nonlinear conduction process in R_p.

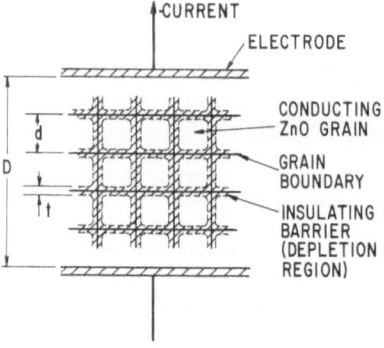

Fig. 2.   "Block model" of a ZnO varistor having grain size d ($\sim 10\,\mu$m) and intergranular layer thickness t (<100A). D is the electrode separation.

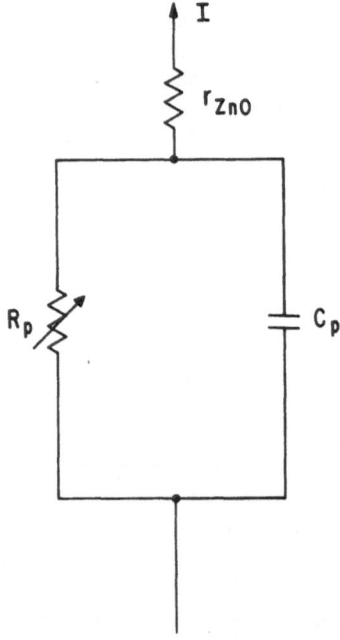

Fig. 3.   Simple equivalent circuit representing a metal oxide varistor as a capacitance in parallel with a voltage-dependent resistor. C_p and R_p are the capacitance and resistance of the intergranular barrier. r_g is the ZnO grain resistance. For low applied voltages R_p behaves as an ohmic loss.

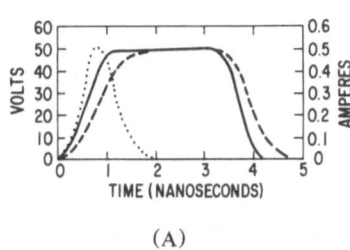

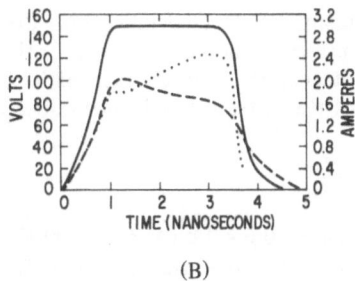

(A)   (B)

Fig. 4.   Voltage response, dashed line, and current response, dotted line, of a varistor chip in a
50 Ω circuit to ~500ps rise-time pulses of magnitude (A) 50 V and (B) 150 V. The
response without the chip in the circuit is shown by the solid line.

FAST RISE-TIME PULSE RESPONSE

The varistor response to ~ 500 ps rise-time pulses is shown in Figs. 4a and 4b. In
Fig. 4a, the pulse voltage is 50 V and only the capacitive charging current is observed.
The decay of the varistor voltage at the end of the pulse is associated with a negative
varistor discharge current (not plotted in the figure). In Fig. 4b, the pulse voltage is
150 V which is above the breakdown voltage of R_p. We note here that the capacitive
charging current is almost hidden by the increased conduction current through R_p and
that the varistor voltage shows a clearly resolved peak and then decreases with increasing time. There is no indication in this curve of any delay in the initiation of the conduction current once the varistor voltage has risen above some minimum value. Hence
the ability of the varistor to "clamp" the incoming pulse, the varistor "turn on" time,
is less than the rise-time of the pulsing equipment, ~5 × 10^{10}s. The initial peaking of
the varistor voltage and its subsequent decrease with time has been denoted the
"overshoot" effect and represents a time-dependent modification of the varistor conduction process, which is inherently very fast. The existence of the overshoot effect has
an important device implication. The protective voltage quoted is usually that determined using standard measuring procedures (pulses of 8 μs rise-time) where the voltage
overshoot, $V_{max} - V_{dc}$, is small. However, for faster rise-time, high current pulses, the
protective level may be appreciably poorer.

One additional comment should be made concerning the ac varistor voltage at low
current levels. The peak (sinusoidal) ac voltage for a given peak ac current is generally
higher then that measured for the same dc current level. The ac/dc voltage shift is usually about 5% (at 60 Hz), although it may be higher or lower depending on the chemical
formulation. The presence of long time polarization currents, which contribute to the ac
power dissipation, may be important here.[12]

ELECTRICAL CONDUCTION

We shall now examine Fig. 1 in more detail. Three regions can be distinguished in
these curves. In the breakdown region, the current is a highly nonlinear function of the
applied voltage for many decades of current and is essentially temperature independent.
The dependence of current upon voltage here is often described by an empirical relation

$$I = C \, V^\alpha$$

where α can achieve values in the range 50 to 100 near 10^{-3} A/cm². At very large currents, the curve exhibits an upturn. This feature is not a property of the breakdown

mechanism, but is associated with the finite resistivity of the grains themselves; the up-turn represents the voltage drop in the grains. In the prebreakdown or leakage region at low voltages, the I-V characteristic is linear. In many cases, the temperature dependence of the current can best be described[13] in terms of an activation energy

$$I = I_0 e^{-e\phi/kt}$$

where e $\phi \sim$ 0.6 - 0.8 eV near 300 K.

An interesting and important feature of the breakdown characteristic is that within reasonable limits it is relatively insensitive to the details of chemical composition and processing. It is of course possible to prepare samples using somewhat arbitrary ingredients in which the nonlinearity is minimal ($\alpha <$ 10) and a clear, sharp breakdown characteristic is not observable.[14] However, for compositions that contain a varistor forming ingredient (heavy elements such as Bi, Pr, Ba and Nd with large ionic radii) and at least one or two varistor performance ingredients (generally transition metal elements such as Co, Mn and Ni) and for sintering temperatures in the range 1100 °C to 1350 °C, the breakdown behavior illustrated in Fig. 1 with $\alpha \sim$ 50 can be achieved.

It can also be shown that the gross ceramic microstructure does not play a major role in the conduction process. In Fig. 5, I-V data are shown for bulk varistor material $\sim$0.15 cm thick, for the as-sintered varistor surface obtained using surface electrodes spaced $\sim$0.10 cm apart, and for a single grain boundary junction.[4,14] The measured voltage is 100 times larger for the bulk and surface electrode data than given on the ordinate scale. The shape of these curves are virtually identical showing that the electrical properties are determined solely by the behavior of the individual grain-grain junctions. The varistor voltage itself is determined by the number of grain boundaries between electrodes. However, we do not mean to imply here that the uniformity of the microstructure is not an important device consideration. As indicated in the idealized microstructure of Fig. 2, between electrode conduction takes place through a multitude of parallel paths. Small perturbations, for example, in the number of grain boundaries in the various parallel conduction paths due to less than optimum processing and sintering

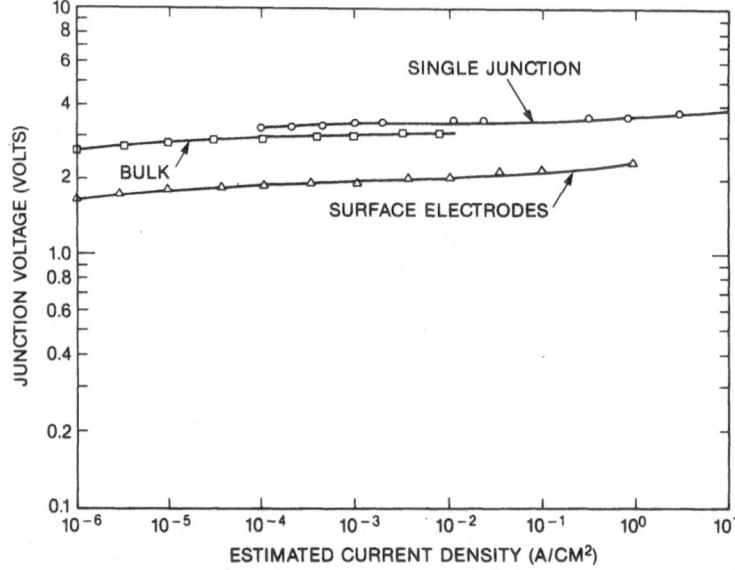

Fig. 5.   Current-voltage characteristics of a commercial varistor material obtained using 1) microelectrodes placed across a single grain junction, 2) electrodes placed 0.1 cm apart on the as-sintered varistor surface, and 3) electrodes placed on the opposite faces of the varistor disk $\sim$0.15 cm thick. The voltages measured in 2) and 3) are 100 times those indicated on the ordinate scale.

procedures can cause the current flow to vary considerably from place to place over the electroded faces of the varistor disk. This behavior can drastically affect varistor performance, especially at high current levels, even in cases where the individual grain-grain barrier electrical characteristics are "optimal."

A number of theoretical models have been proposed that explain grain-grain conduction and other varistor phenomena with reasonable credibility.[4,15–20] They are based on the double depletion layer concept for the region of closest grain-grain contact, which is the generally accepted starting point for any physical model of the varistor junction. The intergranular layer, whose existence derives from the varistor forming and performance additives, is thin (10 Å or even less), and is not specifically described either physically or chemically.

Most of the available theoretical models are phenomenological in their construction in that the physical origin of the surface state and bulk donors that describe the varistor barriers are not considered. For example, in one phenomenological model[20], the surface state density and doping profile are extracted from C-V and I-V data. While it may be possible in this model to assign a set of such parameters to each and every varistor sample and hence "account" for materials and processing variations in terms of these parameters, they are not, of themselves, independently reasoned or justified and are thus of questionable significance. The model of Mahan, Levinson, and Philipp[4], on the other hand, which is illustrated in Fig. 6, does not require any specific form for the doping or surface state density and associates the highly nonlinear conduction with a triggered tunneling process involving hole creation that is intrinsic to such ZnO depletion layer barriers when the voltage exceeds some critical (and universal) value. The agreement of this theory, which contains no arbitrary adjustable parameters, with experiment, is excellent as shown in Fig. 7. The postulated production of holes during breakdown conduction in varistors has recently been directly verified in electroluminescence studies.[21] We further postulate that the overshoot effect described earlier may arise from the time associated with the hole creation process. However, we note that even this model

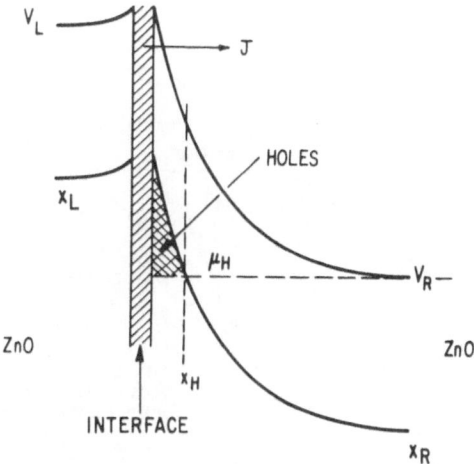

Fig. 6.   The hole model. The holes are created on the forward side of the junction where the hole chemical potential is below the valance band edge. The electron tunneling from the interface is indicated by the arrow labeled J.

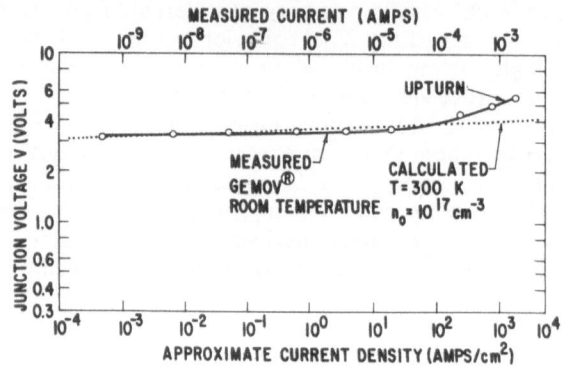

Fig. 7.   Experimental and calculated current-voltage characteristics of a single ZnO varistor grain junction. The observed decrease in nonlinearity at high current densities is due to the series resistance of the ZnO grains. This has not been included in the calculation.

has limited predictive value in that it does not consider (much less clarify) the role of the individual chemical additives in forming and optimizing the grain-grain barriers responsible for varistor action.

A prebreakdown or leakage conduction at lower voltage associated with thermal excitation over the double depletion layer barrier also follows directly from theory. The leakage conduction should be temperature-dependent with an activation energy $e\phi$, which is a measure of the barrier height. This behavior of the leakage conduction is illustrated in Fig. 1. It is found, however, that the magnitude, as well as the temperature dependence of the leakage conduction, is very sensitive to the materials and processing schedules used in the sample preparation as well as to electrical, mechanical, chemical, pressure, environmental (atmosphere), and other stresses that tend to degrade or increase leakage conduction.[22] An example of the effect of dc electrical stress on the I-V characteristics of a ZnO varistor is shown in Fig. 8. The leakage conduction after stress has increased over 4 orders of magnitude and is now polarity-dependent.

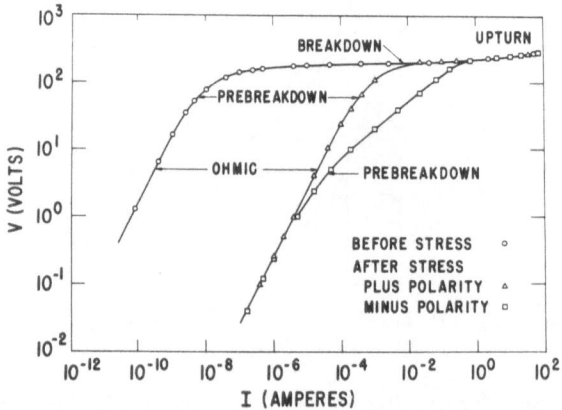

Fig. 8.   The effect of dc electrical stress on the current-voltage characteristics of a ZnO varistor.

Degradation phenomena in ZnO varistors exhibit several general features.[22] First, the rate of degradation (rate of increase of leakage conduction) can be strongly influenced by very small changes in the chemical makeup and/or processing of the varistor, as well as by heat treatments well below the sintering temperature. Second, degradation associated with a diverse set of stresses (electrical, mechanical, environmental, etc.) can often be rapidly reversed by annealing at ~300 °C or lower. Third, and perhaps most striking is the fact that the degradation process mainly affects the varistor leakage and prebreakdown conduction and not the varistor behavior for voltages above the breakdown voltage. This is clearly indicated in Fig. 8. Hence, changes in barrier height, depletion layer thickness, intergranular layer trap density and distribution, etc., which have been associated with the degradation process should be such as to leave unchanged the highly nonlinear breakdown conduction process. The published literature to date does not adequately address this important issue.

Present theory also does not adequately explain conduction in chemically more primitive varistor systems such as ZnO + Bi. Current-voltage data for this material[14] at several different temperatures are shown in Fig. 9. While the curves are not highly nonlinear, the more remarkable feature is that the leakage conduction is insensitive to temperature. According to Eq. 2 this implies that the barrier height $e\phi \sim 0$ and hence the I-V characteristic should be linear, with its magnitude determined solely by the grain resistivity. At one volt applied, this would produce many A/cm^2 of leakage conduction rather than a current density in the 10^{-6} A/cm^2 range as shown in Fig. 9. The temperature dependence of leakage conduction in degraded samples is also often insensitive to temperature and the same argument concerning the magnitude of leakage conduction would apply. To account for this and other problems described earlier, we propose that there is more than one between grain conduction path.[14,23]

In Fig. 10, we show a schematic of the region of grain-grain contact. In the region of closest grain-grain contact, the intergranular layer is very thin, perhaps only a monolayer in thickness.[17,24] At the grain corners, bulk intergranular material of many microns in extent is found.[25] The transition between these extremes appears smooth as illustrated in Fig. 10. We must assume that there are a variety of different between grain conduction paths that operate in parallel. Path (1) through the region of closest grain-grain contact represents conduction over the Schottky barrier and would show a thermally activated temperature sensitive leakage conduction related to the barrier height. Path (2) is through the bulk intergranular material and would be sensitive to the chemical additives used and to applied stress. We speculate that conduction through this path is temperature-insensitive and concerns itself with mechanisms used to describe electronic transport in atomically disordered materials systems. We also require that there is no Schottky barrier formation along path (2), i.e., between the ZnO grain and the intergranular material near the grain corner.

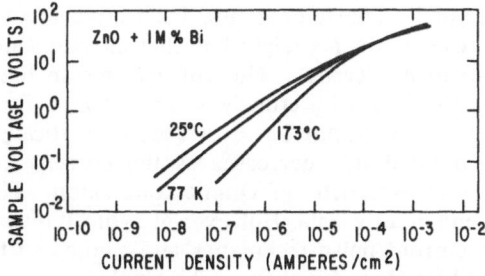

Fig. 9.   Current-voltage characteristics of a varistor material comprised of ZnO plus 1 mole % Bi at temperatures of 25 °C, 173 °C, and 77 K.

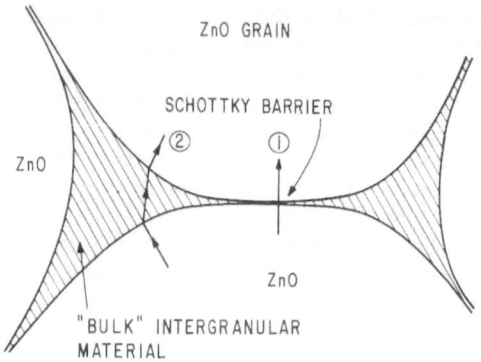

Fig. 10.   Proposed parallel path conduction between grains. Path 1 is in the region of closest grain-grain contact over the Schottky barrier. Path 2 is through bulk intergranular material at grain corners and is temperature insensitive.

GRAIN RESISTIVITY

Intragrain conduction is ohmic and in its simplest representation, Fig. 3, adds a voltage (per unit length) of value

$$V_g = J\rho_g \tag{3}$$

to the breakdown voltage where ρ_g is the grain resistivity. This voltage produces the up-turn shown in Fig. 1 at high currents and is the feature that limits the performance of metal oxide varistors in surge suppression applications. The circuit voltage protection is not the breakdown voltage but is higher because of the voltage drop in the grains themselves.

While pure ZnO is an insulating semiconductor with a $\sim$ 3eV bandgap, sintered ZnO is reasonably conducting. This conductivity probably derives from shallow donor levels in the ZnO associated with oxygen vacancies created in the $\sim$ 1200 °C sintering process.[26] As indicated in Fig. 1, at room temperature $\rho_g \sim 1$ Ω cm and is 10 to 100 times higher at 77 K. The grain resistivity can be controlled to a certain extent by dopants such as Al or Li.[23,27,28] The simple representation of grain resistivity in Fig. 3, however, is only approximate.

At 4.2 K, the grain resistivity of ZnO is $\sim 10^6$ times higher than it is at room temperature[29,30] and, as shown in Fig. 11, grain resistivity effects (upturn) dominate the curve even at low currents.[31] The temperature insensitive breakdown characteristic in these curves can be represented by a straight line that increases slightly in voltage as the current is increased over many decades. The voltage drop in the grains adds a ΔV_g to the breakdown voltage, which can be crudely estimated as a function of current. We note that $\Delta V_g/I = R_g$ is not constant but decreases dramatically as the current is increased. That is, $\alpha = d \ln I/d \ln V$ decreases as the current increases but does not achieve the value $\alpha = 1$ characteristic of Ohmic conduction. This observation is also characteristic of room temperature data. However at room temperature, such data must be obtained using high current pulse techniques and cannot be followed in detail as it can at low temperature where dc techniques can be employed.

A second observation concerning the magnitude of the grain resistivity is also of interest and gives us additional insight into the interpretation of upturn data. The grain

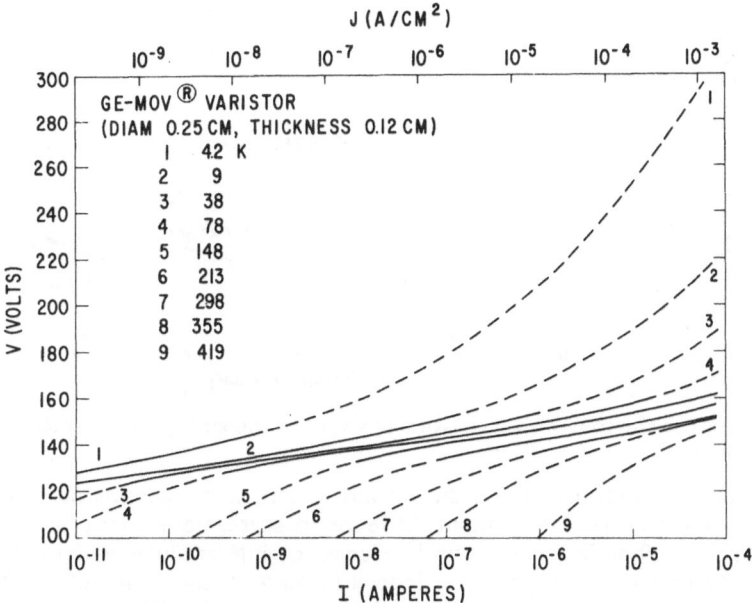

$$J \, (A/CM^2)$$

Fig. 11. The current-voltage characteristics of a small area varistor for temperatures in the range 4.2 K to 419 K. The solid portions of the curves correspond to $\alpha = d \, lnI / dlnV \geqslant 30$.

resistivity can be evaluated by other techniques in addition to the one discussed above where the actual voltage drop in the grains is estimated. The first of these is a high frequency technique.[8] For $f > 10^8$Hz, the intergranular capacitance becomes an effective short circuit for intergranular conduction and the equivalent circuit for the varistor, shown in Fig. 3, can be represented as a resistor r_g whose resistivity is the ZnO grain resistivity. The value of r_g can be determined by high frequency impedance measurements. The second of these is an infrared optical technique.[32] The reflectance of ZnO has a sharp minimum in the infrared whose wavelength position depends on the free electron density and hence on the ZnO grain resistivity. Both of these techniques give comparable values for the grain resistivity which are lower than those obtained from upturn measurements by perhaps a factor of 5 or more.[8]

A possible explanation for the higher value of resistivity obtained from the upturn characteristic arises from current constriction in the region of grain-grain contact. As depicted in Fig. 10, breakdown conduction occurs in the region of closest grain-grain contact where the intergranular layer is very thin. Hence the current is constrained to flow through only a small portion of the ZnO grain area at the interface and thus results in a larger effective grain resistivity.[8] We also speculate that as the voltage is increased above the nominal breakdown voltage, an increasing area for grain-grain current flow is made available; that is, conduction takes place in regions where the intergranular layer is thicker.

The grain resistivity can be increased by doping with Li, which presumably introduces deep levels in ZnO that compensate the shallow donors responsible for grain conduction. Doping with Al, Ga or In, on the other hand, reduces the grain resistivity. Doping with Al is now common practice and grain resistivities of $\sim 0.1 \, \Omega$ cm are achieved in commercial varistor materials. While ZnO resistivities below this are achieved in single crystal studies,[29] efforts to reduce the resistivity below $\sim 0.1 \, \Omega$ cm by conventional means in varistor materials have not met with practical success. One of the problems encountered in varistor grain doping is the fact that leakage conduction is also affected by the use of grain dopants such as Al. Thus, for example, the voltage or clamp ratio, $V_{103A} / V_{0.1ma}$, which is one figure of merit for varistor surge protection capability, may have its optimum (lowest) value at an Al doping concentration, which does not minimize the grain resistivity. The connection between Al doping and leakage con-

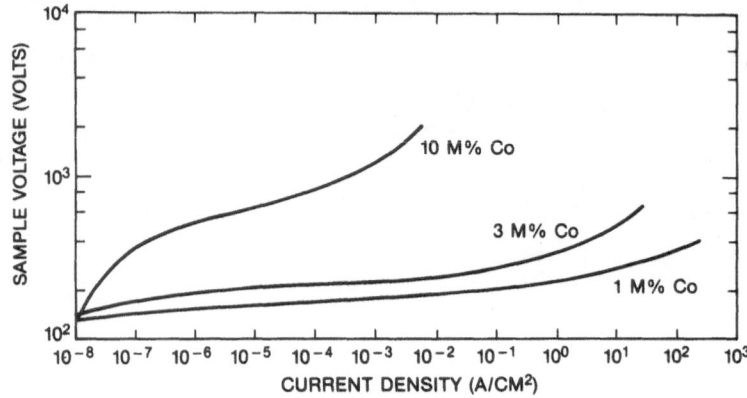

Fig. 12.   The current-voltage characteristics of a varistor containing various amounts of the additive Co.

duction is made somewhat clearer by secondary ion mass spectroscopy (SIMS) depth profiles which show that the element Al tends to segregate at grain boundaries rather than within the ZnO grains themselves.[14] Hence the intergranular material is also doped by Al and apparently such doping does indeed increase leakage conduction. This may be a parallel path effect as discussed earlier and illustrated in Fig. 10.

These same SIMS depth profiles[14] show that the element Co, which is a key ingredient in many commercial varistor mixes, is distributed evenly throughout the varistor material and shows no tendency to segregate at the grain boundary. In fact Co is known to introduce mid-gap levels into ZnO giving such varistors a greenish color.[33,34] In Fig. 12 we show the I-V characteristics of a varistor material containing various amounts of the additive Co. Nominally varistor devices contain ~1 mole % Co. As the level of Co is increased, the upturn occurs at lower and lower currents indicating that this additive does indeed affect the grain resistivity. This result[14] was unexpected. However in concurrent theoretical studies, Mahan predicted this result.[14] He showed in an extension of his earlier work on intrinsic defects in ZnO[26] that the addition of mid-gap donors, in sufficient quantity, dramatically lowers the free electron concentration of sintered ZnO. Capacitance and microelectrode single grain measurements further detail the effect of Co doping on grain resistivity.[14]

We also note in Fig. 12 that the breakdown voltage increases with the level of Co doping. This behavior follows our general observation that additives such as Li and Co, which tend to increase the grain resistivity, also increase the breakdown voltage, while additives such as Al that decrease the grain resistivity tend to lower the breakdown voltage. While the grain size may be affected by such additives, we believe there are other ·factors that may also influence the bulk breakdown voltage. Further speculation on this matter, however, is beyond the scope of this paper.

REFERENCES

1. M. Matsuska, "Nonohmic Properties of Zinc Oxide Ceramics," Jpn. J. Appl. Phys., 10:736 (1971).

2. L. M. Levinson and H. R. Philipp, "The Physics of Metal Oxide Varistors," J. Appl. Phys., 46:1332 (1975).

3. L. M. Levinson and H. R. Philipp, "ZnO Varistors for Transient Protection," IEEE Trans. Parts Hybrids Packaging, PHP-13:338 (1977).

4. G. D. Mahan, L. M. Levinson, and H. R. Philipp, "Theory of Conduction in ZnO Varistors," J. Appl. Phys., 50:2799 (1979).

5. E. C. Sakshaug, J. S. Kresge, and S. A. Miske, "A New Concept in Station Arrestor Design," IEEE Trans. Power App. Syst., PAS-96:647 (1977).

6. "Transient Voltage Suppression Manual," J. C. Hey and W. P. Kram, eds., General Electric Semiconductor Products Dept., Auburn, N.Y. (1978).

7. L. M. Levinson and H. R. Philipp, "AC Properties of Metal Oxide Varistors," J. Appl. Phys. *47*:3116 (1976).

8. L. M. Levinson and H. R. Philipp, "High Frequency and High Current Studies of Metal Oxide Varistors," J. Appl. Phys., *47*:1117 (1976).

9. L. M. Levinson and H. R. Philipp, "Low Temperature AC Properties of Metal Oxide Varistors," J. Appl. Phy., *49*:6142 (1978).

10. H. R. Philipp and L. M. Levinson, "Short Time Pulse Response of ZnO Varistor Grain Boundaries," Advances in Ceramics, *1*:394 (1981).

11. H. R. Philipp and L. M. Levinson, "ZnO Varistors for Protection Against Nuclear Electromagnetic Pulses," J. Appl. Phys., *52*:1083 (1981).

12. H. R. Philipp and L. M. Levinson, "Long Time Polarization Currents in Metal Oxide Varistors," J. Appl. Phys., *47*:3177 (1976).

13. H. R. Philipp and L. M. Levinson, "High-Temperature Behavior of ZnO Based Ceramic Varistors," J. Appl. Phys., *50*:383 (1979).

14. H. R. Philipp, G. D. Mahan, and L. M. Levinson, "Advanced Metal Oxide Varistor Concepts," Final Report ORNL/Sub/84-17457/1, under Subcontract 86X-17457C for ORNL for the DOE under Contract No. DE-AC05-840R21400 (1984).

15. J. D. Levine, "Theory of Varistor Electronic Properties," Crit. Rev. Solid State Sci., *5*:597 (1975).

16. J. Bernasconi, H. P. Klein, B. Knecht, and S. Strassler, "Investigation of Various Models for Metal Oxide Varistors," J. Electron. Mater., *5*:473 (1976).

17. W. G. Morris, "Physical Properties of the Electrical Barriers in Varistors," J. Vac. Sci. Technol., *13*:926 (1976).

18. J. Bernasconi, S. Strassler, B. Knecht, H. P. Klein, and A. Menth, "Zinc Oxide Varistors: A Possible Mechanism," Solid State Commun., *21*:867 (1977).

19. P. R. Emtage, "The Physics of Zinc Oxide Varistors," J. Appl. Phys., *48*:4372 (1977).

20. P. L. Hower and T. K. Gupta, "A Barrier Model for ZnO Varistors," J. Appl. Phys., *50*:4847 (1979).

21. G. E. Pike, S. R. Kurtz, P. L. Gourley, H. R. Philipp, and L. M. Levinson, "Electroluminescence in ZnO Varistors: Evidence for Hole Contributions to the Breakdown Mechanism," J. Appl. Phys., *57*:5512 (1985).

22. H. R. Philipp and L. M. Levinson, "Degradation Phenomena in Zinc Oxide Varistors: A Review," Advances in Ceramics, *7*:1 (1984).

23. K. Eda, "Electrical Properties of $ZnO-Bi_2O_3$ Metal Oxide Heterojunction - A Clue of a Role of Intergranular Layers on ZnO Varistors, in Materials Research Society Symposia Proceedings: Grain Boundaries in Semiconductors," H. J. Leaamy, G. E. Pike, and C. H. Seager, eds., Elsevier, New York (1982).

24. D. R. Clarke, "The Microstructural Location of the Intergranular Metal-Oxide Phase in a Zinc Oxide Varistor," J. Appl. Phys., *49*:2407 (1978).

25. W. G. Morris, "Electrical Properties of $ZnO-Bi_2O_3$ Ceramics," J. Am. Ceram. Soc., *56*:360 (1973).

26. G. D. Mahan, "Intrinsic Defects on ZnO Varistors," J. Appl. Phys., *54*:3825 (1983).

27. T. Miyoshi, K. Maeda, K. Takahashi, and T. Yamarzaki, "Effects of Dopants on the Characteristics of ZnO Varistors," Advances in Ceramics *1*:309 (1981).

28. W. G. Carlson and T. K. Gupta, "Improved Varistor Nonlinearity Via Donor Impurity Doping," J. Appl. Phys. *53*:5746 (1982).

29. G. Heiland, E. Mollwo, and F. Stockmann, "Electronic Processes in Zinc Oxide, in Solid State Physics," F. Seitz and D. Turnbull, eds., Academic, New York, *8*:191 (1959).

30. P. W. Li and K. I. Hagemark, "Low Temperature Electrical Properties of Zn-Doped ZnO," J. Solid State Chem., *12*:371 (1975).

31. H. R. Philipp and L. M. Levinson, "Low-Temperature Electrical Studies on Metal Oxide Varistors - A Clue to Conduction Mechanisms," J. Appl. Phys., *48*:1621 (1977).

32. H. R. Philipp and L. M. Levinson, "Optical Method for Determining the Grain Resistivity in ZnO-Based Ceramic Varistors," J. Appl. Phys., *47*:1112 (1976).

33. D. S. McClure, "The Distribution of Transition Metal Cations in Spinels," J. Phys. Chem.Solids, *3*:311 (1957).

34. H. R. Philipp and L. M. Levinson, "Tunneling of Photoexcited Carriers in Metal Oxide Varistors," J. Appl. Phys., *46*:3206 (1975).

INFLUENCE OF MICROSTRUCTURE AND CHEMISTRY ON

THE ELECTRICAL CHARACTERISTIC OF ZnO VARISTORS

Tapan K. Gupta

Aluminum Company of America
Alcoa Laboratories
Alcoa Center, PA   15069

INTRODUCTION

The ZnO voltage limiter or varistor[1] has an outstanding nonlinear current-voltage (I-V) characteristic and, as such, is used extensively to suppress the transient voltage surges encountered in the lightning and switching surges.[2] The advancement in the state-of-the-art of the varistor in recent years[3-9] makes it impossible to describe all the elements of this development within the scope of this paper. The objective of this paper is to concentrate on the microstructure and chemistry of the ZnO varistor and their effects on electrical properties. However, in order to appreciate this correlation, it is first necessary to briefly describe the current-voltage (I-V) characteristics of the varistor.

Figure 1 illustrates the well-known I-V characteristic of the ZnO varistor. It is an ohmic resistor up to a current density of ~ 10 A/cm^2 and a field strength of 10^2-10^3 V/cm. This is defined as the prebreakdown linear region. Then, if the electric field strength is incrementally increased, the device becomes non-ohmic (nonlinear) and conducts current densities up to 10^3-10^4 A/cm^2 before a second ohmic region is reached (only the beginning of this region is shown in the figure). The degree of nonlinearity is determined by the flatness of the nonlinear region; the flatter the I-V curve in this region, the better the device. Since the nonlinear coefficient α is defined by: $\alpha = d\ell nI/d\ell nV$, it is clear from Figure 1 that α decreases with increasing current density. For practical applications, α is typically calculated over a region of interest of currents and voltages such that

$$\alpha = \frac{\log I_2/I_1}{\log V_2/V_1} \tag{1}$$

where I_2 and I_1 are the currents at V_2 and V_1 with $V_2 > V_1$. For good varistors, the nonlinear region extends over six to seven orders of magnitude of current densities.[10] The onset of nonlinearity occurs at a voltage which has been conveniently defined[10] as the voltage at 0.5 mA/cm^2 and is designated as $E_{0.5}$ in Figure 1.

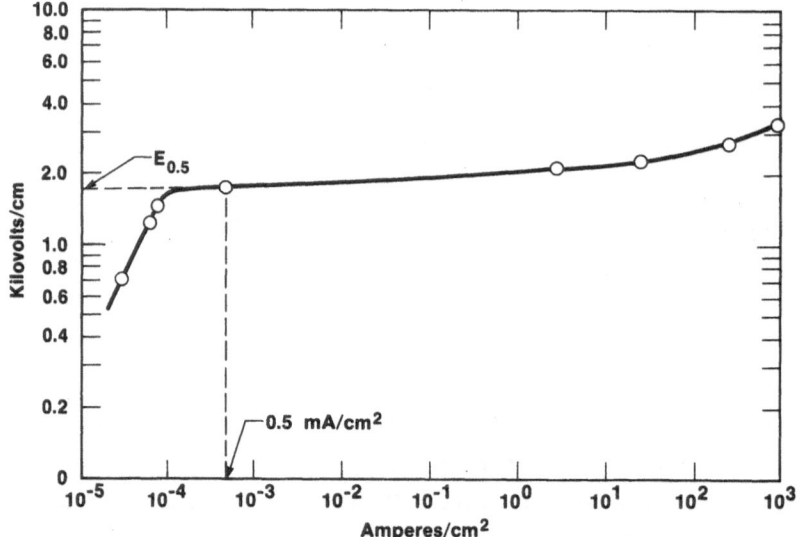

Figure 1. Current—voltage (I–V) characteristic of a typical ZnO varistor from low to high current densities (Ref. 28).

The low current linear region, the intermediate current nonlinear region, and the high current linear region all serve important functions in the design and operation of the surge protector. The low current linear region determines the watt loss during steady application of an external voltage. The nonlinear region determines the clamping voltage upon application of a transient surge with the onset of clamping at $E_{0.5}$. The high current region ($>10^3$ A/cm^2) becomes increasingly important for protection from lightning surges because most lightnings are comprised of currents of high magnitude. However, the linear relation between the current and voltage in this region causes the voltage to go up with current, making the device increasingly less attractive with high magnitude of current densities. It is, therefore, desirable that the flat region be extended and the high current linear region be started as far out in current densities as possible. This is an area of continuing interest to the manufacturers of surge protectors. This paper will discuss how these three regions are closely related to the microstructure and chemistry of the device. This discussion relies partly on existing literature and partly on unpublished data.

MICROSTRUCTURE AND CHEMISTRY

The ZnO varistor is a multiphase polycrystalline ceramic made by conventional sintering of a mixture of oxides, the major component being the ZnO powder. The electrical properties of the ZnO varistor are a sensitive function of the number and type of oxide additions comprising the bulk composition of the device. Commercial varistors can contain five or more additions of other oxides such as Sb, Bi, Co, Mn, Cr, etc. The dependence of the electrical properties on the bulk varistor composition and the complex phase equilibria which must necessarily occur in multicomponent systems has prompted extensive investigations into the microstructural/chemical nature of ZnO varistors.[1,3,11-21]

The microstructural investigations can generally be classified into two broad categories. Those in the first category have emphasized a general characterization of the microstructural features and phase chemistry of the varistor.[1,3,11,15-18,20,22-24] This characterization includes the number, crystallography, and composition of the phases present; the morphology and distribution of these phases in the bulk varistor and usually a correlation between the electrical properties and the salient microstructural features. This category encompasses the early studies on ZnO based binaries[11,15,24] and multicomponent varistors[1] from which the initial concepts on the nature of the electrical barrier were developed. Specifically, the semiconducting ZnO grains were thought to be totally isolated in the bulk device by a highly insulating second phase film. This concept continues to be a subject of considerable controversy. Later microstructural evaluations in this category[16,17,20,22] have focused on the multicomponent varistors which are considerably more complex than the simple binaries.[16-18] Furthermore, the sensitivity of the nonlinearity coefficient, α, and of the electrical properties as a whole have been revealed by systematic variation of the microstructure through control of processing parameters such as composition and thermal treatment.[17,22,23] All told, these microstructural studies have yielded an invaluable understanding of the phase equilibria in ZnO based varistors and have identified the major dopants which control the electrical properties of the ZnO grains and of the grain boundaries. These studies have relied heavily on X-ray diffraction techniques,[1,15,17,20,22,24] optical or scanning electron microscopy (SEM), and electron probe microanalyis.[1,3,15,18,20,22-24] Transmission electron microscopy (TEM) was limited to replication methods[1,11] or dissolution/extraction methods.[13,17,18]

The inability of the general microstructural evaluations to explicitly determine the nature of the electrical barrier at the ZnO boundaries has prompted a second, but limited, class of microstructural studies. These studies have primarily focused on the specific details of the ZnO boundaries[12,14,19,21,25] and therefore, have employed microanalytical techniques with high spatial resolution for both structure and elemental analysis. These techniques have included Auger electron spectroscopy[14] and thin foil transmission electron microscopy,[12,13,19,21] along with related techniques such as direct lattice imaging[12,13,19,21] and scanning/transmission (STEM) microanalysis.[19,21] While a Bi-rich diffused layer at the ZnO boundaries has been identified in both a ZnO-Bi_2O_3 binary[19] and a multicomponent varistor,[14] the remaining studies have failed to clarify the role of various phases on the overall electrical properties of the varistor. However, it became increasingly clear from these studies, that an insulating layer surrounding the ZnO grains was not required for the varistor action and that the nonlinear characteristic resulted from a direct contact of the ZnO grains.[9] A positively charged depletion layer within the grain, compensated by a layer of negative surface charge at the adjacent grain boundary has emerged as the most likely defect configuration of the barrier.[7,26]

The major findings of the microstructural and chemical analysis are summarized in Figure 2. A typical microstructure is shown in Figure 3. The four basic compounds formed are: zinc oxide, spinel, pyrochlore, and several Bi-rich phases. Their location is also indicated in the figure. There are other minor phases which are not readily detectable by conventional techniques. The chemical formulation of the sintered products are complex, and their complexity is further compounded by the nature of the doping elements that are invariably present in each phase. For example, the major dopants in ZnO phase is Co and Mn with a high concentration of Bi and Sb adjacent to grain boundary. The phases at the triple point are rich in Bi, Zn, Sb, and Cr. The spinel and pyrochlore phases are uniformly doped with Mn, Cr, and Co. The effects of these dopants on various phases still remain to be clarified.

MICROSTRUCTURAL/ELECTRICAL CORRELATION

From the accumulated knowledge of microstructure and chemistry, combined with the electrical characteristics of the varistor, a microstructural/electrical model emerges for the ZnO varistor. It is now recognized that the basic building block of the ZnO varistor is the ZnO grain formed as

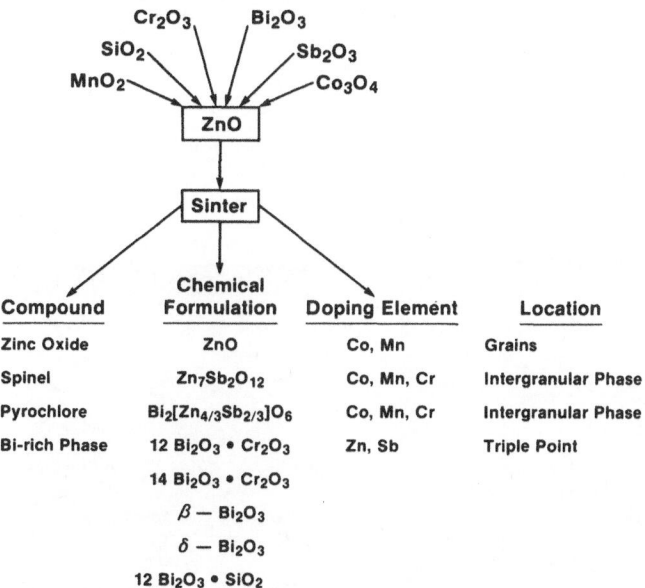

Figure 2. Various crystalline phases, their chemical formulations, and dopants in a typical ZnO varistor.

a result of sintering. During this process, various chemical elements are distributed in such a way in the microstructure that the near grain boundary region becomes highly resistive, $\rho_{gb} \sim 10^{12}$ Ω-cm and the grain interior becomes highly conductive, $\rho_g \sim 1$–10 Ω-cm (Figure 4a), resulting in a sharp drop in resistivity[1,10] from grain boundary to the grain (Figure 4b) within a distance of ~ 500–1000 Å, known as the depletion layer.[5] Thus, at each grain boundary, there exists a depletion layer on either side of the grain boundary[5,7] extending into the adjacent grains. This depletion layer is rich in Bi and Sb whereas the grain interior is rich in Co, Mn, Ni, etc. The varistor action arises as a result of the presence of this depletion layer within the grains. It is a bulk property of the ceramic. Additionally, the presence of two depletion layers adjacent to each grain boundary makes the ZnO varistor insensitive to polarity changes. In this respect the varistor can be described as a back-to-back diode. Furthermore, since the region near the grain boundary is depleted of electrons, a voltage drop appears across the grain boundary upon application of an external voltage. This is known as the barrier voltage (v_{gb}) and is of the order of ~ 2–4 volt/grain boundary.[27] This voltage is comprised of a resistive (R) and a

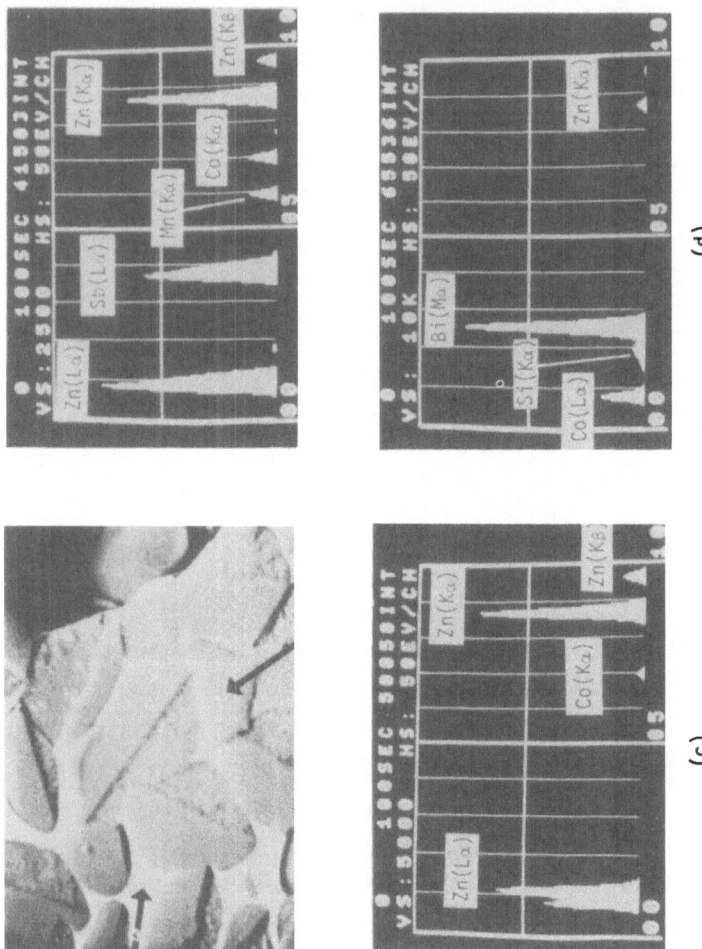

(c)

(d)

Figure 3.   Energy Dispersive X-Ray Analysis (EDAX) on a ZnO ceramic sample with NiO (Sample D-8).
(a) SEM of polished and etched surface (Mag. 2000X); (b), (c), and (d) EDAX data from spinel,
ZnO grain, and bismuth silicate phase, respectively.

capacitive (C) component as shown schematically in the equivalent electrical circuit of Figure 4c.[1,5] Thus, when an AC voltage stress is applied to a ZnO varistor in the pre-breakdown region, the leakage current that flows through the device is entirely of grain boundary origin, and this current again consists of a resistive component and a capacitive component. The roles of these current components in determining the long range stability of the varistor have been discussed elsewhere.[28]

Referring back to Figure 1, we can now cite some correlations between microstructure and electrical characteristic: the low current prebreakdown linear region ($<10^{-4}$ A/cm^2) has been identified to be controlled by the grain boundary resistance and capacitance (see Figure 4c); on the other end of the I-V curve, the high current linear region ($>10^3$ A/cm^2) has been identified to be controlled by the impedance of the grain.[10] The intermedi-

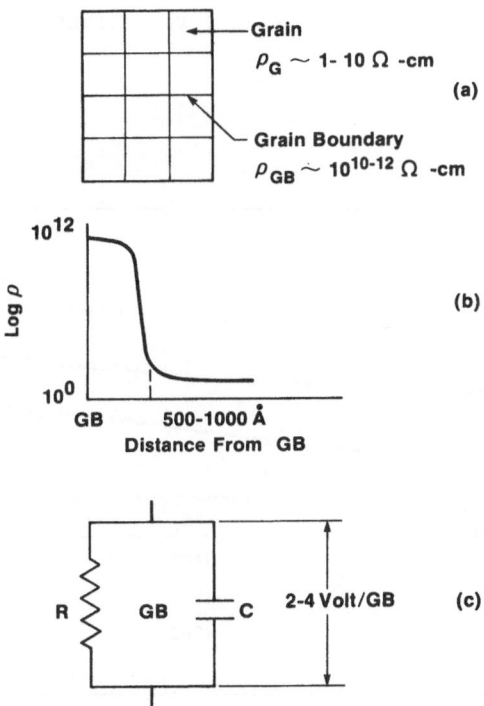

Figure 4. Schematics of microstructure and electrical characteristics. (a) Grain vs. grain boundary resistivity; (b) Resistivity profile at the depletion layer; and (c) Equivalent electrical circuit at the grain boundary (Ref. 28).

ate nonlinear region--the region of major importance for a variety of applications--is indirectly controlled by the resistivity differential between the grain boundary and the grain. If the grain boundary resistivity is enhanced, the prebreakdown linear region is shifted to the left, whereas a decrease in grain resistivity pushes the high current linear region to the right. It then follows that by simultaneously increasing the resistivity of the grain boundary and decreasing the resistivity of the grain, the non-linear region or the flat portion of the I-V curve can be extended over a wider range of current densities, a concept that will be addressed in the next section.

Finally, if the barrier voltage is assumed to be the significant voltage drop in the microstructure, the magnitude of $E_{0.5}$ in Figure 1 can be readily evaluated from the microstructure through the relation

$$E_{0.5} = v_{gb} N_g t \qquad (2)$$

where N_g is a number of grains/cm and t the thickness in cm. Thus, the varistor grain size (GS) manifests its effect through the parameter $N_g \sim (GS)^{-1}$. As in conventional sintering, the grain size in ZnO varistor can be controlled by changing the sintering temperature and time. This then allows one to control the values of $E_{0.5}$ through the phenomena of grain growth during sintering. Figures 5 and 6 illustrate the effect of sintering tempera-

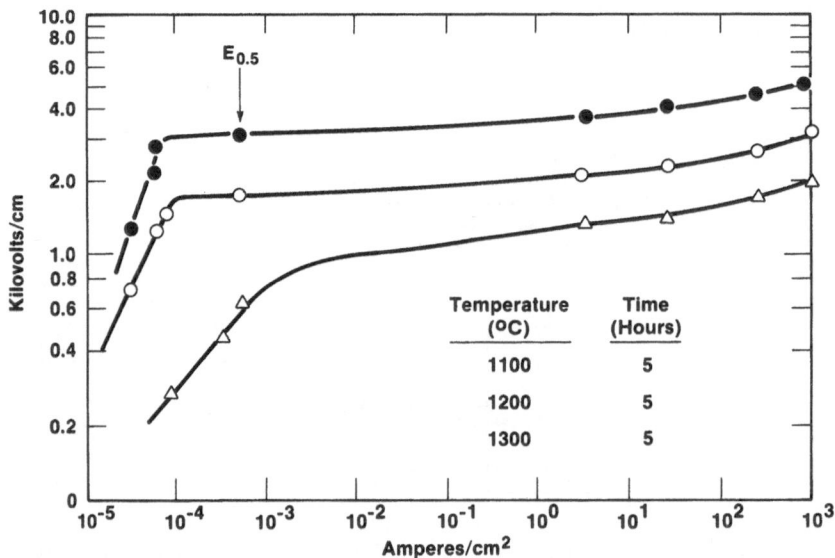

Figure 5. Current-voltage (I-V) characteristics of ZnO varistors made from same composition but sintered at different temperatures for 5 hours (Ref. 2).

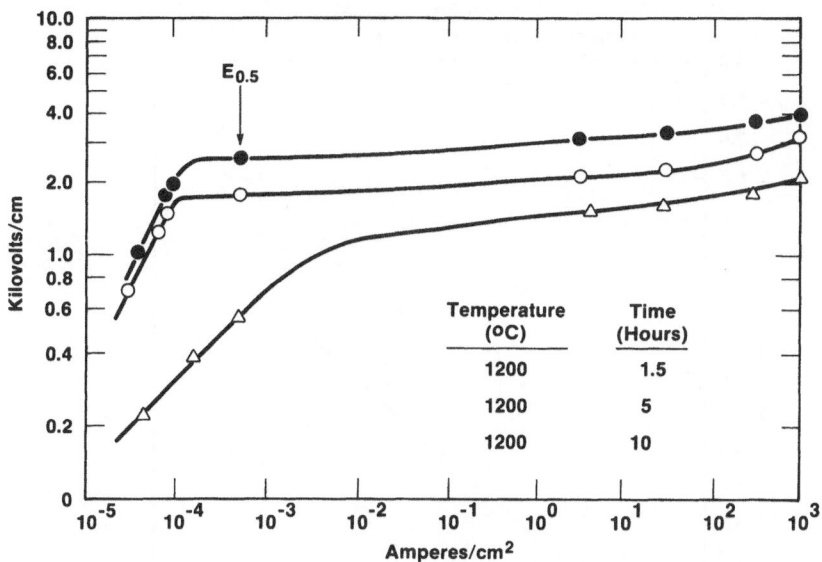

Figure 6. Current-voltage (I-V) characteristics of ZnO varistors made from same composition but sintered at 1200°C for various times (Ref. 2).

ture and time, respectively, not only on the values of $E_{0.5}$ but also on the overall characteristics of the I-V curve, the discussion of which is beyond the scope of this paper. We merely attempt here to point out the effect of grain size on $E_{0.5}$ derived from Figures 5, 6, and similar plots.[2] This correlation is shown in Figure 7. For a limited range of grain size and $E_{0.5}$, the data can be fitted linearly to yield a slope of ~ 2.2 volt/grain boundary, which is a measure of v_{gb} according to Eq. (2). This agrees with the literature data cited[27] previously.

CHEMICAL/ELECTRICAL CORRELATION

So far, we have illustrated the effect of microstructure on the electrical characteristic of the ZnO varistor. Chemistry has an equally important effect on the varistor property and to illustrate this effect, we concentrated on the high current linear region (also known as the "upturn" region) of the I-V curve in Figure 1. We notice a progressively decreasing nonlinearity with increasing current densities. This decrease in nonlinearity at high current densities mitigates the effectiveness of the practical surge protector and, therefore, a means of extending the onset of this voltage rise to higher current densities is always desired.

Since the "upturn" has been recognized to be the grain resistivity limited,[10] the approach taken is that of decreasing the resistivity of the

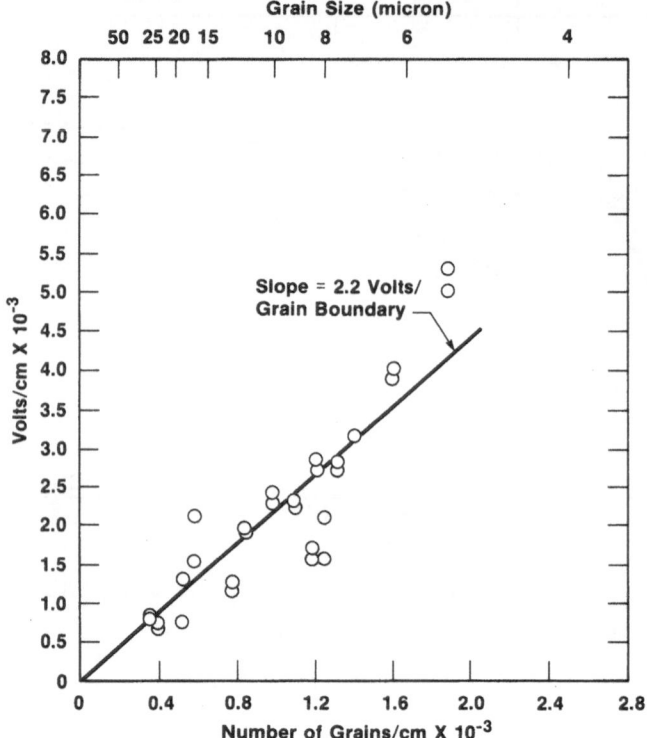

Figure 7. Plot of grain size vs. voltage for estimating the voltage drop at the grain boundary from the slope.

ZnO grain by appropriate doping. The solution becomes obvious by examining the dynamic "apparent resistivity" of the varistor, shown in Figure 8, which is obtained from the I-V curve combined with the knowledge of the resistivities of the grain boundary and the grain. The plot allows the construction of a series of iso-resistivity lines, parallel to one another with the resistivity decreasing from left to right, from that of the grain boundary to the grain. From this simple interpretation of the I-V curve, it is argued that the onset of "upturn" can be delayed to high current densities; i.e., the flatness of the nonlinear region can be extended to higher current densities by decreasing the resistivity of the ZnO grains.

The trivalent dopants are known to act as donors, thus decreasing the resistivity of ZnO ceramics.[29,30] The donor effect on the high current resistivity in the upturn region has been investigated,[10,31] and it was shown that the donor ions such as Al^{3+} and Ga^{3+} can indeed delay the onset

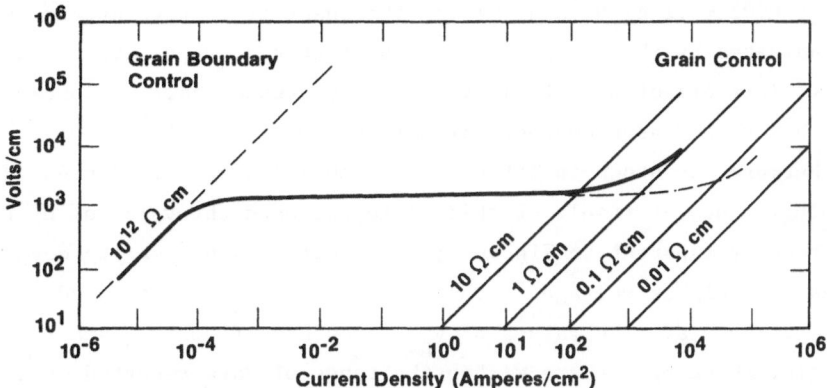

Figure 8.   The current-voltage (I-V) characteristic of a ZnO varistor in
relation to the microstructure of the device: the grain boundary
controlled region at low current densities and the grain controlled region at high current densities.   Also, superimposed are
the iso-resistivity lines representing the slope of the I-V curve
(Ref. 10).

of voltage upturn to higher current densities.   The details of these results
are presented elsewhere.[10]   Suffice is to illustrate here the effect of
Al_2O_3 doping on ZnO varistor I-V characteristics (Figure 9).   The dramatic

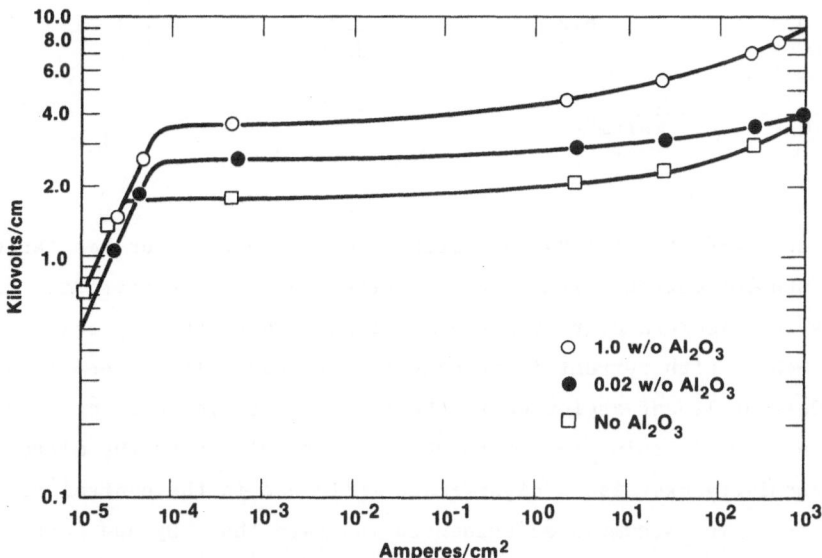

Figure 9.   Effect of Al_2O_3 addition on the flatness of the I-V curve in the
high current region (Ref. 10).

effect of f 0.02 wt.% Al_2O_3 addition on the onset of voltage upturn is
clearly indicated by the extended flatness of this curve to higher current
densities. This effect was attributed to an increased carrier concentration
and as such, an increased conductivity of the ZnO grain.[10]

Furthermore, as shown in Table I, the ZnO grain size is decreased with
Al_2O_3 doping. The net result of this is to increase the value of $E_{0.5}$,
which is readily apparent in Figure 9; the onset of nonlinearity occurs at
an increasingly higher voltage as the amount of Al_2O_3 is increased. Also
shown in the table are the values of v_{gb} calculated from $E_{0.5}$ and the grain
size. Again, these values are within the range of data reported in the
literature[27] and confirms the validity of Eq. (2). The effects of microstructure and chemistry are thus elegantly demonstrated by means of Al_2O_3
doping in ZnO varistor.

TABLE I

EFFECT OF Al_2O_3 ADDITION ON THE VARISTOR GRAIN SIZE AND GRAIN VOLTAGE
FOR ZnO SAMPLES SINTERED AT 1200°C FOR 2 HOURS

w/o Al_2O_3	Average Linear Intercept ZnO Grain Size (μm)	$E_{0.5}$* (volt/cm)	v_{gb} volt/grain
0	15.4	1180	2.7
0.005	12.2	1880	2.3
0.02	8.7	2350	2.0
0.10	8.2	2535	2.1
1.0	7.0	3590	2.5

*Voltage at 0.5 mA/cm^2.

SUMMARY

The current-voltage (I-V) characteristic of ZnO varistor has three
regions: the low current linear region where the I-V characteristic is
ohmic; the intermediate current nonlinear region where the I-V characteristic is non-ohmic with current increasing faster than voltage; and finally,
the high current linear region where the I-V characteristic is again ohmic.
It has been shown in this paper with numerous examples that the microstructure and chemistry exercise a significant influence in the control of all
these regions. The accumulated evidences indicate that, by and large, the
low current linear region is grain boundary resistivity limited, and the
high current linear region is grain resistivity limited; the difference of
these resistivities limits the region of nonlinearity which can extend over
six to seven orders of magnitude of current densities. The high grain

boundary resistivity is caused by the depletion of electrons in this region
as a result of chemical doping with ions such as Bi and Sb. The donor
doping within the grain, on the other hand, can increase the concentration
of the charge carrier and lead to a reduction in the grain resistivity, as
is shown in the example of Al_2O_3 addition to ZnO varistor. Furthermore,
since the grain boundary region is highly resistive, a voltage drop appears
across the grain boundary, with grains sustaining little or no voltage. The
magnitude of this voltage drop is in the range of 2-4 volt/grain boundary,
so the device voltage can be readily controlled by tailoring the grain size
of the varistor. The latter can be accomplished by adjusting the temperature and time of sintering and/or by chemical doping.

REFERENCES

1. M. Matsuoka, "Non-ohmic Properties of Zinc Oxide Ceramics," Jap. J.
    Appl. Phys., 10, pp. 736-746 (1971).

2. A. Sweetana, T. K. Gupta, W. G. Carlson, R. Grekila, N. Kunkle, and
    J. Osterhout, "Gapless Surge Arrestors for Power Systems Applications,"
    EPRI, EL-3166, Vol. 1, Final Report, September 1983.

3. W. G. Morris, "Electrical Properties of ZnO-Bi₂O₃ Ceramics," J. Am.
    Ceram. Soc., 56, pp. 360-364 (1973).

4. L. M. Levinson and H. R. Philipp, "The Physics of Metal Oxide
    Varistors," J. Appl. Phys., 46, pp. 1332-1341 (1975).

5. P. R. Emtage, "The Physics of Zinc Oxide Varistors," J. Appl. Phys.,
    48, pp. 4372-4384 (1977).

6. K. Eda, "Conduction Mechanism of Non-ohmic Zinc Oxide Ceramics,"
    J. Appl. Phys., 49, pp. 2964-2972 (1978).

7. P. L. Hower and T. K. Gupta, "A Barrier Model for ZnO Varistors,"
    J. Appl. Phys., 50, pp. 4847-4855 (1979).

8. G. D. Mahan, L. M. Levinson, and H. R. Philipp, "Theory of Conduction
    in ZnO Varistors," J. Appl. Phys., 50, pp. 2799-2812 (1979).

9. R. Einzinger, "Metal Oxide Varistor Action--A Homojunction Breakdown
    Mechanism," App. Surf. Sci., 1, pp. 329-341 (1978).

10. W. G. Carlson and T. K. Gupta, "Improved Varistor Nonlinearity via
    Donor Impurity Doping," J. Appl. Phys., 53, pp. 5746-5753 (1982).

11. M. Matsuoka, T. Masuyama, and Y. Iida, "Voltage Nonlinearity of Zinc
    Oxide Ceramics Doped with Alkali Earth Metal Oxide," Jap. J. Appl.
    Phys., 8, p. 1275 (1969).

12. D. R. Clarke, "The Microstructural Location of the Intergranular Metal
    Oxide Phase in a Zinc Oxide Varistor," J. Appl. Phys., 49, p. 2407
    (1978).

13. A. T. Santhanam, T. K. Gupta, and W. G. Carlson, "Microstructural Evaluation of Multicomponent ZnO Ceramics," J. Appl. Phys., 50 (2), pp. 852-859 (1979).

14. W. G. Morris, "Physical Properties of the Electrical Barriers in Varistors," J. Vac. Sci. Technol., 13 (4), p. 926 (1976).

15. J. Wong, "Nature of Intergranular Phase in Non-ohmic ZnO Ceramics Containing 0.5 Mol.% Bi_2O_3," J. Am. Ceram. Soc., 57, pp. 357-359 (1974).

16. J. Wong, P. Rao, and E. F. Koch, "Nature of an Intergranular Thin-Film Phase in a Highly Non-ohmic Metal Oxide Varistor," J. Appl. Phys., 46 (4), p. 1827 (1975).

17. M. Inada, "Crystal Phases of Non-ohmic Zinc Oxide Ceramics," Jap. J. Appl. Phys., 17 (1), pp. 1-10 (1978).

18. M. Inada, "Microstructure of Non-ohmic Zinc Oxide Ceramics," Jap. J. Appl. Phys., 17 (4), pp. 673-677 (1978).

19. W. D. Kingery, J. B. Van der Sande, and T. Mitamura, "A Scanning Transmission Electron Microscopy Investigation of Grain Boundary Segregation in a $ZnO-Bi_2O_3$ Varistor," J. Am. Ceram. Soc., 62 (3-4), p. 221 (1979).

20. J. Wong, "Microstructure and Phase Transformation in a Highly Non-ohmic Metal Oxide Varistor Ceramic," J. Appl. Phys., 46 (4), pp. 1653-1659 (1975).

21. P. Williams, O. L. Krivanek, and G. Thomas, "Microstructure-Property Relationship of Rare Earth Zinc Oxide Varistors," J. Appl. Phys., 51 (7), p. 3930 (1980).

22. M. Inada, Formation of Non-ohmic Zinc Oxide Ceramics," Jap. J. Appl. Phys., 19 (3), pp. 409-419 (1980).

23. J. Wong, "Sintering and Varistor Characteristics of $ZnO-Bi_2O_3$ Ceramics," J. Appl. Phys., 51 (8), p. 4453 (1980).

24. J. Wong and W. G. Morris, "Microstructure and Phases in Non-ohmic $ZnO-Bi_2O_3$ Ceramics," Ceram. Bull., 53, p. 816 (1974).

25. D. R. Clarke, "Grain Boundary Segregation in a Commercial ZnO-Based Varistor," J. Appl. Phys., 50 (11), p. 6829 (1979).

26. T. K. Gupta and W. G. Carlson, "A Grain Boundary Defect Model for Instability/Stability of ZnO Varistor," to be published in J. Mat. Sci.

27. T. K. Gupta and W. G. Carlson, "Barrier Voltage and Its Effect on Stability of ZnO Varistor," J. Appl. Phys., 53 (11), pp. 7401-7409 (1982).

28. T. K. Gupta, "60 Hz AC Characteristic of ZnO Varistor Below Breakdown Voltage," 1984 Conference on Electrical Insl. and Dielectric Phenomena, IEEE Elec. Insl. Soc., October 21-25, pp. 437-447 (1984).

29. G. Heiland and E. Mollwo, in <u>Solid State Physics</u>, Vol. 8, Ed. by
    F. Seitz and D. Turnbull, Academic Press, New York, p. 215.

30. R. A. Swalin, in <u>Thermodynamics of Solids</u>, Wiley, New York, 1962,
    Chapter 15.

31. T. Miyoshi, K. Maeda, K. Takahashi, and T. Yamazaki, "Effects of
    Dopants on Characteristics of ZnO Varistors," in <u>Advances in Ceramics</u>,
    Vol. 1, Ed. by L. M. Levinson and D. Hill, Am. Ceram. Soc., 1981,
    p. 309.

INFLUENCE OF CHEMICAL COMPOSITION ON THE BARRIER HEIGHT IN ZnO VARISTORS

Marija Trontelj, Drago Kolar and Viktor Kraševec

Jožef Stefan Institute, University of Ljubljana
Ljubljana, Yugoslavia

INTRODUCTION

ZnO-based varistors exhibit highly nonlinear electrical characteristics, which are due to the electrical properties of the grain boundary regions.

To describe ohmic conduction observed in ZnO-based ceramics at very low currents, a simple model can be adopted in which the varistor conduction results from electrons in the ZnO grains which are thermally exited over a potential barrier in the grain boundary. The potential barrier height was estimated to be 0.7 eV around room temperature for GE-MOV varistors [1] and 0.5 - 0.7 eV for Li and Al doped varistors, depending on the type and concentration of the dopants [2].

Typical varistor compositions contain six to seven oxides, some of them added to create microstructure and which are believed to be electrically inactive. If TiO_2 is added to the standard mixture, varistor grain growth is enhanced [3,4], but its influence on electrical properties has not, to our knowledge, been reported.

The aim of this work was to investigate the influence of dopants such as Na^+, K^+, Ti^{4+} and B^{3+} on the potential barrier height, and to determine the distribution of dopants in the material.

EXPERIMENTAL

Two basic varistor formulations were prepared by mixing oxides of reagent grade purity:

A - ZnO + 0.5 mol % of Bi_2O_3, Co_3O_4, Sb_2O_3, Mn_2O_3 and ZrO_2
B - ZnO + 0.5 mol % of Bi_2O_3, Co_3O_4, Mn_2O_3

Formulation A was doped with 0.2 mol % of Na_2O, K_2O, B_2O_3 or Al_2O_3 in the form of water soluble compounds. To formulation B, TiO_2 was added as an oxide (1.7 wt %) or in the form of a prereacted compounds $Bi_{12}TiO_{20}$ (5 wt % of the compound).

Powder mixtures were pressed into pellets 8mm in diameter and 1mm thick. Sintering was carried out at $1270^{o}C$ in air. For electrodes Ag-based paste, fired at $540^{o}C$ was used.

The current voltage characteristic to 10 mA/cm^2 was measured on an Tektronix 547 oscilloscope. At higher currents the characteristic was obtained by using an instrument constructed in our laboratory, producing normal 8/20 μ sec impulses. The temperature dependence of the resistivity was measured in a thermostatted chamber with a MEGOHM bridge, type 1644-A, General Radio.

The resistivity was measured at 50V on varistors A and at 10V on varistors B.

Specimens for TEM-EDS analyses were prepared from thin sections by the usual ion-beam milling technique. Examinations were carried out on a JEM-2000 FX electron microscope with an EDS attached to it. Suitable boundaries in foils several ten nm thick were analysed using a fixed focused electron beam with an approx. 10nm diameter. The boundaries were considered clean if they did not contain visible traces of a second phase on tilting the boundary by $\pm 30^o$ at a magnification of 400.000, when under observation by TEM and electron difraction. In order to distinguish whether Bi and Ti are associated with the boundary or are incorporated into the ZnO matrix, analyses of ZnO grains with approximately the same thickness were also performed.

The distribution of boron was determined by the neutron induced autoradiographic method. The method is based on registration of radiation damage tracks produced in a solid state nuclear track detector by the passage of charged particles from the ^{10}B (n, α) ^{7} Li nuclear reaction, during the irradiation of the specimen by neutrons [5]. The specimen, a ZnO varistor doped with 0.2 mol % B_2O_3, was irradiated at a neutron fluence of 8.10^{12} n/cm^2. After irradiation the detector foil was etched in 6.25 N NaOH solution at $70^{o}C$ for 30 minutes and photographs were taken.

RESULTS

Chemical and microstructural analysis

Using the TEM-EDS technique it was found that in TiO_2-doped varistor material approximately half of the clean grain boundaries examined contained Bi and Ti. Other elements present in the ED spectrum (Zn, Mn and Co) are associated with the bulk grains. In visible traces of a second phase in the grain boundaries only Bi was detected. Some boundaries are free of Bi and Ti.

Fig. 1 shows the microstructure of a B-doped sample and a typical distribution of boron, obtained by autoradiographic analysis. Dark areas correspond to high boron concentrations. It should be noted that the images of boron-rich particles are somewhat larger due to the unsharpness of the image, which was about 5µm. On encircling areas bordered by black spots, a picture resembling the microstructure is obtained. Clean areas are of approximately the same size as ZnO grains. It was estimated that a few ppb of boron can be detected in a matrix such as ZnO-varistor material, so it can be concluded that ZnO grains do not contain boron. It is mainly incorporated in the intergranular phase, and very likely at grain boundaries.

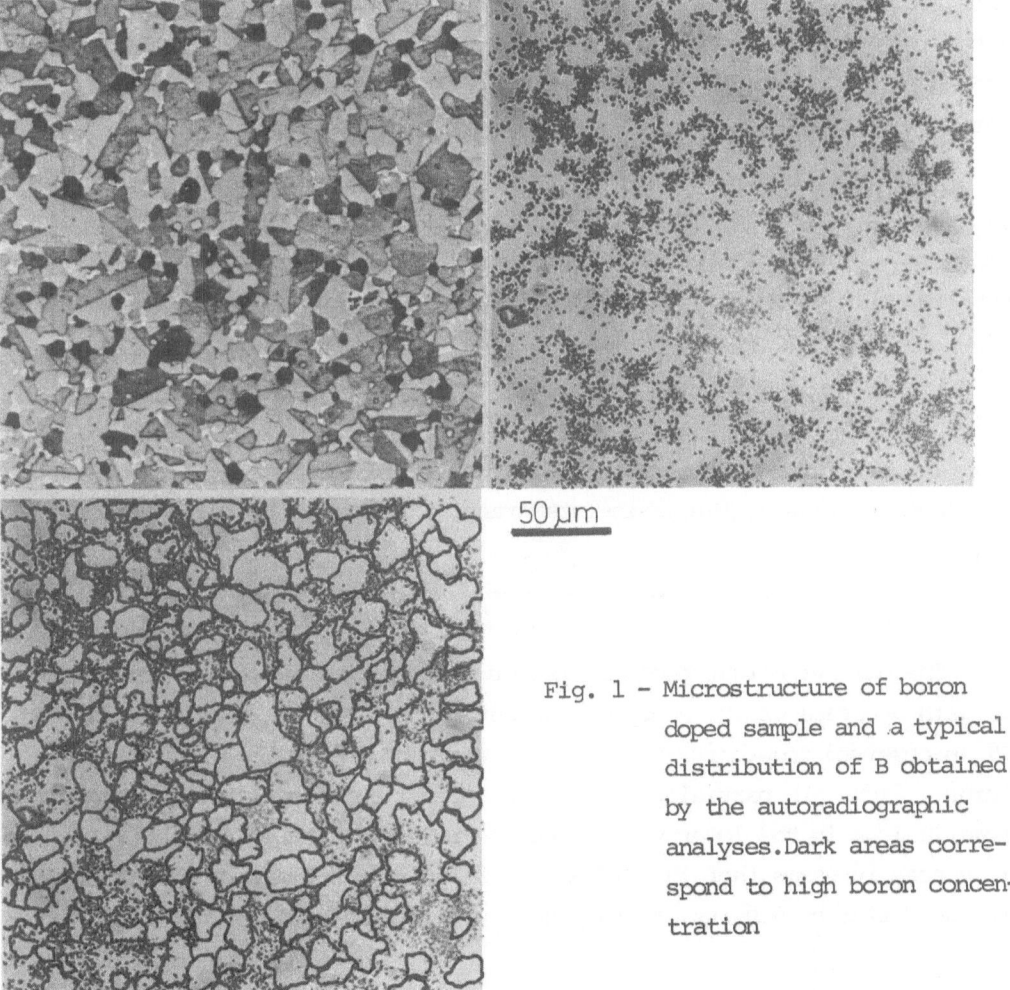

50 µm

Fig. 1 - Microstructure of boron doped sample and a typical distribution of B obtained by the autoradiographic analyses. Dark areas correspond to high boron concentration

Electron microprobe examination of a polished specimen showed no evidence of an increased concentration of K_2O or Na_2O at grain boundaries. Also, the inhibiting effect of K_2O on grain growth was not observed, as was the case in pure ZnO doped with K_2O [6,7]. Most probably the inhibiting effect is affected by other dopants.

Electrical properties

The temperature dependence of ohmic conduction observed at low currents in ZnO-based varistors can be described by the equation.

$$\varrho = \varrho_o \exp{(\phi_B/kT)}$$

where ϱ is the resistivity of the sample, ϱ_o is a constant, ϕ_B is the barrier height, k is the Boltzman constant and T is the temperature in K.

Plotting resistivity vs T^{-1}, an activation energy is obtained from the slope. This activation energy is a measure of the grain - intergranular layer barrier height [1]. The temperature dependence of the resistivity for the five varistor compositions is shown in Fig. 2. Barrier heights calculated from the slopes of the lines are between 0.45 and 0.52 eV for undoped and Na, K and B doped varistors. The barrier height for the Ti doped varistor is 0.29 eV.

In Figs. 3 and 4 voltage - current characteristics are presented. Na_2O and K_2O do not alter U-I the characteristic relative to the undoped varistor. The addition of B_2O_3 increases voltage per 1 mm. The same is observed when monovalent dopants and B_2O_3 were added in pairs. The presence of Al_2O_3 increases the leakage current.

DISCUSSION

The barrier height is related to the donor density and to the density of electron traps at the grain-intergranular interface by the equation:

$$\phi_B = \frac{q^2 N_T^2}{2\varepsilon\,\varepsilon_o\,Nd}$$

The barrier height becomes smaller with increase of donor concentration or with decrease of the acceptor concentration. In TiO_2 doped varistors Ti is observed together with Bi at the grain boundaries. At high temperatures Ti ions are partially reduced to a lower valence state. During cooling some Ti ions in the lower valence state are frozen in, even in air. In this case it seems that the difference in barrier heights can be attributed to the different density of trapping levels at the grain boundaries.

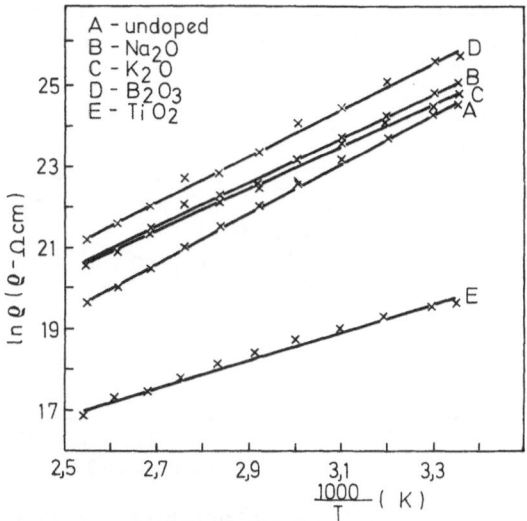

Fig. 2 - The temperature dependence of the resistivity for doped varistors

This explanation is supported by the observation that in TiO_2 doped material the leakage current is much more sensitive to the cooling regime than in the varistor material without TiO_2. Fig. 4 shows the influence of annealing and quenching on the leakage current of TiO_2 - doped material. High conductivity after quenching is ascribed to the large amount of lower valence Ti-ions frozen in.

According to the results it seems that boron does not take an active part in conduction at low voltages. It was shown that trivalent dopants increase the leakage current when built in the lattice[2]. The electrical inactivity of boron is consistent with the results of the analyses of its distribution, which showed that boron is not incorporated in the ZnO grains.

CONCLUSION

Among the dopants whose influence on electrical and microstructural properties were investigated in this work, only TiO_2 showed a lowering of the potential barrier height. It was concluded that its effect is due to its absorbtion on the grain - intergranular phase boundaries. While an increased concentration of K or Na at grain boundaries was not detected,

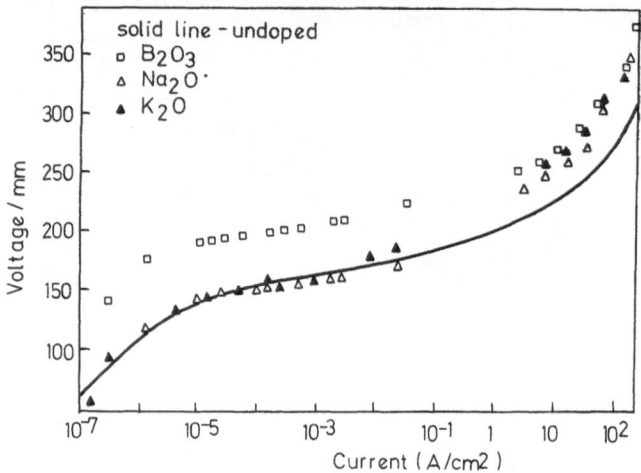

Fig. 3 - Voltage-current characteristics for doped varistors

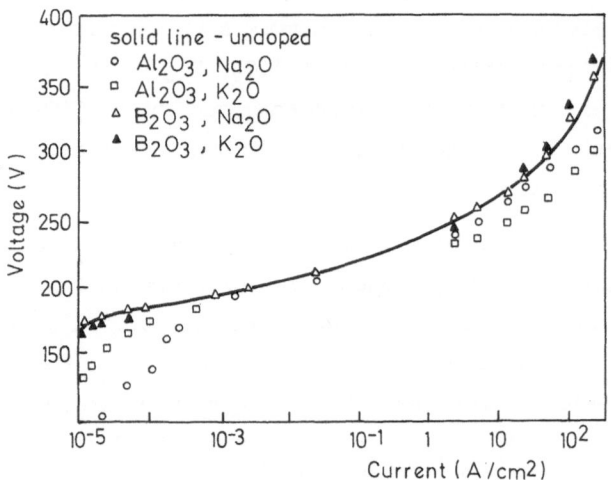

Fig. 4 - Voltage-current characteristics for doped varistors

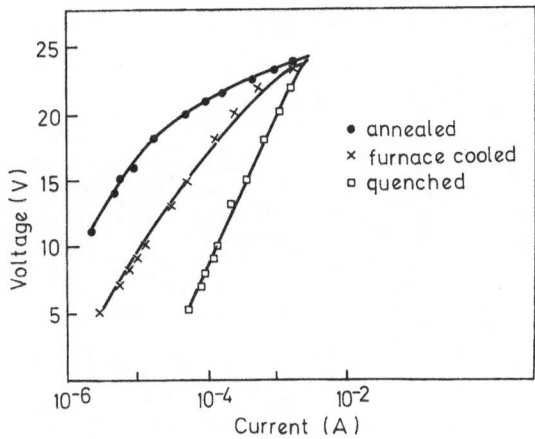

Fig. 5 - The influence of heat treatment on the leakage current of Ti-doped material

B is mainly incorporated in the intergranular phase.

ACKNOWLEDGMENTS

The authors wish to thank A.C.Brown from JEOL, U.K. for TEM-EDS analyses and R.Ilić in this institute for autoradiographic analyses. Technical assistance was given by M. Gec. This work was supported by the Research Council of Slovenia.

REFERENCE

1. M. R. Philipp, L. M. Levinson, J. Appl. Phy. 50, 383 (1979)
2. T. Miyoshi, K. Maeda, K. Takahashi, T. Yamazaki, Advances in Ceramics 1, 309, Columbus: The Am. Cer. Soc. (1981)
3. L. J. Bowen, F. J. Avella, J. Appl. Phys., 54, 27 (1983)
4. M. Trontelj, D. Kolar, Advances in Ceramics 7, 107, Columbus : The Am. Ceram. Soc. (1983)
5. R. L. Fleischer, P. B. Price, R. M. Walker, Nuclear Tracks in Solids, Univ. of California Press, Berkeley,(1975)
6. T. K. Gupta, J. Am. Ceram. Soc. 54 (8) (1971) 413
7. M. Trontelj, D. Kolar, J. Mat. Sci., 12 (1978) 1832

POLYCRYSTALLINE H₃O⁺-β/β″-ALUMINA: A DESIGNED COMPOSITE FOR STEAM ELECTROLYSIS

P.S. Nicholson, M.Z.A. Munshi, and G. Singh
Ceramic Engineering Research Group
Department of Metallurgy and Materials Science
McMaster University
Hamilton, Ontario, Canada

M. Sayer and M.F. Bell
Physics Department, Queen's Univ.
Kingston, Ontario, Canada

ABSTRACT

The synthesis of polycrystalline H_3O^+-β/β''-Al_2O_3 ceramics is described. Design principals are detailed wherein the appropriate strength and conductivity are developed in a precursor K/Na-β/β''-Al_2O_3 ceramic based on microstructural refinement and mixed-alkali/percolation principals respectively. The resulting polycrystals are evaluated in steam electrolysis cells at 100° and 300°C.

1. INTRODUCTION

Single crystal hydronium β''-alumina (0.84 $H_2O\cdot0.84MgO\cdot5Al_2O_3\cdot2.8H_2O$) was reported by Farrington and Briant (1979) to be a fast ion conductor with a proton conductivity of 10^{-5} $(\Omega cm)^{-1}$. These encouraging results were obtained on single crystals or powdered materials and this paper reports the first fabrication, characterisation and use of polycrystalline H_3O-β''/β-Al_2O_3.

β''-Al_2O_3 is a refractory phase and requires 1600°C for sintering to theoretical density. The H_3O^+ β''-Al_2O_3 must be ion-exchanged to the H_3O^+ analogue. The ceramic is destroyed by the requisite expansion of $Na\beta''$-Al_2O_3. It is necessary to produce a refractory form of β''-Al_2O_3 which does not experience the large expansion stresses on exchange or has sufficient strength to withstand them. The production of dense single-phase polycrystalline K-β''-Al_2O_3 would be the most desirable route but K-β''-Al_2O_3 is metastable and sintering inevitably results in K-β-Al_2O_3.

Na-β''/Al_2O_3 ceramics sinter more efficiently than the K analogues (Tennenhouse) (1979). This paper details the development of a mixed-alkali β''/β-Al_2O_3 ceramic composite with sufficient strength to withstand the necessary ion exchange and sufficient conductivity to perform efficiently as a hydrogen-conducting electrolyte.

2. β-ALUMINAS AND H₃O⁺ β/β″-LATTICE DESIGN

The β-Al_2O_3 family of compounds are constructed from spinel-blocks of oxygen and aluminum ions in fcc packing with intervening (conduction) planes containing monovalent or divalent ions and oxygen (05) ion "props". In β-Al_2O_3 there are two spinel blocks/unit cell and in β''-Al_2O_3 there are three. The upper oxygen layer of the spinel block is mirrored across the alkali

517

plane in β–Al₂O₃ but the 3-block structure of β″–Al₂O₃ eliminates this mirror symmetry. The β–Al₂O₃ structure is hexagonal and the β″–Al₂O₃ structure is rhombehedral. The monovalent (or divalent) ions of the conduction planes reside in interstices between the lower and upper oxygen planes and the size of these interstices is governed by the position of these planes relative to each other. The "shapes" of the interstices important to this work are shown in Figure 1. Clearly the

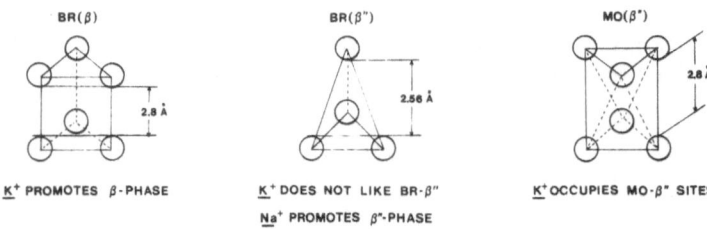

Fig. 1. Schematic of sites in β–β″–Al₂O₃

Beever-Ross site (BR) in the β–Al₂O₃ structure is the largest site and during the synthesis of a mixed alkali (K⁺ = 2.8 Å diameter; Na⁺ = 1.96 Å diameter) β″/β-alumina calcine from the oxides, K⁺ will promote the formation of β–Al₂O₃ and Na⁺ the formation of β″–Al₂O₃ phases. If K⁺ is forced to enter the β″–Al₂O₃ structure, it will tend to the mid-oxygen (MO) site. This preferential site occupancy is postulated as one reason for the so-called mixed-alkali effect (Foster 1979). The relative concentration of alkali ions, the number of conduction planes/unit cell and the defects balancing nonstoichiometry (oxygen interstitials on the conduction plane of β–Al₂O₃ and $(Mg^{2+})'_{Al}$ or $(Li^+)''_{Al}$ in β″–Al₂O₃) will lead to the β″–Al₂O₃ phase having the higher conductivity (Table I). The very low mobility of H_3O^+ in the β–Al₂O₃ structure is possibly due to hydrogen-bonding to 0(5) interstitials. Clearly the β–Al₂O₃ phase is undesirable for the fast conduction of H_3O^+.

Table I. Conductivity Values at 25°C (ohm cm)⁻¹ for β– and β″ –Aluminas

Na –β = 10⁻³	Na –β″ = 10⁻²
K –β = 10⁻⁵	K –β″ = 10⁻⁴
H₃O–β = 10⁻¹¹	H₃O–β″ = 10⁻⁴ –10⁻⁵

H_3O^+ –β″–Al₂O₃ is therefore the desirable phase and, as H_3O^+ and K⁺ have approximately the same diameter, K–β″–Al₂O₃ the desirable precursor. As this cannot be sintered, a sinterable mixed-alkali (Na⁺/K⁺) composition must be developed with sufficient conductivity to facilitate ion-exchange and steam electrolysis/fuel cell performance and sufficient strength to withstand the H_3O^+–K⁺–Na⁺ ion-exchange steps requisite to its synthesis. Knowing that K⁺ will promote the β–Al₂O₃ phase and Na⁺ the β″–Al₂O₃, the volume fraction of β″–Al₂O₃ will be determined by the [Na⁺] and must be optimised with respect to maximum ceramic conductivity and minimum ion exchange stresses.

The question of ceramic conductivity must be approached at two levels, the atomistic and the microstructural. As K⁺ exchanges for Na⁺ in the Na⁺–β″–Al₂O₃ its residence on MO sites will progressively interfere with the Na⁺ mobility. In a typical microdomain of the Na–β″–Al₂O₃ conduction plane, each (BR) site is surrounded by 3 (MO) sites which are each shared between 2 (BR) sites. When larger K⁺ ions can group in threes (Figure 2) around a vacant Na⁺ (BR) site (Foster et al. 1981), the strain energy of the system is relaxable by motion of the K⁺ ions towards

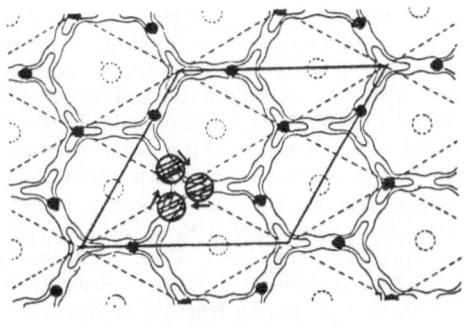

⊜3K⁺(MO)CLUSTERED AROUND A VACANT (BR)

→DIRECTION OF STRESS RELAXATION

Fig. 2. Schematic of the conduction plane of β''–Al_2O_3

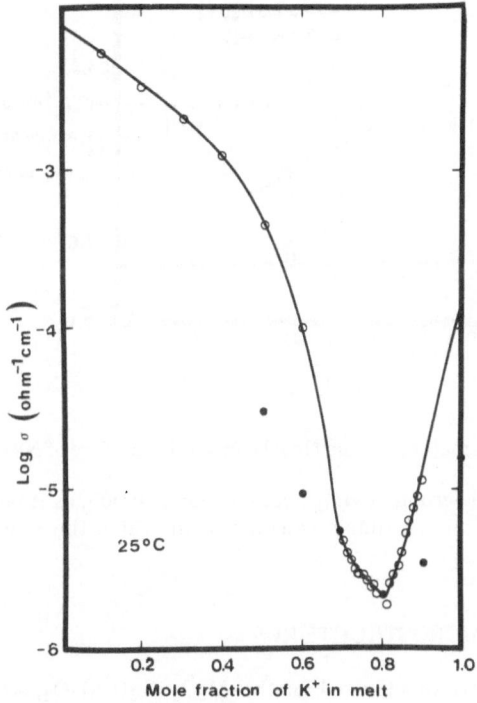

Fig. 3. Conductivity vs. [K⁺] in ion exchange $KNO_3/NaNO_3$ melts at 300°C for 6N3

the BR vacancy. The maximum number of such "clusters" will exist when the $[K+]:[Na+ + K+]$ ratio is ~0.75 and, in view of the relative stability of such complexes, a minimum conductivity (and a maximum activation energy) would be expected for a K^+ mole fraction of 0.75. Ion exchanged Na/K–β/β''–Al_2O_3 results tend to support this conclusion (Figure 3). At ~0.8 K^+–mole fraction in ion-exchange melts, the mole fraction K^+ in the solid is also ~0 8 and a deep conductivity minimum was observed (the ceramic in question, (6N3), is the Na^+–β''–/K^+–β–Al_2O_3 precursor designed and developed in the studies here reported). Clearly K^+ mole fractions approaching 0.8 must be avoided in the β''–Al_2O_3 phase if the ceramic is to have a reasonable conductivity. At a microstructural level, the mixed alkali ceramic will contain a resistive phase (K–β–Al_2O_3) and a conductive phase (Na–β''–Al_2O_3) and percolation considerations suggest the resistive phase must not exceed 0.65 volume fraction of the microstructure (Ast) (1974). The two dimensional nature of conduction in β–aluminas renders a fraction of the conducting β''–Al_2O_3 phase non-conducting (misoriented grains and blocked grain boundaries), and geometric modelling and microscopic observation suggest that 0.33 of the β''–Al_2O_3 is resistively aligned moving the percolation threshold to 0.4 volume-fraction resistive phase ($= K$–β–Al_2O_3 + misaligned Na β''–Al_2O_3) (Bell et al. (1983)). From the conductivity viewpoint therefore the ceramic must contain ≤ 0.4 volume fraction of β–phase (i.e. $f(\beta) \leq 0.4$). The composition required is therefore $(Na_{0.6} K_{0.4})_{1.62} Mg_{0.63} Al_{10.33} O_{17}$ (6N3) and the C_0-lattice parameter of the Na–β''–Al_2O_3 component of this composite must be expanded to that of H_3O^+–β''–Al_2O_3 by ion exchange. The requisite expansion is illustrated in Figure 4. Were the

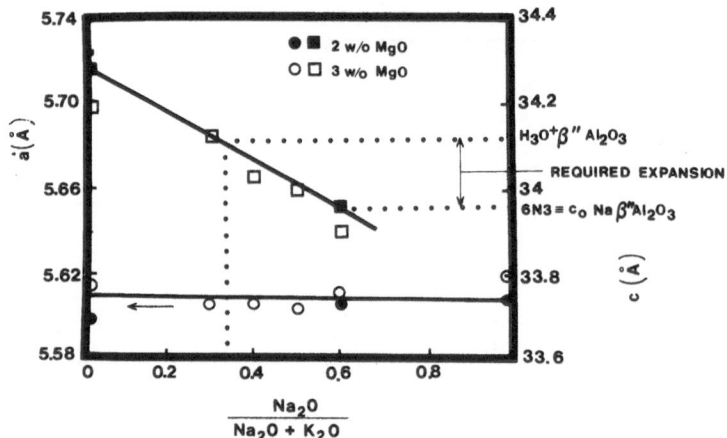

Fig. 4. Lattice parameters versus Na_2O content for sinters of Na–K–β–β''–Al_2O_3

body 100% Na–β''–Al_2O_3, the accompanying strain would be 0.005; as it is only 60% Na–β''–Al_2O_3, the strain involved is 0.003 and a simple calculation indicates the requisite ceramic breaking strength is ≥ 300 MPa.

3. 6N3 SYNTHESIS AND MICROSTRUCTURES

Working phase diagrams were constructed for the Al_2O_3–$K_2O.Al_2O_3$, Al_2O_3–$K_2O Al_2O_3$ (3 wt% MgO) and Al_2O_3–$(KNa)_2O.Al_2O_3$ (3 wt% MgO) systems. The data for these diagrams was gathered via TGA, DTA and sintering/microstructure/X-ray diffraction studies. The resulting diagrams are shown in Figures 5(a), (b) and (c). The phase boundaries and the associated compositions and temperatures indicated in these diagrams are only tentative. The field of coexistence of the β- and β''–Al_2O_3 phases is of major importance. In the Al_2O_3–$K_2O.Al_2O_3$ system, β''–Al_2O_3 has almost no stability (Figure 5a). This result follows from the K^+–BR–in–β–Al_2O_3 site preference discussed earlier. The addition of MgO promotes the β''–Al_2O_3 phase stability and the coexistence dome temperature maximum is increased. (Figure 5(b)) Realising that +1600°C temperatures are

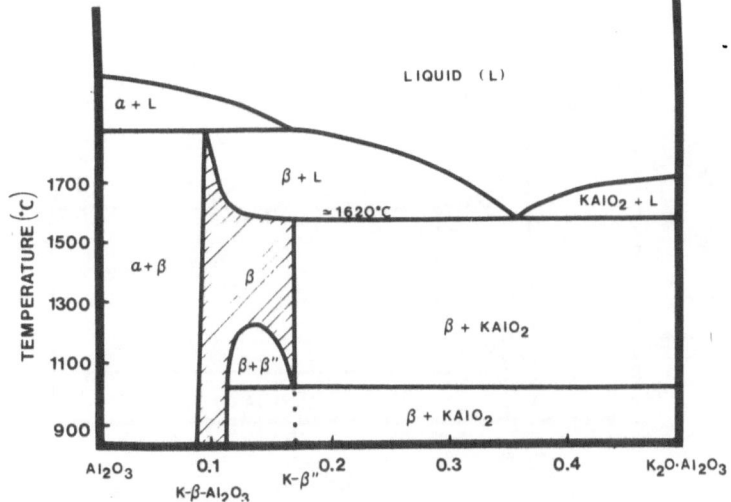

Fig. 5a. Working Al_2O_3–$KAlO_2$ phase diagram

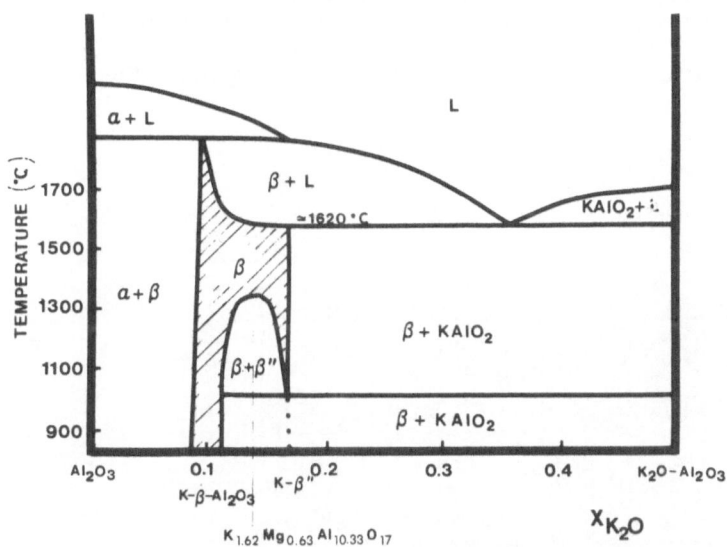

Fig. 5b. Working Al_2O_3–$KAlO_2$ (3 w/o MgO) phase diagram

required for sintering, the coexistence dome maximum must be raised to those temperatures. The substitution of Na_2O for K_2O to the 0.6:0.4 level necessary for 6N3, pushes the dome to these temperatures and results in a ceramic of 0.6 Na–β''–Al_2O_3:0.4 K–β–Al_2O_3. (Figure 5(c)) The β''–Al_2O_3 component of this microstructure will be susceptable to soda loss at sintering temperatures and the resulting increase of f(β) is undesirable. Sintering must therefore be accomplished in a very short time on a calcine of very low f(β). Utilising the liquid-phase/phase field to enhance sintering, the sintering path for 6N3 is superimposed on the phase diagram in Figure 5(c). A calcine with a ~ f(β) < 0.2 is sintered at 1610°C for 1-2 minutes resulting in a dense ceramic (> 95% theoretical density) with an f(β) ≤ 0.4.

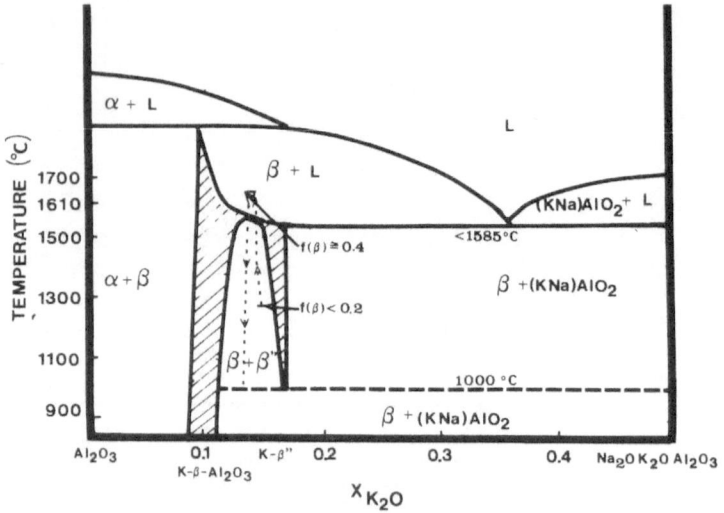

Fig. 5c. Working Al_2O_3–$KAlO_2$–MgO–Na_2O phase diagram with 6N3 heating-path included.

The short sintering times ensure limited grain growth and a fine-grained ceramic results (Figure 6). In this way the natural flaw size in the ceramic is very small (< 2 μm) ensuring its high strength (> 300 MPa). Scanning transmission electron microscopy (STEM) further details the fine grained structure of individual β–Al_2O_3 and β″–Al_2O_3 grains (Figure 7) and intergrowths

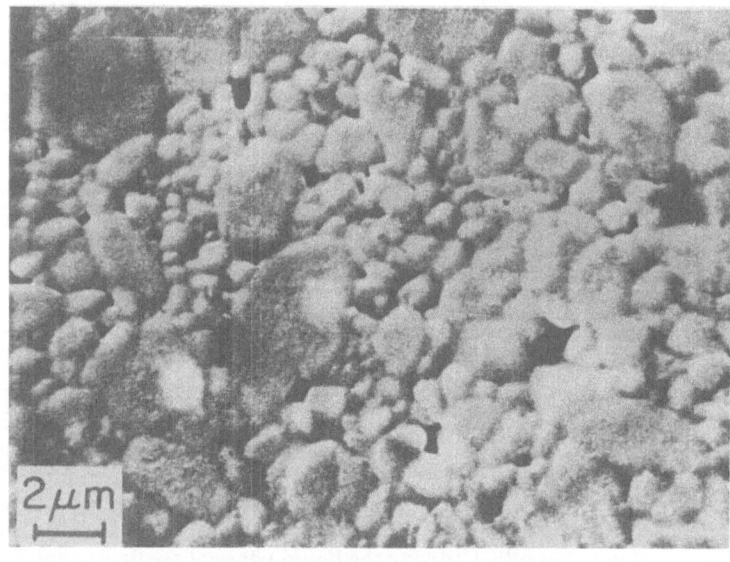

Fig. 6. SEM Microstructure of 6N3

thereof (Figure 8). The selected area microdiffraction pattern of the β–Al_2O_3 and β″–Al_2O_3 phases are shown inserted in Figure 7. Owing to the close coincidence of the hexagonal and orthorhombic

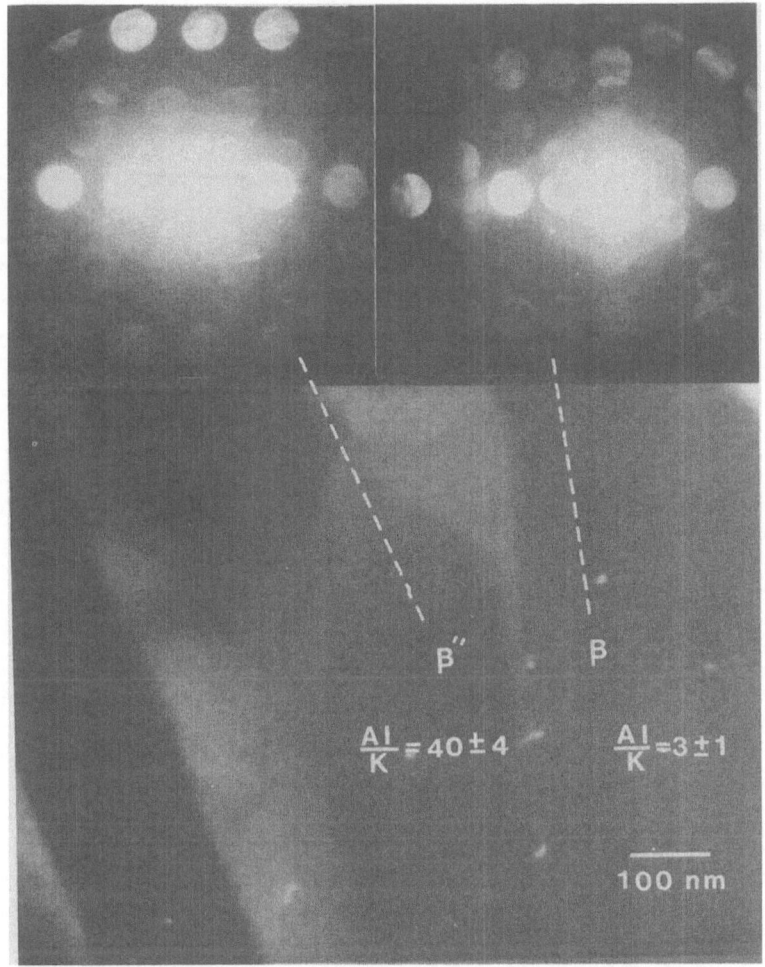

Fig. 7. Individual β and β″ grains in 6N3 with associated micro-electron diffraction patterns

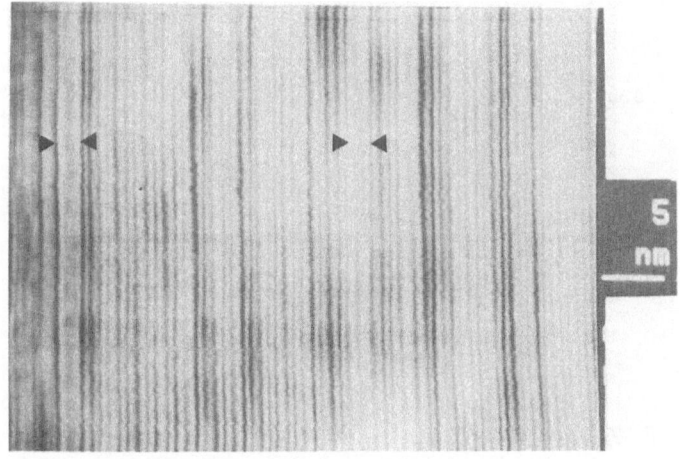

Fig. 8. Intergrowths of β and β″ Al_2O_3 in 6N3 (β–Al_2O_3 between arrows).

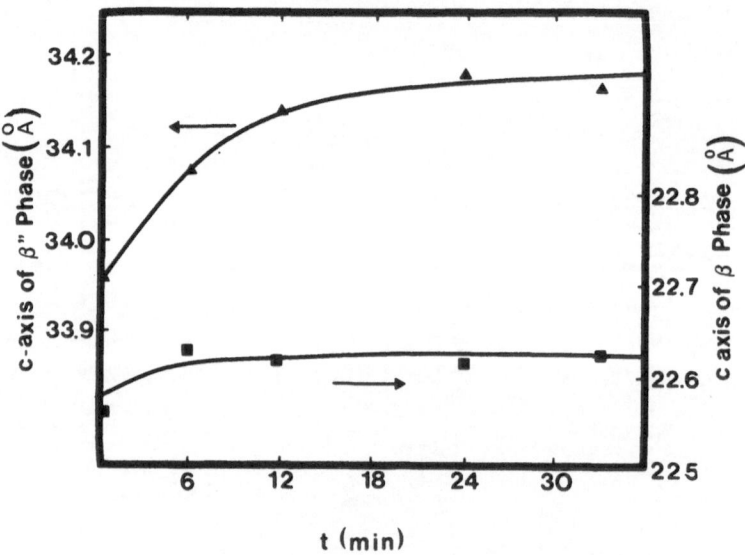

Fig. 9. Variation of c-axis of β-Al$_2$O$_3$ and β″-Al$_2$O$_3$ phases of 6N3 as a function of immersion time in NaCl/KCl melts at 800°C

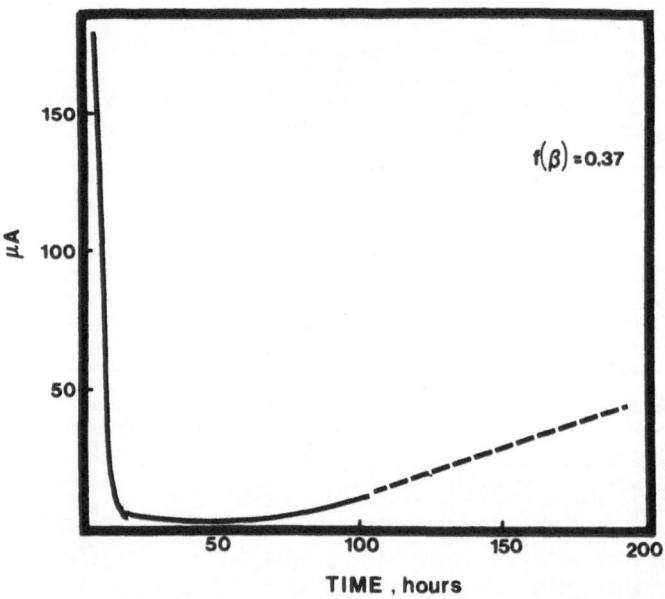

Fig. 10. Current-time curves for field assisted ion exchange to H6N3

reflections of β–Al$_2$O$_3$ and β″–Al$_2$O$_3$ respectively, high index reflections were necessary to identify the two phases. The rhombedral β″–Al$_2$O$_3$ reflection utilised was (02.13) (referred to the hexagonal system). The hexagonal β–Al$_2$O$_3$ crystallites were identified via the 65.15° angle between the (02.13) reflection and the (70.4) reflection. The residence phase of the alkali ions was investigated via STEM/EDAX. The results were only qualitative in view of the instability of the material in the electron beam. The high mobility of Na$^+$ and its lower atomic number made it virtually impossible to count but the slower and heavier K$^+$ was countable in times short enough to preclude significant damage (50 secs). Two probe diameters were used (30 nm and 9 nm) and the Al:K ratio was 3±1 in β–Al$_2$O$_3$ and 40±4 in β″–Al$_2$O$_3$. These values qualitatively confirm the microstructural mixture of K–β–Al$_2$O$_3$ and Na–β″–Al$_2$O$_3$. The STEM lattice image (Figure 8) shows two β–Al$_2$O$_3$ blocks in a β″–Al$_2$O$_3$ matrix. The extinction fringe spacings in the β″–Al$_2$O$_3$ are 1.1 nm, i.e. one third of the rhombedral lattice parameter of 3.3 nm. The 2.2 nm extinction fringes of the hexagonal lattice identify the β–Al$_2$O$_3$ in the microstructure.

4. 6N3 ION EXCHANGE TO K–β″–Al$_2$O$_3$ (K6N3)

As shown in Figure 4, the requisite expansion of the Na–β″–Al$_2$O$_3$ component of the 6N3 microstructure can be accomplished if the Na$_2$O fraction of the phase is reduced to ~0.4 and its K$^+$ content increased to 0.6. Using the data of Kummer,(1972) melts of 0.64 mole fraction K$^+$ are required for this level of ion exchange. In fact, there is an advantage to taking the β″–Al$_2$O$_3$ phase to K–β″–Al$_2$O$_3$ in a melt of K$_2$O mole fraction = 1. Two alternative ion exchange procedures have been used i.e. KNO$_3$/NaNO$_3$ mixtures and KCl/NaCl mixtures. The latter exchange medium has the advantage of higher temperature (800°C vs. 300°C) and shorter times (48 hours vs. 72 hours). Additional advantages are the avoidance of mixed-alkali effects and lower ion-exchange-induced stresses. As a result, the Na–β″–Al$_2$O$_3$ phase can be ion exchanged to K–β″–Al$_2$O$_3$ non-destructively in reasonable times. The nitrate ion-exchange route is more conventional and may be less corrosive of the ceramic although no detrimental effects of the chloride exchange process have been observed. XRD of the K6N3 shows little change of the β–Al$_2$O$_3$ phase (expected as it is K–β–Al$_2$O$_3$) and the predicted expansion of the β″–Al$_2$O$_3$ phase (Figure 9).

5. H$_3$O ION EXCHANGE OF K6N3 TO H6N3

In view of the relative conductivities of H$_3$O$^+$, K$^+$ and Na$^+$ in β″–Al$_2$O$_3$, a field-assisted ion-exchange process was used. Two electrolytes were examined (dilute H$_2$SO$_4$ 25°C: dilute acetic acid 80°C) and the pH of the acids was maintained constant by bubbling H$_2$ over both Pt electrodes. If the β″–Al$_2$O$_3$ phase is mixed-alkali, the relative mobilities of the Na$^+$ and K$^+$ induce a localised mixed-alkali effect ahead of the H$_3$O$^+$ penetration front (Bell et al. (1985)). This results in a marked decrease in current passage with time (Figure 10). If the ceramic contains the K$^+$–β″–Al$_2$O$_3$ phase exclusively and 80°C is used, the exchange is completed in much less time (72 hours vs. 336 hours). This process can be further accelerated by partial H$_3$O$^+$ exchange followed by introduction of the ceramic into a steam-electrolysis cell with a drop of mercury at the cathode. The Hg acts as a sink for residual alkali as it exits from the ceramic to be replaced by H$_3$O$^+$ from the ionised steam. The coherence of the exchanged layer (and therefore the lack of exchange stress development) is illustrated in Figures 11(a) and (b). The H$_3$O$^+$ was first driven off a partially-exchanged sample by heating at 800°C and it was fractured and SEM fractographs taken. The exchanged layer can be seen on the left side of the sample in Figure 11(a) and on the right hand side of the higher-magnification picture shown in Figure 11(b).

6. H6N3 USE IN STEAM ELECTROLYSIS CELLS

The H$_3$O$^+$–β/β″–Al$_2$O$_3$ (H6N3) ceramic has been used in steam electrolysis cells operating at 100°C and 300°C. In both cases, discs of H6N3 were sealed to glass or α–Al$_2$O$_3$ tubes with RTV silicone sealant and porous-platinum electrodes were utilized. Schematics of the two cells are shown in Figures 12(a) and (b). In the 300°C cell, the steam generator and the electrolysis cell reside in neighboring cavities in a large aluminum block heated to 300°C. The steam is maintained at 300°C by heating tape wrapped around the feed pipe between the two units. The characteristics of the two cells are summarised in Table II.

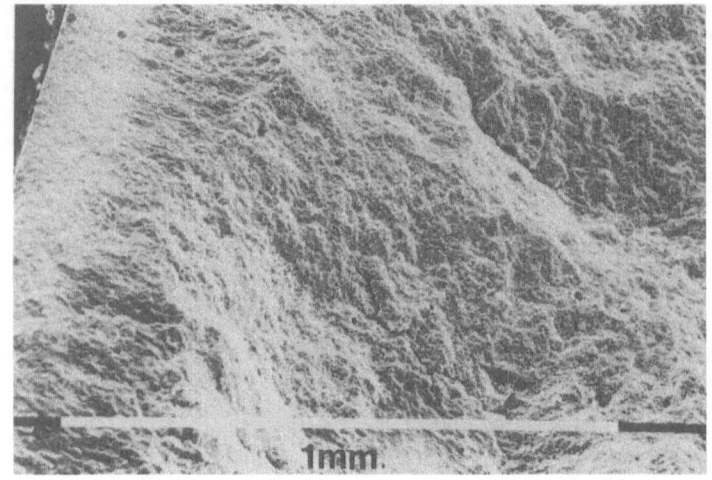

Fig. 11a. Fracture surface of partially H_3O^+ exchanged H6N3

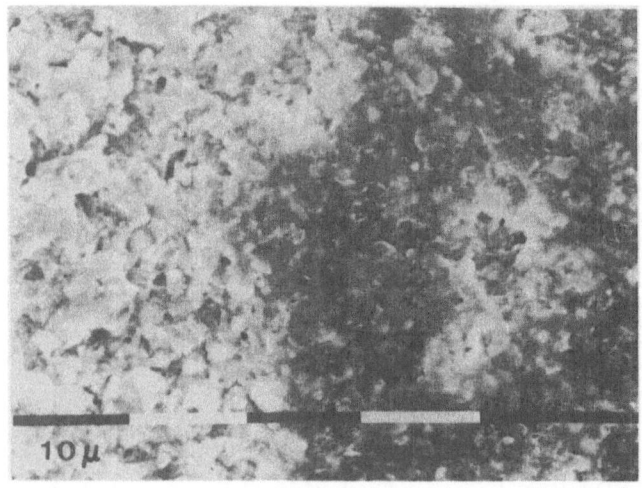

Fig. 11b. Close-up of the interface between H6N3 and K6N3

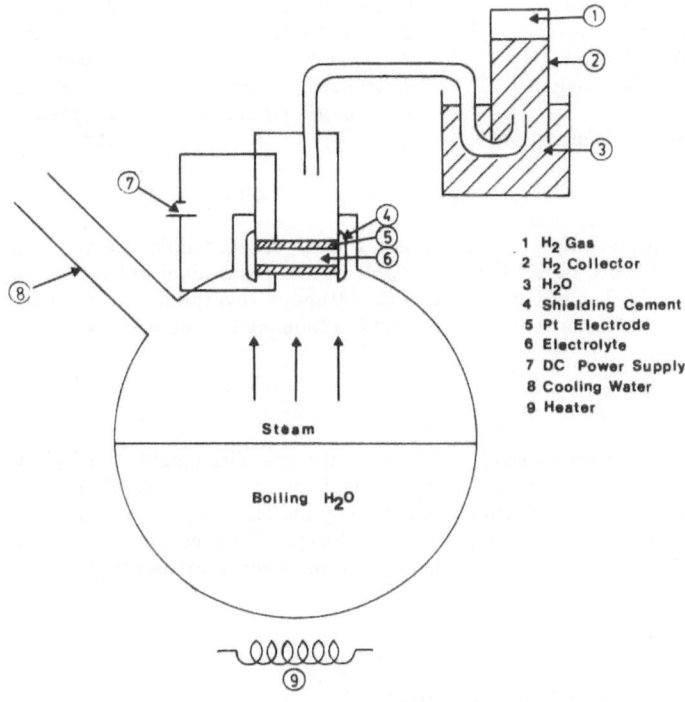

1 H₂ Gas
2 H₂ Collector
3 H₂O
4 Shielding Cement
5 Pt Electrode
6 Electrolyte
7 DC Power Supply
8 Cooling Water
9 Heater

Fig. 12a. Schematic representation of 100°C steam electrolysis unit

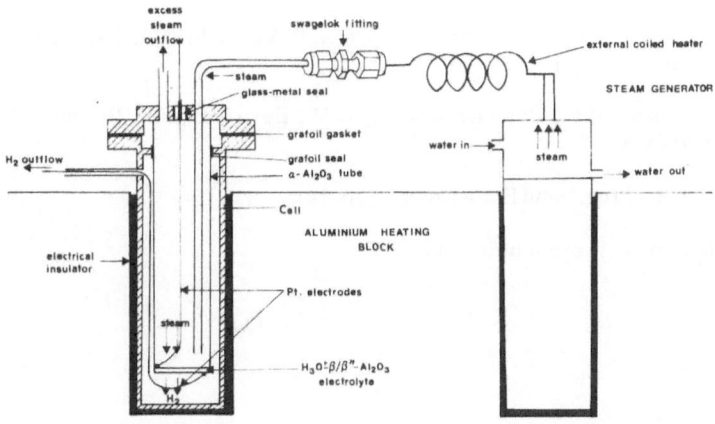

Fig. 12b. Schematic representation of 300°C steam electrolysis unit

Table II. Cell characteristics

Temperature	100°C	300°C
Current	4.1 – 4.6 mA	15 – 17 mA
Electric Power	0.58 – 0.65 W	2 – 2.3 W
Gas Volume (H_2)	27 – 30 cm^3/cm^2/hr	45 – 50 cm^3/cm^2/hr
Total Volume (H_2)	70000 cm^3	500 cm^3
Cell Dimensions	Thickness – 1 mm	Thickness – 2.5 mm
	Diameter – 1.1 cm	Diameter – 1.8 cm
Hours of Operation	2500	12*

* seal failed

The 300°C cell performance characteristics are substantially below those expected by extrapolation the <100°C data. This is possible due to electrode problems but more likely due to partial dehydration of the electrolyte at 200°C. Although reversible, this dehydration will require cell steam pressures > 1 atmosphere if the 300°C efficiency is to be realised.

7. SUMMARY

By combining the unsinterable (K–β″–Al_2O_3) and the unexchangeable (Na– β″–Al_2O_3) in a mixed-phase/mixed-alkali composite ceramic of chemical and phase composition designed to meet the strength and conductivity requirements dictated by induced stresses and mixed alkali/percolation principles, an H_3O^+–conducting β/β″–Al_2O_3 polycrystalline ceramic has been developed. Its performance in steam electrolysis cells up to 300°C has been demonstrated.

REFERENCES

Ast, D.E. (1974). Phys. Rev. Lett., 33, p. 1042.

Bell, M.F. (1), Sayer, M., Smith, D.S., and Nicholson, P.S. (1983). Solid State Ionics, 9/10, p. 731.

Bell, M.F., Sayer, M. and Nicholson, P.S. (1985) (to be published).

Farrington, G.C., and Briant, J.L. (1979). Fast Ion Transport in Solids, eds. P. Vashishta, J.N. Mundy and G.K. Shenoy (North-Holland Amsterdam), p. 395.

Foster, L.M. (1979) Fast Ion Transport in Solids, eds. P. Vashishita, J.N. Mundy and e.K. Shenoy (North Holland Amsterdam), p. 249.

Foster, L.M. Anderson, M.P., Chandrashekhar, G.V., Burns, E., and Bradford, R.B. (1981). J. Chem. Phys. 75 (5) 2412.

Kummer, J.T. (1972). Prog. Solid State Chem., 7, p. 141.

Tennenhouse, E.J., private communication.

THE DEVELOPMENT OF FIBER REINFORCED

GLASSES AND GLASS CERAMICS

Karl M. Prewo

United Technologies Research Center
East Hartford, CT 06108

INTRODUCTION

Historically it can be said that the creation of new types of materials has opened major new avenues to achieving advanced engineering systems and designs. Over the past two decades we have witnessed revolutionary changes in a broad spectrum of applications due to the development of fiber reinforced composites. From sporting goods to industrial and aerospace applications, the availability of these materials has freed the designer from the constraints of metals technology and permitted the development of higher performance systems. This progress has not taken place without difficulty. It has required a whole new way of designing to take into account the tailorability and anisotropy of material properties as well as a whole new way of thinking about material toughness and reliability. These advanced composites, such as carbon fiber reinforced epoxy, exhibit tensile failure strains of less than 1-2% yet they are crack growth resistant and extremely reliable if used properly. It is this successful experience of the past 20 years which has opened up the current real possibility for the success of fiber reinforced ceramics. Combined with a general difficulty in achieving the structural use of monolithic ceramics and the significant effort to find high temperature materials to replace heavy, costly metals has made the current time ideal for such an undertaking.

In this paper it will be shown that fiber reinforced glasses and glass ceramics can make an important contribution to this field of fiber toughened ceramics. The current status of these materials will be described and important aspects of their development traced. It will be clear that, while the potential for the use of these materials is now high, their development has many historical roots.

COMPOSITE SYSTEMS AND THEIR PROCESSING

Fiber reinforced composites of many types have been developed. Using resin, metal and ceramic matrices broad classes of materials have been created that can be conveniently divided on the basis of their matrix types and their potential use temperatures, Fig. 1. It is interesting that of these systems, the glass and ceramic matrix composites offer the greatest range of utility both on a temperature basis as well as based on other environmental considerations such as oxidation resistance, erosion, and chemical attack from acids etc.

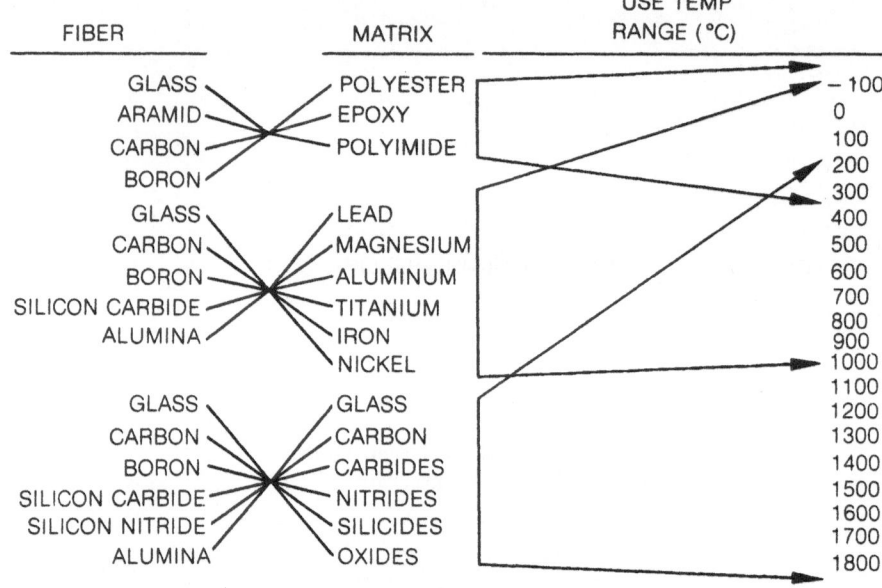

Fig. 1 Fiber reinforced composite systems

From a composite fabrication point of view, the glass matrix composites, as compared to the other ceramic candidates, probably offer the greatest commercial potential from an ease of densification, low cost and also achievement of high performance aspects. The following attributes are important in this sense.

- Glasses can be created with a broad range of chemistries to control fiber-matrix chemical interaction.

- Glasses can be created with a wide range of thermal expansion coefficients to tailor them to nearly match those of reinforcing fibers.

- The low elastic modulus of glasses (50-90 GPa) permits high modulus fibers to provide true reinforcement.

- The ability to control the viscosity of glasses and to flow them easily under pressure permits the physical densification of fiber reinforced composites without mechanical damage to the fibers. It will be shown that relatively high fiber contents can be achieved by several techniques.

- The composite densification process can be rapid since glass matrix flow is all that is required.

Because of the fact that glass can be treated as a thermoplastic material, many of the processes developed for fiber reinforced glasses can be made to emulate those previously used for polymer matrix systems. Sambell, Phillips and Bowen[1] described in detail the development of a procedure for the fabrication of carbon fiber reinforced glasses using a process closely resembling that for polymer matrix systems. Collimated fibers were wound on mandrels after having been infiltrated with a slurry of glass powder. The resultant tapes could then be cut into plys and densified under pressure and high temperature to achieve nearly full density microstructures. Resultant composite flexural strengths were found to depend on bonding conditions and also on the volume percent of fiber reinforcement, Figure 2. While strength increased with fiber content in a manner expected from the rule of mixtures, no further increases, and in fact decreases in strength were noted at 40% fiber and above.

Another approach to precursor tape making was reported[2] where both carbon and glass fibers were cowound to provide a precursor tape. During hot pressing the glass fibers fused together to form the composite matrix. In yet another approach to making precursor tapes, Fitzer, Schlichting and colleagues[2-4] demonstrated that the use of a liquid hydrolyzable metalalkoxide to infiltrate precursor fiber tows provided a superior approach leading to a uniform composite microstructure[2-4], Fig. 3. The hydrolyzed matrix precursor plus fibers is then hot press densified. SiO_2, SiO_2-TiO_2 and Al_2O_3 matrices were all created by this technique.

Other approaches to composite fabrication which do not require the use of precursor tapes and which simulate polymer matrix composite practice have also been reported[5]. Matrix transfer molding, i.e. the injection of hot matrix into a prealigned fiber array, has permitted the fabrication of shapes otherwise not possible. The tube shown in Fig. 4 contains fibers reinforcing it in both the circumferential and radial directions and could not be made by hot pressing. Similarly tubes and other complex shapes can be made using chopped fiber molding compounds[5] that contain both fiber and matrix and can be injected or compression molded into shaped dies at high temperatures.

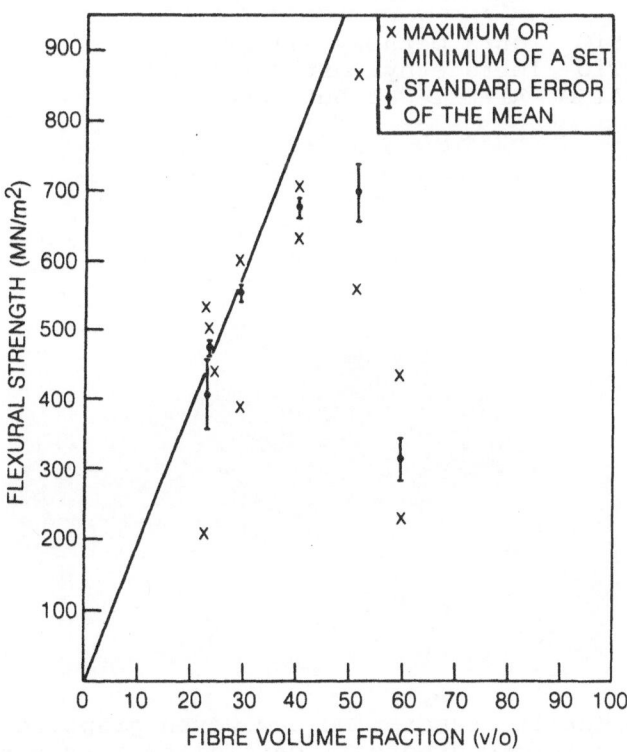

Fig. 2 Flexural strength as a function of volume fraction of fiber in aligned continuous carbon fiber, Ref. 1

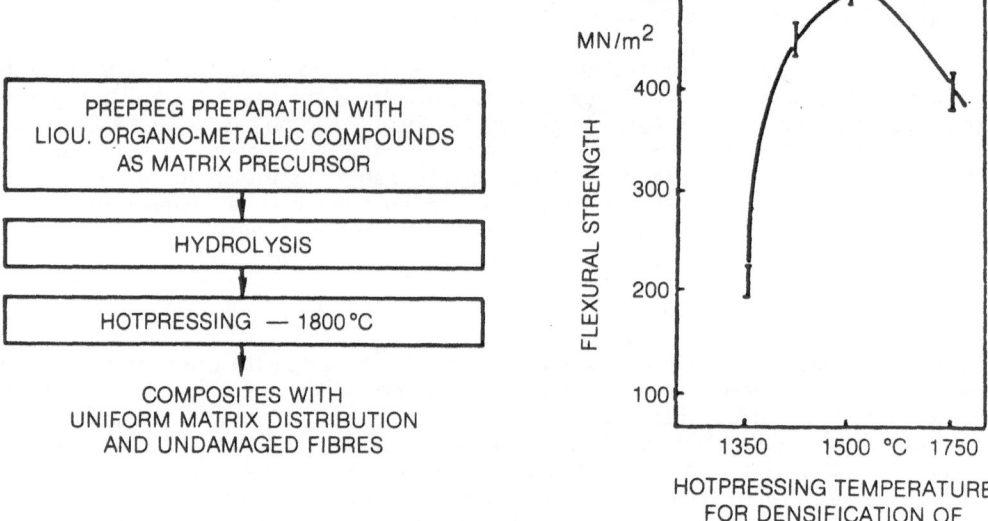

Fig.3 Preparation of composites using metalalkoxides
(Ref. 2, 3, 4). The following starting matrix
materials were used.

For SiO   Tetra methoxy silane
For TiO   Tetra ethyl ortho titanate
For Al O   Al-tri-sec butylate

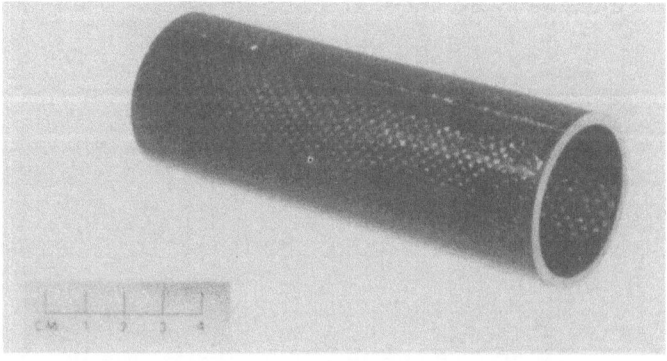

Fig. 4 Matrix transfer molded woven graphite fiber
reinforced borosilicate glass matrix tube

While the above processes were originally demonstrated with glasses, they can also be readily practiced using glass ceramics. The key here is that composite densification takes place while the matrix is in the substantially glassy and viscous form. After full densification the glass can be crystallized under controlled heat treatment conditions to achieve a matrix with superior toughness and high temperature strength.

As a final note, it should be pointed out that the exceptional processability of glasses as matrices has led to some rather ingeneous concepts for the low cost continuous processing of composites [6].

COMPOSITE PROPERTIES

While metal fibers have for many years formed the basis for the reinforcement of glasses and ceramics[7-9] their relatively high density and low thermal stability have prevented any major applications. In contrast the development and availability of advanced high performance fibers such as carbon, boron and silicon carbide has led to composites with exceptional performance potential. It is the characteristics of these latter systems which will be used to illustrate the important aspects of this class of materials.

Carbon Fiber Reinforced Composites

Carbon fibers offer the highest structural performance potential of any of the reinforcements. Available in a wide variety of elastic moduli and at relatively low cost, they offer the opportunity to create economically viable systems right now. Early evidence obtained using a SiO_2 matrix resulted in composites possessing excellent toughness and crack growth resistance as evidenced by non catastrophic failure in bending[10], Fig. 5. Despite being porous, these composites exhibited excellent strength of 350 MPa at both RT and 800°C as well as being unaffected by a water quench from 1200°C. Further evidence for the ability to achieve improved composite toughness for carbon fiber reinforced borosilicate glass, glass ceramic, MgO and Al_2O_3 was demonstrated for discontinuous fibers, however, only in the case where these fibers were carefully aligned or were continuous could composite strength exceed that of the matrix[11,12,13]. The improved toughness could be attributed to low fiber-matrix interface strength which prevented matrix cracks from propagating into the carbon fibers. Similarly, however, the maintenance of a low fiber-matrix strength causes a low composite interlaminar shear strength[14], Fig. 6. This is a significant point in that it causes one to chose between composite "off axis" strength and composite toughness, both of which are considered important performance parameters.

The ability to fabricate a broad range of carbon fiber reinforced glasses and glass ceramics has been demonstrated by several investigators and resultant composite mechanical properties have been characterized for numerous types of test conditions[15-20]. Through the use of very high elastic modulus pitch based carbon fibers composites with exceptionally high elastic moduli were obtained[18,20] Fig. 7. In contrast, using discontinuous carbon fibers of sufficient length permitted the development of composites whose failure strain can exceed 0.5%, whose elastic modulus is less than that of the glass matrix and whose tensile stress-strain behavior can be contrasted with that of resin matrix composites[21], Fig. 8. The very nonlinear stress-strain curve shape for this latter glass matrix system was attributed to the microcracking of the matrix at its failure strain of 0.1-.2% and hence a decrease in composite stiffness with increasing strain. This is unlike the epoxy matrix composite where the matrix failure strain exceeds that of the fibers. The increased compliance of the glass matrix system with increasing strain was also shown to be extremely beneficial to the glass matrix composite during bend testing where, unlike in pure tension, the glass matrix composite flexural beam load carrying capacity exceeded that for the epoxy matrix composite by nearly 50%.

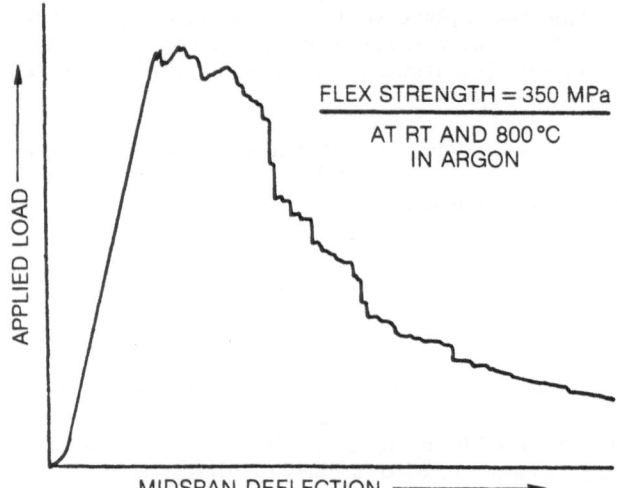

FLEX STRENGTH = 350 MPa

AT RT AND 800°C
IN ARGON

Fig. 5 Applied load vs. deflection for carbon fibers reinforced
SiO at room temperature. Water quenching from 1200°C
has no effect on strength, Ref. 10

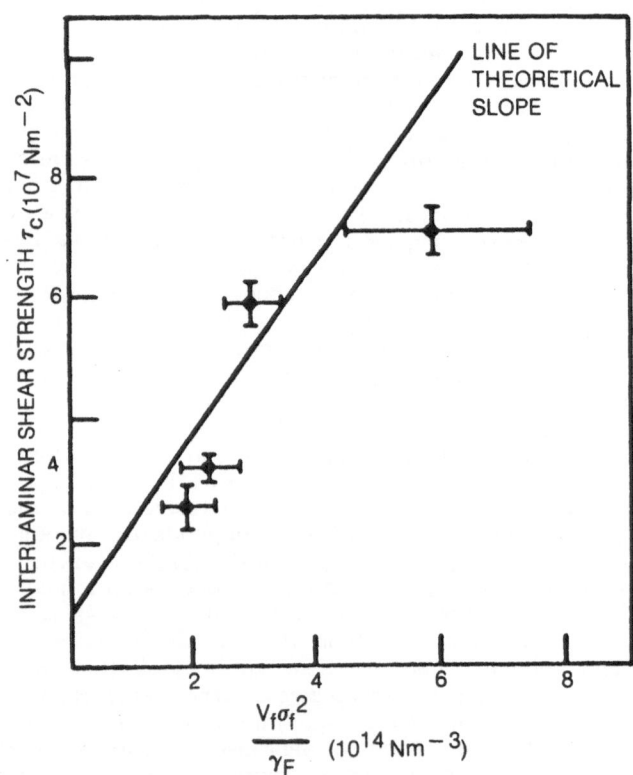

Fig. 6 Composite fracture energy (γ_F) due to fiber pull-out
and composite shear strength, Ref. 14
σ (σ_f = fiber strength, V_f = fiber content)

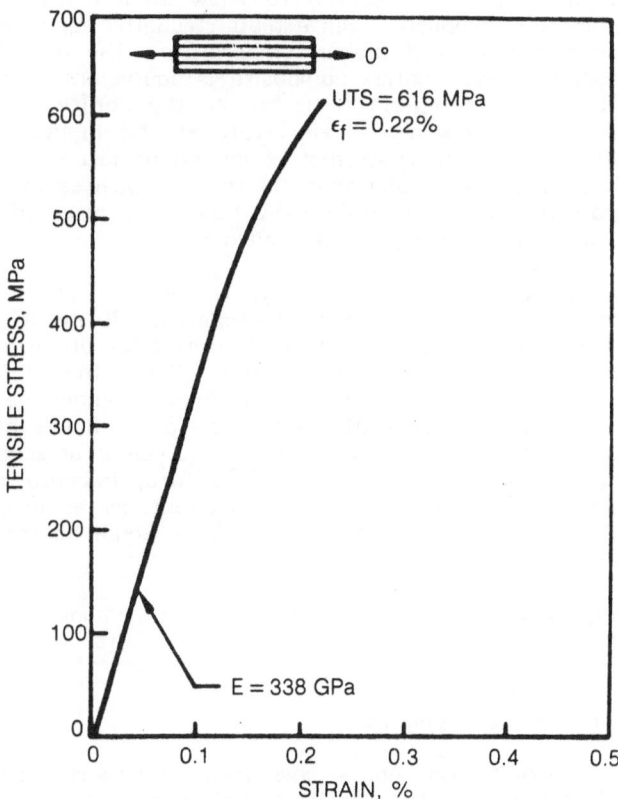

Fig. 7 Tensile stress-strain curve for unidirectional 54 v/o
P-100 carbon fiber reinforced borosilicate glass matrix
composite (Ref. 18, 20)

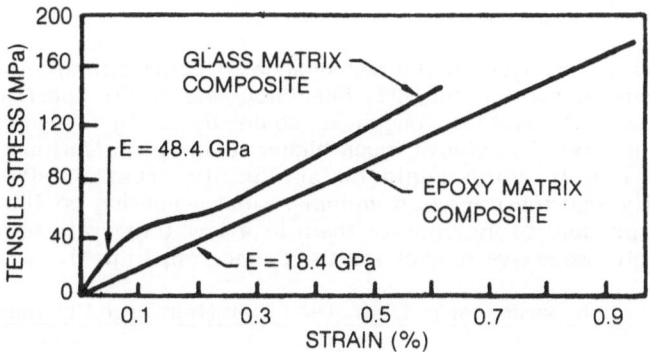

Fig. 8 Tensile stress-strain curves for discontinuous carbon
fiber reinforced borosilicate glass and epoxy. The
fibers are in a 2-D random array. The epoxy matrix composite contains 20 v/o fibers and the borosilicate glass
matrix contains 30 v/o fibers, (Ref. 21)

Further comparison with polymer matrix systems, over a wide test temperature regime,[16], Fig. 9 served to show that continuous carbon fiber reinforced glass matrix composites can exhibit strengths equivalent to their resin matrix counterparts but at much higher temperatures. The fact that the apparent 600°C strength of the glass matrix composite is equivalent to that of the resin matrix material at room temperature is due to the ability of the glass matrix system to redistribute loads more effectively at the higher temperatures. At temperatures above 600°C the glass matrix specimens just deformed and did not fracture at all. Glass ceramic and silica matrix composites extend this strength retention region to over 1000°C,[1,2,18] however, only in a nonoxidizing environment where fiber stability is not limiting.

While most of the emphasis over the years has been on the development of carbon reinforced glass for its mechanical properties, it should not be forgotten that it is a system that can also be extremely useful for other reasons. As in the case of carbon reinforced resins, the carbon fibers impart lubricity to the composite surface and the glass matrix can impart higher hardness and wear resistance[23,24]. The combination of glass and cabon fibers also results in a material with exceptional dimensional stability equivalent or superior to even the most dimensionally stable glasses[18,19,20,25,26]. Also, because of the range of different glasses, carbon fibers and fiber distributions achievable, a range of desired thermal expansion coefficients can also be achieved through tailoring of the composites.

All too frequently the non structural aspects of composite performance are neglected when instead they could become the very first reasons for composite application.

Oxide Fiber Reinforced Composites

Both Al_2O_3 and SiO_2 type fibers have been used to reinforce glasses in the hope of achieving systems with excellent high temperature oxidative stability. Several different types of alumina fibers were used to reinforce high silica glass matrices with the result that modest levels of strength were achieved[27], Fig. 10 and these levels could be maintained up to 1000°C. Through qualitative observations of composite fracture surfaces, however, it was found that composite fracture toughness was much less than that of carbon fiber reinforced composites and this difference was associated with the formation of a much stronger fiber-matrix bond in these systems. Fracture surfaces exhibited only very short lengths of fiber pull out and cracks propagate through these composites much more readily.

A quantitative analysis of the dependence of composite fracture toughness on fiber reinforcement content for SiO_2 fiber/SiO_2 and Al_2O_3 fiber/Al_2O_3 composites showed that overall system toughness could be 2 to 3 times that of the unreinforced matrix[2]. To achieve much higher levels of performance it was shown that an interfacial region could be artifically created between fiber and matrix[20,30]. Through the use of an aluminum metal coating on 100μ diameter SiO_2 fibers, it was possible to incorporate them in a low temperature glass matrix and achieve a notch insensitive impact resistant material, Fig. 11.

Silicon Carbide and Boron Large Diameter Fiber Reinforced Composites

The availability of high strength fibers of boron and silicon cabide produced by chemical vapor deposition has also been actively pursued as an approach to achieving high performance composites. These filaments have been available for nearly as long as carbon fibers but because of their greater cost and their somewhat less convenient and less flexible composite fabrication possibilities they have not received quite as much attention. Boron fiber reinforced glass[6], was shown to provide composites of exceptionally high strength, stiffness and

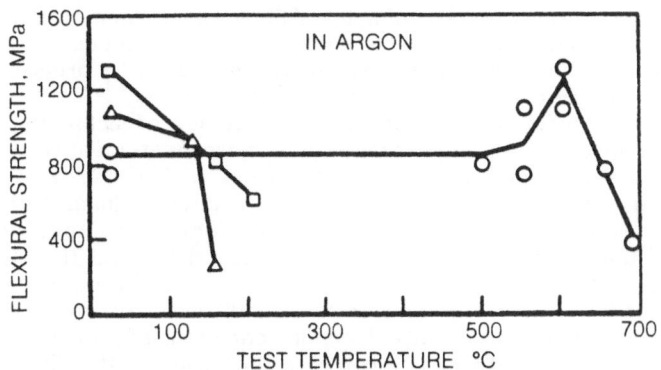

Fig. 9 HM unidirectional carbon fiber reinforced composite
flexural strength comparison for tests performed at temperature in inert argon atmosphere, (Ref. 16)
o-borosilicat glass matrix composite
□-polyether sulfone resin matrix composite
△-polysulfone resin matrix composite

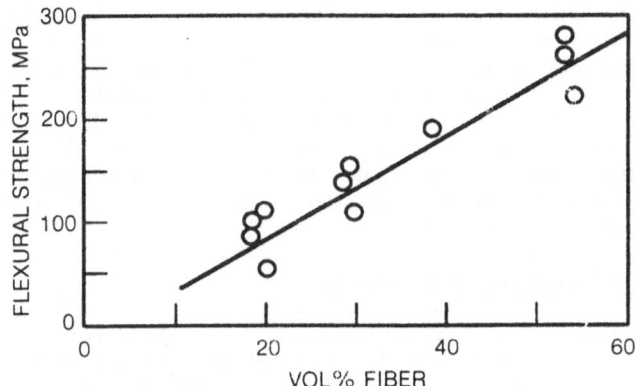

Fig. 10 Room temperature 0° strength of silica matrix composites
as a function of Al_2O_3 fiber content, Ref. 27, 28

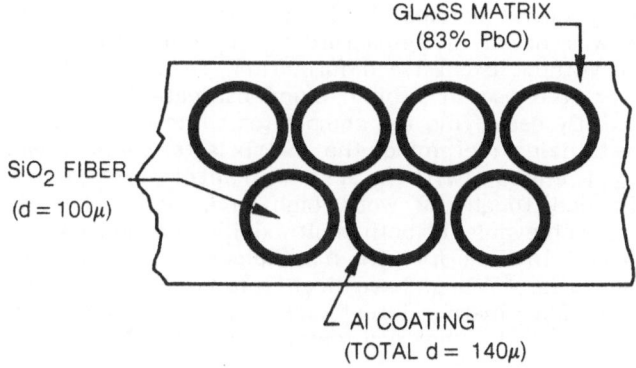

Fig. 11 Metal coated fibers used to create impact resistant
inorganic composites, Ref. 29, 30

toughness. Borosilicate glass reinforced with 30% by volume boron fibers exhibited flexural strengths of 1120 MPa and 980 MPa at room temperature and 540°C respectively. Also, no loss in strength was experienced after 100 hrs. exposure at 540°C. SiC monofilaments, however, were used to create composites with considerably higher use temperature possibilities[33,34]. The use of a cordierite matrix was found to introduce the complexity of achieving the proper heat treatment condition . Post fabrication composite heat treatment principally controlled matrix coefficient of thermal expansion (CTE) and was capable of drastically altering both composite strength and fracture mode. For processing conditions which resulted in composite matrices having a CTE greater than that of the SiC fibers, the composite fracture mode was brittle and the strength low, Fig. 12. This was attributed to a strong mechanical bond between fiber and matrix since, on cooling from the fabrication conditions, the matrix contracted enough to load the fiber-matrix interface in compression. However, by heat treating to create a matrix with CTE much less than that of the fiber, on cooling the fiber-matrix interface is loaded in tension so that, even if a chemical bond is formed during composite fabrication, it is fractured and disrupted during cooling after densification. The resultant composites are nearly 300% stronger than those with the high CTE matrix and the failure mode is now extremely fibrous and tough[33], Fig. 12.

Using a newer form of large diameter SiC fibers that were deposited on a carbon core, borosilicate glass matrix composites were fabricated with high strength and toughness up to test temperatures at which the matrix began to soften[34]. In this case the matrix CTE was less than that of the fiber and the as produced fiber surface was carbon rich. Both of these factors combined in producing a low fiber-matrix bond strength and tough, strong composites, Fig. 13. In a manner similar to that noted in Fig. 9 for cabon fiber reinforced glass composites, the load carrying capabilities of these flexure specimens increased significantly with increasing test temperature and peaked at a temperature (600°C) at which the matrix could deform significantly to redistribute stresses. At higher temperatures the matrix was too compliant and the specimens only deformed and did not fracture at all.

Silicon Carbide Yarn Reinforced Composites

A major increase of interest in the development of fiber reinforced ceramics and glasses can be attributed to the development of a high performance silicon carbide type yarn by Professor Yajima and his coworkers[35]. Available under the name Nicalon with an average tensile strength and elastic modulus of 2060 MPa and 193 GPa respectively this fiber has a unique nonstoichiometric chemistry that makes it particularly suited to the development of high strength glass and glass ceramic matrix composites[34,36,37,38]. Initially, composites with excellent strength were achieved using glass matrices[34,36], Fig. 14 and again an increase in specimen strength was noted at temperatures at which the matrix permitted stress redistribution without excessive deformation. To achieve the highest levels of strength, however, the use of lithium aluminosilicate (LAS) matrices proved most advantageous[37]. By densifying the composites while the matrix is in a glassy state and then crystallizing (ceraming) the matrix afterwards it was possible to fabricate easily yet end up with a very refractory composite. Resultant composite strength and toughness were high and, as in the case of fiber reinforced polymers, achievable in both multiaxially and uniaxially aligned fiber specimens, Fig. 15 and 16. The fracture morphologies of these composites were exceptionally fibrous, Fig. 17, and have been related to the presence of a low strength carbon rich fiber-matrix interfacial region created during composite fabrication and attributable to fiber and matrix chemistry[38].

This low fiber-matrix interfacial strength, while desirable from a toughness point of view, is particularly deleterious to composite off axis strength. The data in Table I summarizes some of the properties of the SiC/LAS composite system

538

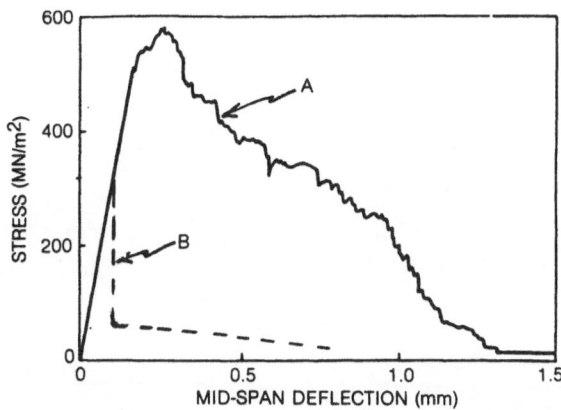

Fig. 12 Flexural stress vs. mid span deflection for SiC large
diameter fiber reinforced cordierite, Ref. 33
A-tough composite with matrix CTE < fiber CTE
B-brittle composite with matrix CTE > fiber CTE

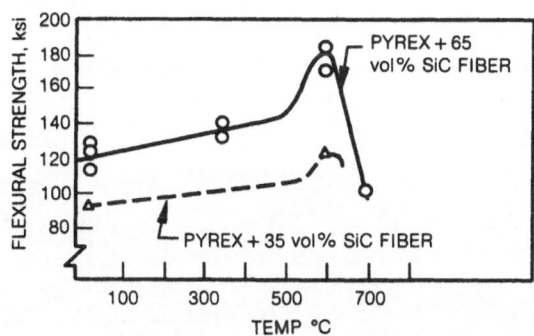

Fig. 13 0°flexural strength in air for SiC large diameter fiber
reinforced borosilicate glass, Ref 34

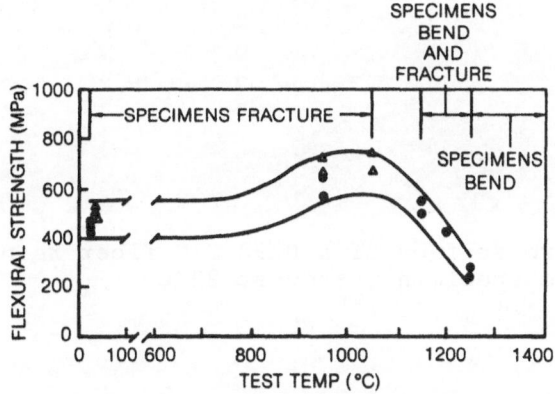

Fig. 14 0° flexural strength of SiC yarn reinforced 96% silica
glass matrix composites tested in argon (Ref. 34, 36)

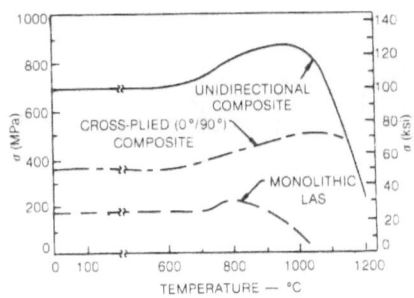

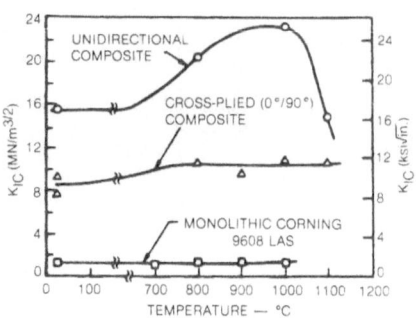

Fig. 15 Three point flexural
        strength in argon of
        SiC reinforced LAS
        glass ceramic, Ref. 37

Fig. 16 Prenotched three point
        bend test determined
        fracture toughness in
        argon of SiC reinforced
        LAS glass ceramic

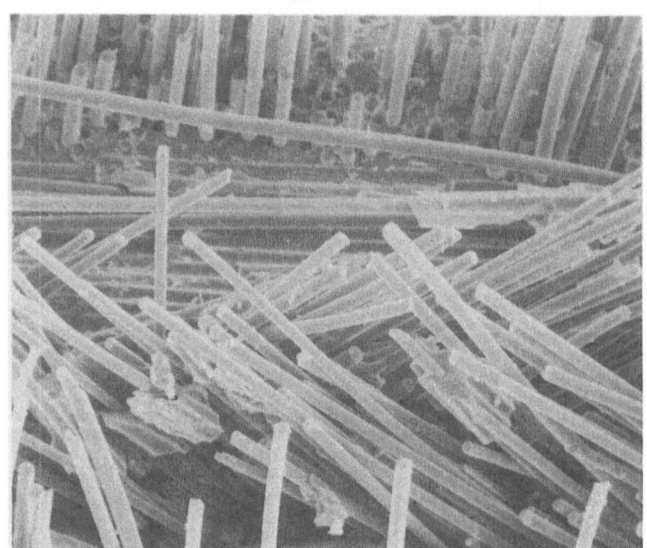

Fig. 17 Fracture surface of a 0/90 SiC fiber reinforced LAS
        tensile specimen tested at 22°C

Table I

Average Properties of Continuous
Nicalon SiC Reinforced
Glass Ceramic (Ceramed LAS)

Unidirectional

$E_0T = 130$ GPa $\qquad$ $E_{90}F = 45$ GPa

$\sigma_0T = 690$ MPa $\qquad$ $\sigma_{90}F = 25$ MPa

$\epsilon_{f_0}T = 0.9\%$ $\qquad$ $\epsilon_{f_{90}}F = 0.06\%$

$CTE_0 = 2.8 \times 10^{-6}{}^{\circ}C^{-1}$ $\qquad$ $CTE_{90} = 1.1 \times 10^{-6}{}^{\circ}C^{-1}$

Vol% Fiber = 46 v/o

$\rho = 2.5$ gm/cm^3

T = tensile test $\quad$ F = flexural test

0/90 Cross Ply

$E_0T = 76$ GPa

$\sigma_0T = 410$ MPa

$\epsilon_{f_0}T = 0.9\%$

$CTE = 2.3 \times 10^{-6}{}^{\circ}C^{-1}$

and show that, for unidirectionally reinforced composites, transverse (90°) strength is less than 5% of axial (0°) strength.

Further testing of these composites in tension has shown that their strength can, in part, be related to the in situ Nicalon fiber tensile strength, and that, when fiber strength and failure strain are great enough to permit matrix failure to occur without overloading the fibers, a nonlinear tensile stress-strain curve results[39], Fig. 18. Also shown in this figure is the tensile stress-strain curve for a Nicalon reinforced epoxy matrix composite[40] which is perfectly linear to failure due to the fact that the epoxy matrix has a higher failure strain than the Nicalon fibers. Repeated mechanical tensile cycling of the LAS matrix composite at increasing values of strain, Fig. 19 results in the observation that composite elastic modulus decreases markedly with increasing strain. This is attributed to the progressive microcracking of the matrix and accompanying decrease in its contribution to the composite stiffness. Eventually the final composite elastic modulus of 88 MPa can be attributed almost solely to the reinforcing fibers. Both the nonlinear shape of this SiC/LAS composite's tensile stress strain curve and also its decreasing stiffness with strain make this composite, just as in the case of the discontinuous carbon reinforced glass composite, Fig. 8, highly tolerant of stress concentrations. A summary of composite tensile and flexural strengths for both the carbon and Nicalon reinforced composites is presented in Table II where it can be seen that in the case of flexure, in which a nonuniform stress state is applied, the glass and LAS matrix composites are significantly stronger than their epoxy matrix counterparts.

While matrix microcracking is the factor which causes this unique stress-strain behavior and accompanying stress gradient tolerance, it also has implications for composite fatigue performance and environmental stability. Composite fatigue resistance is found to be quite good, however dependent on whether the applied stress exceeds the matrix cracking point[41]. Similarly, composite environmental stability is related to this same point[39]. When applied stresses exceed those necessary to cause matrix cracking the surrounding test environment can attack the fiber-matrix interface and change composite fracture morphology from fibrous to relatively brittle. The data in Fig. 20 were obtained by flexure testing in air. At 700°C and above composite strength decreased

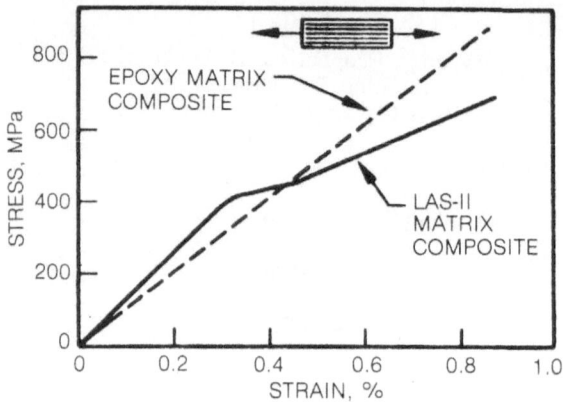

Fig. 18 Tensile stress-strain comparison for unidirectional
SiC reinforced composites, Ref. 39, 40

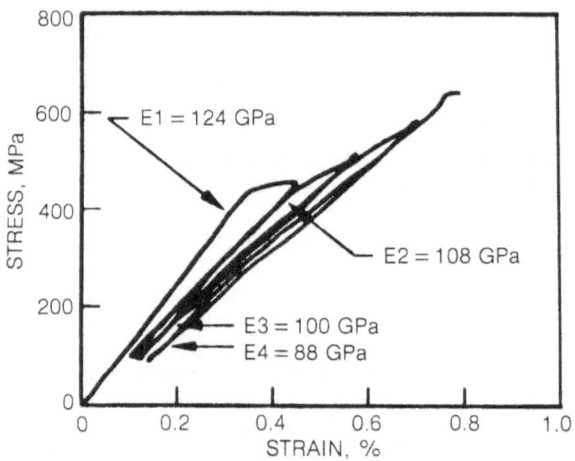

Fig. 19 Cycled tensile stress-strain curves for SiC reinforced
glass ceramic LAS-II, Ref. 39 showing decreasing
elastic modulus with increasing tensile strain

Table II

Composite Strength Comparison

Fiber	Carbon	Carbon	Nicalon	Nicalon
Volume %	20	30	50	44
Orientation	Discon- tinuous	Discon- tinuous	0°	0°
Matrix	Epoxy	Borosilicate	Epoxy	LAS*
Tensile Strength (MPa)	180	150	875	670
Failure strain (%)	0.96	0.63	0.84	0.90
Flexural strength (MPa)	268	400	1240	1380

*Unceramed

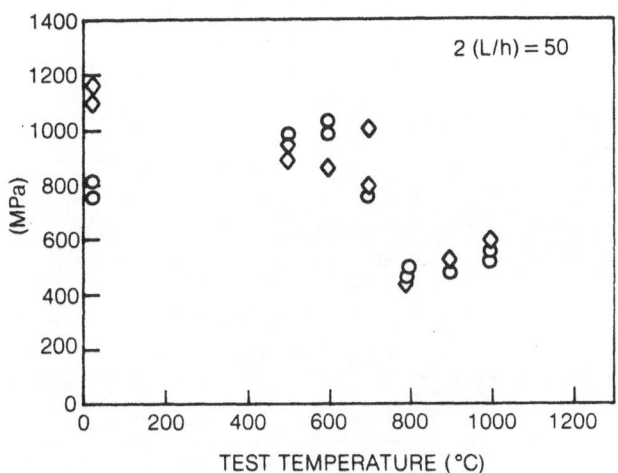

Fig. 20 Flexural strength of 0°-SiC fiber reinforced LAS-II (ceramed) tested in air, Ref. 39

significantly in comparison with the data presented previously in Fig. 15. The testing in air at these temperatures has lowered composite strength to the stress level at which matrix cracking begins to occur. The air then infiltrates the composite and attacks the formerly low strength carbon rich fiber-matrix interface in such a way as to cause an embrittlement of the fracture process,[38][39] Fig. 21.

SUMMARY

From the above presented review of glass matrix composite development it can be seen that a broad range of material combinations has already been explored. Numerous suitable reinforcing fibers are available, matrix compositions have been identified, and fabrication processes have been demonstrated. Also, in all cases it has been found that these composites must be treated as three component systems, i.e. fiber, matrix and fiber-matrix interfacial region. It is this last region of transition which appears to control the fracture process in these composites and hence their relative toughness. The two different types of potential fracture sequences are shown in Fig. 22 for the case where the in-situ matrix failure strain ϵ_m is less than that of the reinforcing fibers ϵ_f. In case A, Fig. 22 total composite failure has occurred when the matrix has failed. A strong fiber-matrix interface has permitted the first cracks, formed in the matrix, to propagate straight through the entire composite. This would typically occur at the matrix failure strain, ϵ_m, and composite failure would be given by the following expression where σ'_f is the fiber strength at a strain equal to matrix failure

$$\sigma_c = \sigma_m V_m + \sigma'_f V_f$$

$$\sigma_c = \epsilon_m [E_m V_m + E_f V_f].$$

A plot of composite tensile strength as a function of fiber volume fraction, V_f, for this type of failure mechanism is shown as line A in Fig. 23 for the case where the fiber elastic modulus E_f is greater than the matrix modulus E_m.

If the fiber-matrix interfacial strength is weak enough to prevent cracks in the matrix from propagating into the fiber, it is possible to achieve significantly stronger composites above a critical fiber content V*. In this case B, Fig. 22, matrix cracking can occur throughout the composite without causing immediate

Fig. 21 Fracture surface of a 0°-SiC fiber reinforced (ceramed)
tested in three point flexure in air at 900°C. The flat
fracture region is closest to the tensile side (bar -
700μm), Ref. 39

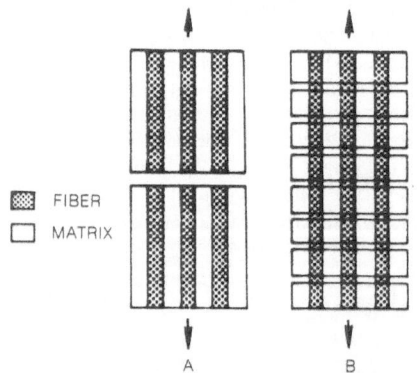

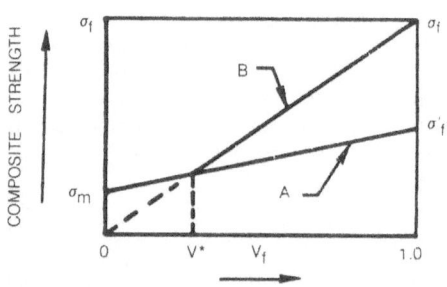

Fig. 22 Fracture types for fiber
reinforced composites.
(A) Matrix and fiber fail
simultaneously in a
strongly bonded composite,
(B) Matrix failure occurs
first in a composite with
low fiber-matrix inter-
facial strength

Fig. 23 Dependence of glass
matrix composite
strength on fiber
content for cases
A and B where matrix
failure occurs first
V_f = volume fraction
fiber

failure. Instead composite failure will occur at the in-situ fiber failure strain and composite ultimate axial strength is given by

$$\sigma_c = \sigma_f V_f$$

In the simplest case there is no effective contribution of matrix to this expression because of the general deterioration of the matrix at its failure strain and σ_f is some complex measure of in situ fiber strength.

It should be noted that composite elastic modulus is affected by this particular sequence of events. For case A the composite elastic modulus can be given by

$$E_o = E_m V_m + E_f V_f$$

For case B this same formula holds, however, only until matirx failure has taken place. Subsequent to matrix failure the composite effective elastic modulus is decreased approximately to the expression

$$E_o = E_f V_f$$

Thus, for a high strength "B" type composite the total stress-strain curve must be nonlinear as shown in Figures 8 and 18.

Major questions still exist as to the significance of this nonlinear stress-strain behavior and whether the added increment of composite strength denoted by the region between curves A and B of Figure 23 can be used in actual engineering applications. Clearly, in the case where the surrounding environment can attack the reinforcing fiber or the important interfacial region, the answer is no, however, in cases where the environment is not very aggressive it may indeed be possible to take full advantage of the strengths associated with a type B fracture. As has been shown, in some cases these composites may even prove to out perform their resin matrix counterparts, Table II.

The following points may be concluded as being important for fiber reinforced glass matrix composites.

- Fiber, matrix and interface must all be tailored to achieve strong, tough composites.

- Low fiber-matrix interfacial strength also implies low off axis strength.

- Nonlinear stress-strain curves and an elastic modulus that decreases with increasing strain will have to be accounted for in successful designs if composites are to see use strain levels above ε_m, the matrix failure strain.

- Flexure tests by themselves are inadequate describe composite stress-strain behavior or strength.

- Environmental stability and composite structural life prediciton will be the key issues for successful composite implementation.

- Non structural composite performance such as tribology and dimensional stability may provide the keys to first composite usage at minimum risk of failure.

It is anticipated that the potential for fiber reinforced glass matrix composites is very significant and that the precise compositions of those materials to be most important are yet to be developed.

REFERENCES

1. R. A. Sambell, D. C. Phillips, D. H. Bowen, "The Technology of Carbon Fibre Reinforced Glasses and Ceramics", Harwell Report AERE-R-7612, Feb. 1974.

2. E. Fitzer, "Fiber Reinforced Ceramics", Proc. Intl. Symp. of Factors in Densification and Sintering of Oxide and Non Oxide Ceramics, 1978, Japan, p. 618-673.

3. M. Sahebkar, J. Schlichting and P. Schubert, "Possibility of Reinforcing Glass by Carbon Fibers", Berichte de Deutschen Keramischen Gesellschaft, Vol. 55, 1978 #5, p. 265-268.

4. J. Schlichting, "Ceramics from Metalloxides", Science of Ceramics, Proceedings of the Tenth Intl. Conference, German Cermaic Society, 1980.

5. G. K. Layden and K. M. Prewo, Advanced Fabrication of SiC Fiber Reinforced Glass Ceramic Matrix Composite, ONR Final Rept R82-915534-1, April 1982.

6. A. C. Siefert, Fiber Ceramic Composites and Method of Producing Same, U.S. Patent 3,575,789, April 20, 1971.

7. I. W. Donald, P. W. McMillan, "Review-Ceramic Matrix Composites", Jl. Mat. Sci, 11, 1976, p. 949.

8. I. W. Donald, P. W. McMillan, "The Influence of Internal Stresses on the Mechanical Behavior of Glass-Ceramic Composites", Jl. Mat. Sci, 12, 1977, p. 290.

9. J. P. Lucas, L. E. Toth, W. W. Gerberich, "A Novel Technique for Producing Five Metal Fibers for Enhancing Mechanical Properties of Glass Matrix Composites", Jl. Am. Cer. Soc., Vol. 63, No. 5, 1980, p. 280.

10. I. Crivelli-Visconti and G. A. Cooper, "Mechanical Properties of a New Carbon Fiber Material", Nature, Vol. 221, Feb. 1969, p. 754-755.

11. R. A. Sambell, D. Bowen, D. C. Phillips, "Carbon Fiber Composites with Ceramic and Glass Matrices, Part 1, Discontinuous Fibers Jl. Mat. Sci, 7, 1972, p. 663.

12. R. A. Sambell, A. Briggs, D. C. Phillips, D. H. Bowen, "Carbon Fiber Composites with Ceramic and Glass Matrices" - Part 2, Continuous Fibers Jl. Mat. Sci, 7, 1972, p. 676.

13. D. C. Phillips, R. A. Sambell, D. H. Bowen, "The Mechanical Properties of Carbon Fiber Reinforced Pyrex", Jl. Mat. Sci, 7, 1972, p. 1454.

14. D. C. Phillips, "Interfacial Bonding and Toughness of Carbon Fiber Reinforced Glass and Glass Ceramics", Jl. Mat. Sci, 9, 1974, p. 1847.

15. S. R. Levitt, "High Strength Graphite Fibre-LAS", Jl. Mat. Sci, 8, 1973, p. 793.

16. K. M. Prewo and J. F. Bacon, "Glass Matrix Composites - I, Graphite Fiber Reinforced Glass", Proc. Second Intl. Conf. on Composites, edited by B. Noton, AIME, 1978.

17. K. M. Prewo, J. F. Bacon and E. R. Thompson, "Graphite Fiber Reinforced Glass", Proceedings of AIME Conf. Advanced Fibers and Composites for Elevated Temperatures, edited by I. Ahmad and B. Noton, 1979.

18. K. M. Prewo and E. R. Thompson, "Research on Graphite Reinforced Glass Matrix Composites", NASA Contract Report 165711, May 1981.

19. K. M. Prewo, J. F. Bacon and D. L. Dicus, "Graphite Fiber Reinforced Glass Matrix Composites", SAMPE Quarterly 10, (4), 42 (1979).

20. K. M. Prewo, E. J. Minford, "Graphite Fiber Reinforced Thermoplastic Matrix Composites for Use at 1000°F", SAMPE Jl. Vol. 21-2, March 1985.

21. K. M. Prewo, "A Compliant, High Failure Strain Fibre Reinforced Glass Matrix Composite", Jl. Mat. Sci. 17, 3549, 1982.

22. K. R. Linger and A. G. Pratchett, "Carbon Fiber Composites for Intermediate Temperatures", Composites, July 1977, p. 139.

23. V. D. Khanna, et al., "Friction and Wear of Glass Matrix-Graphite Fiber Composites", Proc. Mechanical Behavior of Metal Matrix Composites, edited by J. Hock, AIME, 1983.

24. E. Minford and K. Prewo, "Friction and Wear of Graphite Fiber Reinforced Glass Matrix Composites", to be published in the Journal Wear, 1985.
25. K. M. Prewo, "Development of a New Dimensionally and Thermally Stable Composite", Proceedings of the Special Topics in Advanced Composites Mtg, El. Segundo, Calif. 1979.
26. K. M. Prewo and E. J. Minford, "Thermal Stable Composites - Graphite Reinforced Glass", Proceedings of SPIE - the Intl. Society for Optical Engineers, Vol. 505, Aug. 1984.
27. J. Bacon, K. Prewo, R. Veltri, "Glass Matrix Composites - II - Alumina Reinforced Glass", Proc. 1978 Intl. Conf. on Composite Materials, Toronto Canada 1978, Pub. by AIME.
28. E. R. Thompson and K. M. Prewo, "Glass Reinforced by Graphite, Silicon Carbide and Alumina Fibers", AIAA Paper 80-0756-CP, Proc. 21st Structures, Structural Dynamics & Materials Conference, May 1980.
29. A. C. Siefert, Impact Resistant Inorganic Composites, U.S. Patent 3,869,335, March 4, 1975 (applied April 1969).
30. A. C. Siefert, Method of Making Impact Resistant Inorganic Composites, U.S. Patent 3,702,240, Nov. 7, 1972 (applied April 1969).
31. A. C. Siefert, Fiber Ceramic Composites and Method of Producing Same, U.S. Patent 3,575,789 April 20, 1971 (applied Dec. 1966).
32. A. C. Siefert, Fiber Reinforced Ceramics, U.S. patent 3,607,608, Sept. 21, 1971 (applied Jan. 1966).
33. J. Aveston, "Strength and Toughness in Fiber Reinforced Ceramics", Proc. Conf. on Properties of Fibre Composites", IPC Science and Technology Press, 1971.
34. K. M. Prewo and J. J. Brennan, "High Strength Silicon Carbide Fibre Reinforced Glass Matrix Composites", Jl. Mat. Sci, 15, 1980, p. 463.
35. S. Yajima, K. Okamura, J. Hayashi and M. Omori, "Synthesis of Continuous SiC Fibers with High Tensile Strength", Jl. Am. Ceramic Soc., Vol. 58, No. 7-8, 1976, p. 324.
36. K. M. Prewo and J. J. Brennan, "Silicon Carbide Yarn Reinforced Glass Matrix Composites", Jl. Mat. Sci, 17, 1982, p. 1201.
37. J. J. Brennan and K. M. Prewo, "Silicon Carbide Fibre Reinforced Glass-Ceramic Matrix Composites Exhibiting High Strength and Toughness", Jl. Mat. Sci, 17, 1982, 2371.
38. J. J. Brennan, "Interfacial Characterization of Glass and Glass-Ceramic Matrix-Nicalon SiC Fiber Composites", Proc. 21st University Conference on Ceramic Science, Penn. State Univ. July 1985.
39. K. M. Prewo, "Tension and Flexural Strength of SiC Fiber Reinforced Glass Ceramics", to be published Jl. Am. Cer. Soc.
40. J. R. Strife and K. M. Prewo, "Silicon Carbide Fibre Reinforced Resin Matrix Composites", Jl. Mat. Sci, 17, 1982, 65.
41. E. J. Minford and K. M. Prewo, "Fatigue Behavior of SiC Fiber Reinforced LAS Glass Ceramic", Proc. 21st University Conference on Ceramic Science, Penn State Univ. July 1985.

INTERFACIAL CHARACTERIZATION OF GLASS AND GLASS-CERAMIC MATRIX/NICALON SiC FIBER COMPOSITES

John J. Brennan

United Technologies Research Center
East Hartford, CT

ABSTRACT

The strong and tough composite systems consisting of Nicalon SiC yarn reinforced lithium aluminosilicate (LAS) glass-ceramics, with and without Nb_2O_5 additive as a means to form a NbC reaction layer around the SiC fibers during fabrication, have been characterized through the use of TEM replica and thin foil analysis and scanning Auger microprobe (SAM) analysis of composite fracture surfaces. From these studies, it has been found that the chemistry of the fibers within a few hundred angstroms of the fiber/matrix interface has changed significantly from as-received fibers in that a carbon rich zone of 100-400Å in thickness has formed at the fiber surface and that aluminum has diffused into the fibers from the glass-ceramic matrix.

From SAM analysis of Nicalon fiber surfaces from extremely weak and brittle glass and glass-ceramic matrix composites, this carbon rich layer is either nonexistent or much reduced in carbon content. It appears, therefore, that the formation of this carbon rich interfacial zone in the LAS matrix composites under study leads to quite weak bonding at the fiber/matrix interface that directly contributes to the high toughness observed for these materials.

INTRODUCTION

The ceramic composite systems based on the reinforcement of glass and glass-ceramic matrices with Nicalon* SiC fibers have been under study at United Technologies Research Center (UTRC) for the past several years.[1-3] The basic goal of this work was to develop a high strength, high toughness, low density ceramic matrix composite that exhibited a use temperature of at least 1000ºC. This goal was essentially met with the development of the lithium aluminosilicate (LAS) glass-ceramic matrix/Nicalon fiber composite system.[3] However, during the course of this investigation it was found that certain fabrication conditions and/or matrix compositions resulted in quite weak and brittle composites with fracture characteristics quite different than that usually observed for this class of composite. For example, Fig. 1 shows the

---

*Nippon Carbon Co. Ltd., Tokyo, Japan

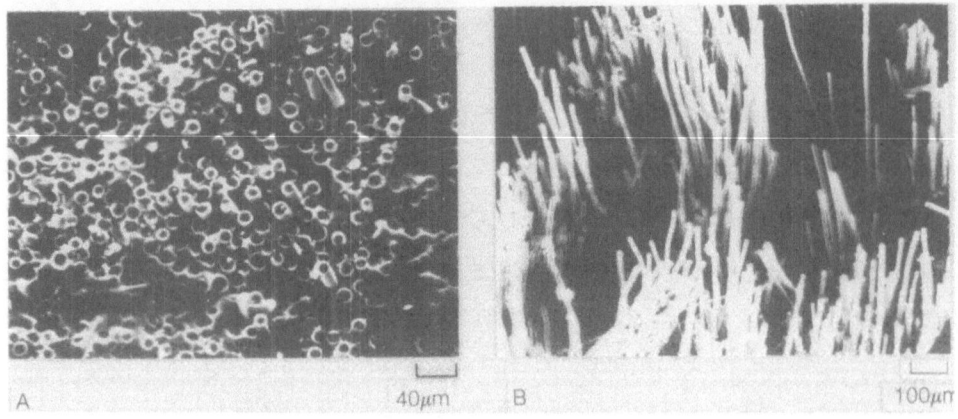

Fig. 1. Fracture surfaces of LAS matrix/Nicalon SiC fiber
composites, (A) weak [σ = 31 ksi (214 MPa)] and
(B) strong [σ = 119 ksi (820 MPa)]

fracture surfaces of two different LAS matrix/Nicalon SiC fiber composites;
one is a very weak [31 ksi (214 MPa)] and brittle material while the other is
strong [119 ksi (820 MPa)] with a very tough and fibrous mode of fracture. In
order to gain an understanding of the factors controlling the type of fracture
behavior observed in these composite systems, a comprehensive scanning transmission electron microscope (STEM) analysis of replicas of polished composite
cross-sections as well as ion beam thinned composite sections and scanning
Auger microprobe (SAM) analysis of composite fracture surfaces was
undertaken. These analytical methods allowed the phase assemblages and
microchemical composition of the fiber/matrix interfacial regions to be
determined to a fine degree of spatial resolution. The results of this study are
discussed in the following sections of this paper.

MATERIALS

The SiC fiber used throughout this study is that produced by Nippon
Carbon Co. in Japan and currently distributed in this country by Dow Corning
Corp., Midland, Michigan. The fibers are obtained on spools of continuous
length (~500 m) tows of 500 fibers/tow with an average fiber diameter of
~15 µm. The average tensile strength and elastic modulus of this fiber, as
measured at UTRC, is 2400 MPa (350 ksi) and 193 GPa (28 x 10^6 psi),
respectively.

The LAS matrix materials used in this study are designated LAS-I,
LAS-II, and LAS-III. LAS-I is the matrix very similar in composition to Corning
9608, except that ~3 wt% ZrO_2 replaces the ~3 wt% TiO_2 used as a nucleating
agent in 9608. LAS-II is identical to LAS-I except that it contains an addition
of ~5 wt% Nb_2O_5 for NbC reaction barrier formation. LAS-III also contains 5
wt% Nb_2O_5 and, in addition, is formulated to be much more refractory than
either LAS-I or LAS-II. The glass matrix material evaluated in composite form
is Corning 1723 aluminosilicate. All of the matrix materials were received
from Corning Glass Works as glassy powder of 8-12 µm average particle size.

COMPOSITE FABRICATION

The glass and glass-ceramic matrix composites were fabricated by
passing the SiC yarn through an agitated slurry of glass powder, water, and an
acrylic binder, onto a rotating drum. The resultant tape was then cut into the

appropriate dimensions and fiber orientations, heated in air to 600°C to remove the binder, and stacked in a graphite hot-press die to the desired thickness. The composites were hot-pressed at 1000 psi (6.9 MPa) pressure at temperatures of 1000-1350°C and times of 15-30 min, depending on the matrix composition. Samples to be studied in the "ceramed" or matrix crystallized state were subjected to a post hot-pressing heat treatment at temperatures of 900-1150°C, again depending on the matrix composition. For the LAS matrix composites, the ceraming step crystallized the major portion of the matrix into either the β-quartz-silica solid solution $[(Li_2O,MgO) \cdot Al_2O_3 \cdot nSiO_2(n \gtrless 2)]$ or β-spodumene-silica solid solution $[(Li_2O,MgO) \cdot Al_2O_3 \cdot nSiO_2 (n > 3.5)]$ phase. The remaining matrix consists of a continuous glassy grain boundary phase plus small crystals of mullite $(3Al_2O_3 \cdot 2SiO_2)$ and zircon $(ZrSiO_4)$.

COMPOSITE INTERFACIAL ANALYSIS

The main analytical tools utilized to study the glass and glass-ceramic matrix/Nicalon SiC fiber composite interfacial areas were the scanning transmission electron microscope (STEM)* and the scanning Auger electron micropobe (SAM).** The STEM was utilized to study replicas of polished composite cross-sections and ion beam thinned composite sections. The SAM was utilized to study fiber surfaces, both loose fibers and fibers lying in the plane of a composite fracture surface, and matrix troughs from which fibers had been pulled away. Both of these analytical methods were indispensible for the study of fiber/matrix interfaces due to their fine spatial resolution and, in the case of the SAM, the ability to detect low atomic number elements.

TEM Replica Analysis

Typical TEM replicas of transverse cross-sections of the LAS-I matrix/Nicalon SiC fiber composite system are shown in Fig. 2. The blocky crystals that can be seen in both samples are mullite $(3Al_2O_3 \cdot 2SiO_2)$ that crystallizes out of the glass on cooling from hot-pressing. The sample on the left (A) exhibits a partially crystallized matrix while the one on the right (B) is almost completely crystallized to the β-quartz-silica solid solution LAS phase. One occasionally also finds small elongated crystals of zircon $(ZrSiO_4)$ within the matrix that form during cooling from fabrication. It is also obvious from Fig. 2 that a thin ring exists around the fibers at the fiber/matrix interface. This ring was not present on the fibers to begin with and is never

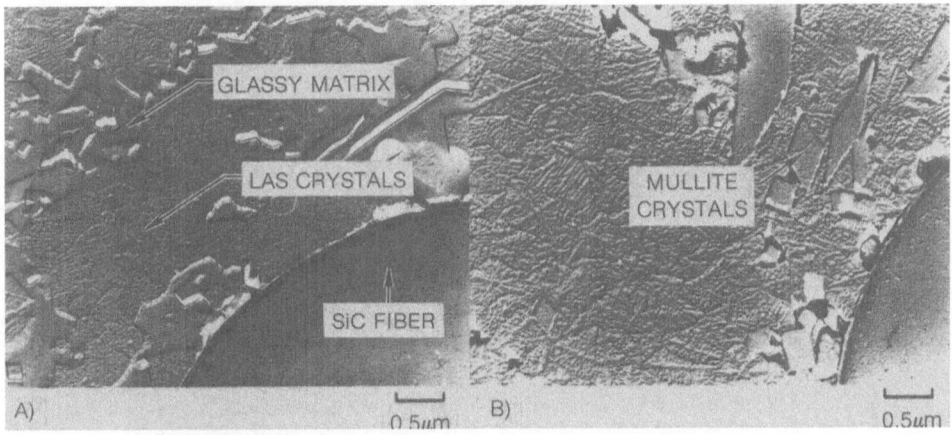

Fig. 2. TEM replicas of LAS-1 matrix/Nicalon SiC fiber composite

*Phillips EM 400T
**Perkin Elmer Physical Electronics 600H

observed when these fibers are incorporated into resin matrix composites. Extraction of this ring material by overetching the sample with HF was not successful, as it was in identifying the crystalline phases within the glass-ceramic matrix.

Figure 3 shows a TEM replica of an LAS-II matrix/Nicalon SiC fiber composite in the as-pressed condition. In this condition, the matrix is predominantly glassy, with a few mullite and β-quartz-silica solid solution LAS phase crystals. At the fiber/matrix interface, there now exists a second reaction layer next to the matrix. From X-ray diffraction analysis of crushed samples and by STEM extraction replica elemental distribution maps, as shown in Fig. 4 for a fiber that is emerging from the matrix at a low angle, this layer

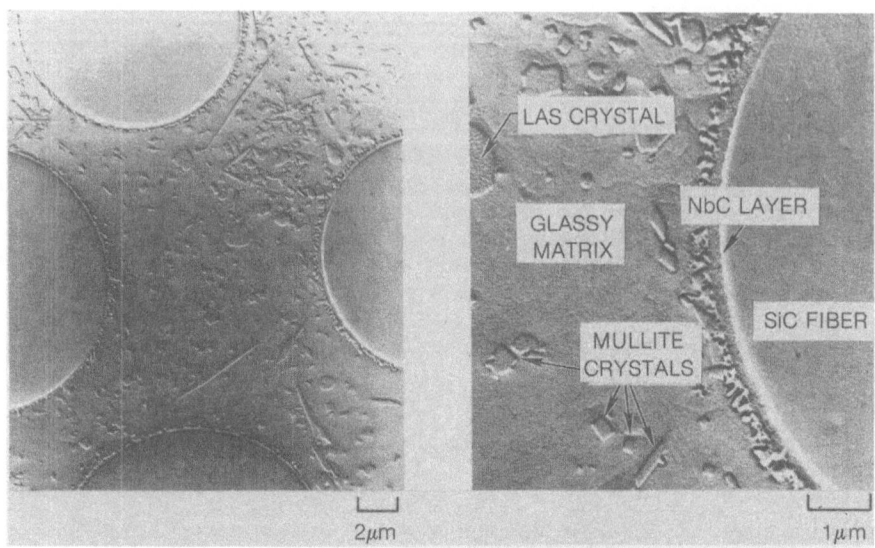

Fig. 3. TEM replica of as-pressed LAS-II matrix/Nicalon SiC fiber composite

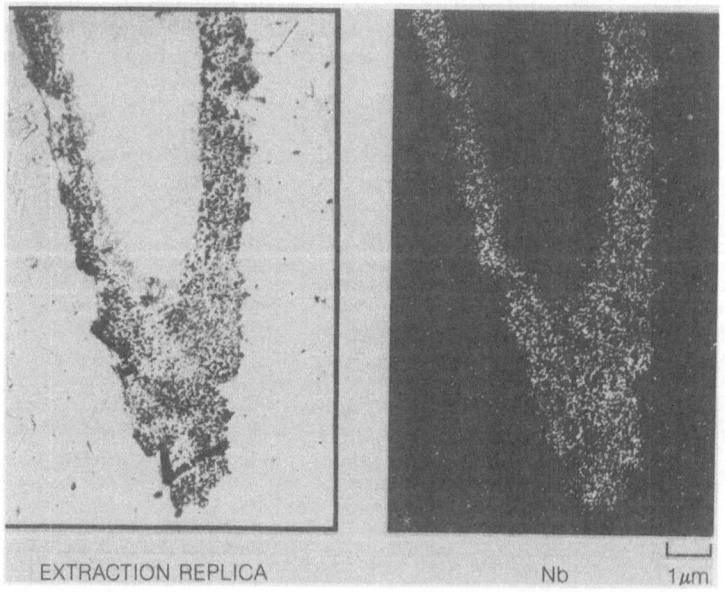

Fig. 4. STEM element map of as-pressed LAS-II matrix/ Nicalon SiC fiber composite

has been identified as NbC. The 5% Nb_2O_5 added to the LAS composition has reacted with the SiC fibers during hot-pressing to form this layer. Other studies at UTRC have shown that Ta_2O_5 additives react in the same manner.

TEM Thin Foil Analysis

In order to characterize the LAS matrix/Nicalon SiC fiber composite interfacial area in greater detail, thin foils were prepared by ion beam thinning. During this process, it was found that the LAS matrix and Nicalon fibers generally did not thin at the same rates and that the weak fiber to matrix bonding made the preparation of good foils rather difficult. The greatest success was achieved with composites in which the fibers were lying essentially parallel with the foil surface. The TEM thin foil analyses were done on LAS-III matrix composites.

Figure 5 shows TEM thin foil micrographs of two different areas in a LAS-III matrix composite in the as-pressed condition. It can be seen that there exists a light colored interfacial zone between the SiC fibers and the LAS matrix. The dark particles along the interfacial zone/matrix interface are NbC crystals. The small dark spots in the matrix also contain Nb while the larger, lighter colored crystals are mullite. Diffraction patterns taken from the SiC fiber near the interface show it to be microcrystalline in nature, while the interfacial zone appears to be essentially amorphous. Most of the matrix adjacent to the NbC layer is glassy in nature.

Energy dispersive X-ray (EDX) analysis of the interfacial region shown in Fig. 5B was performed in a stepwise fashion and, as shown in the sequence of EDX results in Fig. 6, indicate that Al and a small amount of Mg are diffusing into the SiC fiber to a distance of at least 1000Å from the light colored interfacial zone. The distances measured from the interface must be interpreted with the knowledge that the section of the fiber thinned is not necessarily from the fiber mid plane but may be a section near the fiber surface. In the latter case, what appears to be 1000Å from the interface may in reality be considerably less.

The interfacial zone (point 4) appears to contain mostly Si, with some Nb, Al, and Mg present as well. The much more jagged nature of the EDX curves for the interfacial zone in contrast to the fiber or matrix EDX is because a very long counting time (approximately 15 times the other EDX curves) was necessary to obtain significant amounts of Si and Nb, indicating that these elements were actually present in very small amounts with the major

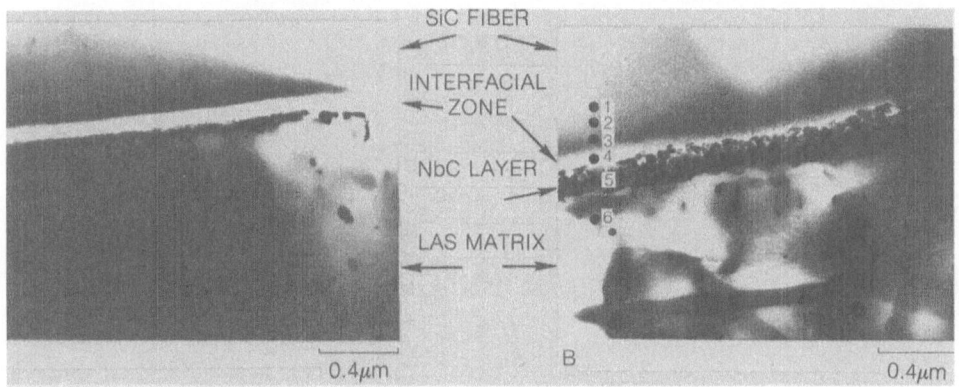

Fig. 5. TEM thin foil micrographs of LAS-III matrix/Nicalon SiC fiber composite

chemical phase present being undetectable (i.e. atomic number <11) using this method. Unfortunately, the electron energy loss spectroscopy (EELS) unit incorporated into the Phillips EM 400T STEM at UTRC has not performed up to expectations, so that low atomic number elemental detection using this instrument has not been possible.

From TEM investigation of fractured composite surfaces, it was found the Nicalon fibers often had small regions of a very thin film of material adhered to them. Figure 7 shows this film along with its diffraction pattern. The rather diffuse ring pattern for this film was indexed and appears to correlate quite well with that of carbon (ASTM Card 6-0675). Some of these

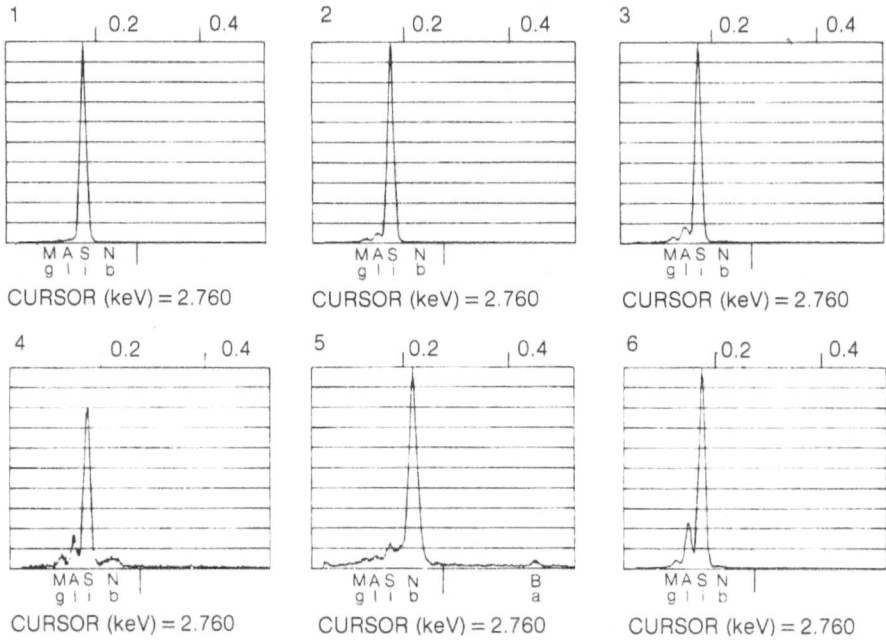

Fig. 6 Energy dispersive x-ray spectroscopy (EDX) analysis of fiber/ matrix interfacial area of Fig. 5B

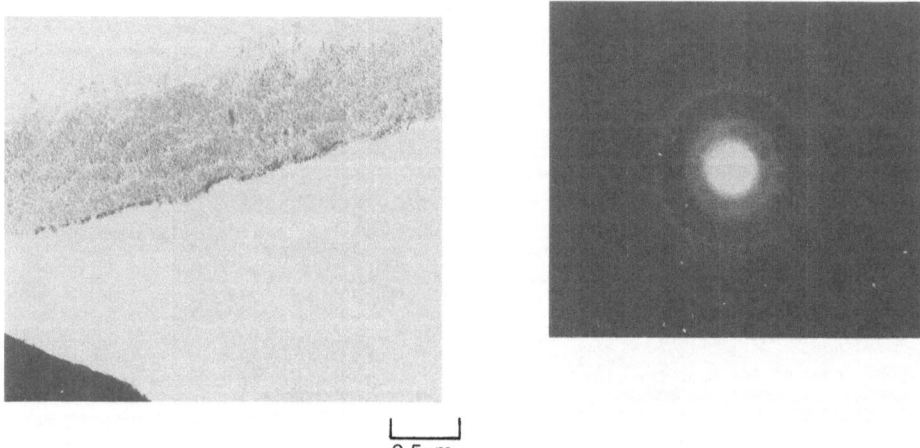

0.5μm

Fig. 7. Thin film on surface of Nicalon SiC fiber from fracture surface of LAS-III matrix composite

films contain small amounts of Si and/or NbC particles, as well. It appears that this film is at least part of the light colored interfacial zone that was found from thin foil studies to exist at the fiber/matrix interface, between the NbC layer and the SiC fiber. It is obviously rather weakly bonded to both the matrix and fiber and therefore is the main contributing factor to the ability of these composites to deflect matrix cracks which gives rise to the observed high fracture toughness.

Scanning Auger Microprobe (SAM) Analysis

This analytical tool was utilized to investigate the surface and near-surface regions of Nicalon SiC fibers, both before and after incorporation into composite form, and the matrix interfacial region of a composite fracture surface. The SAM system used has a minimum electron beam spot size of less than 500Å with a typical Auger electron emission depth of 10-30Å. In this study the analyses were performed in the spot mode using a beam diameter of nominally 3000Å in order to provide a more integrated area of analysis. This instrument has excellent sensitivity for most low atomic number elements and has been used to detect elements down to atomic no. 3 (Li). Depth profile information was obtained by in situ sputtering with an Ar ion beam. Since the rate of material removal has been calibrated using a Ta_2O_5 standard, the depth profiles obtained for the Nicalon fibers and the LAS matrix must be taken as approximations.

A Nicalon SiC fiber that had been passed quickly through a bunsen burner flame to remove the vinyl acetate sizing was depth profiled from its surface inward in the SAM. Figure 8 shows the SAM depth profile for this fiber. From this data it appears that the surface of the as-flamed Nicalon fibers is high in oxygen and low in carbon. To verify that this is the true chemistry of the fibers and not just created by the oxidizing bunsen burner flame, the sizing was also removed in Ar and in N_2. Both of these fibers, as well as a more recent fiber lot received from Nippon Carbon Co. with no sizing applied, showed oxygen rich surfaces and overall fiber chemistry similar to that shown in Fig. 8 when depth profiled in the SAM. While most of the Nicalon fibers investigated to date using SAM analysis have given a bulk fiber chemistry of 50-55 at% C, 35-40 at% Si, 6-10 at% O, and ~2-3 at% N; some fibers from the same tow would give markedly different results with one fiber being in the above chemistry range while another would indicate carbon content of ~60 at% and Si content of 25-28 at%. The SAM analysis has been found at UTRC to be very repeatable; thus, the Nicalon fibers actually vary in chemistry, even within the same fiber tow.

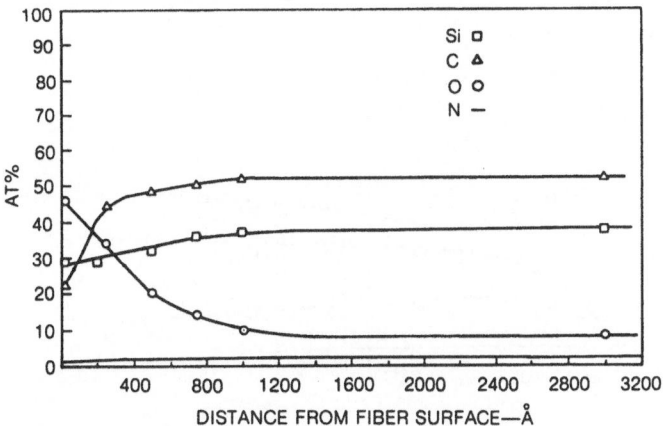

Fig. 8. Scanning Auger analysis of Nicalon SiC fiber after bunsen burner flame treatment to remove sizing

When a strong glass or glass-ceramic matrix/Nicalon SiC fiber composite is fractured in the plane of the fiber orientation, many of the fibers lying along the fracture surface break free of the matrix, as shown for an LAS-III matrix composite in Fig. 9. Careful selection of those fibers that are free of matrix debris can then be done for SAM analysis. Also, the matrix troughs from which fibers have pulled away from during composite fracture can also be SAM depth profiled.

A typical SAM analysis of a fiber from an LAS-I matrix composite is shown in Fig. 10. It can be seen that the surface composition of the Nicalon fiber in the composite has changed from oxygen rich to extremely carbon rich. This carbon layer appears to be on the order of 100Å of pure carbon plus another 200Å of high carbon that is grading into the usual fiber composition. A small amount of Al was also found inside of the carbon layer within the fiber to a depth of ~1500Å from the fiber surface. Figure 11 shows the SAM depth profile of a matrix trough in an LAS-I matrix composite from which a fiber has been removed. A small amount of carbon diffusion into the matrix is detected up to a depth of ~300Å. No Li or Mg was detected in the matrix even though these elements are present in LAS-I in amounts of ~5 and 2 at%, respectively.

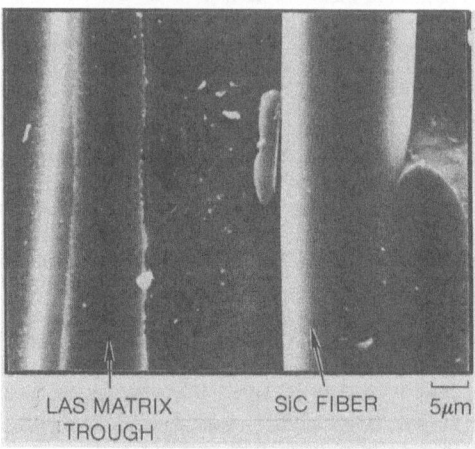

Fig. 9 SEM of LAS-III matrix/Nicalon SiC fiber composite longitudinal fracture surface

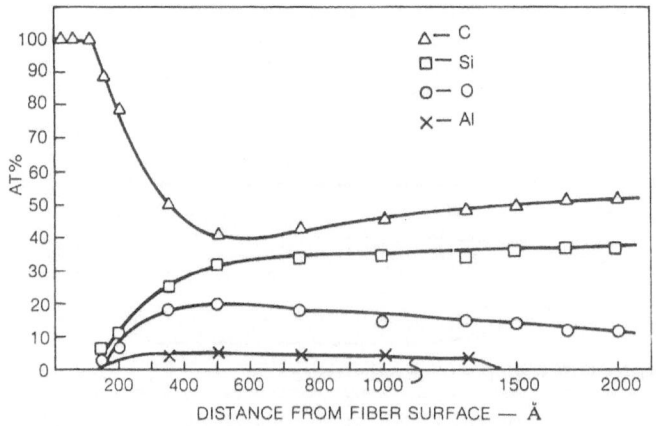

Fig. 10. Scanning Auger analysis of Nicalon SiC fiber from LAS-I matrix composite fracture surface

The SAM analysis for a Nicalon fiber from an LAS-III/SiC fracture surface is shown in Fig. 12. This analysis shows that, similar to the fiber from a LAS-I matrix composite, a very carbon rich surface now exists on the fiber that is on the order of 300-500Å in thickness. Also, a small but measurable amount of aluminum (~3-5 at%) was found between ~400-1400Å from the fiber surface, substantiating the TEM thin foil observations. The fiber composition appears to stabilize out to the bulk fiber composition at a depth of ~1600-1800Å from the fiber surface. As discussed previously, the distance given in these depth profiles must be taken as approximate due to the nature of

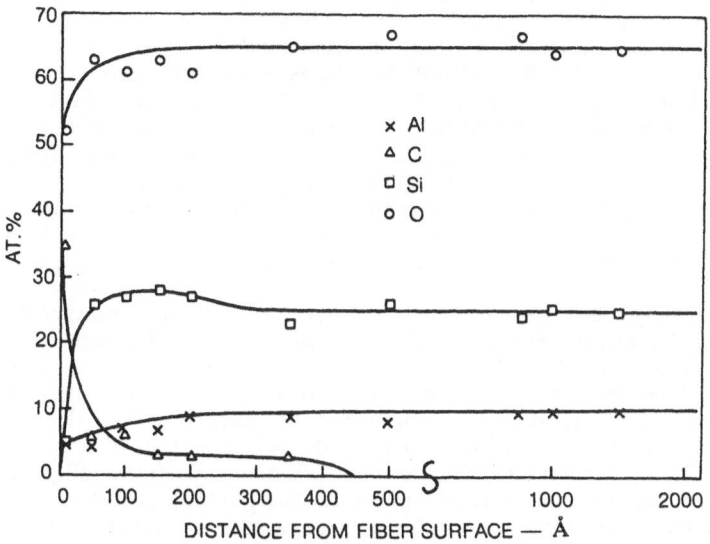

Fig. 11. Scanning Auger analysis of matrix trough from LAS-I/ Nicalon SiC fiber composite fracture surface

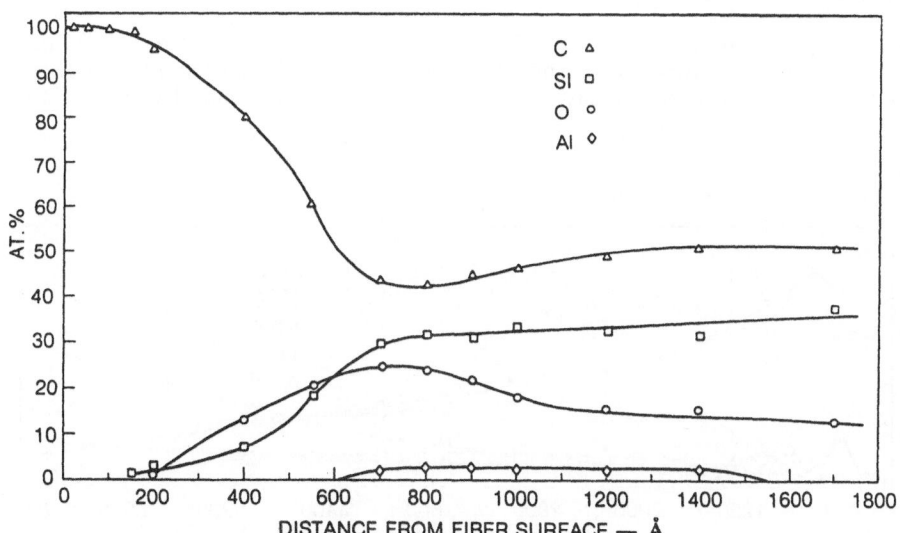

Fig. 12. Scanning Auger analysis of Nicalon SiC fiber from LAS-III matrix composite fracture surface

the analysis. No real difference was noted for the SiC fiber surface chemistry between fibers from as-pressed or heat-treated LAS-III matrix composites. Also, no evidence was found for diffusion of Li, Mg, Nb or other matrix constituents into the SiC fibers.

Figure 13 shows the SAM analysis of the LAS-III matrix trough from which the Nicalon fiber has lifted away. In contrast to the LAS-I matrix trough analysis (Fig. 11), the LAS-III matrix trough analysis is rather complicated because of the many elements present. From Fig. 13, it is apparent that there exists a very high concentration of carbon at the fiber/matrix interface that declines gradually as the sputtering depth increases, but is still present even to a depth of 6500Å. The niobium content gradually increases from essentially zero right at the interface to a maximum of ~30 at% at a depth of 3000Å. It then parallels the carbon in essentially a 1:1 ratio that would be expected for NbC. The Si content increases very gradually from essentially zero at the fiber/matrix interface to ≈16% at a depth of 6500Å. The oxygen content jumps between 5 and 20 at% near the interface and then increases in a steady manner from ~5 at% at 1400Å to ≈50 at% at 6500Å. The Li content was essentially zero to ~1600Å, then held steady at 10-12% to a depth of 5000Å, and then rather rapidly declined to 1-2% by 6500Å. Aluminum was not found until a depth of ~5000Å, increasing to ~5% by 6500Å. No Mg or Zr were identified up to the last sputtering depth of 6500Å.

Neglecting the 5 wt% Nb_2O_5 and 2 wt% ZrO_2 in the matrix composition, the starting overall elemental atomic composition of the matrix is approximately 5% Li, 2% Mg, 7% Al, 24% Si, and 62% O. On cooling from hot-pressing, however, the presence of mullite and zircon crystals can alter this overall glassy matrix composition somewhat by lowering the Al, Zr, Si, and O content. The SAM analysis obviously did not encounter either a mullite $(3Al_2O_3 \cdot 2SiO_2)$ or zircon $(ZrSiO_4)$ crystal. Accounting for a slight loss in Al, Si, and O from mullite and zircon crystallization, it appears that at a depth of 6500Å from the fiber/matrix interface, the matrix composition is approaching that expected for the bulk matrix, except for the low Li and no indication of Mg. At 2 at%, the Mg may be difficult to detect. The Li appears to concentrate near but not at the interface and somewhat parallels the Nb concentration. The glassy matrix in which the NbC particles appear to be imbedded, as seen in the TEM thin foil study, may have a rather Li rich composition.

It is apparent from the SAM studies on the LAS-I and III matrix/Nicalon SiC fiber composites that the carbon rich layer at the fiber/matrix interface is the same as the layer previously seen from STEM studies and also identified as

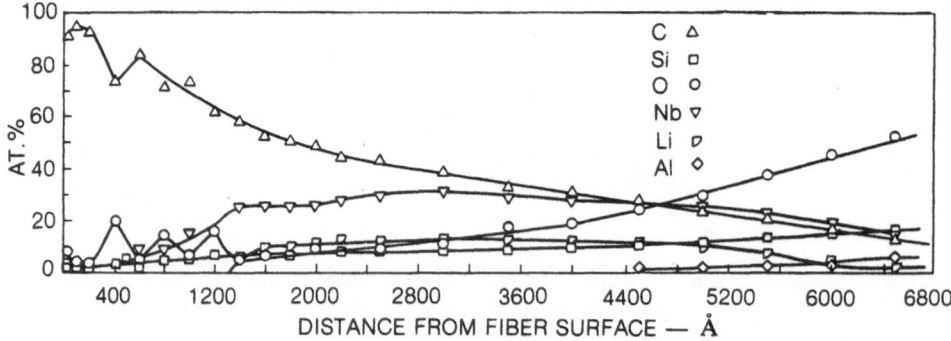

Fig. 13. Scanning Auger analysis of matrix trough from LAS-III Nicalon SiC fiber composite fracture surface

carbon. SAM analysis of the thin film adhering to a fractured fiber, as was shown in Fig. 7, also verified that this film consists primarily of carbon. The formation of this carbon layer with its resultant weak bonding to the matrix apparently accounts for the crack blunting and deflection that leads to high fracture toughness in these composites.

This carbon rich interfacial layer has also been found in other strong and tough glass matrix composites utilizing Nicalon fibers, such as those with Corning 1723 aluminosilicate glass matrices. In this case, the carbon interfacial layer is very thin, being ~25Å in width. It has also been found in a high silica glass matrix (Corning 7930) composite that was quite strong and tough (Fig. 14), but was not present in a weak and brittle composite utilizing

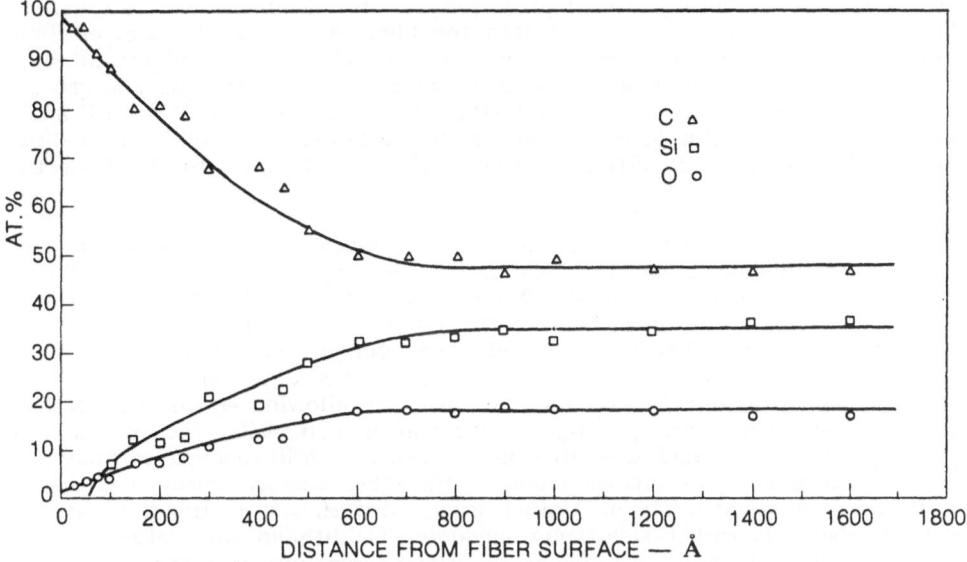

Fig. 14. Scanning Auger analysis of Nicalon SiC fiber from a strong 7930 glass matrix composite fracture surface

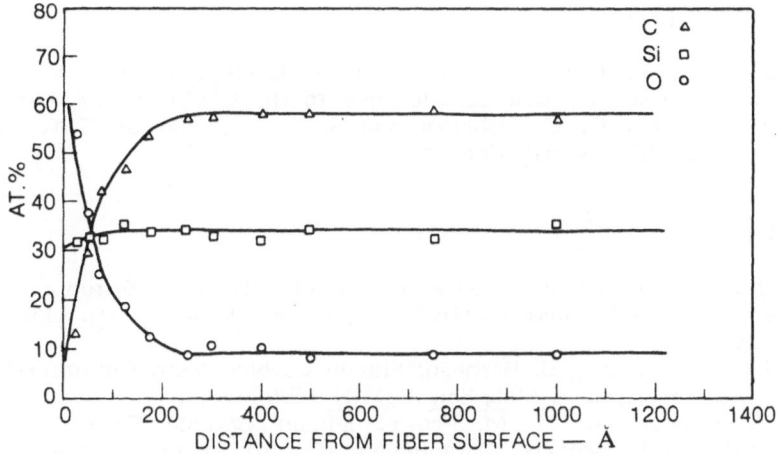

Fig. 15. Scanning Auger analysis of Nicalon SiC fiber from a weak 7930 glass matrix composite fracture surface

this matrix that was processed under different conditions (Fig. 15). A composite utilizing an LAS matrix with Nicalon fibers that was quite weak and brittle also evidenced a much reduced carbon content at the fiber/matrix interface.

CONCLUSIONS

It has been found from TEM thin foil and scanning Auger microprobe (SAM) analysis that the chemistry of Nicalon SiC fibers within a few hundred angstroms of the fiber surface undergoes a rather profound change when the fibers are incorporated into LAS matrices. Among the changes noted is that a thin (~100-400Å) surface layer forms on the fibers that is extremely carbon rich, being almost totally devoid of Si and O. This layer gradually increases in Si and O as a function of depth into the Nicalon fibers. As-received fibers have a slightly oxygen rich surface. The weakly bonded interfacial layer appears to consist of carbon on one side (nearest the LAS matrix) and grade to Si, C, and O on the other side next to the Nicalon SiC fibers. In LAS-II and III matrix composites, this layer is also surrounded by NbC particles In addition, aluminum appears to be diffusing from the LAS matrices into the fiber to a depth of ~600Å.

From SAM analysis of Nicalon fiber surfaces from extremely weak and brittle glass matrix composites, this carbon rich layer is either nonexistent or much reduced in carbon content. It appears, therefore, that the formation of this carbon rich interfacial zone in the LAS and glass matrix composites is responsible for the high toughness of these composites, in that it is quite weakly bonded to both the matrix and the fiber thus allowing crack deflection to occur along the fiber/matrix interface but also allowing enough load transfer to occur so that strengthening by traditional composite theory can take place. Exactly how and why this layer forms is still unknown. Apparently oxygen and silicon are diffusing out of the fiber surface during hot-pressing while some Al is diffusing in. Carbon is also diffusing from the fiber into the LAS-III matrix in rather significant amounts. No lithium was detected in the fibers but it could be present since Li in small amounts is rather difficult to pick up by SAM analysis. It is apparent from this investigation that the particular chemistry and/or microstructure of the polymer derived Nicalon SiC fibers is responsible for the type of interface formed in the glass-ceramic and glass matrix composites.

ACKNOWLEDGEMENTS

The author would like to thank Mr. G. McCarthy, and Drs. V. Patarini and B. Laube of UTRC for their contributions to the STEM and SAM analyses, respectively, and Drs. R. C. Pohanka and S. G. Fishman of ONR for the primary support of this investigation.

REFERENCES

1.   K. M. Prewo and J. J. Brennan, High Strength Silicon Carbide Fiber-Reinforced Glass Matrix Composites, J. Mat. Sci. 15:463-468 (1980).

2.   K. M. Prewo and J. J. Brennan, Silicon Carbide Yarn Reinforced Glass Matrix Composites, J. Mat. Sci. 17:1201-1206 (1982).

3.   J. J. Brennan and K. M. Prewo, Silicon Carbide Fibre Reinforced Glass-Ceramic Matrix Composites Exhibiting High Strength and Toughness, J. Mat. Sci. 17:2371-2383 (1982).

FATIGUE BEHAVIOR OF SILICON CARBIDE FIBER REINFORCED

LITHIUM-ALUMINO-SILICATE GLASS-CERAMICS

Eric Minford and Karl M. Prewo

United Technologies Reseach Center
East Hartford, CT

ABSTRACT

The mechanical fatigue behavior of Nicalon silicon carbide-type fiber reinforced lithium-alumino-silicate glass-ceramic matrix composites (SiC/LAS) was examined as a function of matrix composition and fiber ply lay-up. Both tensile and flexural fatigue behavior were determined at room temperature in air. The results are discussed in terms of the effect of the non-linearity in the stress-strain behavior on the fatigue life of these composites as well as the effects of applied stress state. Differences in fracture surface appearance are related to composite matrix composition and ply lay-up.

INTRODUCTION

The utilization of high performance ceramic matrix composites in structural applications requires their complete thermal and mechanical characterization so as to provide the designer with sufficient information to make intelligent use of these composites. Characterization of this type has been the subject of on-going corporate and Office of Naval Research funded programs at United Technologies Research Center. As part of these programs, the Nicalon silicon carbide-type fibers have been used to reinforce glasses and glass-ceramics since 1976.[1-6] During these studies, it has been shown that exceptional levels of strength and toughness can be achieved. It has also been observed, however, that matrix microcracking occurs prior to fiber failure in most applied stress states. Unlike the situation in resin matrix composites this matrix microcracking significantly alters composite stiffness and controls the shape of the composite's stress-strain curve.

EXPERIMENTAL PROCEDURE

Materials

The composites tested consisted of 40-50 volume percent of Nicalon silicon carbide-type multifilament yarn in a lithium-alumino-silicate glass-ceramic matrix. Two different matrix compositions, designated LAS-II and LAS-III, were examined which differ primarily in trace element concentrations. Composites tested were either unidirectional, 0°, composites, in which all fibers are in parallel alignment, or cross-ply, 0°/90°, composites, in which alternating plys of fibers are perpendicularly oriented with the composite through thickness center being a plane of symmetry. Further discussion of composite fabrication can be found elsewhere.[3]

Flexural Fatigue

Four point flexural fatigue testing was performed in air at room temperature using a ratio of minimum to maximum flexural stress, R, of ~0.1. The flexural stresses were calculated using an elastic beam formula and are thus the maximum applied flexural stress or the 'outer fiber' flexural stress. The inner span was 0.75 inches and the outer span was 2.5 inches. The spans were chosen such that for the specimen thicknesses used the ratio of the maximum applied flexural stress to the maximum applied shear stress $(\sigma_{max}/\tau_{max})$ was approximately 45. The test specimens were all unidirectonally reinforced SiC/LAS-II and applied load vs. mid-span deflection traces were obtained for all tests. Tests were run at a rate of 0.5 cycles per second and were terminated at a maximum of 100,000 cycles. Any specimens which survived 10^5 cycles without failure were then flexure tested to determine residual strength. Failure was non-catastrophic and was therefore defined as a significant increase in sample deflection caused by degradation of the composite elastic modulus. Nominal sample dimensions were 0.25" wide by 0.08" thick.

Tensile Fatigue

Tensile fatigue testing was performed in air at room temperature using a ratio of minimum to maximum tensile stess, R, of ~0.1. Parallel sided specimens were used with a 1" gage length. Test specimens were both unidirectional and cross-ply SiC/LAS-II and SiC/LAS-III. Applied load vs. tensile extension traces were obtained for all tests. Tests were run at a rate of 5-10 cycles per second and were terminated at a maximum of 100,000 cycles. Any specimens which survived 10^5 cycles without fracture were then tensile tested to determine residual strength. A circular hole, approximately 0.0625" in diameter, was machined in the center of the gage section of several of the tensile specimens of the cross-ply reinforced SiC/LAS-II and SiC/LAS-III composites. Nominal sample dimensions were 0.4" wide by 0.08" thick.

RESULTS

Flexural Fatigue

The results of room temperature four point flexural fatigue of 0° SiC/LAS-II are shown in Fig. 1. Three unfatigued specimens were tested to obtain the initial strength and load-deflection behavior. The initial flexural strength ranged from 100 to 118 ksi and all showed first failure on the compressive side. Samples fatigued at 100 and 109 ksi failed during fatigue testing, while those fatigued at 71.1 to 103 ksi survived the 100,000 cycle exposure. The residual strength of the survivors, calculated using an elastic beam formula, was comparable to or greater than the initial strengths. In all fatigue specimen cases composite failure was intiated on the compression side of the specimen in a manner similar to the unfatigued samples. Specimens did not fracture into two pieces but instead retained substantial load bearing capacity after test.

Tensile Fatigue

Unidirectional Composites. Results are shown in Fig. 2 for uniaxial tension-tension fatigue tests of both 0° SiC/LAS-II and 0° SiC/LAS-III. Two unfatigued samples of 0° SiC/LAS-II and six unfatigued samples of 0° SiC/LAS-III were tested to obtain the initial strength and load-extension behavior. The

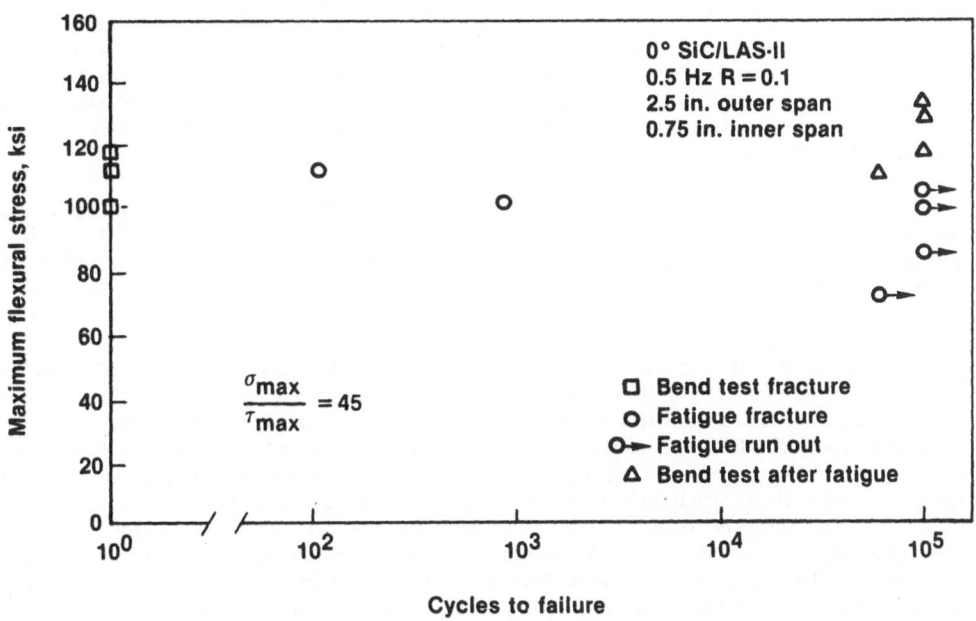

Fig. 1. Room temperature four-point flexural fatigue of 0°
SiC/LAS-II composites.

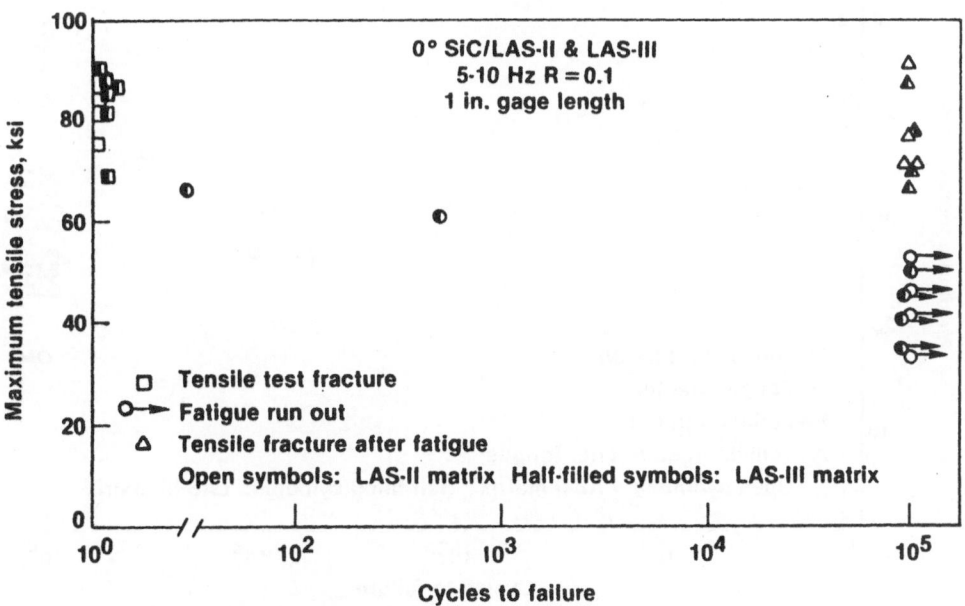

Fig. 2. Room temperature uniaxial tensile fatigue of 0° SiC/
LAS-II and 0° SiC/LAS-III composites.

initial tensile strength ranged from 68.2 to 89.8 ksi and all samples failed in a fibrous mode with some shear parallel to the fibers. Samples fatigued at 61.0 and 66.7 ksi failed during fatigue testing, while those fatigued at 32.4 to 51.8 ksi survived the 100,000 cycle exposure. The residual strength of the survivors was comparable to the initial strengths. Fatigue specimens failed in a similar manner to those tested monotonically.

Cross-Ply Composites. In Fig. 3 the results are shown for uniaxial tension-tension fatigue tests of both 0º/90º SiC/LAS-II and 0º/90º SiC/LAS-III. Eleven unfatigued samples of 0º/90º SiC/LAS-II and four unfatigued samples of 0º/90º SiC/LAS-III were tested to obtain the initial strength and load-extension behavior. The initial tensile strength ranged from 40.8 to 60.8 ksi for the LAS-II matrix composites and from 35.8 to 41.8 ksi for the LAS-III matrix composites and all samples failed in a somewhat fibrous mode with some interlaminar shear. LAS-II matrix samples fatigued at greater than ~35 ksi failed during fatigue testing, while those fatigued at stresses less than ~35 ksi survived the 100,000 or more cycles. In the case of the LAS-III matrix composites, those samples fatigued at greater than ~26 ksi fractured during fatigue, while those fatigued at lower stress levels survived the 100,000 cycles. One LAS-III sample failed after 311,800 cycles at 25.4 ksi. The residual strength of the survivors was comparable to the initial strength for both matrices. The fatigue specimens failed in a more fibrous manner, particularly in the case of the LAS-III matrix composites, and exhibited less shear.

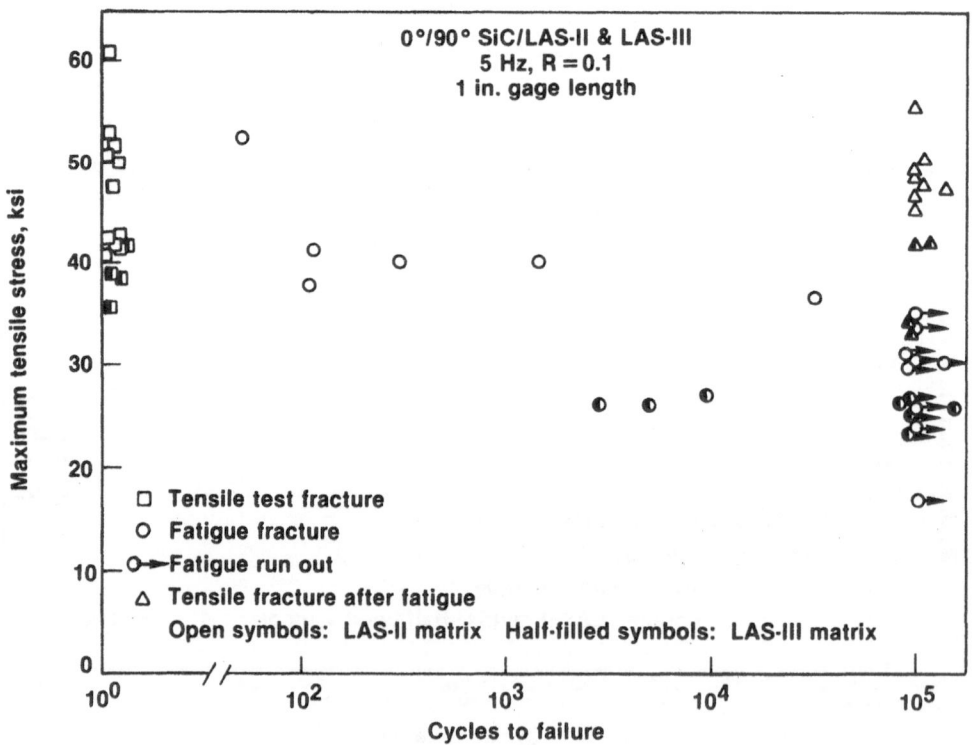

Fig. 3. Room temperature uniaxial tensile fatigue of 0°/90° SiC/LAS-II and 0°/90°SiC/LAS-III composites.

Table 1. Effects of Centrally Located Circular Holes on the Uniaxial Tensile
Strength of Cross-Ply Reinforced SiC/LAS Composites

Specimen ID	Matrix	Hole[+]	Tensile Strength (ksi)	Tensile Modulus (Msi)	Strain to Failure (%)
3095-1C	LAS-II	No	42.5	11.4	0.777
3095-2C	LAS-II	Yes	41.3*	12.4	0.745
3095-3C	LAS-II	Yes	42.6*	18.1	0.626
3123-3C	LAS-III	No	18.0	9.82	0.267
3123-4C	LAS-III	Yes	21.5*	9.91	0.412
3123-5C	LAS-III	Yes	18.7*	8.89	0.325

+ Hole diameter approximately 0.0625"

* Net section stress: $\sigma = \dfrac{P\ max}{(w-d)xt}$, w = sample width, d = hole diameter and t = sample thickness

The tension-tension fatigue behavior of 0°/90° SiC/LAS-II and 0°/90° SiC/LAS-III specimens with centrally located 0.0625" diameter circular holes was also examined. As is shown in Table 1 the net section strength of these composites was unaffected by the presence of the holes. In Table 2, where the initial and residual strengths are compared for notched and unnotched samples which survived 10^5 cycles, it can be seen that there is less than 10% difference in the fatigue behavior. The notched sample data is included in Fig. 3.

DISCUSSION

In order to better understand and compare the flexural and tensile fatigue results shown in Figures 1-3, it is necessary to examine the nature of the stress-strain behavior of these SiC/LAS composites. The stress-strain behavior of both unidirectional and cross-ply reinforced SiC/LAS composites is shown schematically in Fig. 4. For unidirectional composites the stress-strain curve is initially linear with a modulus, E_1, predictable using rule-of-mixtures. At some point matrix microcracking begins to limit the matrix contribution to composite stiffness and the behavior departs from linearity. By convention, a measure of the onset of this departure from linearity has been chosen to be the stress obtained with a 0.02% off-set in strain, designated $\sigma_{0.02\%}$, Fig. 4. The situation is the same for cross-ply composites with the exception that the 90° plys fail at low stress levels resulting in an additional transition in the stress-strain behavior and a reduction in modulus from E_1 to E_2. The offset stress is still defined in the same way. It should be noted that, even after extensive matrix microcracking has taken place, the composites still retain additional load bearing capacity and can exhibit failure strains of greater than 0.7%.

In order to compare the fatigue behavior of composites tested in different applied stress states (flexure and tension), different matrices (LAS-II and LAS-III) and different reinforcment geometries (unidirectional and cross-ply) it is necessary to normalize the strengths and applied fatigue stresses. This has been done by taking the ratio of the residual strength relative to the initial strength for each composite (σ_R/σ_I) and the maximum fatigue stress relative to the 0.02% off-set stress for each composite ($\sigma_C/\sigma_{0.02\%}$). In this way not only are variations in properties due to differences in test type, matrix composition, and reinforcement geometry normalized, but also composite plate to plate variability as well.

Table 2. Effects of Centrally Located Circular Holes on the Tensile Fatigue Behavior of Cross-Ply Reinforced SiC/LAS Composites

LAS-II Matrix

	Unnotched Strength (ksi)			Notched Strength (ksi)	
Spec. 1D	Initial	Residual	Spec. 1D	Initial	Residual
2649B-2	50.8	48.4	3097-2	47.6	45.2
2649B-3	50.8	48.9	change		-5.0%
2654B-1	46.8	55.2			
2716A-1	46.8	47.1			
3093-3	52.2	49.8			
3094-3	42.0	46.7			
3094-4	42.0	47.5			
Average	47.3	49.1			
Change		+3.8%			

LAS-III Matrix

	Unnotched Strength (ksi)			Notched Strength (ksi)	
Spec. 1D	Initial	Residual	Spec. 1D	Initial	Residual
3345-5	39.0	33.9	3205-4	41.8	41.8
3204-5	35.8	41.8	3208	38.6	33.3
Average	37.4	37.9	Average	40.2	37.6
Change		+1.3%	Change		-6.5%

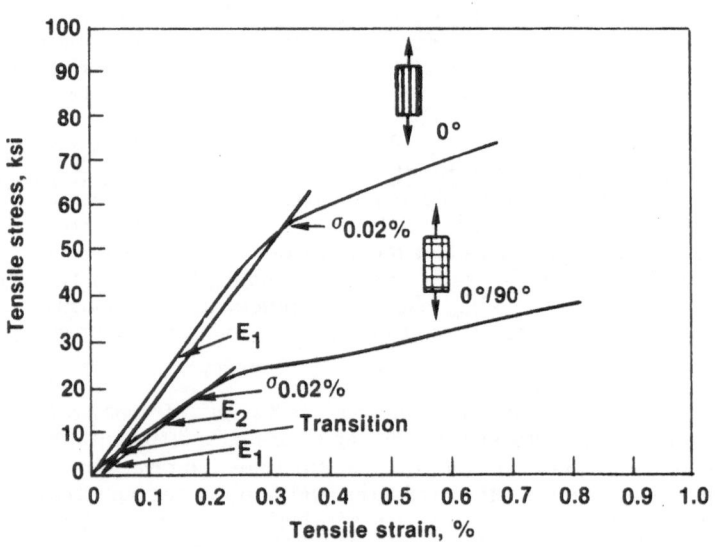

Fig. 4.   Tensile stress-strain curves for 0° and 0°/90° SiC/LAS at room temperature.

In Fig. 5, where the room temperature flexural and tensile fatigue results for SiC/LAS composites are presented in terms of normalized stresses, it can be seen that the fatigue behavior of the SiC/LAS composites, although appearing quite similar in the standard fatigue life plots (Figures 1-3), is, in fact, quite different depending on applied stress state and ply lay-up. In Fig. 5a, where the flexural fatigue results for 0° SiC/LAS-II are presented, there appears to be the suggestion of fatigue induced strengthening with σ_R/σ_I increasing with increasing fatigue stress ($\sigma_C/\sigma_{0.02\%}$). More likely this behavior is an artifact of the flexure test in which interpretation of results can be complicated by such factors as non-tensile failure modes (compressive and shear failures are often observed) and the shift of the neutral axis within the composite beam as a result of matrix microcracking in the high tensile stress region. This forms a more compliant surface region and yields an over-estimate of the outer fiber tensile stresses at failure when simple elastic beam calculations are used.

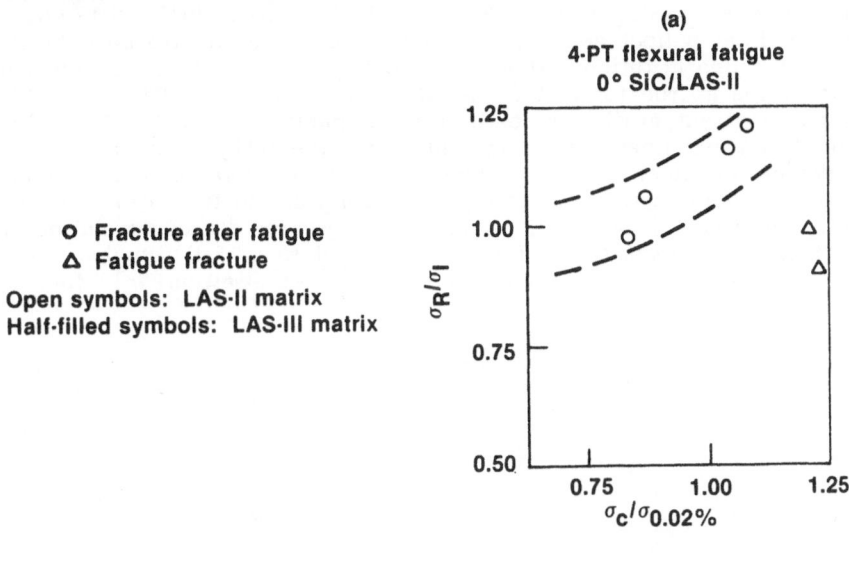

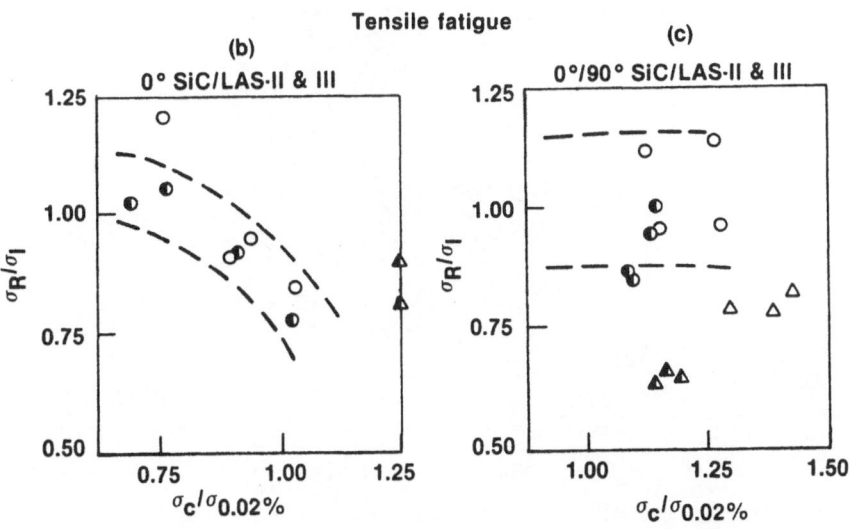

Fig. 5.   Normalized stress representation of flexural and tensile fatigue behavior of SiC/LAS composites.

In Fig. 5b, where the tensile fatigue results for 0° SiC/LAS-II and SiC/LAS-III are presented, a much different type of behavior is observed. Cumulative damage-type behavior is seen where the residual strength (σ_R/σ_I) decreased as fatigue stress ($\sigma_C/\sigma_{0.02\%}$) increased. This type of behavior could be a result of the opening and closing of matrix cracks during stress cycling as reported by Marshall and Evans.[7]

Lastly, in Fig. 5c for the tensile fatigue of cross-ply SiC reinforced LAS-II and LAS-III there appears to be no change in σ_R/σ_I as a function of $\sigma_C/\sigma_{0.02\%}$. This could be related to some form of fiber-matrix decoupling for these composites which reduces fiber damage during fatigue.

Further evidence of a difference in unidirectional and cross-ply composite fatigue behavior is obtained by examining the fracture morphologies of SiC/LAS composites tested in monotonic tensile loading and in tensile fatigue as shown in Fig. 6. In Fig. 6a, for 0° SiC/LAS-III, there is little difference in the degree of fiber pullout and appearance between a tensile tested composite and one which failed in fatigue other than a decrease in the tendency for shear failure. The same is true for 0° SiC/LAS-II. In Fig. 6b, for 0°/90° SiC/LAS-III and to a lesser extent in Fig. 6c for a notched specimen of 0°/90° SiC/LAS-II, much more extensive fiber pullout is evident for the fatigue specimens. This greater degree of fiber pullout would imply that more extensive matrix deterioration or fiber-matrix decoupling is occuring during the fatigue of these cross-ply composites. In addition, the somewhat greater degree of pull-out for the LAS-III matrix composites in fatigue compared to LAS-II samples may be related to a lower interlaminar shear strength observed elsewhere[4,5,6] for these composites.

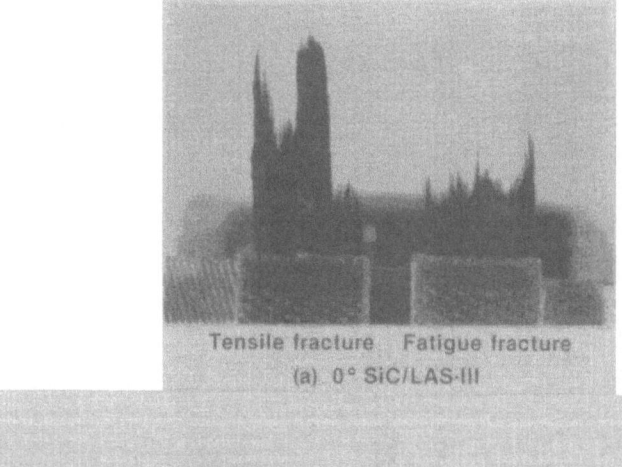

Tensile fracture   Fatigue fracture

(a) 0° SiC/LAS-III

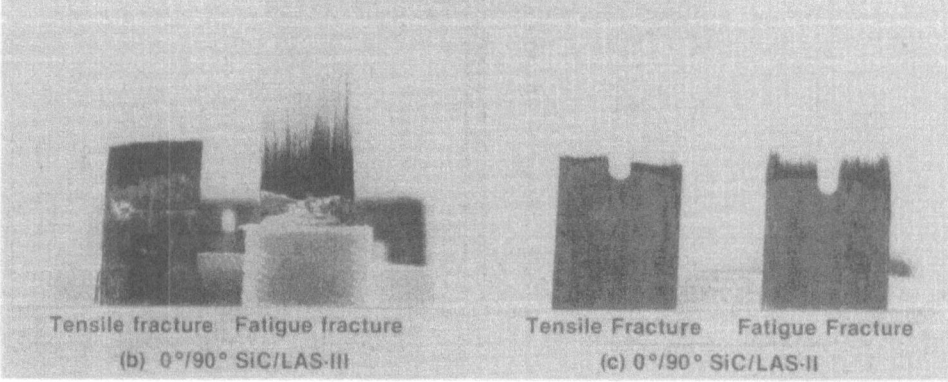

Tensile fracture   Fatigue fracture   Tensile Fracture   Fatigue Fracture

(b) 0°/90° SiC/LAS-III   (c) 0°/90° SiC/LAS-II

Fig. 6.   Fracture appearance of tension and tensile fatigue samples of SiC/LAS.

Fig. 7. Prenotched 0°/90° SiC/LAS-III specimen cracked during fatigue testing at $\sigma_{max} \simeq 27$ ksi after 5031 cycles. Sample failed at 5068 cycles.

A result of the fibrous failure mode of the cross-ply composites is considerable matrix crack tolerance. In Fig. 7 a pre-notched tensile fatigue specimen is shown in which matrix cracks developed during the fatigue cycling. Even though the cracks are nearly fully developed through the composite matrix this composite was able to survive over 30 additonal cycles to approximately 27 ksi (20% above the proportional limit) before failure occurred.

CONCLUSIONS

It has been shown that the complete characterization of Nicalon silicon carbide fiber reinforced LAS glass-ceramic matrix composites is an extremely complex issue. This study, limited to just room temperature fatigue behavior, has identified significant differences in composite performance depending on test geometry, ply lay-up, and matrix composition.

For unidirectionally reinforced composites both monotonic flexural strength and 10^5 cycle fatigue limit stress were much greater than similar quantities measured in uniaxial tension. It is suggested that this is due to the fact that in the tensile stress region of the flexural specimens matrix micro-cracking causes a nonlinear distribution of stress through the specimen thickness. This change in stress distribution leads to an overestimation of the failure stress when elastic beam calculations are used. Another factor which can complicate the interpretation of flexural results is the possibility that failure of the composite beam can initiate on the compressive surface of the beam in a compressive failure mode or in the interior of the beam via an interlaminar shear mechanism. Composite failure by either of these alternative modes can lead to an underestimate of composite tensile behavior.

An additional feature observed during flexural fatigue testing of unidirectional SiC/LAS composites was the suggestion of fatigue induced strengthening, while tensile fatigue tests on similar composites gave evidence of cumulative damage type behavior. It is implied that the strengthening observed in flexure is an artifact of the test technique (i.e. flexure as opposed to tension) and is a consequence of the change in stress distribution within the beam as discussed earlier.

The tensile fatigue behavior of cross-ply reinforced SiC/LAS composites differs from that of the unidirectional composites in that there is apparently no dependence of residual strength on maximum fatigue stress level. This coupled with the change in the degree of fiber pullout appear to indicate that the damage is more uniformly distributed in the cross-ply composites.

In general, the fatigue test results indicate that considerable caution must be exercised in using flexure test results to predict tensile behavior and vice versa. The extremely complex nature of stress distributions in flexure specimens, particularly during fatigue above the proportional limit, makes analysis of results very difficult. Additionally, caution must also be exercised in using data from uniaxial composites to predict the behavior of multiaxially reinforced material. More extensive analysis of the nature of residual stresses in brittle matrix composites will likely be required.

ACKNOWLEDGEMENTS

The authors would like to acknowledge the support of Drs. R. Pohanka and S. Fishman of the Office of Naval Research, through Contract N00014-81-C-0571, for the work described.

REFERENCES

1. K. M. Prewo and J. J. Brennan, "High-Strength Silicon Carbide Fiber-Reinforced Glass-Matrix Composites", J. Mater. Sci, 15, 463-468 (1980).

2. K. M. Prewo and J. J. Brennan, "Silicon Carbide Yarn Reinforced Glass Matrix Composites", J. Mater. Sci., 17, 1201-1206 (1982).

3. J. J. Brennan and K. M. Prewo, "Silicon Carbide Fiber Reinforced Glass-Ceramic Matrix Composites Exhibiting High Strength and Toughness", J. Mater. Sci., 17, 2371-2383 (1982).

4. K. M. Prewo, "Advanced Characterization of SiC Fiber Reinforced Glass-Ceramic Matrix Composites", Interim Report, Contract N00014-81-C0571, Project No. NR420-002/4-16-81(260), June 1983.

5. K. M. Prewo and G. K. Layden, "Advanced Fabrication and Characterization of SiC Fiber Reinforced Glass-Ceramic Matrix Composites", Interim Report, Contract N00014-81-C-0571, Project No. NR420-004/4-23-82(260), April 1984.

6. K. M. Prewo, G. K. Layden, E. J. Minford, and J. J. Brennan, "Advanced Characterization of Silicon Carbide Fiber Reinforced Glass-Ceramic Matrix Composites", Interim Report, Contract N00014-81-C-0571, Project No. NR420-004/4-23-82(260), June 1985.

7. D. B. Marshall and A. G. Evans, "Failure Mechanisms in Ceramic-Fiber/Ceramic-Matrix Composites", J. Am. Ceram. Soc., 68 [5] 225-31 (1985).

FIBRE REINFORCED COMPOSITES VIA THE SOL/GEL ROUTE

E. Fitzer and R. Gadow

Institut für Chemische Technik
Universität Karlsruhe
D-7500 Karlsruhe, FRG

1. INTRODUCTION

The outstanding physical, mechanical and chemical high temperature properties of modern refractory oxide materials make them to favorite candidates for application in advanced high performance thermal engines. The low cost raw materials if compared with those for super alloys support this tendency. However, the problems arising from the main disadvantages of all oxide ceramics such as low flexural strength and insufficient fracture toughness especially at lower temperatures must have been solved before the intended broad applications can be realized.

One method for improving the strength and toughness of ceramic materials is the dispersion hardening by the zirconia phase transition (1). Another one is the alloying with binder phases showing plastic deformation at elevated temperatures (2). A third perhaps more promising possibility for the development of structural materials with high flexural strength and fracture toughness is the reinforcement of refractory oxides with fibres and whiskers of high tensile strength (3).

1.1 The Technique of Fibre Reinforcement

The technique of fibre reinforcement is best known from composites with polymer matrix, and has started in the fourties with the so-called fibre glass, glass fibre reinforced unsaturated polyesters (UPE) and is culminating now-a-days with the so-called "advanced composites" already in technical application (4). These are epoxy, polyimides and high temperature polymers reinforced with polyaramide and carbon fibres.
The fibres have diameters between 5 and 15 μm. We distinguish principally between two types of composites :

<u>Type 1)</u> Short fibre reinforcement resulting in composites with predominantly isotropic mechanical properties, and fibre contents mostly not exceeding 40 v%. These composites are fabricated by extrusion or molding of a plastic deformable mixture of solid fibres and a molten matrix or a liquid matrix precursor. In case of whisker reinforcement volume fractions up to 60 % are achievable.

<u>Type 2)</u> The reinforcement with continuous fibres is a means by which "tailored" composites with controlled anisotropic properties and fibre volume contents up to maximum 70 v% can be obtained. Also in this case, a liquid matrix or matrix precursor (for instance curable prepolymer) is needed as binder for the pre-arranged fibre skeleton.

In translation of these experiences to fibre reinforced oxides, we need either meltable oxide powders as matrix precursor which can be densified by melting techniques or liquid precursors which can be transferred by hydrolysis/polymerization into solid gels and by subsequent heat treatment into solid glasses or ceramics (see Fig.1).

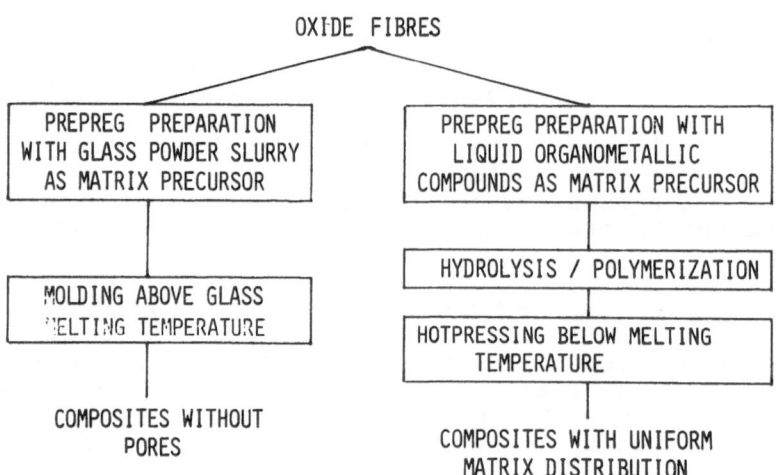

Fig.1 – Alternative methods for preparation of fibre reinforced oxide composites

1.2 Glass Powder-Melting Technique

An example of the first fabrication method is the work of SAMBELL (5) in which the fabrication of UD-composites with C-fibres in various glasses is described. The author used a glass powder slurry as matrix precursor and a hot pressing technique up to 1400°C.

In our institute similar techniques were applied for composites of CVD-B and SiC-fibres in low melting PbO glasses as matrix. Although UD composites with good translation of the fibre strength (70 % in case of 30 v% fibres) were achieved

(σ_B = 925 MN/m^2, ILSS = 38 MN/m^2, impact strength = 38·10^{3}kJ/m^2) the homogeneity of the fibre distribution was insufficient (see Fig.2). The good adhesion between fibre and matrix corresponds to the suitable wetting behaviour measured separately (6) (Fig.3).

Also, the results of PREWO (7) with wet spun polycarbo-silane-based SiC fibres (NICALON) in glass matrix as well as with carbon fibres in borosilicate glass matrix should be mentioned. In the case of the last mentioned combination, composites with low thermal expansion coefficient (3.25·10^{-6} K^{-1}) and the highest values for flexural strength of UD-composites (1324 MPa at 873 K, 71 v% C-fibres, K_{ICmax} = 22.1 MN/m$^{3/2}$) have been achieved (9).

The applicability of these fabrication method is controlled by the thermal, mechanical and chemical stability of densification and thus limited to relatively low melting matrices.

1.3 Sol/Gel Technique

The second fabrication process is the so-called sol/gel route, which applies liquid matrix precursors. The sol/gel technique was first developed in 1939 already for preparation of glass coatings (9) and became known during the last decade for preparation of bulk glasses without the high melting temperatures of the conventional glass production technology (10). The advantages and disadvantages of the sol/gel route as compared with the conventional road are compiled in Table 1 (11).

The sol/gel technique is also applied in ceramics for replacing conventional clay minerals as binder material as well as for fabrication of uranium oxide in technology of nuclear fuels. There exist good compilations on techniques and further applications in the proceedings of the first (12) and the second (13) "International Workshop on Glasses and Ceramics from Gel".

In principle, these techniques consist in preparation of homogeneous non crystalline xerogels from metal alkoxides and inorganic salts as starting materials and subsequent heat treatment to transform these gels into solid glasses and ceramics. This concept was studied intensively by Rustum ROY and his co-workers starting from 1948 (14). At about 1971 a clear picture on the mechanism and on combinations of various alkoxides to form multicomponent oxide glasses was available (15).

One of us published in the fifties the successful use of TEOS as a temporary binder for the preparation of sintered MoSi$_2$ materials from silicide powder by the slurry casting technique. In that application the high shrinkage during gel and glass formation from the alkoxide was used for densification without pressure application. The formed SiO$_2$-glass binder between the grains was then removed by heat treatment in hydrogen above 1600 °C in order to increase the amount of shrinkage and thus densification of the product (Fig.4) (16).

Fig.2 - Inhomogeneneous distribution of CVD-SiC-fibres in a glass matrix prepared by the fusion method

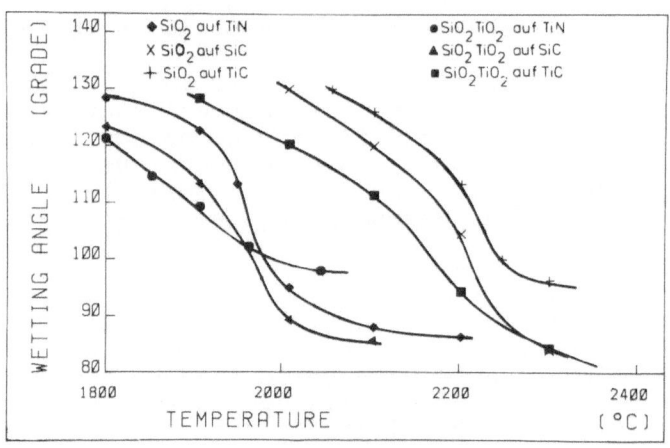

Fig.3 - Wetting angle of SiO_2 glass melts on various refractory coatings on carbon

Table 1 - Advantages and Disadvantages of the Sol/Gel Method Compared with Sintering and Melting

Advantages	Disadvantages
(1) better homogeneity from raw materials	(1) large shrinking during processing
(2) higher purity	(2) residual fine pores
(3) lower temperatures of preparation (mostly temperatures less than 1300 K compared with 1600 ... 2200 K for conventional melting	(3) residual carbon and hydroxyl groups

574

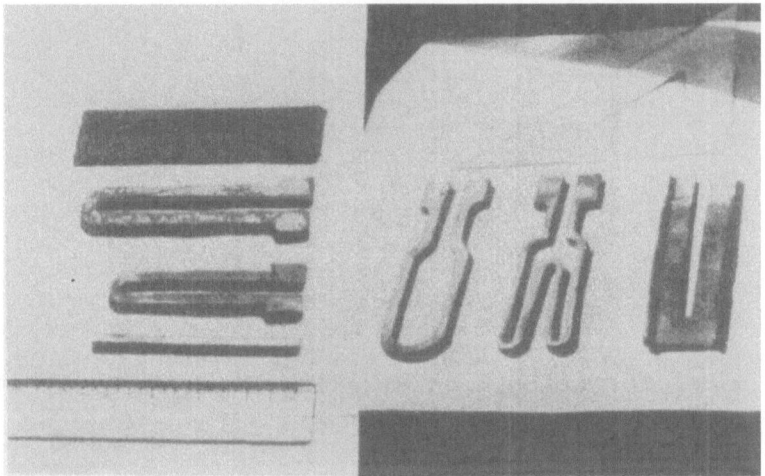

Fig.4 - Use of TEOS as temporary binder in slip casting powder
metallurgy of silicides (heating elements)

    a) comparison of final sinter shrinkage of $MoSi_2$-cylin-
ders prepared by conventional cold pressure densification (left) and pressureless slip casting technique (right)

    b) sintered $MoSi_2$ products of complex geometry fabrica-
ted by pressureless slip casting technology (16)

     The experiments described in the present paper, namely pre-
paration of fibre reinforced oxides via the sol/gel route, used
this alkoxide not only as temporary binder but as precursor for
the matrix between the refractory fibres. The shrinking densification is insufficient and final hot pressing became inevi-
table in all cases (17).

2. CHEMISTY OF SOL/GEL DERIVED OXIDES AND GLASSES

2.1 Gel-Formation from Alkoxides

A sol from metal alkoxide is defined as a liquid mixture or a solution of a single or of various mixed alkoxide compounds. It can contain different catalysts, dissolved salts, dissolving intermediaries and additives which influence the viscosity. Water, alcohols and various other polar organic solvents are suitable as solvents. In a simple way, one can say, that an alkoxide compound opposed to air humidity is yet a sol because of beginning hydrolysis. Sometimes, a sol is described as colloidal dispersion of a hydrous oxide in which at least one dimension of the particles is between 1 nm and 1 μm (18).

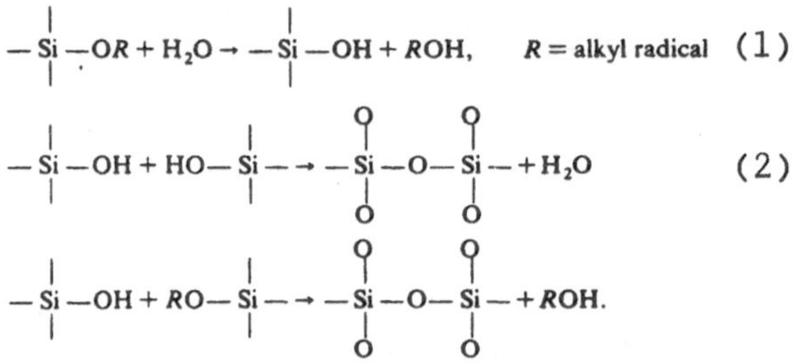

$$-Si-OR + H_2O \rightarrow -Si-OH + ROH, \qquad R = \text{alkyl radical} \quad (1)$$

$$-Si-OH + HO-Si- \rightarrow -Si-O-Si- + H_2O \qquad (2)$$

$$-Si-OH + RO-Si- \rightarrow -Si-O-Si- + ROH.$$

Fig.5 – Sol formation from alkoxides by hydrolysis (1) and condensation (2)

Solid gels are formed from such liquid sols by further hydrolysis and subsequent dehydration and polymerization . In principle one has to distinguish between the following steps:

1. hydrolysis,

2. condensation between hydrolyzed and non-hydrolyzed molecules as shown in Fig. 5,

3. further condensation reaction and completing hydrolysis,

4. polymerization to a solid gel under decomposition and evaporation of alkoxy and silanol groups.

This transformation can be followed by the increasing viscosity. A special example is shown in Fig. 6a for TEOS with varied H_2O-contents (19), causing geling at various temperatures.

The oxide content of the hydrolysis product increases strongly with increased amount of water added to the sol but levels out with a water/alkoxide ratio of 10 to 1 as shown in Fig.6b (20). At such high water contents of the sol the resulting gel consists of isolated spherolitic particles instead of a bulk structure.

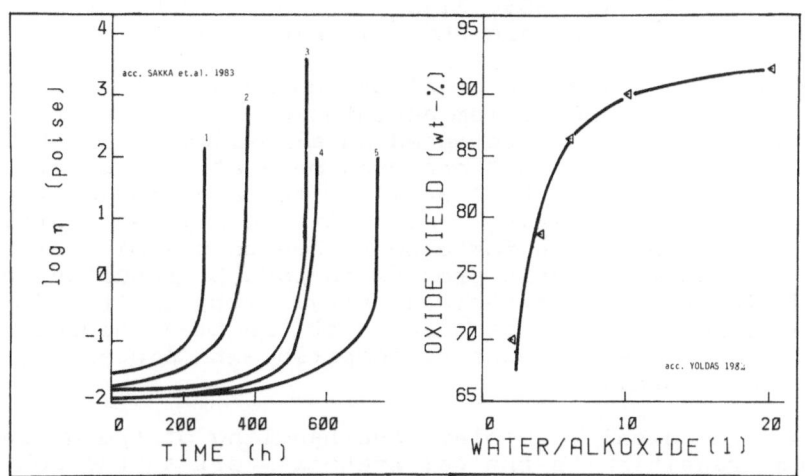

Fig.6 - Influence of H₂O amount in the sol during geling:
   a) curve 1-3 HCl catalyzed TEOS with increasing
      H₂O content; curve 4 and 5 NH₄OH catalyzed TEOS (19)

   b) oxide yield after calcination as function of in-
      creasing H₂O content during hydrolysis (20)

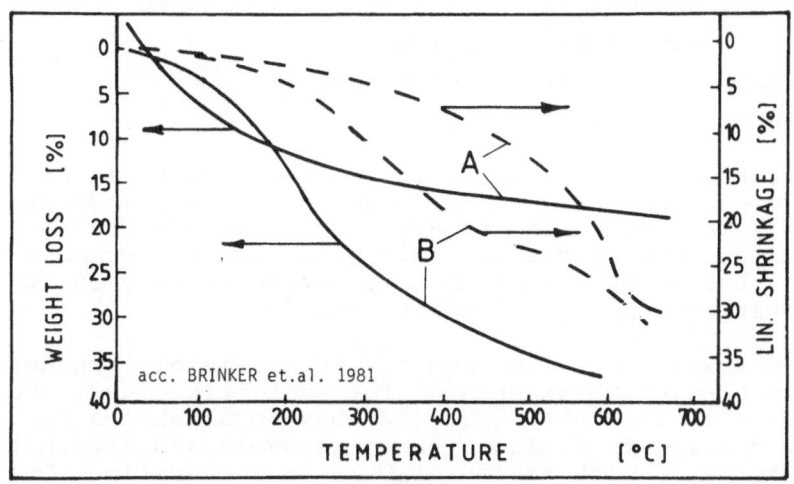

Fig.7 - Modes of sol preparation for the same final glass
        product strongly influence shrinkage and weight loss
        during gelation and calcination (21)

Also the nature of the solvent, the pH and catalysts present in the sol influence the condensation/polymerization reactions. As example weight losses and shrinkage behaviour during heat treatment up to 700 °C for the same multicomponent silicon glass (wt% 66 SiO_2, 18 B_2O_3, Al_2O_3, 6 Na_2O, 3 BaO) prepared by varied sol/gel processes are shown in Fig.7 (21).

In process A a mixture of alkoxy-compounds and aqueous solutions of sodium and barium acetate were used, in process B, however, all components were added as alkoxides in alcoholic solution. Gelation in case B occurred only after exposure of the gel to approximately 80 % humidity for a period of 2 to 5 days. Also with increasing temperature, continued hydrolysis is found to be lower than for case A. Due to the minimal amount of water introduced by this process in case B, resultant gel contained high amounts of residual alkyl groups. Therefore, during densification by subsequent heat treatment higher weight loss can be explained in case B, compared with process A, as shown in the figure 7.

From a technical viewpoint, the handling of the geling sol during transition to the gel state as well as the achieve-ability of controlled density and microstructure of the final xerogel is most important. This gel structure will distinctively control the further densification by heat treatment and sintering.

2.2   Gel to Glass Transition

During the gel to glass conversion both chemical (calcination and cross linkages), and structural (sintering) transformations take place which can be summarized as follows:

1. physical desorption of water from micropore walls,,
2. thermal decomposition from residual organic groups,
3. Condensation polymerization,
4. Formation and collapse of micropores
5. Viscous sintering.

The structural transformation is represented in Fig.8 for a borosilicate glass (66 SiO_2, 18 B_2O_3, 7 Al_2O_3, 6 Na_2O, 3 BaO (21). The degree of the BET surface area decreases while the hardness increases. This glass can be formed via the sol/gel technique at a maximum temperature of only 600°C, whereas melting temperature needs 1600 °C.

Sol/gel science and technology is intimately connected with the system $Al_2O_3 \cdot SiO_2$ as studied for synthetic clays. Recently HOFFMANN , ROY and KOMARNENI (22) have published a study on mullite formation from single phase xerogels and alternatively diphasic xerogels with varied Al_2O_3/SiO_2 composition. The term xerogel is understood as a gel structure from which all unbound water has been extracted. The single phase xerogel was prepared from a solution of tetraethoxysilane (TEOS) and aluminium nitrate 9-hydrate in pure ethanol, diphasic xerogels were derived from a mixture of aqueous silica in boehmite sols.

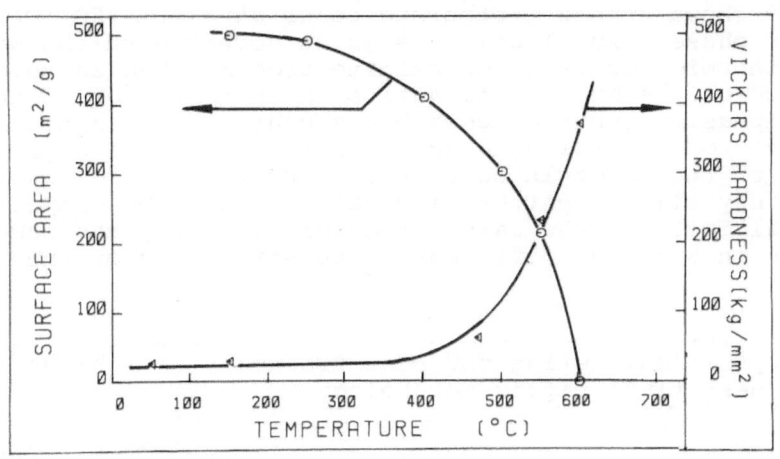

Fig.8 - Structural transformation during heat treatment of xerogels (borosilicates (21))

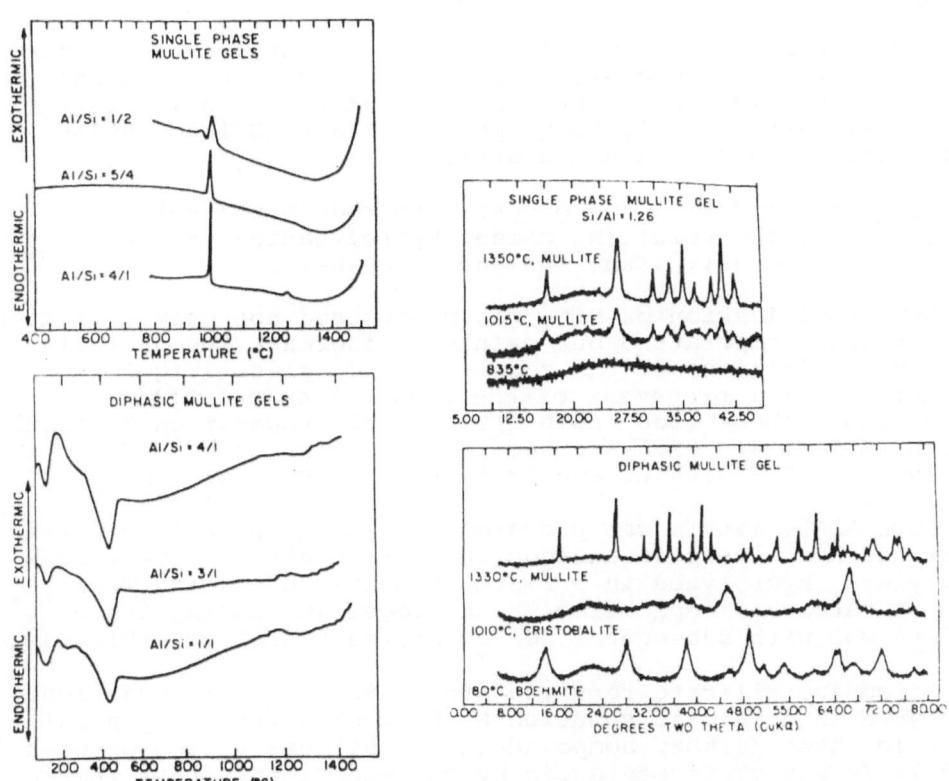

Fig.9 - Mullite formation from single phase and diphasic xerogels followed by DTA and X-ray diffraction (22)

Fig.9 shows strong exothermic peaks at about 960 °C for the single phase xerogel which is in agreement with the well known exothermic formation of mullite from metakaolinite. There is no trace of the mullite formation exotherm at 960 °C in the case of diphasic precursor gels but a continuous reaction occurs over a range of several hundred degrees. The final crystalline structure of the xerogels heat treated up to 1350 °C shows, however, very similar pattern indicating, that the composition and crystallinity of the calcined oxide is nearly the same in both cases in spite of different phase structures of the xerogels.

This result can be taken as guideline for the optimization of sol preparation, geling and heat treatment to achieve primarily best final matrix morphology.

3. MATERIALS AND PROCESSES FOR THE EXPERIMENTAL PREPARATION OF FIBRE REINFORCED HIGH MELTING GLASSES AND REFRACTORY OXIDES

3.1 Matrix Precursor

The alkoxides used in this experimental study are compiled in Table 2. For the pure SiO_2-glass matrix tetraethoxysilane with HCl as catalyst and C_2H_5OH as solvent was used (50 v% part TEOS, 40 v% part ethanol, 10 v% parts 1 % aqu. HCl. Gelation occurs after 5 to 7 h in humid air).

TiO_2 modified SiO_2 was prepared to reduce the softening temperature of the resulting oxide. Hydrolyzation was performed simultaneously with TEOS and $Ti(OC_2H_5)_4$.

GeO_2 modification of SiO_2 glass was used not only to reduce the softening temperature but mainly to increase the thermal expansion coefficient of the resulting SiO_2 glass (Fig.10) (23) needed for improvement of the physical compatibility with various fibres (see Table 2). The sol preparation by simultaneous TEOS/TEOG hydrolysis was later simplified by $GeCl_4$ additives to TEOS without any further catalyst additions (24).

Pure Al_2O_3 matrix was prepared from Al-isopropylate precursor and hydrolysis in analogy to TEOS or alternatively from Al-butylate, hydrolyzed in a 2-step-process, beginning with NH_4OH as catalyst, separating the hydrogel and adding it to butylate-sol with subsequent HCl catalyzed hydrolysis (Fig.11).

Zirconium silicate resulted from TEOS and $ZrCl_4$ addition, hydrolyzed in ethanolic solution without catalyst. In general the ratio water /alkoxy compound in the sol was varied between 1 and 2. It was confirmed again by our experiments,that the most important step during hydrolysis/polymerization from a multicomponent sol to a homogeneous gel is the controlled co-precipitation/polymerization so that a selective hydrolysis of mostly one compound is avoided.

Table 2 - Sol/Gel-Derived Oxide Matrices

SOL/GEL-based OXIDE MATRICES	USED PRECURSOR	GEL-FORMATION PROCESS	COEFFICIENT OF THERMAL EXPANSION ($10^{-6} K^{-1}$)	
SiO_2	Tetraethoxysilane (TEOS)	HCl catalysis in ethanolic solution	0.35 ... 0.5	
$TiO_2 - SiO_2$	TEOS + $Ti(OC_2H_5)_4$ TEOS + $Ti(iso-OC_3H_7)_4$	Simultaneous hydrolysis HCl catalysed in ethanolic solution	1.5 ... 1.8	$(10 \, TiO_2-SiO_2)$
$GeO_2 - SiO_2$	TEOS + Tetraethoxy-germane (TEOG)	HCl catalysis in ethanolic solution	2.5 5.8	$(5 \, GeO_2-SiO_2)$ $(10 \, GeO_2 SiO_2)$
	TEOS + $GeCl_4$	ethanolic solution without additional catalyst	3.0 6.3	$(5 \, GeO_2.SiO_2)$ $(10 \, GeO_2-SiO_2)$
$B_2O_3 - SiO_2$	TEOS + B_2O_3 hydr.	HCl catalyst in ethanolic solution	1.2	$(10 \, B_2O_3-SiO_2)$
$Al_2O_3- SiO_2$	TEOS + $Al(sec-OC_4H_9)$	aqueous ethanolic solution	5.6 4.3	(Mullite) $(2 \, SiO_2-3 \, Al_2O_3 \, glass)$
Al_2O_3	$Al(iso-C_3H_7)_3$	HCl catalysis in ethanolic solution	8.4	
	$Al(sec-OC_4H_9)_3$	2-step(NH_4OH/HCl) catalysis	8.9 ... 9	
$ZrO_2 - SiO_2$	TEOS + $ZrCl_4$	ethanolic solution without catalyst	1.8 ... 2.4	$(12 \, ZrO_2-SiO_2)$

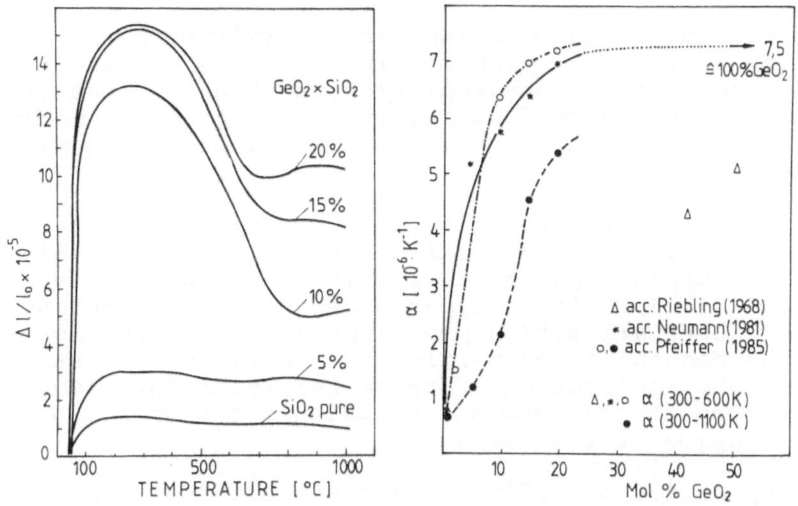

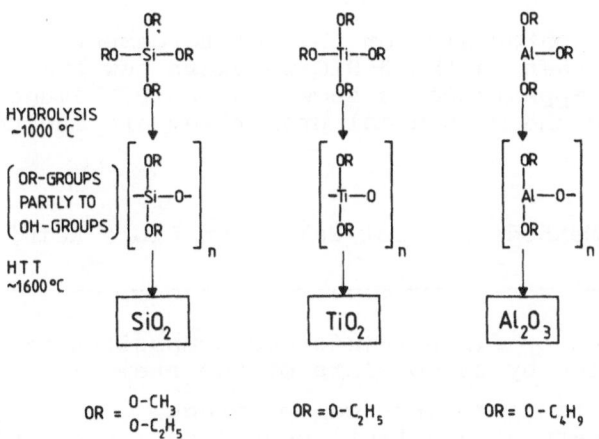

Fig.10 - Increase of reversible thermal expansion behaviour of SiO_2-glass by GeO_2 additions (24)

Fig.11 - Representative alkoxides and resulting oxide matrices

3.2 Reinforcing Fibres and Whiskers for Composites with Oxidic Matrix

The suitability of reinforcement components for oxidic composites is controlled by their chemical and physical compatibility. From chemical viewpoint high melting oxide fibres are most candidate materials. However, silicon carbide is also known to be compatible with high melting oxides. The compatibility of carbon fibres with oxides can only be expected if reaction barriers are applied, for instance silicon carbide layers. Refractoriness of the reinforcement components is mostly combined with high modulus (see Fig.12). As the YOUNG's modulus of the fibres will control in most cases the strength of the composites, it is an important selection criterion for the reinforcement component. The fused SiO_2 fibres with relatively low modulus are included in our experiments because of their good compatibility with modified SiO_2 glass matrices.

More detailed information on the reinforcement elements are compiled in Table 3. High modulus C-fibres are known to remain unchanged in their physical properties after heat treatment up to temperatures above 2000 °C, in contrast to the behavior of HF type C-fibres. Therefore, only HM C-fibres were used in this study.

The thermal stabilities of the commercially available SiC-fibres derived from polycarbosilane as well as the α-Al_2O_3 fibres were studied experimentally. The results are given in Figs. 13 and 14. Both fibre types undergo temperature and time controlled recrystallization processes by which the microstructure and the mechanical properties of the fibres are strongly influenced. The application of these reinforcement elements is thus limited by the maximum hot pressing temperatures needed for densification of the matrix.

The deposition of diffusion and reaction barrier on carbon fibres, if performed accurately, does not influence decisively mechanical properties of the fibres (25). SEM image of SiC-coated C-fibres with partially removed layer and mechanical data are given in Fig. 15.

New promising materials for the reinforcement of ceramic composites can be seen in the ß-SiC-whiskers now available in the market. Their appearance is shown in the REM image in Fig.16. Literature data on the mechanical properties are included in Table 3.

3.3 Preparation Methods for Sol/Gel Based Fibre Reinforced Composites

The processing of fibre reinforced composites via the sol/gel route as applied by us consists of two steps :

(i) the preparation of fibre gel prepregs
(ii) thermal treatment and final densification by hot pressing.

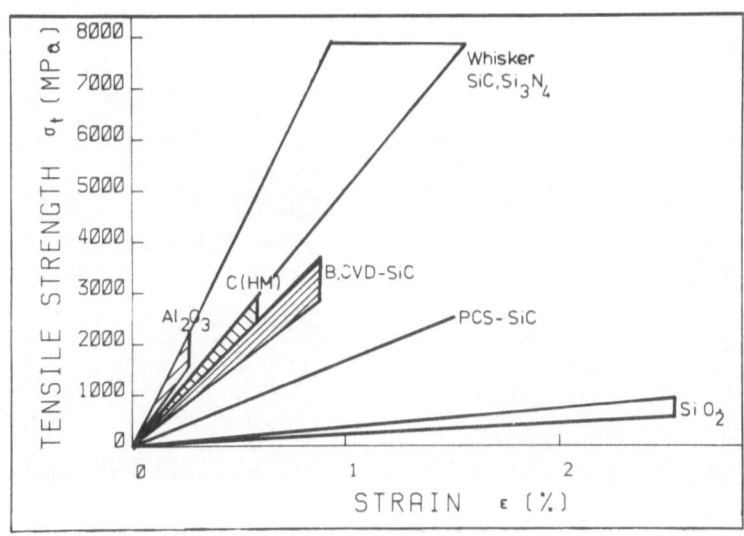

Fig.12 – Stress-strain diagram of refractory reinforcement composites

Table 3 – Mechanical and physical properties of reinforcement components

	ENDLESS FIBRES					WHISKER
	C-FIBRE TYPES		SIC-FIBRES		α-AL₂O₃-FIBRE	β-SIC WHISKER
	THORNEL 400	M 40	CVD-SIC ON W	PCS-SIC	TYPE FP	
DIAMETER (μM)	7.5	7.0	100	9...15	15....25	0.05... 1.5
TENSILE STRENGTH σ_T (MPA)	2800	3200	3700	1900	1400	3000....14000
STRAIN TO FAIL. ε (%)	1.25	0.85	0.95	1.0	0.4	1.0 - 1.5
YOUNG's MODULUS (GPA)	230	380	380...420	190	400	400....700
DENSITY (G/CM³)	1.8	1.9	3.45	2.58	3.95	3.19
THERM. EXPANSION COEFFICIENT (10⁻⁶ K-1)	-0.5 ... 0.2 (=) 10 ... 15 (⊥)		4.2 - 4.5 (=) 4.2 - 5.2 (⊥)	3.1	7.2 - 8.6	2.4 - 2.5

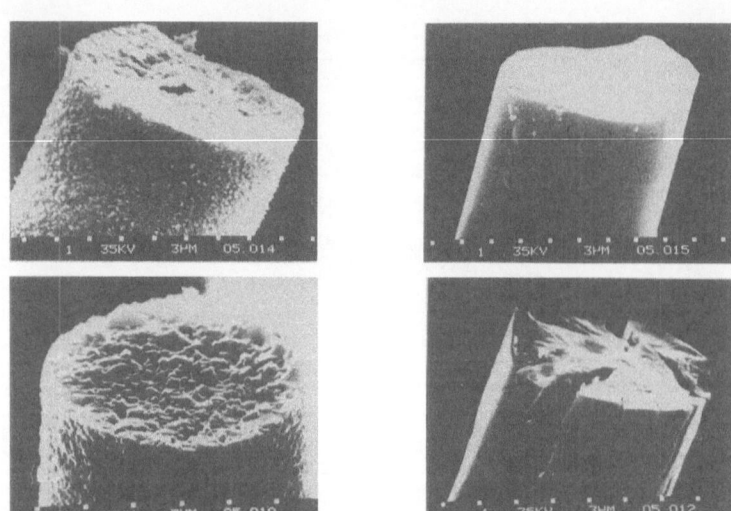

Fig.13 - SEM fracture surfaces of SiC and Al₂O₃ fibres
as received (upper row) and after 100 h heat treatment
at 1600 K in air (lower row)
a) Polycarbosilane derived (PCS)   b) α-Al₂O₃ fibre
SiC fibre (NICALON)

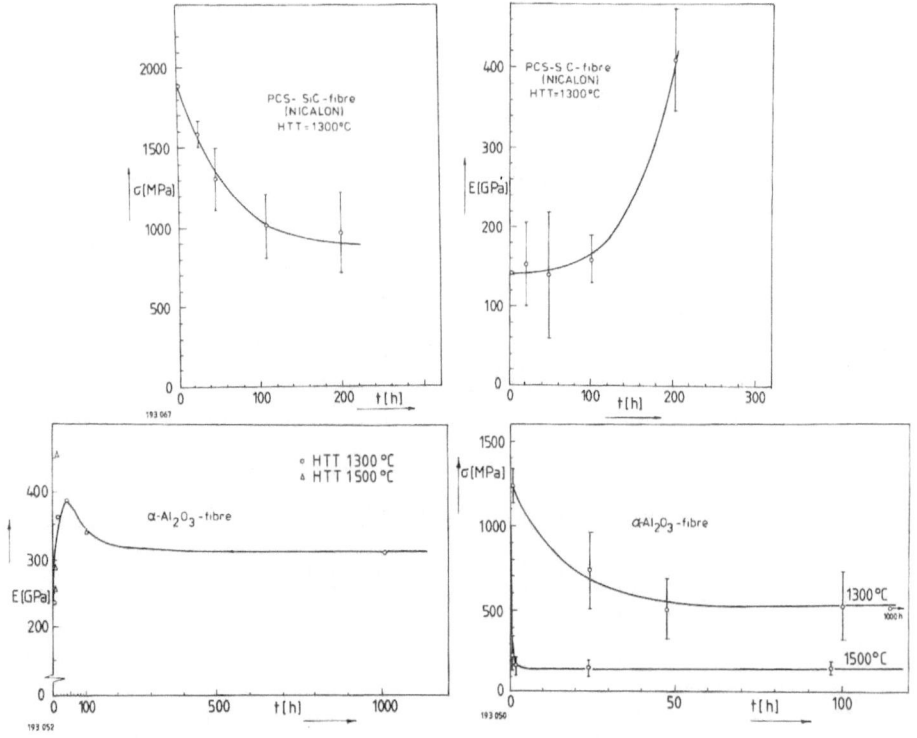

Fig.14 - Influence of heat treatment on mechanical properties;
upper row: PSC-SiC fibres, HTT 1300°C,
lower row: α-Al₂O₃ fibres, HTT 1300 and 1500°C

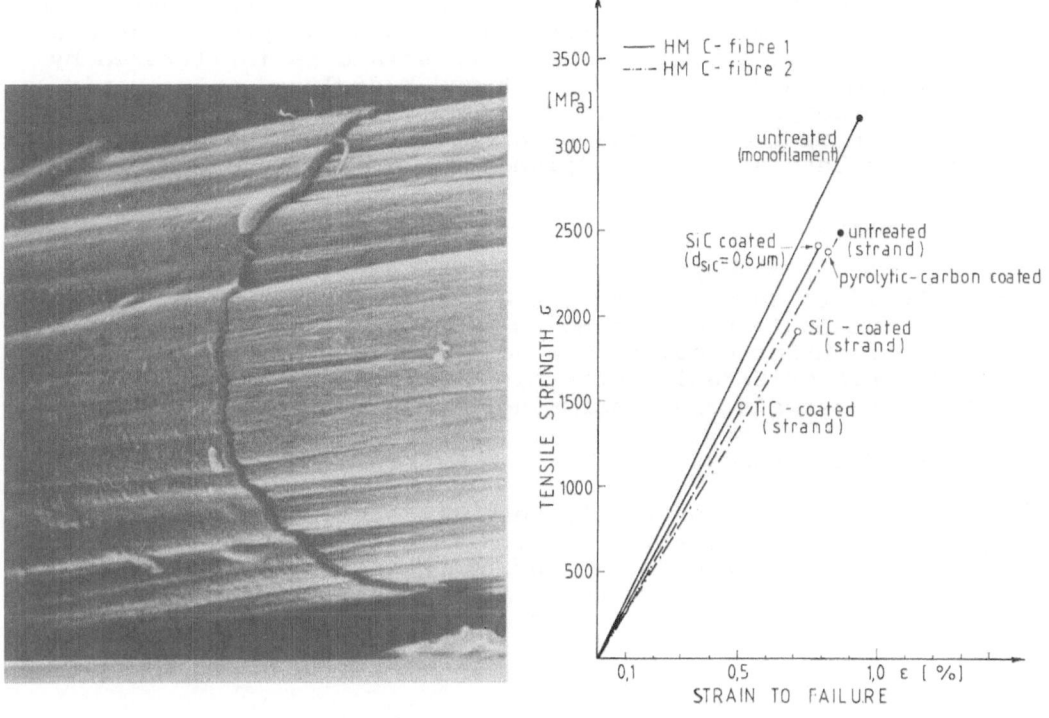

Fig.15 - C-fibres coated with refractory carbides

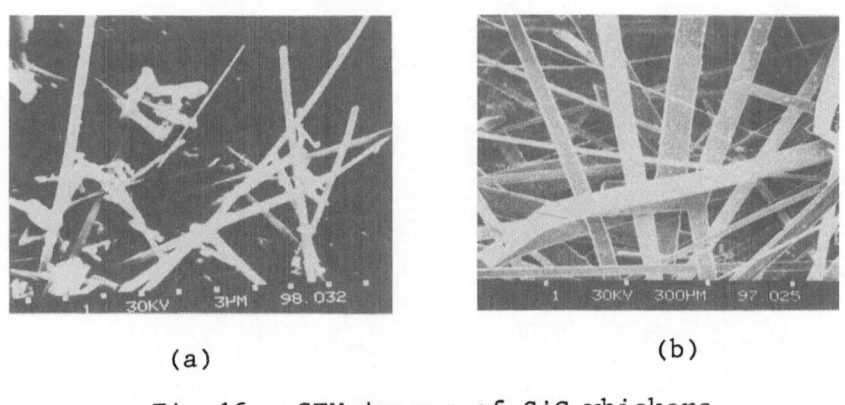

(a)   (b)

Fig.16 - SEM-images of SiC whiskers
         β - SiC (part a)
         α - SiC (part b)

a) Strand dipping method

For systematic studies in laboratory scale the preparation of gel impregnated fibre strands was found to be the best method to obtain sample materials with good mechanical properties and acceptable reproducibility. The fibre strand is infiltrated by sol with a minimum solvent content and hydrolyzed only by air humidity to complete geling. The advantage of this strand dipping process is the homogeneous distribution of the gel in the interspaces between the monofilaments.

b) Wet winding process

The wet winding process can follow the well-known procedure where the fibre is drawn through a partially hydrolysed sol bath. The gel content in the prepreg is mainly controlled by the tensile stress during winding. The advantage of this process is the precise fibre orientation, the disadvantage, however, limited gel content between the inner monofilaments of the fibre bundle.

c) Dry winding process

The dry winding of the fibres with subsequent pressure impregnation after evacuation results in a homogeneous gel distribution between the fibres. Shrinkage cracks during drying (see Fig.17) can be refilled by repeated impregnation/curing cycles.

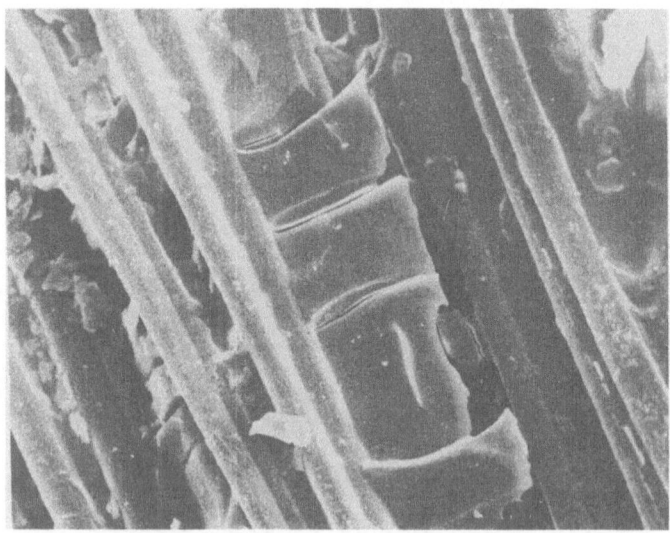

Fig.17 - SEM micrograph of xerogel matrix precursor
         with shrinkage cracks

As mentioned before, hot pressing was found to be necessary in all cases for densification of the xerogel matrix with pores and cracks. The static hot pressing method in graphite molding tools and inductive heating (100 kW) is shown in Fig.18, left part. Right part of Fig. 18 gives the temperature cycles for densification of the various composites and the pressure periods.

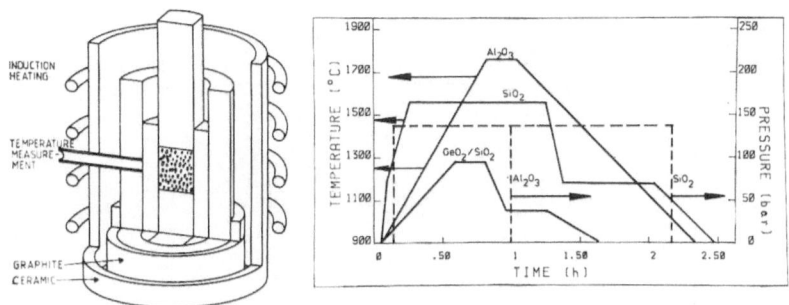

Fig.18 - Applied hot pressing techniques
   a) hot pressing tool
   b) temperature cycles for various oxides
      and the pressure periods

In more detail, the densification of pure SiO_2 and Al_2O_3 matrices from xerogels is shown in Fig. 19a, compared with the densification of solid powder precursors. Fig. 19b shows the mullite formation and densification derived from simultaneous hydrolysis of varied mixtures of methoxysilane and aluminium butylate. The densification depends on the composition and occurs in different steps. The first step can be explained by densification of the lower melting components (region I to III). The mullite formation is seen in region IV, whereas mullite densification occurs subsequently in regions V and VI.

The hot pressing conditions (temperature and time) on oxides in a graphite mold are limited by the chemical reactivity between the two materials. In the case of Al_2O_3 densification, chemical reaction occurs above 1700 °C causing high porosity and low strength of the sintered material as shown in Fig. 20 (26).

4. EXPERIMENTAL RESULTS ON VARIOUS FIBRE/MATRIX COMBINATIONS OF SOL/GEL DERIVED OXIDIC COMPOSITES

4.1 Fibre Reinforced SiO_2-glasses

SiO_2-glass is the highest temperature resistant glass and its formation from TEOS is the best studied of all alkoxide derived oxides. Because of the high hot pressing temperatures up to

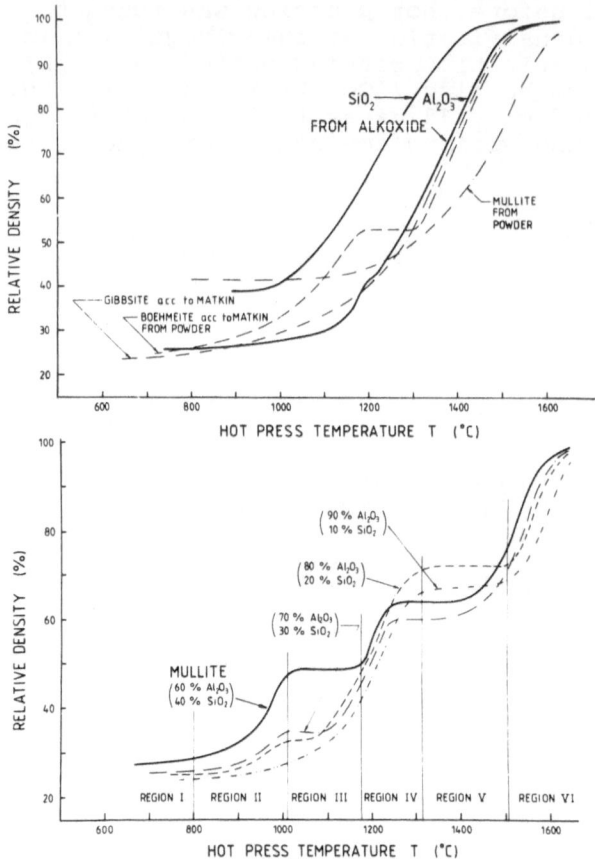

Fig.19a – Densification of oxide matrix during hot pressing of hydrolyzed matrix precursor (top)
  19b – preformed oxide powder, stepwise mullite formation and densification from hydrolyzed SiO_2 and Al_2O_3 precursor (bottom) (26)

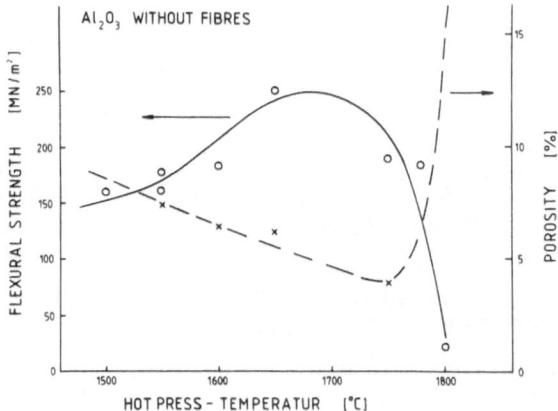

Fig.20 – Flexural strength and porosity of pure Al_2O_3 from Al-butylate depending on hot pressing temperature

1700 K (see Fig.18b) only carbon fibres can be expected to remain mechanically uneffected, whereas Al_2O_3-fibres and PCS-SiC-fibres are sensitive against heating above 1600 K already (compare Fig.14). For such temperature sensitive fibres modified SiO_2 matrices with lower softening temperatures had to be applied (7-10 % TiO_2 and 5-15 % GeO_2) (see Table 2).

4.1.1 Fibre Reinforced Pure SiO_2 Glass

Non-coated carbon fibres will react chemically with SiO_2-glass at temperatures above 1500 °C to form volatile carbon oxides and residual silicon carbide. Carbon fibres coated with high melting carbides or nitrides are chemically compatible with SiO_2-glass at least during short time preparation of the composites. Such coatings are also necessary to achieve good wetting during processing and in consequence good adhesion between fibre and matrix in the final composite.

Carbon fibres were coated by vapour deposition with TiC, TiN and SiC (see Fig.15). These refractory coatings were proved to guarantee sufficient wetting during processing by systematic measurements with the sessile drop method (Fig.3). The UD-composites were prepared by the dry winding method with subsequent impregnation.

The fibre matrix adhesion can be recognized by the degree of pullout as shown in Fig.21. This adhesion in the final composite is best in case of SiC-coatings and worst in case of TiC-coatings. The extremely heavy pullout of TiC coated fibres in SiO_2 glass is shown in Fig.22, whereas the fracture surface of SiC coated fibres shows nearly no pullout (see Fig.23) (6).

The precalculated strength of unidirectionally reinforced composites according to the rule of mixture is given in Fig.24 as straight upper line. The measuring data of flexural strength reach these precalculated values, indicating that in best cases a translation of fibre strength within the composites up to 100 % has been achieved. With fibre contents above 50 v% the strength itself and thus also the degree of translation of fibre strength decreases because of insufficient filling of all interspaces between the too tight package of the fibre bundles.

This disadvantageous geometrical effect is also found for TiN coated C-fibres as shown in Fig. 25. The overwhelming qualitative effect on the adhesion by different coatings can clearly be recognized. Metal coated fibres (in this case Ta coatings) behave even worse than uncoated carbon fibres.

Bulk specimen of pure SiO_2glass have a low YOUNG's modulus and show also low strain to failure values because of the minor strength. The improvement of both, strength and elongation, by increased volume fractions of carbon fibres is given in Fig.26. The elongation of the pure matrix at room temperature is increased by factor 5, indicating a toughening effect. Special attention should be drawn to the step-like fracture behaviour even in case of only small fibre contents.

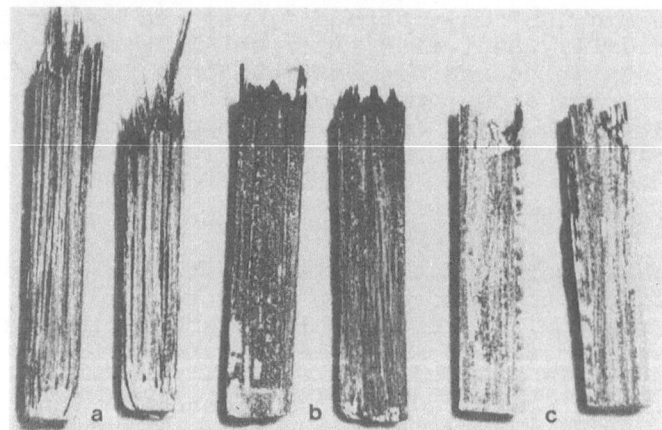

Fig.21 - Influence of various refractory surface layers on
C-fibres under fibre pullout during fracture of
SiO$_2$ composites (6)
   a) 40 v% TiC-coated C-fibres
   b) 35 v% TiN-coated C-fibres
   c) 35 v% SiC-coated C-fibres

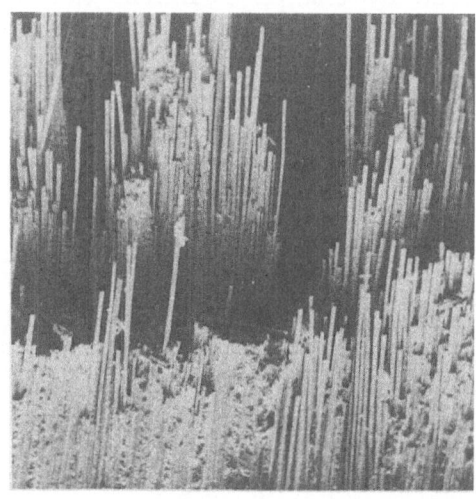

Fig.22 - Extremely high pullout of TiC-coated C-fibres from
SiO$_2$ matrix

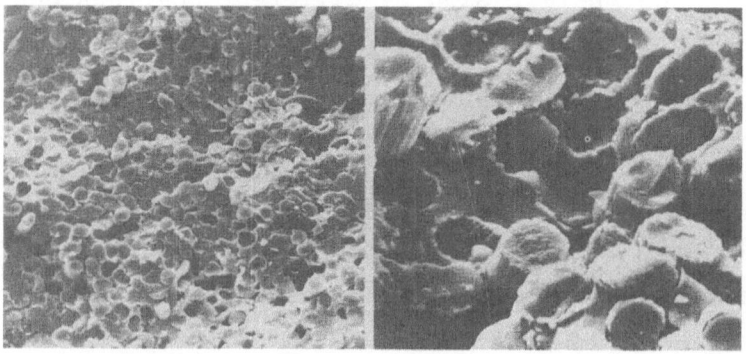

Fig.23 - SEM fracture surface of SiC-coated C-fibres/SiO$_2$com-
posites with only minor fibre pullout

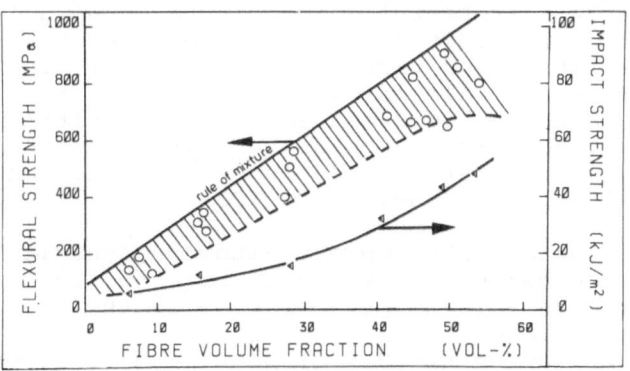

Fig.24 – Effect of volume content of SiC–coated C–fibres on flexural and impact strength of SiO$_2$ glass composites.

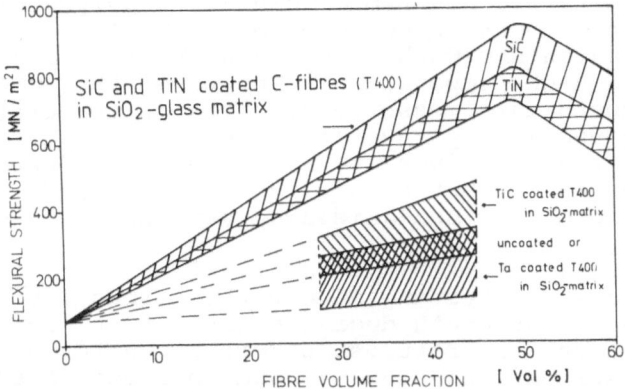

Fig.25 – Influence of the nature of coating in the reinforcing effect of C–fibres in SiO$_2$ glass.

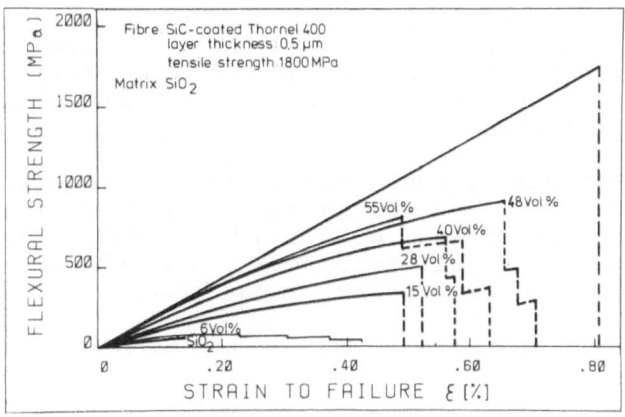

Fig.26 – Stress–strain diagram for C–fibres (SiC–coated) with different fibre contents.

The slightly different fracture surfaces with increasing volume content of SiC-coated carbon fibres are shown in Fig.27 (part a: 25 v% fibres and part b: 35 v% fibres). The complete wetting and filling of all interspaces in case of 25 v% fibres in SiO_2 glass shown in Fig.28 causes a clear brittle fracture without pullout, whereas in composites with increased fibre content the only thin matrix bridges between the monofilaments cause a partial pullout (compare Fig.23).

Carbon fibre contents improve also the dynamic behaviour of SiO_2 glass composites. As shown in Fig.29, 10^5 cycles can be expected at a dynamic loading of 65 % of the static flexural strength, whereas fracture will occur after 100 cycles in the case of dynamic loading at 80 % of the flexural strength.

Silica glass is known to exhibit extreme brittleness at room temperature. The toughening effect by increased C-fibre content can be recognized from the impact strength values, included in Fig.24 with maximum values up to 50 MN/m.

Also in CFRP (carbon fibre reinforced polymers) impressive mechanical improvement is achieved by increased carbon fibre content. The advantage with SiO_2-glass as matrix can be expected in an improved high temperature strength of the composite. Flexural strength data measured at elevated temperatures up to 1000 °C are given in Fig. 30. It can be seen that the room temperature strength values are not only maintained up to higher testing temperatures, but even slightly increased. These tests have been performed in non-oxidizing atmosphere.

In the case of testing in air the carbon fibres burn out. Fig. 31 shows the strength decrease after 50 h heat treatment in air at various temperatures. Literature data on C-fibre reinforced borosilicate glasses (5) are included. A suitable protection against fibre burn out was achieved with a coating consisting of a fibre free SiO_2-layer (0,5 mm thickness) prepared by a post treatment via sol/gel process. The flexural strength after 50 h in air is maintained up to 1000 °C heat treatment temperatures (6).

The fracture behaviour of such a composite with a non-reinforced outer layer is shown in Fig.32a and is typical for flexural fracture initiated in the surface on tensile stress side. Fig.32b shows such a crack in the fibre free outer layer which will initiate the fracture of the composite. Such cracks open also the entrance for oxidizing gases and can initiate burn out of the fibres.

One may recognize, that carbon fibre reinforced SiO_2 glass composites show promising mechanical properties but their sensitivity against oxidation is inhibited only partially so far. As a consequence, oxidation resistant fibres would be preferrable in high temperature composites.

Fig.27 - Effect of C-fibre content in SiO$_2$ glass composites on the pullout 25% fibres (l), 35% fibres (r)

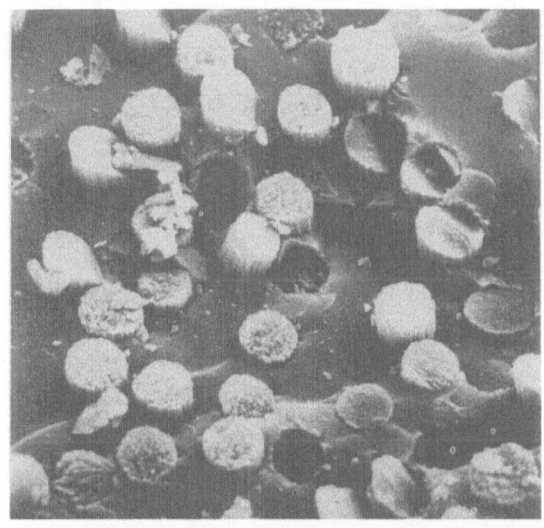

Fig.28 - Complete surrounding of all C-fibre monofilaments by SiO$_2$ glass in composites with 25 v% fibres

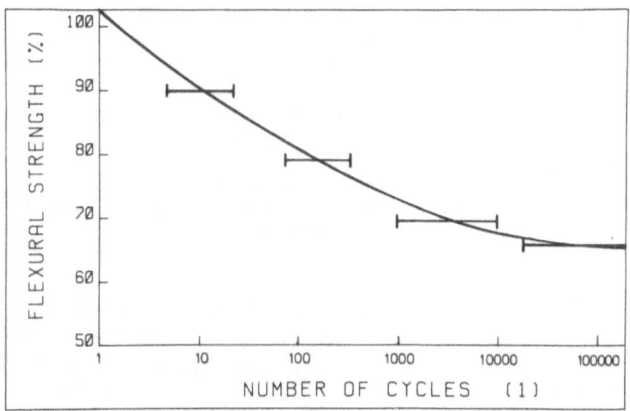

Fig.29 - Test results on cyclic loading of SiC reinforced SiO$_2$ glass at room temperature

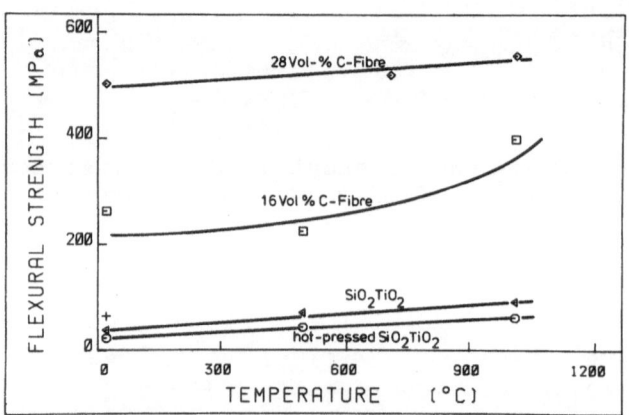

Fig.30 - High temperature flexural strength of C-fibre reinforced SiO$_2$ glass, measured in non-oxidizing atmosphere

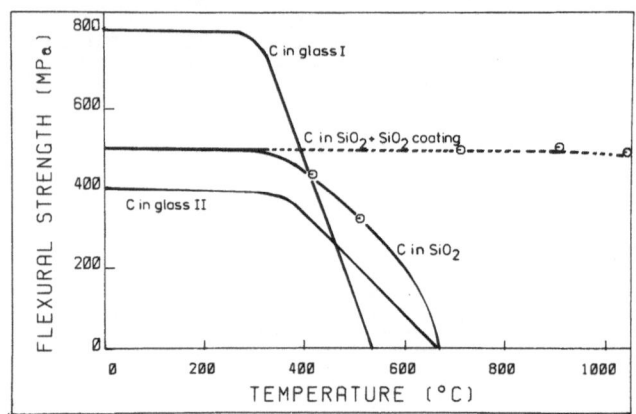

Fig.31 - Strength reduction of SiO$_2$-composite reinforced
with 30 v% of SiC-coated carbon fibres after
50 h oxidation in air (6) (in glass I and II acc.
SAMBELL (5))

(a) 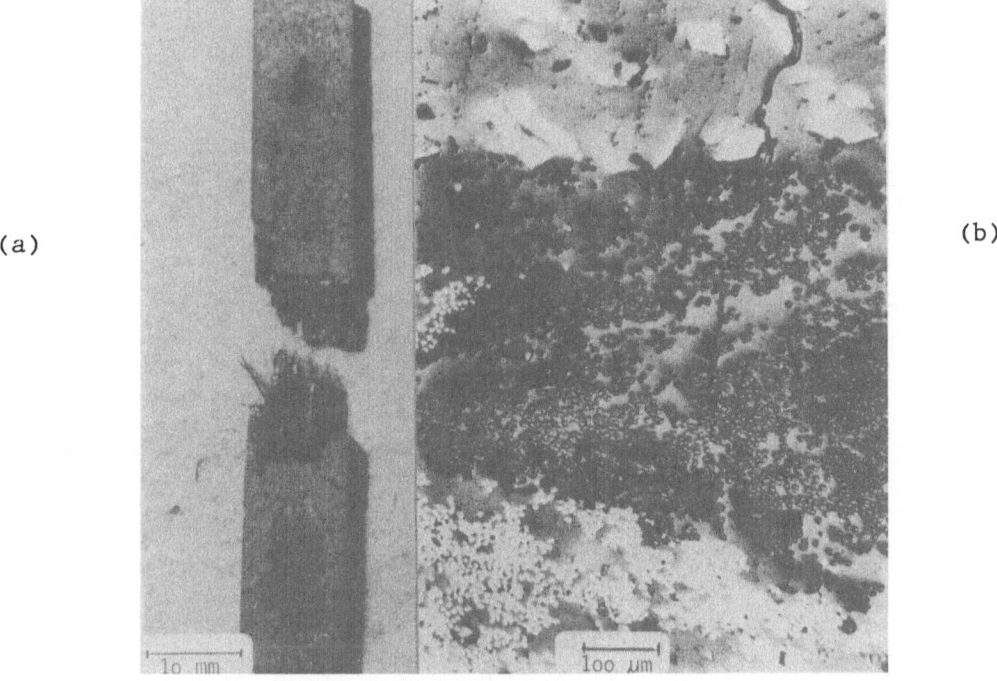 (b)

Fig.32 - C-fibre reinforced SiO_2 glass composite with fibre free
         outer layer as protection against oxidation
         a) sample after flexural fracture at room temperature
         b) cross section with fracture initiating crack in
            the fibre free surface layer

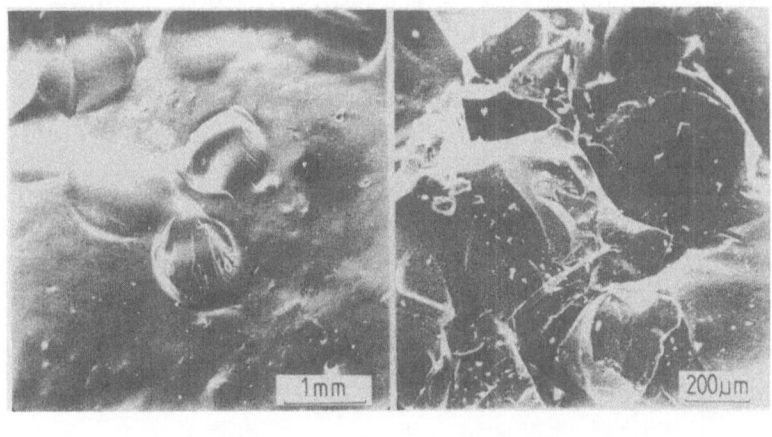

(a)   (b)

Fig.33 - Fracture surface of SiO_2 rods :
         a) in SiO_2/7 % TiO_2 glass
         b) in borosilicate glass

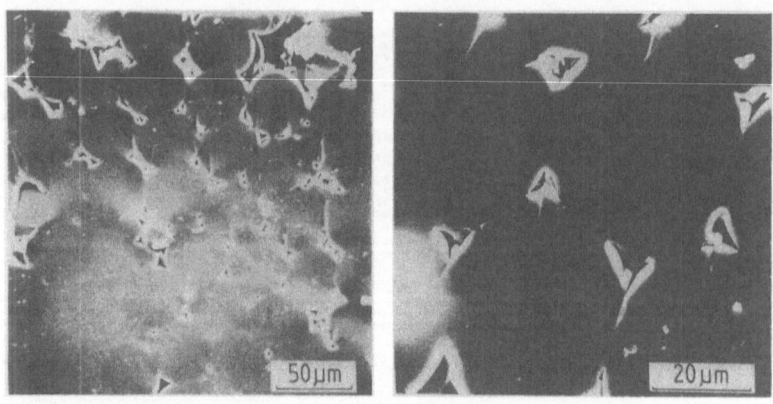

Fig.34 – Cross section of the SiO$_2$ fibre reinforsed SiO$_2$ glass with unfilled
interspaces between the monofilaments.

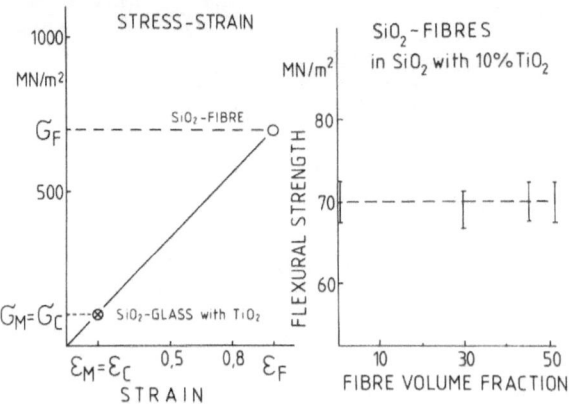

Fig.35 – Mechanical properties of the composites SiO$_2$ – fibres in TiO$_2$
modified SiO$_2$ glass.

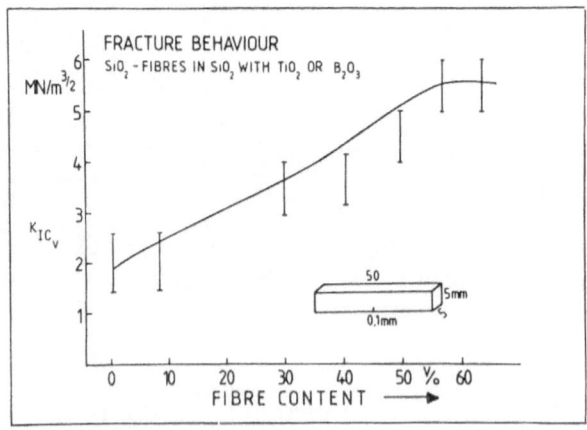

Fig.36 – Critical stress intensity factor in SiO$_2$ fibre reinforced SiO$_2$
glass composites.

4.1.2 SiO$_2$-Fibres in SiO$_2$-Glass

For SiO$_2$-fibres/SiO$_2$-matrix composites the hot pressing temperature of the matrix must be considerably lower than the softening temperature of the SiO$_2$-glass fibres. In pre-tests SiO$_2$-rods with 1 mm thickness were inbedded in TiO$_2$ and B$_2$O$_3$ modified SiO$_2$-glass (see Fig.33). Chemical compatibility and physical maintainance of the fibre form were observed in the TiO$_2$ modified glass, whereas the silica fibre reacts with the boron modified glass during hot pressing.

For our composite fabrication only fibre bundles with mono-filaments of 40 µm were available. As can be seen from Fig.34, we did not succeed in obtaining pore free bulk structure of the composites. The interfibrellar spaces remained partly un-filled.

The mechanical properties of such SiO$_2$/SiO$_2$ composites are shown in Fig.35. No strengthening effect was achieved with increasing fibre contents. However, the fracture behavior was improved (Fig.36). This retarding effect on crack propagation by fibre additions can not be interpreted on the basis of the rule of mixture. It offers an additional advantage in the application of fibre additions to brittle matrix.

4.1.3 Fibre Reinforcement of Modified SiO$_2$-Glass by α-Al$_2$O$_3$ and PCS/SiC-Fibres

As shown in Figs. 13 and 14 both α-Al$_2$O$_3$ as well as the PCS-SiC fibres are sensitive against heat treatment above 1500 K. With 10 mol% GeO$_2$ additions to SiO$_2$ glass an optimum hot pressing temperature of 1400 K can be applied. GeO$_2$ modifications have the additional advantage to increase the thermal expansion be-haviour of SiO$_2$-Glass, for instance to α-values of 2-$3 \cdot 10^{-6}$ K^{-1} with 10 % GeO$_2$ additions (see Fig.10). We did not succeed, how-ever, to prepare SiO$_2$-glasses with further increasing GeO$_2$ con-tent because of the high vapour pressure of this component re-sulting in porosity and increased brittleness of the glass. Because of this only partial improvement in direction of physical compatibility we have still to expect high tensile stresses on the fibres after cooling from hot pressing temperature because of the higher shrinkage of the fibres (α Al$_2$O$_3$ ca $8 \cdot 10^{-6}$ K^{-1}. PCS/SiC ca $3 \cdot 10^{-6}$ K^{-1}).

The stress-strain behavior of UD-composites with 40 % of both fibres is shown in Fig.37. The data for the pure oxide ma-trix are included for comparison. One recognizes only moderate reinforcing effect in both cases. The flexural strength values of 250 and 320 MN/m^2 indicate translation of fibre strength below 50 %. Obviously, the internal stresses built up during cooling and some sensitivity of the fibres against heat treat-ment at these lower temperatures already can be taken for ex-planation.

Most dangerous for the fracture initiation are the micro-cracks formed in the matrix as shown in the SEM image (Fig.38).

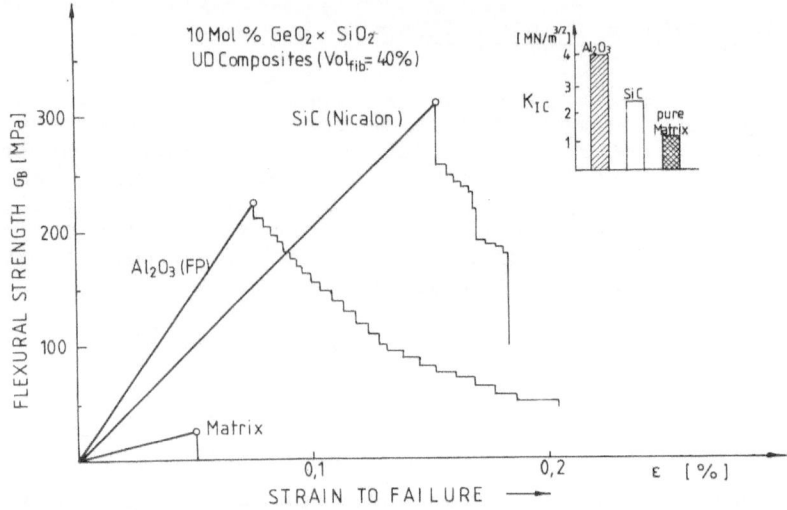

Fig.37 - Reinforcement of GeO$_2$ modified SiO$_2$-glass by 40 v%
α-Al$_2$O$_3$ and PCS/SiC-fibres

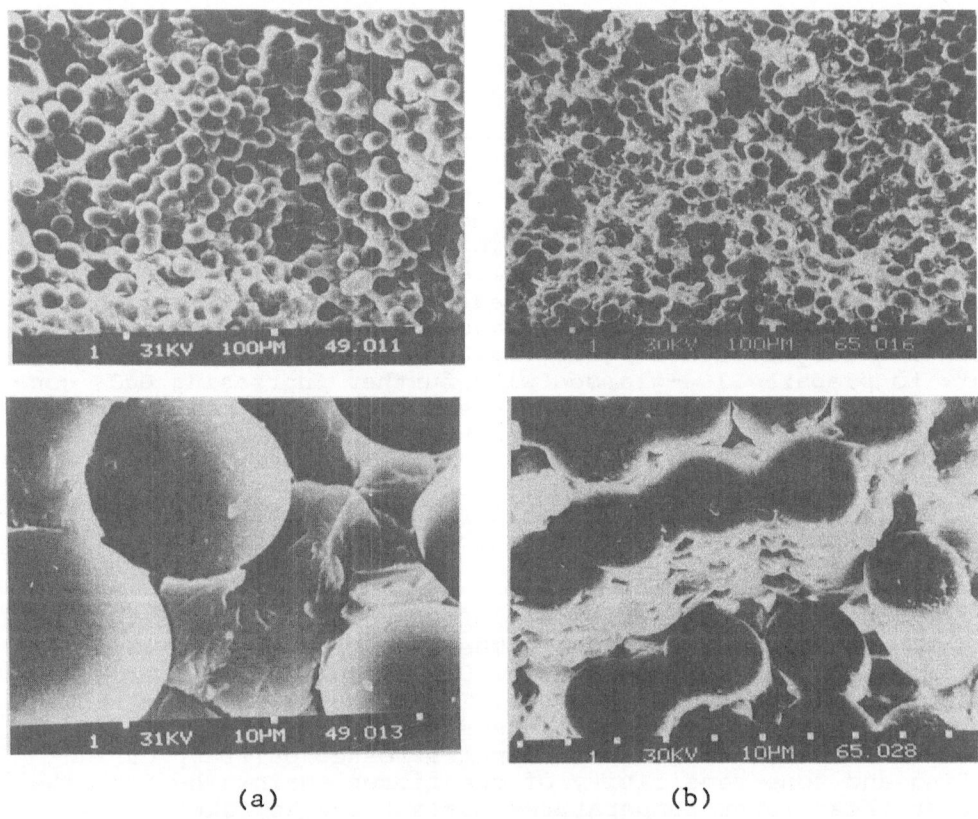

(a)   (b)

Fig.38 - Surface fracture of composites with 10 GeO$_2$·SiO$_2$
matrix and UD reinforcement by 40 v% of fibres
a) PCS/SiC-fibres
b) α-Al$_2$O$_3$-fibres

So far as the toughening effect by the fibre addition is concerned, the Al$_2$O$_3$ fibres are superior to the SiC fibres as shown by the K$_{IC}$-data included in Fig.37.

Both fibres differ also in respect to their chemical compatibility. The wetting between SiC-fibres and glass matrix seems sufficient as shown in Fig.38. Between matrix and Al$_2$O$_3$ fibres a surface reaction can be recognized which can support the adhesion but can also influence the toughness of the composite.

4.1.4 SiO$_2$-Glass Reinforced by SiC Whiskers

The very fine β-SiC whiskers distributed were mixed with the viscous sol of TEOS with GeCl$_4$ additions and hydrolyzed in humid air. It was found advantageous to apply ultrasonic densification of the mixture consisting of higher contents of whiskers and the solidifying hydrogel. It was tested before, that the applied hot pressing temperature around 1500 K does not influence the crystalline structure of the whiskers.

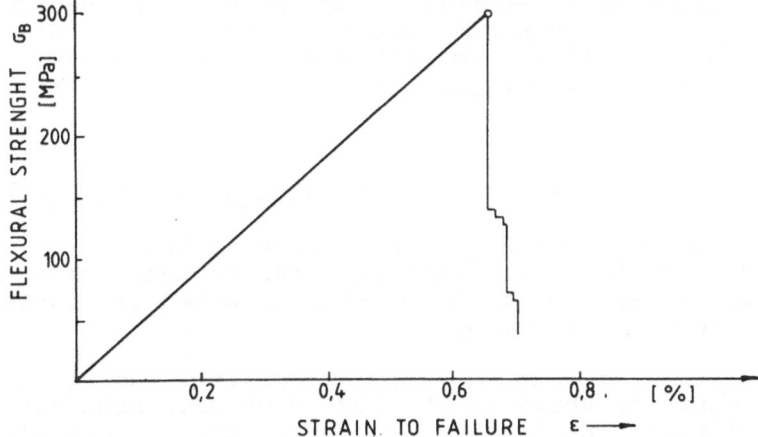

Fig.39 - GeO$_2$ modified SiO$_2$-glass reinforced with 60 v% of β-SiC whiskers

Fig. 39 shows stress/strain behaviour of composites with 60 v% of whiskers. Final flexural strength of 300 MPa and an elongation of 0.7 % were achieved. It is very difficult to calculate translation of whisker strength into the composite because the whisker diameter, whisker length and also whisker strength are nonuniform (see Fig.15 and Table 3). If we assume 100 % translation of whisker strength and use the calculation according the rule of mixture divided by 6 for isotropic reinforcement, an average whisker strength of 3000 MN/m^2 would result. This value is obviously too low. A more realistic value between 4000 and 4500 MN/m^2 would indicate a translation of whisker strength of 60 to 70 %. Also this assumption leads to the conclusion of an acceptable adhesion of the fibres in the isotropically reinforced matrix. Fig. 40 shows that the whiskers are distributed quite homogeneously. There are, however, remaining pores to see in the fracture surfaces. This indicates, that the densification process is not yet optimized. No toughening behaviour was measured so far experimentally.

Fig.40 - Fracture surface of modified 10 GeO_2-SiO_2-glass
with 60 v% of SiC-whiskers

4.2   Fibre Reinforced Alumina and Mullite

Al_2O_3 is known as a high strength ceramic material. If derived from aluminium sec-butylate, hot pressing around 2000 K
was needed to achieve minimum porosities below 5 v%. With in
situ formation of mullite from alkoxides (TEOS and Al
$(sec-C_4H_9)_3$ hot pressing temperatures can be lowered to 1900 K
(Fig. 41).

4.2.1   Carbon Fibre Reinforcement of Alumina and Mullite

Also in the case of Al_2O_3 and mullite matrices, C-fibres
with SiC coatings have been used as reinforcement component.
For these later experiments M 40 C-fibres were used instead of
the THORNEL 400 type (Fig.43).

The flexural strength of Al_2O_3 composites increases only
with fibre contents above 10 v%. 500 MN/m^2 have been achieved
with 40 v% fibres. The importance of SiC-coatings can clearly
be recognized. With fibre additions below this critical 10 %
value a strength decrease of these aluminium oxide composites
was found (see Fig. 32).

The fracture toughness is considerably increased in case
of good adhesion between fibre and matrix (SiC-coated C-fibres)
with K_{IC} values up to 10 $MN/m^{3/2}$, but not with uncoated fibres.

A similar beneficial effect of SiC-coated fibres additions
was found for the reinforcement of sol/gel based mullite derived from a mixture of TEOS and aluminium butylate. Obviously,
because of the lower hot pressing temperatures and better densification behaviour of mullite matrix, final flexural strength
up to 800 to 900 MN/m^2, similar as with SiO_2 glass matrix has
been achieved (Fig.44). Coating of the carbon fibre was found
not to be so essential for the K_{IC}-values. It is assumed, that
some SiC layers are formed between the carbon fibre surface
and the SiO_2-component of the matrix during the hot pressing
step.

Nevertheless, there is only a limited bonding between
fibres and matrix, and obviously circumferential micro-cracks
are formed which improve strength and toughness by stopping and
deviating crack propagation within the matrix. From fracture

surface in Fig.45 it can be seen that the delamination and pullout of the fibres will influence the toughness and fracture behaviour of the samples.

4.2.2   Aluminium Oxide and Mullite Reinforced by Al_2O_3-fibres

The most compatible reinforcement fibres for Al_2O_3 based matrices is Al_2O_3 itself. The only disadvantage for such a combination is the sensitivity of this fibre against heat treatment during hot pressing above 1000 °C. Unfortunately, densification of the calcined xerogels used by us, was not achieved at

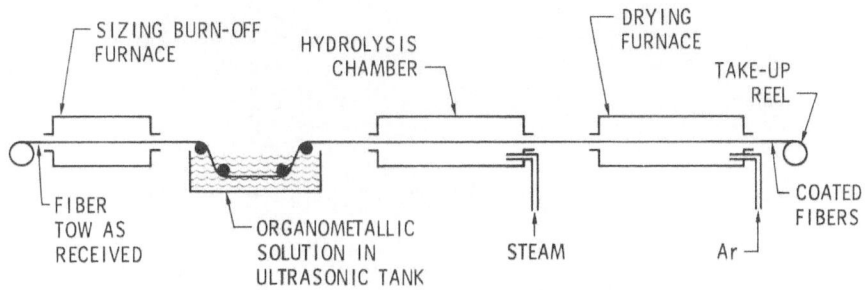

$$Si(OC_2H_5)_4 + 2H_2O \rightarrow SiO_2 + 4C_2H_5OH\uparrow$$

$$Si(OC_2H_5)_4 \xrightarrow{\Delta} SiO_2 + 2C_2H_5OH\uparrow + 2C_2H_4\uparrow$$

Fig.41 - Experimental optimization of hot pressing temperatures for C-fibre reinforcement of mullite and its pure components

these low temperatures (see Fig.41). The composites have passed 1 h at 1700 °C for Al_2O_3 and 1600 °C for mullite matrix (26).

It was not expected, therefore, that the strength will increase by the fibre addition. This is confirmed experimentally in Fig. 46a for both, Al_2O_3 and for mullite matrices. However, the fracture toughness shows some promising improvement (see Fig.46b).

It is difficult to explain the toughening effect, because in case of Al_2O_3 fibre reinforcement of these alumina based matrices no pullout was observed in the fracture surface (see Fig.47) contrary to the fracture behaviour of C-fibre reinforced mullite (see Fig.45).

SiC COATED C-FIBERS IN SiO$_2$ MATRIX

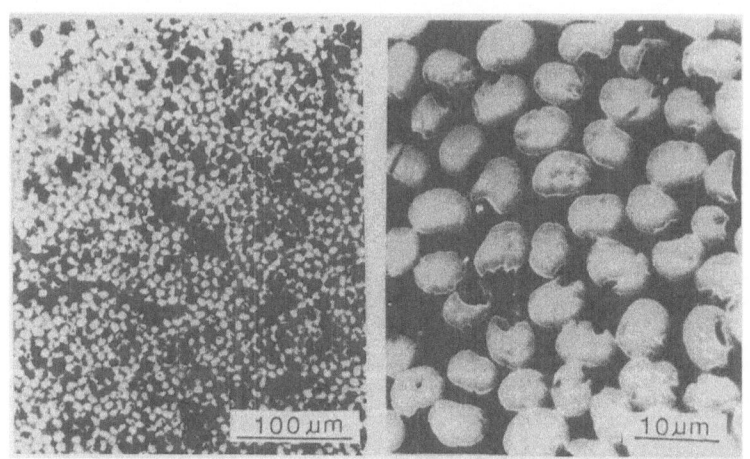

Fig.42 - Micrographs of SiC coated C-fibre   (type THORNEL 400)
reinforced SiO$_2$ glass (upper row) and
Al$_2$O$_3$-composites (C-fibre type M 40) (lower row)

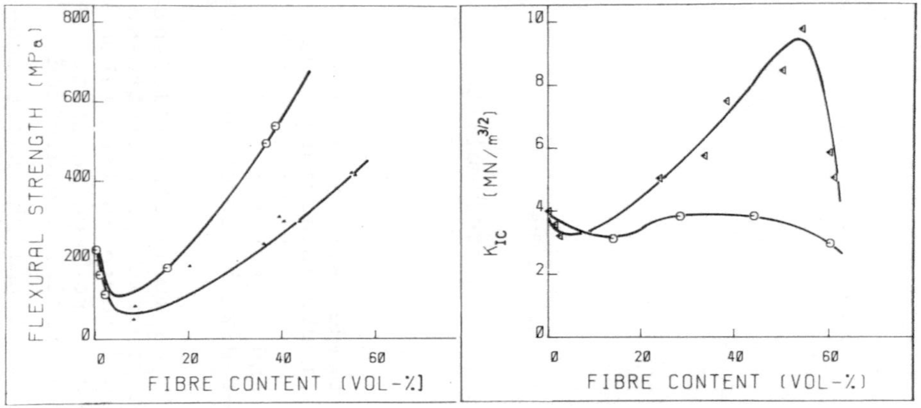

Fig.43 - Strength and toughness
of SiC coated C-fibre
reinforced Al₂O₃ com-
posites depending on
fibre volume content

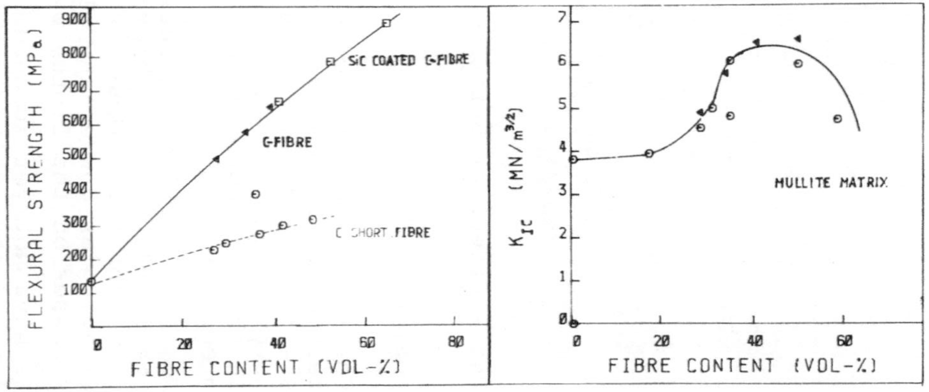

Fig.44 - Mechanical properties
of C-fibre reinforced
mullite composites,
depending on fibre
volume content
a) flexural strength
b) fracture toughness

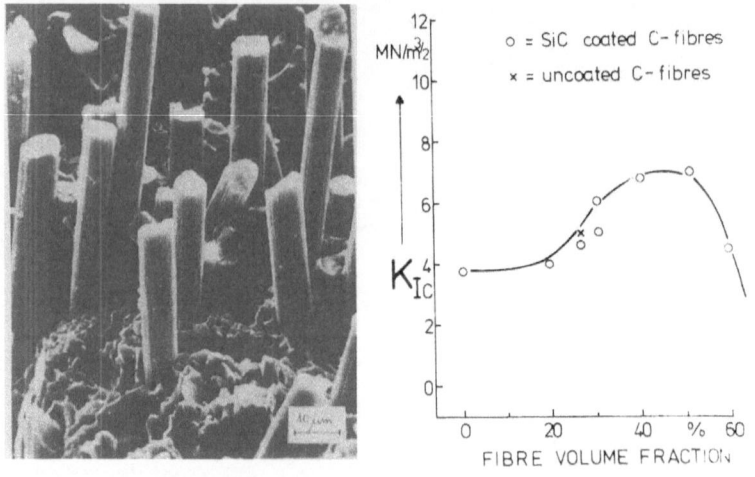

Fig. 45 - Effect of C-fibre additions to sintered mullite

CONCLUSIONS

1. For the preparation of glass and oxide matrices in fibre
   reinforced composites the sol/gel route is advantageous
   similar as in bulk glass fabrication because of the much
   lower densification temperatures and higher flexibility
   in chemical composition, if compared with powder or slurry
   precursor.

2. For high melting matrices (SiO_2/Al_2O_3/mullite) sufficient
   densification was achieved by hot pressing post treatment
   only. For this procedure the sol/gel route is less fibre
   damaging as powder and slurry precursors because shear
   stresses on the fibres by powder particles are avoided.

3. We have not yet succeeded to use the total shrinkage tendency (dehydration, calcination, sintering) for complete
   densification of the final oxide matrix. The feasibility
   of such pressureless densification via sol/gel technique
   was demonstrated before in powder metallurgy of silicides.

   It becomes obvious, that isotropic shrinkage may be the
   precondition. The anisotropic shrinkage in composites,
   perpendicular to the fibre direction hinders a more complete densification by pressureless techniques during
   heat treatment. It should be applicable, however, for
   whisker reinforcement.

3. What remains to do in future work :

   (a) further modification of the sol precursor composition
       and of the process parameters to improve densification
       at lowest possible temperatures,

   (b) improvement of the thermal stability of chemical compa-
       tible fibres and whiskers.

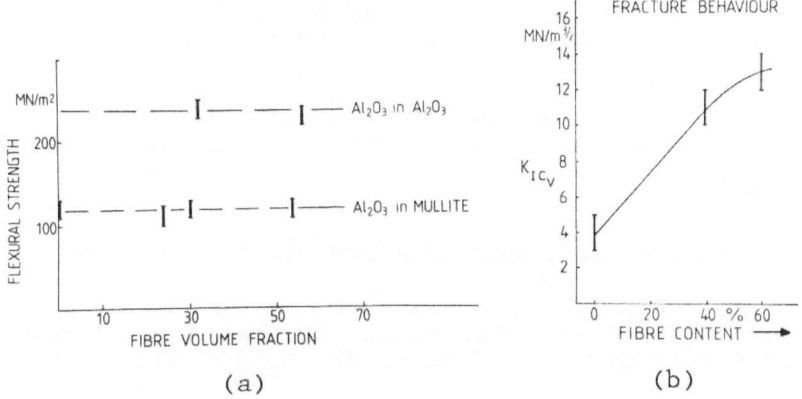

Fig.46 - Mechanical properties of Al_2O_3-fibre reinforced
(a) Al_2O_3 and (b) Mullite

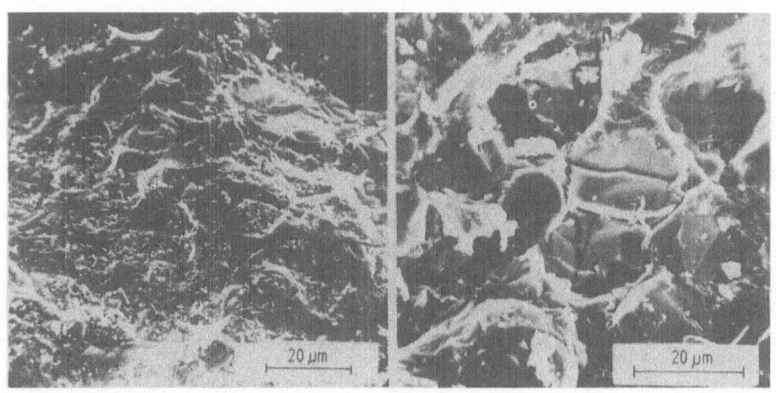

Fig.47 - Fracture surfaces of Al_2O_3 (a) and Mullite (b)
reinforced with Al_2O_3-fibres

6. REFERENCES

1) CLAUSSEN, N.
Z. Werkstofftech. 13, 138-147 (1982)

2) STEINBRECH, R.; KNEHANS, R.
in "Hochtemperaturreaktionen Keramischer Werkstoffe",
24.-25.11.1983, Stuttgart, R.F. PABST ed., MPI Stuttgart 1984

3) FITZER, E.
"Fibre Reinforced Ceramics and Glasses", Proc. of Int. Symp.
Densification and Sintering of Oxide and Non-oxide Ceramics,
Tokyo 1978, 618-673

4) LUBIN, G.
Handbook of Fiberglass and Advanced Plastics Composites

5) SAMBELL; R.; PHILLIPS, D.; BOWEN, D.
Proc. Int. Conf. Carbon Fibres 18.-20. Feb.1974, London
Plastics & Polymers Conf. Suppl. Nr. 6, 105-113

6) SCHUBERT, P.
Ph.D. Thesis, Faculty of Chemistry, Universität Karlsruhe 1977

7) PREWO, K.; BRENNAN, J.J.
J. Mat. Science 17 (1982) 1201-1206, 2371-2383

8) PREWO, K.; THOMPSON, E.R.
NASA Contractor Rep. 165711, Research on Graphite Reinforced
Glass Matrix Composites, May 1981

9) GEFFCKEN; W.; BERGER; E.
Deutsches Reichspatent 736 411, Jenaer Glaswerk Schott u.Gen.
(1939)

10) SCHROEDER, H.
Thin Film, (1969), 87 ff

11) MACKENZIE; J.D.
J. of Non-Cryst. Solids 48 (1982), 1-10

12) GOTTARDI, V. ed.
"Glasses and Glass Ceramics from Gels", Proc. Int. Workshop
on Glasses and Glass Ceramics from Gels, Padova, Italy,
October 8-9, 1981, printed in J. of Non-Cryst. Solids 48
(1982), Nos. 8,10

13) SCHOLZE; H. ed.
Proc. of the 2nd Int. Workshop on Glasses and Glass Ceramics
from Gels, Würzburg, Germany, July 1-2, 1983, printed in J.
of Non-Cryst. Solids 63 (1984) Nos. 1,2

14) ROY, R.
J. Am. Cer. Soc., 39 (4) 145-46 (1956)

15) DISLICH; H.
Angew. Chem. Int. Ed. Engl. 10 (6) 1971, p. 363 ff.

16) FITZER, E.
Österr. Patent 217721 (1.7.1954) DBP 1124166 (8.3.1955)

17) FITZER, E.; SCHLICHTING, J.
High Temperature Science 13, 149-172 (1980)

18) SEGAL, D.L.
J. of Non-Cryst. Solids 63 (1984) pp 183-192

19) SAKKA, K.; KAMIYA, K.
J. of Non-Cryst. Solids 48 (1982), 31-46

20) YOLDAS, B.E.
J. of Non-Cryst. Solids 51 (1982), 105 ff

21) BRINKER, C.J.; MUKHERJEE, S.P.
J. Mat. Science 16 (1981) 1980-88

22) HOFFMAN; D.W.; ROY, R.; KOMARNENI, S.
J. Am. Cer. Soc. Vol. 67, Nr. 7 (1984) 468-71

23) SCHLICHTING, J.; NEUMANN, S.
J. of Non-Cryst. Solids 48 (1982) 185-194

24) PFEIFFER, A.
Thesis, Institut für Chemische Technik, Universität
Karlsruhe 1985

25) SCHOCH, G.
Thesis, Institut für Chemische Technik, Universität
Karlsruhe 1983
19th Carbon Conf. 1984, Bordeaux, Abstracts Vol., 162-163

26) YASUDA, E.; SCHLICHTING, J.
Z. Werkstofftechnik 9, 310-315 (1978)

FIBER-MATRIX INTERACTIONS IN CARBON FIBER/CEMENT MATRIX COMPOSITES

D. C. Cranmer and D. J. Speece

Materials Sciences Laboratory
The Aerospace Corporation
P. O. Box 92957
Los Angeles, CA 90009

ABSTRACT

An understanding of the nature of the interface between fiber and composite matrix is essential for development of ceramic matrix and metal matrix composites. It has been shown that the use of organometallic-based coatings on carbon fibers aids processing and improves properties in metal matrix composites. The present paper deals with the use of similarly coated carbon fibers in various cement matrices. It describes the method of applying the coatings and the differences in the fiber-matrix interfacial region due to the use of different coatings. The tensile properties are discussed and related to the differences seen at the interface and differences in processability.

INTRODUCTION

There is a large and growing activity in the development of advanced fiber-reinforced ceramics and glasses for use in a variety of structural applications. The properties of these composites are influenced by the interactions between the filament and matrix materials. A strong fiber-matrix bond can result in a strong, but still brittle composite while a weak bond can result in a material with poor properties. To optimize the properties and to be able to make design trade-offs (strength vs. toughness, e.g.), a more complete understanding of the interactions between the matrix and fibers is required. This paper reports on the first results of our investigation into fiber-matrix interactions in a ceramic matrix system.

The purpose of this investigation is to examine the fiber-matrix interactions in carbon fiber-cement systems where an oxide coating of SiO_2, Al_2O_3, Al_2O_3-SiO_2 (3:2), or ZrO_2 has been deliberately introduced at the interface. The carbon fiber-cement system was chosen for several reasons. The most important is the potential for development of a zero or near-zero thermal expansion coefficient material. This is made possible by the combination of the negative axial thermal expansion of the carbon fiber and the low modulus of the cement. Also, this system requires no elevated temperature processing since the cements chosen cure at room temperature. Finally, the cements can be readily modified by the addition of organometallics and/or particulates. The coatings were chosen to provide differing degrees of chemical compatibility with the matrix cements.

There are a variety of techniques available for coating carbon fibers including chemical or physical vapor deposition but one technique which has been shown to be effective in metal matrix composites involves coating the fiber with an oxide material of 1000 – 2000 Å thickness (1) using organometallic precursors. The coating adheres to the fiber and provides a surface which the molten metal will wet during further processing. The exact coating used depends on the matrix metal desired. The organometallic is applied in an ultrasonic bath and hydrolyzed to form the oxide. It is then heat treated to drive off the organic components and consolidate the oxide. A schematic of the coating process is shown in Figure 1. Coatings of carbides or nitrides can also be deposited by this technique.

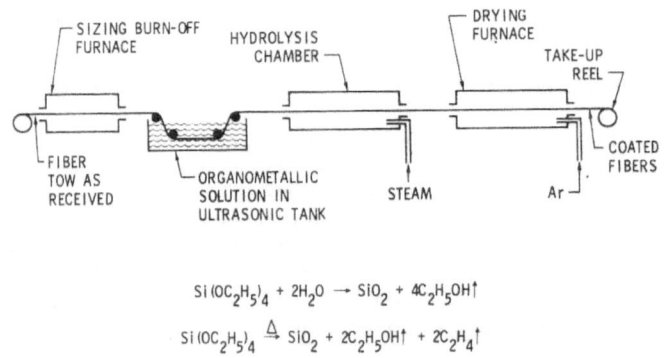

$$Si(OC_2H_5)_4 + 2H_2O \rightarrow SiO_2 + 4C_2H_5OH\uparrow$$

$$Si(OC_2H_5)_4 \xrightarrow{\Delta} SiO_2 + 2C_2H_5OH\uparrow + 2C_2H_4\uparrow$$

Figure 1.   Schematic of Fiber Coating Process and Typical Reaction for SiO_2 Coating.

EXPERIMENTAL PROCEDURE

Several 15 meter lengths of carbon fiber tows[a] were double coated with SiO_2, Al_2O_3, $SiO_2-Al_2O_3$, and ZrO_2. The organometallic precursors were tetraethylorthosilicate (TEOS), aluminum isopropoxide, and zirconium n-propoxide. The precursors were applied to the tows in an ultrasonic bath, hydrolyzed in a steam bath, and heat treated to 575 °C. Lengths of coated and uncoated tows were cut into ~ 15 cm segments and infiltrated with an epoxy resin or one of two porcelain cements[b] in an ultrasonic bath. The major constituent of the cements was either zircon or silica. All of the matrices were cured at room temperature.

---

[a]T300-1K, Union Carbide Corp.

[b]Sauereisen 29, and 31, Sauereisen Cement Co.

The tensile strengths of the various infiltrated tows were measured by attaching metal tabs to the ends, mounting in a tensile tester, and pulling at a crosshead speed of 0.0042 cm/s until complete failure. The fracture surfaces were examined in a scanning electron microscope to determine the failure modes and origins.

RESULTS AND DISCUSSION

Fracture strengths were calculated from the breaking loads and areas measured in the SEM at the fracture origin. The results are given in Table 1. It is evident from the data for the resin matrix materials that there has been no degradation of fiber properties due to the application of the coatings.

Table 1. Tensile Strength (MPa) of Various Composite Tows

Fiber	Matrix Resin	Zircon	Porcelain
Unsized	---	172 ± 46	----
Sized	3130 ± 205	232 ± 83	340 ± 118
SiO_2 Coated	2680 ± 532	111 ± 28	416 ± 136
Al_2O_3 Coated	3060 ± 375	16 ± 11	124 ± 56
SiO_2-Al_2O_3 Coated	3090 ± 193	45 ± 32	218 ± 111
ZrO_2 Coated	2780 ± 435	87 ± 50	110 ± 52

Note: Unreinforced Cement is ~ 2.8 MPa

Figures 2 - 7 show the tows infiltrated with the zircon base cement for the unsized, sized, SiO_2-, Al_2O_3-, SiO_2-Al_2O_3-, and ZrO_2-coated materials, respectively. From the chemical compatibility viewpoint, it would be expected that the SiO_2- and ZrO_2-coated tows would show the greatest degree of bonding with the cement, while the remainder would be less well bonded. Examination of the figures leads to several observations. The first is that the fiber with no sizing or coating (Figure 2) has very poor infiltration of the cement into the tow compared to the remaining materials. Second, the distribution of the fibers in the cement as well as the overall diameter of the infiltrated tow varies considerably with the applied coating. The best distribution and smallest diameter occur in the sized (Figure 2), SiO_2 (Figure 4), and ZrO_2 (Figure 7) materials. The Al_2O_3 (Figure 5) and Al_2O_3-SiO_2 (Figure 6) materials showed clumping of the fibers and large diameters. A unique observation about the Al_2O_3 coated tows can be seen in the high magnification micrograph of Figure 5. Although each fiber is surrounded by cement, there appears to be no contact of the fiber and the matrix, i.e., there is no apparent fiber-matrix bonding. Such a condition would lead to fracture and instantaneous release of the individual fibers during tensile testing, resulting in very poor strengths. Such a result is seen in Table 1, where the strength of the Al_2O_3 coated material is the lowest of the zircon cement materials.

Table 2 compares the measured tensile strengths of the T300/zircon cement tows with the values expected from the rule-of-mixtures. The volume fraction of fibers was measured from the micrographs and strains estimated from the load versus time chart acquired during the tensile tests. For all of the materials except the Al_2O_3 coated tows, the rule-of-mixtures appears to provide a reasonable prediction of the tensile strength. The rule-of-mixtures calculations assume complete bonding of the fiber and the matrix.

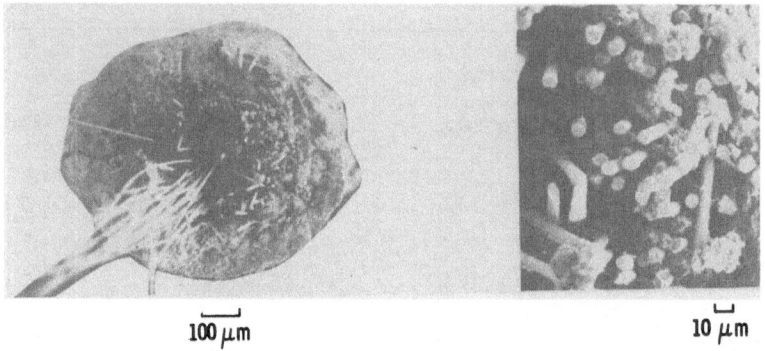

100 μm   10 μm

Figure 2.   Carbon Fiber Reinforced Cement Composite with Uncoated Fibers

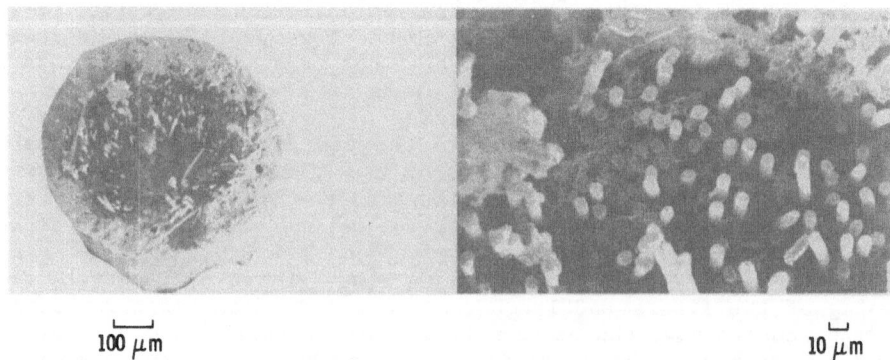

100 μm   10 μm

Figure 3.   Carbon Fiber Reinforced Cement Composite with Sizing Only

100 μm   10 μm

Figure 4. Carbon Fiber Reinforced Cement Composite with SiO_2 Coating

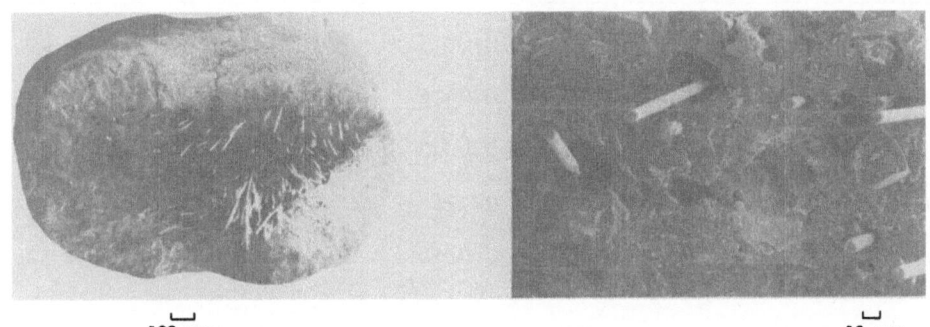

100 μm   10 μm

Figure 5. Carbon Reinforced Cement Composite with Al_2O_3 Coating.

100 μm   10 μm

Figure 6. Carbon Fiber Reinforced Cement Composite with Al_2O_3 – SiO_2 Coating ·

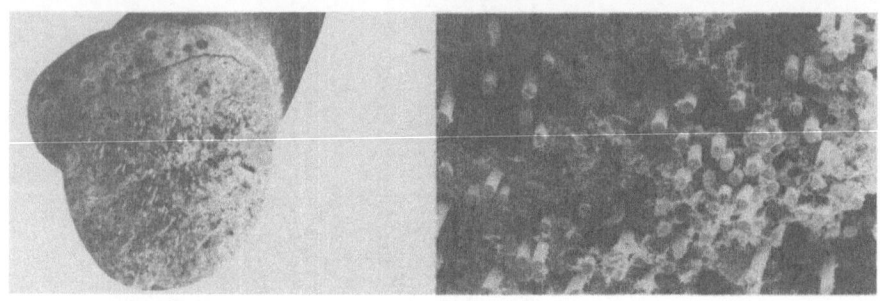

100 μm   10 μm

Figure 7.   Carbon Fiber Reinforced Cement Composite with ZrO_2 Coating.

Table 2.   Tensile Strength of Infiltrated Tows
Versus Rule-of-Mixtures Calculations

	V_F	$\varepsilon_{Matrix\ Break}$	σ_{ROM}(MPa)	σ_{ACTUAL}(MPa)
Sized	0.11 ± 0.03	0.0093 ± 0.0011	241 ± 78	231 ± 83
Unsized	0.13 ± 0.02	0.0077 ± 0.0016	232 ± 49	171 ± 46
SiO_2	0.07 ± 0.02	0.0063 ± 0.0013	114 ± 41	111 ± 26
Al_2O_3	0.04 ± 0.03	0.0072 ± 00.24	75.8 ± 59.8	16.7 ± 11.2
Al_2O_3-SiO_2	0.04 ± 0.02	0.0041 ± 0.0010	42.5 ± 25.1	45.2 ± 32.5
ZrO_2	0.07 ± 0.03	0.0042 ± 0.0011	70.0 ± 36.3	86.7 ± 50.6

CONCLUSIONS

Differences in mechanical properties in the carbon fiber-cement system due to the presence of different coatings have been seen. The difference can be related to the ability to infiltrate the carbon tows and to the degree of bonding between the fibers and matrix. For the zircon cement system, the rule-of-mixtures appears to provide a reasonable prediction of the tensile strengths and yields an indication of the degree of bonding enhancement due to the coatings.

ACKNOWLEDGEMENTS

This work was supported by The Aerospace Corporation MOIE program. The authors would like to thank Dr. H. Katzman for his helpful technical discussions.

REFERENCES

1.   H. A. Katzman, "Carbon Reinforced Metal-Matrix Composites," U.S. Patent #4, 376, 803, 15 March 1983.

MORPHOLOGICAL AND MECHANICAL CHARACTERIZATION OF CERAMIC COMPOSITE

MATERIALS

J.L. Chermant, M. Gomina, and F. Osterstock

Equipe Matériaux-Microstructure du L.A. 251, ISMRa
Université, 14032 CAEN Cedex, France

ABSTRACT

Fiber-reinforced CVD-ceramic composites, C-SiC, have been tested in
three point bending for three orientations of the notch prior to the
applied stress. For the two main orientations, load-unloading sequences
have been performed to investigate the rupture parameters using the Sakai
and al's method. The very good mechanical behaviour of these materials
are found to be due to the energy dissipative effect of the reinforcement. For the weaker orientation an empirical method is proposed to
measure the compliance at any point of the P(h) curve.

I - INTRODUCTION

The use of thermomechanical ceramic composites has come into an
industrial stage because to the often tough - and stress oriented -
applications in which they are used in advanced technologies. Thus, these
composite materials have many potential applications in fields such as :
aeronautic, biology, off-shore, sports, ... Among the various composites
which have been investigated those being reinforced with fibers are the
most interesting because they enable one to adapt the structures to the
direction of the applied load.

The way these composites are processed and the resulting microstructure make them resemble to coarse grained microstructures. As a
consequence designers are often faced with problems linked to the signi-
fication and the applicability of mechanical parameters as they are
defined most generally for homogeneous and isotropic materials. For
example, linear elastic fracture mechanics (LEFM) appears to be well
fitted to investigate the fracture behaviour of fine grained polycrystalline ceramics. This is due to the fact that the microstructure, especially grain and porosity sizes, is small in comparison with the specimen
size. This is no longer the case with CVD filled up fiber composites.
First, and as it will be pointed out later, the microstructure is coarse,
and as consequence the mechanical response of the microstructure is
anisotropic. Secondly, the fact that these materials have a coarse microstructure will need other or enlarged crack initiation and propagation
criteria.

In this way, fracture criteria already used with metallic specimens exhibiting gross yielding can be thought to be adapted to coarse grained ceramics.

Some attempts have been made to homogenize the properties of composites (1)(2). However it is difficult to quantify the resulting effect of the interfaces in the homogenized material. Analytical models, based on the elastic constants of the components of the composite, have been proposed in order to explain the specific shape of crack propagation in some composites (3)(4). They do not, however, give a measure of the change of the mechanical parameters as a function of damage. One possible approach is to investigate the overall experimentally measured parameters. Recent models tend to work-out the load-displacement record, P(h), of notched specimens as a whole in order to investigate the rupture behaviour of composites (5)(6)(7). These load-displacement curves are investigated taking into account non-linear aspect, due to the coarse grained structure.

The aim of this paper is to apply the previously cited models to investigate the rupture behaviour of laminate fiber composites as a function of the stress orientation and crack propagation with respect to the three principal axis of the composite.

II - MATERIALS AND EXPERIMENTAL PROCEDURE

Materials

The materials are carbon fiber-silicon carbide matrix composites. Woven clothes of PAN precursor carbon fibers (diameter: 8µm to 10µm) are piled-up and then filled-up with SiC using a CVD process (8). The optical micrograph (Fig. 1) shows the woven shape of the fibers. Results of metallographical investigations, using image analysis (9)(10), show that small pores, 15µm in mean size, are located between the fibers of a bundle. Larger pores (60µm) are located at the crossing of the fiber bundles.

The radial distribution of the fibers within a bundle has also been investigated. A comparison with known piling-up models has been made. It has been shown that the mean number of contacts is higher in our material than for a random distribution (1.5 against 0.7 respectively for a volume

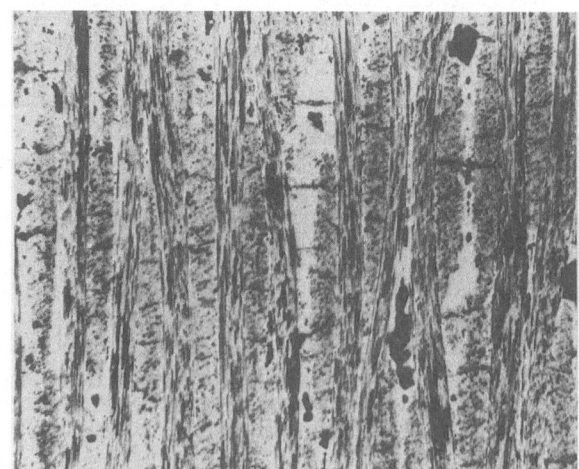

1 mm

Fig. 1. Optical micrograph of a C-SiC materials.

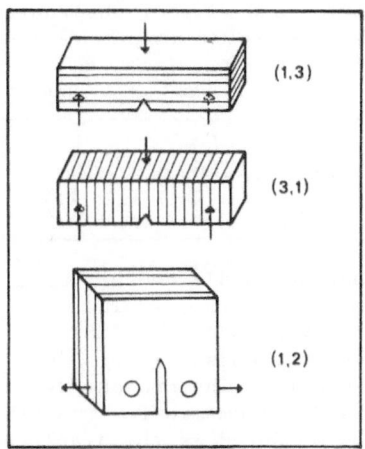

Fig. 2. Schematic of the test specimens used.

fraction of fibers of 0.3). This is an important information to get when considering the load transmission.

Three specimen orientations have been tested, as shown on figure 2. Directions (1) and (2) are equivalent and are in the plane of the woven carbon clothes. The denomination (i,j) means that the direction of tensile stress is "i" when the specimen is loaded in the direction "j".

The density of the specimens has been measured using the mass over volume ratio. Its value lies between 2.2 and 2.4. The graph in figure 3 illustrates the influence of the porosity on the elastic properties for the material with the (3,1) orientation.

Rupture tests

All the specimens have been processed and ground by the Société Européenne de Propulsion (SEP, Bordeaux, France). Table I gives the orientations and sizes of the specimens which were used.

The notches have been introduced using diamond saws (copper discs

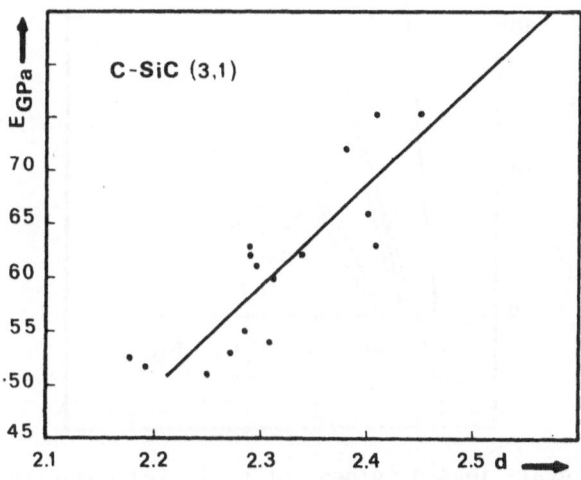

Fig. 3. Change of the elastic modulus, E, as a function of density, d.

Table I : Orientation and size of the rupture test specimens

Orientation	L mm	B mm	W mm
(3,1)	50	5	10
(1,2)	100	10	20
(1,3)	70	7.5	15

with electrolytically deposited diamond grains). They were made with saw-thickness ranging from 0.1 to 0.5 mm. The controlled rupture tests have been run in three point bending on an INSTRON 1185 universal testing machine, with a span over thickness ratio, L/W = 4. The load was measured with a cell of 10kN capacity. The crosshead-speeds were 0.05 mm/min for the (1,3) orientation and 0.1 mm/min for the others. The deflection of the bending specimen was measured as the displacement of the upper knife

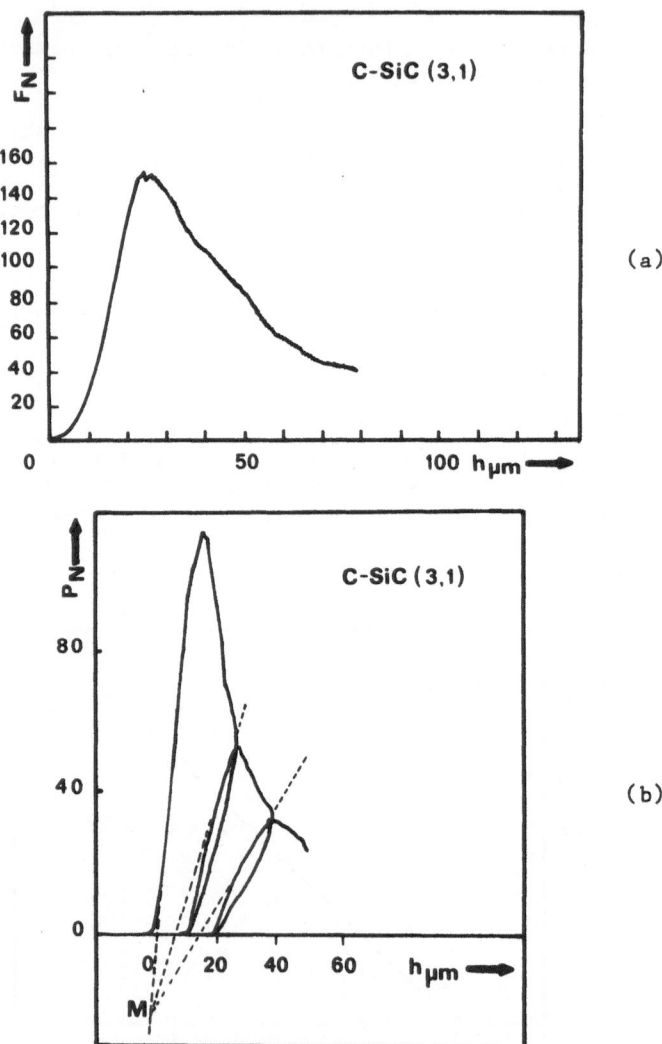

(a)

(b)

Fig. 4. Experimental load-displacement curve (a) and with several load-unloading loops (b) for C-SiC (3,1) composites.

with respect to the steel plate supporting the lower knives. An inductive transducer and an amplification system allowing an accuracy of 1μm
were used. The area measurements on the load-displacement curves were
made using a planimeter.

III - RESULTS

1. Orientation (3,1)

Figure 4 (a) presents the load-displacement plot, P(h), of a (3,1)
specimen. The maximum load is reached after only a small part of deviation from linearity. During loading-unloading cycles an important amount
of residual deformation appears. Working-out these plots within the scope
of linear elastic fracture mechanics (LEFM) as for homogeneous fine-
grained bulk materials is thus not recommended.

We propose thus to measure the crack length on each point of the
P(h) curve using the compliance determined according to the following
method. We have noticed, with specimens of this orientation, that the
tangents to the re-loading part passing through the unloading point on
the P(h) record overcross all at one and the same point, M (see figure
4b). This point corresponds to a shift of the origin on the load-axis.
Compliance is measured from the slope of the straight line joining this
new reference point, M, to the one considered on the P(h)-curve. Figure 5
shows the compliance calibration curve of these specimens. It has been
plotted as a function of E*C(a) vs. relative notch length, in order to
take the porosity into account. It allows to associate the crack length
(notch + extension) to the value of the compliance at each point of the
P(h)-curve. The effect of that crack length is considered to be equivalent to the real damage of the material (main and secondary cracks). For
such a material with a coarse microstructure, the evaluation of a stress
intensity factor, K, as it is used for monolithic polycrystalline ceramics, needs the experimental determination of the dimensionless factor

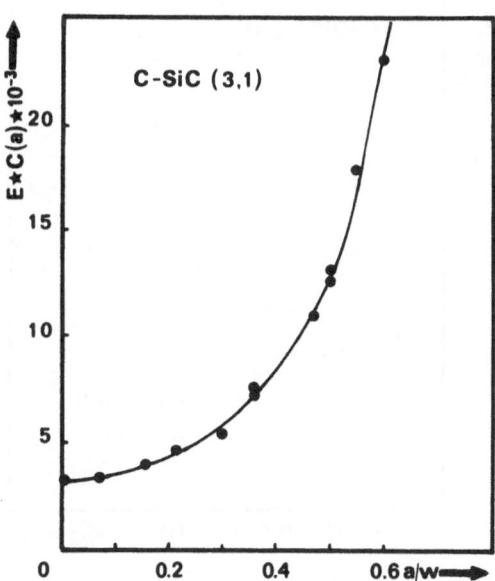

Fig. 5. Experimental change of E*C(a) as a function of the relative notch
depth (a/W).

Y(a/W), which is used in calculating K as $K = \sigma Y \sqrt{a}$. This factor presumes a single crack with an elastic stress intensification field. This is not realy the case in our materials.

The correction factor, Y, can be derived from compliance measurements. In three point bending it is given by :

$$Y(a/w) = (\frac{W}{L}) \left[\frac{1}{(a/W)} \frac{d(E*C)}{d(a/W)} \frac{2b}{9(1 - \nu^2)} \right]^{1/2} \quad [1]$$

Numerical treatment of the values of E*C from figure 5 gives the following polynomial expression best fitted to our material :

$$Y(a/W) = 8.18 - 53(a/W) + 510.8(a/W)^2 - 157.4(a/W)^3 + 57.3(a/W)^4 \quad [2]$$

The variation of Y as a function of a/W is shown in figure 6 and is compared to the theoretical expression proposed for monolithic polycrystalline dense ceramics (11). The values of K_c and G_c have been calculated using expression [2], on several points of the load-displacement curves. Results are given in table II.

Figure 7 represents G and R curves for two initial notch lengths. In such a representation, the point of instability of the specimen is defined by the point of tangency of the two curves, i.e. G = R and dG/da $\geqslant$ dR/da, which means that the rate of energy release is higher than the rate of crack extension resistance. The critical values are respectively $G_c = 70$ J.m^{-2} and 60 J.m^{-2} for a/W = 0.4 and 0.5. Since G_c varies with initial crack length, the LEFM criteria are not fulfilled. Consequently we have used J-integral measurements following the procedures outlined by J.D. Landes and J.A. Begley (12) with the aim to verify the validity of the above presented G_c-results. In addition to the strain energy release rate, G, used for the creation of new surfaces, the value of J also

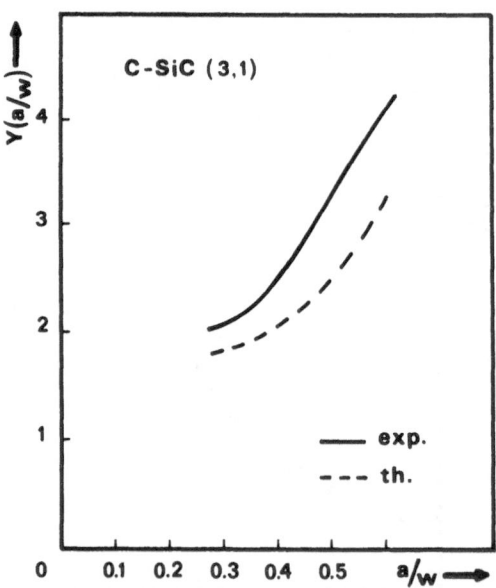

Fig. 6. Variation of the Y polynomial with the relative notch depth, a/W. Curve in solid line corresponds to the experimental curve for the composite and that in dotted line corresponds to the theoretical curve for monolithic ceramics.

Table II : Values of the various rupture parameters related to the orientation (3,1)

$\dfrac{a_0}{w}$	$\left(\dfrac{a}{w}\right)_{Fmax}$	$\left(\dfrac{a}{w}\right)_{Rmax}$	F_{max} N	R_{Fmax} J/m^2	K_{Fmax} $MPa\sqrt{m}$	F_{Rmax} N	R_{max} J/m^2	K_{Rmax} $MPa\sqrt{m}$
0.30	0.37	0.45	152.0	48	1.80	132.0	77	2.3
0.40	0.45	0.47	115.2	65	2.10	113.6	90	2.5
0.44	0.50	0.54	92.8	61	1.83	91.0	64	2.0
0.49	0.50	0.56	92.8	54	1.83	88.0	64	2.0
0.50	0.53	0.57	93.6	57	1.90	88.8	65	2.0

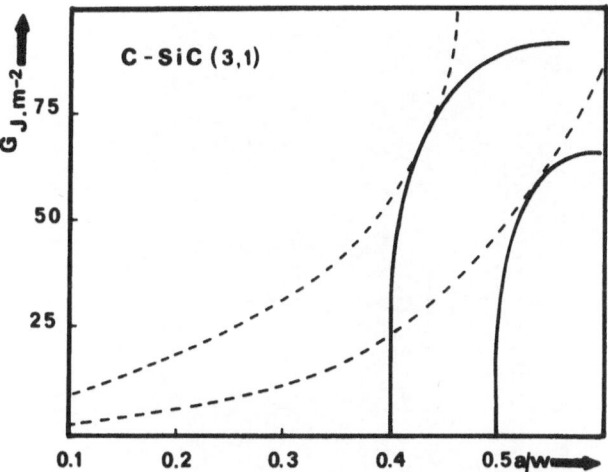

Fig. 7. R and G curves as a function of the notch depth, a/W, for C-SiC (3,1) composites (R in solid line, G in doted line).

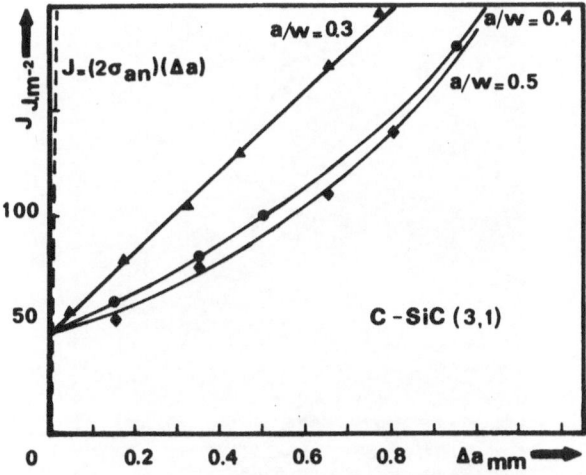

Fig. 8. Change in J curve as a function of the crack displacement, Δa, for three values of notch depth, a/W, for C-SiC (3,1) composite.

comprises the other terms of energy consummed at the tip of the propagating crack. Figure 8 shows the variation of J as a function of crack
extension Δa, for three values of initial crack length, a_o/W.

The critical value of J, J_c, corresponding to the transition from
crack blunting to the crack extension, is 50 J/m^2.

Optical observations of the crack extension path on the specimen
surface shows that the notch extends by crossing over the pores which are
in the neighbourhood of the notch plane. Numerous microcracks, which
either are branches or join the main crack, are observed in front of the
main crack which zigzags around the notch plane. Fractographic observation with a scanning electron microscope (Jeol 840) on the fracture
surface confirms that the crack meets several pores (Figure 9). The crack

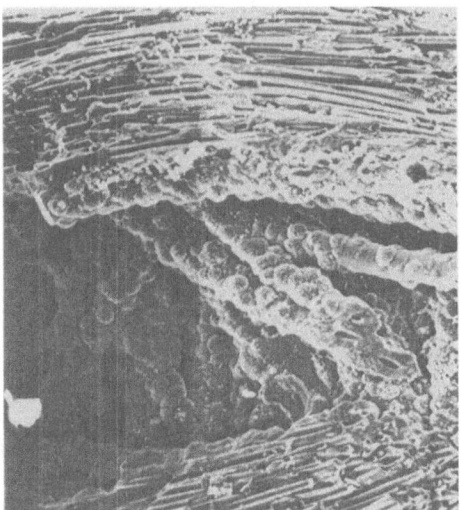

10 µm

Fig. 9. Interaction of the fracture path with pores.

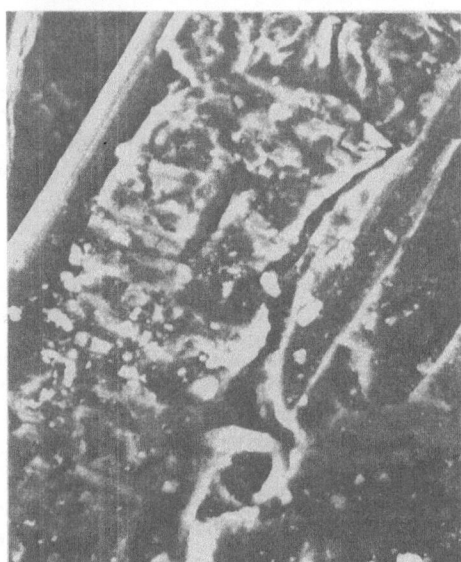

10 µm

Fig. 10. Microfractography showing the fracture of the SiC matrix and the
debonding of the fiber.

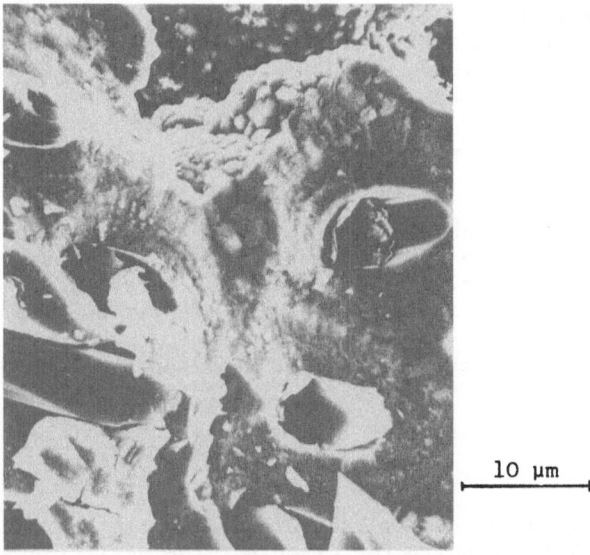

Fig. 11. Microfractography showing the pull-out of the SiC fibers.

propagation path not only produces fiber-matrix decohesion (figure 10) but also rupture of fiber bundles (figure 11).

2 - Orientations (1,3) and (1,2)

For these two orientations, the empirical method proposed by D. Sakai and al. (7) was used to work-out the load-displacement curves. In the scope of this method several energetic parameters (R, G, J, anelastic energy) are identified on the P(h) curve on which unloading-reloading cycles have been performed. The principles of this method are sketched on figure 12. The work furnished by the external forces to extend the crack

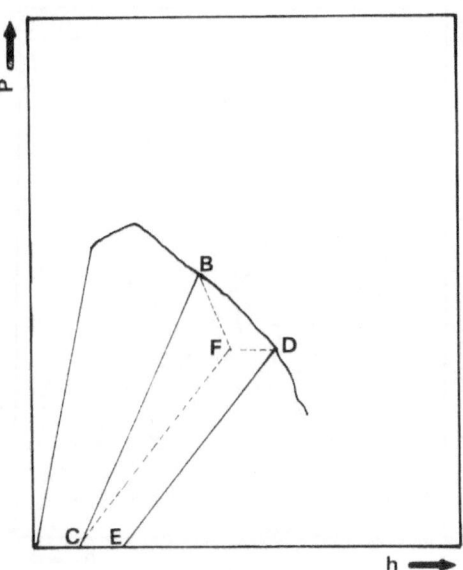

Fig. 12. Principle of the empirical method proposed by M. Sakai and al. for calculation of R, G, ∅.

from point B to D (i.e. a surface creation of total area 2ΔA) is represented by the area BCED. It corresponds also to the resistance of the
material which opposes to the crack extension. So the crack extension
resistance is given by :

$$R = \frac{\text{area BCDE}}{\Delta A}$$

The distance, Δh, between points C and E corresponds to the residual
deformation observed from unloading from points B and D. If the material
would exhibit no anelastic deformation, points C and E would be the same
and point D would, in fact, be F (DF = Δh). The area BCF represents thus
the amount of energy G, comsummed for the creation of the only area 2ΔA.
As a consequence the point CEDBF represents Ø, the rate of anelastic
energy spent during the creation of supplementary surfaces :

$$G = \frac{\text{area BCF}}{\Delta A} \; ; \; \emptyset = \frac{(\text{area BCDE}) - (\text{area BCF})}{\Delta A}$$

In fact the unloading and reloading plots are not straight lines. In
polycrystalline graphite for example, the deviation from linearity would
be due to plastic deformation in front of the crack tip during unloading
and reloading. The J-integral takes this supplementary work into account
in the evaluation of the work necessary for the creation of new surfaces.
The quantity J is obtained by measuring the area $\overgroup{\text{BCF}}$ (figure 13) :

$$J = \frac{\text{area } \overgroup{\text{BCF}}}{\Delta A}$$

The parameters R,G,Ø and J measured on the P(h) plot are related to
the anelastic crack propagation in the material. The critical value of
the strain energy release-rate, G_c^o is obtained in the absence of any
anelastic energy consomption (Ø = 0). In these conditions the total spent

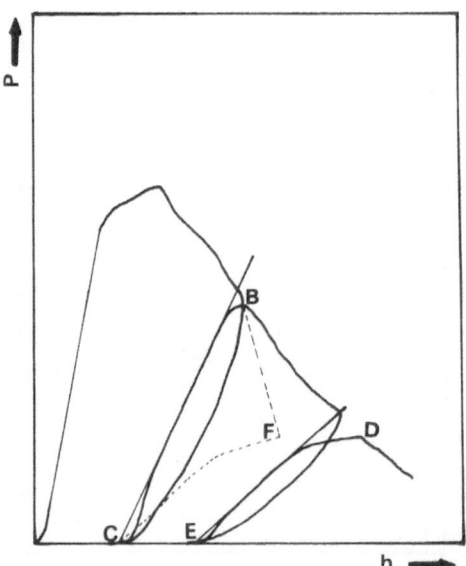

Fig. 13. Principle of the empirical method proposed by M. Sakai and al.
for calculation of J.

energy is used to create new surfaces :

$$\lim_{\emptyset \to 0} R = \lim_{\emptyset \to 0} G = \lim_{\emptyset \to 0} J_c = G_c^o$$

The work of rupture of the material is given by the mean of the crack propagation resistance, R :

$$2\gamma_r = \frac{\int_o^S R(a)dA}{S}$$

where S represents the total surface of crack extension.

Figures 14a and 14b show respectively the P(h) plots for orientations (1,2) and (1,3). The evolutions of the parameters R, G, J and $\emptyset$ as a function of crack extension, ΔA, are given in figures 15a and 15b.

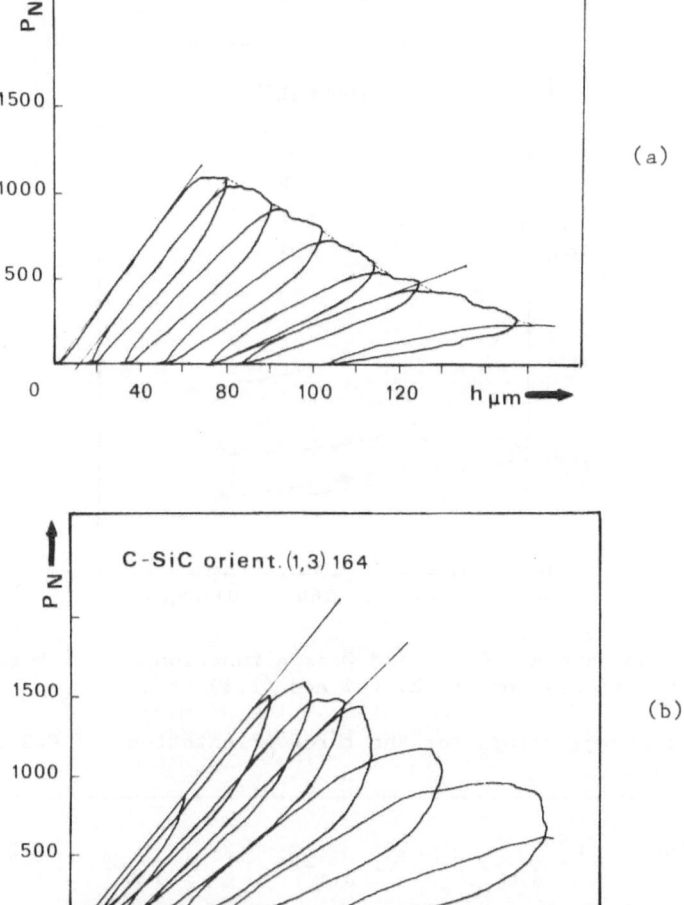

Fig. 14. Parts of the load-load point displacement curves for the orientations (1,2) (a) and (1,3) (b).

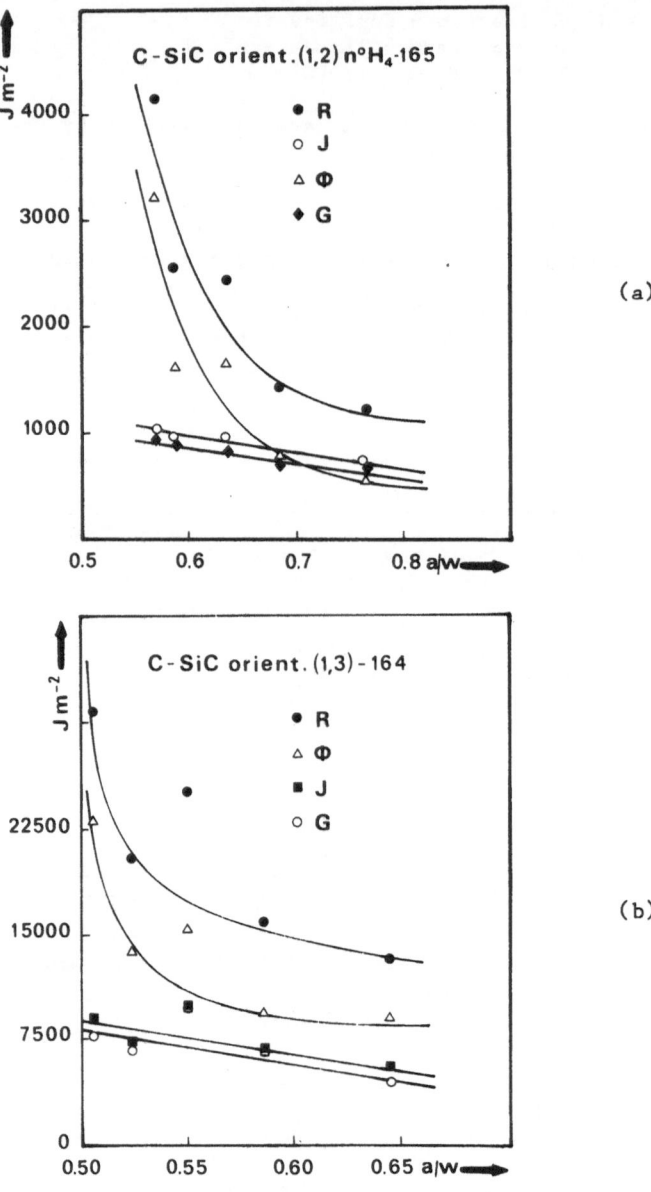

Fig. 15. Parameters R, G, J and $\emptyset$ as a function of crack extension, Δa, for orientations (1,2) (a) and (1,3) (b).

Table III : Energy values for the three orientations of C–SiC composites.

Orientation	G_c^o $J.m^{-2}$	$2\gamma_r$ $J.m^{-2}$	$2\gamma_{el}$ $J.m^{-2}$	$2\gamma_{an}$ $J.m^{-2}$
(3,1)	50–60	–	–	–
(1,2)	640	2040	800	1240
(1,3)	4140	17470	5950	11520

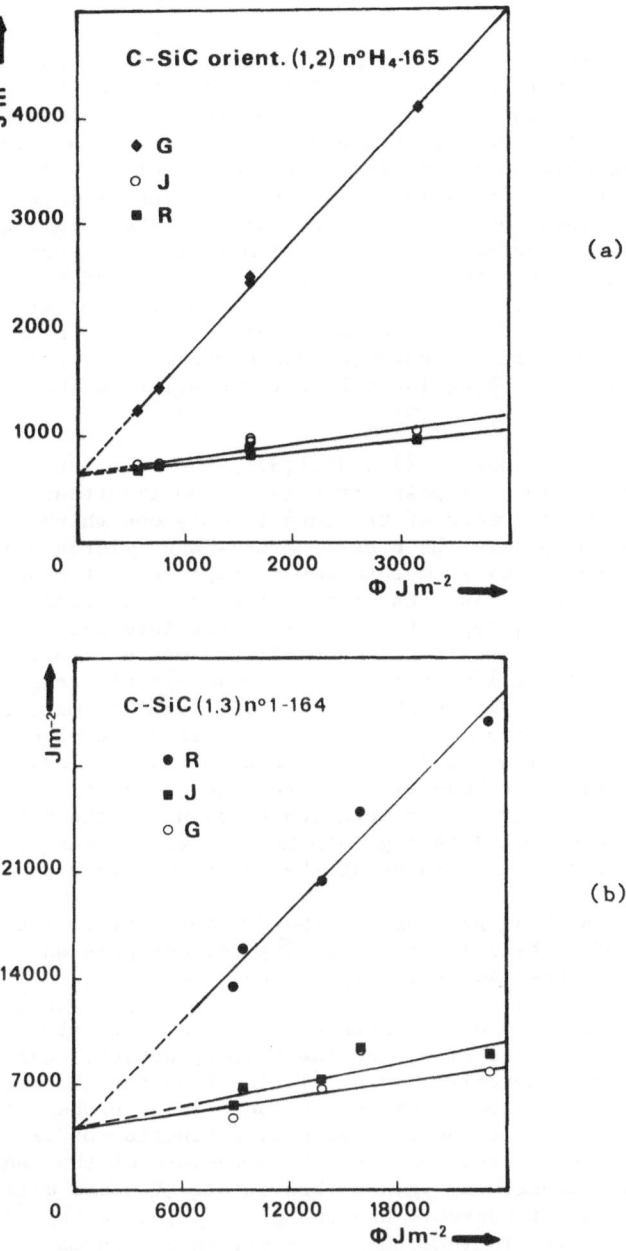

Fig. 16. Plots of the parameters R, J, G as a function of the anelastic term, $\emptyset$, for orientation (1,2) (a) and (1,3) (b).

Figures 16a and 16b show respectively the variation of the parameters R, G and J as a function of $\emptyset$. The critical values G_c^o for the orientation (1,2) and (1,3) are respectively 640 J.m^{-2} and 4140 J.m^{-2}. The high value of the work of fracture (1040 J.m^{-2}) in the case of orientation (1,2) comes from the important contribution of the anelastic term (617 J.m^{-2}) which represents 60% of the total amount. For orientation (1,3), the proportion of anelastic energy is 66% (i.e. 5760 J.m^{-2}) of the work of fracture γ_r (8735 J.m^{-2}). All the results for the three orientations investigated are summerized in table III.

627

DISCUSSION

The method of compliance measurement we propose is based on the observation that the tangents of the reloading plot converge to one and the same point on the load axis. The results of the application of this principle to the rupture of graphites for electrical siderurgical furnaces confirm this approach (13). The values of G_c = 50 to 60 J.m^{-2} measured on the C-SiC composites are among the highest measured as compared to values obtained for monolithic polycrystalline ceramics. The difference between the experimental Y(a/W) polynomial and those given for monolithic ceramics shows that the latter is not adapted to composite materials with a coarse microstructure. This is due to the stress concentration effect of the fiber bundles whose orientation is perpendicular to the plane of the crack. However this experimental polynomial correction factor allow one to define a K-factor similar to those used for homogeneous materials.

For orientations (1,2) and (1,3), the most interesting observation from a technological point of view is the importance of the anelastic energy term. The work of fracture is only one third of the total energy needed to propagate the crack. Macro- and microscopic observations suggest three anelastic mechanisms to explain that amount :

- observable microcracking at the onset of crack propagation on the crack tip. This type of damage is taken into account when the crack length is evaluated from the compliance measurement. The extension of this microcracking zone previously to macroscopic crack extension corresponds to the "plastic zone" observed for metallic materials.

- the deviation of the main crack from the plane of the initial notch (orientation 1,3) under the effect of a local residual stress introduced by fiber bundles or sharp edged pores for example. Investigation of the rupture of orientation (1,3) using the model of M. Ichikawa and S. Tanaka (14) taking into account the shear component, yielded strain energy release rates at the onset of crack propagation of 3000 J.m^{-2} (15).

- the work of pulling-out the C-fibers out of the SiC matrix or the friction of fibers in regard during reloading is an almost inexistant phenomena in the case of monolithic ceramics.

Some comments on the method of M. Sakai et al., as it can be used for fiber composites with a coarse microstructure, can also be made. Its advantage over the criterion of T. Ichikawa and S. Tanaka is to give at one and the same time crack initiation, crack propagation and anelastic values. The ratio of the two later as a function of crack propagation and stress direction can be used as a measure of the anisotropy of the composites. Concerning the method itself (coarse microstructure), the load-displacement curve is broken up in various energy terms. The physical meaning given to them has, however, only a formal sense. The ratio of anelastic over elastic terms can be expected to vary with specimen size and eventually with relative notch length. This is interesting and in some way offers much development to that method. Specimens which exhibit such a behavior are far from the validity criteria of LEFM. Decreasing and increasing the specimen sizes and looking for the relative amounts of the anelastic terms should yield a K_{IC} value by extrapolation beyond the largest specimens. The method of M. Sakai et al. can thus be extended to give propagation and damage values for various sizes and loading directions of structural parts.

ACKNOWLEDGMENTS
This work has been performed under a contract MRT-CNRS-DRET, AIP-ASP n°19.84.54.

REFERENCES

1. Ichikawa H., Chou T., Taya M., J. Mat. Sci., 17 : 832-842 (1982).
2. Murphy M.C., Outwater J.O., Proc. of the 28th Annual Technical Conference, Reinforced Plastics/Composites Institute, Washington D.C.. The Society of the Plastic Industry Inc., Feb. 1973, p. 17-A.
3. Ichikawa T., Tanaka S., Int. J. Fract., 22 : 125-131 (1983).
4. Bilby B.A., Cardew G.E., Howard I.C., Fourth International Conference on Fracture, 3 : 197-200 (1977).
5. Shah S.P., Sixth International Conference on Fracture, 1 : 495-514 (1984).
6. Swain M.V., Rose L.R.F., Sixth International Conference on Fracture, 1 : 473-494 (1984).
7. Sakai M., Urashima K., Inagaki M., J. Amer. Ceram. Soc., 66 : 868-874 (1983).
8. Christin F., Naslain R., Bernard C., Proc. 7th Int. Conf. CVD, edited by Sedwick T.O. and Lydtin M., The Electrochemical Society, Princeton, p. 499 (1979).
9. Coster M., Chermant J.L., "Précis d'Analyse d'Images" Ed. CNRS (1985).
10. Gomina M., Chermant J.L., Coster M., Acta Stereol., 2 Suppl. I : 179-184 (1983).
11. Gross B., Srawley J.E., "Stress intensity factors for single-edge-notch specimens in bending of combined bending and tension by boundary collocation of a stress function", Tech. Note NASA-TN-D-2603, (Janv. 1965).
12. Landes J.D., Begley J.A., ASTM - STP N° 632, p. 57-81 (1977).
13. Gomina M., Unpublished results.
14. Ichikawa M., Tanaka S., Int. J. Fract., 22 : 125-131 (1983).
15. Gomina M., Chermant J.L., Osterstock F., Proc. IInd Conference on "Creep and Fracture of Engineering Materials and Structures", Edited by Wilshire B. and Owen D.R.J., Pinebridge Press, Vol. I, p. 541-550 (1984).

TOUGHNESS ANISOTROPY OF A SiC/SiC LAMINAR COMPOSITE

Y.M. Pan*, M. Sakai*, J.W. Warren+, and R.C. Bradt*

*Dept. of Mat. Sc. and Eng., Univ. of WA, Seattle, WA 98195
+Refractory Composites, Inc., Whittier, CA 90606

ABSTRACT

The toughnesses of a 2-D laminar SiC/SiC composite consisting of a 40 v/o 8H/S weave Nicalon fabric and a CVI beta SiC matrix were measured at room temperature for the three principal orthogonal directions. The inter-laminar toughness was 1.30 MPa $m^{\frac{1}{2}}$ and those across the fabric were 9.12 and 12.56 MPa $m^{\frac{1}{2}}$. Other mechanical properties exhibited similar anisotropy. The results are compared with the toughnesses of other SiC materials and a C/C laminar composite of similar macroscopic construction.

INTRODUCTION

Recent advances of glass and carbon fiber/polymer matrix composites and also ceramic fiber/metal matrix composites to the status of practical engineering materials have focused a new research emphasis to develop advanced ceramic fiber/ceramic matrix composite systems with potentially much higher application temperatures. While these composite materials may not initially achieve the low temperature toughnesses of some polymer and metal matrix systems, there currently exists an aura of enthusiasm that substantial toughness increases beyond those commonly observed for monolithic ceramics are possible, and even highly probable. Much of this excitement is the consequence of a recent report that composites consisting of short SiC fibers in a glass matrix exhibit exceptionally high toughnesses (1). Successful processing of ceramic-fiber-containing ceramic matrix composites by chemical vapor infiltration methods has further demonstrated the feasibility of consolidating ceramic composites with more exotic matrix materials (2-5). Of course, CVD, or CVI techniques are not new to composites as these methods have been utilized in C/C composites for some time (6); however, the processing of carbide and nitride matrices by this technique does provide for some exciting ceramic composites, opportunities and challenges. Warren (5) has demonstrated that SiC matrices can be successfully deposited within continuous fiber structures by using a silane gas system. This paper reports on the mechanical properties of one of those composites, a laminate utilizing a woven fabric. Emphasis is directed toward the toughness anisotropy.

EXPERIMENTAL

The SiC/SiC composite that was tested in this study consisted of a Nicalon ceramic grade fabric with an 8 H/S weave of uncoated fibers. The composite was manufactured to contain 40% of the woven fabric by first layering and molding the fabric and then infiltrating it using a silane gas system at <1200°C to yield a beta-SiC matrix. The final composite density was 2.16 gm/cc. Scanning electron micrographs of fracture surfaces revealed a dense sheath of SiC surrounding each of the individual fibers and occasional minor granular SiC deposits where the weaves crossed. Considerable porosity remained in the body. Plates of this composite were then diamond sawed and diamond ground to test specimens with dimensions of approximately 0.50 x 0.75 x 5 cm. No subsequent thermal treatments were applied to the specimens.

Fracture measurements were made of these composites for the three crack propagation orientations as shown in Figure 1. The nomenclature applied here is the flatwise orientation for the crack plane across the fabric layers with the crack growth direction parallel to the molding direction, perhaps the most likely orientation for high stress application of these materials. The term edgewise is applied to the same orientation crack plane, but for the crack propagation direction at a right angle to the fabric layers, approaching the fabric layers on edge. The interlaminar orientation is that for the crack plane extending between the laminated fabric layers. For the straight-through-notched (STN) specimens and the chevron-notched specimens (CHN) the flatwise and edgewise orientations were tested on specimens that were utilized as-received; however, for interlaminar testing, the compact tension and bend-test specimens were constructed using sections that were diamond sawed from the as-received bars then glued together. While this is not a desirable practice and is certainly less than optimal, the thickness dimension of the manufactured structure necessitated this approach. Fortunately, failures never occurred at the glued sections of these specimens. The notches were diamond sawed in the specimens using a 0.3 mm thickness blade and for the case of the straight-through-notch, sharpened using diamond paste and a razor blade. The dimensionless crack length ($\alpha_o = a_o/W$) was kept between 0.3 and 0.4 for accuracy of the stress intensity calculations (7,8).

The "K_{Ic}" (STN) was calculated using the stress intensity formulas for the bend bar and compact tension specimens as reported in ASTM 399. The "K_{Ic}" (CHV) was evaluated from:

$$"K_{Ic}" \text{ (CHV)} = (\frac{P_{max}}{B\ W^{\frac{1}{2}}})\ Y_{min} \tag{1}$$

where P_{max} is the maximum fracture load of the composite specimen during steady state crack propagation and Y_{min} is the minimum value of the shape factor. Work-of-fracture values for both the (STN) and (CHV) specimens were determined by integrating the areas under the load-displacement curves and dividing by the total projected fracture surface areas. Modulus of rupture values were determined in three point bending of unnotched specimens while elastic moduli were estimated from the load displacement curves accounting for the test machine compliance as appropriate. Fracture surfaces were examined by scanning electron microscopy.

Because of the complicated fracture processes which occur in the frontal process zone and also the wake of this SiC/SiC fabric reinforced composite, the usual toughness, "K_{Ic}", is not adequate to characterize the nonlinear fracture process. Crack propagation in brittle ceramics and ceramic composites has been previously studied on the basis of energy principles (9,10) for which graphical methods have been outlined to determine the nonlinear energy fracture resistance parameters, including: the nonlinear fracture toughness (G_c) the crack-growth resistance (R_c) and the irreversible energy dissipation ($\emptyset_{ir}$). The energy toughness ($\tilde{G}_c$) is the net potential

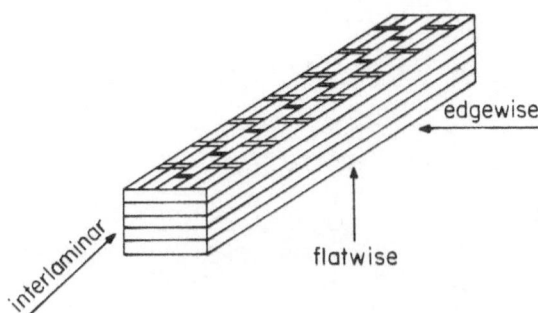

Fig. 1. Testing orienta-
tions of the SiC/SiC
laminar composite.

energy available for crack extension (11) while the crack-growth resistance
(R_c) is defined by the sum of ($\tilde{G}_c$) and ($\emptyset_{ir}$), i.e., the total energy consumption during a unit incremental increase of crack area. The dimension-
less load-displacement method for determing these fracture resistance parameters has outstanding advantages, especially for the highly nonlinear fracture of ceramic/ceramic composites. The technique has been applied in the
present study of this SiC/SiC composite for determination of (R_c).

The dimensionless load (P_r) and dimensionless loadpoint displacement
(u_r) can be defined as:

$$P_r = P_c(a)/K_c BW^{\frac{1}{2}} \tag{2}$$

and

$$u_r = u(a)E'/K_c W^{\frac{1}{2}} \tag{3}$$

where $P_c(a)$ and $u(a)$ are, respectively, the load and the corresponding loadpoint displacement given as functions of crack length (a). K_c, B, and W are
the toughness, thickness and width of the specimen, respectively. Using the
fact that (P_r) and (u_r) for a perfectly linear elastic material yields a   .
universal relation given by the following expressions:

$$P_r = 1/Y(a/W), \tag{4}$$

and

$$u_r = \lambda(a/W)/Y(a/W), \tag{5}$$

the aforementioned nonlinear fracture parameters can be determined by the
deviation of the experimentally observed ($P_r - u_r$) curve from the universal
linear elastic curve (1/Y vs. λ/Y). The details of the application of this
technique are reported elsewhere (10). The graphical method gives the
following physical meaning to the nonlinear fracture resistance parameters:

$$\tilde{G}_c = K_{Ic}^2/E', \text{ and} \tag{6}$$

$$R_c = K_{Ic}^2/E' + \emptyset_{ir}, \text{ where} \tag{7}$$

$$\emptyset_{ir} = \frac{\partial\Gamma_{micro}}{\partial A} + \frac{\partial U_{in}}{\partial A} + \frac{\partial U_{re}}{\partial A} \; . \tag{8}$$

Here ($\partial\Gamma_{micro}$) is the total fracture surface energy of induced microcracks
for the increment (∂A) of the primary macrocrack surface area. The energies
(U_{in}) and (U_{re}) are, respectively, the inelastic energy dissipation such as
the frictional energy associated with the fiber pull-out and the residual
elastic energy stored in the process zone wake.

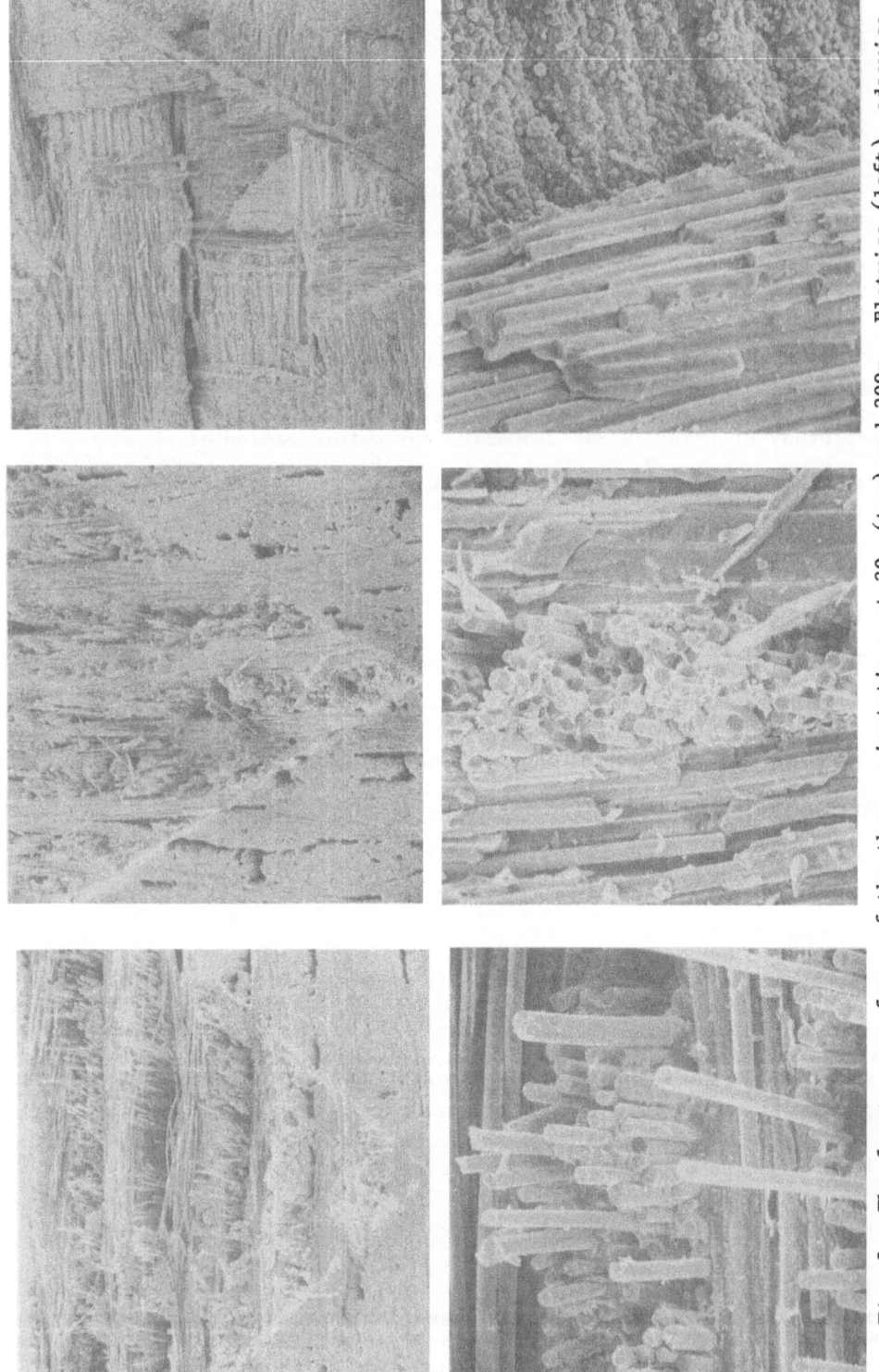

Fig. 2. The fracture surfaces of the three orientations at 20x (top) and 200x. Flatwise (left), edgewise (center), and interlaminar (right).

RESULTS AND DISCUSSION

Table I summarizes the properties for the three orientations of this composite. Figure 2 illustrates both the macro- and micro- fracture surfaces of the three orientations. The toughness values in Table I are enclosed in quotes as it is recognized that they may not be true K_{Ic} values for there exists uncertainty as to the form and location of the crack tip in composite materials of this type. It is for this latter reason that the anisotropic analysis of Sih, et al (12) was not applied either. In addition to their impressive values, there are several other features of the toughnesses that are appropriate to consider. For the cross fabric orientations the chevron-notch method yields a higher toughness than the straight-through notch. This is because the toughness is calculated at (P_{max}, Y_{min}) after some crack extension and at a time when the frontal process zone and its following wake have had the opportunity to develop substantially (fully?). This is a consequence of rising R-curve behavior. Little or no contribution from the wake can be expected for a straight-through notch test where the initiation load is used to calculate the toughness. Since considerable fiber bridging across the new crack surfaces occurs in the wake region of this composite, a substantial wake effect may be expected and in fact does occur as discussed later.

Although both the elastic modulus and modulus of rupture values exhibit the trends and correspondence expected from the toughness values, the work-of-fractures do not. The flatwise orientation has about a 70% higher work-of-fracture than the edgewise orientation. This is because the orientation of the fabric sheets normal to the cracking direction promotes extensive delamination cracking. This process has been previously addressed in detail by Sakai, et al (13) as it directly applies to toughening processes in laminar structures.

It is evident that the interlaminar toughness and other properties are inferior to those of the cross fabric values; however, they are quite comparable to similar properties for many monolithic SiC ceramic bodies. The reason for the lower interlaminar modulus, strength and toughness values can be readily discerned in Figure 2 where it is evident that essentially no fiber contribution to the bonding exists for the interlaminar test orientation. In fact, although both the flatwise and edgewise specimens were always held together by the fibers after testing, the interlaminar samples readily cleaved into two completely separate pieces.

The dimensionless (load) versus (load point displacement) plot is shown for the flatwise orientation in Figure 3. The edgewise orientation was similar, but that of the interlaminar orientation was not interesting. Figure 3 shows the feature which is directly related to the fracture surfaces in Figure 2 and also the (R_c) trend which is depicted in Figure 4. The experimental (P_r-u_r) curve in Figure 3 is in excellent agreement (coincidence) with the theoretical linear elastic curve during the initial stages of testing, however, substantial non-linear behavior dominates after the initial stages of crack growth. This is consistent with the R_c behavior shown in Figure 4 and also the observations during testing that linear elastic behavior persists until the sound of fiber breakage is audibly detected. This composite behavior relates directly to that of the woven fibers to which the initial load is transferred prior to crack extension (14,15).

The rapid increase of R_c values during crack extension into the latter

Table 1. Mechanical Properties of the SiC/SiC Composite

	Flatwise	Edgewise	Interlaminar
E(GPa)	117.6±8.2	111.6±6.1	17.6±4.2
MOR (MPa)	351.5±17.2	342.6±47.0	17.7±1.4
"K_{Ic}" (CHV) (MPa $m^{\frac{1}{2}}$)	9.12±0.36	12.56±0.64	1.30±0.94
"K_{Ic}" (STN)	8.46±0.38	8.64±0.32	1.30±0.02
γ_{wof} (CHV) (KJ/m^2)	3.53±0.32	2.14±0.04	0.08±0.01
γ_{wof} (STN)	3.40±0.28	2.05±0.17	0.14±0.01

part of the specimen ligament is not that expected for a material which develops a large frontal process zone. For example, Sakai and Bradt (10) have observed rapid decreases in the R_c values when the frontal process zone contacts the specimen end (surface), just the reverse of the behavior in Figure 4. A logical interpretation of the behavior in Figure 4 is that the frontal process zone is not the primary contributor to this composite's toughening process, but rather it is the wake region behind the advancing crack front which is the critical feature. This is quite reasonable, for as the crack propagates the frontal process zone is actually decreasing as it encounters the specimen back surface, but the extent of the wake region is increasing in size. The R_c value is similarly rapidly increasing. The natural conclusion is that the two are directly related.

Many phenomena may occur in the wake region for different types of materials; however, for this composite a natural process is the bridging of the new surfaces by fibers which are only partially pulled out. That this wake enhancing mechanism occurs is evidenced by the fact that the flatwise and edgewise orientation specimens never actually broke into two distinct pieces, but were always held together by the partially pulled out fibers bridging the wake region. Frictional processes from anisotropic grains have also been observed to produce similar effects (16). It must be concluded that the high toughness values and the rapidly increasing R_c values are directly related to the fiber strengths, their partial pullout and the integrity that the process imparts to the composites. The fiber pullout and bridging processes associated with the wake region during fracture are a major factor in the properties of this composite.

It is worthwhile to compare the anisotropic properties of this SiC/SiC composite with other similar composites and also commercial monolithic SiC bodies. Since the composite's density is low, its elastic moduli are only modest compared to dense SiC; however, the strengths in the cross fabric directions are quite good by any standards, even though the MOR for interlaminar fracture is low. The work-of-fracture values are outstanding for the flatwise and edgewise orientations. Relative to the toughnesses, only some C/C composites have similar laminated structures for direct comparison. A typical C/C composite with woven fabric and both CVD and pitch matrix carbon has been measured to have toughnesses of 5.25, 7.13, and 0.79 MPa $m^{\frac{1}{2}}$ for the same orientations as in Table I (17). The toughness anisotropy in the two composites is quite similar, but the C/C material is not as tough as this SiC/SiC composite. Monolithic SiC bodies have been frequently studied and have toughnesses comparable to the interlaminar fracture in many instances. Abe, et al (18) have summarized a number of results illustrating that the porous recrystallized SiC bodies usually have room temperature toughnesses between 1 and 2 MPa $m^{\frac{1}{2}}$, while the fully dense SiC bodies (hot pressed or

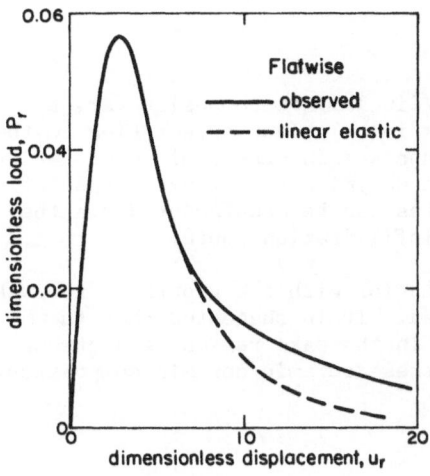

Fig. 3. The dimensionless load/dimensionless displacement plot for the flatwise orientation.

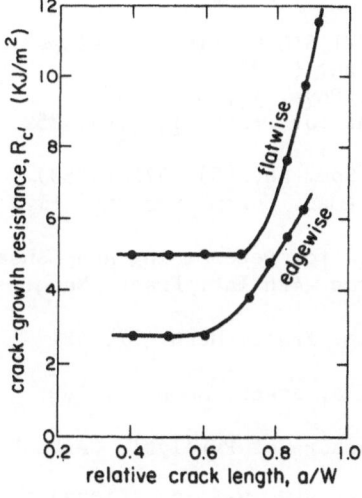

Fig. 4. The R_c curves for the flatwise and edgewise orientations.

sintered) are about 4 MPa m$^{\frac{1}{2}}$. The various types of 'siliconized' bodies are between these two extremes. It is evident that this SiC/SiC composite has a toughness comparable to monolithic SiC bodies, even in its weakest orientation, and is at least 2-3 times as tough as those materials in its tougher orientations.

CONCLUSIONS

The toughness anisotropy of this SiC/SiC laminar composite varies nearly an order of magnitude from the weak interlaminar orientation to the tough cross fabric orientation. The toughnesses in excess of 10 MPa m$^{\frac{1}{2}}$ are among the highest reported for any brittle ceramics and provide expectations that even tougher ceramic/ceramic composites can be manufactured via the continuous fiber (fabric)/chemical vapor infiltration route.

The correspondence of a rising R_c behavior with the growth of the wake indicates that the two are directly related. It is suggested that the fiber bridging of the newly created crack in the wake region is a strong contributing factor to the toughening of these ceramic/ceramic composites.

ACKNOWLEDGEMENTS

The authors acknowledge the suggestions of E. Anderson regarding this SiC/SiC composite and the partial support of M.S. and R.C.B. by The Dept. of Energy, Cont. No. 86A-00209C, Ceramic Technology for Advanced Heat Energy Program, Work Breakdown Struct. Elem. FM-3.4.

REFERENCES

1.   J.J. Brennan and K.W. Prewo, J. Mat. Sc. 17, 2371 (1982).
2.   M. Dauchier, P. Lamicq, and J. Mace, Rev. Int. hautes Temp. Ref. Fr. 19, 285 (1982).
3.   E. Fitzer, D. Hagen, and H. Strohmeier, p. 525, "Proc. 7th Int. Conf. CVD", Electrochem. Soc., Princeton (1979).
4.   A.J. Caputo and W.J. Lackey, "Fabrication of Fiber Reinforced Ceramic Composites by CVI", Oak Ridge National Laboratory Report, ORNL/TM-9235, Oct. (1984).
5.   J.W. Warren, "Fiber and Grain Reinforced CVI SiC Matrix Composites", Amer. Cer. Soc. Conf., Cocoa Beach, FL., Jan. (1985).
6.   E. Fitzer and W. Huttner, J. Phys. D: Appl Phys. 14, 347 (1981).
7.   D. Minz, R.T. Bubsey, and J.E. Srawley, Int. J. Fract. 16, (9), 359 (1980).
8.   M. Sakai and K. Yamasaki, J. Amer. Ceram. Soc. 66, (5), 371 (1983).
9.   M. Sakai, K. Urashima, and M. Inagaki, J. Amer. Ceram. Soc. 66. (12) 868 (1983).
10. M. Sakai and R.C. Bradt, "Graphical Methods for Determining Non-Linear Fracture Parameters", (to be published) Proc. 4th Int. Fract. Mech. Ceramics, VPI (1985).
11. J. Eftis, D.L. Jones, and H. Liebowitz, Eng. Fract. Mech. 17, 481 (1975).
12. G.C. Sih, P.C. Paris, and G.R. Irwin, Int. J. Fract. Mech. 1, 189 (1965).
13. M. Sakai, R.C. Bradt, D.B. Fischbach, "Fracture of Pyrolytic Carbon" (to be published in J. Mat. Sc.).
14. D.K. Hale and A. Kelly, p. 405, Vol. 2, Ann. Rev. Mat. Sc. (1972).
15. R.W. Rice, Cer. Eng. Sc. Proc. 2, 7-8, 661 (1981).
16. M. Sakai, R.C. Bradt, and A.S. Kobayashi, (to be published).
17. Y.M. Pan and R.C. Bradt (to be published).
18. H. Abe, H. Chandan, and R.C. Bradt, Bull Amer. Cer. Soc. 57, (6) 587 (1978).

WHISKER REINFORCED CERAMIC COMPOSITES*

T. N. Tiegs and P. F. Becher

Oak Ridge National Laboratory
P. O. Box X
Oak Ridge, Tennessee 37831

INTRODUCTION

The reinforcement of ceramics, such as alumina and mullite, with short, discontinuous SiC whiskers (~0.5 μm diam. × ~30 μm length) has been shown to significantly improve their fracture toughness and strength.[1-4] Equally important is the fact that the excellent fracture toughness and strength are maintained up to temperatures of at least 1000°C. As a result, these ceramic composites are extremely attractive for possible high-temperature structural applications.

The primary toughening mechanism observed in the alumina and mullite matrices has been crack deflection by the SiC whiskers although there is also a contribution from whisker pull-out. Both the fracture surface observations and effect of whisker volume concentration on toughness are consistent with such mechanisms. For the case of alumina, whisker reinforcement has also been shown to substantially increase the resistance to slow crack growth, which should increase the reliability and lifetime of ceramics under service.[4]

Experimental Procedure

The ceramic powder–SiC whisker composites were fabricated by conventional ceramic processing techniques. The ceramic powders and SiC whiskers are mixed together in a liquid medium, dried, and then hot-pressed or pressureless sintered. Dispersing the whiskers properly is a major variable in achieving a composite having optimum properties because the whiskers tend to agglomerate strongly. The difficulty of dispersion increases as the volume content of whiskers in the composite is increased. Use of an ultrasonic homogenizer[†] and/or ball milling provides considerable improvement in achieving a satisfactory dispersion of the SiC whiskers in the alumina or mullite powders.

---

*Research sponsored by the Advanced Materials Development Program, Office of Transportation Systems, U.S. Department of Energy, under contract DE-AC05-84OR21400 with the Martin Marietta Energy Systems, Inc.

†Model PT45/80, Brinkmann Instrument Co., Westbury, New York.

639

The powder-whisker mixtures were hot-pressed in a graphite die and heated in a graphite resistance furnace at temperatures from 1500°C to 1850°C under a vacuum of approximately 10^{-5} torr. Densities achieved were greater than 98% of theoretical for composites containing <50 vol % SiC whiskers.

Fracture toughness measurements were obtained using both the applied moment double cantilever beam (AMDCB) and the multiple indent flexure strength (MIFS)[5] techniques. The latter technique used flexure bars having nominal dimensions of $2.2 \times 2.5 \times >20$ mm, with polished tensile surfaces that contained 3 Vicker's DPH indents to produce controlled flaws in the region of maximum tensile stress. The fracture strengths were determined by four-point flexure on flexure bars having the same dimensions as the MIFS specimens, but with 180 grit diamond ground tensile surfaces.

An important point to keep in mind is that during uniaxial hot-pressing, the longitudinal axes of the whiskers tend to orient normal to the original pressing axis. On the other hand, within a plane normal to the original pressing axis, the longitudinal axes of the whiskers are randomly oriented. The result is that the whiskers have a sheet-type rather than a uniaxial texture, resulting in some anisotropy in the mechanical properties.[1,2] The optimum properties are achieved when a crack propagates perpendicular to the whisker axes. These will be the property values reported in the present paper.

The thermal shock resistance of the alumina-SiC whisker composites was determined by inserting flexure bars ($2.2 \times 2.5 \times >20$ mm) into a resistance heated tube furnace at temperature, soaking for 10 to 15 min at a given temperature and then dropping into a boiling water bath. The test bars were then broken in four-point flexure at room temperature and the fracture strengths determined.

SiC Whisker Reinforced Alumina Composites

Typical room temperature mechanical properties for SiC whisker reinforced alumina at various whisker concentrations are summarized in Fig. 1. As shown, the fracture toughness and strength are significantly improved over monolithic alumina. In addition these properties remain relatively constant up to temperatures of at least 1000°C. Further discussions of the whisker reinforced alumina composite properties can be found in References 1 through 4.

This previous work reported results on composites made with SiC whiskers fabricated by ARCO Chemical Company.* Since SiC whiskers are now available from both domestic and foreign sources, a series of Al_2O_3-20 vol % SiC whisker composites was made to determine the properties of composites using these alternate whiskers. The composites were hot-pressed under identical conditions to samples containing ARCO whiskers (1850°C, 41 MPa). The results of the mechanical property testing are summarized in Table 1. As shown, the mechanical properties were significantly different for the composites containing the whiskers from various sources.

Ceramographic examination of polished sections from composites made with the Tateho and Tokai Carbon whiskers indicated that the SiC whiskers were apparently intact. However, examination with a scanning electron microscope of a fracture surface revealed a relatively smooth texture with

---

*ARCO Chemical Co., Greer, South Carolina.

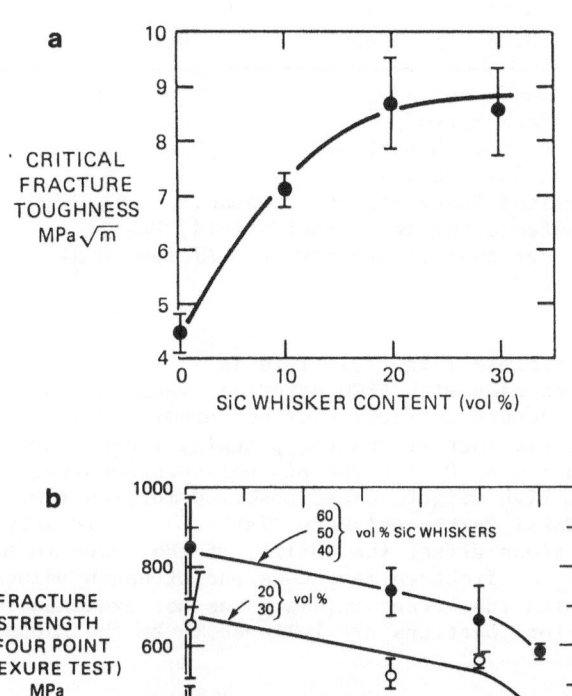

Fig. 1. (a) Fracture toughness of alumina–SiC whisker composites as a function of whisker content. (b) Fracture strength of alumina–SiC whisker composites as functions of temperature and whisker content.

Table 1. Summary of mechanical testing of
Al_2O_3–20 vol % SiC whisker composites
from various whisker sources

SiC whisker source	Density (Mg/m^3)	Flexural strength (MPa)[a]	Fracture toughness $(MPa \cdot m^{1/2})$[b]	Weight loss during hot pressing (%)
ARCO[c]	3.82	650	8.3	0.8
Tokai Carbon[d]	3.79	270	7.2	5.17
Tateho[e]	3.74	370	4.2	1.53
Versar[f]	3.72	340	[h]	2.32
LANL[g]	3.83	430	9.1	4.85

[a]Four-point bend test.
[b]Multiple indent method.
[c]ARCO Chemical Co., Greer, S.C.
[d]Tokai Carbon Co., Japan.
[e]Tateho Chemical Industries Co., Japan.
[f]Versar Manufacturing Inc., Springfield, Va.
[g]Los Alamos National Laboratory, Los Alamos, N.M.
[h]Not determined.

no SiC whiskers visible (Fig. 2). This is in direct contrast to Al_2O_3–SiC whisker composites made with ARCO material, where the whiskers are easily visible on the fracture surface. Further examination with an electron microprobe suggested that at the hot-pressing temperature (1850°C) a reaction between the Al_2O_3 and the SiC whiskers occurred. This is also indicated by the high weight losses observed for the composites made with the Tateho and Tokai Carbon whiskers (Table 1). Evidently, the degradation reactions affect the ability of the crack to be deflected, resulting in the low fracture toughness and strength values. While the composite made with the Versar whiskers was not examined in such detail, similar degradation reactions are believed to be the cause of the poor mechanical properties.

The degradation reactions observed between the alumina and some SiC whiskers appear to be minimized at lower hot pressing temperatures. Additional hot pressings at ORNL have been made at temperatures <1600°C with no apparent degradation of the whiskers. Other researchers[6] have incorporated Tateho SiC whiskers into partially-stabilized-zirconia by hot-pressing for ~10 min at a temperature of 1450°C with no apparent degradation. While differences in ceramic powders is well known, variations in the thermal stability of SiC whiskers are not. Obviously, composites made with the various ceramic powders and SiC whiskers must have the processing tailored for those particular starting materials.

The composite made with the whiskers from Los Alamos National Laboratory (LANL) exhibited high fracture toughness (however with signifi- cant scatter), but relatively low strength. Examination of fracture sur- faces showed the whiskers to have a wide distribution of diameters and lengths, but were in good condition with no apparent degradation. The fracture surfaces also revealed large flaws (<300 μm diameter) which had nickel and iron associated with them. Presumably, these inclusions result from the catalyst balls used in the whisker growth process. The whiskers were processed to minimize the catalyst ball content, however some evi- dently were not removed. The high weight losses observed during the hot- pressing of the composite with the LANL whiskers may also be attributable

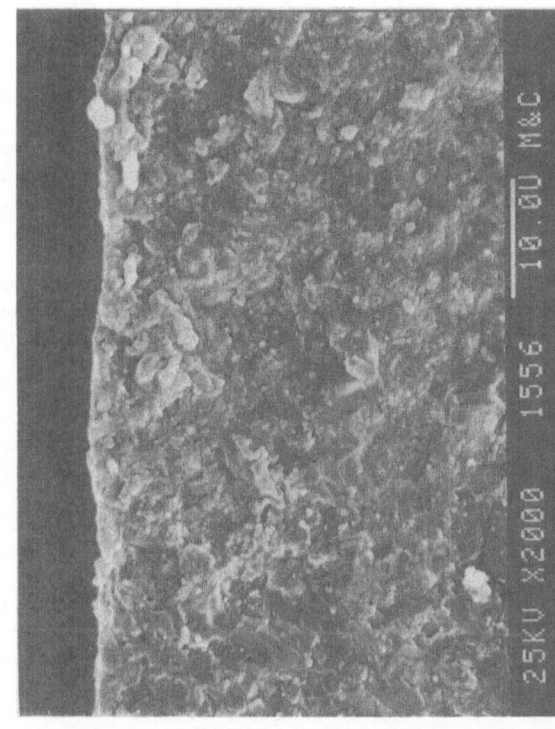

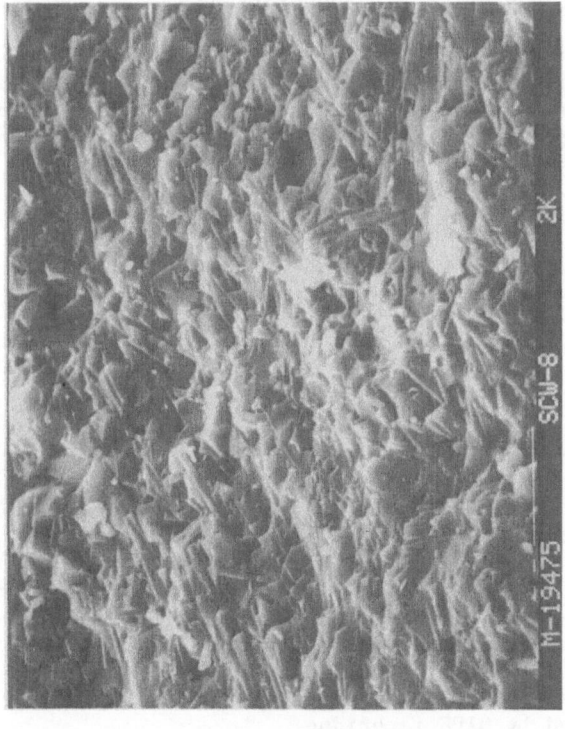

Fig. 2. (a) Typical fracture surface of Alumina-SiC whisker composite with no whisker degradation. (b) Fracture surface of alumina-SiC whisker composite showing whisker degradation. 10 μm bar in Fig. 2b applicable to both surfaces.

to the presence of some catalyst balls, however we have no specific data to confirm this.

Alternate Ceramic Matrix Composites

In addition to alumina, SiC whisker reinforcement of other ceramic matrices was examined. These alternate matrices include mullite, mullite-zirconia, alumina-zirconia and alumina-chromia. We were successful in fabricating composites from these matrices with the exception of the alumina-chromia. Alumina and chromia form a complete solid solution and this material is attractive for certain applications because of its lower thermal conductivity as compared to pure alumina. However, during fabrication at elevated temperatures, reduction of the chromia to metallic chromium by the SiC whiskers occurs. The resulting microstructure showed that all of the whiskers were affected by this reaction.

The mechanical properties for the composite with alternate matrices were determined at room temperature and are summarized in Table 2. Typically, monolithic mullite has fracture toughness and strength values of 2.2 MPa·m$^{1/2}$ and 200 MPa, respectively. As shown in Table 2, significant increases in the toughness and strength of mullite are achieved by reinforcement with SiC whiskers.

The mullite-zirconia and alumina-zirconia matrices offer the possibility of combining transformation toughening with whisker reinforcement for further improvements in mechanical properties. As indicated in Table 2, the incorporation of dispersed zirconia particles results in some improvement in both the toughness and fracture strength for the mullite matrix, whereas the alumina with dispersed zirconia particles shows improvement in the strength, but some decrease in toughness for comparable whisker concentrations.

These preliminary results suggest that whisker reinforcement can indeed be used in conjunction with other toughening mechanisms that can lead to the development of very strong and tough ceramic composites. Such ceramics would be especially attractive for structural components in advanced heat engines and gas turbines.

Table 2. Summary of mechanical properties on SiC whisker reinforced ceramic matrices

Matrix material	Reinforcement (vol %)[a]	Flexural strength[b] (MPa)	Fracture toughness (MPa·$^{1/2}$)
Mullite	10	420	3.6[c]
Mullite	20	420	4.7[c]
Mullite-Zirconia	20	440	5.6[c]
Alumina	20	650	8.3[c,d]
Alumina-Zirconia	20	750	7.8[d]

[a]All composites used ARCO Chemical Co. SiC whiskers.
[b]Measured by four-point flexure.
[c]Measured by AMDCB technique.
[d]Measured by MIFS technique.

Thermal Shock of Alumina-SiC Whisker Composites

Another important consideration in the use of these SiC whisker reinforced alumina composites for structural applications is their thermal shock resistance. Because of the high fracture strength and toughness of these composites they would be expected to have good thermal shock resistance, but since this property is a complex function of several material properties, tests were conducted to determine if indeed these composites had good thermal shock resistance. An alumina-20 vol % SiC whisker composite hot-pressed to full density was comparted to literature data[7] on alumina with no SiC whisker reinforcement.

The changes in flexural strength as a function of the temperature drop from the furnace to the boiling water bath are illustrated in Fig. 3. As shown the alumina-20 vol % SiC whisker composite shows virtually no decrease in flexural strength with temperature differences up to 900°C. Alumina, on the other hand, shows a significant decrease in flexural strength with a temperature change >400°C. Other researchers have shown similar results from thermal shock tests on alumina[8]. Plans are to increase the temperature difference up to 1400°C to determine if any strength degradation occurs.

Because the thermal shock resistance was excellent for the single quench, some specimens were subjected to 10 quench cycles from 800°C into the boiling water bath. As shown in Fig. 3, some strength degradation was observed: an average strength of 540 MPa versus >610 MPa for the unshocked and singly shocked composites. While this indicated some fatigue effects, the thermal shock resistance of the alumina-20 vol % SiC whisker composite is considered to be excellent.

Conclusions

SiC whisker reinforcement has been shown to significantly improve the mechanical properties of alumina and mullite. For example, typical fine-grained alumina with ground surfaces has fracture toughness, K_{Ic}, and flexural strength values of 4 to 5 MPa·$m^{1/2}$ and 350 to 450 MPa, respectively. We have demonstrated that, by the addition of 20 vol % SiC whiskers to the alumina, corresponding values of 8 to 8.5 MPa·$m^{1/2}$ and 650 MPa are obtained. In addition, other studies have found that these toughness and strength values are essentially maintained up to temperatures of at least 1000°C.[4] Analysis has indicated that crack deflection by the whiskers is the major toughening mechanism[1,4].

By increasing the fracture toughness, whisker-reinforced ceramics become less sensitive to flaw size. Previous work[4] has shown that this results in improvement in resistance to slow crack growth for alumina-20 vol % SiC whisker composites. The present work further enhances the potential of the SiC whisker reinforced alumina by showing excellent thermal shock resistance up to temperature changes of 900°C.

The final properties of the composites depend not only on the volume fraction of whiskers and the processing, but on the starting materials as well. While this is well known for ceramic powders, great differences in the behavior of commercially available SiC whiskers has not been previously reported. Results from an evaluation of whiskers from several commercial sources with respect to final composite properties show that SiC whisker-alumina incompatability at elevated fabrication temperatures can result in severe degradation of the mechanical properties. Processing of the whisker reinforced ceramic composites therefore must be tailored to specific powder-whisker combinations.

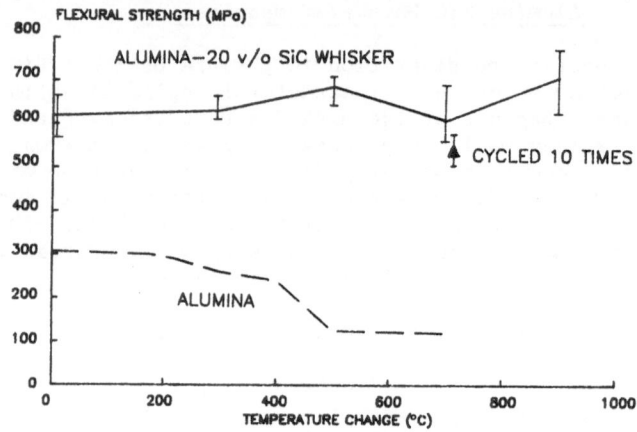

Fig. 3. Thermal shock behavior of alumina-20 vol % SIC whisker composite and monolithic alumina. All data from single quench rest with expectation of data ( ) with 10 quench cycles . Data for alumina taken from Ref. 7.

The mechanical property gains of the whisker reinforced composites over the monolithic ceramics are impressive. However, recent results suggest that further improvements in mechanical properties can be attained by combining whisker reinforcement with transformation toughening through the addition of zirconia particles to the composite.

REFERENCES

1. P. F. Becher and G. C. Wei, "Toughening Behavior in SiC-Whisker-Reinforced Alumina," Comm. Am. Ceram. Soc. 67(12), 259-60 (1984).
2. G. C. Wei and P. F. Becher, "Development of SiC-Whisker-Reinforced Ceramics," Am. Ceram. Soc. Bull. 64(2), 298-304 (1985).
3. T. N. Tiegs and P. F. Becher, "Particulate and Whisker Toughened Alumina Composites," pp. 479-485 in Proc. Auto. Tech. Dev. Contractors' Coord. Meet., 22nd, 1984, Vol. P-155, Society of Automotive Engineers, Inc., Warrendale, Penn., March 1985.
4. P. F. Becher, T. N. Tiegs, J. E. Ogle, and W. H. Warwick, "Toughening of Ceramics by Whisker Reinforcement," Presented at Fourth International Symposium on the Fracture Mechanics of Ceramics, June 19-21, 1985, Blacksburg, Virginia.

5. R. F. Cook and B. R. Lawn, "A Modified Indentation Toughness Technique," <u>Comm. Am. Ceram. Soc.</u> 11, C-200-1, 1983.
6. N. Claussen, K. L. Weisskopf and M. Ruhle, "SiC Whisker Reinforced TZP," Presented at Fourth International Symposium on the Fracture Mechanics of Ceramics, June 19-21, 1985, Blacksburg, Virginia.
7. P. F. Becher, "Transient Thermal Stress Behavior in ZrO_2-Toughened Al_2O_3," <u>J. Am. Ceram. Soc.</u> 64(1), 37-39, 1981.
8. H. P. Kirchner, <u>Strengthening of Ceramics</u>, Marcel Dekker, Inc. New York, p. 63-65, 1979.

WHISKER-REINFORCED ZIRCONIA-TOUGHENED CERAMICS

N. Claussen and G. Petzow[*]

Technische Universität Hamburg-Harburg
2100 Hamburg 90, FRG
[*]Max-Planck-Institut für Metallforschung
7000 Stuttgart, FRG

ABSTRACT

Dense and homogeneous composites with matrices of either tetragonal zirconia polycrystals, or ZrO_2-containing Al_2O_3, mullite or cordierite, and with SiC whiskers as reinforcement were fabricated by conventional P/M processing including hot-pressing. The toughness was increased in all cases with crack deflection being the main toughening mechanism. The room-temperature strength of tetragonal zirconia polycrystals was reduced due to the high thermal mismatch stresses with the SiC whiskers. However, the tetragonal (t) - to - monoclinic (m) transformation in ZrO_2 can be utilized to adjust the thermal mismatch between SiC whiskers and those zirconia-containing ceramics which have large coefficients of thermal expansion.

INTRODUCTION

Zirconia-toughened ceramics (ZTC) exhibit excellent mechanical properties at low and intermediate temperatures (1,2). Record strength values of > 2500 MPa for tetragonal zirconia polycrystals (TZP) have been reported (3), values which 10 years ago would have been unthinkable for a bulk polycrystalline ceramic material. However, at elevated temperatures (~ 700°C), the loadbearing capability of such materials is severely limited because of reduced transformation thoughening and the onset of creep. The latter is due to the fine grain size (except for partially-stabilized ZrO_2, PSZ) and the presence of a glassy intergranular phase. Of the various possibilities suggested (4) for overcoming these deficiencies, whisker reinforcement appears to be the most promising. Whiskers are a better choice than fibers because of their ability to survive the severe hot-pressing conditions necessary for incorporation into ceramic matrices (5).

Another attraction of introducing whiskers into ZTC is the prospect of increasing the room temperature strength of some of the high-toughness materials (especially TZP and PSZ) where the strength is not controlled by the largest flaw but by the mean critical stress required to initiate the transformation (6,7). The Ce-TZP ceramics (8) with a toughness of ~ 15 MPa $\sqrt{m}$ and strengths of ~ 500 MPa would, for example, represent an ideal matrix for whisker reinforcement. This is shown in schematic form in Fig. 1 where the strength, σ_F, of ZTC can be seen to be limited by the critical stress, σ_c, (I) on the high-toughness side and by the flaw size (II) on the low-toughness side. The numbers in the diagram represent the approximate range of values anticipated.

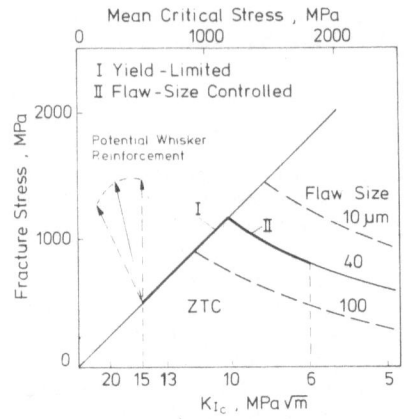

Fig. 1. Schematic diagram of strength-toughness relation in ZTC ceramics. I, yield-limited strength, $\sigma_F \propto \sigma_c$. II, Flaw-size controlled strength, $\sigma_F \propto 1/\sigma_c \propto K_{Ic}/\sqrt{a_c}$

Furthermore, additional toughening may be expected from interactions between transformation toughening and whisker reinforcement, e.g., from stress transfer into regions further away from the crack plane, from enhanced stress-assisted nucleation along the whiskers, from suppression of stress singularities as long as the matrix material transforms adjacent to the whiskers, and from other such mechanisms.

EXPERIMENTAL PROCEDURE

For most experiments in the present work, SiC whiskers of grade SCW-1 ("separated and dispersion improved") from Tateho Chem. Ind., Kariya, Japan, were used. The property data given for these whiskers by the supplier are a composition of ~ 95 % β-SiC, an impurity content (%) of 0.15 Mg, 0.25 Ca, 0.31 Al, 0.04 Fe, a range of diameters from 0.06 to 0.2 μm and a range of lengths from 10 to 40 μm. Because the lot also contained larger SiC particles (> 5 μm), the whiskers were first ultrasonically dispersed in isopropanol and deagglomerated in a tumbling mixer with plastic balls (3 mm Ø) for 24 h. Thereafter, most of the fragments > 5 μm were eliminated by sedimentation. In two exploratory experiments were Al_2O_3 whiskers (Thermokinetics, Washington, D.C.) and Si_3N_4 whiskers (SNW-1, Tateho) incorporated into TZP and cordierite.

Table 1. Ceramic powders used for ZTC/whisker composites

Ceramic	Type	Supplier
Al_2O_3	A16	Alcoa, USA
Mullite	Dynamullit (fused)	Dynamit-Nobel, FRG
Cordierite	experimental	Schott Glaswerke, FRG
$m-ZrO_2$	SC 20	MEL, U.K.
$t-ZrO_2$	TSK 3, mole % Y_2O_3	Toyo Soda, Japan

The ceramic matrix powders used for the ZTC/whisker composites are given in Table 1. The cordierite is an experimental, fully crystallized powder which is readily sinterable. The other materials are well-characterized standard grades.

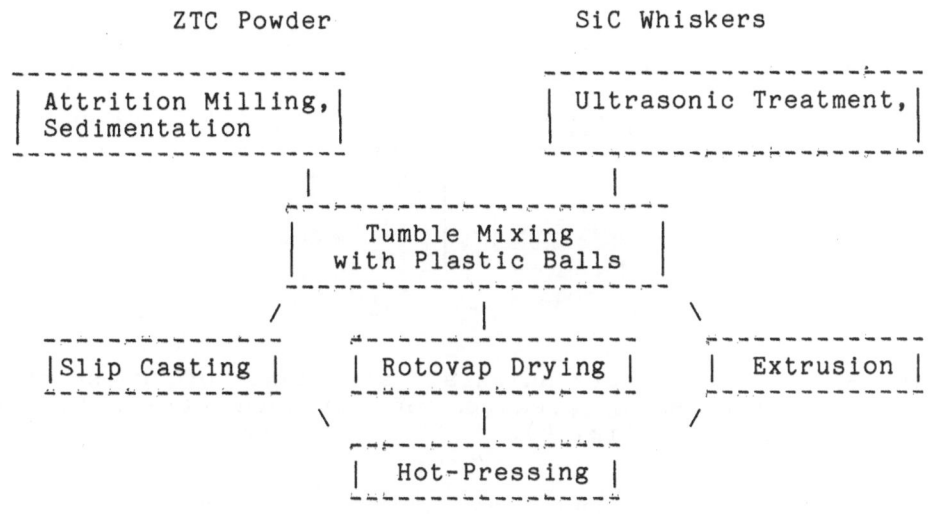

Fig. 2. Processing of ZTC/whisker Composite

The processing of the composites is summarized in Fig. 2. All matrix compositions with and without ZrO_2 additions were attrition milled for 2 h in isopropanol using TZP balls for the Al_2O_3 and cordierite composites and Al_2O_3 SiO_2 balls for the TZP and mullite composites. The whisker powder slurries were then tumble mixed with 3 mm plastic balls. In most cases, the mixtures were dried in a rotovap dryer followed by uniaxial hot-pressing in BN-washed graphite dies. The hot-pressing conditions were 30 min. at 1550°C for Al_2O_3, 30 min. at 1600°C for mullite, 10 min. at 1300°C for cordierite and 10 min. at 1450°C for TZP composites. In a few experiments, extrusion through a hypodermic syringe with a 1 mm orifice was tested with a plasticized mullite/whisker mixture as a means of achieving uniaxial alignment of the whiskers (9). In this case, packages of aligned extruded rods were hot-pressed (Figs. 3 and 4).

The hot-pressed discs (35 mm in dia., 10 mm thick) were cut into rectangular bend bars (30 x 30 x 2.5 mm). The fracture toughness was measured by the ISB technique (10) in 4-point bending. Controlled flaws were introduced by Vickers indentation with loads of between 10 and 1250 N. The fracture strength was measured in 4-point bending (20/7 mm), the orientation being such that the tensile surface was perpendicular to the HP direction. Six specimens were tested for each set of conditions.

The fracture surfaces were analyzed by SEM, and the microstructure was examined by TEM and STEM.

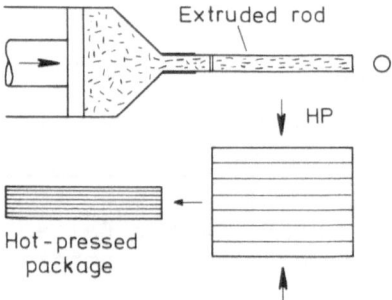

Fig. 3.   Schematic of uniaxial alignment of whiskers by hot-pressing packages of extruded ceramic whisker rods (cf. Fig. 4).

RESULTS AND DISCUSSION

The processing conditions used resulted in almost theoretically dense composites with very homogeneous distribution of the whiskers. The preferred long-axis orientation of the whiskers was on planes perpendicular to the hot-pressing direction (whisker-plane). The overview optical micrograph in Fig. 5 shows a whisker plane in a 10 vol. % ZrO_2 mullite composite with 20 vol. % SiC whisker. The TEM overview of a 30 vol. % SiC whisker/TZP material (Fig. 6)

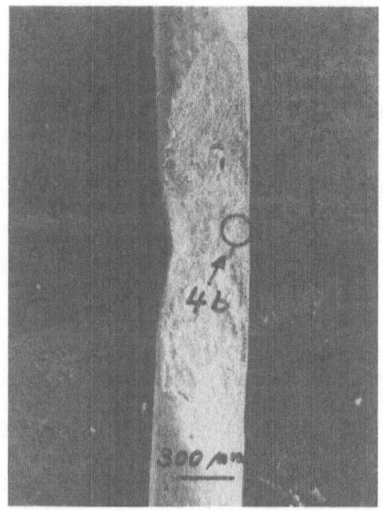 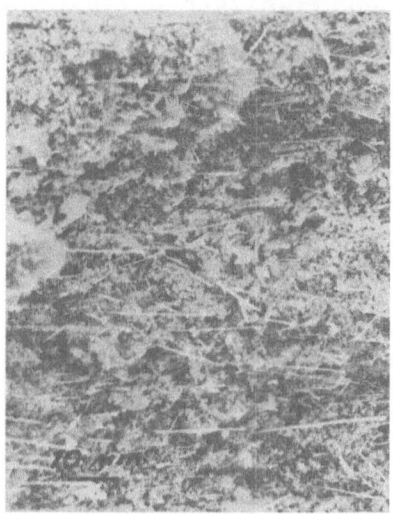

Fig. 4.   Mullite/10 vol. % SiC whisker composite extruded
          through a 1mm orifice of a hypodermic syringe.
          a) Dried green rod split with a needle.
          b) SEM fractograph of the green extruded rod showing
          the whisker alignment in the extrusion direction.

also demonstrates the homogeneity of the whisker distribution
(11). Fracture of whiskers was rarely observed. Reactions
between the SiC whiskers and the ZTC ceramics studied could
not be detected, as was also the case for Al_2O_3 whiskers in
TZP. Si_3N_4 whiskers, however, reacted severely with ZrO_2-
containing materials.

Fig. 5.   Microsection of a 10 vol. % ZrO_2 mullite composite
          with 30 vol. % SiC whiskers showing the "whisker
          plane" (see text).

Results obtained with the various ZTC/SiC whisker systems are
discussed in the following paragraphs.

a) ZrO_2-Al_2O_3/SiC whisker composites

Table 2. lists strength and toughness data obtained at
room temperature. For comparison, data from Ref. 12 are also
included. The latter values, however, were obtained with
different materials under different hot-pressing conditions.

Table 2. Strength and toughness of ZrO_2-Al_2O_3/SiC whisker
composites

No.	Composition, vol.% Matrix	Whisker	K_{Ic} MPa$\sqrt{m}$	Strength MPa
1	Al_2O_3		4.7*	520
2	Al_2O_3	+ 20 SiC	8:5*	650*
3	Al_2O_3+15 t-ZrO_2		6:2	1080**
4	(Al_2O_3+15 t-ZrO_2)	+ 20 SiC	13:5+	~700
5	Al_2O_3+40 m-ZrO_2		2	-
6	(Al_2O_3+40 m-ZrO_2)	+ 20 SiC	8.5	880
7	m-ZrO_2		< ↑	-
8	m-ZrO_2	+ 20 SiC	9:0++	n.d.

* data from Ref. 12, ** sample HIPped at 1600°C for 10 min.
+ annealed at 1500°C in Ar for 24h,
++ determined from indentation crack lengths

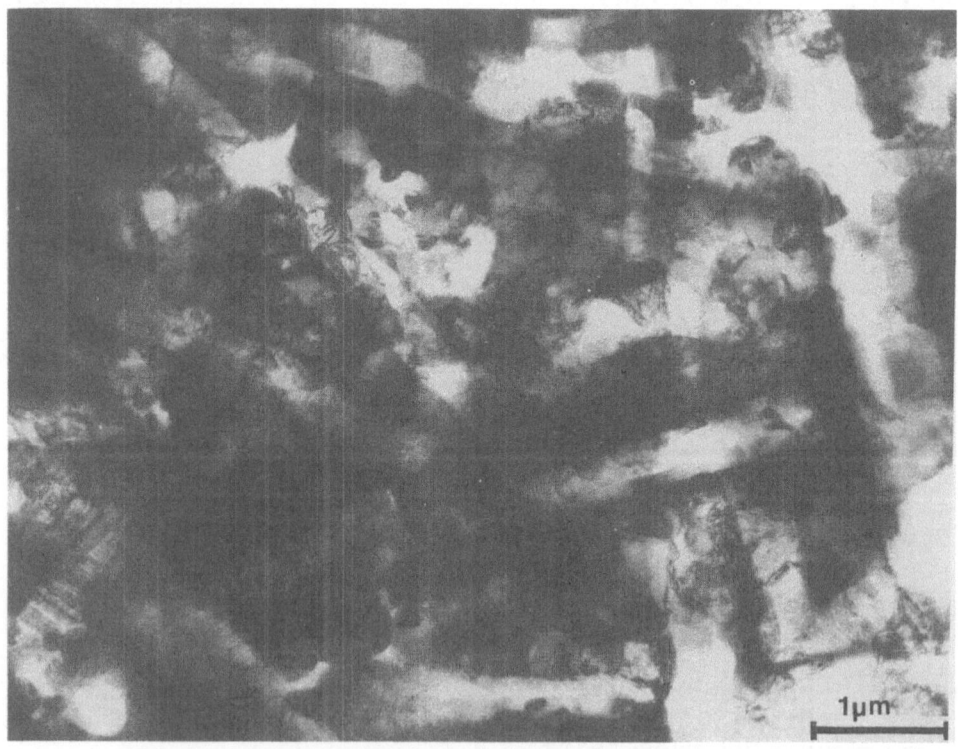

Fig. 6. TEM overview of the "whisker plane" in a TZP /
30 vol. % SiC whisker composite.

The addition of t-ZrO$_2$ to Al$_2$O$_3$ is more effective in increasing the strength than the toughness, a feature which is due to the refined microstructure (13). On the other hand, whisker reinforcement results in remarkable toughness enhancement, although the strength increase is less pronounced. This fact can be explained by the high tensile matrix stresses due to the thermal mismatch between Al$_2$O$_3$ (α = 8.5 MK^{-1}) and SiC (α = 4.7 MK^{-1}). An average value for these stresses (σ_m) may be estimated from (14)

$$\sigma_m = \frac{(\alpha_m - \alpha_w)\ (Tg-T)\ E_w\ w}{1 + w\ (E_w/E_m - 1)} \tag{1}$$

where E_m ($\sim$ 400 GPa) and E_w ($\sim$ 580 GPa) are the matrix and whisker Young's moduli, w is the whisker volume fraction and T_g ($\sim$ 1200°C) the temperature below which interfacial relaxation no longer takes place. At room temperature, σ_m = 485 MPa for w = 0.2. In addition, the effect of a larger critical flaw size due to some whisker clustering has to be considered.

The combination of the two toughening methods (No 4 in Table 2) yields a toughness increase ($\sim$ 9.0 MPa $\sqrt{m}$) which is much more than the sum of the two single toughness increments (see Table 2, No 2: 3.8 MPa$\sqrt{m}$ and No 3: 1.5 MPa$\sqrt{m}$). Hence, the existence of a positive interaction appears to be evidenced by the results.

The same arguments as for the No 2 composite apply to the strength of the No 4 composite. Further increases in strength should be possible by reducing the thermal mismatch stresses, e.g., by introducing Al$_2$O$_3$ whiskers or by utilizing the phase transformation.

The latter method was tested with composite No 6 (Table 2). While a whisker-force Al$_2$O$_3$ with 40 vol.% m-ZrO$_2$ (No 5) was heavily cracked due to the large volume change ($\sim$ 5 %) associated with the zirconia transformation, the same system when used as matrix with 20 vol. % SiC whiskers exhibited both high K_{Ic} and strength. Since the two-phase matrix is not transformation toughened, the enhanced performance, especially in strength, has to be attributed to the reduced thermal mismatch stresses in the matrix. Even composites consisting of m-ZrO$_2$ as the only matrix phase could, with 20 vol. % SiC whiskers, be fabricated with high toughness (No 8).

The control of the thermal mismatch stresses by the tetragonal-monoclinic transformation is explained in Fig. 7. This schematic diagram shows the temperature dependence of the average residual matrix stresses without (v_o) and with (v_1, v_2) two different volume fractions of thermally transformable ZrO$_2$. The matrix stresses in the composite without transformable ZrO$_2$, i.e., containing only retained t-ZrO$_2$, are tensile at room temperature for $\alpha_m > \alpha_w$., while the stresses at v_1 are just balanced and those at v_2 are compressive. In order to estimate σ_m at $T < M_f$ (martensite finish temperature, $\sim$ 700°C), the extent of the transformational volume expansion has to be considered in Eq. (1):

$$\sigma_m T = \frac{[(\alpha_m - \alpha_W)(T_g - T) - \varepsilon^T v] E_W \, w}{1 + w (E_W / E_m - 1)} \qquad (2)$$

where ε^T is the dilational transformation strain. The balancing volume fraction v_1 is obtained from

$$v_1 = \frac{\Delta\alpha \, \Delta T}{\varepsilon^T} \qquad (3)$$

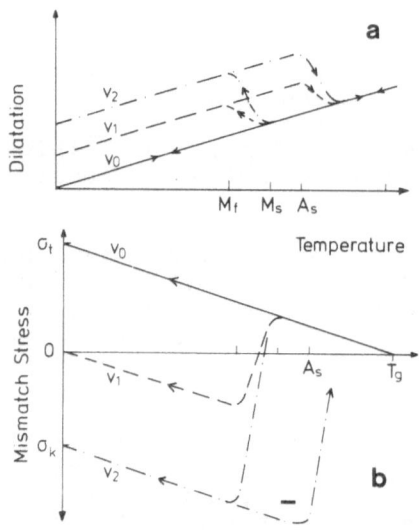

Fig. 7. a) Schematic dilato-meter curve of a whisker composite without (v_0) and with two different volume fractions of thermally trans-formable ZrO_2 (v_1, v_2).
b) Average residual matrix stresses in the whisker composite for $\alpha_m > \alpha_W$. M_S and M_f are the martensite start and finish temperature, A_S the austenite start temperature.

With $\Delta\alpha = 3.5 \ MK^{-1}$, $\Delta T = 1200°C$ and $\varepsilon^T = 0.015$, the volume fraction necessary to achieve zero average matrix stress at room temperature is ~ 0.3. With Eq. (2) the residual stress in the No 6 composite is -190 MPa, hence compressive.

SEM evaluation of the fracture surfaces suggests crack deflection as the main whisker toughening mechanism. Pull-out was rarely observed.

The dilatometer curves of the No 5 and No 6 composites (Fig. 8) show that the presence of the whiskers has little effect on the M_S and A_S temperatures, while the composite strain is reduced by ~ 60 %.

(b) ZrO_2-mullite/SiC whisker composites

Room temperature strength and toughness data (9) are given in Table 3.

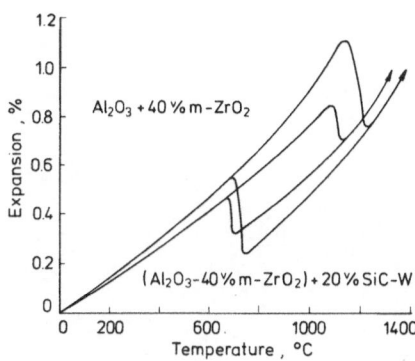

Fig. 8. Dilatometer diagram of Al_2O_3 - 40 vol. % in ZrO_2 without and with 20 vol. % SiC whiskers, measured in air.

Table 3. Strength and toughness of ZrO_2-mullite/SiC whisker
          composites (9).

No	Composition, vol. % Matrix	Whisker	K_{Ic}, MPa$\sqrt{m}$	Strength, MPa
1	Mullite		2.8	244
2	Mullite + 10ZrO_2	3.5	4.4	
3	Mullite	+ 20 SiC	4.4	452
4	(Mullite + 10ZrO_2)	+ 20 SiC	5.4	580
5	(Mullite + 10ZrO_2)	+ 30 SiC	6.7	551
6	Mullite*	+ 10 SiC	4.1	277
7	Mullite*	+ 20 SiC	4.6	407

* extruded (cf. Fig. 4)

In mullite composites, whisker and ZrO_2 toughening appear to be additive. There is no obvious additional interaction effect.   Since the thermal mismatch is negligible between mullite and SiC, the strength and K_{Ic} are improved to the same degree. The relatively low strength of the aligned whisker composites (Nos. 6 and 7 in Table 3, Fig. 9), may be explained by the fact that stress singularities at strongly bonded whiskers can lead to premature failure and that crack deflection is less effective as a mechanism if the crack plane is perpendicular to the whiskers. At high temperature, when whisker pull-out may become effective due to interface softening, the aligned composites may exhibit superior strengths.

Fig. 9. Microsection of a mullite/10 vol. % SiC whisker composite showing the "whisker plane" (cf. Fig. 3 and 4).

The SEM fractographs in Fig. 10 show progressively increasing evidence for crack deflection in the sequence from a) mullite - 10 % ZrO_2, to b) mullite/20 vol. % SiC whisker to c) mullite - 10 vol. % ZrO_2/20 vol. % SiC whiskers. As with the Al_2O_3 whisker composites, very little whisker pull-out could be detected.

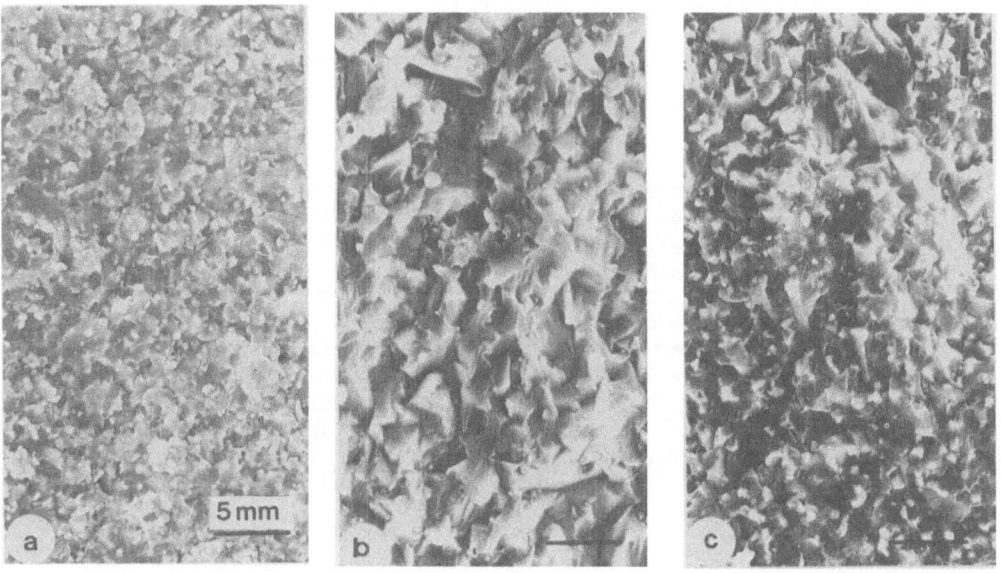

Fig. 10. SEM fractographs of a) mullite - 10 vol. % ZrO_2, b) mullite/20 vol. % SiC whiskers and c) mullite-10 vol. % ZrO_2/20 vol. % SiC whiskers.

(c) ZrO_2-cordierite/SiC whisker composites

Table 4 gives some preliminary K_{Ic} and strength results obtained at room temperature and at 1000°C.

Table 4. Strength and toughness of ZrO_2-cordierite/SiC whisker composites (ZrO_2 + 3 Mole % Y_2O_3).

No	Composition, vol. % Matrix	Whisker	K_{Ic}, MPa√m RT	Strength, MPa RT	1000°C
1*	Cordierite		2.2	180	170
2	Cordierite + 20 ZrO_2		n.d.	190	160
3*	Cordierite	+ 20 SiC	3.7	260	245
4	(Cordierite + 20 ZrO_2)	+ 20 SiC	n.d.	380	300

* Hot pressed at 1250°C for 30 min; data from Ref. 15

The cordierite/ZrO_2 composites did not show stress-induced transformation, i.e. only t-ZrO_2 was detected on the fracture surfaces. The favorable thermal mismatch becomes obvious from the greatly improved strength of the whisker composites. The higher strength of composite No 4 when compared to that of No 3 is partly due to the enhanced densification behaviour associated with ZrO_2 dispersions in cordierite (16) (density of No 3: ~ 96 % TD; of No 4: > 99 % TD).

(d) TZP/SiC whisker composites

The TZP matrix consisted of tetragonal grains, typically < 0.5 μm in size, this microstructure being a result of the short time and low temperature of the hot-pressing. No crystalline phases were detected other than t-ZrO_2 and SiC. An amorphous phase at the whisker matrix interface, 1-5 nm thick, was observed in TEM (17). In some cases, glassy pockets had formed adjacent to the whiskers (Fig. 11).

Table 5. Strength and toughness of TZP/SiC whisker composites (17)

No	Composition, vol. % Matrix	Whisker	K_{Ic}, MPa√m RT	Strength, MPa RT	1000°C
1	3Y-TZP		6.8	1150	160
2	3Y-TZP	+ 20 SiC	10.2	600	340
3	3Y-TZP	+ 30 SiC	11.0	590	380

The toughness values and strengths of 3Y-TZP with 0, 20 and 30 vol. % SiC whiskers are given in Table 5. The matrix shows the characteristic properties of high-strength TZP with relatively low toughness. The latter may be attributed to the small size of the t-ZrO_2 grains which prevents their transforming on stress loading. Thus, the increase in fracture toughness of the whisker composites is assumed to originate from the whiskers where crack deflection appears to be the dominant toughening mechanism.

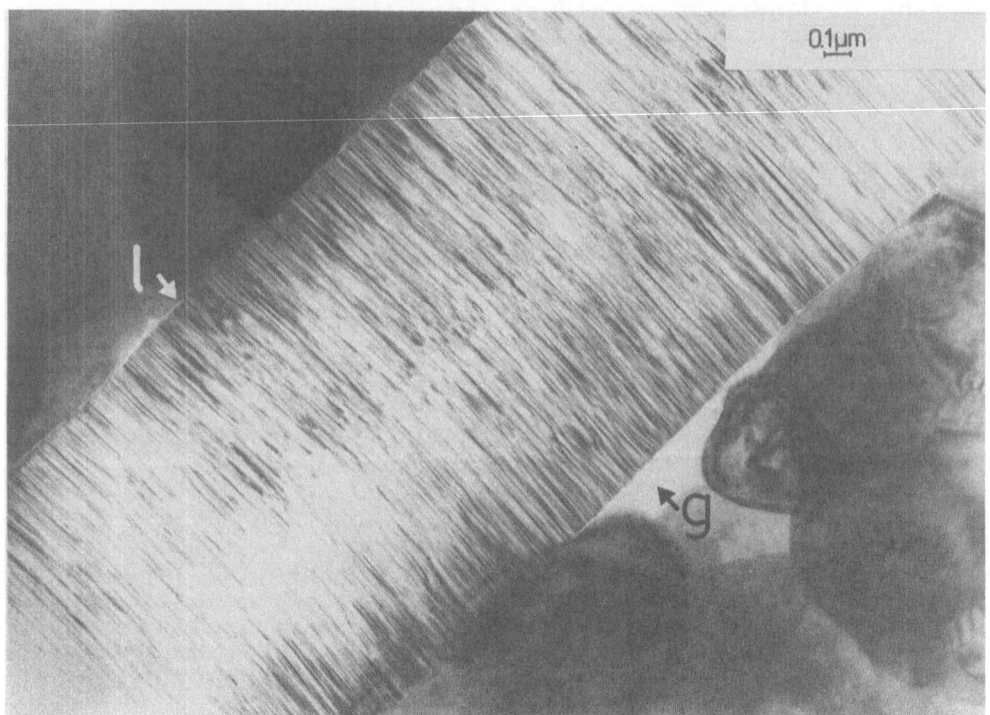

Fig. 11. Dark-field TEM of SiC whisker/TZP interface showing
a very thin glass layer (l) and glass pockets (g).

Whisker debonding and pull-out at room-temperature was rarely
observed, probably due to the strong interfacial bonding, and
so did not contribute much to the toughening. The bond
strength may result from chemical bonding, thermal mismatch
strains and interfacial roughness. Only in cases where the
whiskers were nearly parallel to the fracture plane did some
debonding take place. In all cases, the matrix fracture was
predominantly intergranular.

The fracture strength is reduced to almost half the TZP
matrix strength by the incorporation of whiskers because of
the high tensile matrix stresses due to the thermal mismatch
($\alpha_{TZP} = 11$ MK^{-1}).

The situation is markedly different at high temperatures. At
1000°C, the strength of the whisker composite is nearly twice
that of the TZP matrix. The thermal mismatch stresses are
much lower than at room temperature and the interfacial layer
has softened to the extent that some whisker pull-out (Fig.
12) can take place.

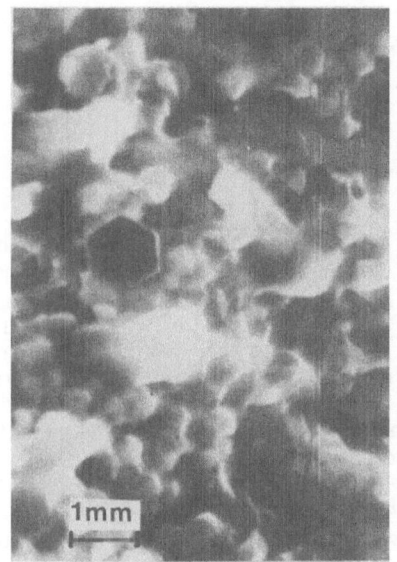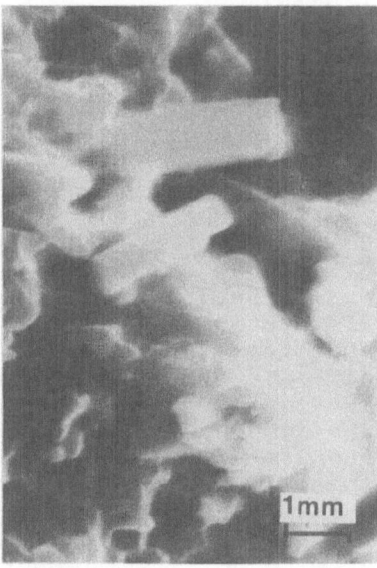

Fig. 12.   SEM micrographs of 3Y-TZP / 30 vol. % SiC whisker
composite fractured at 1000°C showing some whisker
pull-out.

CONCLUSIONS

(a) Almost fully dense ZTC composites reinforced with up to
    30 vol. % homogeneously dispersed SiC whiskers were
    fabricated by hot pressing at relatively low tempera-
    tures.

(b) The thermal mismatch stresses strongly affect the
    strength of the composites, e.g. the room-temperature
    strength of TZP is reduced, while that of mullite and
    cordierite based composites is increased by the incor-
    poration of SiC whiskers.

(c) The thermal mismatch stresses in TZP- and Al_2O_3-based
    composites can be reduced by adding an appropriate amount
    of thermally transformable ZrO_2.

(d) The fracture toughness of all ZTC ceramics studied was
    enhanced by the incorporation of SiC whiskers. The
    dominant toughening mechanism appears to be crack deflec-
    tion (18).

(e) An additional toughness contribution may result from the
    interaction of the stress-induced transformation and the
    whisker reinforcement.

(f) The strength at 1000°C of TZP and cordierite was conside-
    rably enhanced by the incorporation of SiC whiskers.

ACKNOWLEGMENT

The authors thank R.J. Brook, M.L. Mecartney and K.-L. Weiss-kopf for helpful discussions and critical review of the manuscript.

REFERENCES

1. A.H. Heuer and L. Hobbs (eds.), Science and Technology of Zirconia, Advances in Ceramics, vol. 3, The American Ceramic Society, Columbus, OH, (1981).

2. N. Claussen, M. Rühle and A.H. Heuer (eds.), Science and Technology of Zirconia II, Advances in Ceramics, vol. 12, The American Ceramic Society, Columbus, OH, (1984).

3. K. Tsukuma, K. Ueda, and M. Shimada, Strength and Fracture Toughness of Isostatically Hot-Pressed Composites of Al_2O_3 and Y_2O_3 - Partially - Stabilized ZrO_2, J. Am. Ceram. Soc. 68: C-4 (1985).

4. N. Claussen, Strengthening Strategies for ZrO_2-Toughened Ceramics at High Temperatures, Mat. Sci. Eng., 71: 23 (1985).

5. P.F. Becher and G.C. Wei, Toughening Behavior of SiC-Whisker Reinforced Alumina, J. Amer. Ceram. Soc., 67: C 267 (1984).

6. M.V. Swain, Strength - Toughness Relationships for Transformation - Toughened Ceramics, to be published in Fracture Mechanics of Ceramics, vol. 7/8.

7. A.G. Evans and R.M. Cannon, Toughening of Brittle Solids by Martensitic Transformations, to be published in Acta Met.

8. K. Tsukuma and M. Shimada, Strength, Fracture Toughness and Vickers Hardness of CeO_2-Stabilized Tetragonal ZrO_2-Stabilized Tetragonal ZrO_2 Polycrystals (Ce-TZP), J. Mat. Sci., 20: 1178 (1985).

9. H.Y. Liu, Ph.D.Work, Max-Planck-Institut, Stuttgart, (1985).

10. P. Chantikul, G.R. Anstis, B.R. Lawn and D.B. Marshall, A Critical Evaluation of Indentation Techniques for Measuring Fracture Toughness: II Strength Method, J. Am. Ceram. Soc., 64: 539 (1981).

11. N. Claussen, K.L. Weisskopf and M. Rühle, Mechanical Properties of SiC Whisker Reinforced TZP, to be published in Fracture Mechanics of Ceramics, vol. 7/8.

12. T.N. Tiegs and P.F. Becher, this issue

13. N. Claussen, Microstructural Design of Zirconia - Toughened Ceramics (ZTC), p. 325, in Ref. 2.

14. D.C. Phillips, Fibre Reinforced Ceramics, in: Handbook of Composites, A Kelly and S.T. Mileiko, eds., vol. 4, p. 373, Elsevier (1983).

15. T.K. Kang, K.L. Weisskopf and G. Petzow, unpublished results.

16. K. Niezcery, Ph.D.Work, Max-Planck-Institut, Stuttgart, (1985).

17. N. Claussen, K.L. Weisskopf and M. Rühle, to be published in J. Am. Ceram. Soc.

Si_3N_4-SiC WHISKER COMPOSITE MATERIAL

Ryozo Hayami, Kazuo Ueno, Isao Kondou, Nobuyuki Tamari
and Yasuo Toibana

Government Industrial Research Institute, Osaka

1-8-31, Midorigaoka, Ikeda-shi, Osaka 563 Japan

INTRODUCTION

Brittleness is the weakest point of ceramics when they are used as structural component such as a diesel engine or a gas turbine. One of the ways to improve the toughness of ceramics is the "fiber-rein-forcement". Recent studies have revealed the potentiality of ceramic composite as an advanced material. The fibers used in the FRC (Fiber Reinforced Ceramics) can either sustain the applied stress or absorb the fracture energy through being pulled out and deflecting crack propa-gation.

Among several candidates of the fibers for FRC, SiC whisker seems to have an outstanding position because of its high tensile strength and high thermal and chemical stabilities. In this work, SiC whisker reinforced Si_3N_4 was fabricated and some mechanical properties were evaluated. This FRC showed good electrical conductivity and electric discharge machining was applied to the FRC.

EXPERIMENTAL

Chemicals   Alpha silicon nitride powder (Toshiba Ceramic Co., Grade A) has an average particle size of 1.01 μm and its composition is: Si 59.2, N 37.9, C 0.92, O 1.82, Fe 0.007, Ca 0.007, Al 0.005, Mg 0.001 wt%. Characteristics of β-silicon carbide whisker (Tokai Carbon

Table 1. Characteristics of Silicon Carbide Whisker

Diameter	0.1 - 1.0 μm (almost 0.2 - 0.5 μm)
Length	50 - 200 μm
Aspect ratio	50 - 300
Density	3.17 g/cm^3
Bulk density	0.03 - 0.05 g/cm^3
Heat-resistivity	1600°C in air
Tensile strength	3 - 14 GPa
Elastic modulus	400 - 700 GPa

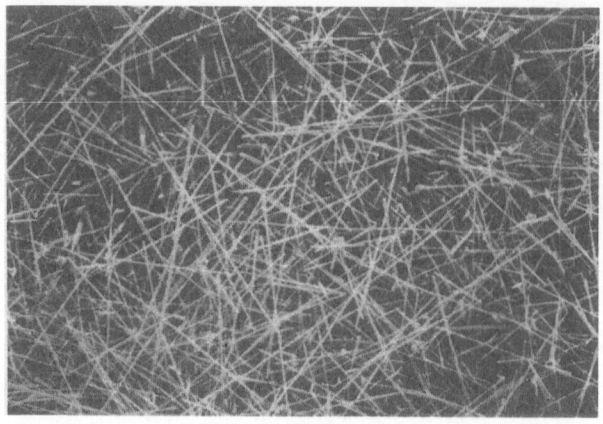

Fig. 1. SiC whisker observed by SEM (x500).

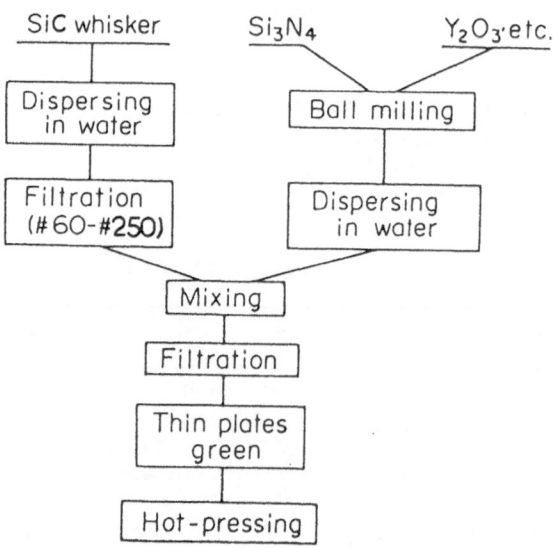

Fig. 2. Fabrication process for SiC whisker-
Si$_3$N$_4$ FRC.

Co.) are shown in Table 1.   Figure 1 is a SEM image of the whisker.
As sintering ades, 10-25 mol% Y_2O_3 and La_2O_3 are added to Si_3N_4 powder.

Fabrication Process   Figure 2 shows the fabrication process for SiC(w)
(silicon carbide whisker)-Si_3N_4 FRC.[1]   Entangled lumps of the whisker
should be untied in order to get homogenious composite microstructure.
SiC(w) is dispersed in water by mechanical stirring for 15-30 min.
The remaining lumps are removed through #60-#250 mesh sieves.   Granular
silicon carbide and other impurities are removed through the decantaion
of the dispersion. Si_3N_4, previously doped with Y_2O_3 and La_2O_3 by ball-
milling in ethanol, is ultrasonically dispersed in water.   Then the
SiC(w) and Si_3N_4 dispersions are mixed under stirring, and filtered
through a Buchner funnel, which gives the mixture sheet with the thickness of 1-3 mm.   Thus obtained thin green plates are cut into the fitting
shape for hot-pressing die and dried in an oven.   Then they are
piled up to the amount enough to get a desired size, and hot-pressed
at 1800°C under 34.3-39.2 MPa for 15-90 min. by using the graphite
die coated with BN powder on the inner surface.   The sintering atmosphere
is assumed to be N_2+CO.

Characterization   The microstructure of the FRC are examined by an
optical microscope, SEM, and X-ray diffractometer.   Three points bending
strength is measured with the samples which have the shape of about
3 x 3 x 40 mm and the rounded edges, under the conditions of 0.5 mm/min.
cross head speed at the temperature from ambient to 1300° in air.

RESULTS AND DISCUSSION

Observation of the microstructure of FRC

     Figure 3 shows the microstructure of the FRC containing 30 wt%
SiC(w).   As can be seen, the whisker, looks brighter in the composite
texture, seems to keep its needle-like shape even after the hot-pressing.
This means that the whisker is stable and does not react with Si_3N_4 and
additive oxides(Y_2O_3, La_2O_3) during the hot-pressing.   When Y_2O_3 and
Al_2O_3 were used as sintering ades, the whisker seemed to melt partially

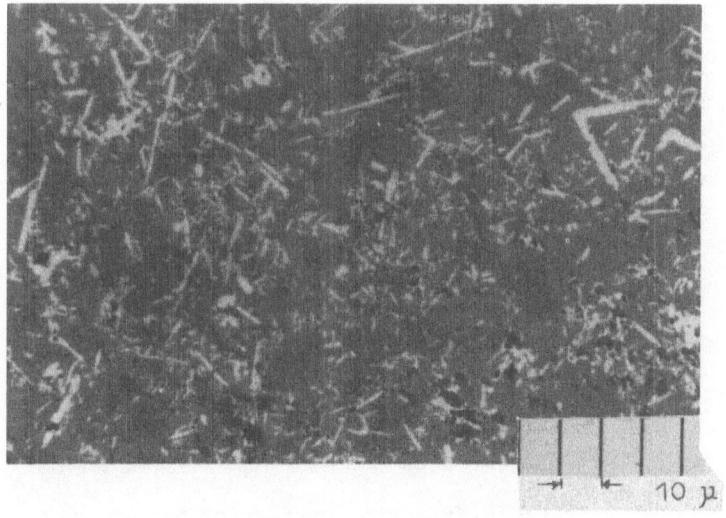

Fig. 3. Microstructure of Si_3N_4-SiC whisker FRC.

into the matrix Si_3N_4 phase. Therefore, Al_2O_3 doping is not suitable for this FRC. The system Y_2O_3 + La_2O_3 did not so remarkably attack the whisker as the Al_2O_3 containing system did. In this case, the features of the boundary between the whisker and the matrix could not be elucidated by SEM examination.

From microscopic observation, it was concluded that almost all of whisker aligned two-dimensionally perpendicular to the pressing direction. Filtration into thin green plates and the one dimensional shrinkage during the hot-pressing may cause the orientation of the whiskers. The orientation is favorable for fiber reinforcement toward the stress applied in one or two directions, but will introduce aniso-tropies in mechanical, electrical, and thermal properties.

X-ray diffraction pattern of the FRC is shown in Figure 4. The sample contained 30 % whisker. The starting α-type Si_3N_4 has trans-formed to β-type as much as over 80 % after the hot-pressing for 30 min. or longer. Complete change into β -type was recognized after the hot-pressing for 90 min.
Besides Si_3N_4 and SiC(w),there were detected several products of the reaction among the doped oxides, Si_3N_4, and SiO_2 on the surface of Si_3N_4. Y_2O_3 + La_2O_3 system forms Y-Si-O-N and La-Si-O-N oxynitride phases in grain boundary as shown in Figure 4. As yttrium containing oxynitride, N-melilite ($Si_3N_4 \cdot Y_2O_3$), J-phase($Y_4Si_2O_7N_2$), and H-phase (N-apatite, $Y_5(SiO_4)_3N$) were detected. And as lanthanum containing phase, $2Si_3N_4 \cdot La_2O_3$ was detected. These crystalline phases in the grain boundary are believed to be effective to improve the refractoriness of Si_3N_4 ceramics.

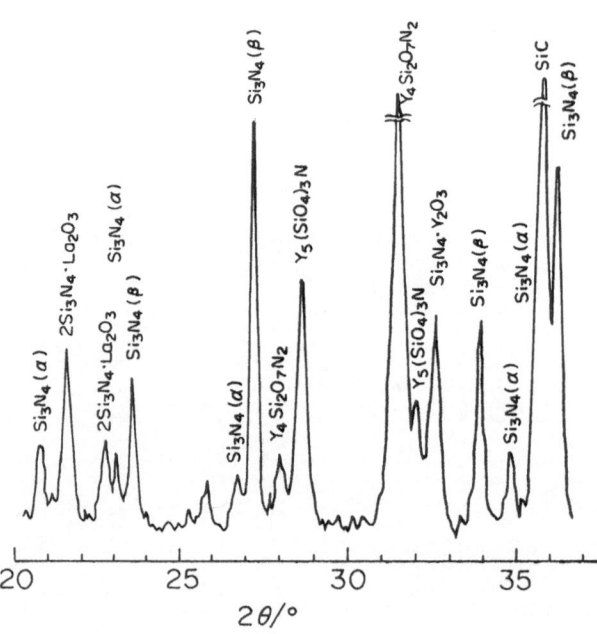

Fig. 4. X-ray diffraction pattern of SiC whisker
-Si_3N_4FRC. SiC whisker: 30 %

Influence of the sintering time

As needle-like whiskers prevent the shrinkage of green body, the sintering rate of the FRC is lower than that of the Si_3N_4 itself. Therefore, it is critical to choose the sintering time which is long enough to get the full-dense body and, at the same time, is short enough in order to minimumize the reacton between the whisker and the matrix. The FRC, containing 30 % SiC(w), were hot-pressed for 15, 30, 60, and 90 min. Figure 5 shows the relative density and the bending strength at room temperature and 1300°C. From the variation of the relative densities, 60 min. hot-pressing was long enough to get the nearly full-dense body. Hot-pressing for less than 30 min. left some porosities in the FRC.

On the other hand, strengthes at room temperature were 590-680 MPa, being not so much influenced by the hot-pressing time. At 1300°C, in contrast to this, strength of the FRC hot-pressed for less than 30 min. was 100 MPa lower than that of hot-pressed for 60 min. The FRC hot-pressed for 90 min. showed slightly lower strength than that of 60 min. hot-pressing.

From these results following are suggested:

(1) Most suitable hot-pressing time is 60 min. for this FRC. The hot-pressing for less than 30 min. is not sufficient to eliminate the porosity completely. Longer sintering over 90 min. deteriorates the whisker owing to the reaction with the matrix.

(2) The FRC sintered for 15 min. had the porosity more than 5 %, nevertheless its room temperature strength did not differ so much from that of the full-dense body. In general, strength of ceramics is very sensitive to inner defects and cracks. Strength (σ) of Si_3N_4 at room temperature degrades exponentially as a function of the porosity (p), as proposed by Ryshkewitch:

$$\sigma = \sigma_0 e^{-bp},$$

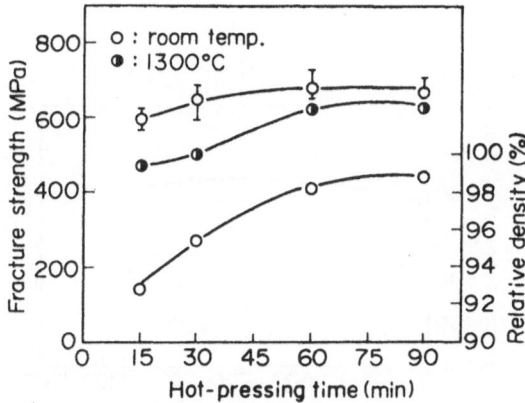

Fig. 5. Relative density and strength of SiC whisker-Si_3N_4 FRC.

where σ_0 is the σ at p=0 and b is a constant. Our FRC does not follow this relation, meaning that the strength becomes insensitive to the porosity by the fiber reinforcement.

(3) When the FRC was hot-pressed over 60 min., only about 50 MPa degradation from the room temperature strength was observed at 1300°C. Si_3N_4 ceramics, having the same composition as the matrix of the FRC, showed the strengthes of 1050 MPa at room temperature and 735.5 MPa at 1300°C. This corresponds about 30 % degradation. Therefore, it is concluded that SiC(w) in the composite is effective to decrease the drop in the strength at high temperatures.

Figure 6 shows the temperature dependence of the strength of the FRC, containing 30 % SiC(w) and hot-pressed for 60 min. Up to 1300°, the FRC had nearly constant strength. The strength degradation in Si_3N_4 at higher temperatures above 1300°C can be attributed to the grain boundary sliding due to the softening of the boundary phase. The elasticy of the matrix (E_M) goes down at higher temperature, but the elastic modulus of the whisker (E_F) remains without significant decrease. That is, E_F becomes much higher than E_M at these temperatures. As pointed out by Fitzer, there comes the possibility for fiber reinforcement, when the condition $E_F > E_M$ is attained. At ambient temperature one cannot expect above effect because $E_F \cong E_M$.

Influence of the amount of SiC whisker

Figure 7 indicates the strengthes at ambient and 1300°C for the FRC containing 10-50 wt% whisker. For these composites, since the SiC whisker having 30-50 % shorter mean length was used, the values were tend to be lower than that described in the previous chapter. In the cases of the FRC containing over 40 % whisker, much oxides were added to the matrix Si_3N_4 to facilitate the sintering.

The strengthes at room temperature were about 590-680 MPa, being independent to the whisker content. Table 2 shows the compositions and the densities of the FRC. Although the amount of the additives was increased, rather porous ceramics were fabricated when the amount of the whisker exceeded 40 %. Nevertheless, there were no strength degradation with increasing porosity, indicating the effect of fiber reinforcemnt.

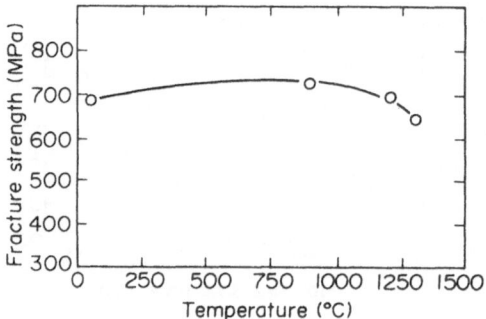

Fig. 6. Temperature dependence of the strength of SiC whisker-Si_3N_4 FRC, containing 30 % SiC(w).

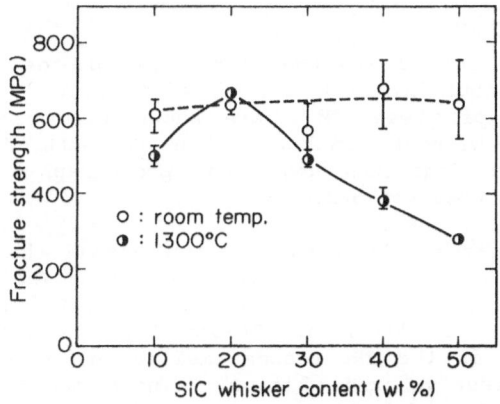

Fig. 7. Strengthes at room temperature and 1300°C for FRC
containing 10 - 50 % SiC whisker.

On the other hand, the strength at 1300°C decreased as increasing
the amount of the whisker. This may be attributed to the increased
additive oxides, which will leave the glassy phases in the grain boundary. In this case, the whisker might be pulled off from the matrix
under the stress, that is, could not carry the applied load, because
of the softening of glassy boundary between the whisker and the Si_3N_4
grains.

Fracture energy of the FRC

Figure 8 shows the effect of the content of SiC(w) on the fracture
energy of the FRC. Values relative to that of the Si_3N_4 ceramics are
plotted in the diagram. Fracture energy increased with increasing
the content of the whisker. The energy for the FRC with 30 % SiC(w)
is about 2.5 times larger than that of Si_3N_4 ceramics.

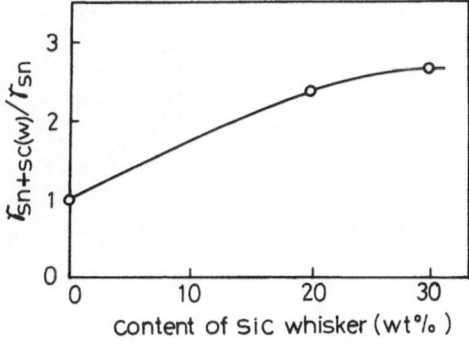

Fig. 8. Effect of whisker content on the fracture
energy of SiC whisker-Si_3N_4 FRC.

Fractography of the FRC

Figures 9, 10, and 11 are the microscopic photos of the fractured surface for the FRC containing 10, 20, and 40 % SiC(w).[5] The compositions and the densities of these FRC's are shown in Table 2. Figure 12 is that of the Si_3N_4 without whisker. The strength for the fracture of each samples are indicated under the photograph. Followings are suggested from these observations.

The fracture mirror, the planar region around the fracture origin offers various informations about the material. The mirror size is thought to be proportional to the brittleness. Starting from the clear, round one on the Si_3N_4 as shown in Figure 12, the size and shape of the fracture mirror in the FRC become smaller and more indistinct with increasing the content of the whisker. There scarcely was observed the mirror in the FRC with 40 % SiC(w), and the strength was higher than those for the FRC with 10 or 20 % SiC(w).

Eveness in and near around the mirror decreased with increasing the whisker content, meaning that the whisker changed the course of the crack propagation (bowing) and increased the fracture energy as descrived in the previous chapter.

Fig. 9. Fractured surface for the
FRC with 10 % SiC(w).
(Strength=638 MPa)

Fig. 10. Fractured surface for the
FRC with 20 % SiC(w).
(Strength=613 MPa)

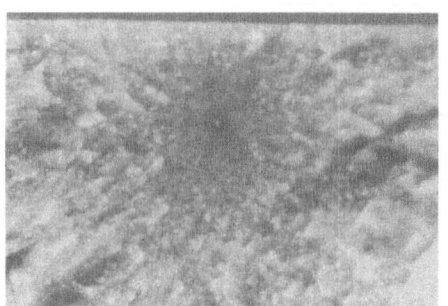

Fig. 11. Fractured surface for the
FRC with 40 % SiC(w).
(Strength=747 MPa)

Fig. 12. Fractured surface for the
Si_3N_4 ceramics.
(Strength=790 MPa)

Table 2. Compositions and densities of SiC whisker-Si_3N_4 FRC with various content of the whisker

Composition(wt%)		Matrix (mol%)			Density (g/cm^3)	Relative density (%)
SiC	Matrix	Si_3N_4	Y_2O_3	La_2O_3		
10	90	90	5	5	3.40	98.8
20	80	85	7.5	7.5	3.45	98.3
30	70	85	7.5	7.5	3.30	94.8
40	60	80	10	10	2.99	84.9
50	50	75	12.5	12.5	3.12	88.4

The prevention effect to crack propagation by the whisker was clearly observed in the samples with controlled surface flaw introduced by means of a Knoop indentor in advance to bending measurement. Figure 13 shows the fractured surfaces of the composite with 30 % SiC(w) and Si_3N_4 ceramics, both having about 200 μm length flaw.

The fractured surface of the FRC was rather rough, while the Si_3N_4 surface was very smooth. The introduction of flaw reduced the strength from 637 to 294 MPa, to about one half, in the case of the FRC. For the Si_3N_4 ceramics, significant down to about one fourth, from 882 to 225 MPa, was observed. Apparently, SiC whisker had improved the toughness of Si_3N_4 ceramics.

Reliability of the FRC

In almost all of FRC samples, the fracture was found to originate from the lumps of SiC whisker. If these lumps are removed completely from the raw whisker, the strength and the reliability of FRC are expected to be much improved. In order to confirm this presumption, mesh filter having various opening sizes were used to refine the whisker. Weibull's analysis was done with eight strength data for each sample containing 30 % SiC(w).

(a)

(b)

Fig. 13. Fractured surfaces of (a) FRC containig 30 % SiC(w), and (b) Si_3N_4 ceramics, both having controlled surface flaw.

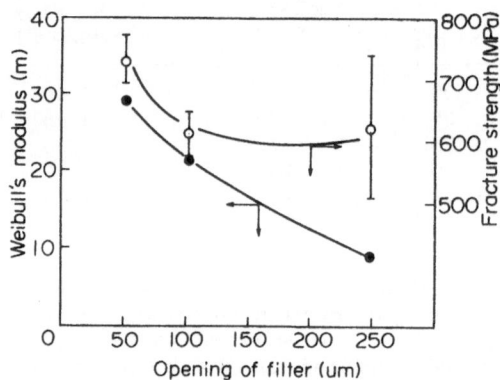

Fig. 14. Relation of strength and Weibull's modulus
to filter opening.

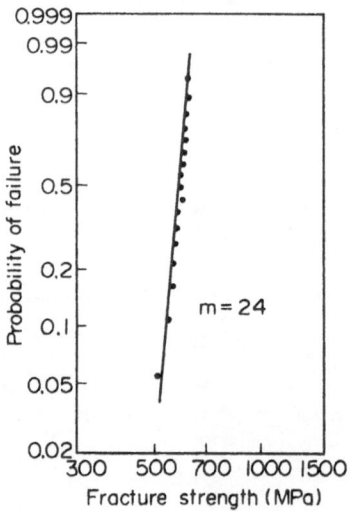

Fig. 15. Weibull's plots for 30 % SiC whisker-Si_3N_4
FRC.

Figure 14 shows the relations of strength and Weibull's modulus to filter openings. The finer the size of filter opening was, the higher the strengthes were. Weibull's modulus became significantly higher as finer the filter was. The FRC made by using usual 105 μm filter showed the modulus of m≅23. This value is much more exellent than the values of m=10-15 for ordinal Si_3N_4. When a finer filter having 53 μm openings was used, the highest value, m≅29 was obtained, which ranks with the values for metals.

Eight test pieces for one sample is not enough to Weibull's analysis. In order to get as many pieces as possible from one sintered body, a large size FRC (47 x 42 x 10 mm, 30 % SiC(w) filtered with 88 μm opening sieve) was fabricated. Eighteen test pieces were cut out from the FRC.

The modulus m=24 was calculated from the data as shown in Figure 15. Now we cannot explain clearly the reason why thus fabricated FRC shows such high reliability, but it is assumed that the filtering of SiC(w) gave the FRC nearly same size defects, which resulted in very narrow strength distribution.

Electrical conductivity of the FRC

Silicon nitride ceramics is an electric insulator, having the resistance of 10^{12} - 10^{13} Ω cm. The FRC, on the other hand, showed good electric conductivity as shown in Table 3. When the whisker content was less than 30 %, the resistance was tended to increase. The 10 % SiC(w)-FRC showed about 30 Ω cm.

As Si_3N_4 itself is an insulator, electrical conductivity should be attributed to SiC, a kind of semiconductor. Incidentally, ordinal SiC ceramics have larger resistance than that of the FRC, perhaps on account of barrier layer at grain boundaries, and is severely influenced by species of additive, sintering time and temperature. In the FRC, SiC whiskers may contact each other strictly after the hot-pressing.

Electric discharge machining

Precise machining of ceramics having complex shaped is one of the key technologies when ceramics are used as an engineering material. As described above, electrical conductivity of the FRC gives an possibility for electric discharge machining.

Figure 16 shows several samples submitted to the machining. Complex forms could be cut out from the FRC block. One can select either fast or slow processing speed by controlling the current, pulse interval

Table 3. Electrical resistance of hot-pressed Si_3N_4, SiC, and SiC whisker-Si_3N_4 FRC at room temperature

Ceramics	Electrical resistance (Ω cm)	
Si_3N_4	10^{12} - 10^{13}	β-type
SiC	1 - 10^5	"
SiC(w)-Si_3N_4	1	30 % whisker
"	0.56	40 "
"	0.42	50 "

and polarity during the discharging.   Remarkable feature is very small consumption of the electrode.

Fig. 16. Samples cut out from the Si_3N_4-SiC whisker FRC block by electric discharge machining.

REFERENCES

1. Y. Toibana and K. Ueno, USP 4,507,224. Mar.26,1985.
2. K. Ueno and Y. Toibana, Yogyo-kyokai-shi(J.Cer.Soc.Japan), 91,409-414, (1983).
3. E. Ryshkewitch, J.Am.Cer.Soc., 36, 65-68(1953).
4. E. Fitzer, Proc. of International Symposium of Factors in Densification and Sintering of Oxide and Non-oxide ceramics, 1978, Japan, 618-673.
5. R. Hayami, Y. Toibana and K. Ueno, Zairyou Kagaku, 19, 335-340 (1983).

MECHANICAL PROPERTIES OF SiC FIBER-REINFORCED REACTION-BONDED Si_3N_4

COMPOSITES

Ramakrishna T. Bhatt

AVSCOM – Propulsion Directorate
NASA Lewis Research Center
Cleveland, OH 44135

INTRODUCTION

Because of their lightweight, excellent oxidation resistance, high temperature strength, environmental stability, and nonstrategic nature, silicon-based ceramics are candidate materials for high performance advanced gas turbine and diesel engines. However, the use of these materials is severely limited because of their inherent flaw sensitivity and brittle behavior. Past studies [1,2] have shown that reinforcement of ceramics by high strength, high modulus, continuous length ceramic fibers should yield stronger and tougher materials. Glass matrix composites [3] have clearly demonstrated the feasibility of obtaining strong and tough materials. These newly developed composites, however, are presently limited in temperature capability by matrix properties, interfacial reactions, and by thermal instability of the fibers above about 1000°C.

In contrast, the present study reports the development of a strong and tough ceramic matrix composite which should have potential for use in advanced engines operating at temperatures above 1200°C. In this study, silicon carbide fiber-reinforced reaction-bonded Si_3N_4 (SiC/RBSN) composites were fabricated and their physical and mechanical properties determined. The study was performed to obtain basic understanding about the reinforced effects of fibers, fiber interaction with the matrix, and fiber strength stability under processing conditions. Wherever possible, metallographic and fractographic analyses were also performed to study microstructure and failure mechanisms.

EXPERIMENTAL

Fiber

The SiC fibers used in this study were obtained from AVCO Specialty Materials Division. These fibers were produced by chemical vapor deposition (CVD) from methyl trichlorosilane onto a heated carbon monofilament which was pulled continuously through the deposition reactor [4]. Different surface coatings were deposited onto the SiC fibers by introducing hydrocarbon gas or a mixture of hydrocarbon gas and silane vapor near the exit port of the reactor. A schematic diagram of the fiber cross section is shown in Fig. 1(a). The fiber consists of a SiC

sheath with an outer diameter of 142 μm surrounding a pyrolytic
graphite-coated carbon core with diameter of 37 μm. The SiC sheath is
entirely comprised of columnar β-SiC grains growing in a radial
direction with {111} preferred direction [4]. The fiber used for
composite fabrication contained a surface coating consisting of an
overlayer with high silicon/carbon ratio on top of an amorphous carbon
layer (cf. Fig. 1 (b)). The total thickness of the coating is ~2 μm.
This type of fiber, labelled SCS-6 by the manufacturer, was originally
developed for metal matrix composites. The average room temperature
tensile strength of the as-received SCS-6 fibers was 3.8 GPa. (gauge
length = 50 mm).

Ceramic Powder

High purity silicon powder was used for preparation of the
reaction-bonded Si_3N_4 matrix. The impurity content of the as-received
powder is shown in Table I. This powder contained a high volume
fraction of large particles (>20 μm) which are even larger the interfiber
distance of the desired composites. To reduce particle size and to
promote reactivity during nitridation, the silicon powder was attrition
milled for several hours in a Si_3N_4 container using Si_3N_4 balls and
heptane solvent. After milling, the excess solvent was evaporated from
the powder. The powder was then dried and stored for use in composite
fabrication. The chemical analysis, average surface area, and particle
size of the powder before and after attrition milling are shown in Table
II. It is obvious from the table that after attrition milling there was
a significant increase in the oxygen and carbon contents and essentially
no increase in iron content. The surface area of the powder increased
from 1 m^2/g to 10 m^2/g while the average particle size decreased from 6
μm to 0.4 μm.

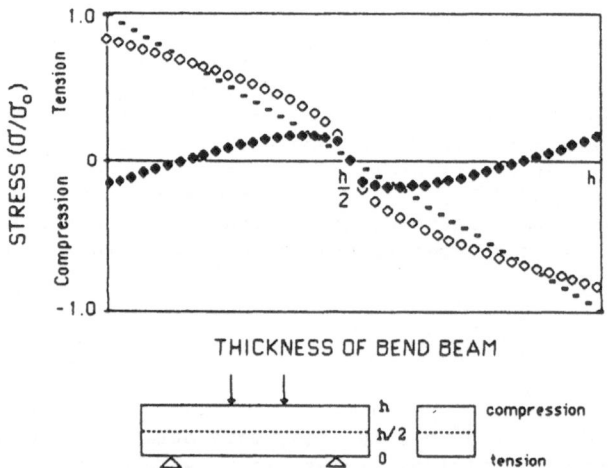

Figure 1. (a) Schematic cross section of CVD SiC fiber.
          (b) The composition profile of carbon-rich coating on the
              surface of SiC fiber (AVCO SCS-6 fiber).

TABLE I - TRACE IMPURITY ANALYSIS OF AS-RECEIVED SILICON POWDER

Elements	Wt. Percent
Aluminum	0.17
Calcium	0.02
Chromium	0.08
Iron	0.53
Manganese	0.08
Titanium	0.03
Vanadium	0.02
Zirconium	0.02

TABLE II - CHEMICAL ANALYSIS, SURFACE AREA, AND AVERAGE
PARTICLE SIZE OF SILICON POWDER

Material	Oxygen Wt %	Carbon Wt %	Nitrogen Wt %	Iron Wt %	Surface Area m^2/gm	Average Particle Size μm
As-Received Silicon Powder	0.43	0.025	0.004	0.60	1.644	6.0
Attrition Milled	1.20	0.31	0.07	0.60	10.216	0.4

Composite Fabrication

The SiC/Si_3N_4 matrix composite material was consolidated by a method similar to that employed for fiber-reinforced metal matrix composites [5]. The details of the fabrication process will be reported elsewhere [6].

To describe the fabrication method briefly, the SiC fiber was wound with desired spacing on a circular drum and coated with a slurry consisting of attrition milled silicon powder, an additive (for enhancing nitridation), a fugitive polymer binder, and a solvent. After carefully drying the fiber mat, it was cut into strips of required dimensions. These strips were stacked up in a graphite die and hot pressed in vacuum or in a nitrogen environment at a suitable combination of temperature and pressure to produce handleable green compacts. Using this method, two types of composite green compacts were produced. One contained ∿ 23 volume fraction and the other ∿ 40 volume fraction SiC fibers. The volume fraction of fibers in the green compact was varied by controlling fiber spacing or by adjusting the thickness of the silicon slurry. Before further processing, the green compacts are weighed and their densities and porosity measured by a mercury porosimeter.

The green compacts were then transferred to a horizontal nitridation furnace consisting of a recrystallized Al_2O_3 reaction tube with stainless steel end caps. The compact loading and furnace description were the same as reported in ref. [7]. The nitriding gas (N_2 or N_2 + 4 percent H_2) of commercial purity was purified according to the procedure followed in ref. [7] and then flowed through the furnace before, during, and after nitridation. The nitriding temperature and nitriding schedule for the composite were similar to that employed for monolithic reaction-bonded silicon nitride [8]. Upon completion of nitridation, the composite plates were reweighed and their densities measured. The nitrided composite materials were examined by X-ray diffraction using the technique developed by Gazzara and Messier [9] to determine the amount of $\alpha-Si_3N_4$, $\beta-Si_3N_4$, and residual silicon

Specimen Preparation

The as-fabricated nitrided composite plates contained excess matrix layer on their top and bottom surfaces. This matrix layer was removed by grinding the composite plate on SiC papers. Specimens for mechanical testing were prepared from the composite plates by cutting and grinding with a diamond impregnated abrasive wheel. For three-point bend tests, typical specimen dimensions were 50 mm x 6.4 mm x 1.2 mm, and the span to height ratio (L/h) was 35. For four-point bend tests, typical specimen dimensions were 63.5 mm x 6.4 mm x 1.2 mm for (L/h) = 45 and 31.8 mm x 6.4 mm x 1.2 mm for (L/h) = 15. Maximum bend stresses were calculated assuming a homogeneous microstructure for the composites. For tensile tests, specimens of dimensions 127 mm x 12.7 mm x 1.2 mm were adhesively bonded with glass fiber reinforced epoxy tabs at the specimen ends, leaving 50 mm as the test gauge length. Tensile and bend tests were performed in an Instron machine at a constant cross-head speed of 1.26 mm/min. Axial strains were measured with a clip gauge attached to the 25 mm center portion of the specimen gauge length. For each volume fraction, the strengths of five specimens were measured. Whenever possible, broken composite specimens were examined by light and scanning electron-microscopy to determine the mode of failure.

RESULTS

Microstructure and Density

A typical cross section of a composite specimen with the excess outer matrix layers is shown in Fig. 2. In all composite specimens, mechanical polishing resulted in erosion of the matrix faster than the fiber. Because of this, a high magnification photograph of the fiber/matrix interface could not be taken. The composite matrix displayed density variations and significant porosity. Average density and porosity data in the green state and after nitridation for the 23 and 40 volume percent SiC fiber/RBSN and for the unreinforced RBSN prepared under similar conditions are listed in Table III. The composites showed nearly 40 volume percent porosity. X-ray analysis of composites after nitridation revealed a small amount of excess silicon (6-8 percent) in addition to $\sim$60 percent αSi_3N_4 and $\sim$ 32 percent Si_3N_4.

Composite Properties

Axial bend and tensile strength. The room temperature ultimate strength results for the 23 and 40 volume percent SiC/RBSN composites are shown in Table IV. Also included in the table for comparison purposes are the four-point bend strength data of unreinforced RBSN fabricated under similar processing conditions. The table also shows the influence of (L/h) ratio on four-point bend strength. Part of the scatter in strength data seen in the table can be attributed to a small variation in fiber volume fraction (±3 percent). In general, the data indicates that ultimate fracture strengths of the composites were significantly higher than those of the unreinforced RBSN matrix and that composite strengths increased in fiber volume fraction.

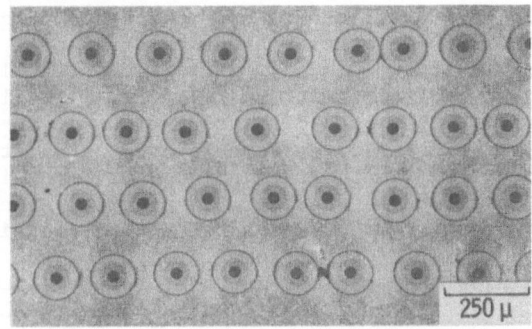

Figure 2.-A typical cross section of SiC/RSBN
composite showing fiber distribution.

TABLE III.   DENSITY AND POROSITY DATA FOR RBSN AND SiC/RBSN
COMPOSITES

Fiber Vol. Fraction	Before Nitridation		After Nitridation	
	Composite Density gms/cc	Matrix Porosity Percent	Composite Density gms/cc	Matrix Porosity Percent
0	1.56	35	1.98	37
23±3	1.70	54[a]	2.19	39[a]
40±2	1.90	51[a]	2.36	40[a]

a. Matrix porosity calculated from composite density and from
theoretical density for CVD SiC fiber (3.0 gm/cc) and from density for
silicon (2.4 gm/cc) or for Si_3N_4 (3.2 gm/cc).

TABLE IV.   ROOM TEMPERATURE STRENGTH OF RBSN AND SiC/RBSN
COMPOSITES

TESTS	AXIAL STRENGTH: MPa		
	0% Fiber	23±3% Fiber	40±2% Fiber
4-Point Bend (L/h=15) [a]	107±26[c]	539±48[c]	616±36[c]
4-Point Bend (L/h=45)	–	675±42	868±32
3-Point Bend (L/h=35)	–	717±80	958±45
Tensile [b]	–	352±73	536±20

a   L/h refers to span to height ratio of test specimen.   h~1.2 mm
b   Tested at 50 mm gauge length
c   Standard deviation for five tests

Typical load-displacement behavior for a composite in a three-point bend test at room temperature is shown in Fig. 3. This specimen contained 20 volume percent SiC fibers. In general, the load displacement curve showed two distinct regions. Initially, load increased linearly with displacement up to a stress level of 240 MPa. Above this stress level, the load-displacement curve was no longer linear. This nonlinear behavior appears to originate with the onset of matrix cracking normal to the fiber reinforcement direction. In Fig. 4, a photograph of the tensile surface of a specimen stressed beyond the linear region reveals the formation of multiple matrix cracks. These cracks did not exist at any stress level below the nonlinear region and initiated on the specimen surface where the tensile stress was maximum. As the sample was stressed to higher stress levels, more matrix cracks appeared at nearly regular spacing. For a given fiber loading, both four-point and three-point bend specimens displayed similar average matrix crack spacings. The crack spacing was observed to vary with the matrix density, processing temperature, and fiber volume fraction. Data for the average matrix crack spacing and the composite stress at which the matrix first cracks are given in Table V for the 23 and 40 volume percent SiC/RBSN. With continued stressing, the composite reached its ultimate stress and broke noncatastrophically due to random fracture of the fibers.

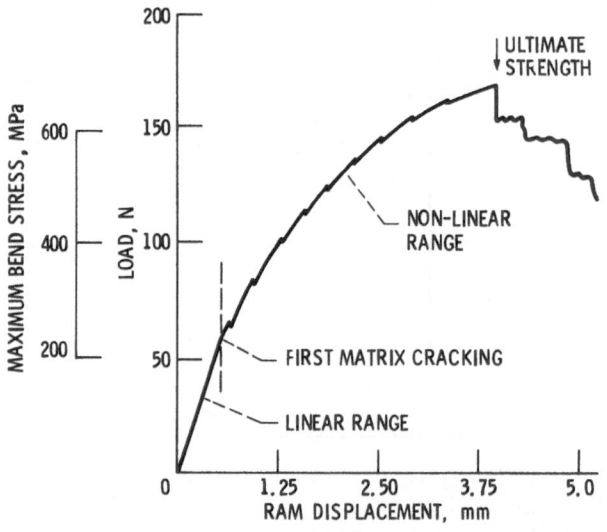

Figure 3.-The load-deflection behavior in 3-point bending for 20 vol% SiC fiber/RBSN composite.

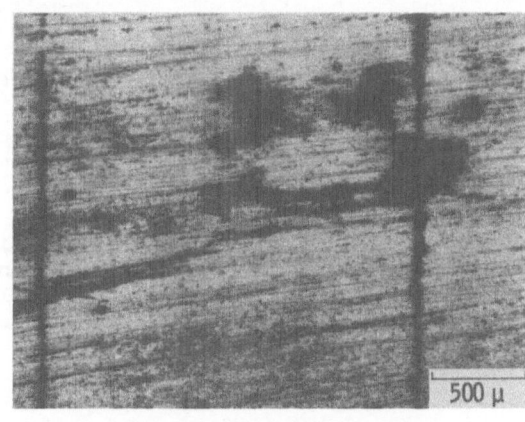

FIBER
DIRECTION

500 μ

Figure 4.-A typical 3-point bend specimen stressed
above the non-linear range showing matrix cracks
normal to fibers on tensile surface.

TABLE V. MEAN MATRIX CRACK SPACING AND FIRST MATRIX CRACKING
STRESS FOR SiC/RBSN COMPOSITES

Fiber Fraction Percent	Matrix Crack Spacing mm	Comp. Stress at Which Matrix First Cracked MPa
23±3	2.0±0.3[a]	237±25[b]
40±2	0.9±0.2	293±15

a Standard deviation for 30 cracks on five bend specimens
b Standard deviation for 5 specimens measured in 3-point bend

The stress-strain behavior for a typical 20 volume percent
SiC/RBSN composite specimen tested in tension at room temperature is
shown in Fig. 5. This figure is similar to that shown in Fig. 3 except
for a second linear region after the nonlinear region. The deviation
from linearity occurred at 220 MPa which is slightly lower than the
stress level seen in Fig. 3. Above this stress level, audible noises
were heard. It was not clear, however, whether these noises
corresponded to matrix or fiber cracks. When the composite reached its
ultimate strength, it broke noncatastrophically. In most tested
specimens, fibrous fracture without complete separation of the specimen
was observed. A specimen which was broken into two pieces, Fig. 6,
displays the fibrous fracture. The matrix around the fiber fell apart
due to high energy release at ultimate fracture.

Fiber/Matrix Interface

To evaluate the fiber-matrix interfacial bonding, a composite
specimen was cut transversely to the fiber direction and tested in a
three-point bend mode. The specimen containing 23 volume percent SiC
fiber showed a transverse bend strength of 76 MPa. The fracture
surface, shown in Fig. 7, revealed that failure of the composite
invariably occurred through a mechanism of interfacial splitting. The
interfacial splitting occurred between the carbon rich surface coating
of the fiber and the matrix, indicating that the interfacial bonding
was weak.

DISCUSSION

The results obtained in this study clearly demonstrate that reinforcement of RBSN by high strength, high modulus, large diameter (CVD) SiC fiber can yield a stronger and tougher (higher strain to failure) material than unreinforced RBSN. The mechanisms by which these improved occur can be analyzed from composite theory.

In general, when a ceramic matrix composite containing uniaxially aligned continuous ceramic fibers is stressed in tension in a direction parallel to the fiber, the composite extends elastically until one of the components fractures. The stress in the composite varies linearly

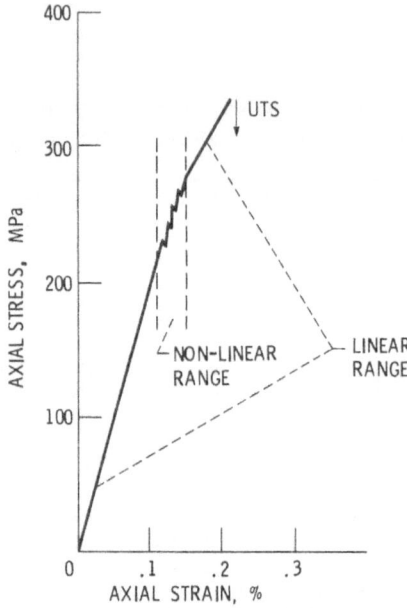

Figure 5.-The tensile stress-strain behavior for 20 vol% SiC fiber/RSBN composite showing linear and non-linear ranges.

Figure 6.-A fractured tensile specimen showing fibrous fracture and fiber pull out.

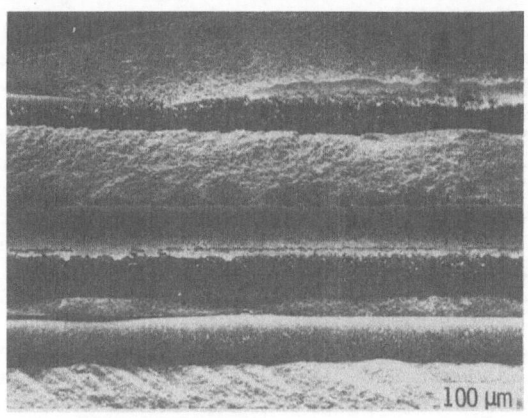

Figure 7.-Fracture surface of a 23 vol% SiC/RSBN
composite tested in transverse flexure showing interfacial splitting between fiber and matrix.

with strain, and at any strain level the composite stress σ_c is given
by

$$\sigma_c = \varepsilon_c E_c = \varepsilon_c \ (E_f V_f + E_m V_m) = \sigma_f V_f + \sigma_m V_m \tag{1}$$

where σ is stress, ε is strain, E is modulus, V is volume fraction, and
c, f, and m refer to composite, fiber, and matrix, respectively. The
stress in each component will then be proportional to its modulus. The
implicit assumptions of this equation are that the strain in the fiber,
matrix, and composites are the same and that there is sufficient
fiber-matrix bonding for load transfer.

   In SiC/RBSN composites, the matrix is the weaker phase and cracks

before the fibers. Therefore at matrix failure $\varepsilon_c = \varepsilon_m^f$ where

ε_m^f is the matrix fracture strain. Substituting ε_m^f for ε_c in Eq. [1] and
rearranging, the matrix fracture strain can be calculated.

$$\varepsilon_m^f = \frac{\sigma_c^f}{E_f V_f + E_m V_m} \tag{2}$$

where σ_c^f represents the composite stress at which the matrix first
cracks. The matrix modulus E_m varies, however, with the volume
fraction of porosity, but can be estimated from the equation

$$E_m = 300 \ \exp \ (-3P) \ GPa \tag{3}$$

where P is volume fraction of porosity [10]. For $P \sim 0.39$ typically
measured for the RBSN matrix (cf. Table III), the estimated value of E_m

was 93 GPa. For the calculation of ε_m^f the values of $E_f = 390$ GPa

[11], $E_m = 93$ GPA, and the appropriate values of σ_c^f corresponding to

683

two different volume fractions were used (cf. Table V). The calculated matrix fracture strain values for the 23 and 40 volume percent SiC/RBSN composites were 0.15 percent and 0.14 percent, respectively. These matrix strain values are comparable with the value of 0.11 percent obtained for unreinforced RBSN. This indicates that reinforcement of RBSN by large diameter SiC fibers did not significantly affect matrix failure strain. This is in contrast to the microcrack bridging effect [1] seen in small diameter SiC fiber or graphite fiber reinforced glass composites [3]. The implication of this finding is that the CVD SiC fibers have a diameter approximately equal in size or larger than the size of critical strength-controlling flaws in the RBSN.

Although the microcrack bridging mechanism is probably not operative with the large diameter SiC fibers, from Eq. 2 there are other methods available for increasing σ_c^f-, the composite stress at which the first matrix crack appears. One method is by decreasing matrix porosity which will not only increase E_m (cf. Eq. 3), but also increase ε_c^f [10]. Another method is by increasing the volume fraction of fibers.

Assuming RBSN matrix with E_m = 206 GPa and ε_m^f = 0.15 percent (i.e., porosity 20 percent), the composite stress at which matrix cracks first occur was calculated for 50 fiber percent SiC/RBSN to be σ_c^f = 447 MPa which is significantly higher than 300 MPa, the typical strength of RBSN with 20 percent porosity. This analysis shows that, even without microcrack bridging , reinforcement of RBSN by high modulus, high strength, large diameter SiC fibers can yield a stronger material whose properties can be tailored to suit a variety of design strength requirements.

After the first macrocrack in the matrix is formed, the load carried by the matrix will be transferred to the fiber and Eq. 1 is no longer valid. As the composite is loaded further, more matrix cracks appear and this process continues until the matrix effectively no longer bears any load. In this stage, the matrix is severly cracked while fibers are still intact. Thus the fibers impart the composite with higher toughness as manifested by noncatastrophic failure upon matrix cracking and a higher total strain capability. In contrast, in unreinforced RBSN, once a crack initiates, the material fails catastrophically at low strain. In some cases, however, multiple cracking of the matrix may not be desirable from a practical point of view because it may result in exposure of the fibers to the environment which at high temperature may promote severe loss in fiber strength. Therefore after initiation of the first matrix crack, usefulness of the composite may be lost.

With continued stressing, the fiber bundle will elongate until its failure stress is reached. Because SiC fiber is elastic up to failure, a second linear region is observed in the stress-strain curve. In fact, the ultimate tensile strength of the composite is a measure of the bundle strength of the fibers. When the matrix is completely cracked and carries no load Eq. 1 reduces to

$$\sigma_c^u = \sigma_f^u V_f \tag{4}$$

where σ_c^u is the ultimate composite stress and σ_f^u represents the bundle strength of fibers. Using the UTS values reported in Table IV, the bundle strength calculated from Eq.. 4 was 1.43 GPa. This value is less than the value 2.0 GPa, measured for the room temperature bundle strength of untreated, as-received fibers at a gauge length of 50 mm. It is also less than the 2.41 GPa value measured for the average room temperature tensile strength of SiC fibers heat treated in the nitriding furnace under similar processing conditions as that of the composites. These results suggest fiber strength degradation may have occurred during processing of the composite.

Interfacial Shear Strength

The shear strength between the SiC fiber and the RBSN matrix can be estimated from the equation

$$\tau = \frac{D \ \sigma_c^f}{2.98 x V_f \ \left[1 + \dfrac{E_f V_f}{E_m E_m}\right]} \tag{5}$$

where σ_c^f is the composite stress corresponding to the first matrix crack, x is the mean separation between the matrix cracks, D is the fiber diameter, and the other terms have their usual meaning [1, 12].

Using the values of x and σ_c^f from Table V, the calculated mean interfacial strength was 10 MPa. This low value of shear strength is consistent with interfacial splitting observed on the fracture surface of the transverse flexural test specimen. A possible mechanism for low is the radial shrinkage of the fiber away from the matrix which can occur on cooling from the processing temperature due to a difference in fiber-matrix thermal expansion.

SUMMARY OF RESULTS

A fabrication method for reinforcing a powder-derived ceramic matrix with ceramic fibers have been developed. Using this method, 23 and 40 percent SiC fiber-reinforced RBSN composites have been fabricated. Initial room temperature properties for this composite have been evaluated. The important findings are as follows:

1. Room temperature strength measurements show that the ultimate axial tensile and flexural strengths increased with volume fraction of fibers and were significantly higher than unreinforced RBSN matrix of comparable porosity.

2. The composite matrix crazed before final fracture. The composite stress at which matrix first cracks increased with increasing volume fraction of fiber and was significantly greater than the fracture strength of unreinforced RBSN of the same porosity. This effect is primarily due to the higher modulus of CVD SiC fiber.

3. Due to their large diameter, the CVD SiC fibers did not measurably affect the matrix failure strain; however, due to their high strength and strength retention during processing, the CVD fiber allowed the composite to display extensive matrix cracking and high ultimate fracture strain, an indication of a tough material.

4. The composite displayed weak fiber-matrix bonding which resulted in low interfacial shear and low transverse flexural strength. Microstructural examination of the cross section of the composite indicated little evidence of the reaction between the fiber and matrix.

5. Physical property measurements indicate that the composites have density values of 2.19 – 2.36 gm/cc and contain 40 percent matrix porosity.

REFERENCES

1. J. Aveston, G. A. Cooper, and A. Kelly., "The Properties of Fiber Composites" Conference Proceedings, National Physical Laboratory (IPC Science and Technology Press Ltd, 1971) Paper I, p. 15.

2. J. Aveston and A. Kelly, "Theory of Multiple Fracture of Fibrous Composites," (1973), J. Mat. Sci., (8), p. 352.

3. J. J. Brennan and K. M. Prewo, "Silicon Carbide Fiber Reinforced Glass–Ceramic Matrix, Composites Exhibiting High Strength and Toughness," (1982), J. Mat. Sci., (8), p. 2371.

4. F. W. Wawner, A. Y. Feng, and S. R. Nutt, "Microstructural Characterization of SiC (SCS) Filaments," (1983), SAMPE Q., (4), p. 39.

5. R. A. Signorelli, "Metal Matrix Composites for Aircraft Propulsion Systems," (1976), Proceedings of the 1975 International Conference on Composite Materials, edited by E. Scala, E. Anderson, I. Toth, and B. R. Norton, The Metallurgical Society AIME, New York, (1) p. 411.

6. R. T. Bhatt, "Fabrication of Continuous Fiber Reinforced Ceramic Composites," to be published.

7. T. P. Herbell, T. K. Glasgow, and N. J. Shaw, "Reaction Bonded Silicon Nitride Prepared from Wet Attrition Milled Silicon," (1980), NASA TM-81428.

8. J. A. Mangels, "Strength–Density–Nitriding Cycle Relationships for Reaction-Sintered Si_3N_4 in Nitrogen Ceramics," (1977), p.569, edited by F. L. Riley, Noordhoff, Leyden, Netherlands.

9. C. P. Gazzara and D. R. Messler, "Determination of Phase Content of Si_3N_4 by X-Ray Diffraction Analysis," (1977), Am. Ceram. Soc. Bull. 56 (9) p. 777.

10. A. J. Moulson, "Reaction-Bonded Silicon Nitride, Its Formation and Properties," (1979), J. Mat. Sci., (14) p. 1017.

11. J. A. DiCarlo and W. Williams, "Dynamic Modulus and Damping of Boron, Silicon Carbide, and Alumina Fibers," (1980), NASA TM-81422.

12. A. C. Kimber and J. G. Keer, "On the Theoretical Average Crack Spacing in Brittle Matrix Composites Containing Continuous Aligned Fibers," (1982), J. Mat. Sci. letters (1) p. 353.

SURFACE ENERGY AS AN INDICATOR OF INTERFACIAL MECHANICAL RESPONSE

H. T. Godard and K. T. Faber

Department of Ceramic Engineering
The Ohio State University
Columbus, Ohio 43210

ABSTRACT

Contact angles between silicon and a variety of silicon carbide-based fibers have been measured and correlated to the mechanical behavior of the fiber-matrix interface. Infiltration by the molten silicon into the fibers and/or reaction with excess carbon to form silicon carbide gives rise to low contact angles and high toughness interfaces. It is suggested that the high toughness of the interface is morphology-controlled.

INTRODUCTION

The performance of ceramic-matrix composites is partly determined by the individual properties of the fiber and matrix phases, but largely by their interaction. Interactions may arise from thermal expansion or elastic mismatch, giving rise to residual stresses, or from chemical reaction between fiber and matrix. Chemical bonding is necessary for load transfer from the matrix to the fibers, resulting in enhanced strength in the fiber direction and strength retention in the transverse direction. Weaker bonding is required for crack deflection along the interface and subsequent fiber pull-out resulting in enhanced fracture toughness. Often the interface is characterized as "weak" or "strong" by inspection of the fracture surface or the stress-strain response of the composite. If the "brush-type" fracture surface characteristic of fiber pull-out is observed, the fiber-matrix interface bonding is said to be "weak". It would be useful to first, quantify the mechanical behavior of the interface and second, be able to predict which systems would provide toughening based on interfacial energies.

Attempts to relate the interfacial energy to the work of adhesion for ceramic-metal interfaces have recently been reviewed by Cannon.[1] Of significance is work by Klomp who compared the work of adhesion to the interfacial energy between Al_2O_3 and a number of FCC metal substrates.[2] In these materials, the interfacial energy followed the same trend as the work of adhesion, but the systems studied were restricted to those having contact angles greater than $90°$.

The present work describes an investigation of interfacial energies via contact angle measurements as related to the strength and toughness of fiber/matrix interfaces between silicon and silicon carbide fibers.

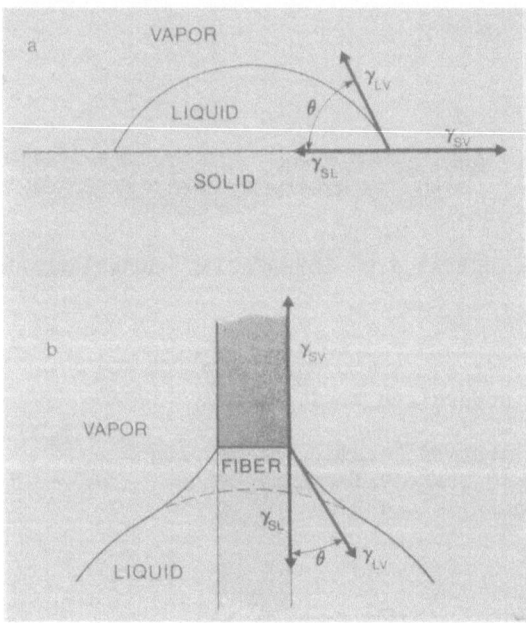

Fig. 1. Two configurations for measurement of surface energies. (a) Sessile
   drop on a flat substrate; (b) Sessile drop on a cylindrical
   substrate.

Silicon/silicon carbide was chosen as a model system for the high-performance
silicon-based ceramic matrix composite systems so avidly studied today.   The
Si/SiC system has been well characterized in previous studies and provides a
sound background for this study.   Modifications to widely used interfacial
energy measurement techniques for single fiber/matrix configurations are
described herein along with the methods of mechanical characterization of the
fiber/matrix interface.   A correlation between the interfacial fracture
toughness and the interfacial energy is attempted for use in composite
design.

Thermodynamic Treatment

   The determination of solid-solid interfacial energies is very difficult.
It is somewhat easier to determine the liquid-solid interfacial energy.   One
method for assessing the latter is by means of a sessile drop whereby the
contact angle between a liquid drop and solid substrate is measured.[1,2]
(Figure 1(a)) The Young equation describes the relationship between the
various interfacial energies:

$$\cos \theta \; = \{ \gamma_{sv} - \gamma_{sl} \} / \; \gamma_{lv} \tag{1}$$

where γ is the surface energy, the subscripts, s,v and l refer to the solid,
vapor, and liquid species, respectively, and θ is the contact angle at the
edge of the drop, measured from the liquid-solid interfacial plane.   The
solid-vapor interfacial energy, γ_{sv} can be determined from "zero-creep"

688

measurements and γ_{lv} from the drop shape.[3] The contact angle can be measured in situ or from photographs of the drop assuming the substrate is flat and no reactions occur. From the Young equation, the surface energy of the liquid-solid interface can be determined. This technique has been used to determine surface energies in numerous ceramic-metal systems.[4-8] It is not without drawbacks, however. Large drops require corrections for the effect of gravity, particularly if the contact angles are obtuse.[9] The surface energies of the solid and liquid are greatly affected by the atmosphere, making the use of literature values somewhat problematic.

Often, a system is described by the contact angle alone, where   is used to describe the apparent energetics of the interfaces. A liquid-solid interaction is then described by the terms, non-wetting, wetting or spreading for $\theta > 90^\circ$, $< 90^\circ$ or $= 0^\circ$, respectively.

Although these techniques were designed for hemispherical drops, the method is extended here to a single fiber/matrix system for conditions where the "matrix" acts as the sessile drop. Shown in Figure 1(b), a vertical fiber surrounded by a liquid droplet can be used for contact angle measurements.

Mechanical Response

Quantitative information about the interfacial mechanical properties in a fibrous composite can be obtained by use of any number of techniques, the choice of which depends on the the strength of the interface. For those systems in which no bonding occurs and the interfacial shear stress is purely frictional, a hardness indentation technique recently proposed by Marshall is suitable.[10] A Vickers indentor acts as a point load source, along the fiber axis, depressing the fiber relative to the matrix. (Figure 2(a)) If the fiber protrudes from the bottom of the matrix, the interfacial shear stress, τ, can be calculated from:

$$\tau = F/[2\pi Rt] \tag{2a}$$

where F is the load applied, R is the fiber radius, and t is the sample thickness. Alternatively, the technique can be extended to long fibers which only slip over a short proportion of their length. In this case the amount of displacement is determined from the indentation markings in the matrix and the geometry of the indentor, and,

$$\tau = F^2/4\pi^2 (b - a) \cot\psi R^3 E_f \tag{2b}$$

where a, b and ψ are defined in Figure 2(a) and E_f is the elastic modulus of the fiber.

This is a powerful technique for studying the fiber-matrix interface, particularly in a single fiber system, but is only useful for small interfacial stresses. For example, to produce a minimum displacement sufficient to mark the matrix (b = R in Eqn. 2b), SiC fibers loaded below the threshold for crack initiation (1.5 N) require the interfacial shear stress to have a maximum value of 1 MPa for this technique to be applicable.

For those systems in which chemical bonding is significant, the fiber cannot be depressed and cracks may be propagated through fiber, interface and matrix alike. (Figure 3) In these instances, methods which have been used to determine the work of adhesion are more appropriate. Hardness indentations may again be used (Figure 2(b)); a radial crack is propagated parallel to the interface. [11] The relative toughness of the interface may be deduced by comparing the crack lengths in the matrix, the fiber and the interface where the fracture toughness, K_c, is related to indentation crack length by the

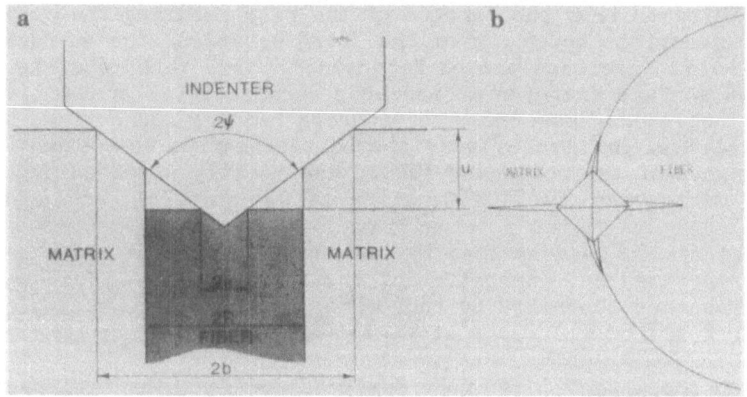

Fig. 2. Schematic of hardness indentation techniques for measuring (a) fiber/matrix frictional stress, and (b) interfacial fracture toughness.

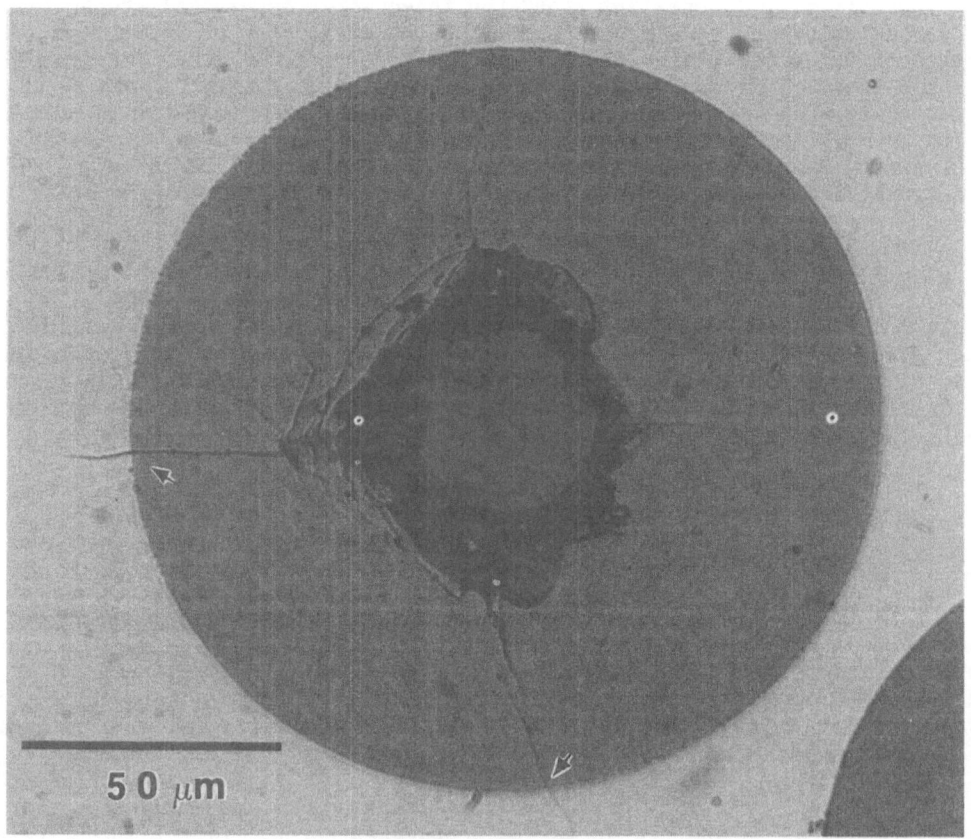

Fig. 3. A Vickers indentation (29.4 N) placed in a SiC fiber. Note, by arrows, that the cracks propagate through interface and into silicon matrix.

690

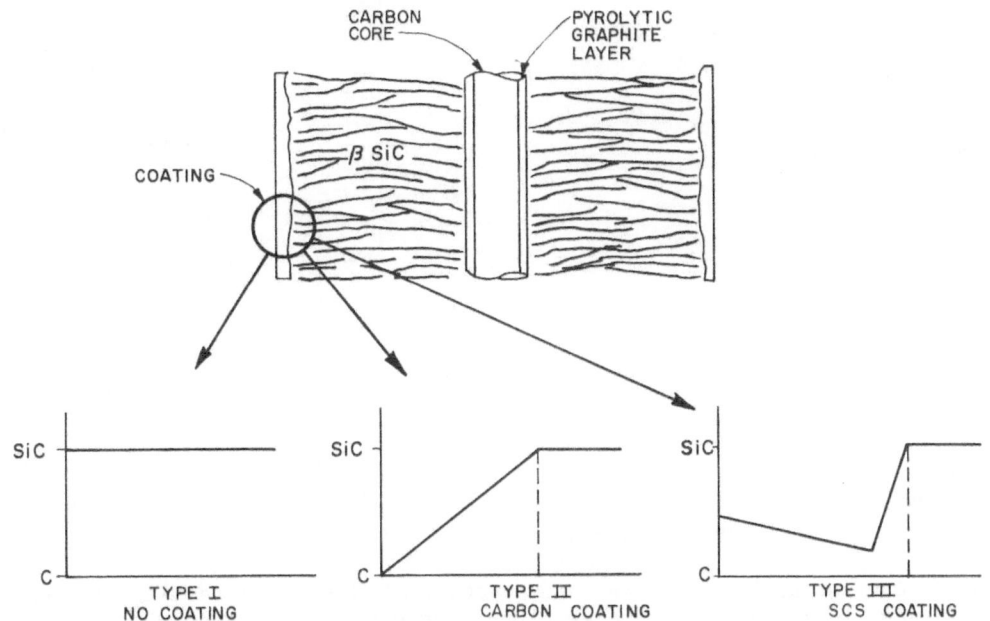

Fig. 4. Cross-sectional schematic of SiC fiber with graphs of relative SiC/C ratios versus distance from external surface for three surface treatments: (a) stoichiometric SiC (b) carbon (c) graded SiC.

following relation: [12]

$$K_c = 0.016 \ (E/H)^{1/2} \ P/c^{3/2} \tag{3}$$

where H is the hardness, c is the radial crack length and P is the peak indentation load.

EXPERIMENTAL PROCEDURE

Silicon carbide filaments[a] with either a stoichiometric SiC surface, a carbon surface, or a silicon-poor, graded, surface were used for study. (Figure 4) Hot-pressed α-SiC plates[b] were ultrasonically polished and machined to hold the SiC filament vertical. A retaining ring of hexagonal BN was placed around the fiber and three or four millimeter-sized pieces of high purity silicon.[c] Several short lengths of filament were placed along with a piece of silicon on a small BN plate for mechanical property studies. These were positioned on an Al_2O_3 plate in a tube furnace equipped with a viewport. The furnace was operated under gettered pre-purified H_2 at 1417-1418°C. Oxygen partial pressure was measured by a ZrO_2-5 w/o CaO cell in a region of the furnace at 700 ± 10°C to remain in the ionic conductivity regime for this electrolyte.[13] Assuming a H_2/H_2O gas buffer system and a quasi-steady state condition the room temperature mixing ratio was calculated and used to derive the partial pressure at the sample temperature.[14] The O_2 partial pressure was varied from 10^{-17} to 10^{-19} atm. Contact angles were taken from enlarged photographs of the samples at temperature.

[a] AVCO, Specialty Materials Division, Lowell, MA.
[b] NC-203, Norton Company, Worcester, MA.
[c] Cerac Inc., Milwaukee, WI.

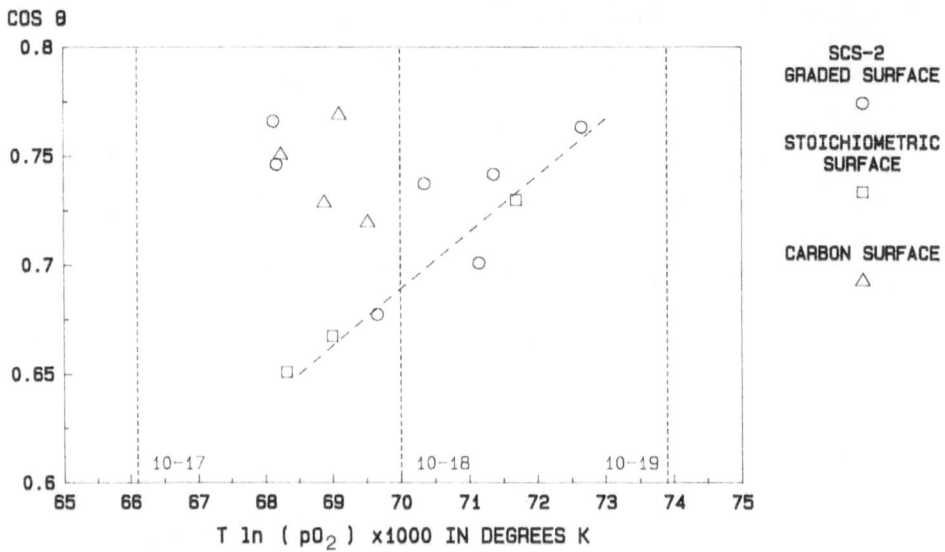

Fig. 5. Contact angle, θ, versus the product of the oxygen partial pressure and temperature for three fiber surfaces.

Upon cooling, the samples were diamond polished perpendicular to the fibers for mechanical property measurements. Polished drops were mounted for indentation measurements to determine the interfacial shear stress and interfacial toughness.

RESULTS AND DISCUSSION

Surface Energy

Data derived from contact angle and oxygen partial pressure measurements are plotted in Figure 5. Contact angles were measured with a precision of 3 degrees and ranged from 39° to 50°. The contact angles measured were advancing angles as the meniscus was observed to "wick" up the fibers at the melting temperature and "equilibrate". In a purified argon atmosphere, the water content, derived from the measured oxygen partial pressure and the H_2/H_2O equilibrium, was found to be constant at 3 ppm over the temperature range of the oxygen sensor. The oxygen partial pressures presented in Figure 5 are, thus, considered to be an upper limit.

The data in Figure 5 fall into two groups. The graded-SiC surface and the stoichiometric-SiC surface generally demonstrate an increase in contact angle with increasing oxygen partial pressure. Fibers having a carbon surface, over the limited range of oxygen partial pressures studied, show no relationship between contact angle and oxygen partial pressure, but consistently demonstrate lower contact angles than their stoichiometric SiC counterparts. As the measured sample temperatures varied over 6°, cos θ is plotted versus $-Tln(pO_2)$.

The measured contact angles agree within experimental error with those of previous investigations on flat SiC substrates. [15,16] The decrease in contact angle with increasing oxygen partial pressure is consistent with results obtained by Kingery and Humenik, who attributed this to a change in γ_{1v} with oxygen partial pressure.[6] It is possible to explain changes in measured contact angles in this system by γ_{1v} alone, suggesting γ_{sv} and γ_{s1}

are relatively invariant with pO_2. Variations in γ_{1v} of 18% can account for a change in θ of 10° in the range of contact angles measured.

Microscopy of the interfaces revealed an interpenetrating microstructure indicative of a diffusion-controlled process. (Fig. 6 and 7) The thickness of the interface for the graded and stoichiometric fibers could not be correlated to the oxygen partial pressure or to the contact angle. Both groups, however, had thinner interfaces than the carbon surface fibers. The width of the interfacial region was significantly larger than the coating thickness for both the graded and carbon surface fibers.

The micrographs of the interface microstructure show that the assumptions for interfacial energy measurements (a smooth surface and no interfacial reaction) are violated. It is well established that carbon reacts to form SiC in the presence of molten silicon.[17] Likewise, coarsening of SiC in the presence of silicon is also well documented, rate-limited by carbon transport at the Si-SiC interface.[18] The observed microstructure is consistent with these diffusion-controlled processes and attack along the grain boundaries of the CVD SiC, which are normal to the fiber axis.

Correlation with Mechanical Properties

In each of the systems studied, regardless of fiber coating or atmosphere, the interface that resulted was too strong to measure interfacial stresses by depressing the fiber. (Figure 3) Nucleation and growth of radial cracks from the fiber through the interface and into the matrix always occurred before any displacement of the fiber. Propagation of radial cracks along the interface to determine the interfacial fracture toughness, however, was possible using loads of 0.5 to 1.5 N. In most instances, depending on the precise placement of the indentation, the radial cracks propagated through the interface into the region of least fracture toughness, the low modulus silicon matrix. (Figure 8(a)) Indentations along the interface indicate that interactions with the "finger-like" microstructure of the interface inhibits crack propagation. Results are listed in Table 1 and demonstrated in Figure 8(b). The results suggest that the lower contact angles do indeed correlate with higher toughness interfaces.

The interpenetrating microstructure and the dissolution of the fiber into the silicon produce a strong interface capable of stopping or deflecting cracks, as evidenced by the inability to produce fiber slippage upon indentation and by the facility with which cracks cross the interface radially. The difficulty of crack propagation along the interface cannot unequivocally be attributed to the magnitude of interfacial energy alone. Since the geometry of the reaction interface is complex, crack deflection processes in this reaction zone may significantly increase the toughness and reduce the indentation crack lengths.[19] Crack deflection may increase the toughness by as much as a factor of two, and may be the main mechanism for the apparent toughness of the interface.

CONCLUSIONS

Silicon contact angles have been measured on SiC fibers and found to range from 39 to 50°, in agreement with literature values for SiC plates.

A strong interpenetrating interface is formed between the fiber and matrix with mechanical properties similar to the matrix. Fiber pull-out is not expected in this system due to the large resistance to crack propagation along the interface. Although trends in contact angle and fracture toughness have been noted, toughening mechanisms along the interface may also be operative.

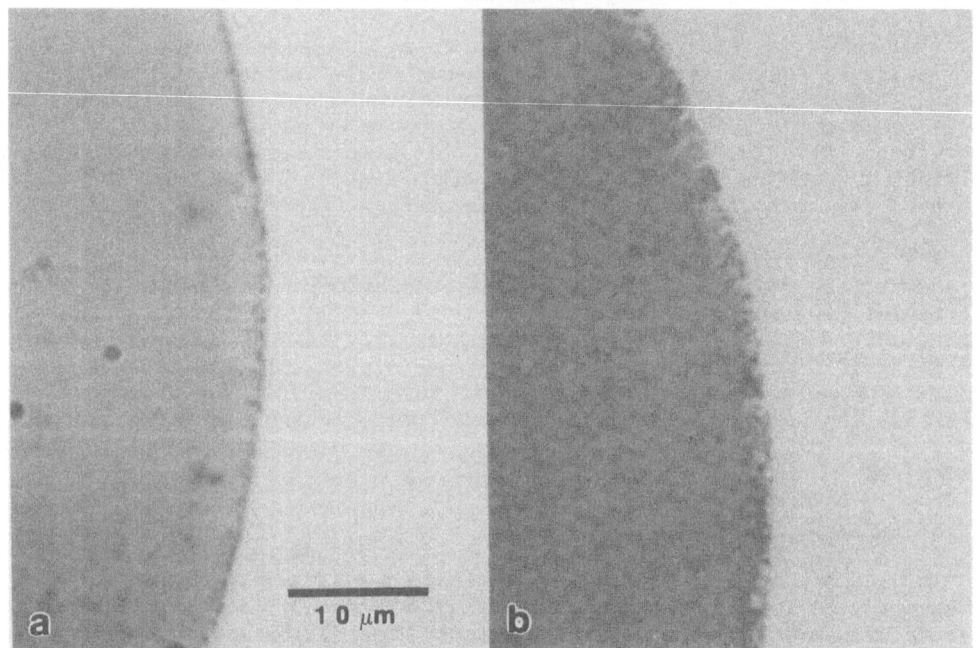

Fig. 6. Interface of stoichiometric SiC/Si in (a) high pO$_2$ atmosphere, and (b) low pO$_2$ atmosphere.

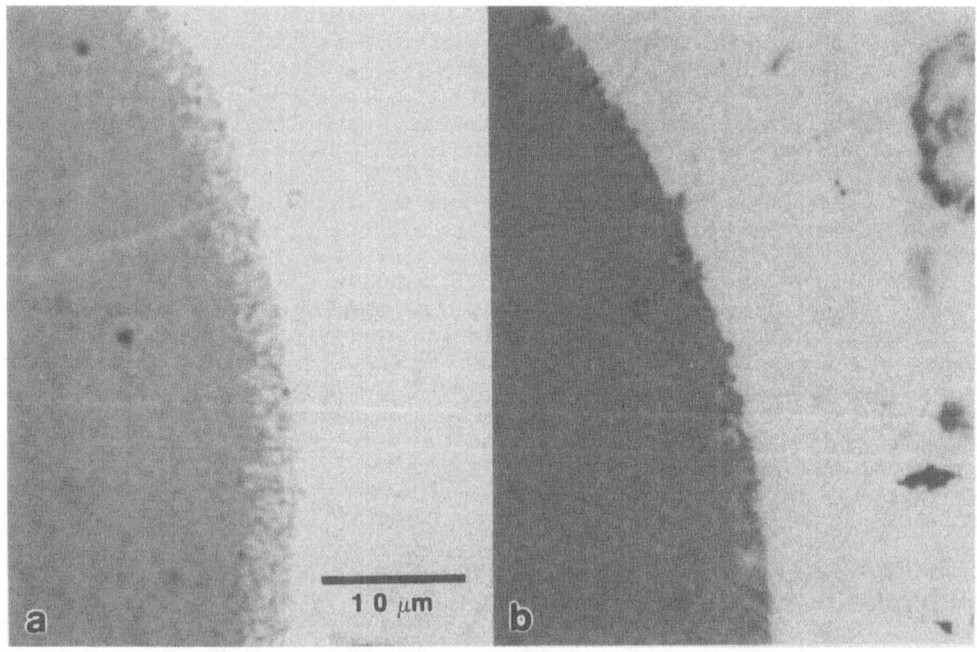

Fig. 7. SiC/Si interface with fiber having (a) carbon surface, and (b) graded SiC surface.

694

Table 1. Fracture Toughnesses of the Si/SiC System

	SiC Surface $\theta = 49°$	SiC Surface $\theta = 40°$	C Surface $\theta = 40°$
K_c (matrix)	0.7 MPa√m	0.7 MPa√m	0.7 MPa√m
K_c (fiber)	4.0	4.0	4.0
K_c (interface)	2.8	$\geq$4.0	3.5 - 4.0

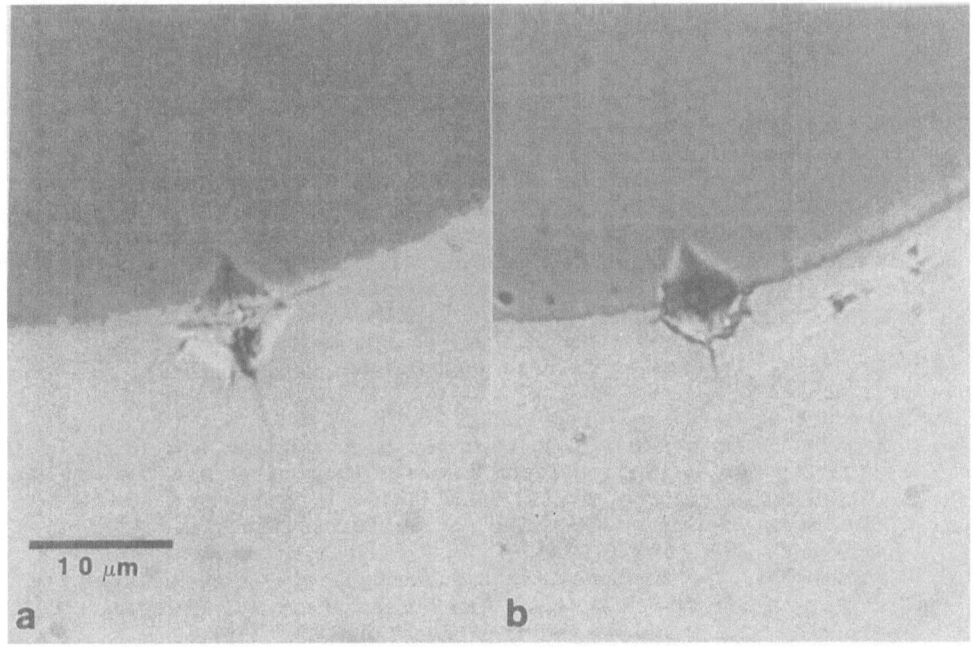

Fig. 8. Indentations produced with 0.5 N load at a graded SiC/Si interface where (a) crack propagates into silicon, and (b) where crack propagates along interface.

695

Extension of this work to lower oxygen partial pressures, surface coatings to produce interfaces which accommodate fiber pull-out, and chemical characterization is anticipated.

ACKNOWLEDGMENTS

This research was supported by the General Electric Company, Evendale, Ohio. The authors thank D. W. Readey for critically reviewing the manuscript and S. Guruswamy, M. Shaw and C. Brown Jr. for helpful discussions.

REFERENCES

1. R. M. Cannon, Factors Controlling Interfacial Bond Strengths of Ceramic-Metal Interfaces, 87th Meeting of the American Ceramic Society, Cincinnati, OH, 1985.
2. J. T. Klomp, Sci. Ceram. 5:501 (1970).
3. L. E. Murr, Interfacial Energetics in Metal-Metal, Metal-Ceramic, Metal-Semiconductor, and Related Solid-Solid and Liquid-Solid Systems in: "Surfaces and Interfaces in Ceramic-Metal Systems" J. A. Pask and A. G. Evans, eds. Plenum Press, New York (1981).
4. M. Humenik, Jr. and W. D. Kingery, Metal-Ceramic Interactions: III, J. Amer. Ceram. Soc. 37 [1] 18 (1954).
5. N. Eustathopoulos, Energetics of Solid/Liquid Interfaces of Metals and Alloys, Int. Met. Rev. 28 [4] 189 (1983).
6. W. D. Kingery and M. Humenik, Jr., Surface Tensions at Elevated Temperatures, I, J. Phys. Chem. 57:359 (1953).
7. W. D. Kingery, Metal-Ceramic Interactions: IV, J. Amer. Ceram. Soc. 37 [2] 42 (1954).
8. B. C. Allen and W. D. Kingery, Surface Tension and Contact Angles in Some Liquid Metal-Solid Ceramic Systems at Elevated Temperatures, Trans. Met. Soc. AIME 215 [2] 30 (1959).
9. G. L. Mack, The Determination of Contact Angles from Measurements of Dimensions of Small Bubbles and Drops I: The Spheroidal Segment Method for Acute Angles and II: The Sessile Drop Method for Obtuse Angles, J. Phys. Chem. 40 [2] 159 (1936).
10. D. B. Marshall, An Indentation Method for Measuring Matrix-Fiber Frictional Stresses in Ceramic Composites, J. Amer. Ceram. Soc. 67 [12] C259 (1984).
11. P. Dokko and A. G. Evans, unpublished work.
12. G. R. Anstis, P. Chantikul, B. R. Lawn and D. B. Marshall, A Critical Evaluation of Indentation Techniques for Measuring Fracture Toughness: I, J. Amer. Ceram. Soc. 64 [9] 533 (1981).
13. J. W. Patterson, Conduction Domains for Solid Electrolytes, J. Electrochem. Soc. 3 [7] 1033 (1971).
14. P. Deines et al., Temperature-Oxygen Fugacity Tables for Selected Gas Mixtures in the System C-H-O at One Atmosphere Total Pressure, Bull Earth and Min. Exp. Sta. Penn. State Univ. 88: (1974).
15. T. J. Whalen and A. T. Anderson, Wetting of SiC, Si_3N_4, and Carbon by Si and Binary Si Alloys, J. Amer. Ceram. Soc. 58 [9-10] 396 (1975).
16. Y. V. Naidich and G. M. Nevodnik, Wettability of the Surface of SiC Single Crystals by Molten Metals, Izv. Akad. Nauk SSSR Neorg. Mat 5 [12] 2066 (1969).
17. W. P. Minnear, The Reaction of Carbon with Liquid Silicon, 14th Biennial Conference on Carbon, Pennsylvania State University, 1979.
18. W. P. Minnear, Interfacial Energies in the Si/SiC System and the Si+C Reaction, J. Amer. Ceram. Soc. 65 [1] C10 (1982).
19. K. T. Faber and A. G. Evans, Crack Deflection Processes - I, Acta Metall. 31 [4] 565 (1983).

PROSPECTS FOR ULTRA-HIGH-TEMPERATURE CERAMIC COMPOSITES

William B. Hillig

General Electric Corporate Research and Development
PO Box 8, Schenectady, NY 12301

INTRODUCTION

In the search to improve the efficiency and performance of existing devices and systems, there is a constant demand for structural materials that are stronger, more reliable, and have higher temperature capability. Other attributes are also often sought such as low cost and a lesser dependence on scarce elements. This paper is concerned with the question of how far can this search continue before nature finally intervenes.

In the last decade there has been a major international competitive effort to develop a high-performance ceramic turbine engine for automotive, truck, and power gas turbine use. The relative sensitivity, compared with metals, to flaws and accidental damage of the preferred silicon carbide and silicon nitride materials has been a continuing challenge. This sensitivity manifests itself in low fracture toughness values and low Weibull moduli for tensile strength, again relative to metals. However, transformation-toughened ceramics based on zirconia dispersions can be made that exhibit remarkable toughness values over a temperature range of several hundreds of degrees. Furthermore, the recent availability of high-temperature fibers such as the polymer-derived silico-carbon continuous filaments, or the rice-hull-derived silicon carbide whiskers have rekindled efforts to achieve "tough," i.e., damage-tolerant, ceramic materials through the use of fiber reinforcement. Results to date have been encouraging, and they allow speculation that such ceramics will become the ultimate high-temperature structural materials, particularly in heat engines. This paper examines, in part, the factors that are likely to determine the useful upper temperature bounds for composite, as well as monolithic ceramic materials. The factors are then considered in total to estimate the rankings and the maximum use temperatures of the ten highest melting oxide and the ten highest melting nonoxide compounds.

GENERAL CONSIDERATIONS

For most applications, materials must meet several requirements simultaneously, such as melting temperature, minimum strength, and environmental resistance. Accordingly, no single property can rank the adequacy of a material for some intended purpose. For high-temperature heat engines, the first requirement is that the material not melt, vaporize, dissociate, or chemically degrade in its normal environment. In addition there are generally mechanical requirements, such as strength, stiffness, and creep resistance. Engine weight and inertial force considerations often put a premium on low density. These properties will be considered for the various high-temperature substances.

Because of the large number of materials involved, the approach is to broadly treat classes of materials and properties. Unfortunately little high quality thermochemical data are available for most of the specific materials, and the theoretical basis for extrapolating mechanical properties to very high temperatures has received little study. No attempt has been made to examine

697

critically the data cited in this paper for any specific material. Such data has been derived mainly from various general sources.[1-5]

Other important issues, not covered here, are the practical considerations of material availability, material scarcity, cost, and fabricability. Although properties such as thermal shock resistance are undoubtably important, the necessary data are lacking. Flaw or damage tolerance is recognized as being an extremely important parameter, which not only depends on the intrinsic physical properties of the material, but also on the microstructure. However, it is debatable whether damage tolerance can be adequately measured by K_c as is often supposed. Furthermore, in the case of composites, the properties of the separate constituents, as well as the details of the heterogeneous structure, and the interface properties, affect the flaw sensitivity. Hence, neither flaw sensitivity nor fracture toughness is treated here as a parameter quantifiable on an a priority basis.

The three laws[6] of high-temperature chemistry are

1. Everything reacts with everything at high temperatures.

2. Everything reacts faster at high temperatures.

3. The products may be anything.

On this basis, pure monolithic materials may be advantageous relative to composites, although reaction with the environment is still possible. Ease of fabrication is probably another advantage. However, the potential for enhanced "toughness," plus the potential for tailoring, controlling and stabilizing the microstructure, as well as the benefit in inhibiting creep, suggest that ceramic composites will emerge as the preferred structural high-temperature materials. In addition to requiring that the constituents be compatible (in contradiction to the above laws), the other firm restraint is that the reinforcement phase be elastically stiffer than the matrix phase in which it is immersed. Finally, because only oxides (except for a few precious metals) are stable against environmental oxidation, only the refractory oxides will be considered as matrices or as monolithic ceramics.

MELTING TEMPERATURES

The classes of materials considered, ranked in decreasing order of the melting point of the most refractory representative of each class are carbides, nitrides, metals (including intermetallics), oxides (including mixed oxides), borides, silicides, and sulfides, selenides, and tellurides. Table 1 gives the distribution of the melting temperatures for each class and summarizes the number of substances remaining solid when the temperature is raised to that listed.

Table 1: Distribution of Melting or Dissociation
Temperatures for Various Classes of Substances

Temp. Range °C	C	N	Met	Ox	B	Si	S	Cum. Total	Nonoxide + Oxide Combinations
3751-4000	2	0	0	0	0	0	0	2	0
3501-3750	1	0	0	0	0	0	0	3	0
3251-3500	2	1	1	1	1	0	0	9	8
3001-3250	2	3	3	0	3	0	0	20	19
2751-3000	3	3	1	5	2	0	0	34	161
2501-2750	9	4	2	3	11	3	0	66	513
2251-2500	4	2	3	18	13	4	2	112	2295
2001-2250	7	2	9	20	15	15	20	200	7191
1750-2000	6	1	22	54	4	8	20	315	21614
Max.Temp.	3985	3385	3370	3320	3250	2510	2450	---	---
Material	TaC	HfN	W	ThO_2	HfB_2	Ta_5Si	CeS	---	---

Code: C=Carbides, N=Nitrides, Met=Metals+Intermetallics, Ox=Simple+Mixed Oxides, B=Borides, Si=Silicides, S=Sulfides+Selenides+Tellurides.

The right-hand column in the table indicates the number of oxide matrix composites mathematically possible in combination with nonoxide reinforcements. In a protective environment, all of the materials given in the column labeled Cumulative Total could be candidates as monolithic materials. In an oxidizing atmosphere, only the oxides and the precious metals would be possible, unless protective oxide coatings could be applied to the materials of the other classes. Although it is obvious that more composites can be made from pairs of consituents, than there are constituents, it seems noteworthy that more than 7000 hypothetical composite systems can even be identified that exist at least up to 2000 °C, for which the continuous phase is oxidation-resistant. Other considerations will significantly reduce the number of realistic possibilities.

STABILITY WITH RESPECT TO VOLATILIZATION

Gas evolution can result in substantial material removal or degradation. Such volatilization can be illustrated by such reactions as

(a) $MgO(s) ---> MgO(g)$ (sublimation)

(b) $SiO_2(s) ---> SiO(g) + (1/2)O_2$ (disproportionation)

(c) $Cr_2O_3(s) + (3/2)O_2 ---> 2CrO_3(g)$ (oxidation)

(d) $SiO_2(s) + SiC(s) ---> CO + SiO(g)$ (reduction)

(e) $B_2O_3(l) + H_2O(g) ---> 2HBO_2(g)$ (hydration)

Monolithic structures, as well as the matrix phase of a composite, are more vulnerable to volatilization loss than is the reinforcement phase of a composite that is totally encased by the matrix. Accordingly, the gas pressure generated by the reinforcement can be greater than that tolerated for the continuous phase, although internal gas pressures in excess of the ambient probably would result in long-term degradation.

Vapor pressure offers a potentially useful yardstick for estimating gas phase transport behavior. However, quantitative measures of vulnerability to volatilization cannot be given with much precision because the rate of material loss will depend sensitively, among other factors, on the velocity and on the composition of the surrounding gas phase— in the case of volatilization from the external surface. Interfacial and bulk diffusion will determine the rate of material loss in the case of a driving force for internal gas generation. Unfortunately, there is a paucity of data relating to the vapor pressure of refractory oxides. Such data are difficult to obtain. Furthermore, as indicated above, vaporization often involves other chemical reactions and is therefore dependent on the concentrations in the environment of the species involved in the reaction. Thus, the measured pressures cannot always be interpreted simply.

Figure 1 gives the vapor pressures of those high melting oxides contained in a compilation[7] of the measurements from several investigators. The figure also gives the dissociation pressure for several nitrides as calculated from thermodynamic tables.[8,9] A considerable linear extrapolation of the tabulated data for the standard free energy of HfN was required. The reader is advised to treat the absolute values presented with caution. The present purpose is merely to estimate structural use limits of high-temperature materials.

Common experience indicates that a vapor pressure, or dissociation pressure as great as 0.001 torr presents no long-term problem with respect to the degradation of a structural material. On the same basis, 1 torr is probably excessive. Therefore, some intermediate value, such as 0.01 torr, can be taken as as an upper bound for marginal utility. Table 2 uses that criterion to define the temperature limit and compares that limit with the melting point for the case of various refractory oxides.

The values for the volatilization limit are subject to the uncertainties and arbitrariness discussed above. However, if the data and arguments are at all credible, the matrix (and monolithic ceramic) candidates should be limited to temperatures below their melting points by an amount that depends on the desired useful life of the structure. Among the above oxides, ZrO_2 has the highest estimated effective use temperature. However, its high permeability with respect to oxygen transport, makes any nonoxide reinforcement materials subject to oxidative attack.

Table 2: Comparison of Volatilization Limit vs. Melting Temperature

Oxide	Melting Temperature	Volatilization Limit (Pressure = .01 Torr)
ThO_2	3220 °C	2100 °C
MgO	2825 °C	1550 °C
ZrO_2	2765 °C	2500 °C
BeO	2570 °C	2050 °C
SrO	2455 °C	1800 °C
La_2O_3	2265 °C	2000 °C

INTERNAL CHEMICAL REACTIVITY

Intrinsic thermal dissociation of a material produces solid, liquid, or gas products. Similar products can result from chemical reactions between the constituents of a composite system. Assuming the matrix to be an oxide, considerations of relative mechanical stiffness, and of the bond strength between the reinforcement and the matrix (in the case of fiber reinforced composites) make the nonoxide materials attractive candidates for the reinforcement phase. With the exception of the precious metals, these are all reducing agents. Thus, reactions between the (reducible) oxide matrix and the (oxidizable) reinforcement phase are to be expected.

The free energy of formation of many of the high melting nonoxide compounds reduces the free energy driving force for such redox reactions. The degree by which it is diminished is revealed by comparing the Gibbs free energy change in Kcal per gram atom of oxygen for the oxidation of the compounds with that for the oxidation of the constituent elements.

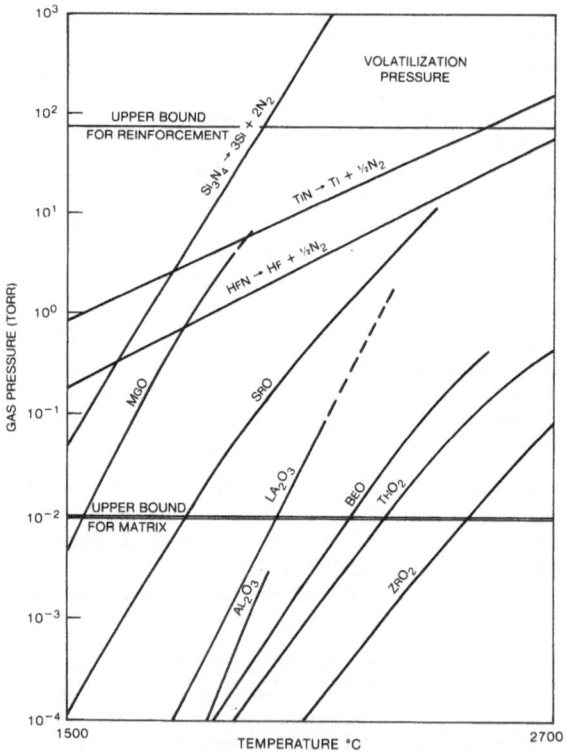

Figure 1. Gas pressure generated by vaporization or dissociation of various oxides and nitrides.

Table 3: Effect of Compound Formation on Oxidation Driving Force

Temp (K)	$\Delta G/[O]$											
	B	BN	N	Si_3N_4	Si	SiC	C*	Ti	TiB_2	TiN	TiSi	TiC
1500	72	63	--	63	73**	65	58	81	64	62	71	60
2000	64	58	--	60	62**	62	68	70	55	55	61	57
2500	56	53	--	--	64#	62	78	61	47	51	55	55

*) C oxidized to CO
**) Si oxidized to SiO_2
#) Si oxidized to SiO

As the table shows, compound formation between the nonmetallic constituents has little effect on the driving force. Furthermore, even the formation of a stable refractory compound by reaction with a metal such as Ti produces little benefit with respect of redox behavior. Thus, as a first approximation one can treat the nonoxide compounds as if they were simply mixtures of the constituent elements. This implies that for oxide matrices to be chemically stable when in intimate contact with the reinforcement phase, the oxide should preferably occur in only a single valence state and should not tolerate deviations from stoichiometry. Thus, for example, aluminum oxide offers advantage over titanium oxide.

Next we consider a general approach for estimating the stabilities of composites comprised of a matrix oxide and a nonoxide reinforcement, i.e., in predicting which way the following (unbalanced) reaction will go

$$MO_w + JX_a \quad = \quad JO_y + MX_b + XO_z \tag{1}$$

The symbols M and J refer to the metal species, and X represents N, B, C, Si, etc. Reaction (1) can be expressed as the combination of two half-reactions:

$$JX_a + O_2 \quad = \quad JO_y + XO_z \tag{2}$$

$$MX_b + O_2 \quad = \quad MO_w + XO_z \tag{3}$$

for which the free energy change is abbreviated $D(JX_a)$ for reaction (2) and $D(MX_b)$ for (3). If the compound XO_z is unstable, as is the case when X = nitrogen, these reactions must be modified accordingly. Plots of D, the driving force for the oxidation of a nonoxide compound of interest, can be made as a function of temperature, in analogy to the familiar plots shown in Figs. 2a and 2b of the free energy of formation of carbides and nitrides from the elements. Examples of the corresponding plots of D are given for the cases of the carbides and nitrides in Figs. 3a and 3b. Note that the rankings of stabilities in Figs. 3a,b are different from those in Figs. 2a,b. Reaction (1) can now be predicted to proceed from left to right if $D(MX_b)$ lies below $D(JX_a)$, otherwise the reverse reaction is favored.

Normalizing the half-reaction free energy with respect to oxygen allows the overall course of the reactions to be compared even when the compounds do not follow the usual valency rules. However, some modification may be required where the overall reaction indicates the net consumption of a gaseous species that may not be initially present in the reacting system. These D plots are also useful for predicting which of several possible end products would be favored, such as in a reaction of an oxide with SiC, in which case either a silicide or a carbide could result.

Finally, for purpose of illustration, we consider the stability in the temperature range 2000-3000 K of two possible composites comprised of the highest melting oxide ThO_2 in combination with the second highest melting nitride HfN. It is unusual that critically compiled free energy data are available[8] for essentially all of the relevant substances, viz., HfO_2, ThO_2, HfN, and ThN. The data for the oxides are complete over the entire solid range; however, the data on the other substances cover only a portion of the temperature range. Hence, extrapolation was required. The reaction of concern is

$$ThO_2 + HfN \quad = \quad HfO_2 + ThN$$

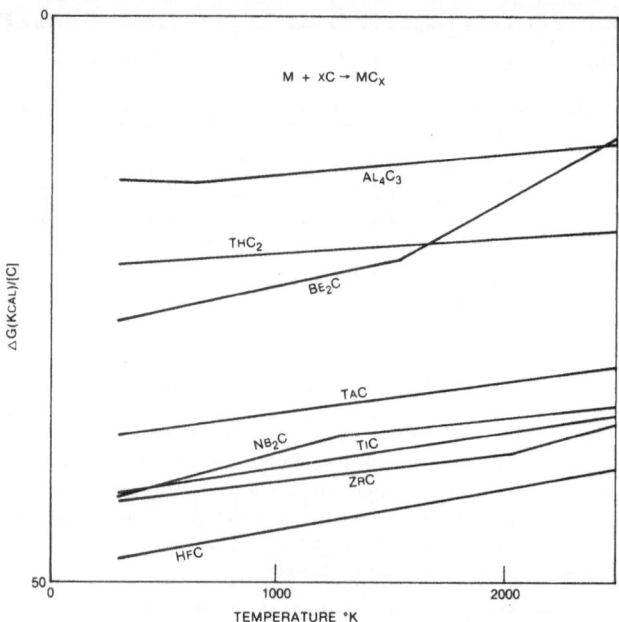

Figure 2a. Free energy of formation of carbides as a function of temperature.

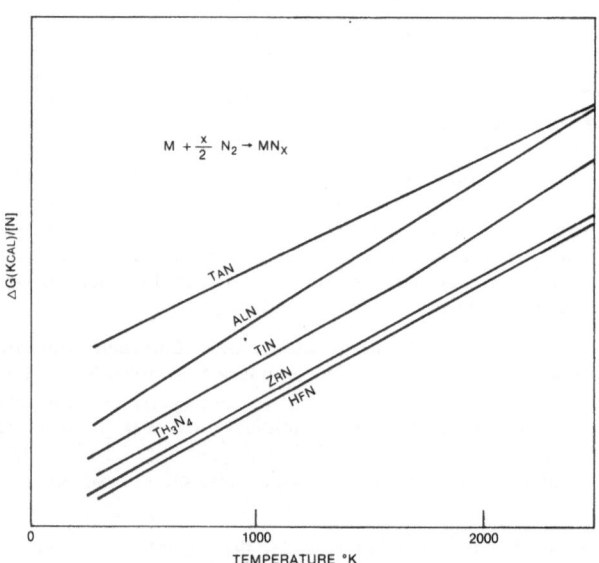

Figure 2b. Free energy of formation of nitrides as a function of temperature.

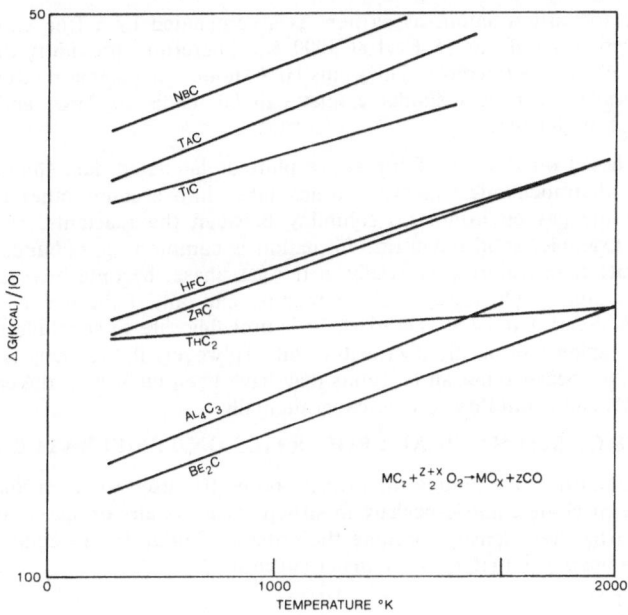

Figure 3a. Free energy of oxidation of carbides as a function of temperature.

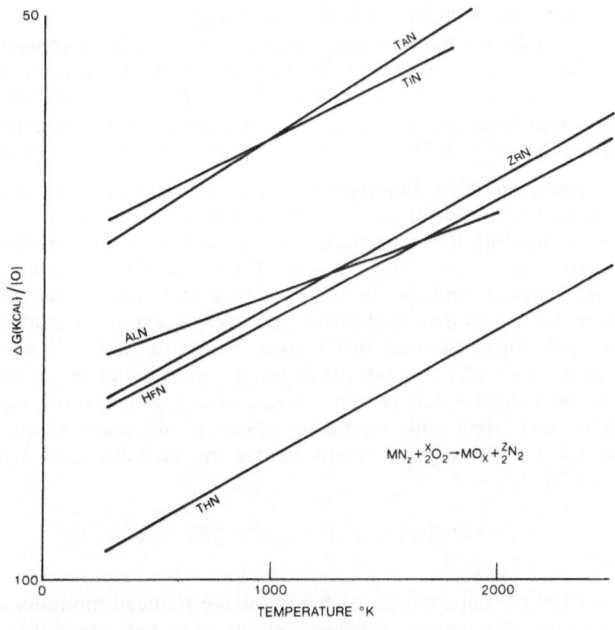

Figure 3b. Free energy of oxidation of nitrides as a function of temperatures.

At 2000 K the forward reaction, as written, is accompanied by a free energy increase of 22 Kcal, which decreases only to 15 Kcal at 3000 K. Therefore, providing that the dissociation pressure of the nitrides is tolerable, ThO_2 plus HfN should be a stable mixture. Lack of the required data does not permit a similar estimate to be made for ThO_2 and the high melting borides, carbides, or silicides.

Predictions based on the use of the above plots or tabulated data should be confirmed by more complete thermochemical analysis, which takes into account other factors such as the composition of the gas environment, solubility between the reactants, and other competing reactions. For example, solid oxynitride formation is common; compounds not usually found at more moderate temperatures, especially in the gas phase, become increasingly prominent at very high temperatures. Obviously such analysis is valid only if the major reactions have been properly identified and if reliable data exist. It is probably safe to conclude that if a calculation indicates that reaction will occur, it probably will. However, if the analysis concludes that no reaction will occur, because not all reactions may have been identified, the conclusion is not always dependable, and should be verified experimentally.

DEPENDENCE OF MECHANICAL PROPERTIES ON TEMPERATURE

Strength and stiffness are among the primary properties used to select materials. In rotating or reciprocating machines, and especially in structures to be air- or space-borne, these properties, normalized by their density, become the criteria of interest. In addition the creep resistance of high-temperature materials is a major concern.

Elastic Properties

The dependence of the elastic constants on temperature has been studied more extensively in the cryogenic regime than at temperatures approaching the melting point. These constants are temperature independent near 0 K, decrease with increasing temperature, and for many materials decrease linearly above their Debye temperatures. This behavior is supported by theoretical thermodynamic models[10] based on lattice dynamics. These models generally consider that the change of the mean (or rms) vibrational frequency is proportional to strain. However, the interatomic force laws, such as given by the Born-Madelung potential power function, indicate a more complicated behavior at large vibrational amplitudes, i.e., at high temperature. The thermodynamic model predicts the linear decrease in the elastic constants with increase of temperature. It also predicts[10] that the thermal expansion coefficient to be temperature-independent in the same regime, whereas the coefficient increases with increasing temperature in most materials at high temperatures. Thus, whether the temperature dependence of the elastic constants remains theoretically linear up to the melting point is open to question.

However, one can empirically inquire whether some kind of reduced law might reveal a common behavior, such as a plot of E/E_o vs T/T_m where E is the Young's modulus at a temperature T, E_o is the limiting low temperature value, and T_m is the absolute melting temperature. Figures 4a, 4b, and 4c show such results[11] for the single crystal C_{11} tensile elastic constant, the C_{44} shear constant, and also in polycrystalline materials for the shear constant G and for E. All materials for which data could be found over an extended temperature range tended to display an essentially linear decrease in C_{11} over the entire range. However, the relative decrease is greater in the case of the halides than for the metals and the oxides. The C_{44} and G moduli show a similar behavior except that the rate of decrease is more marked, especially for Al_2O_3, CeO_2, ZrO_2, and MgO with increasing effect in the order given. The data for the Young's modulus for a wide range of mostly oxides and carbides show a more consistent expressible empirically by

$$E/E(298K) = 1-(T/T_m)^3 \quad 298 < T < T_m \tag{4}$$

This equation describes the apparent upper bound of the reduced modulus and suggests that its use would lead to somewhat generous values for the predicted modulus. The form of Eq. 4 comes from analysis of the data at low temperature. Its validity close to T_m, where the calculated modulus vanishes, is open to question. On the other hand, E is essentially proportional to the shear modulus G and to the bulk modulus K in accordance with

$$E = 9GK/(3K + G) \tag{5}$$

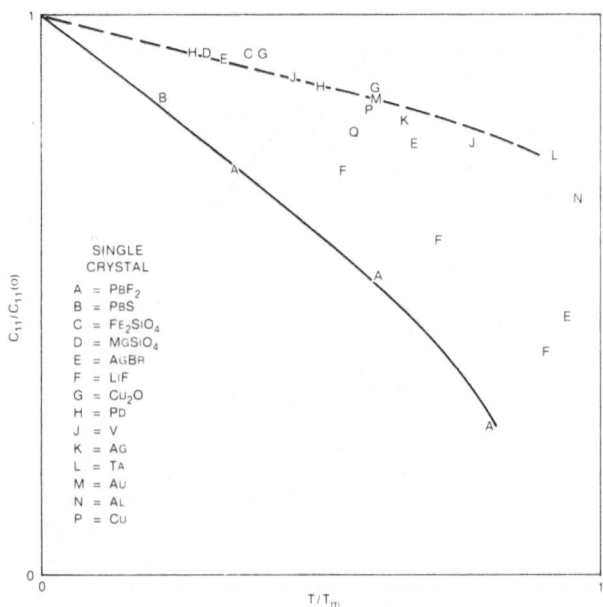

Figure 4a. Dependence of relative tensile modulus on relative temperature.

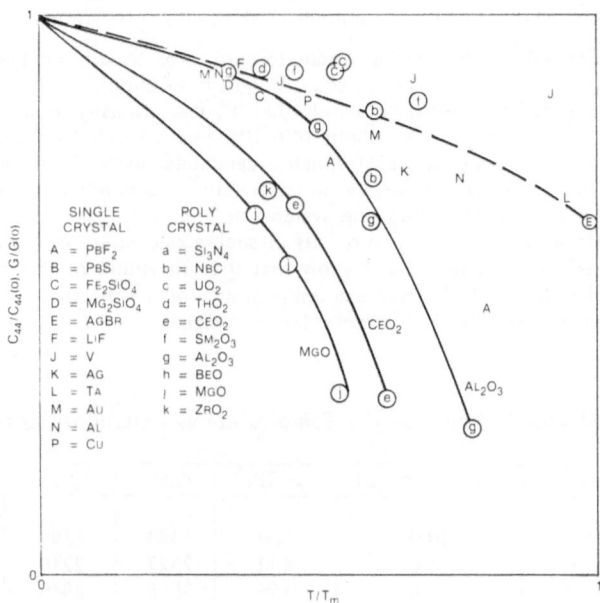

Figure 4b. Dependence of relative shear modulus on relative temperature.

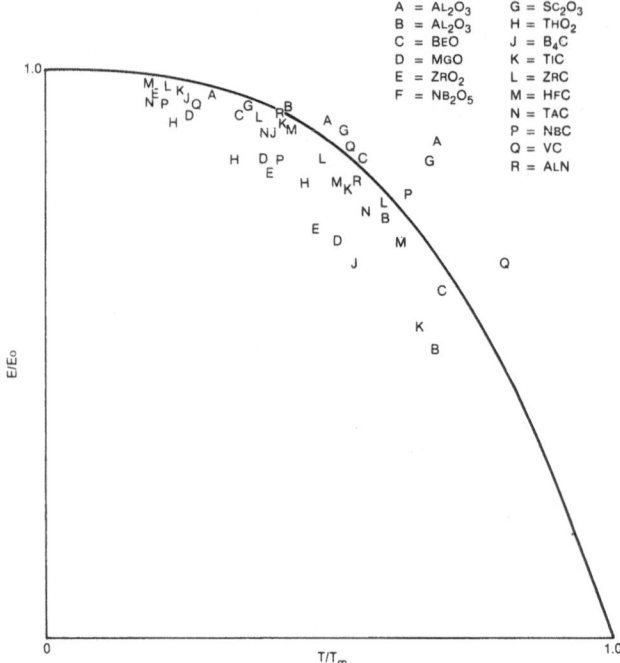

Figure 4c.   Dependence of relative young's modulus on relative temperature.

As discussed above, the shear modulus G for the oxides and other materials as well, drops off sharply in the vicinity of T_m. Such a sharp dropoff is not evident in the case of C_{44} constant for the metals. Thus, the use of Eq. (4) near T_m may not be justifiable for all classes of materials. Subject to revision as better information becomes available, we will use Eq. (4) as a guide for estimating E of oxide and nonoxide compounds at elevated temperature.

If the refractory materials of the present discussion are to be used in applications currently served by superalloys, their stiffnesses on an equal weight basis should at least be comparable. A typical nickel-base superalloy is useful to about 1000 °C, has a density of about 8.3 g/ml and an elastic modulus at that temperature of about 140 GPa, i.e., a specific stiffness E/density of 1.7×10^6 m. The maximum temperature at which a candidate material still has the requisite stiffness can be estimated by using this specific stiffness value as a standard in combination with Eq. (4). In the case of a candidate composite system the appropriate law of mixtures for that composite must be introduced as well. The results of such a calculation for various monolithic high-temperature materials given in Table 4 show that the imposition of this specific stiffness criterion means that the allowable effective use temperature for the material can be sustantially below its melting point, particularly for the more dense materials.

Table 4:   Calculated Maximum Use Temperature as Determined by Stiffness

Material	Density (g/ml)	E_o(GPa)	T_m(K)	T_{max}(K)
ThO_2	10.0	180	3493	1390
Al_2O_3	4.0	490	2327	2210
MgO	3.6	350	3098	2910
BeO	3.0	380	2843	2710
AlN	3.3	270	3033*	2810*
HfC	12.7	410	4100	3200

* AlN decomposes congruently at 1 atm at 2572 K

706

Strength

Although extremely important, if not decisive, for many applications, the (tensile) strength of brittle materials is an elusive quantity with which to deal merely on the basis of composition and temperature. Nevertheless, if we assume that yielding does not occur and that the mechanism of failure does not change with temperature, then we can formulate a simple model for the decrease in strength with increasing temperature. We postulate that when fracture does occur, each "atom" pair at the leading edge of the fracture must be subjected to a force just exceeding the ultimate strength of that bond. We approximate the force F between atoms to have the form:

$$F = F_o \sin (bu) \qquad (6)$$

in which F_o is the force needed to separate the atom pair, and b is a constant such that when the interatomic displacement u reaches a particular value, the force reaches its maximum value. The derivative of F with respect to the interatomic displacement is the restoring force constant. Thermal expansion produces a finite displacement u, and also produces a decrease of the force constant as well as of the net maximum force needed to separate the atoms. It is reasonable to assume that E is proportional to the force constant, and that the force that must be applied to separate the atoms is proportional to the applied stress needed to cause fracture, i.e., the strength of the material. This leads to the expectation that

$$(\delta S)/S_o = \sqrt{(2(\delta E)/E_o)}. \qquad (7)$$

This relationship states that at any given temperature, the decrease in relative strength will be much greater than the decrease in relative stiffness. Figure 5 gives a plot of the right-hand side of Eq. (7) versus the left side for data on two different types of Al_2O_3 single crystal filaments. Although there is scatter, the data follow the trend suggested by this simple model. More study is needed to test whether such a trend is generally obeyed, preferably using systematic data obtained by the same investigator of both the elastic properties and the strength on unambiguously brittle materials. Coupling Eq. (7) with Eq. (4) suggests that strength controlled by brittle failure should decrease proportionally to about the 3/2 power of the reduced temperature, and more specifically that the strength at $0.5\ T_m$ should be about half of the room temperature value.

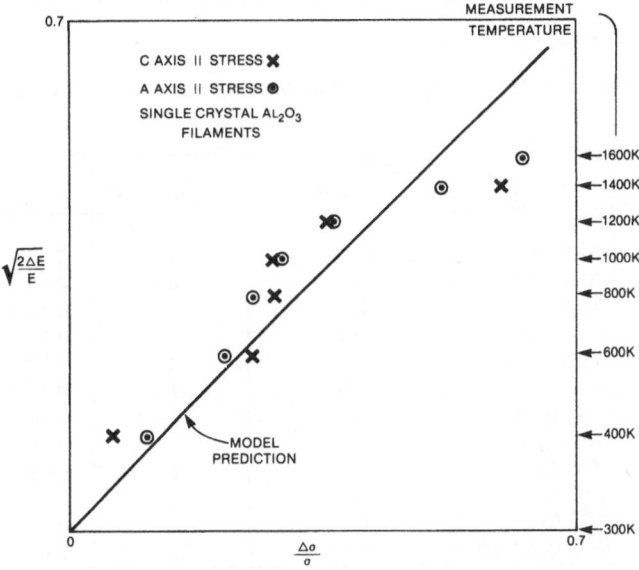

Figure 5. Relationship between decrease in strength and decrease in stiffness due to temperature.

Creep Resistance and Yielding

Dimensional stability or resistance to long-term deformation, by creep, is basic to the consideration of the use of any material to be used under substantial stress. Numerous models of steady state creep rate $\dot{e}$ have been formulated and are broadly expressible[12] by

$$\dot{e} = (ADGb/kT)((b/d)^p)(S/G)^n \qquad (8)$$

in which A, p, and n are constants characteristic of a particular mechanism, G is the shear modulus, b the Burger's vector, d the mean grain size, S the applied stress, and D the diffusion coefficient $D = D_o \exp(-Q/RT)$ for which D_o is the frequency factor, and Q the activation energy for the type of diffusion specified by the particular model. The terms k, R, and T have their usual meanings. All models invoke dislocation movement and/or some mechanism for relative grain boundary motion. An extensive recent review[12] summarizes the constants and corresponding models for a wide range of ceramic materials. The various mechanisms depend on factors such as the details of the microstructure, stress configuration, and impurities.

Creep rates of about 10^{-7}/sec are probably close to the maximum tolerable level for long-term usage of highly stressed structural materials for which working strains can be expected to be at least 0.0001 in general. In the case of fiber-reinforced composites, the fiber strain may be ten times greater. In view of the many possible mechanisms it is unfortunately not feasible, when taking a broad perspective, to use the wealth of detailed information and understanding that is available. Instead we shall resort to a rule-of-thumb approach for estimating the upper use temperature based on creep resistance considerations.

Table 5 summarizes creep information on a variety of representative refractory materials for which data[3] were conveniently available. The table focuses on the lowest temperature available data for creep rates approximating the toleration limit given above. Elastic strain values are used to quantify the driving force for creep.

The creep rate of single crystals depends sensitively upon their orientation with respect to the applied stress and may account for the high relative temperature needed in the case of single crystal filament melt grown Al_2O_3 compared with single crystal MgO to effect comparable rates of creep. Inspection of the T/T_m column shows that creep rates of the order 10^{-7} sec can occur when the relative temperature reaches an average value of about 0.6. The upper bound of T/T_m for the polycrystalline materials is about 0.65. Because creep mechanisms are dependent on purity, microstructure, etc., suitable material control can probably result in a further increase in

Table 5: Low Temperatures at Which Creep Rate Limit is Obtainable

Material	Strain Rate $\times 10^7$ (1/sec)	Strain $\times 10^4$	T/T_m
Al_2O_3 (sx)	14	6	0.83
Al_2O_3 (px)	5	0.5	0.63
MgO (sx)	5	1	0.51
Cr_2O_3 (px)	0.1	0.1	0.64
ThO_2 (px)	0.2	2	0.50
UN (px)	0.08	1	0.44
TiC (px)	0.3	1	0.56
HfC (px)	20	0.4	0.66
ZrC (sx)	0.7	5	0.45
ZrC (px)	0.4	4	0.56
SiC (px)	0.01	5	0.60
SiC (CVD)#	0.15	50	0.67

Code: sx = single crystal; px = polycrystalline solid
#CVD = chemical vapor deposited material[13]

the relative temperature for the onset of excessive creep. Therefore, somewhat arbitrarily, we suggest provisionally that $0.7\,T_m$ defines the probable upper bound for the use of polycrystalline ceramic materials as monolithic structures or as reinforcement fibers. However, the creep rate of composite systems can be significantly reduced by the reinforcement. Therefore, matrix materials can function satisfactorily to somewhat higher temperatures. Accordingly we have arbitrarily selected $0.85\,T_m$ as defining the upper use temperature for matrix materials that are used in conjunction with particulate or fiber reinforcement.

An analysis[14] of creep behavior in a composite consisting of a parallel array of short fibers has shown that the creep rate is inversely proportional to $c\,(a/L)^2$ where c is a constant that increases with increasing volume fraction of fibers, and the ratio a/L is the (fiber radius)/(fiber length). For a volume fraction of 0.3 the value of c is about 5, so that for the hypothetical case of fibers having an aspect ratio of 200, the strain rate could be reduced to about 0.0001 of its unreinforced value. This process, however, transfers load to the fibers and implies the availability of strong fibers, which can carry the stress normally supported by the matrix.

Perfect Crystal Behavior

The recent development of relatively low cost methods to make SiC and Si_3N_4 whiskers in quantity has reopened interest in the use of these and other whiskers as possible fiber reinforcements. Perfect whiskers, grown with a single screw dislocation cannot creep because they lack the necessary dislocation structure. Cottrell[15] has treated the work required to nucleate a dislocation loop in a solid, and Kelly[16] has suggested that thermal activation may assist in lowering the stress needed to initiate such a dislocation. Once this occurs then the whisker becomes the equivalent of a polycrystalline fiber. Following Cottrell, the energy barrier U^* is given by

$$U^* = B\,X\,(Ln(X) - 1) \qquad (9)$$

where $X = r/w$ (the loop radius/the dislocation width) and X is related to the applied shear stress S by

$$S = C\,(Ln(X) + 1) \qquad (10)$$

in which $C = G\,b/4\pi\,w$, the symbols having the meanings defined above. Further analysis[17] of these relationships shows that temperature has little effect on the critical shear stress S^* needed for nucleation and that this stress is given closely by

$$S^* = p\,(1 - \nu)/e \qquad (11)$$

where ν is the Poisson ratio, e is the base of the natural logarithm, and p is the maximum shear stress that can develop between adjacent atoms, i.e., the ultimate intrinsic shear strength. This ultimate strength can be expected to decrease with increase of temperature, just as discussed above for the intrinsic tensile strength. However, because S^* remains of the same order as p, there appears to be little likelihood that spontaneous nucleation of dislocations will occur prior to the normal brittle failure of the whisker. Thus, if whiskers such as those of SiC can remain free of local stress-raising imperfections that might result from chemical attack, these fibers have the potential of being effective to very high temperature in strengthening the composites and inhibiting creep.

INTEGRATION OF THE SEPARATE CRITERIA

Having considered separately some of the individual criteria that must be satisfied for a material to function satisfactorily as a high-temperature structural component, we finally turn to applying these together to gain a "realistic" perspective of the potential for such materials. For this purpose we examine the ten highest melting oxides and nonoxides in Tables 6 and 7 in which the criteria are stated in kelvins. Assuming that the latter are to be used as reinforcements with the oxides serving as the matrices, we impose the additional condition that the stiffness (per unit weight) of the reinforcements be at least twice that of the matrices.

One may readily question the reliability of the estimates in the above tables. The criteria used are probably somewhat optimistic in estimating the property values retained at any given temperature. The minimum temperatures T_{min} given should be interpreted as provisional upper bounds for the use temperature of the various materials. As more reliable information and models become available, the above conjectures can be replaced by values providing greater confidence. Furthermore, as one scrutinizes the individual materials, it is obvious that other, if less quantifiable, criteria must be included. For example, Table 6 indicates that the oxides having the greatest potential decrease in the order $CaO > BeO > ThO_2 > ZrO_2 > UO_2 > HfO_2 > Al_2O_3$. This sequence does not take into account the strength of these materials. Clearly, a meaningful basis for estimating the strengths of these materials in regimes where plastic deformation occurs is needed. Furthermore, one must take into account the facts that CaO is subject to moisture attack at low temperature, that BeO, ThO_2, and UO_2 present health hazards, that HfO_2 is not available in quantity, and so forth. Each new consideration can serve to lower further the upper bound temperatures.

Table 6: Combined Criteria for Oxides

Oxide	T_m	Creep		E/ρ	Volatility	T_{min}	Rank
		$.85\,T_m$	$.7\,T_m$				
ThO_2	3493	2970	2445	2290	2375	2290	3
HfO_2	3117	2650	2180	2020*	--	2020	6
UO_2	3113	2645	2180	2050	--	2050	5
MgO_2	3098	2635	2170	2870	1825	1825	9
ZrO_2	3037	2580	2125	2195	2775	2195(2125)	4
CaO	2882	2450	2015	2430	--	2430(2015)	1
BeO	2843	2415	1990	2710	2325	2325(1990)	2
SrO	2727	2320	1910	1750	2075	1750	10
Al_2O_3	2327	1980	1630	2210	--	1980(1630)	7
Cr_2O_3	2300	1955	1610	2015	--	1955(1610)	8

* Assumes E is the same as that reported for ZrO_2
  Values in parentheses in T_{min} column refer to use as monolithics
  if that use results in a lower estimated use temperature.

Table 7: Combined Criteria for Nonoxides

Mat'l	T_m	$.7\,T_m$	E/ρ	Volatility	T_{min}	Rank
TaC	4258	2980	Fails	--	Fails	--
HfC	4163	2915	Fails	--	Fails	--
NbC	3886	2720	2365	--	2365	6
C	3825	2670	3685	~3500	2670	2
ZrC	3803	2660	2715	--	2660	3
HfN	3660	2560	--	--	2560	(5)
TiC	3530	2470	3035	--	2470	4
HfB_2	3523	2463	2345	--	2345	7
TaN	3360	2350	1855	--	1855	9
TaB_2	3310	2315	--	--	2315	(8)
SiC	3100	NA	2850	2780	2780	1

Of the materials listed in Table 7 only C and SiC are currently available as fibers. It is noteworthy that these two materials also emerge as having the highest upper bound use temperature. Carbon is the only strong filamentary material that actually increases in strength with increase of temperature. However, both C and SiC are expected to be highly reactive towards most oxides. This suggests that coatings may be required in the use of these fibers. The use of the other materials will depend on their future availability in suitable form.

The cursory examination of the ten highest melting members of the two classes suggests that structural composites that can function up to the range 2000-2400 K are conceivable. Much basic data and modeling are needed to provide a firmer basis for such a conclusion, and would undoubtably provide a more efficient approach to identifying systems of high potential, than would the individual study of the multitude of possible composite systems.

ACKNOWLEDGMENTS

I wish to thank my colleagues, and in particular Drs. Robert L. Fleischer, Krishan L. Luthra, and Farhad N. Mazandarany, for many helpful discussions and sharing of useful information.

REFERENCES

1. Campbell, I.E. and Sherwood, E.M., "High Temperature Materials and Technology," John Wiley and Sons, Inc., New York, 1967.

2. Samsonov, G.V. and Vinitskii, I.M., "Handbook of Refractory Compounds," transl. Shaw, K., IFI/Plenum, New York, 1980.

3. "Engineering Property Data on Selected Ceramics," Vol. I, Nitrides 1976; Vol. II, Carbides, 1979; Vol. III Single Oxides, 1981; Metals and Ceramics Information Center, Battelle Columbus Laboratories, Columbus, OH.

4. Lynch, J.F., and Duckworth, W.H., "Engineering Properties of Ceramics," Air Force Materials Laboratory Technical Report AFML-TR-66-52, Battelle Memorial Institute, Columbus, OH, 1966.

5. J.L. Margrave, Ed., "Refractory Materials," Vol. 7, L.E. Toth, Ed., "Transitions Metal Carbides and Nitrides," Academic Press, New York, 1971.

6. J.L. Margrave, "High Temperature Thermodynamnics," Ref. 1, p. 23, quoting A.W. Searcy, "Proceedings of the International Symposium on High Temperature Technology," McGraw-Hill Book Co., New York, 1961.

7. Goldsmith, A., Waterman, T.E., and Hirschorn, H.J., eds., "Handbook of Thermo-Physical Properties of Solid Materials," Pergamon Press, New York, 1961.

8. a. Barin, I. and Knacke, O., "Thermochemical Properties of Inorganic Substances," Springer-Verlog, Berlin/Heidelberg, 1973.

   b. Barin, I., Knacke, O., and Kubeschewski, O., Supplement to the above, 1977.

9. JANAF Thermochemical Tables, NSRDS-NBS37, U.S. Government Printing Office, Washington, 1971.

10. Huntington, H.B., "The Elastic Constants of Crystals," Vol. 7, Solid State Physics, Seitz, F. and Turnbull, D., eds., Academic Press, New York, 1958.

11. a. Hearmon, R.F.S., "The elastic constants of crystals and other anisotropic materials," Landolt-Bornstein, New Series III, Vol. 2, p. 16ff, Springer-Verlag, Berlin/Heidelberg, 1984.

    b. Ibid, Vol. 3, p. 20ff.

    c. Ibid, Vol. 18, p. 62ff.

12. Cannon, W.R. and Langdon, T.G., J. Mater. Sci. *18*, 1 (1983).

13. Carter, C.H., Jr., Davis, R.F., and Bentley, J., J. Am. Cer. Soc. *67*, 732 (1984).

14. Hillig, W.B., "The Tensile Properties of Fiber-Reinforced Composites II. Time Dependent Response to Tensile Stress," General Electric Report 71-C-322, Schenectady, NY, 1971.

15. Cottrell, A.H., "Dislocations and Plastic Flow in Crystals," Oxford, 1953.

16. Kelly, A., "Strong Solids," Oxford, 2nd Ed., 1973.

17. Foreman, A.J., Jaswon, M.A., and Wood, J.K., Proc. Phys. Soc., *64B*, 156 (1950).

MICROSTRUCTURAL ENGINEERING OF CERAMICS

FOR HIGH-TEMPERATURE APPLICATION

M. H. Lewis

Centre for Advanced Materials Technology
University of Warwick, U.K.

INTRODUCTION

The meaning of 'high'-temperature in relation to ceramic applications is normally interpreted in terms of the current requirement to operate energy-conversion systems at temperatures beyond that achievable with metallic alloy components. Thus in gas-turbines a temperature increment of 200–400°C above the ~1000°C limit set by superalloy properties is desireable for improved efficiency. The 1000–1400°C temperature interval is also that in which severe problems are introduced in using ceramic components under stress, especially in oxidising or corrosive environments.

In the lower temperature regime the use of ceramic composites or ZrO_2 transformation-toughened oxides has much promise in overcoming such problems as susceptibility to thermal shock, via their enhanced fracture-toughness. Microstructural stability required for these toughening mechanisms is not normally retained above ~1000°C where there remains a need for refinement of microstructure of monolithic ceramics with appropriate intrinsic properties. It is in this area that the nitrides, oxynitrides and carbides of silicon offer the most significant advantages, conferred by high-covalency Si-N and Si-C bonding with typical tetrahedral-network crystal structures. Although these structures have decomposition temperatures well in excess of 1500°C there are significant problems in the 1000–1500°C temperature range originating from a fabrication process which normally involves liquid-phase sintering. It is now well-established that the transformation in properties above ~1000°C in the nitride and oxynitride ceramics is due to liquid residues which provide sites for creep-cavity nuclei and easy paths for diffusion of metallic ions to the initially-protective surface SiO_2 oxidation film.

The purpose of this paper is to review the development and refinement of microstructure in ceramics prepared via liquid phase sintering and post-sintering heat-treatment to extend their temperature of application to ~1400°C. The review is focussed on monolithic ceramics based on oxynitride phases in the Si-Al-O-N system.

THE OXYNITRIDE CERAMICS

Crystallography of Major Phases

The nitride ceramics which have been most intensively studied utilise βSi_3N_4 as a major phase. The tetrahedral, SiN_4, network structure is retained in the derivative solid-solution β' $(Si_{3-x}Al_xO_xN_{4-x})$[1] which is the basis of 'Sialon' ceramics in which Al and O substitute for Si and N respectively (Fig.1). β and β' at the substitution level normally used, have similar intrinsic properties (thermal expansion, oxidation resistance, decomposition temperature, hardness and resistance to dislocation motion etc.). Hence the real benefits of ceramic 'alloying' lie in the ability to more easily control the volume and composition of the sintering liquid and residual phases.

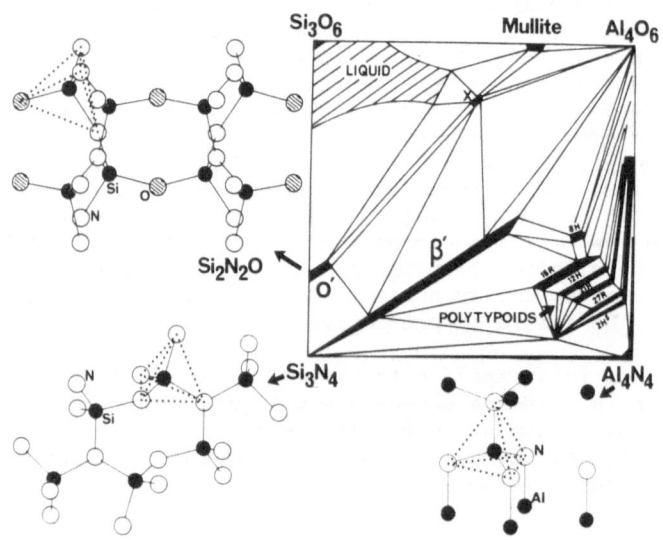

Fig. 1. Crystal structures of oxynitride phases in the Si-Al-O-N system and their relative positions in the behaviour diagram for a 1700° isotherm (based on Ref. 1.).

A related structure is the oxynitride phase Si_2N_2O which also exists as a substituted derivative, O', but with very limited Al+O solid solution (Fig.1). In Si_2N_2O one of the nitrogen atoms in SiN_4 coordination tetrahedra is replaced by oxygen. These are selected in an ordered manner with pseudo-planar arrays of SiN_3 units joined by Si-O-Si bonds, the oxygen atoms forming a singular (100) plane in the orthorhombic symmetry lattice (Fig.1). This singular plane, with a greater proportion of ionicity in bonding, is expected to introduce a greater anisotropy and impairment to properties compared to β and β' crystals. However, although single-phase data is not available, ceramic surfaces or bulk-ceramics containing a large Si_2N_2O volume fraction exhibit high hardness and resistance to deformation comparable with the high-covalency nitrides. The excellent oxidation resistance of this phase and its compatibility with β or β' support its contention as a high-temperature ceramic phase.[3]

The Al-rich high covalency structures are based on the hexagonal wurtzite AlN crystal which again has a tetrahedral AlN_4 structure unit. The series of polytypoid phases (15R, 21R etc. - Fig.1) are derived from AlN where substitution of Si and O is accompanied by stacking-modulations parallel to the hexagonal basal plane. The series of six phases may be identified by the metal/non-metal atom ratio M_nX_{n-1}, where n has values 4,5,6,7 and 9, and the additional non-metal atoms, $_{4,5}X_5$ are associated with layers in preferred octahedral oxygen coordination.[4,5] Ceramics containing polytypoids as a major phase have been infrequently studied; the problem being the relation between phase composition and liquid-sintering aid rather than of intrinsic properties.

Remaining oxynitrides in the Si-Al-O-N system, the oxygen-rich X-phase and the spinel AlON are not appropriate as major phases in high-temperature engineering ceramics in view of the similarity in intrinsic properties to the pure silicate or oxide phases. Derivatives of the nitrogen-rich oxynitrides may be stabilised by an additional element, for example Ca or Y within interstices in the tetrahedral SiN_4 network stabilise the αSi_3N_4 structure as $\alpha' \left| M_x(Si,Al)_{12}(O,N)_{16} \right|$[6] which is a potential ceramic phase in parallel with β'. Solubility of metallic elements in the β' or O' phases is, by comparison, negligible.

System Selection

An established requirement for fabrication of silicon nitride and oxynitride ceramics is the use of liquid-forming additives to catalyse the sintering process. The liquid phase should satisfy the following requirements;

(i) It should have a convenient melting temperature, well below that for decomposition of the major -phase ($\sim$1800°C) and exhibit partial solubility for the latter.

(ii) The liquid phase should form a binary tie-line with the major phase and, for high-temperature application, have a composition which may be controllably crystallised to high-temperature oxide or oxynitride phases.[7] βSi_3N_4 ceramics were initially hot-pressed with an MgO addition which was later identified as one component of a Mg-Si-O-N eutectic sintering liquid.[8] Trivalent oxides, such as Y_2O_3 were later favoured in generating superior high-temperature properties either because of higher-viscosity glass residues or yttrium oxynitride crystallisation products.[9] The latter suffer from oxidation-induced microstructural instability at intermediate temperatures[10] and are difficult to sinter to near-theoretical density without the application of pressure.

'Sialon' ceramics may be hot-pressed using the relatively high-melting liquid in the Si-Al-O-N system (Fig.1). However the best densities and properties are achieved with eutectic liquids of the type M-Si-Al-O-N. In early hot-pressed ceramics a 'transient' Mg-Si-Al-O-N liquid was success-

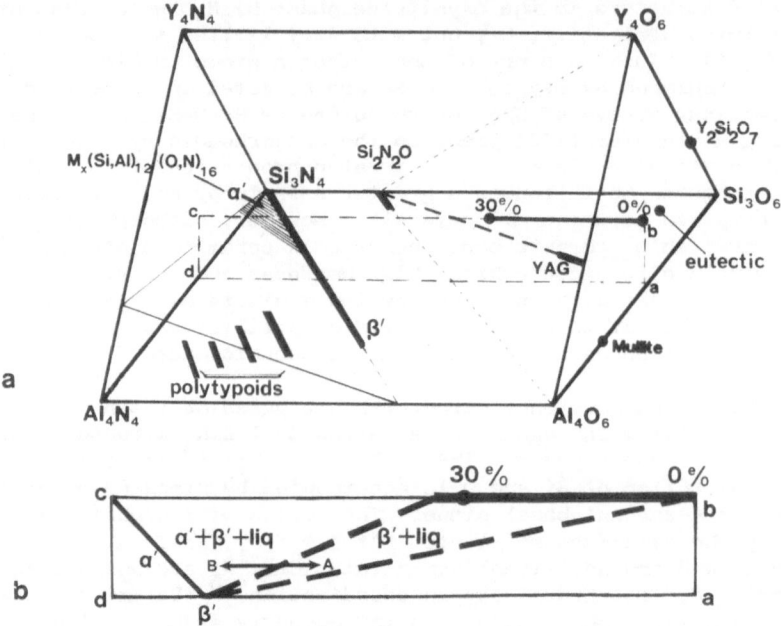

Fig. 2. (a) A Janecke prism representation for the Y-Si-Al-O-N system showing the relation between major ceramic phases and sintering liquid/
residual glass compositions. The prism section abcd, sketched in
(b), illustrates the emergence of the α′ phase at high N/O ratio
in a β′ ceramic.

fully used, with control of overall composition near to the monophase β′
via AlN additions[11] to compensate for excess surface SiO_2 on the main component – αSi_3N_4. More recently Y_2O_3 has become the favoured[12] addition in
relation to the pressureless-sintered series of ceramics known commercially as Syalon* ceramics. These have essentially bi-phase microstructures,
the evolution of which may be described with reference to Fig.2.

The Janecke prism (Fig.2) is a convenient illustration of the relation
between initial stoichiometric ceramic components, liquid sintering compositions and final ceramic phases. A β′ Syalon ceramic is fabricated from
mixtures of the terminal components via an overall reaction :

$$\alpha Si_3N_4 + SiO_2 + Y_2O_3 + (Al_2O_3 + AlN) \rightarrow \beta' Si_{3-x}Al_xO_xN_{4-x} + \text{Y-Si-Al-O-N}$$

surface   liquid

SiO_2 is derived from the surface oxidation film on αSi_3N_4 and AlN is
normally added as one of the near-neighbour polytypoids and is used as a
sensitive control of O/N ratio, substitution level (x) in β′ and hence of
O/N ratio in the final liquid phase. The liquid is based on the ternary
$Y_2O_3-Al_2O_3-SiO_2$ eutectic which may dissolve up to 30 equivalent percent
(e/o) of nitrogen (Fig.2). The use of Y-Si-Al-O-N liquids in place of
Y-Si-O-N liquids is important in reducing the eutectic temperature from
~1600°C to ~1300°C, to accelerate the α→β′[13] solution-recipitation and
densification kinetics. It is the basis for an increasing number of
commercial pressureless-sintered β′ ceramics. The key to a successful
high-temperature ceramic is to maximise N/O ratio in the final liquid,
which increases its viscosity and, more important, controls the liquid
crystallisation products. For example, experiments on bulk-synthesised

*Lucas-Cookson-Syalon, U.K.

716

glasses[14] show a change from yttrium disilicate $Y_2Si_2O_7$) + mullite ($Al_6Si_2O_{13}$) mixtures in pure silicate liquids to yttro-garnet ($3Y_2O_3.5Al_2O_3$-YAG) + Si_2N_2O mixtures in high-nitrogen liquids. This is consistent with the shift in tie-line (Fig.2) and is mirrored in the ceramic by the emergence of YAG as a crystallisation product at high N/O ratio. The absence of Si_2N_2O in the latter is explained by the solubility of its components in β' evidenced by a reduction in its substitution level.[12,13]

Similar principles may be applied to other systems of the general form M-Si-Al-O-N, for example, when M=Mg the oxide spinel $MgAl_2O_4$ is the dominant liquid crystallisation product.[15] However, divalent ions M^{2+} are not favoured if crystallisation is incomplete, because of a reduction in the viscosity of minor glassy residues with severe impairment to high-temperature properties.

The use of M-Si-Al-O-N eutectic sintering liquids may be extended to ceramics containing alternative major phases. Eutectic temperatures below 1600°C are essential in the case of Si_2N_2O which decomposes via the reaction :

$$3Si_2N_2O \rightarrow Si_3N_4 + 3SiO + N_2$$

Near theoretical densities have been achieved using low O/N eutectic liquids in the Y-Si-Al-O-N system[3] in which the major phase is low-substitution Si_2N_2O (O'). It has been possible to crystallise the liquid residue mainly as $Y_2Si_2O_7$ and the refinement of composition necessary for complete crystallisation of oxide phases is currently being explored with a view to its use as a high-temperature ceramic.

Polytypoid ceramics have been fabricated by hot-pressing using Mg-Si-Al-O-N liquids of small volume fraction. This is a further example of the use of transient liquids due to the high solubility of Mg and other liquid components in the polytypoid structures. Attempts to extend pressureless-sintering to the polytypoids using Y-Si-Al-O-N liquids are complicated by the appearance of other phases due to the remote positioning of major phase and sintering liquid (Fig.2).

Ceramics containing α' phase have not been developed for high-temperature use to the same degree of refinement as β' ceramics. They may be sintered using similar Y-Si-Al-O-N liquids, since Y is an α'-stabilising element.[6] α' progressively replaces β' as a solution-reprecipitation product with increased N/O ratio in the sintering mixture (in the direction A $\rightarrow$ B within a Janecke prism section - Fig.2).

Fabrication routes

A survey of the main fabrication routes for β' ceramics (Fig.3) shows a starting point of mixing various oxide and nitride additives with αSi_3N_4 via wet milling (using Al_2O_3 or Si_3N_4 balls) to ensure homogeneity and further reduce particle size. For Si_2N_2O or polytypoid ceramics αSi_3N_4 is normally replaced by the presynthesised major phase but may be formed in situ by reaction, e.g. $\alpha Si_3N_4 + SiO_2 \rightarrow 2Si_2N_2O$.

In earlier development work the milled and dried powders were hot die-pressed, with subsequent component or specimen shaping by diamond machining (route 1-b). With the current requirement for more convenient and economic 'near-net shaping' during sintering, routes 2a and 2b involving pressureless-sintering[12] have assumed greater importance. However, the occurrence of large-scale porosity during the 'burn-out' stage of injection moulding for large components, has emphasised the value of hot-isostatic pressing (HIP) either at a post-pressureless-sintering stage or from an encapsulated green pressing (route 1a).[19,20]

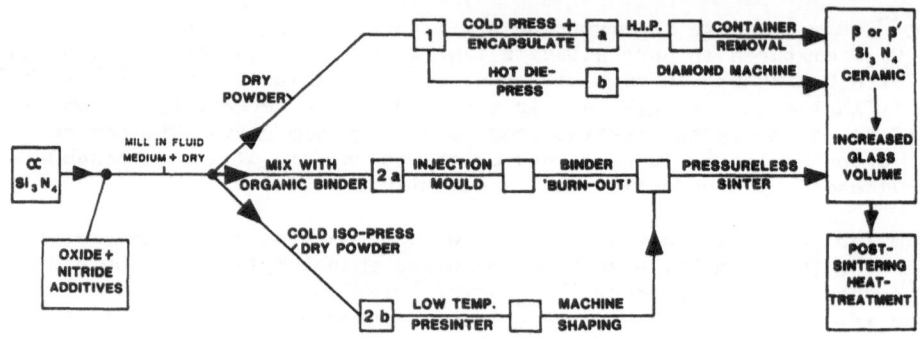

Fig. 3. A survey of possible fabrication routes for oxynitride ceramics.

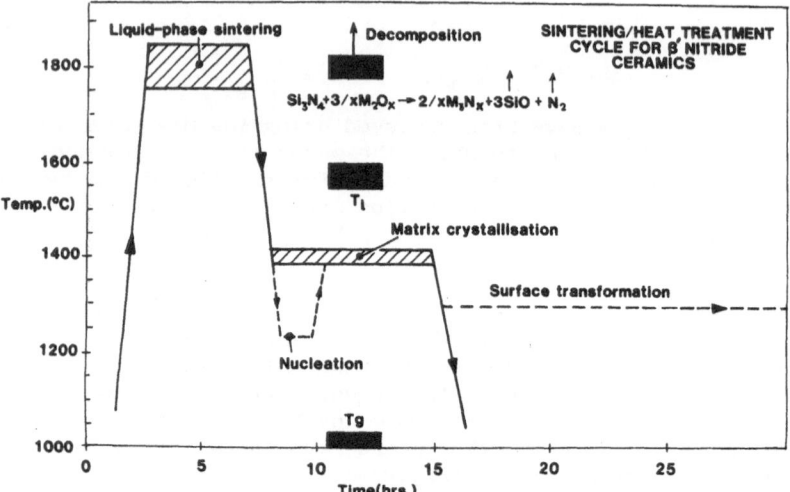

Fig. 4. The complex sintering/heat-treatment schedule required to optimise
ceramic microstructure for high-temperature application.

In general, pressureless-sintering requires a relatively large liquid
volume ($\sim$10%) and for high-temperature application should preferably be
followed by a matrix-crystallisation heat-treatment.[12]   The use of low
liquid volumes (<5%) and chemistries specifically tailored for the encapsulated HIP route has not yet been fully explored.

A typical sintering/post-sintering heat treatment schedule for
optimising the microstructure of a pressureless-sintered β' ceramic is
illustrated in Fig.4.   Sintering temperatures, and hence densification
kinetics are optimised just below that for the most prominant decomposition
process viz. active oxidation via the additive oxides with evolution of SiO
and N_2 (Fig.4).   Matrix liquid crystallisation is achieved between the
M-Si-Al-O-N glass transition T_g and T_ℓ for the main crystallisation phase.
This is preferably a 2-step 'nucleation' and 'growth' sequence similar to
glass-ceramics.   This and a subsequent surface transformation for application at the highest temperature are further discussed in following sections.

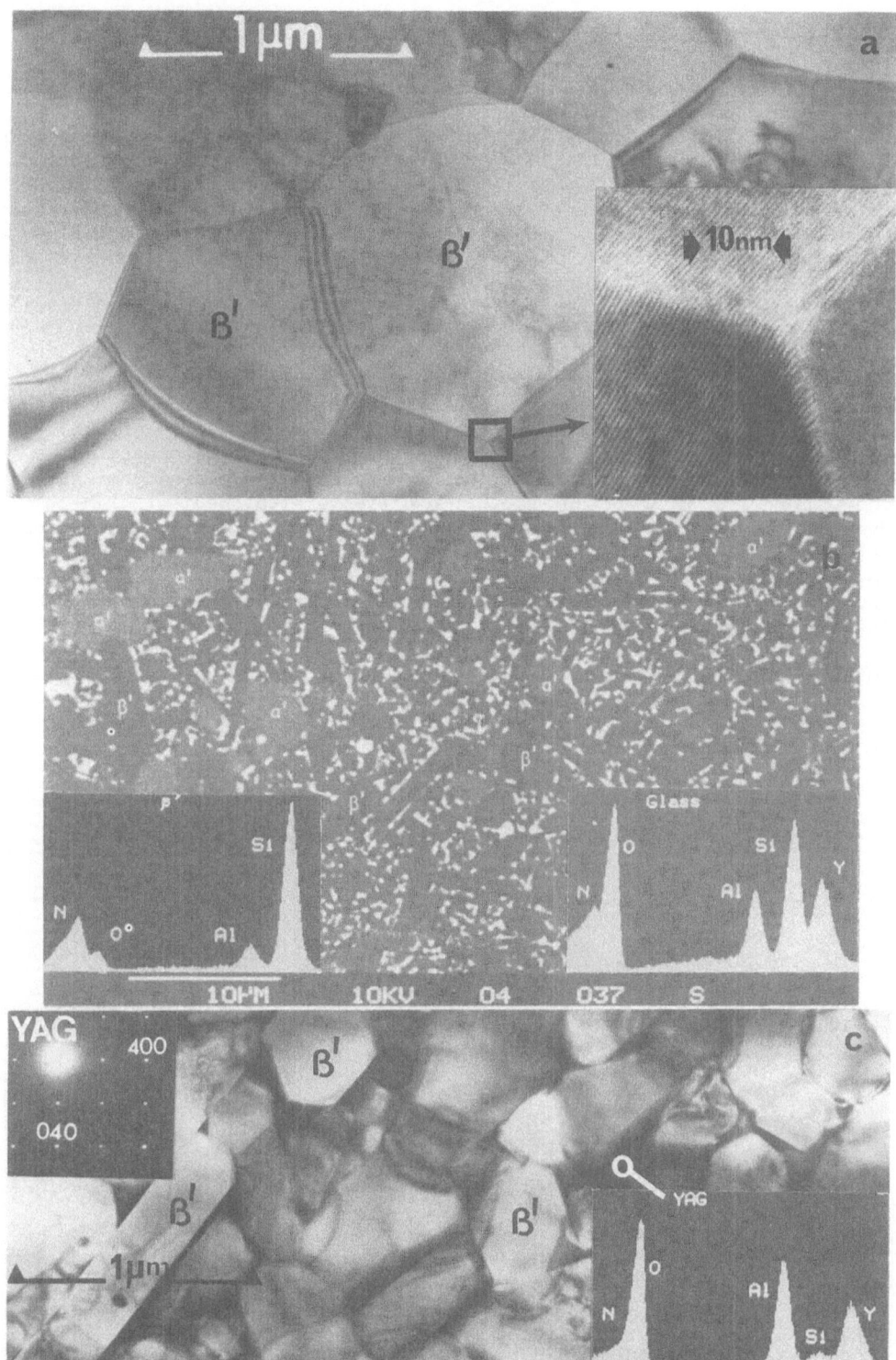

Fig. 5. Comparison of microstructures in hot-pressed monophase β´ ceramic
(a) and pressureless-sintered bi-phase β´ ceramic in the 'as-
sintered' (b) and matrix-crystallised states (c).

β′ ceramics

The simplest microstructures are the earlier monophase β′ ceramics produced by hot–pressing with a transient liquid phase. An example is reproduced here for comparison with the more recently developed pressureless–sintered bi–phase ceramics (Fig.5). β′ grains have a distinct non–faceted morphology produced by impingement at interfaces which do not have a marked energy–anisotropy.[11] With careful composition control (using AlN additions to balance excess SiO_2) these microstructures contain a negligible grain–junction residual glass content and were the first examples of β or β′ ceramics exhibiting pure diffusion–creep without cavitation.[21] The comparative morphological isotropy and larger grain sizes than bi–phase ceramics impairs fracture–toughness and hence MOR.

During the solution–recipitation mechanism in pressureless–sintered ceramics β′ crystals have greater freedom to grow with anisotropic and faceted morphology within the comparatively large liquid matrix (Fig.5). β′ crystals are in mutual contact only over part of their periphery and the intercrystalline liquid normally has a near eutectic glass–forming composition.[12,13] An important factor in ultimate attainment of high–temperature performance is the saturation of nitrogen content in the glass via AlN (or polytypoid) addition. Spatially–resolved direct analysis of glass nitrogen content is obtainable only with electron energy–loss spec–troscopy[22] (ELS) or 'windowless' X–ray energy–dispersive spectrometry (EDS – Fig.5b). However, a simple indication of high nitrogen levels is the appearance of the α′ phase above a critical polytypoid addition in Y–Si–Al–O–N ceramics[23] (Fig.5b). Microstructure changes accompanying polytypoid additions are summarised in Fig.6. The increase in Al+N/Si+O in the system

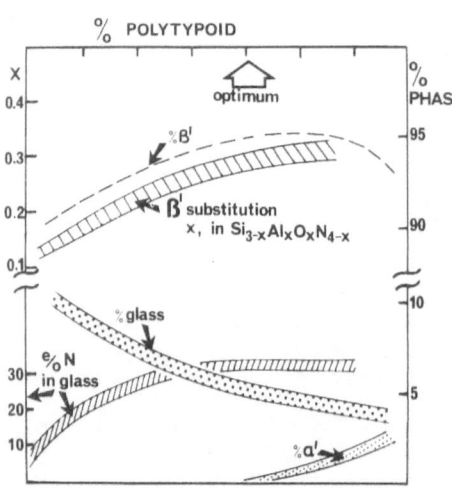

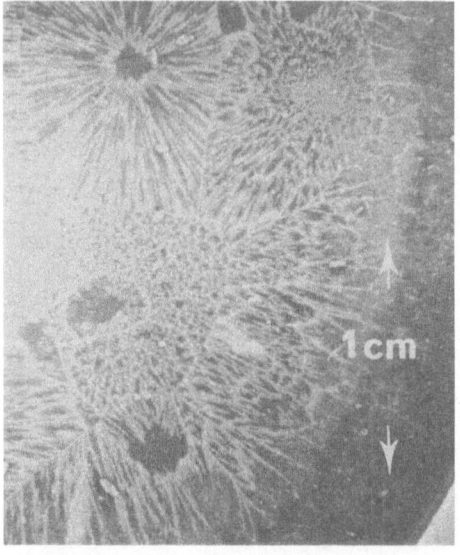

Fig. 6. A summary of constitutional changes with varying N/O ratio in β′ Syalon ceramics.

Fig.7. Optical micrograph of course cellular matrix crystallisation in a β′ Syalon ceramic without prior nucleating treatment.

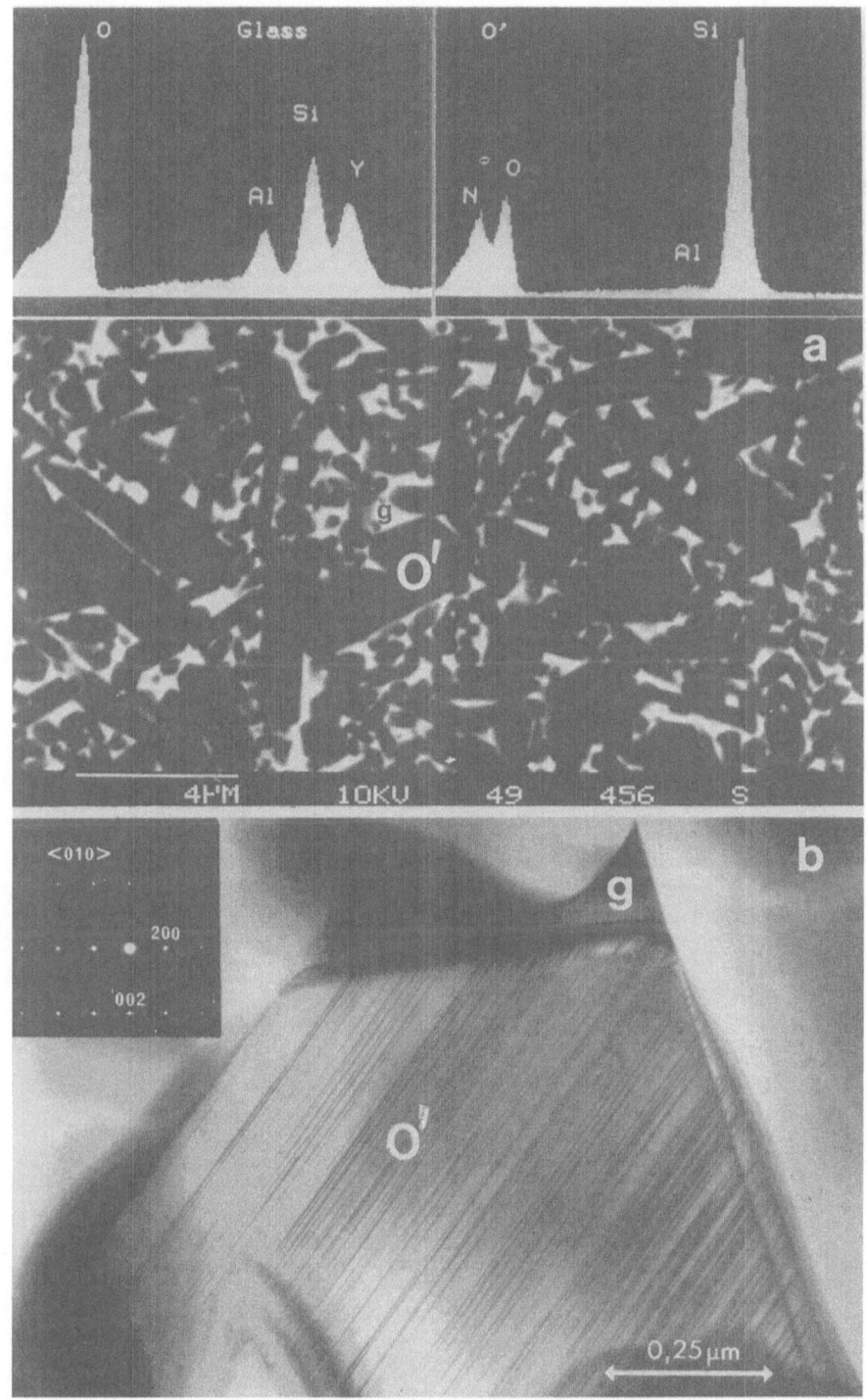

Fig. 8. (a) Si_2N_2O ceramic microstructure, showing the morphological
similarity with β´ ceramics when sintered with a Y–Si–Al–O–N
eutectic liquid. The TEM (b) reveals a high stacking fault
density parallel to the plane of oxygen atoms.

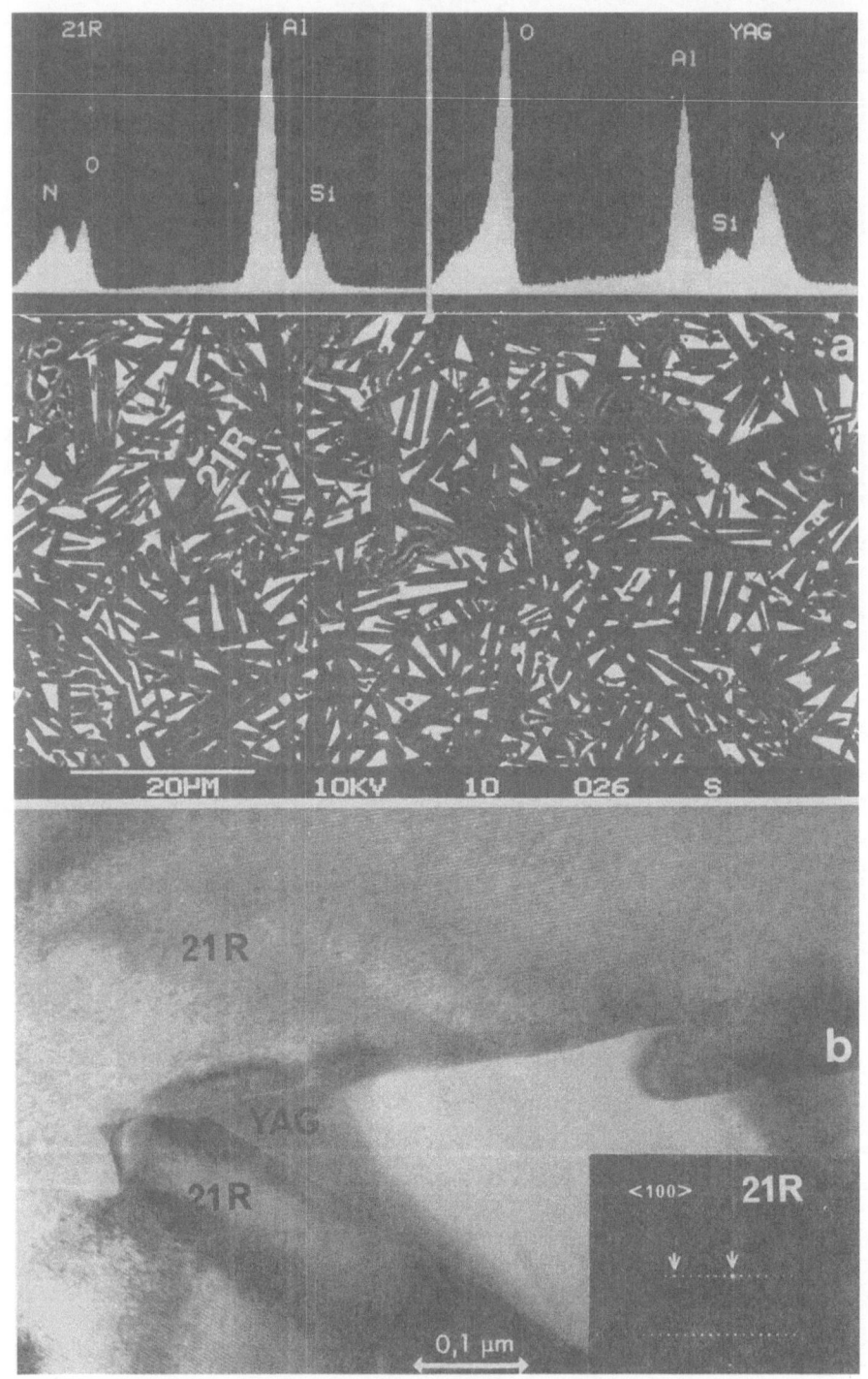

Fig. 9. A polytypoid ceramic microstructure showing course tabular 21R
crystals within a matrix of YAG. (a) SEM, (b) TEM.

reduces liquid volume and increases β'Al+O substitution level which, in a conservative system, amplifies the N/O ratio within the glass matrix.

The attainment of a bi–phase β'+YAG microstructure by total crystalli-sation of the matrix is dependent on high–nitrogen content and high purity of the glass. A prior nucleation heat–treatment (Fig.4) is also beneficial in avoiding low density heterogeneous nucleation on surfaces and internal defects with the development of large cellular growth units and associated intercellular segregation or cavitation due to the glass–crystal volume change (Fig.7 is an extreme example of this phenomenon). Long nucleation heat–treatments at temperatures close to Tg should be avoided, due to the increased probability of crystallisation of metastable phases. These are kinetically preferred within the high–viscosity matrix and hence normally have similar compositions to the parent glass. Such phases, which have been studied in Mg–Si–Al–O–N and Y–Si–Al–O–N glasses,[14,24] generally decompose above 1000–1100°C, well below the temperatures for intended ceramic application.

Following a conveniently short heat–treatment (a few hours between 1200–1400°C) crystallisation is nearly complete with isolated glass remaining in the more acutely–angled β'–β' junctions. This is normally of subcritical size for creep–cavity nucleation. Long heat–treatment, associated with creep deformation at 1300–1400°C, results in a morphological change characterised by the loss of β' facet planes due to equilibration of β'/YAG/β' junctions under conditions of solid–solid interfacial energy isotropy (Fig.10).

Alternative oxynitride ceramics

Silicon oxynitride (Si_2N_2O) ceramics sintered with Y–Si–Al–O–N eutectic liquids are microstructurally similar to β' ceramics (Fig.8). Low energy crystal/liquid facet planes in this structure result in a pseudo–hexagonal prism morphology bounded by (100) (110) and (1$\bar{1}$0) planes in the orthor-hombic lattice.[3] The distinction between the morphologically similar β' and Si_2N_2O ceramic microstructures (Fig.5 and 8) may be made via EDS or transmission electron microscopy; the singular (100) plane is a stacking-fault plane in Si_2N_2O crystals (Fig.8b) which have a relatively large oxygen content (Fig.8a inset). Variation in O/N content of the matrix glass to achieve a preferred oxide crystalline phase, such as YAG, is more difficult in this system due to the limited O' substitution level (Fig.1). Yttrium disilicate ($\beta Y_2Si_2O_7$) is the main crystallisation product.[3]

Polytypoid ceramics contain insufficient surface and dissolved Si+O to generate a ternary Y–Si–Al–O–N eutectic liquid when sintered with Y_2O_3+Al_2O_3 additions. The polytypoid phase grows via solution–reprecipi-tation as tabular crystals within an (Y+Al)–rich liquid, based on a eutectic near YAG. In the relatively low viscosity liquid polytypoid crystals grow rapidly and the matrix liquid spontaneously crystallises on cooling, normally as YAG (Fig.9).[18] The relatively course microstructure and lack of theoretical density generates inferior mechanical properties. Attempts to control polytypoid crystal growth and matrix crystallisation via SiO_2 additions results in a multiphase ceramic due to the absence of a binary tie–line between polytypoid and Y–Si–Al–O–N liquid phases.

HIGH–TEMPERATURE PROPERTIES

Oxynitride Ceramics other than those containing β' as a major phase have not yet achieved an equivalent degree of microstructural refinement for high–temperature application. Hence this survey of deformation and fracture mechanisms is based principally on research on β' ceramics.

Creep Deformation

A prominent feature of creep-curves for both hot-pressed monophase and pressureless-sintered, matrix crystallised, bi-phase ceramics is a long period of transient creep. This is due to the superposition of a primary visco-elastic, partially recoverable, creep mechanism and a micro-structural change. The primary creep may be removed by long prior heat-treatment at 1300–1400°C during which residual glass is crystallised (Fig.10). In high N Y-Si-Al-O-N bi-phase ceramics glassy residues continue to crystallise as YAG. In hot-pressed, nearly monophase ceramics glassy residues are stabilised by metallic additive and impurity ions (normally Mg and Ca) such that their out-diffusion to a SiO_2 surface 'sink' accelerates the final crystallisation step in oxidising atmospheres.[21,25] The final crystallisation product in Si-Al-O-N ceramics may be partially β', due to its solubility for the components of the Mg-depleted glass. An alternative is Si_2N_2O but this is not detected by diffraction.

Measurements of creep parameters, such as stress-exponents (n) or activation-energies (Q) in the equation for strain-rate $\dot{\varepsilon} = \text{const. } \sigma^n \exp(-Q/kT)$ are meaningless in the period of long transient creep. The equation refers to the true steady-state (linear) creep rate which is only reached after $\gg$ 100 hours. Under these conditions both types of ceramic exhibit stress exponents near unity, indicative of diffusional creep (Fig.11), with activation energies believed to be that for intergranular diffusion.[21,23] The terminal value for β'-YAG ceramics is higher ($\sim$1000 kJ mol^{-1}) than for monophase β' ceramics ($\sim$850 kJ mol^{-1}) due either to the necessity for diffusion of Si+N etc. along β'-YAG interfaces, in series with β'-β' interfaces, or to the influence of differing grain boundary impurities (Mg and Y). The perfection of crystallisation in these systems is such that a creep mechanism of viscous-flow or 'solution-reprecipitation with transport via a glassy intergranular phase[26] is unrealistic. Values of Q are much higher than reported for other βSi_3N_4 ceramics or for transport mechanisms in glasses.

The most important advance presented by these refined high-temperature ceramic microstructures is the demonstration of a diffusional creep mechanism and suppression of creep cavitation when junction glass residues are below a critical size for cavity nucleation under hydrostatic tension. The absence of creep cavitation is confirmed microstructurally and is consistent with the similarity in creep rates for bend and compressive loading and the absence of stress exponents n>1, typically observed in βSi_3N_4 ceramics (Fig.11). Thesholds for cavitation are a function of stress and temperature, as well as residual glass volume; the most recent β'+YAG ceramics are capable of non-cavitating diffusion-creep to 400 MPa and 1375°C but require oxidation-protection at these temperatures[23] (Fig.11).

There is now evidence for 3 main regimes of creep behaviour, mainly dependent on residual glass volume in β' ceramics:

(i) Prior to matrix crystallisation in bi-phase ceramics, at low stress and high temperatures, there is a free viscous flow of the glass phase between β' grains, without cavity nucleation.

(ii) If matrix crystallisation is incomplete (or in hot-pressed monophase ceramics with grain-junction residues) hydrostatic tension in assymetric glass pockets, due to β'-β' grain – boundary sliding, causes cavity nucleation. Cavity growth contributes to creep strain in addition to grain-boundary sliding.

(iii) For complete matrix crystallisation or in ideally monophase ceramics Coble creep, by interfacial or grain boundary diffusion, is the dominant rate-controlling mechanism.

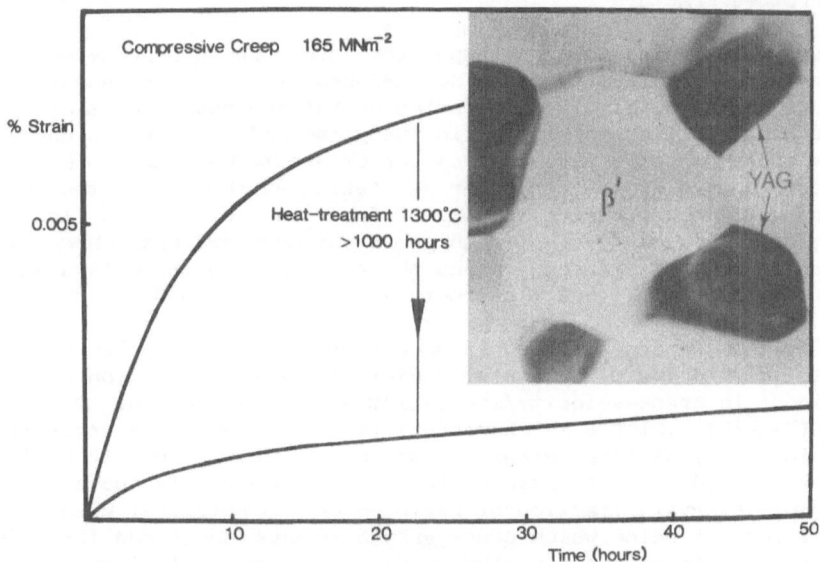

Fig. 10. Creep curves for a β'-YAG ceramic showing the reduction in
transient creep after long heat-treatment. The morphology
change in YAG (inset) is typical of specimens heat-treated
under creep conditions for 1000 hours.

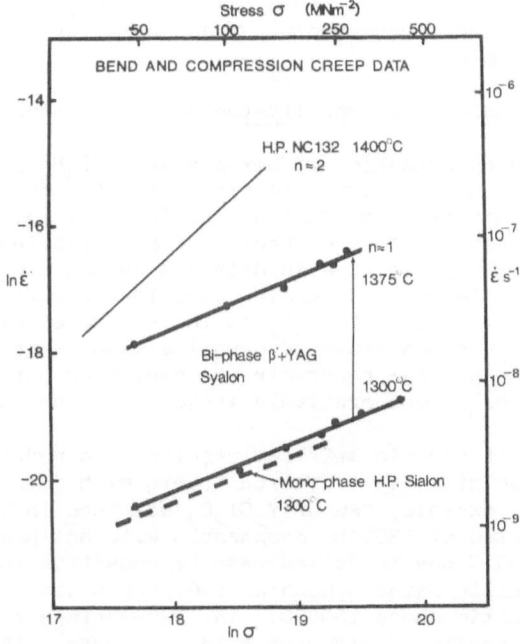

Fig. 11. Logarithmic strain rate vs stress plot under steady-state creep
conditions for mono and bi-phase β' ceramics, with comparative
data for a commercial hot-pressed ceramic (NC132) which undergoes creep cavitation. Stress exponent n≃1 is typical of diffusional creep in completely crystalline ceramics.

High--Temperature Fracture

An appropriate guide to high--temperature creep--rupture behaviour of
β' ceramics may be obtained from the notched--beam fracture toughness (K_{1c})
-- temperature relation (Fig.12). Ceramics containing phase--junction
glass residues exhibit a peak in K_{1c} in the glass softening range and an
optically--visible region of sub--critical crack growth on fracture surfaces.
This is due to a zone of intergranular cavitation at the tip of the macrocrack and an extra contribution to work of fracture from viscous flow of
glass bridging the crack together with grain pull--out and crack--branching.
A completion of crystallisation removes the K_{1c} peak and there is a small
reduction from the totally brittle--fracture level to ~1500°C.

This division in fracture behaviour correlates with a marked difference
in susceptibility to creep--rupture by a similar crack--propagation mechanism.
This is evident in crack--velocity/stress intensity data for the two microstructures (Fig.13). Plot A is characteristic of incompletely crystallised
ceramics which, only at large glass volumes may exhibit a low--stress intensity (K_1) 'threshold' effect, related to creep mechanism (i) above. Plot B
typifies complete crystallisation of residues or YAG matrices, with a threshold stress intensity below which crack growth is suppressed and there is
evidence for diffusional creep at the macro--crack tip. This correlates
with creep mechanism (iii) above. At high stresses crack growth has been
modelled by a coupled crack--surface/grain boundary diffusional transport
mechanism.[27,28].

Although it is unreliable to convert double--torsion data (Fig.13) to
quantitative creep--rupture data the trends may be predicted for the
different microstructural states (Fig.14). It is clear that the only
acceptable state for high--temperature application is that of complete
crystallinity, generating stress--rupture plots characterised by a high--stress
crack--blunting 'threshold'.

The Influence of Oxidising and Impurity--Containing Atmospheres

Initial research on oxidation of hot--pressed Si_3N_4 and β' Sialon
ceramics demonstrated the importance of viscosity of a passive SiO_2--rich
oxidation layer in controlling oxidation kinetics. Of special interest
was the indirect rate--controlling mechanism of metallic ion out--diffusion
from grain--boundary glassy residues in determining the SiO_2 layer
viscosity.[29,30] This, in turn, determines the inward diffusion rate of
oxygen or the dissolution rate of Si_3N_4 in the silicate layer (Fig.15a).
It is clear that pressureless--sintered ceramics with large volume glass
matrices which provide a large reservoir and easy transport path for
metallic ions will exhibit comparatively large oxidation rates.[30]

Crystallisation of ceramic matrices results in a reduced reactivity
of the SiO_2 -- metallic silicate diffusion couple with marked reduction in
oxidation rates. For example, YAG or $Y_2Si_2O_7$ matrices in Syalon ceramics
have low oxidation rates at 1300°C, comparable with hot--pressed β' monophase
ceramics. Above 1300°C the oxidation rate is sensitive to matrix crystal
chemistry; most metallic oxides, such as YAG, react with SiO_2 to form a
ternary silicate eutectic above 1300°C. The sub--surface reversion of matrix
to the liquid state greatly accelerates oxidation rate[31] (Fig.16). This is
a special problem in the preferred β'--YAG Syalon ceramics but has been
resolved by controlled transformation of a surface layer to $\beta' + 0'$
(substituted Si_2N_2O) via reaction of the type :

$$\beta' Si_3N_4 + SiO_2 \rightarrow 0'(Si_2N_2O)$$

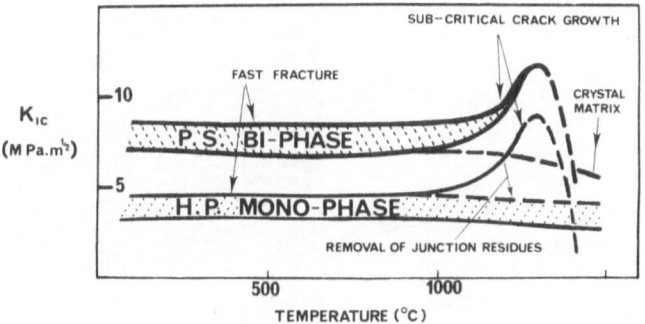

Fig. 12. Fracture-toughness (K_{1c}) – temperature data for β´ ceramics before and after matrix crystallisation (bi-phase β´-YAG) and following removal of junction glass residues (monophase ceramics).

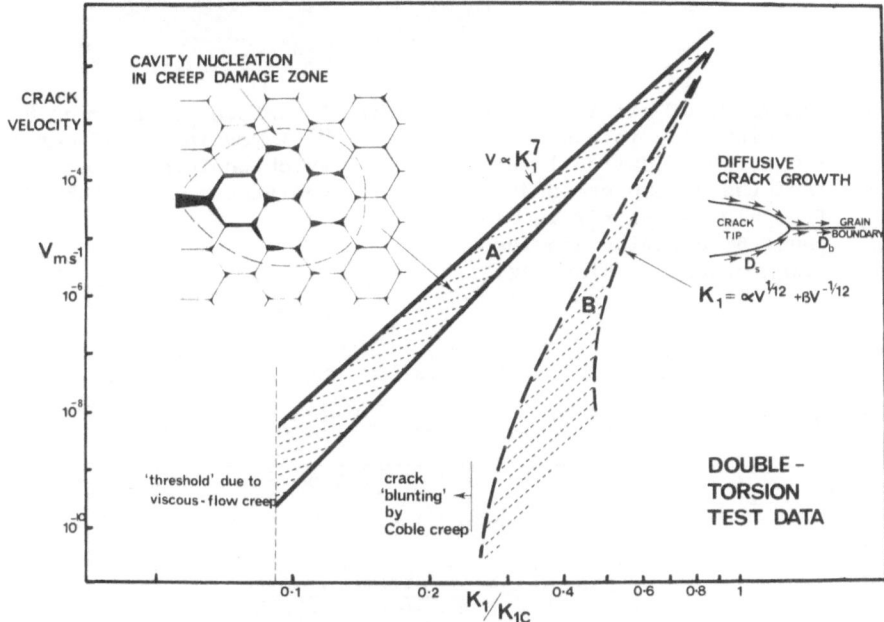

Fig. 13. A survey of high-temperature (1300–1400°C) slow crack growth data illustrating the change in growth mechanism for β´ ceramics with glassy residues (A) to microstructures which are strictly mono-phase or have fully crystalline matrices (B).

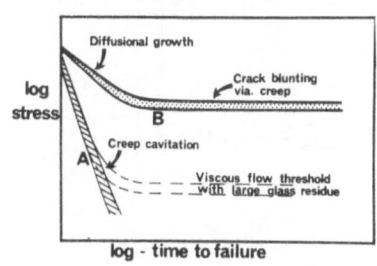

Fig. 14. Schematic creep-rupture diagram based on type A and B crack growth data of Fig. 13.

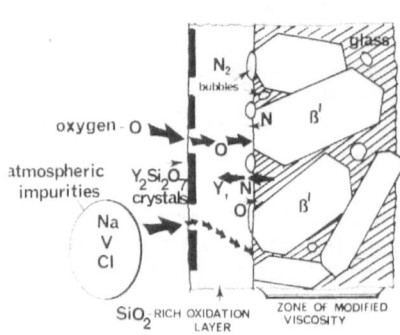

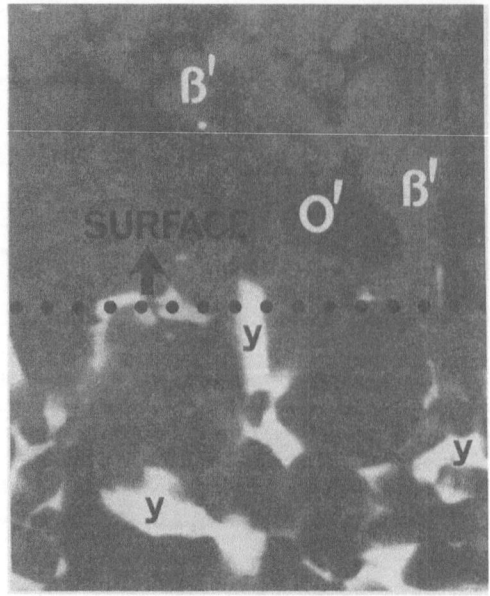

Fig. 15(a). Schematic of metallic ion diffusion into a SiO$_2$-rich oxidation layer from the atmosphere and from a glass matrix in bi-phase β´ ceramics accelerating oxidation kinetics.

Fig. 15(b). A surface transformed β´+ O´ layer in β´-YAG ceramic which confers oxidation-resistance above 1350°C.

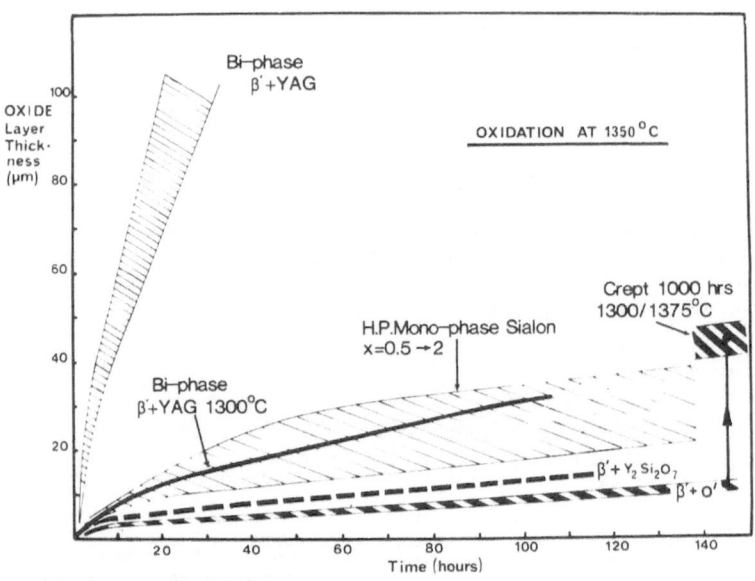

Fig. 16. A comparison of oxidation kinetics for β´ ceramics with different phase combinations within near-surface microstructures.

following out-diffusion of matrix yttrium ions.[32] The β´+O´ surface layer
(Fig.15b) has extremely limited oxidation kinetics (Fig.16) and has been
used in extending creep data in oxidising atmospheres to nearly 1400°C
Fig.11). In the absence of a surface transformation layer creep-rupture
lifetimes are negligible above 1350°C due to penetration of the YAG+liquid
reversion reaction into the bulk ceramic.

An additional problem in, for example, turbine applications is the
influence of impurities within a hot-gas atmosphere on oxidation kinetics.
Sodium or chlorine in a marine environment or metallic fuel impurities are
expected to reduce the viscosity of SiO_2 oxidation layers and hence accelerate oxidation even when out-diffusion of intergranular metallic ions is
suppressed (Fig.15a). This environmentally-induced mechanism has received
little attention and could be one of the most significant limitations to
application at temperatures above 1200°C.

CONCLUSION

It is clear that to achieve the flexibility in near-net shaping,
afforded by the pressureless-sintered fabrication route, in combination with
good high-temperature properties it is necessary to tailor initial compositions to generate a completely crystalline ceramic during a post-sintering
heat-treatment. In addition it may be necessary to accommodate the problem
of high-temperature oxidation by surface transformation or diffusion-barrier
coating. This processing sequence, summarised in Fig.4, has evolved to a
degree of complexity comparable with metallic alloys which may, in severe
applications, be replaced by ceramic alloys.

ACKNOWLEDGEMENTS

The material reviewed in this paper is derived from research programmes
supported principally by Lucas-Cookson-Syalon Ltd., the Wolfson Foundation
and Rolls-Royce Ltd.

The assistance of Mrs. P.M. Kirby in preparation of this manuscript
is gratefully acknowledged.

REFERENCES

1. K. H. Jack, J.Mater.Sci. 11 1135 (1976).
2. I. Idrestedt and C. Brosset, Acta.Chem.Scand. 18 1879 (1964).
3. M. H. Lewis, C. J. Reed and N. D. Butler, Mater.Sci.Eng., 71 87 (1985)
4. D. P. Thompson, in 'Nitrogen Ceramics', ed. F. L. Riley (Noordhoff,
     Leyden) 129 (1977).
5. N. D. Butler, R. Dupree and M. H. Lewis, J.Mater.Sci.Letters 3 469
     (1984).
6. K. H. Jack in 'Progress in Nitrogen Ceramics', ed. F. L. Riley
     (Martinus Nijhoff) 45 (1983).
7. G. G. Deeley, J. M. Herbert and N. C. Moore, Powder Metall. 8 145 (1961).
8. P. Drew and M. H. Lewis, J.Mater.Sci., 9 261 (1974).
9. G. E. Gazza, Bull.Amer.Ceram.Soc. 54 778 (1975).
10. F. F. Lange in 'Progress in Nitrogen Ceramics', ed. F. L. Riley
     (Martinus Nijhoff) 467 (1983).
11. M. H. Lewis, B. D. Powell, P. Drew, R. J. Lumby, B. North and A. J.
     Taylor, J.Mater.Sci., 12 61 (1977).
12. M. H. Lewis, A. R. Bhatti, R. J. Lumby and B. North, J.Mater.Sci.,
     15 103 (1980).
13. M. H. Lewis and R. J. Lumby, Powder Metall. 26 73 (1983).
14. G. Leng-Ward and M. H. Lewis, Mater.Sci.Eng., 71 101 (1985).
15. M. H. Lewis, A. R. Bhatti, R. J. Lumby and B. North, J.Mater.Sci.,
     15 438 (1980).

16. J. Dodsworth and D. P. Thompson, 'Special Ceramics 7', Proc.Brit. Ceram.Soc. 51 (1981).

17. R. K. Ball, M. H. Lewis, A. Szweda and E. Butler, Mater.Sci.Eng., 71 137 (1985).

18. N. D. Butler and M. H. Lewis, unpublished research, Univ. of Warwick (1983).

19. H. T. Larker in 'Progress in Nitrogen Ceramics', ed. F. L. Riley (Martinus Nijhoff) 717 (1983).

20. G. Bandyopadhyay, K. W. French and C. L. Quackenbush, Proc. 21st Automative Technology Development Meeting (Dearborn, Mich., Nov. 1983).

21. B. S. B. Karunaratne and M. H. Lewis, J. Mater.Sci. 15 449 (1980).

22. S. M. Winder and M. H. Lewis, J. Mater.Sci. Letters 4 241 (1985).

23. M. H. Lewis, S. Mason and A. Szweda, in Proc.Conf. on Non-Oxide Technical Ceramics, Limerick (to be published – Elsevier 1985).

24. S. Wild, G. Leng-Ward and M. H. Lewis, J.Mater.Sci. 16 1815 (1981).

25. B. S. B. Karunaratne and M. H. Lewis, J.Mater.Sci. 15 1781 (1980).

26. R. Raj, R. L. Tsai, J-G. Wang and C. K. Chyung in 'Deformation of Ceramics II', ed. R. E. Tressler and R. C. Bradt (Plenum Press) 353 (1984).

27. M. H. Lewis and B. S. B. Karunaratne in 'Fracture Mechanics of Ceramics, Rocks and Concrete' (ASTM-STP745) 13 (1981).

28. T. J. Chuang, J.Amer.Ceram.Soc., 65 93 (1982).

29. D. Cubicciotti and K. H. Lau, J.Amer.Ceram.Soc. 61 512 (1978).

30. M. H. Lewis and P. Barnard, J.Mater.Sci. 15 443 (1980).

31. M. H. Lewis, B. S. B. Karunaratne, J. Meredith and C. Pickering in 'Creep and Fracture of Engineering Materials and Structures', ed. B. Wiltshire and D. Owen (Pineridge Press) 365 (1981).

32. M. H. Lewis, G. R. Heath, S. M. Winder and R. J. Lumby in 'Deformation of Ceramics II', ed. R. E. Tressler and R. C. Bradt (Plenum Press) 605 (1984).

TAILORING OF THE THERMAL TRANSPORT PROPERTIES AND

THERMAL SHOCK RESISTANCE OF STRUCTURAL CERAMICS

D. P. H. Hasselman

Department of Materials Engineering
Virginia Polytechnic Institute and State University
Blacksburg, VA 24061

ABSTRACT

A survey is presented of the approaches which can be taken in the modifications of the thermal transport properties and thermal shock resistance of structural ceramics. Experimental data are presented which indicate that the thermal conductivity and thermal diffusivity as well as their dependence on temperature, are strongly affected by alloying and impurity elements and structural defects, and the presence of second phase dispersions including porosity and microcrack formation. Time-dependent effects such as crack closure and healing, and changes in crystallinity are shown to play a significant role as well. All these mechanisms, in principle, should allow enhancement of heat conduction as well as thermal insulating ability as governed by design requirements.

Thermal shock resistance can be improved by appropriate modifications of those properties which affect the magnitude of thermal stress and corresponding failure criteria as well as by control over the nature and extent of crack propagation. Microcracking is singled out as a highly effective mechanism for simultaneously improving thermal shock resistance as well as thermal insulating ability.

INTRODUCTION

The increasing trend in many fields of engineering such as energy-conversion, chemical processing and aerospace, towards ever-increasing operating temperatures places complex demands on the development and selection of materials to meet design requirements and long-term satisfactory performance. In general, candidate materials for functions which involve high temperature should exhibit melting points well in excess of the use temperature, as well as high chemical and structural stability. The thermophysical properties of the candidate materials which meet the above property requirements also can play a vital role in governing design and performance criteria. These thermophysical properties, on which this paper will concentrate, can include the thermal conductivity and diffusivity, the specific heat, coefficient of thermal expansion and thermal shock resistance.

For example, candidate materials for heat exchangers should have values of thermal conductivity as high as possible. In contrast, for many other designs, the choice of the optimum materials is based on its thermal insulating qualities, i.e., low thermal conductivity in order to keep heat losses to a minimum. Low thermal conductivity also is essential for heatshields and thermal barrier coatings to assure thermal protection for the underlying component or structure. For heat storage purposes candidate materials should have values of specific heat as high as possible in combination with high density in order to optimize the volumetric heat capacity. In contrast, low values of the volumetric heat capacity are essential for energy conservation for any high-temperature structure for which the recovery of the stored heat is not practical.

Dimensional changes due to the effects of thermal expansion on heating or cooling generally are undesirable. This requires that values of the coefficients of thermal expansion should be as close to zero as possible.

Catastropic thermal shock fracture is regarded as a major problem of candidate materials for engineering applications involving high temperature. Such thermal shock fracture results from the high levels of steady-state or transient heat fluxes to which structures and components at high temperatures are inevitably subjected. The resulting spatially non-uniform temperature distributions can lead to thermal stresses of high magnitude. Because of their nature of atomic bonding and more complex crystal structures candidate materials for high temperature service tend to be highly brittle. Because of such brittleness, failure under the influence of thermal stresses can be highly catastrophic, rendering the structure or component totally unsuitable for continued satisfactory service [1]. For this reason, in the design and selection of materials considerable emphasis must be placed on avoiding failure by thermal stresses. It has been established that materials with high resistance to the onset of crack propagation under the influence of thermal stresses should have values for the appropriate fracture stress and thermal conductivity in combination with low values for the coefficient of thermal expansion and Young's modulus of elasticity [2,3]. In practice, situations can be encountered in which the thermal stresses are of such high magnitude that even in the optimum material, the initiation of thermal stress failure cannot be avoided. Under these conditions the criterion for high thermal shock resistance is based on the ability of candidate materials to undergo rapid crack arrest following the onset of crack propagation [1]. It should be noted that the requirement of high thermal conductivity for high thermal shock resistance conflicts with the requirement of low thermal conductivity for those designs for which energy conservation is of primary importance. Clearly, for designs with conflicting property requirements appropriate trade-offs need to be made.

The purpose of this paper is to review the principles and methods by which the thermo-physical and -mechanical characteristics of structural ceramics can be modified to meet design requirements for specific applications and/or operating conditions.

DISCUSSION

A discussion of the tailoring of the thermo-physical properties of ceramic materials is carried out most effectively in terms of the basic phenomena which underlie the specific property of interest, illustrated by examples encountered in practice or literature data.

Thermal Conductivity, Diffusivity And Specific Heat

The thermal conductivity (K) of a material for uniaxial heat flow in the x-direction is defined by [4]:

$$q = -K \, dT/dx \tag{1}$$

where q is the heat flux per unit area and unit time, T is the temperature and dT/dx is the temperature gradient.

The thermal diffusivity of a solid can be defined in terms of the transient heat conduction equation [4]:

$$K \, \partial^2 T/\partial x^2 = c\rho \partial T/\partial x \tag{2}$$

where K, ρ and c represent the thermal conductivity, density and specific heat, resp. The thermal diffusivity (κ) is defined by the ratio:

$$\kappa = K/\rho c \tag{3}$$

The thermal diffusivity which represents the "diffusion constant" for the propagation of a spatially non-uniform temperature is a more fundamental property of a material than the thermal conductivity, which represents the product of the thermal diffusivity and the volumetric heat capacity.

The specific heat (c) of a material is defined by:

$$c = dQ/dT \tag{4}$$

where Q is the internal energy per unit mass. The quantity $C = c\rho$ is the volumetric heat capacity.

In general, the conduction of heat in solids is controlled by variables at the atomic level and at the microstructural or continuum level.

At the atomic level, the conduction of heat in single-phase materials occurs by three primary mechanisms, namely phonon, electron [5,6] and photon transport [7], the latter mechanism being effective primarily at elevated temperatures. Theoretical calculations of the thermal conductivity for any of these three mechanisms relies on the general equation:

$$K = 1/3Cvl \tag{5}$$

where C is the volumetric heat capacity, v is the carrier velocity and l is the "mean-free-path" between collisions of the phonon or electron or the distance of propagation of a photon between emission and re-absorption. In eq. 5, the quantity 1/3vl represents the thermal diffusivity. For this reason, theoretical estimates for the thermal conductivity rely on separate derivations of the specific heat, the carrier velocity and the mean-free path. From the perspective of the primary objective of this paper, the specific heat and carrier velocity for any given material at any temperature are intrinsic properties not strongly affected by extrinsic variables. Any major changes in the thermal conductivity and diffusivity can be brought about by corresponding changes in the mean-free-path only.

The mean-free-path is controlled by intrinsic variables such as phonons and thermally induced vacancies as well as extrinsic variables such as alloying elements, impurities, vacancies, crystal defects, elastic and optical discontinuities, etc., which can lead to phonon and electron scattering or the absorption of photons.

The total mean-free-path, l_{total} in terms of the intrinsic and extrinsic mean-free-path, l_{int} and l_{ext}, resp., is:

$$1/l_{total} = 1/l_{int} + 1/l_{ext} \tag{6}$$

which indicates that l_{total} in a material in which the mean-free-path is affected by extrinsic variables is always less than l_{int}.

At the microstructural or continuum level, the conduction of heat can be affected by thermal discontinuities such as pores, cracks or other thermal barriers and second phase inclusions, which all serve to modify the internal temperature distributions within the solid and associated effective thermal diffusivity and conductivity. Theoretical studies of these effects have been extensive [8-14].

As an example, the upper (K+) and lower (K-) bounds on the thermal conductivity of a two-phase composite is given by [11]:

$$K+ = K_1 V_1 + K_2 V_2 \qquad (7)$$

and

$$1/K- = V_1/K_1 + V_2/K_2 \qquad (8)$$

where V and K are the volume fraction and the thermal conductivity, resp. and the subscripts 1 and 2 refer to the two phases.

For a matrix with spherical inclusions, the effective thermal conductivity, K_c is [8,9]:

$$K_c = K_o [2K_o + K_d - 2V_d (K_o - K_d)] / [2K_o + K_d + V_d (K_o - K_d)] \qquad (9)$$

where K_o and K_d are the values for the thermal conductivity of the matrix and dispersed phase, resp. and V_d is the volume fraction of the dispersed phase.

For the special case of dilute concentrations of a pore phase, with $K_d = 0$, eq. 9 becomes:

$$K_c = K_o (1 - 3V_d/2) \qquad (10)$$

The effect of cracks on thermal conductivity was studied as well. For a matrix with dilute concentrations of randomly oriented penny-shaped cracks of equal size, the effective thermal conductivity is [14]:

$$K_{eff} = K_o (1 + 8Nb^3/9)^{-1} \qquad (11)$$

where K_{eff} and K_o represent the thermal conductivity of the matrix with or without cracks, resp., N is the crack density and b is the radius of the cracks.

The volumetric heat capacity of a composite, C_c is:

$$C_c = V_1 C_1 + V_2 C_2 \qquad (12)$$

In terms of the objective of tailoring the thermo-physical properties of structural ceramics, a number of general conclusions can be drawn from the above discussion: The value of thermal conductivity or diffusivity governed by the intrinsic mean-free-path represents the theoretical maximum value for a single-phase material. The presence of any compositional or structural lattice defects which cause the creation of an extrinsic mean-free-path will also serve to decrease the effective thermal conductivity and diffusivity. Pores and cracks will result in a similar effect. Increases in thermal conductivity can be achieved by reducing the concentration of lattice defects which control the extrinsic mean-free-path or by second phase dispersions with a value of thermal conductivity higher than the matrix phase. In order to achieve this latter effect, good interfacial adhesion between the matrix and dispersed phase is essential to assure good thermal contact. On the other hand, poor adhesion should create a thermal barrier resistance, which can assist in decreasing the thermal conductivity if so desired. Changes, either upward or downward in the specific heat (per unit mass or unit volume) can be achieved effectively by the addition of a second phase with higher or lower specific heat, as desired.

The above approaches provide a fairly wide latitude in tailoring the thermal conductivity/diffusivity and specific heat to meet specific design requirements, which can be illustrated by experimental data for either thermal conductivity or thermal diffusivity. Thermal conductivity at low to moderate temperatures is measured most conveniently by steady-state methods. At high temperatures, for reasons of experimental convenience, the measurement of the thermal diffusivity by a transient method such as the flash-diffusivity method [15] is preferred. Separate measurement of the specific heat and density permits the calculation of the thermal conductivity from the thermal diffusivity and vice-versa.

The effect of the above variables on heat conduction behavior can be illustrated by a number of examples.

Fig. 1 shows the effect of non-stoichiometry on the thermal diffusivity of TiO_2 measured by Siebeneck et al [16]. The data indicate that the thermal diffusivity decreases rapidly with increasing oxygen deficiency as the result of phonon scattering at the vacant oxygen lattice sites. In general, vacancies are particularly effective in lowering the thermal diffusivity. The effectiveness of impurity or alloying element as phonon scatterers increases with the increasing differences in their atomic mass and the corresponding mass of the host lattice [17]. For vacancies such mass difference reaches a maximum.

The effectiveness of lattice vacancies in decreasing the thermal diffusivity also is demonstrated by the data shown in fig. 2, for the thermal diffusivity of monoclinic and cubic zirconia with 5 wt % MgO. In the monoclinic zirconia the magnesia exists as a grain-boundary phase, whereas in

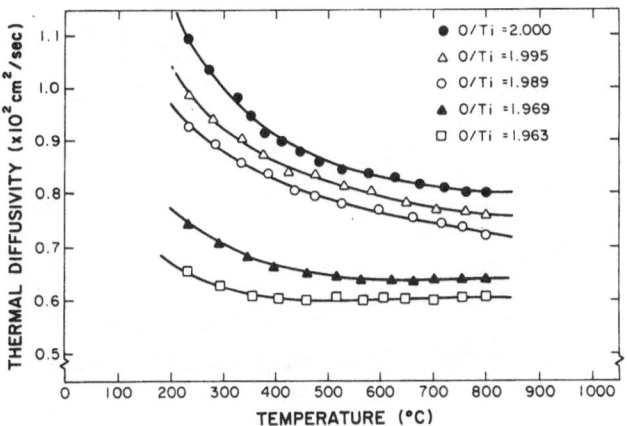

Fig. 1. Effect of non-stoichiometry on the thermal diffusivity of titanium dioxide [16].

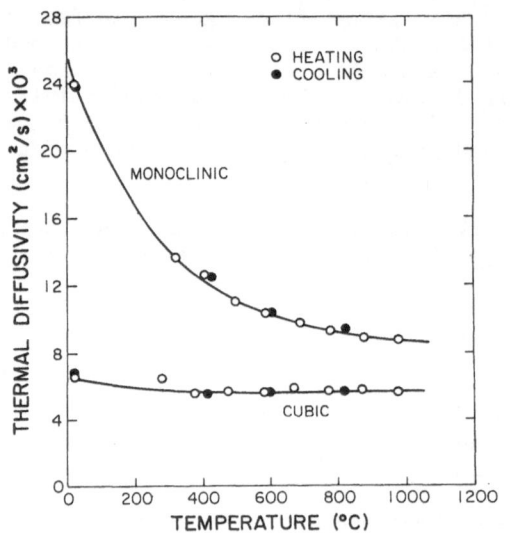

Fig. 2. Comparison of thermal
diffusivity of zirconia
with 5 wt. % magnesia in
the monoclinic or cubic
crystal structure.

the cubic zirconia the magnesia exists in the form of a solid-solution with excess vacancies on the oxygen sub-lattice. At room temperature, the vacancies lower the thermal diffusivity by a factor in excess of three.

Solid-solution alloying of structural ceramics also serves to lower the thermal diffusivity or thermal conductivity significantly. Examples include solid-solutions of alumina and silicon nitride to form sialon [18], as shown in fig. 3, and the alloying of alumina with chromia [19,20].

Impurities have a significant effect on the thermal conductivity. In particular, electrically active impurities in single and polycrystals of such materials as silicon, silicon carbide, aluminum nitride and diamond appear to be particularly effective in lowering the thermal conductivity [21]. As an example of the impurity effect, fig. 4 compares the thermal diffusivity of an industrial grade silicon with about 0.6 wt % B, 0.1 wt % Fe, 0.1 wt % Cu as principal impurities and a purified grade of silicon with

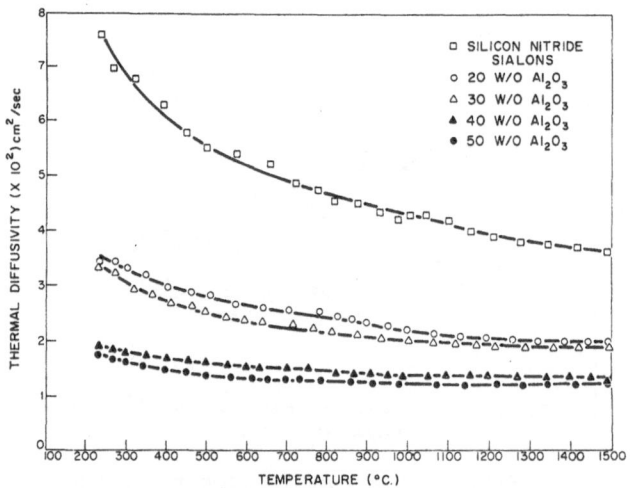

Fig. 3. Thermal diffusivity of sialons with various
compositions [18].

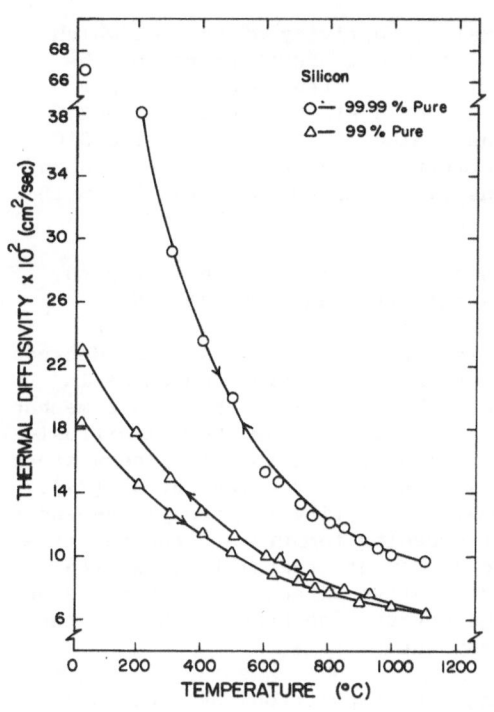

Fig. 4.   Thermal diffusivity of
a pure and impure
silicon [22].

at least 99.99% purity [22]. This latter material clearly exhibits a much
higher thermal diffusivity then the industrial grade silicon. These differences are particularly pronounced at the lower ranges of temperature and
are relatively less at the higher temperatures. This occurs because at the
higher temperatures the thermal diffusivity already is suppressed by
increased phonon-phonon interactions, so that impurities become less effective in lowering the thermal diffusivity even further. Fig. 4 also indicates that on return to room temperature the thermal diffusivity has a
higher value than during the heating part of the cycle. The reason for
this is not known, but could be the result of impurity segregation or other
effects. Regardless of the reason, however, this observation suggests that
the thermal diffusivity of impure material could be a function of prior
thermal history.

Experimental data for the thermal diffusivity at room temperature for
structural polycrystalline silicon carbide measured by this writer and co-
workers have ranged from in excess of 1.0 $cm^2.s^{-1}$ for a high-purity grade
to as low as 0.25 $cm^2.s^{-1}$ for silicon carbide with high level of impurities
and sintering aids added to improve processibility. This range of data
indicates that the heat conduction behavior of silicon carbide can be
tailored over a wide range to meet specific design requirements. Especially,
the lowering of the thermal conductivity and diffusivity is accomplished
very easily.

Increasing the thermal conductivity of structural silicon carbide to
its optimum value for pure silicon carbide is not easily accomplished.
Nevertheless, the development of silicon carbide and other materials such
as aluminum nitride with very high thermal conductivity is critical for
substrate purposes in the electronics industry for high-density electronic
packaging with high energy dissipation per unit volume. In this respect,
the development of polycrystalline silicon carbide with approx. 1 wt % BeO
with high thermal conductivity and increased electrical resistivity must be
regarded as a breakthrough [23]. The mechanism by which the addition of
the BeO leads to these improvements is not quite clear. It cannot be

attributed to the high value of the thermal conductivity of the BeO which is comparable to the value of pure silicon carbide. Furthermore, on the basis of composite theory, 1 wt % BeO would be expected to have a minor effect only. Instead, some effect based on solid state principles must be involved. The rather high value of the dielectric constant of the SiC with BeO, suggests that the BeO serves to introduce a resistive grain boundary layer combined with an increased density of charge carriers within the grains.

Amorphous dielectric materials, because of a limited mean-free-path, generally exhibit much lower thermal conductivity than materials which exhibit long-range crystallinity. For this reason, modification in the degree of crystallinity should provide a mechanism for increasing or decreasing the heat conduction properties of those materials which can exist in either the amorphous or crystalline state. The effectiveness of this mechanism is illustrated in fig. 5 which shows the value for the thermal diffusivity of a series of cordierite specimens ranging from a totally glassy state to the fully crystallized glass-ceramic, as measured by Chyung et al [24]. Especially at room temperature, the thermal diffusivity is strongly dependent on the crystallization treatment. With increasing temperature the relative difference in thermal diffusivity for the different crystallization treatments becomes less because of the greater relative effect of temperature on the length of phonon mean-free-path in the crystalline samples than on the samples with a greater proportion of glassy phase.

Crystallization, coupled with a change in composition, of amorphous silicon carbide fibers as indicated by the data shown in fig. 6, causes a permanent increase in the thermal diffusivity of those fibers infiltrated with CVD-SiC, as observed by Tawil et al [25]. Especially, heating to temperatures as high as 1800°C causes a major increase in the thermal diffusivity at room temperature.

Crystallization of the grain boundary phase in a sialon ceramic also was observed to lead to a significant increase in the thermal diffusivity at room temperature [26]. In general, these observations suggest that materials subject to crystallization or devitrification will exhibit thermal transport properties which can be a function of thermal history.

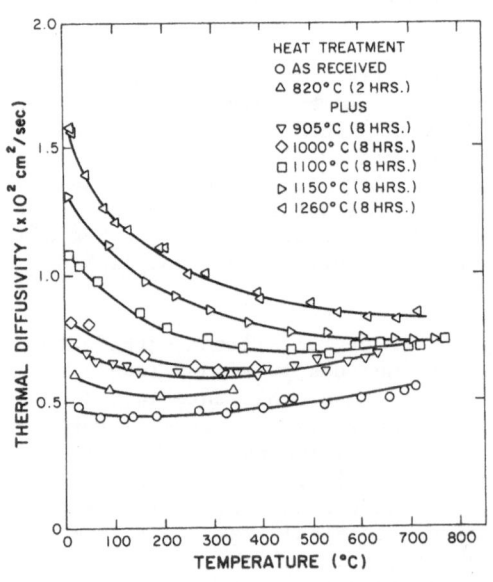

Fig. 5. Effect of crystallization treatment on the thermal diffusivity of a cordierite glass-ceramic [24].

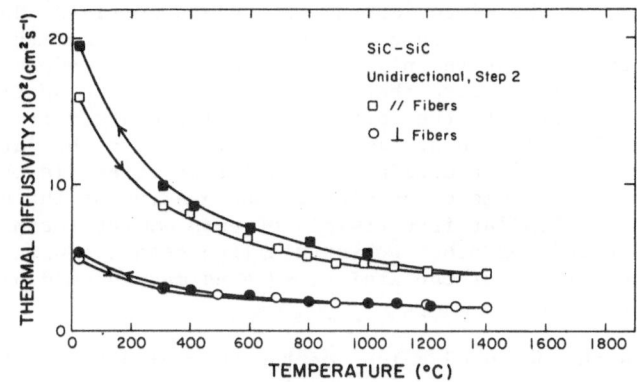

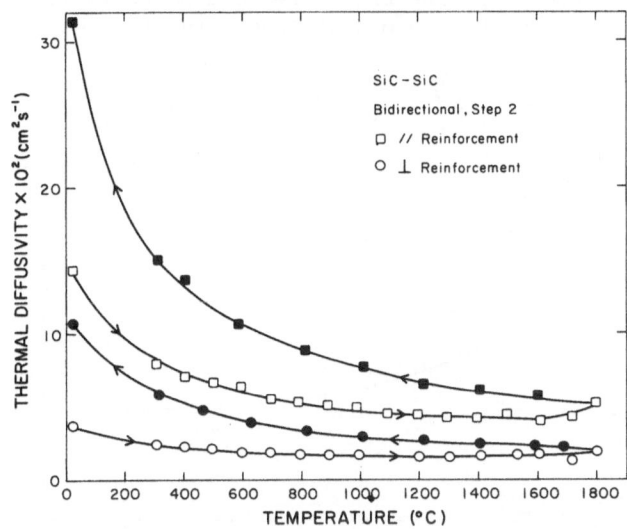

Fig. 6. Effect of crystallization of amorphous silicon carbide
fibers on thermal diffusivity during thermal cycling
of SiC fiber reinforced CVD-SiC, a: heated to 1400°C,
b: heated to 1800°C [25].

A number of examples, illustrative of the effect of variables at the
microstructural level on the thermal diffusivity and conductivity can be
given as well.

Microcracking, as discussed earlier, can have a significant effect on
thermal conductivity or thermal diffusivity. Microcracking in structural
ceramics can result from spatially non-uniform thermal expansion. In
single-phase polycrystalline ceramics spontaneous microcracking can arise
from internal stresses due to the thermal expansion anisotropy of the
individual grains. Microcracking in brittle matrix composites can occur
because of mismatches in the coefficients of thermal expansion of the
individual components. The incidence of microcracking, however, is a function of the grain size, with no microcracking occurring below a specific
value of critical grain size.

Figs. 7a and 7b compare the experimental data for the thermal diffusivity as a function of temperature of a fine-grained microcrack-free iron-
titanate and a coarse-grained heavily microcracked iron-titanate, resp. [27].

These data indicate that the microcracking causes a significant decrease in thermal diffusivity. Also to be noted is that the data for the micro-crack free iron-titanate shows the negative temperature dependence typical for a dielectric material and also that the data obtained on heating and cooling coincide. In contrast, the data for the microcracked iron-titanate show an irreversible behavior on heating and cooling, with the data on heating above those obtained on cooling. This latter effect is thought to be attributable to irreversible crack closure and healing at the higher levels of temperature. Similar irreversible effects on the thermal diffusivity have been observed for other polycrystalline ceramics with high thermal expansion anisotropy of the grains, such as magnesium dititanate and aluminum titanate [28,29].

Fig. 8 compares the measured values of the thermal diffusivity of microcracked composites consisting of a continuous matrix of polycrystalline magnesia with a dispersed phase of silicon carbide, with the values calculated from composite theory for the crack-free composite [30]. The data clearly indicate that microcrack formation lowers the thermal diffusivity significantly.

a.

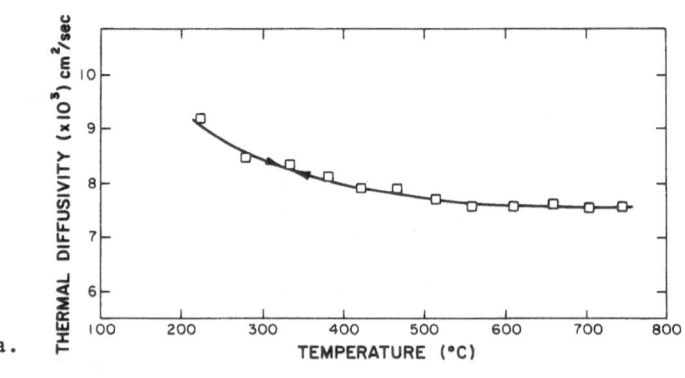

b.

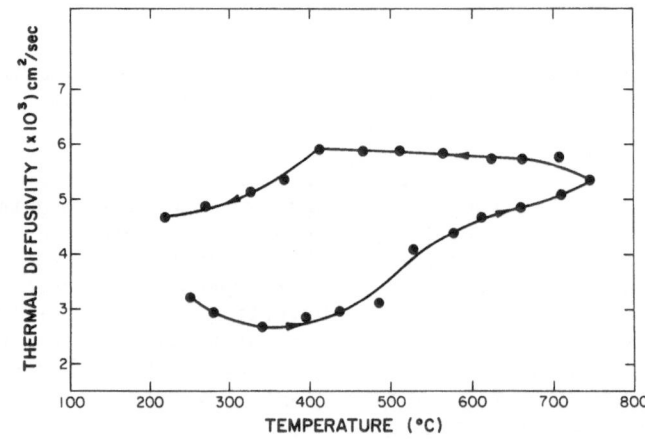

Fig. 7.   Temperature dependence of iron titanate,
          a: fine-grained microcrack-free and
          b: coarse-grained, heavily microcracked [27].

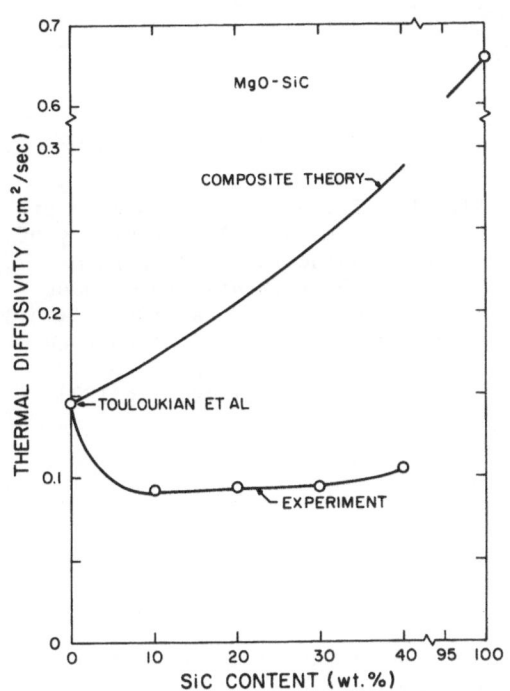

Fig. 8. Comparison of calculated values of thermal diffusivity at room temperature of composites of magnesium oxide without microcracks and experimental data [30].

Fig. 9 shows the data of the temperature dependence of a magnesia matrix with 20 % SiC. On cooling the data exceed those obtained on heating, again as for the data for the iron titanate, attributable to the closure and healing of cracks at the higher values of temperature. In general, microcracking has little effect on the specific heat per unit volume. For this reason the relative effect of microcracking on thermal diffusivity and thermal conductivity should be almost identical. For the same reason, microcracking is a particularly attractive mechanism for lowering the thermal diffusivity for such applications as heat-shields for which high values of the volumetric heat capacity are a critical design requirement.

Second-phase dispersions can be used very effectively to increase or decrease the thermal diffusivity or conductivity. A second phase with zero thermal conductivity such as porosity should be most effective in that respect as a pore phase will not contribute to heat conduction at all, at least over those temperature ranges at which radiative heat transfer is negligible. Generally, it was found that for randomly-oriented pores one percent of porosity decreases the thermal conductivity by approx. three percent, which is greater than the 1.5% predicted by eq. 10 for dilute concentrations of spherical pores. This difference most probably is attributable to the deviation of the pore geometry from the ideal spherical geometry assumed for the derivation of eq. 10.

Although a dispersed pore phase can be used effectively to improve thermal insulating ability, at any given value of density a fibrous microstructure will be far more effective in achieving low values of thermal conductivity than a dispersed pore phase. The reason for this is that fibrous microstructures exhibit poor thermal contact between the fibers which effectively suppresses the contribution of heat flow through the solid. The fibers also inhibit the contribution of convection to the total heat transfer as well. A further advantage of a fibrous microstructure is that it exhibits a low volumetric heat capacity which can significantly reduce the total energy required to reach a given temperature. This latter

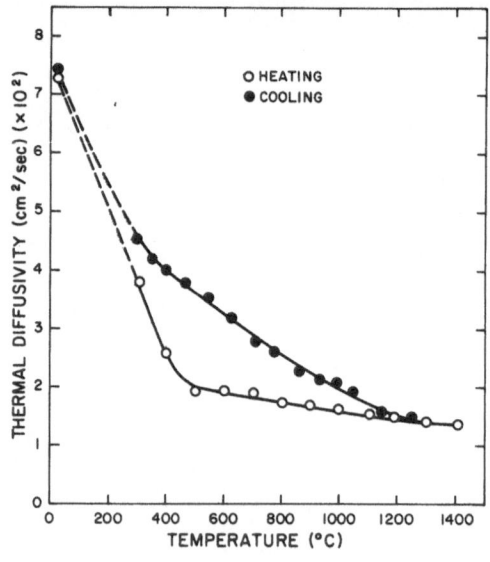

Fig. 9. Temperature dependence of
the thermal diffusivity of
magnesium oxide with a
20 wt. % dispersed phase
of silicon carbide, showing
the effect of microcrack
healing and re-formation
on thermal cycling [30].

aspect is critical for such structures as furnaces and kilns subject to
cyclic heating and for which the energy required can only be partially
recovered if at all. Fibrous insulating materials also have the advantage
of some degree of structural stability and therefore require minimal other
mechanical support.

Excellent thermal insulating properties are exhibited by porous
materials in the form of fine powder [7]. In such materials the conduction
of heat in the solid phase is suppressed due to the very poor particle-to-
particle thermal contact. Furthermore, due to the small dimensions of the
inter-particle spaces, heat conduction in the gas-phase must occur by the
inefficient process of heat conduction rather than by gaseous convection.
For such materials, the composition of the gas phase affects the overall
thermal conductivity which is lowest for a vacuum. One disadvantage of
powder insulations is that because of their lack of structural integrity
they require external support.

Increases in thermal diffusivity or conductivity can be achieved by
the addition of a dispersed phase with a value of thermal conductivity
higher than the matrix phase. As an example of this latter effect, fig.
10 compares the values of the thermal diffusivity of a lithium-alumino-
silicate (LAS) glass-ceramic without and with a uniaxial reinforcement of
approx. 35 vol. % carbon fibers measured along the fiber direction [31].
These data indicate that the carbon fiber reinforcement increases the
thermal diffusivity by a factor in excess of an order of magnitude. Such
an increase is expected because of the high value of thermal conductivity
of the fibers along their axial direction which corresponds to the basal
plane of the graphite crystal structure. As judged by data for pyrolytic
graphite [32,33,34] the thermal diffusivity perpendicular to the basal
plane is some two orders of magnitude lower than parallel to the basal
plane.

Fig. 11 shows the data for the thermal diffusivity of the lithium-
alumino-silicate (LAS) glass-ceramic matrix uniaxially reinforced with
carbon fibers perpendicular to the fiber direction. These data indicate
that the diffusivity in this direction only slightly exceeds the

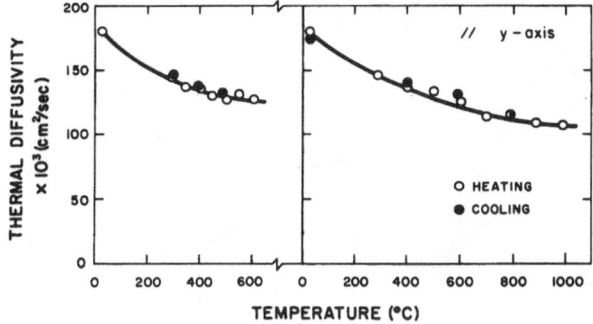

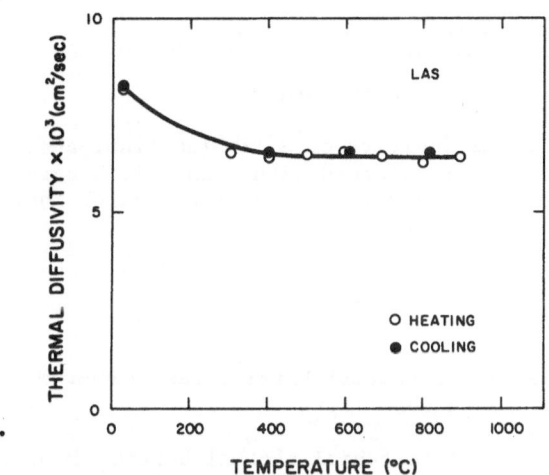

Fig. 10. Thermal diffusivity of a lithium-alumina-silicate
glass-ceramic, a: without and b: with ∿ 35 vol. %
uniaxial carbon fiber reinforcement parallel to
fiber direction [31].

corresponding value of the LAS matrix. The reason for this is two-fold.
Not only do the fibers exhibit a low thermal conductivity perpendicular to
their axial direction, but the LAS matrix being continuous is the controlling
phase in establishing the over-all value of the thermal diffusivity. The
resulting highly anisotropic thermal diffusivity and corresponding thermal
conductivity can be used to advantage for design geometries such as hollow
cylinders for which high axial thermal conductivity is required to reduce
the incidence of thermal stress fracture due to axial temperature variations,
but which at the same time require high thermal insulating ability in the
radial direction for purposes of improved energy efficiency. This combination of design criteria, for instance, is encountered for the cylinder walls
of internal combustion engines.

Fig. 11 shows another interesting feature, namely that on heating and
cooling the thermal diffusivity exhibits an irreversible behavior. Because
this effect is not evident for heat conduction parallel to the fiber direction, it is thought that this irreversibility is related to the existence

743

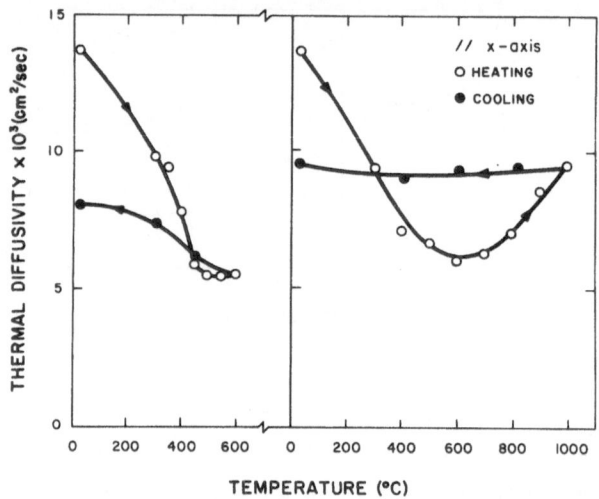

Fig. 11. Thermal diffusivity of a lithium-alumina-silicate
uniaxially reinforced with ∿ 35 vol. % carbon
fibers perpendicular to the fiber direction.

of a thermal-history-dependent thermal barrier resistance at the interface
between the LAS matrix and carbon fibers.

The significance of an interfacial thermal barrier in decreasing the
thermal diffusivity of composites is also indicated by data in fig. 12 for
a soda-borosilicate glass with Ni spherical inclusions, measured by Powell
et al [35]. In these composites there is little or no adherence between the
glass matrix and Ni dispersions. Furthermore, the Ni metal has a coefficient
of thermal expansion which is much larger than the corresponding value for
the borosilicate glass matrix. As a result, on cooling from the temperature
of ∿ 700°C at which these composites were hot-pressed, the Ni dispersions
shrink away from the glass matrix leaving an interfacial gap. The width of
this gap and associated thermal barrier resistance increases with increasing
degree of cooling. This effect, in turn, results in a temperature dependence
of the thermal diffusivity which is strongly positive, not normally encountered for dielectric materials. These observations suggest that at least in
principle, by modifications of the nature of adhesion between the components,
a degree of control can be exercised over the heat conduction behavior and
associated temperature dependence of composites.

As a final example of the tailoring of the heat conduction behavior of
materials by the combination of a number of different mechanisms, fig. 13
shows the data for the thermal diffusivity at room temperature of solid-
solutions of alumina and chromia, without and with a zirconia dispersed
phase [36]. These materials were developed to meet the contradictory
requirements of low thermal conductivity for purpose of minimizing heat loss
in combination with improved thermal shock resistance. This latter mechanism
was achieved by the mechanism of transformation-toughening due to the
tetragonal-to-monoclinic phase transformation of the zirconia dispersed
phase, which serves to enhance crack stability and limit the extent of crack
propagation. The data shown in fig. 13 indicate that the presence of the
chromia causes a significant decrease in the thermal diffusivity. A similar

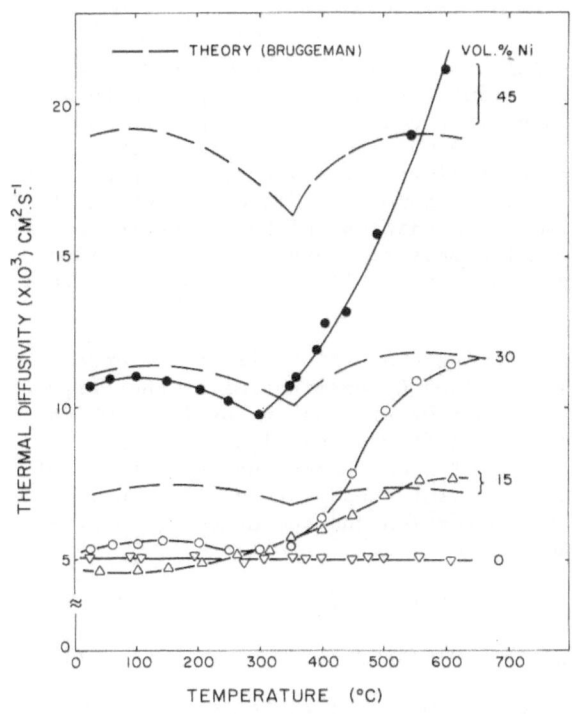

Fig. 12. Thermal diffusivity of a borosilicate glass with nickel dispersion showing the effect of a temperature dependence of an interfacial thermal boundary resistance.

effect is achieved by the zirconia dispersed phase. Furthermore, at the lower chromia content grain-growth during hot-pressing was significantly extensive that the critical grain size for spontaneous microcracking was exceeded. As a result, for the alumina-chromia-zirconia advantage is taken of three different mechanisms for lowering the thermal diffusivity.

The above discussion and examples illustrate that over a fair range of latitude a variety of methods is available by which the heat conduction behavior of structural ceramics can be modified to suit specific design requirements encountered in engineering practice.

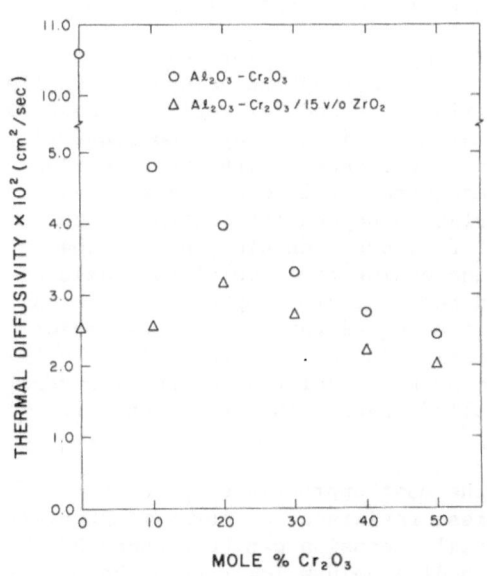

Fig. 13. Thermal diffusivity at room temperature of transformation toughened solid-solutions of alumina and chromia [36].

The resistance of ceramic materials to failure by thermal shock can be improved by taking two distinctly different routes. The resistance to the onset of thermal stress fracture can be improved by the appropriate modification of those properties which affect the magnitude of thermal stress in combination with the improvement in the failure stress or other appropriate failure criteria depending on the specific thermal conditions to which the ceramic is subjected. Alternatively, for conditions of heat transfer of such severity that the onset of thermal stress fracture cannot be avoided, thermal shock resistance can be improved by controlling the mode and extent of crack propagation.

The thermal, mechanical and other properties which affect the onset of thermal stress fracture can be ascertained by examination of the "figures-of-merit" or "thermal stress resistance parameters" on the basis of which the relative thermal stress resistance of different candidate ceramic materials can be judged. A large number of such thermal stress figures-of-merit were derived [37,38]. Some of those for steady-state heat flow, convective transient heat transfer by convection and radiation for opaque and semi-transparent ceramics include:

$$\sigma_f(1-\nu)/\alpha E; \quad \sigma_f(1-\nu)K/\alpha E; \quad \sigma_f(1-\nu)\kappa/\alpha E$$

$$\sigma_f(1-\nu)K/\varepsilon\alpha E; \quad \sigma_f(1-\nu)K/\varepsilon\mu\alpha E \qquad (13)$$

in which σ_f is the tensile fracture stress, ν is Poisson's ratio, K is the thermal conductivity, κ is the thermal diffusivity, α is the coefficient of thermal expansion, E is Young's modulus, ε is the emissivity (with $\varepsilon = 1-r$, where r is the reflectivity) and μ is the absorption coefficient.

The resistance of a ceramic material is directly proportional to the magnitude of the above figures-of-merit. These also indicate that thermal stress resistance is not governed by a single material property but depends on the combined effect of a number of material properties. Furthermore, the relative role of any of these individual properties depends on the specific heat transfer conditions to which the ceramic is subjected. For this reason, a detailed analysis of the thermal environment is essential in order to assess the most effective approaches towards the improvement of the thermal stress resistance of a candidate ceramic for a specific application.

In general, the above figures-of-merit suggest that improvements in thermal stress resistance can be achieved by increasing the values of tensile strength, thermal conductivity, thermal diffusivity and reflectivity, coupled with decreases in the coefficient of thermal expansion, Young's modulus, Poisson's ratio, emissivity and absorption coefficient. With the exception of tensile strength, for a pure single-phase structural ceramic the above properties are governed by interatomic variables beyond the control of the materials technologist. The principles of fracture mechanics suggest that the tensile strength can be increased by increasing the size of the failure-initiating flaws coupled with increases in the fracture toughness, achieved by microstructural control and optimum processing techniques to avoid major internal or surface flaws. However, increasing thermal stress resistance by increasing tensile strength can lead to problems related to the improvement of thermal shock resistance by controlled crack propagation, as will be discussed later.

Composite development appears to be the most appropriate approach towards the improvement of the thermal stress resistance of ceramic materials. Elastic moduli and the coefficient of thermal expansion can be reduced by the introduction of second phases with corresponding values lower than the matrix phase. In part, this approach serves to explain the improvement of the thermal stress resistance of such materials as carbides, nitrides and oxides by the incorporation of a dispersed phase of graphite or boron nitride [39]. As discussed earlier, the thermal conductivity and diffusivity can be improved by the introduction of dispersed phases with thermal conductivity higher than the matrix phase. The incorporation of a ductile metal, which should improve both strength and fracture toughness is responsible for the enhanced thermal shock resistance of cobalt-bonded tungsten carbide, alumina-chromia cermets [40] and metal-modified zirconia [41]. The emissivity and reflectivity can be modified by appropriate metallic or dielectric coatings. Significant reductions in the magnitude of the thermal stresses can be achieved by thermal barrier coatings with very low heat transfer character-istics.

Alternative approaches to improving thermal stress resistance include compressively pre-stressing the ceramic components in those regions subjected to the maximum values of thermal stress [42,43,44]. The introduction of a spatially varying thermal conductivity gradient, with a lower value of thermal conductivity found in those regions at the highest temperatures, results in a lowering of the magnitude of the tensile stresses in the colder parts of the ceramic component [45, 46, 47].

One specific method of improving the thermal stress resistance of ceramics by microstructural control which needs special mention is to intro-duce high density of microcracks [48]. As mentioned earlier, microcrack formation can arise from mismatches in the coefficients of thermal expansion of the individual components of composites or by the thermal expansion aniso-tropy of the individual grains.

The effectiveness of microcracking in improving thermal stress resis-tance will be discussed first from the perspective of avoiding the onset of failure, in terms of the effect of microcracking on the values of a number of the continuum properties which affect thermal stress failure. As will be discussed later, microcracking also is highly beneficial from the point of view of improving thermal shock resistance.

The specific continuum properties which will be considered are Young's modulus, Poisson's ratio, the coefficient of thermal expansion and thermal conductivity. For simplicity, the microcracks will be assumed to consist of non-interacting, randomly oriented penny-shaped cracks of equal size.

For a ceramic which undergoes microcracking of the above type, the effective Young's modulus and Poisson's ratio are [49]:

$$E_{eff} = E_0[1+\{16(10-3\nu_0)(1-\nu_0^2)\}Nb^3/45(2-\nu_0)]^{-1} \tag{14}$$

$$\nu_{eff} = \nu_0[1+\{16(3-\nu_0)(1-\nu_0^2)\}Nb^3/15(2-\nu_0)]^{-1} \tag{15}$$

where E_{eff} and ν_{eff} are Young's modulus and Poisson's ratio of the micro-cracked material, respectively, and E_0 and ν_0 are Young's modulus and Poisson's ratio of the crack-free material, respectively, N is the number of cracks per unit volume and b is the radius of the cracks.

Examination of eqs. 14 and 15 shows that microcracking can cause signi-ficant decreases in Young's modulus and Poisson's ratio, in support of experimental data [50,51].

The effect of microcracks on the coefficient of thermal expansion can be examined on the basis of a theoretical expression for the coefficient of thermal expansion of a composite in terms of the appropriate values for the individual components. As derived by Turner [50] for a two-component com-posite the coefficient of thermal expansion is:

$$\alpha_c = \left[\alpha_1 V_1 B_1 + \alpha_2 V_2 B_2\right] / \left[V_1 B_1 + V_2 B_2\right] \qquad (16)$$

where α, V and B refer to the coefficient of thermal expansion, volume fraction and bulk modulus, resp. and the subscripts c, 1 and 2 refer to the composite and the two components resp. Microcracking is expected to occur in the component with the highest value of the coefficient of thermal expansion as this component on cooling from the processing temperature is expected to be subjected to a state of tensile stress. In direct analogy, in a polycrystalline material with grains with high thermal expansion anisotropy, fracture is expected to occur in the direction within each grain with the highest value of thermal expansion coefficient.

The effect of microcracking on the coefficient of thermal expansion can be ascertained by substitution of the appropriate expression for the effect of microcracking on the bulk modulus of the component subject to microcracking. As derived by Walsh [53], the effective bulk modulus, B_{eff} of a material with randomly oriented microcracks is:

$$B_{eff} = B_o [1 + 16(1 - \nu_o^2) N b^3 / 9(1 - 2\nu_o)]^{-1} \qquad (17)$$

where B_o is the bulk modulus of the crack-free material.

Substitution of eq. 17 for the bulk modulus of the component which undergoes microcracking, indicates that the thermal expansion behavior of a microcracked composite can approach the value of the component with the smaller value of the coefficient of thermal expansion. Similarly, for a heavily microcracked polycrystalline ceramic, the coefficient of thermal expansion approaches the corresponding value for the crystal direction with the minimum value of coefficient of thermal expansion. For some crystal structures this value may approach zero, or may even be negative.

An estimate of the tensile strength of microcracked composites is more difficult to obtain. For failure originated at non-interacting microcracks, tensile strength is not affected by adjacent microcracks. Failure-initiating flaws in structural ceramics generally are much larger than the grain size and corresponding dimensions of the microcracks. In that case, an estimate of the tensile strength of a microcracked ceramic should take into account the effect of microcracks on elastic behavior and the fracture energy and/or fracture toughness. Experimental data for microcracked materials suggest that they exhibit values of tensile strength lower than the corresponding values for non-microcracked ceramics. However, the accompanying decrease in Young's modulus is such that the ratio of tensile strength to Young's modulus (i.e., the strain-at-fracture) increases substantially. Because the avoidance of thermal stress fracture basically requires the successful accommodation of differences in thermal strain, the strain-at-fracture is more critical than either tensile strength or Young's modulus.

The combined effects, then, of the increase in strain-at-fracture and the decrease in both thermal expansion and Poisson's ratio suggest that microcracking represents a very effective mechanism for improving thermal stress resistance. The corresponding decrease in thermal conductivity, however, tends to decrease the resistance to thermal stress failure. This, however, will only be the case where the value of thermal conductivity affects the magnitude of thermal stress. This, for instance, is not the case for convective heat transfer for high values of Biot number [2]. The decrease in thermal conductivity due to microcracking, however, can be of major advantage for those designs for which candidate materials should exhibit high thermal stress resistance in combination with high thermal insulating ability.

Improving thermal shock resistance by way of controlled fracture is subject to those variables which affect crack propagation in thermal stress fields. The pertinent variables which control such crack propagation were established by the present writer [1,54] on the basis of a simple fracture-mechanical analysis of a flat plate with parallel, equally-sized, non-interacting Griffith cracks, prevented from thermal expansion by external constraints in the direction perpendicular to the plane of the cracks. For this model and a material with a fracture energy independent of crack size, the critical temperature difference (ΔT_c) required for crack instability due to cooling of the plate (for conditions of plane stress) is [54]:

$$\Delta T_c = (2\gamma_f/\pi l \alpha^2 E_o)^{1/2}(1+2\pi N l^2) \tag{18}$$

where γ_f is the energy required to create unit area of fracture surface, l is the half-length of the cracks, α is the coefficient of thermal expansion, E_o is Young's modulus of the crack-free plate and N is the number of cracks per unit area.

The value of ΔT_c exhibits a minimum at a value of crack length, l_m:

$$l_m = (6\pi N)^{-1/2} \tag{19}$$

At instability at $\Delta T = \Delta T_c$, for crack with initial length, $l_o < l_m$, the strain energy release rate exceeds the fracture energy. As a result, the crack will propagate in a dynamic mode, with crack arrest for $l_o \ll l_m$, occurring at a value of final crack length, l_f:

$$l_f = (4\pi N l_o)^{-1} \tag{20}$$

It should be noted that l_f is an inverse function of the initial crack length, l_o and crack density but is independent of the fracture energy or toughness. However, by means of the Griffith criterion, l_o can be expressed in terms of the tensile strength, which yields:

$$l_f = \sigma_f^2/4N\gamma_f E \tag{21}$$

From the point of view of selection of the optimum material which undergoes a minimum of unstable crack propagation, eqs. 20 and 21 suggest that the material with the larger initial crack combined with a higher value of crack density is preferred. Alternatively, as suggested by eq. 21, moderate strength combined with high Young's modulus is preferred to the combination of high strength and low Young's modulus. Critical to note is that this latter conclusion is diametrically opposed to the requirement of high tensile strength and low Young's modulus for high resistance to the initiation of thermal stress fracture.

For $l_o > l_m$ crack propagation will occur in the stable mode with the crack length varying with ΔT_c, as described by eq. 18.

The dependence of ΔT_c (eq. 18) and l_f (eq. 20) on initial crack length is shown in fig. 14 for two values of crack density. Thermal stress resistance as measured by the magnitude of ΔT_c can be improved either by decreasing the initial crack size for $l_o < l_m$ or by increasing the crack size for $l_o > l_m$. This latter conclusion, although perhaps surprising at first sight results from the significant reduction in the effective Young's modulus of the plate with increasing crack size. ΔT_c, at any constant value of crack size with $l_o > l_m$, can be increased significantly also by increasing the crack density.

Increasing thermal stress resistance by decreasing the initial crack length for $l_o < l_m$ has the disadvantage that the final value of crack length

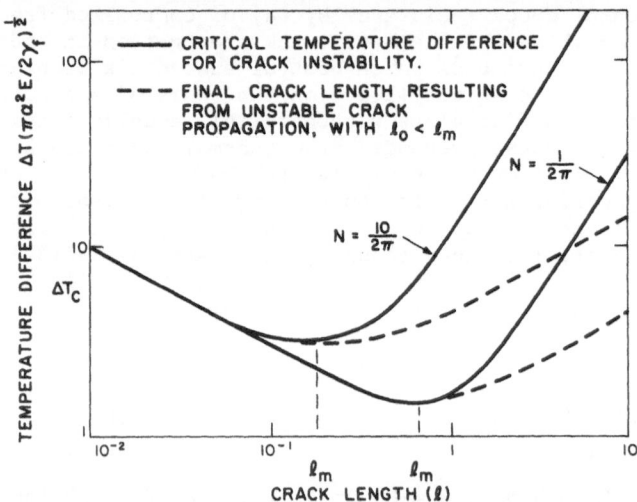

Fig. 14. Crack stability and propagation behavior
for a constrained plate thermally stressed
by cooling over temperature difference ΔT.

l_f which results from unstable crack propagation as indicated by eq. 20,
and also shown by the dotted lines in fig. 14, is inversely proportional
to the initial crack length. For this reason, if the decrease in crack
length still is insufficient to avoid the onset of crack propagation the
resulting fracture will be increasingly catastrophic with associated
increased losses in load-bearing ability and other properties adversely
affected by large cracks. The net result is opposite of the result required.

As an illustration and verification of the validity of the above conclusions, fig. 15 compares the strength at room temperature of two different
polycrystalline aluminas subjected to thermal stress failure by quenching
into water over a range of temperature differences [55]. The value of ΔT
at which the strength exhibits a sharp decrease represents the critical
quenching temperature difference required to initiate propagation of pre-
existing flaws. Comparison of the values of ΔT_c indicates that, as expected,
the initially stronger alumina requires a higher value of quenching temperature difference to initiate crack propagation than the alumina with the
initially lower strength value. However, the strength retained for $\Delta T > \Delta T_c$,
for the initially weaker material is higher than the corresponding value for
the initially stronger material. These data indicate that depending on
whether thermal shock resistance is based on the onset of crack instability
or the resulting extent of crack propagation and associated change in continuum properties, the relative thermal shock resistance of different
materials, in fact, may well be interchanged.

A solution to the problem of extensive unstable crack propagation due
to thermal shock is to find materials which exhibit a fracture energy or
toughness which increases with increasing crack length. For the simple case
of a crack with a fracture energy for the onset of crack propagation, γ_o
and the fracture energy for continued propagation and arrest, γ_f, the final
crack length following unstable crack propagation can be derived to be [56]:

$$l_f = (\gamma_o/\gamma_f)(4\pi N l_o)^{-1} \tag{23}$$

Eq. 23 indicates that low values of l_f require values for γ_o as low as
possible in combination with values of γ_f as high as possible. Low values
of γ_o assure that fracture occurs at a low value of stress which reduces

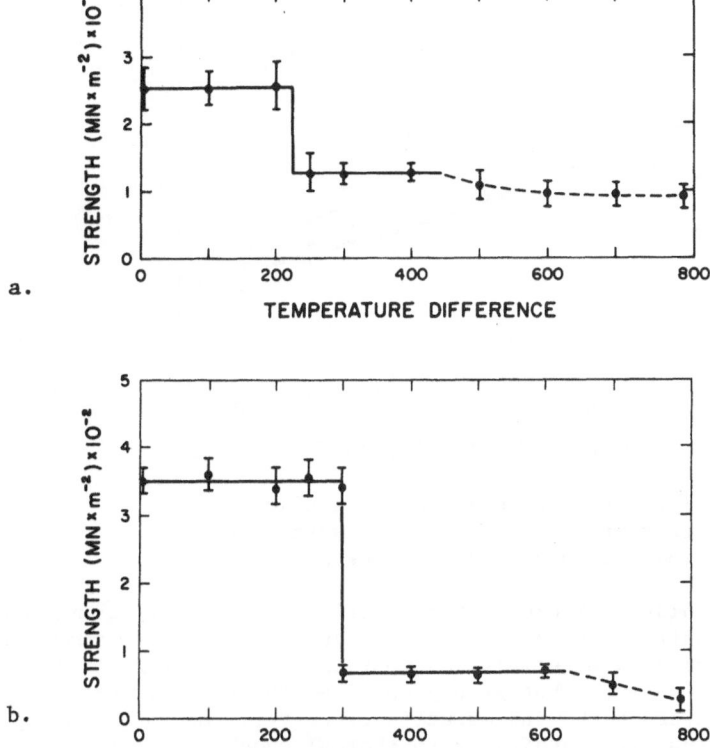

Fig. 15.   Comparison of strength loss behavior of two
polycrystalline aluminas subjected to thermal
shock by a water quench.

the driving force for crack propagation; at the same time a high value of
γ_f assures rapid crack arrest.   Such a crack size-dependent fracture energy
or toughness was thought to be responsible for the thermal shock behavior
of refractories [57], de-stabilized zirconia [58] and a cordierite glass-
ceramic [59].

An alternative to reducing the extent of unstable crack propagation is
to choose or develop materials which exhibit stable crack propagation only.
This can be accomplished by developing microstructures such that $l_o \sim l_m$.
This is the direction taken for many refractories for furnace linings which
are subject to thermal shock of such severity that the onset of thermal
stress failure cannot be avoided.   Such materials, although exhibiting a
rather low strength value exhibit a monotonic increase in crack size and
associated strength loss instead of the catastrophic crack propagation
typical for unstable fracture.   The strength loss behavior of a high-alumina
refractory subjected to a water quench is shown in fig. 16.   The data
indicate the absence of the sharp decrease in strength at ΔT_c evidenced by
the data shown in fig. 16.

A very effective approach to improving thermal shock resistance by
controlled crack propagation is based on the earlier conclusion that by
incorporating high densities of cracks with $l_o \stackrel{\sim}{\sim} l_m$ should result in high
values of ΔT_c.   In effect, this implies that extensive microcracking not
only should result in high resistance to the onset of crack propagation but
also should result in stable crack propagation as well in case the onset of

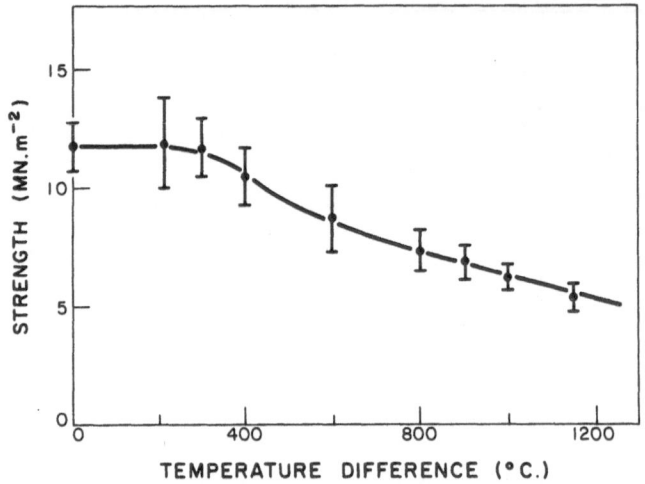

Fig. 16. Strength loss behavior of a high-alumina
refractory subjected to a water quench
indicative of stable crack propagation.

failure cannot be avoided. Microcracked materials are finding increasing
applications in industrial and other applications. The high thermal shock
resistance of microcracked materials was most effectively demonstrated by
Rossi [60] who by time-lapse photography compared the fracture behavior of
a magnesia without microcracks and magnesia extensively microcracked by a
dispersed phase of tungsten, subjected to thermal shock by an oxygen-
acetylene torch. The fracture behavior of the magnesia can be described as
violent. In contrast, the microcracked material retained its geometry without any evidence of thermal shock damage.

In summary, it has been shown how the heat conduction behavior of
structural ceramics can be tailored by compositional as well as microstructural control. Furthermore, guidelines were presented for tailoring ceramics
with improved thermal shock resistance.

ACKNOWLEDGMENT

This review was prepared with financial support provided by the Office
of Naval Research under Contract No. N00014-78-C-0431.

REFERENCES

1. D. P. H. Hasselman, J. Amer. Ceram. Soc. 52 (1969) 600.
2. W. D. Kingery, J. Amer. Ceram. Soc. 38 (1955) 3.
3. D. P. H. Hasselman, Bull. Amer. Ceram. Soc. 49 (1970) 1933.
4. H. S. Carslaw and J. C. Jaeger, Conduction of Heat in Solids, 2nd. Ed.
      Clarendon Press, Oxford (1960).
5. C. Kittel, Introduction to Solid State Physics, 2nd. Ed. John Wiley,
      New York (1962).
6. R. Berman, Thermal Conduction in Solids. Clarendon Press, Oxford (1976).
7. W. D. Kingery, H. K. Bowen, D. R. Uhlmann, Introduction to Ceramics,
      2nd. Ed. John Wiley, New York (1976).
8. L. Rayleigh, Phil. Mag. 34 (1892) 481.
9. J. C. Maxwell, A. Treatise on Electricity and Magnetism, 1, 3rd. Ed.
      Oxford University Press (1904).
10. D. A. G. Bruggeman, Annalen Physik 24 (1935) 636.

11. A. E. Powers, Conductivity in Aggregates, Knolls Atomic Power Laboratory Report-2145, General Electric Corp. (1961).

12. Z. Hashin, J. Comp. Mat. 2 (1968) 284.

13. S. C. Cheng and R. I. Vachan, Int. J. Heat and Mass Transfer, 12 (1969) 249.

14. D. P. H. Hasselman, J. Comp. Mat. 12 (1978) 403.

15. W. J. Parker, R. J. Jenkins, C. P. Butler, G. L. Abbott, J. Appl. Phys. 32 (1961) 1679.

16. H. J. Siebeneck, W. P. Minnear, R. C. Bradt and D. P. H. Hasselman, J. Amer. Ceram. Soc. 59 (1976) 84.

17. B. Abeles, Phys. Rev. 131 (1963) 1906.

18. F. F. Lange, H. J. Siebeneck and D. P. H. Hasselman, J. Amer. Ceram. Soc. 59 (1976) 454.

19. D. W. Lee and W. D. Kingery, J. Amer. Ceram. Soc. 43 (1960) 594.

20. J. E. Matta and D. P. H. Hasselman, J. Amer. Ceram. Soc. 58 (1975) 458.

21. G. A. Slack, J. Appl. Phys. 35 (1964) 3460.

22. M. Srinivasan, L. D. Bentsen, D. P. H. Hasselman, pp. 877-87 in Thermal Conductivity 17. Ed. by J. G. Hust, Plenum Press (1983).

23. M. Ura and O. Asai, F. C. Report Vol. 1, No. 4, Japan Fine Ceramics Ass., Tokyo, Japan (1984).

24. K. Chyung, G. E. Youngblood and D. P. H. Hasselman, J. Amer. Ceram. Soc. 61 (1978) 530.

25. H. Tawil, L. D. Bentsen, S. Baskaran and D. P. H. Hasselman, J. Mat. Sc. (in press).

26. L. D. Bentsen, T. Y. Tien and D. P. H. Hasselman, J. Amer. Ceram. Soc. 67 (1984) C-85.

27. H. J. Siebeneck, D. P. H. Hasselman, J. J. Cleveland and R. C. Bradt, J. Amer. Ceram. Soc. 59 (1976) 241.

28. H. J. Siebeneck, D. P. H. Hasselman, J. J. Cleveland and R. C. Bradt, J. Amer. Ceram. Soc. 60 (1977) 336.

29. H. J. Siebeneck, J. J. Cleveland, D. P. H. Hasselman and R. C. Bradt, pp. 753-62 in Proc. of Symposium on Ceramic Microstructures, Westview Press (1977).

30. L. D. Bentsen, D. P. H. Hasselman, N. Claussen, pp. 369-82 in Environmental Degradation of Engineering Materials in Aggressive Environments. Ed. by M. R. Louthan, R. D. McNitt and R. D. Sisson, Jr. Virginia Polytechnic Institute (1981).

31. D. P. H. Hasselman, R. Syed, R. F. Johnson, M. P. Taylor, K. Chyung, J. Mat. Sc. (in press).

32. S. Nasu, T. Takahashi, T. Kikuchi, J. Nucl. Mat. 43 (1972) 72.

33. T. Tanaka, H. Suzuki, Carbon 10 (1972) 253.

34. M. R. Hull, W. W. Lozier, A. W. Moore, Carbon 11 (1973) 81.

35. B. R. Powell, Jr., G. E. Youngblood, D. P. H. Hasselman and L. D. Bentsen, J. Amer. Ceram. Soc. 63 (1980) 581.

36. D. P. H. Hasselman, R. Syed and T. Y. Tien, J. Mat. Sc. (in press).

37. D. P. H. Hasselman, Am. Ceram. Soc. Bull. 49 (1970) 1933.

38. D. P. H. Hasselman, Ceramurgia 4 (1978) 147.

39. D. P. H. Hasselman, P. F. Becher and K. S. Mazdiyasni, Materials Technology and Testing 11 (1980) 82.

40. M. S. Tacvorian, Soc. Franc. Ceram. Bull. 29 (1955) 20.

41. A. Arias, J. Am. Ceram. Soc. 49 (1966) 339.

42. H. P. Kirchner, R. M. Gruver and R. E. Walker, Am. Ceram. Soc. Bull. 47 (1968) 498.

43. H. P. Kirchner, R. M. Gruver and R. E. Walker, J. Appl. Phys. 40 (1969) 3445.

44. J. Gebauer, D. A. Krohn and D. P. H. Hasselman, J. Am. Ceram. Soc. 55 (1972) 175.

45. K. Satyamurthy, J. P. Singh., M. P. Kamat and D. P. H. Hasselman, J. Am. Ceram. Soc. 62 (1979) 431.

46. K. Satyamurthy, J. P. Singh, M. P. Kamat and D. P. H. Hasselman, Proc. Brit. Ceram. Soc. 80 (1980) 10.

47. K. Satyamurthy, J. P. Singh, D. P. H. Hasselman and M. P. Kamat, J. Am. Ceram. Soc. 63 (1980) 363.

48. D. P. H. Hasselman and J. P. Singh, Bull. Am. Ceram. Soc. 58 (1979) 856.

49. R. L. Salganik, Izv. Akad. Nauk SSSR. Mekh. Tverd. Tela, 4 (1973) 149.

50. E. A. Bush and F. A. Hummel. J. Am. Ceram. Soc. 41 (1958) 189.

51. J. A. Kuszyk and R. C. Bradt, J. Am. Ceram. Soc. 56 (1973) 420.

52. P. S. Turner, J. Res. Natl. Bur. Stand. 37 (1946) 239.

53. J. B. Walsh, J. Geophys. Res. 70 (1965) 381.

54. D. P. H. Hasselman, pp. 89-103 in Ceramics in Severe Environments, Ed. by W. W. Kriegel and H. Palmour, Plenum Press, New York (1971).

55. D. P. H. Hasselman, J. Am. Ceram. Soc. 53 (1970) 490.

56. D. P. H. Hasselman and J. P. Singh, Theoretical and Applied Fracture Mechanics 2 (1984) 59.

57. D. R. Larson, J. A. Coppola, D. P. H. Hasselman and R. C. Bradt, J. Am. Ceram. Soc. 57 (1974) 417.

58. M. V. Swain, J. Mat. Sc. Lett. 2 (1983) 279.

59. C. J. Fairbanks, H. L. Lee and D. P. H. Hasselman, J. Am. Ceram. Soc. 67 (1984) C-236.

60. R. C. Rossi, Bull. Amer. Ceram. Soc. 48 (1969) 736.

CREEP RUPTURE OF SILICONIZED SILICON CARBIDE

S.M. Wiederhorn, L. Chuck, E.R. Fuller, Jr., and N.J. Tighe

National Bureau of Standards
Gaithersburg, MD 20899

ABSTRACT

Creep and creep-rupture of siliconized silicon carbide were
studied in flexure as a function of temperature and applied stress. The
behavior of this material was dominated by the formation of cavities at
the silicon-silicon carbide interface. At temperatures less than
1200°C, and applied stresses less than 300 MPa, cavity densities were
low, and only transient creep was observed. The strength at room
temperature and at the creep temperature were the same for these
conditions. At 1300°C and stresses greater than ~250 MPa, many large
cavities formed on the tensile side of the specimen, effectively
reducing the internal cross-section, and enhancing the rate of creep.
These cavities lined up in rows to form cracks that limited specimen
lifetime. A shift of the plane of zero strain from the geometric
center towards the compressive surface of the specimen was also
observed as a consequence of the creep and cavitation. Cavity
formation during creep caused a significant reduction in the room
temperature strength. By contrast, the strength at elevated
temperatures did not decrease until extensive cavity linkage occurred
at creep times that approached the creep-rupture life.

1. INTRODUCTION

Current interest in the mechanical behavior of composite materials
has its origin in the intended use of these materials in high-
temperature structural applications, such as heat engines and heat
exchangers. Composites usually consist of a continuous matrix, in
which is imbedded a second fibrous or particulate phase that may or may
not be continuous. The strength and toughness of composites at
elevated temperatures depend on the refractoriness of the two phases
that make up the composite, the distribution and morphology of these
phases, and the strength of the interface between the two phases. At
elevated temperatures where creep occurs, the mechanical properties of
the composite usually depend on the behavior of the continuous matrix,
which is normally the less refractory phase. Creep of the matrix leads
to the formation of defects such as cavities, voids and cracks that
eventually limit component strength and cause component failure. Since

the use of composite materials at elevated temperatures requires establishing conditions for lifetime prediction, the role of creep and its contribution to defect generation must be understood.

One ceramic composite that is being considered for use in high-temperature applications is siliconized, silicon carbide [1,2], which consists of a dispersion of silicon carbide particles in a continuous matrix of silicon [3]. At the maximum use temperature, 1200-1300°C, for this composite, silicon is highly ductile [4], whereas silicon carbide is almost completely rigid so that creep of the composite depends primarily on the rheology of the silicon and the way in which the silicon carbide particles affect that rheology. During creep, flow of the silicon between the grains of silicon carbide accommodates the deformation. This flow is also responsible for the generation of internal stresses within the composite [5], and for the occurrence of cavitation [6] when the silicon can not flow rapidly enough. Thus, cavity nucleation, cavity growth and cavity linkage are principally responsible for limiting the lifetime and strength of this composite [7].

In this paper, the creep and creep-rupture behavior of siliconized, silicon carbide are characterized in four-point bending as a function of temperature and stress. It is observed that cavities nucleate early in the creep process. With additional deformation, the nucleation of cavities continues, eventually forming large cracks that lead to failure. As a consequence of cavitation processes, the plane of neutral strain for bending shifts towards the compressive surface of the test specimen. Cavitation also results in a reduction of the room-temperature strength subsequent to the elevated temperature creep.

2. EXPERIMENTAL PROCEDURE

The material selected for study was a fully dense commercial grade of siliconized, silicon carbide. Also known as reaction sintered silicon carbide[1], the material consists of 5 μm grains of silicon carbide in a continuous matrix of silicon [3]. Depending on the billet of material from which the specimen was taken, agglomerates of silicon carbide grains and large pockets of silicon were occasionally observed throughout the billet. Other billets, however, were relatively homogeneous and the agglomerates of silicon carbide were not readily apparent. Similarly, pockets of silicon were not equally apparent in all of the specimens. The volume fraction of silicon in all specimens tested ranged from approximately 30 to 40 percent.

All studies were conducted in air using a four point bending fixture with inner and outer loading spans of 10 and 40mm, respectively. Specimens, 50 by 4 by 3 mm, were machined to final size using a 180 grit diamond grinding wheel, and were usually tested after machining without further mechanical finishing. To observe cavitation, some specimens were polished on the tensile surface, and on an adjacent side surface before testing. After deformation, selected bend bars were sectioned and polished to reveal internal damage as a function of creep strain. By this means, the process of cavity nucleation and growth could be studied. To locate the surface of neutral strain as a function of total strain, two parallel rows of indentations (formed by an indentation load of 2N) were placed in the polished side of the bend bars at a predetermined separation, typically 8 mm. The rows of

---

[1]KX01, made by the Sohio Co. The use of this material does not imply endorsement by the National Bureau of Standards.

indentations lay perpendicular to the longest dimension of the bend bar. By measuring the displacement of these indentations after creep, the amount of permanent strain that occurred through the cross-section could be calculated as a function of distance from the tensile surface, and the point of zero strain could be identified.

The apparatus used for the creep studies has been described previously [8,9]. Briefly, a static load was applied to the specimen by a pneumatically driven ram. The load on each specimen was controlled by regulated air pressure, and was monitored by a load cell attached to the ram. In this way, loads could be controlled to better than one percent of full load. Static loads were maintained until rupture occurred, or until sufficient strain occurred to characterize the early stages of creep. The test apparatus contained three test stations in each furnace, which permitted up to three creep or creep-rupture experiments to be conducted simultaneously. In the present set of experiments, two test stations were usually used for creep and creep-rupture tests, and the third was used as a reference leg to eliminate spurious signals resulting from electrical noise and thermal expansion of the test frame due to room temperature thermal fluctuations. Specimen displacement during creep was measured from the loading points of the specimens using a linear voltage displacement transducer (LVDT). Strains were estimated using the equations developed by Hollenberg et al. [10]. Linear power law creep was assumed in both the tensile and compressive halves of the flexure bars, and the linear equation from the analysis of Hollenberg et al. was used to calculate both the nominal stress and nominal strain. Although this procedure leads to errors in estimating the stress-strain behavior of the material, as a consequence of nonlinear behavior and shifts in the neutral axis, the procedure provided a convenient way of summarizing the experimental data. In the absence of creep data obtained in pure tension and compression, a more complete analysis of the creep behavior of this material in bending was not possible.

To characterize component strength as a function of applied creep stress and time under load, interrupted creep-rupture tests were conducted. In each test, a fixed load was applied to the specimen, and after a predetermined period of time, the specimen was broken and its strength was calculated from the fracture load and the specimen dimensions. For each specimen that broke prematurely under the static load, the rupture time was recorded, as it was for specimens that were intentionally permitted to creep to failure. Loads equivalent to initial stresses of 200, 250 and 300 MPa were used in these studies, and strengths were measured at 1100, 1200 and 1300°C, and at room temperature subsequent to high-temperature creep. For room temperature measurements, the initial applied load was maintained on each specimen as it was cooled to room temperature, at which time the load was increased until failure occurred.

Strength measurements were also made on as-machined specimens, both prior to any heat-treatment and after annealing in a stress free condition at 1100 and 1300°C. Specimens that were only annealed were also broken at both elevated and room temperature. With the exception of the as-machined specimens tested without heat treatment, all specimens were annealed at temperature for ~1 hour before any mechanical testing. Finally most data were collected as data pairs to minimize statistical fluctuations from experimental and material variability. In this procedure, two specimens were placed in the test apparatus and were exposed to identical test conditions during high-temperature creep. At a predetermined time, one of the specimens was broken at temperature, while the other was cooled to room temperature before being broken. To reduce material variation in these tests, the specimens used were nearest neighbors in the original billet from which they were machined. The loading rate for these studies was ~40 MPa/s.

3. EXPERIMENTAL RESULTS

3.1 Microstructure

The microstructure of the material studied, figure 1, indicates that most silicon carbide grains (light grey) were surrounded by the silicon matrix (white). Pockets of silicon and agglomerates of silicon carbide grains can be seen distributed throughout the material, indicating that the microstructure of this specimen is inhomogeneous. For the test conditions used in the present study (1100°C to 1300°C and 200 to 300 MPa) the silicon behaves as a ductile metal [4]. Test conditions lie above the ultimate tensile strength of the silicon and deformation is accompanied by slip and dynamic recrystallization of the silicon. Examination of specimens by transmission electron microscopy after deformation gives some evidence of dislocation activity within the silicon. The silicon matrix consists primarily of silicon grains that span the space between the silicon carbide grains. Evidence of dynamic recrystallization in the form of small-angle grain boundaries was obtained from specimens that were deformed, figure 2. Some deformation of the silicon carbide is observed at points of contact between the silicon carbide grains. The dislocation activity in figure 3, for example, appears to be related to sliding contact damage between grains of silicon carbide.

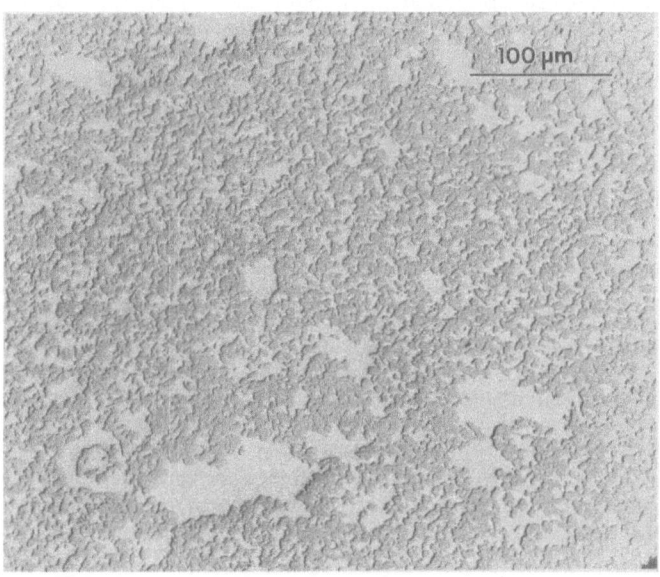

Fig. 1. Microstructure of siliconized, silicon carbide consisting of grains of silicon carbide (light grey phase) in a continuous matrix of silicon (white phase).

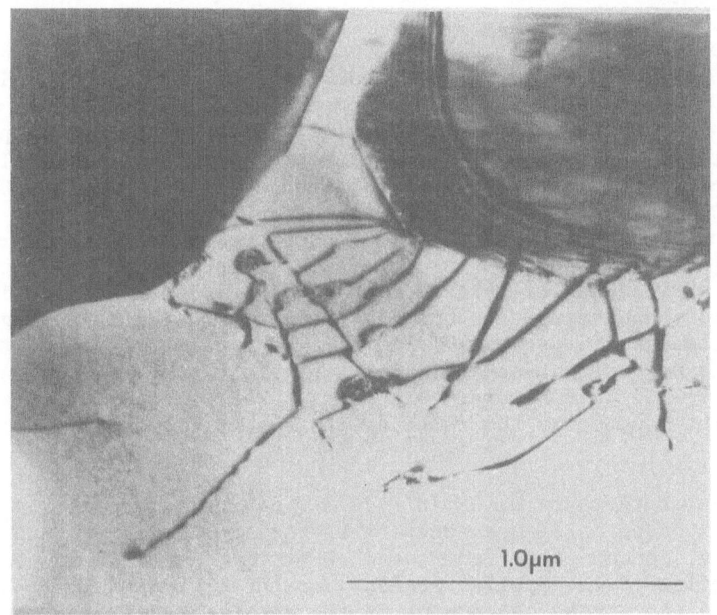

Fig. 2. Electron micrograph of the silicon phase after deformation at
1200°C. The only evidence for dislocation motion within the
silicon phase is remnant small angle grain boundaries, such as
that shown in this figure.

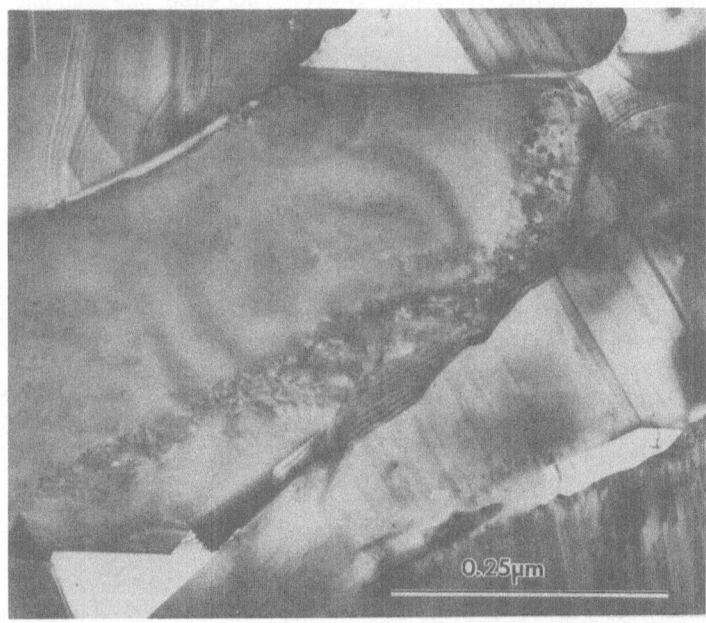

Fig. 3. Evidence for dislocation motion in the silicon carbide grains.
The dislocations appear to have formed as a consequence of
contact between the two grains during the deformation
process.

3.2 Creep Data

Creep curves for this material, are shown in figure 4, for tests conducted at 1300°C and applied stresses of 200 and 250 MPa. The curve for 200 MPa, figure 4a, indicates apparent transient creep for the entire period of the test. The material appears to deform in primary creep, with no indication of steady state or tertiary creep. This curve is also typical of the data obtained at 1100°C and 1200°C in this study. At 1300°C and applied stresses of 250 MPa and 300 MPa, however, the strain rate and the shape of the creep curves are altered dramatically. The creep curves now appear to be typical of the tri-modal curves usually obtained for metals. Closer examination of these curves, however, indicates only primary and tertiary stages of creep. Although the curves in figure 4b have points of inflection, there is no region that approximates a straight line. From the shape of the curves in this figure, it appears that the transient stage of creep is interrupted by the tertiary stage that eventually leads to fracture.

Curves of the type shown in figure 4 can be summarized by plotting the minimum creep rate as a function of the applied stress and temperature. Creep rates from these curves are usually expressed as a power function of the applied stress, and the plot in figure 5a illustrates the stress dependence of the creep rate for the materials studied in this paper. At 1100°C and 1200°C the slope of the curves is approximately 4, which is close to the stress exponent obtained for the creep of silicon alone [4]. This finding suggests that at these temperatures, creep of the composite is controlled entirely by the deformation of the silicon. At 1300°C, the results differ considerably from those at the lower temperatures. Although the data fall on a straight line, the slope of the line, ~13, is considerably greater than would be expected from dislocation models of creep, which suggests that a second mechanism of creep is occurring to increase the stress exponent of the creep rate. The fact that the apparent activation energy for creep is not a constant, but depends on the applied stress, figure 5b, also suggests that more than one process controls the creep of this material.

3.3 Cavitation

The creep data reported in figures 4 and 5 can be understood in terms of the cavities that form as a function of creep strain. Specimens that exhibit only transient creep, figure 4a, also exhibit only slight cavitation, as is shown in figure 6a for a specimen tested at 200 MPa and 1300°C. Nucleation of cavities occurs in early stages of the creep process. Although cavities form at random locations within the specimens, they always form at interfaces between the silicon and silicon carbide. Cavities are not observed within the silicon pockets, but may be located at interfaces bordering these pockets. Cavities seem to have an equal probability of forming in dense agglomerates of silicon carbide, or in less dense regions of the material. Cavities that form at this early stage of deformation appear to have little effect on the strength, or on the creep process.

Cavity formation seems to dominate the creep process in specimens that exhibit tertiary creep, figure 4b. For the specimen used to obtain the data in figure 4b, for example, cavities are no longer distributed in a random array, figure 6b. Instead, they tend to line up, eventually forming creep cracks that result in specimen rupture.

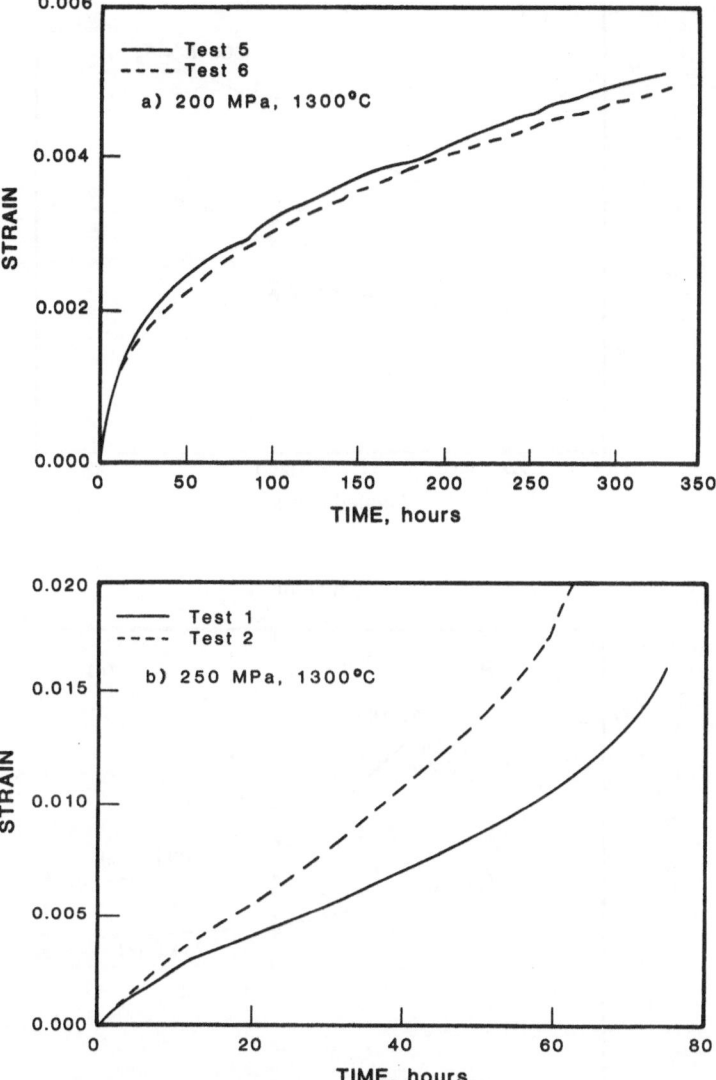

Fig. 4. Creep curves for siliconized, silicon carbide deformed at
1300°C: (a) 200 MPa initial applied load; (b) 250 MPa initial
applied load. The difference between these two curves is a
consequence of cavity formation at the higher applied load,
which enhances the creep rate.

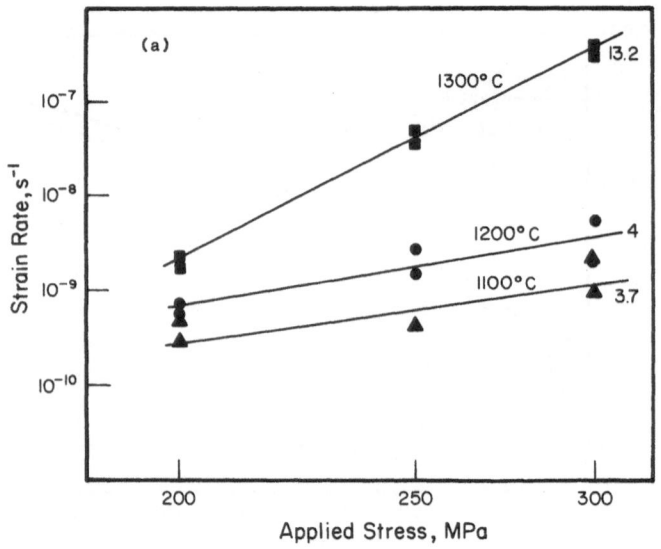

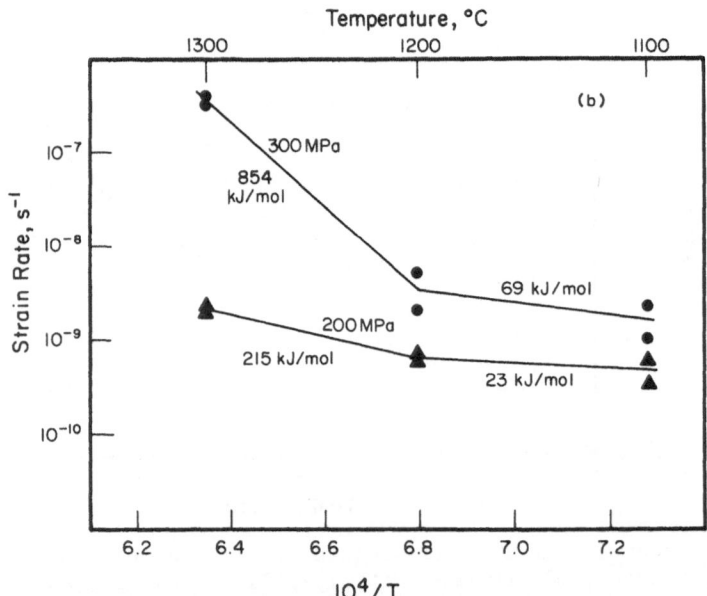

Fig. 5. Summary of creep data: (a) Stress dependence of the creep
rate; (b) Temperature dependence of the creep rate. The
changes in slope in both figures is a consequence of cavity
formation and growth at higher temperatures and applied
stresses.

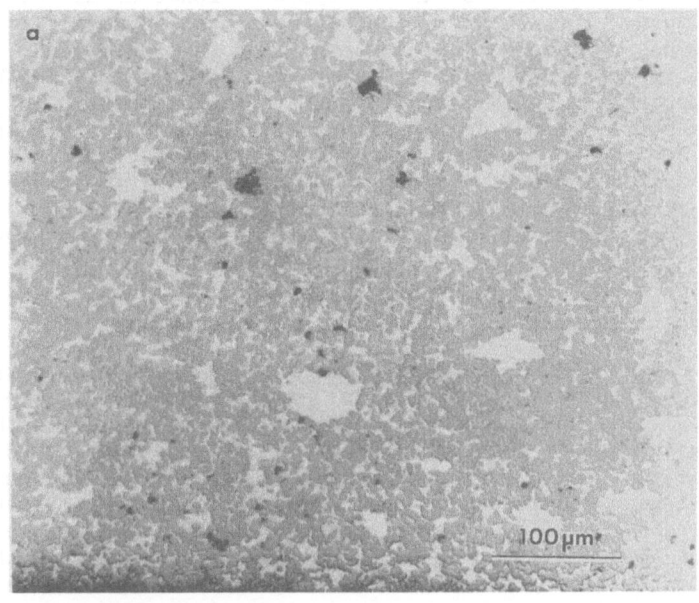

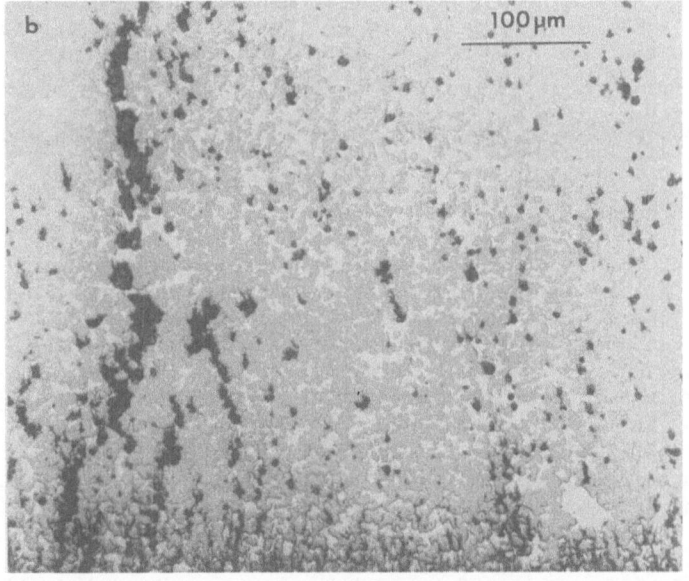

Fig. 6. Cavity formation in siliconized silicon carbide at 1300°C as a
consequence of creep: (a) 200 MPa initial applied stress; (b)
250 MPa initial applied stress. The high density of cavities
at 250 MPa correlates with the increase of creep rate at this
applied stress.

Perhaps the most interesting observation of cavity formation in this study is the generation of cavities on the compressive side of the geometric center of the flexure specimens, figure 7. This finding indicates that creep does not occur with equal facility in tension and compression. As indicated by the analysis of Cohrt et al. [11], Fett [12], and Chuang [13], the position of the neutral stress plane depends on the response of the material to creep in both tension and compression. Ceramic materials usually creep more easily in tension than in compression, which leads to a shift in the position of the neutral plane towards the compressive surface of the flexure specimen. This shift of the neutral axis is enhanced by cavitation, which causes an internal reduction of the cross section, and thus increases the creep rate in the tensile section of the flexure specimens. The shift can be substantial, as is seen in figure 7, where cavities have formed well past the geometric center of the specimen.

3.4 Plane of Zero Strain

The position of the plane of zero strain was determined by measuring the displacement of parallel rows of indentations that were placed in the side surface of the specimens perpendicular to the initial neutral axis. Although, these rows of indentations were no longer parallel after deformation, figure 8, they still lay on straight lines, which supports the usual assumption that planar surfaces remain planar during creep [14]. The residual strain at any distance from the tensile surface can now be measured from the displacement of the two rows of indentations. The strain calculated by this method is shown in figure 9 for a specimen subjected to an initial applied stress of 250 MPa at 1300°C for 15 hours. This analysis suggests a shift of the plane of zero strain from the geometric center to a point that lies at approximately 67% of the distance from the tensile to the compressive surface. As both strain and cavitation increase, the neutral plane moves even closer to the compressive surface. Positions of the plane of zero strain of as large as 85% of the distance from the tensile to the compressive surface have been measured when cavitation and creep have been extensive.

3.5 Strength Measurements

Cavities that are generated during creep not only enhance the creep rate in the tensile section of the flexure specimen, but are also a principal cause of strength degradation in this material after extensive creep deformation at elevated temperatures, figure 10. This strength degradation is particularly apparent in room temperature strength measurements, where substantial reductions in strength are indicated by the interrupted creep-rupture tests conducted at a temperature of 1300°C and a load of 300 MPa, figure 10a. Failure by creep-rupture for these conditions is observed to occur at times ranging from between ~10 and ~30 hours of exposure (see the data points on the 300 MPa stress line in figure 10a). Within this time frame, large cavities generated by creep reduce the room temperature strength from ~500 MPa (measured at 0.1 and 1 hour of exposure) to ~ 400 MPa.

Strengths measured at 1300°C seem to be less sensitive to cavity formation than those measured at room temperature, as can be seen by the fact that most of the strengths measured between 10 and 30 hours of exposure at 1300°C have values (~580 MPa), that are close to those measured at 0.1 and 1 hours of exposure, figure 10a. The difference between the strength measured at room temperature and that measured at 1300° is made more apparent by displaying the data as matched pairs, table 1. Of the five sets of data points obtained as matched pairs,

Compression Side

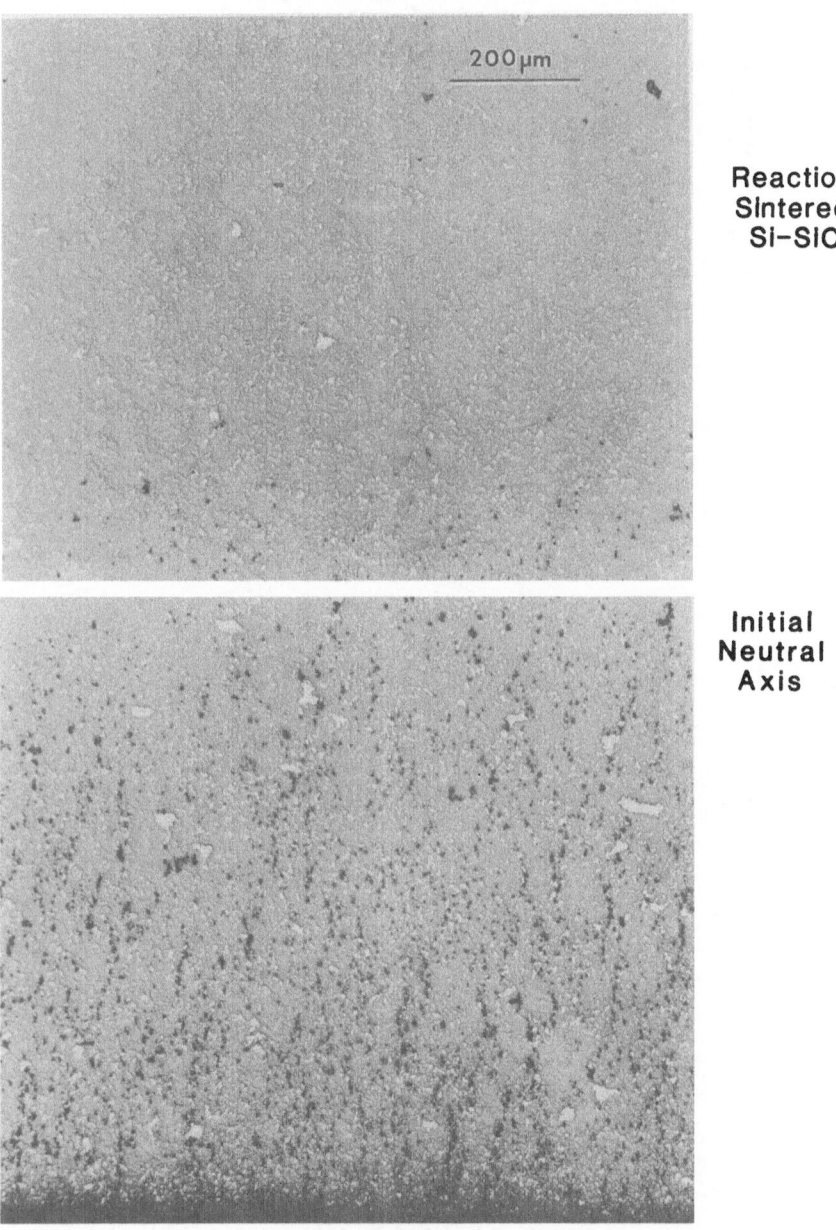

Reaction
Sintered
Si–SIC

Initial
Neutral
Axis

Tensile Side

Fig. 7. Cavity formation within the cross-section of a specimen tested
at 1300 °C and 300 MPa. Note that cavity formation has
occurred on what was originally the tensile side of the
cross-section.

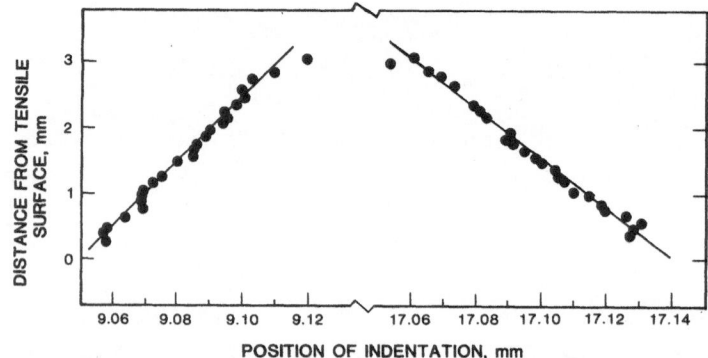

Fig. 8. Position of indentations on a polished side surface of a bend
specimen. The indentations originally lay in two parallel
rows perpendicular to the initial neutral axis and separated
by 8 mm. The spacing of the indentations has changed as a
consequence of the deformation.

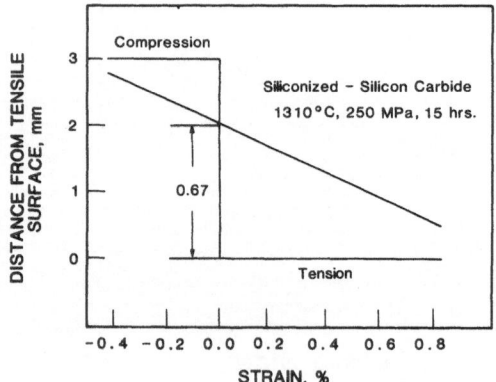

Fig. 9. Strain calculated from data shown in figure 8. Note that the
axis for zero strain has moved towards the compression surface
of the specimen.

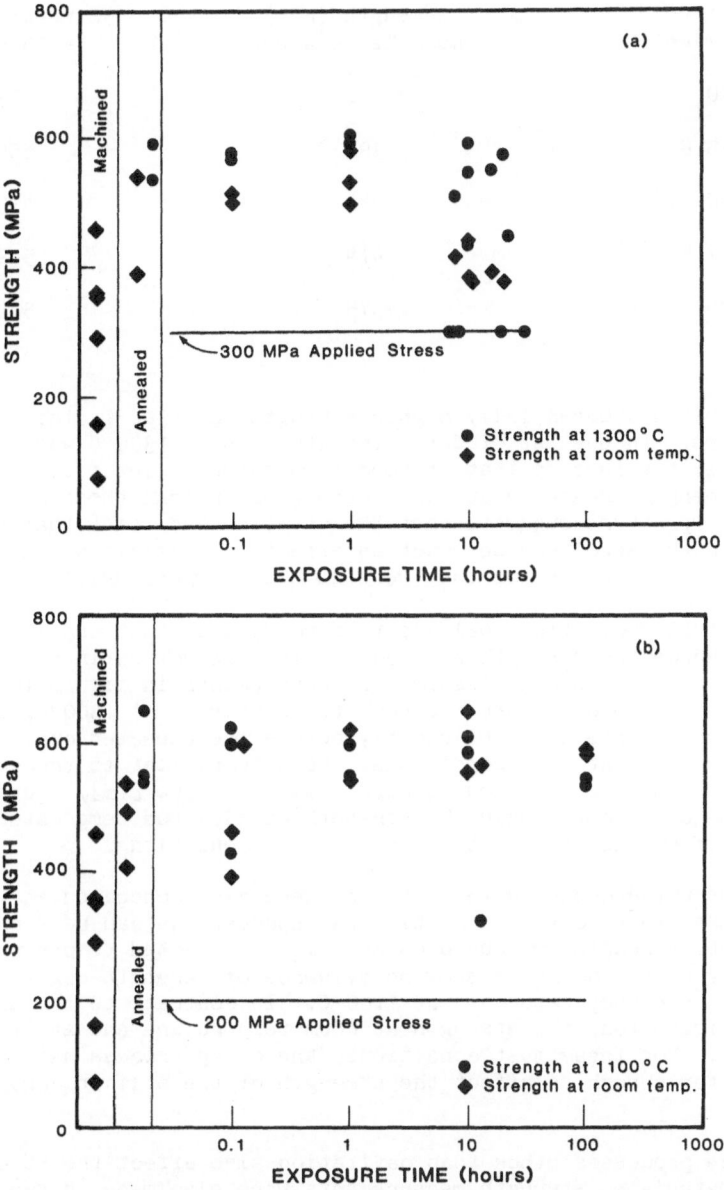

Fig. 10. The strength of specimens tested after exposure to creep at
elevated temperature: (a) 1300°C and 300 MPa; (b) 1100°C and
200 MPa initial stress. Strength measurements were made both
at room temperature and at the temperature of creep. The
initial (as-received) strength distribution is shown on the
left-hand side of the figure. In (a) one specimen broke at
0.01 hours under a static load of 300 MPa at 1300°C. This
data point is off scale in the figure.

Table 1. Strength of siliconized silicon carbide measured after creep
at 1300°C and 300 MPa. Specimens were matched pairs tested
simultaneously in the same creep apparatus.

Exposure Time (hr)	Strength (MPa) Room Temperature	Strength (MPa) 1300°C
10	382	593
15.8	391	550
10	442	433
7.7	414	509
20	375	574

four exhibited substantially higher strengths at 1300°C, than at room
temperature. In one set of data, the strength at 1300°C was
essentially the same as that at room temperature. The fact that most
of the strengths measured at 1300° were greater than those measured
at room temperature suggests that the cavities and voids that were
present in the specimens were not as effective in limiting the strength
at elevated temperature as they were at room temperature.

Eventually cavities coalesce to form defects that are large enough
to also reduce the strength at 1300°C. This reduction in strength is
indicated by the two high-temperature data points in figure 10a that
fall close to those measured at room temperature. At 1300°C, strength
degradation appears to occur shortly before the expected creep-rupture
time under constant load conditions. Results similar to these were
obtained recently on vitreous bonded aluminum oxide [15]. Here too,
creep damage did not impair the strength at elevated temperatures until
shortly before the failure time under creep conditions.

Interrupted creep-rupture studies were also conducted at 1100°C
and 200 MPa, where cavity formation was sparse. As can be seen in
figure 10b, strengths measured on specimens subjected to creep for
periods of up to 100 hours show no evidence of strength degradation,
either at room temperature or at 1100°C. In contrast to the data at
1300°C, figure 10a, the strength at room temperature and at 1100°C were
identical. For these test conditions, the creep process has apparently
not gone far enough to affect the strength of the silicon carbide
specimens.

Since processes other than cavitation also affect the strength of
ceramic materials, strength measurements were also made on two other
sets of specimens: one that had been freshly machined, the other that
had been machined and then annealed at high temperature under zero
load. Shown as on the left hand portion of figure 10 the strength of
the freshly machined specimens ranged from ~70 MPa to ~460 MPa; this is
considerably less than the values measured after exposure to elevated
temperatures. This range of strengths is also considerably less than
that obtained at elevated temperatures on freshly annealed specimens
that were not subjected to creep. Thus, the strength at elevated
temperatures after annealing was approximately equal to that measured
after deformation by creep. Strengths measured at room temperature
after annealing lay at the high end of the strength distribution of the

as-machined specimens and are slightly lower than the room temperature strengths after creep. Because of the small number of data points, the statistical significance of these results remains to be verified.

4.0 DISCUSSION

4.1 Cavitation

A qualitative model of cavitation during creep can be developed from the theoretical work of Drucker [5], who examined the effect of viscous, and plastic intergranular phases on the stress distribution in specimens that were subjected to uniaxial creep[2]. The main effect of a uniaxial compressive stress is to squeeze the intergranular phase from between the rigid grains of the solid, thus setting up localized pressures and stresses within the intergranular phase over and above the stresses applied to the material. Conversely, a uniaxial tensile stress applied to the solid tends to draw the intergranular phase into the space between the grains, setting up a negative pressure within this phase. Drucker considered several types of intergranular phases: a viscous fluid, a perfectly plastic solid and a solid that undergoes power law creep. He concluded that the maximum stress at the grain interface was about twice the applied uniaxial stress. On the tensile side of a flexural specimen, this stress is the driving force for cavity formation. In the present study, the applied stress for cavitation was approximately 300 MPa. Prior to the nucleation of cavities, this stress is reduced somewhat by the creep, but can go no lower than two-thirds of the applied stress if creep is symmetric prior to cavitation. Thus the stress that develops in the test specimen prior to cavitation will be approximately 400 MPa.

Using the equation developed by Young [16], the pressure, P, needed to grow a cavity of radius, R, is given by $P = 2\Gamma/R$, where Γ is the surface tension of the solid. The size of the cavities observed in this paper were of the order of 5 µm. Assuming a surface tension for silicon of approximately 1 J/m^2, an equilibrium pressure of 0.4 MPa is required to maintain cavities of this size. Since this pressure is far less than the estimated pressure at the silicon-silicon carbide interface (400 MPa), these cavities would be expected to grow in response to the applied stresses. The fact that they do not seem to grow larger than 5 µm is probably a consequence of the fact that the cavities are forced to grow in the silicon matrix. Hence their growth rate and their size is limited by the dimensions of the microstructure. The minimum cavity radius for cavity stability can also be determined from the Young equation. If the maximum stress is assumed to be approximately 400 MPa, then a cavity radius of approximately 5 nm is required for cavity nucleation. Defects of this size can be small areas of decohesion, or small inclusions or cavities along the silicon-silicon carbide interface. The search for defects of this size at the interface is a subject for future investigation.

4.2 Strength Degradation

Once cavities grow to a substantial size, they act as sources of fracture at room temperature. This result was first reported by Carroll and Tressler [17], who showed that room temperature strengths

---

[2]Such a model of cavitation-creep was developed for a viscous intergranular fluid by Lange [6].

of specimens decreased as cavities grew in the tensile cross-section of
the specimen. Observations of this paper are consistent with these
results. The room temperature strength of specimens loaded in air at a
stress of 300 MPa and a temperature of 1300°C decreased well before the
failure time from creep-rupture, figure 10a. At elevated temperatures,
however, the strength of such specimens appears to be maintained until
shortly before the creep-rupture failure time in bending.

The difference between the strengths measured at room temperature
and those measured at elevated temperature can be understood in terms
of the toughness, i.e. K_{Ic}, of this material. Since, siliconized,
silicon carbide is a brittle material with a toughness of approximately
4 MPa-m$^{1/2}$ [18], its strength at room temperature is susceptible to
cavities that are generated during the creep process. At high
temperatures, >1000°C, the toughness of this material depends on both
temperature and loading rate, as has been reported by Chuck and Fuller
[18], Hillig et al. [19], and Srinivasin and Kasprzyk [3]. At 1300°C,
the toughness is reported to go through a maximum and can be as high as
20 MPa-m$^{1/2}$ [18]. This five-fold increase in toughness at elevated
temperatures reduces the sensitivity of specimen strength to the flaws
that grow during creep.

A rising resistance to fracture is also possible for this material
at 1300°C. Although cavities nucleate and grow as a function of time,
no reduction of strength at temperature is evident from figure 10a,
which suggests an insensitivity of the fracture strength to the flaw
size. Similar behavior was observed by Wiederhorn et al. [15] on a
grade of vitreous bonded aluminum oxide tested under similar
conditions. Thus, both an increase in toughness and a rising resistance
to fracture probably account for the strength behavior of this material
at elevated temperatures, whereas at low temperature, cavity size and
density are probably the critical determinants of strength.

Composite materials similar to that studied in this paper were
studied earlier by Trantina [20], Carroll and Tressler [17], Cohrt
et al. [21] and by Carroll et al. [22], who showed that the flexural
strength of these materials is strongly affected by the application of
creep loads prior to the strength measurement. When specimens were
tested with the original tensile surface in tension, the strength was
found to be greater than that obtained when the specimens were only
annealed prior to testing. Conversely, when the original tensile
surface was placed in compression during the strength test, a decrease
in the strength was obtained [17, 20-22]. The effect of the creep
stress on the strength of the composite was attributed to changes in
the stress distribution that result from the process of creep. The
strength data of this paper are in qualitative agreement with these
results, although additional data are required to firmly establish the
statistical significance of these results. The shift of the plane of
neutral strain, discussed below, supports the interpretation made by
these authors.

4.3 Neutral Plane Shift

If the creep rate is assumed to be both a power function of the
applied stress, and symmetric in tension and compression, then the
stress distribution within the flexure specimen must change with time
as the specimen is deformed [10-14]. The maximum tensile and
compressive stresses are less than that given by the linear elastic

stress distribution first imposed on the specimen [10, 14]3. The stress distribution is however symmetric through the specimen cross section, figure 11. The distribution is further modified if the creep rate in tension differs from that in compression [11-13]. Then, not only does the stress distribution change with time, but the neutral plane of the stress distribution changes its position and is no longer congruent with the geometric center of the test specimen. An example of the type of distribution expected when such shifts of the neutral plane occur is shown in figure 11. The shift in the neutral plane for strain measured in this paper and the observation of cavity formation on the tensile side of the geometric center of the specimen suggests that creep in this grade of silicon carbide is asymmetric, creep occurring more readily in tension than in compression. This asymmetry in creep behavior may be a result of the creep cavitation that occurs on the tensile side of the specimen.

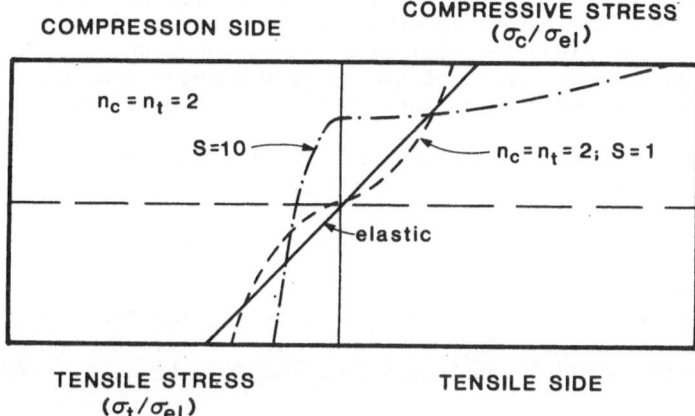

Fig. 11. Theoretical stress distributions after deformation by creep. The linear elastic distribution is maintained during power law creep if creep is symmetric and if the power law exponent is equal to unity, otherwise, the stress distribution after deformation is nonlinear. If creep is not symmetric in tension and compression, then the stress distribution is also not symmetric, and the neutral axis shifts from the geometric center of the specimen. The creep rates in compression and tension are given respectively by the following equations:

$$\dot{\varepsilon}_c = \dot{\varepsilon}_o (\sigma/\sigma_o)^{n_c}; \quad \dot{\varepsilon}_t = \dot{\varepsilon}_o (S\sigma/\sigma_o)^{n_t}$$

where $\dot{\varepsilon}_o$, S, n_c and n_t are empirically determined from the creep data, and σ_o is a normalization constant.

3If the power exponent is unity, then the stress distribution in creep is the same as the linear elastic distribution [10].

5.0 SUMMARY

Cavity formation seems to dominate the mechanical behavior of the fine-grain siliconized, silicon carbide composite discussed in this paper. When this material is subjected to creep deformation at elevated temperatures, cavities probably nucleate from small preexisting defects estimated to be approximately 5 nm in radius. They grow rapidly to a substantial size (5 μm), which accounts for the decrease in specimen strength at room temperature. At elevated temperatures, the strength is considerably higher, most likely as a consequence of an increase in the toughness of the material at elevated temperatures.

Cavity formation and growth also affects the creep behavior of the material. Massive generation of cavities in the tensile cross-section of the flexural test specimen results in an internal reduction of cross-section that enhances the creep rate and results in tertiary creep and eventual failure. As the cavities generate, the neutral planes for stress and strain shift from geometric center towards the compressive surface. Studies of the shift of the neutral plane for strain show that planar sections of the specimen before deformation remain planar after deformation, which is consistent with the usual assumption made for materials subjected to pure bending deformation. The shift of the plane for neutral strain towards the compressive surface suggests that creep of the silicon carbide studied in this paper occurs more readily in tension than in compression.

Acknowledgement: The authors are grateful for the support of the Department of Energy, Fossil Energy Program.

REFERENCES

1. V.J. Tennery, G.C. Wei, and M.K. Ferber, "High Temperature Behavior of Silicon Carbide and Aluminum Oxide Ceramics in Coal and Residual Oil Slags," Cer. Eng. and Sci. Proc. 2, 1171-88 (1981).
2. M.K. Ferber and V.J. Tennery, "Behavior of Tubular Ceramic Heat Exchanger Materials in Basic Coal Ash from Coal-Oil-Mixture Combination," Am. Ceram. Soc. Bull., 63, 898-904 (1984).
3. M. Srinivasan and M. Kasprzyk, "The Effect of Microstructure on the Mechanical Properties of Reaction Sintered Silicon Carbides," Presented at the Fall Meeting of the Basic Sciences Division of the American Ceramic Society, New Orleans, Louisiana, November 1979, Available as a Sohio Company Report.
4. H.J. Frost and M.F. Ashby, Deformation-Mechanism Maps: The Plasticity and Creep of Metals and Ceramics, Pergamon Press, New York (1982).
5. D.C. Drucker, "Engineering and Continuum Aspects of High-Strength Materials," pp. 795-833 in High Strength Materials, V.F. Zackey, ed., John Wiley and Sons, Inc., New York (1965).
6. F.F. Lange, "Non-Elastic Deformation of Polycrystals with a Liquid Boundary Phase," pp. 361-81 in Deformation of Ceramic Materials, R.C. Bradt and R.E. Tressler, Eds., Plenum Press, New York (1975).
7. R.E. Tressler, E.J. Minford and D.F. Carroll, pp. 551-63 in Creep and Fracture of Engineering Materials and Structures, Part I. Edited by B. Wilshire and D.R.J. Owen, Pineridge Press, Swansea, U.K., 1984.
8. N.J. Tighe and S.M. Wiederhorn, "Effects of Oxidation on the Reliability of Silicon Nitride," pp. 403-23 in Fracture Mechanics of Ceramics, Vol. 5, R.C. Bradt, A.G. Evans, D.P.H. Hasselman and F.F. Lange, eds., Plenum Press, New York (1983).

9. S.M. Wiederhorn, B.J. Hockey, R.F. Krause, Jr., and K. Jakus, "Creep and Fracture of Vitreous Bonded Aluminum Oxide," J. Mater. Sci., in press.

10. G.W. Hollenberg, G.R. Terwilliger and R.S. Gordon, "Calculation of Stress and Strains in Four-Point Bending Creep Tests," J. Am. Ceram. Soc. 54, 196-99 (1971).

11. H. Cohrt, G. Grathwohl and F. Thümmler, "Non-Stationary Stress Distribution in a Ceramic Bending Beam During Constant Load Creep," Res. Mechanica, 10, 55-71 (1984).

12. T. Fett, "Stress Distribution in a Bending Beam for Cyclic Loading Under Creep Conditions," Res. Mechanica, in press.

13. Tze-jer Chuang, "Estimation of Power-Law Creep Parameters from Bend Test Data," J. Mat. Sci., in press.

14. I. Finnie and W.R. Heller, Creep of Engineering Materials, McGraw-Hill Book Co., Inc., New York (1959).

15. S.M. Wiederhorn, R.F. Krause, Jr., and K. Jakus, "Strength versus Crack Size in Vitreous Bonded Aluminum Oxide at Elevated Temperatures," Presented at the Fourth International Symposium on the Fracture Mechanics of Ceramics, Blacksburg, VA, June 19-21 (1985).

16. A.W. Adamson, Physical Chemistry of Surfaces, 4th Ed., John Wiley and Sons, Inc., New York (1982).

17. D.F. Carroll and R.E. Tressler, "Time-Dependent Strength of Siliconized Silicon Carbide Under Stress at 1000° and 1100°C," J. Am. Ceram. Soc., 68, 143-6 (1985).

18. L. Chuck and E.R. Fuller, "Fracture Behavior of a Siliconized Silicon Carbide at Elevated Temperatures," Presented at the Fourth International Symposium on the Fracture Mechanics of Ceramics, Blacksburg, VA, June 19-21 (1985).

19. W.B. Hillig, R.L. Mehan, C.R. Morelock, V.J. DeCarlo and W. Laskow, "Silicon/Silicon Carbide Composites," Am. Ceram. Soc. Bull. 54, 1054-56 (1975).

20. G.G. Trantina, "Strengthening and Proof Testing of Siliconized SiC," J. Mater. Sci. 17, 1487-92 (1982).

21. H. Cohrt, G. Grathwohl and F. Thümmler, "Strengthening after Creep of Reaction-Bonded-Siliconized Silicon Carbide," pp. 515-26 in Creep and Fracture of Engineering Materials and Structures, B. Wilshire and D.R.J. Owen, eds., Pineridge Press, Swansea, U.K., (1984).

22. D.F. Carroll, R.E. Tressler, Y. Tsai and C. Near, "High Temperature Mechanical Properties of Siliconized Silicon Carbide Composites," in Tailoring Multiphase and Composite Ceramics, held at The Pennsylvania State University, University Park, Pa., July 17-19, 1985.

HIGH TEMPERATURE MECHANICAL PROPERTIES OF

SILICONIZED SILICON CARBIDE COMPOSITES

D. F. Carroll, R. E. Tressler, Y. Tsai and C. Near

The Pennsylvania State University
Department of Materials Science and Engineering
University Park, PA 16802

INTRODUCTION

Siliconized silicon carbide composites are candidate materials for ceramic components in gas turbines and heat exchangers (1,2). Properties which make these materials suited for these applications are good oxidation resistance, high thermal conductivity, low thermal expansion and adequate mechanical strength (1,3). Siliconized silicon carbide is a two-phase material whose microstructure consists of interpenetrating phases of silicon and silicon carbide. The silicon phase is dispersed throughout the continuous silicon carbide phase in either a continuous or non-continuous matrix, depending upon the amount of free silicon in the material.

Past investigations have revealed a strengthening phenomena in some types of siliconized silicon carbides during interrupted static load tests in bending at elevated temperatures (4-6). This strengthening phenomena has been attributed to either localized flaw blunting processes or macroscopic stress redistribution in the bending beam. Recently, Carroll and Tressler have shown both of these processes to be operative in a specific siliconized silicon carbide (7). Flaw blunting may result in strengthening through localized deformation processes around the crack tips of severe flaws in the material.

Strengthening through macroscopic stress redistribution during static load tests may also occur in a material that exhibits non-linear creep behavior or a different creep rate in tension than in compression. An analysis developed by Cohrt et al. enables the calculation of the stress redistribution process in a bend beam as a function of time (8). Figure 1 summarizes the stress redistribution process in a bending beam for a material that exhibits non-linear creep behavior only (i.e., the creep exponent (n) is greater than one, while the ratio (S) of the creep rate in tension to compression is equal to one). At time t=0, the initial applied stress is linear throughout the beam. As time progresses, the non-linear creep behavior of the material causes a redistribution of the stress in the beam. At $t=t_1$, the stress at the outer tensile and compressive fiber has decreased while the stress around the neutral axis has increased compared to the initial values. If the beam is loaded to failure at $t=t_1$, a strength increase would be observed due to the reduction of the outer fiber tensile stress. If the specimen is unloaded at $t=t_1$, a residual compressive stress would develop on the tensile surface while a residual

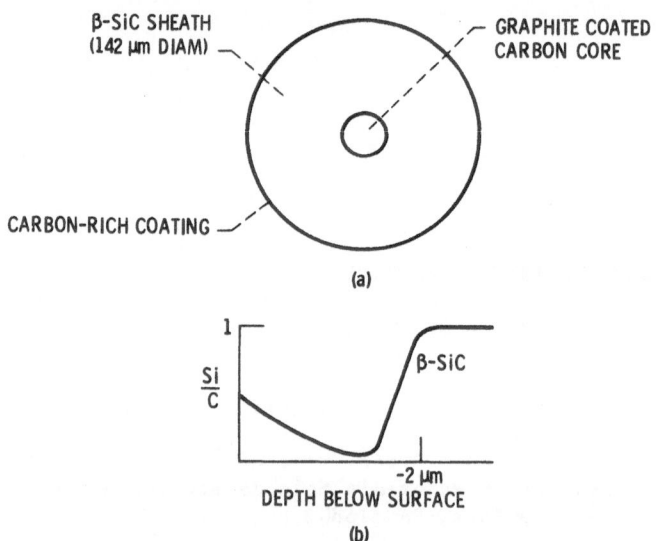

Figure 1. Stress redistribution in a four-point bend beam with time for a
            material which exhibits non-linear creep behavior only (n=2,
            S=1, σ_o = initial outer fiber tensile stress).
            - initial stress distribution at t=0,
            ◇ stress distribution at $t=t_1$
            ◆ residual stress distribution after unloading specimen
              at $t=t_1$

tensile stress would develop on the compressive side.

     The stress redistribution process in a bend beam, when the material
exhibits linear creep behavior (n=1) and a creep rate in tension that is
ten times greater than in compression (S=10), is summarized in figure 2.
Again, at time t=0, the initial stress distribution is linear throughout
the bend beam. At time $t=t_1$, the stress redistribution process has again
caused the outer fiber tensile stress to decrease. In contrast to the
previous example, the outer fiber compressive stress has not decreased.
The outer fiber compressive stress has increased compared to its initial
value. The neutral axis has shifted from h/2 toward the compressive side
of the specimen. If at $t=t_1$, the specimen is loaded to failure, a
strength increase would be observed due to the reduction of the outer fiber
tensile stress. If the specimen is unloaded at $t=t_1$, a residual
compressive stress would develop on both the tensile and compressive
surfaces while a residual tensile stress would develop in the interior of
the bend beam.

     The purpose of this paper is to summarize the strengthening phenomenon
observed in two commercially available siliconized silicon carbides with
significantly different volume fractions of silicon, during static load
tests at elevated temperatures. The relative amount of strengthening in
each material due to localized flaw blunting processes and macroscopic
stress redistribution will also be discussed.

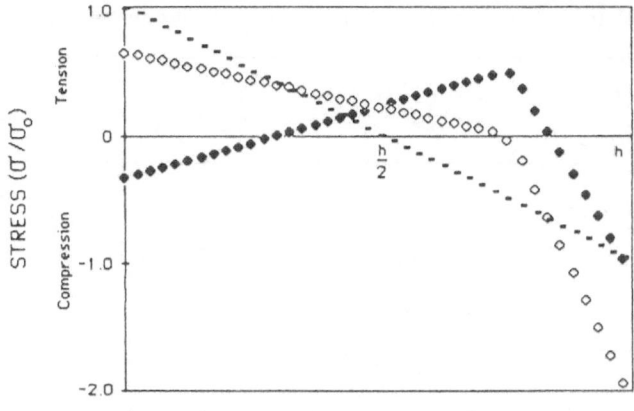

THICKNESS OF BEND BEAM

Figure 2. Stress redistribution in a four-point bend beam with time for a
material which exhibits linear creep behavior and a creep ratio
in tension ten times greater than in compression (n=1, S=10,
σ_o = initial outer fiber tensile stress).
  ▪ initial stress distribution at t=0
  ◇ stress distribution at $t=t_1$
  ◆ residual stress distribution after unloading specimen
    at $t=t_1$

EXPERIMENTAL PROCEDURE

     The two commercially available siliconized silicon carbide materials
investigated in this study were designated A and B.* The material
specifications are summarized in table 1.· Material A is composed of 4-6
micron SiC grains and approximately 29 volume percent free silicon in a
continuous matrix. Material B contains a bimodal distribution of 10 and
100 micron SiC grains with 15 volume percent free silicon dispersed
throughout the continuous SiC network. The densities of material A and B
were 2.95 and 3.08 g/cc, respectively.

     Bend beams of material.A were cut from billets using a diamond saw.
The major faces were ground parallel with a 600 grit diamond wheel. The
approximate specimen dimensions were 25 by 2.5 by 2.5 mm. The edges
forming the tensile plane were bevelled by hand with 600 grit SiC grinding
paper to remove any edge checking flaws. In all cases, the bend beams were
proof tested at 495 MPa to narrow the strength distribution. This proof
stress yielded a survival rate of 70%.

     The as-received material B was precut into specimens with approximate
dimensions of 76.2 by 7.6 by 6.3 mm. The tensile and compressive surfaces
were finished to 320 grit while the other two faces were finished to 150
grit. The edges which formed the tensile plane in these specimens were
bevelled using a 15 micron diamond wheel. Since these specimens exhibited
a narrow starting strength distribution, proof testing was not necessary to

---

* Material A - KX01, Sohio Engineered Materials Co., Niagara Falls, NY
   Material B - NC430, Norton Co., Worcester, MA

Table 1.   Description of material characteristics

Material	SiC Grain Size	Volume % Silicon	Density (g/cc)
A	4-6 μm	29	2.95
B	Bimodal 10 μm 100 μm	15	3.08

apply the technique described by Tressler et al. (4). Smaller specimens were also cut and machined from these specimens to dimensions similar to that of material A.

The fracture toughnesses of these specimens were determined as a function of temperature using the single edge notch beam technique (9). An argon environment was used during the fracture toughness tests at elevated temperatures. The notched specimens were fractured in four-point bending using a crosshead speed of 0.25 cm/min.

The strength distributions of the siliconized silicon carbides were determined as a function of temperature in four-point bending. The inner and outer span dimensions of the knife edges used for the A specimens were 4.76 and 19.05 mm, respectively. The knife edge dimensions used for the B specimens were 12.7 and 50.8 mm, respectively. A crosshead speed of 0.25 cm/min was used for all fracture tests.

In order to observe the strengthening phenomena in these two siliconized silicon carbides under stress, interupted static load tests were conducted at temperatures between 1000 and 1300°C for times up to 50 hours. The applied stress levels during static loading varied from 0.50 to 0.80 σ_f, where σ_f is the average fracture stress of the material at test temperature. A summary of the static loading conditions is shown in table 2. At the end of the static load time, the specimens were fractured at test temperature by rapidly loading from the static load level.

Some of the specimens, after static loading, were not fractured at the end of the static load time, but were unloaded and cooled to room temperature. By fracturing these specimens with either the original tensile or compressive surface in tension, the existence of residual stresses was determined by comparing the resulting strengths to those in the initial distribution. If a residual compressive stress exists on the tensile surface of the specimen and either a residual tensile or compressive stress exists on the compressive surface, strengthening through macroscopic stress redistribution as described by Cohrt et al. has occurred (8). The smaller B specimens were used in this phase of the investigation.

In order to attempt to separate the strengthening contributions of localized flaw blunting and macroscopic stress redistribution, a series of artificial flaw experiments were conducted. Large artificial flaws were introduced into specimens using the controlled flaw technique (10). A 19.6 and 88.2 N Knoop indentation was used to produce artificial surface cracks in the A specimens and the smaller B specimens. The larger indentation load was necessary for the B material so that the artificial flaws would be larger than the inherent flaws in the material. After indenting, the crater was removed by grinding the surface to eliminate residual stresses.

Table 2. Summary of static load conditions used in this investigation

Material	Temperature ($^{\circ}$C)	Static Load Stress	Time of Static Load (hrs)
A	1000	$0.50\ \sigma_f$	2
	1100	$0.80\ \sigma_f$	10
B	1300	$0.50\ \sigma_f$	4
		$0.75\ \sigma_f$	16
			50

The resulting strength distributions of the indented A specimens were determined at room temperature, 1100°C in argon and 1100°C in argon after a 10 hour anneal. During furnace heat-up to 1100°C, a prestress of 0.75 σ_{fI}, where σ_{fI} is the average room temperature strength of the indented specimens, was applied to prevent flaw healing (11). Indented A specimens were then static loaded for 10 hours in argon at an applied stress of 0.80 σ_{fI}', where σ_{fI}' is the average fracture stress of the indented specimens at 1100°C. The strength distributions of the indented B specimens were determined at room temperature and 1300°C in argon. A prestress of 0.50 σ_{fI} was used during furnace heat-up to help prevent flaw healing. Four hour static load tests of the indented B specimens were also conducted at 1300°C using applied stresses of 0.50 and 0.75 σ_{fI}'.

RESULTS AND DISCUSSION

Fracture Toughness

The fracture toughnesses of materials A and B as a function of temperature are summarized in figure 3. At room temperature, where the silicon is not as tough as silicon carbide, material A has the lower toughness due to its higher silicon content. The fracture toughnesses at room temperature for materials A and B are on the average 2.95 and 3.73 MPa.m^{172}, respectively. As temperature is increased, the fracture toughness of material A increases sharply while the toughness for material B decreases slightly. The increase in fracture toughness of material A is due to the free silicon phase which is ductile at these temperatures. The silicon phase (29 volume percent) in material A forms a continuous matrix. The plastic strain capabilities of the ductile matrix are responsible for the increase in fracture toughness. A similar increase in toughness with temperature of material A was reported by Chuck et al. (12). There was no increase observed in the fracture toughness of material B at elevated temperatures even though this material contains 15 volume percent free silicon. The fracture toughness of this material is controlled by the rigid SiC network. During processing, the rigid SiC network is formed during the refire step which partially sinters the SiC structure (13). The molten silicon is then infiltrated into this porous structure reacting with the free carbon to form SiC. The remaining pores are then filled with silicon. The resulting silicon phase is located in pockets throughout the SiC network forming a semi-continuous matrix. The deformation capabilities of the silicon are restricted by the SiC network. Therefore, the fracture toughness does not change dramatically with temperature. The fracture toughness of material B as a function of temperature, reported by Fuller, is shown to remain relatively constant for a displacement rate of 0.05

cm/min (14). However, his results did show a slight increase in toughness with temperatre when a much slower displacement rate, 5×10^{-4} cm/min, was used. This result suggests that the higher displacement rate, 0.25 cm/min, used in the present work does not allow sufficient time for silicon deformation to have an effect on the fracture toughness.

Strengthening Behavior During Static Load Tests

The results of the static load tests as a function of time for materials A and B are summarized in figure 4. Material A was static loaded at 0.80 σ_f (480 MPa) at 1000 and 1100°C for 2 and 10 hours while material B was static loaded at 0.75 σ_f (170 MPa) at 1300°C for 4, 16 and 50 hours. Both materials exhibit an increase in strength with time of static loading. The increase in strength in material A is larger than that observed in material B. For example, after two hours of static loading at

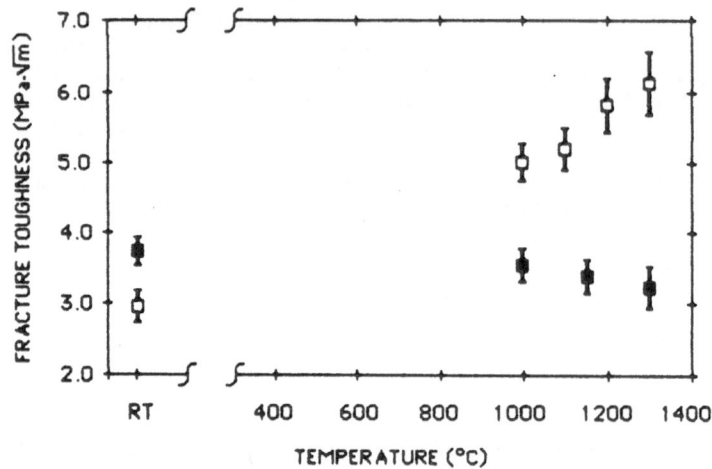

Figure 3.   Fracture toughnesses versus temperature for materials A (□) and B (■) determined using the notched-beam method.

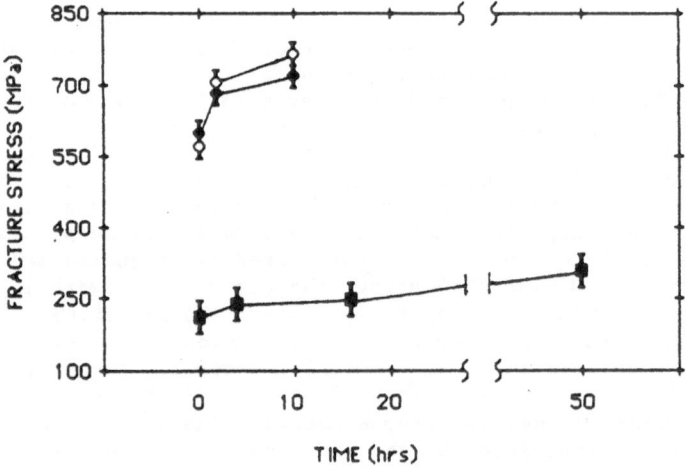

Figure 4.   Fracture stress of material A after static loading at 0.80 σ_f for 2 and 10 hours at 1000 (◆) and 1100°C (◇) and material B after static loading at 0.75 σ_f for 4, 16 and 50 hours at 1300°C (■).

an applied stress of 0.80 σ_f at 1100°C, the average fracture stress of material A increased 19% from 570 to 680 MPa. The average fracture stress of material B after static loading 4 hours at 0.75 σ_f increased 14% from 208 to 238 MPa. The amount of strengthening, initially large for both materials at shorter static load times, tends to approach a limiting value at longer times. Strengthening in material B under stress at elevated temperatures has also been reported by Trantina (5). His results have shown a similar time dependent strength increase at 1200°C under a stress equivalent to 86% of the baseline strength.

A temperature dependence of the strengthening phenomenon is also observed in material A by comparing the results at 1000 and 1100°C. This dependence is more evident after 10 hours of static loading where the average fracture stress at 1000 and 1100°C increased 20 and 27%, respectively, over their initial distributions.

The stress dependence of the strengthening phenomenon during static loading is summarized in figure 5. The results show no observable strength increase after static loading at 0.50σ_{fo} for 10 hours at 1000 and 1100°C for material A and for 16 hours at 1300°C for material B. However, a strength increase was observed after static loading for the same times and temperatures under a higher applied stress of 0.80 σ_f and 0.75 σ_f for material A and B, respectively. Again, material A exhibits a larger strength increase under the higher applied stress for shorter static load times. These results indicate that a time, temperature and stress dependent strengthening process exists in both materials A and B.

Permanent Deformation

The amount of permanent deformation in A and B specimens was determined by measuring the outer fiber strains of specimens which were unloaded after static loading. The permanent strains measured for the various test conditions are summarized in figure 6. The results show that material A experienced more deformation than material B under the higher applied stress levels. The permanent strain in the A specimens after static loading at 1000 and 1100°C for 10 hours at 0.80 σ_f was 1.8 and 2.7%, respectively. The permanent strain in the B specimens after static

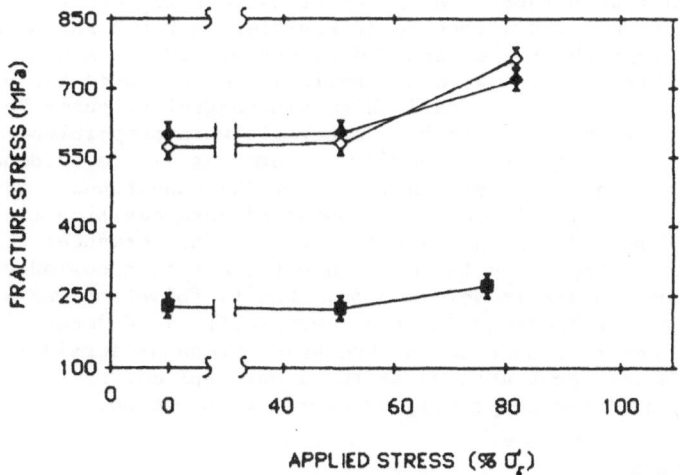

Figure 5. Fracture stress of material A after static loading at 0.50 and 0.80 σ_f for 10 hours at 1000 (◆) and 1100°C (◇) and material B after static loading at 0.50 and 0.75 σ_f for 16 hours at 1300°C (■).

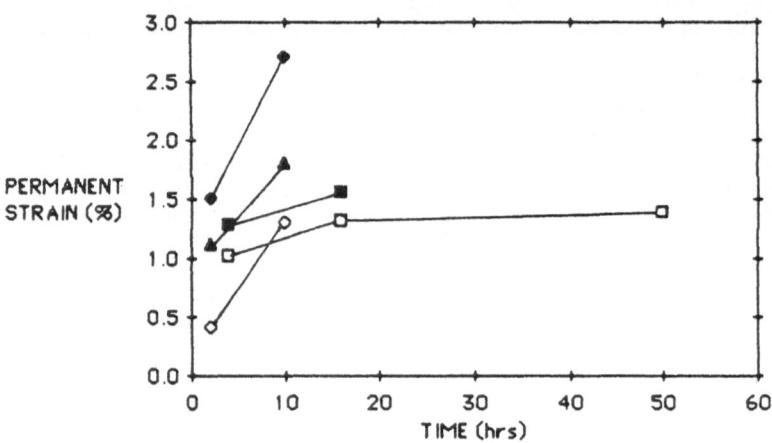

Figure 6. Permanent strains in material A after static loading at 0.50 and
0.80 σ_f for 2 and 10 hours at 1000 and 1100°C and material
B after static loading at 0.50 and 0.75 σ_f for 4, 16 and 50
hours at 1300°C.
◇ material A, 0.50 σ_f, 1100°C
▲ material A, 0.80 σ_f, 1000°C
◆ material A, 0.80 σ_f, 1100°C
□ material B, 0.50 σ_f, 1300°C
■ material B, 0.75 σ_f, 1300°C

loading at 0.75 σ_f for 16 hours at 1300°C was 1.55%. The larger amount
of deformation in material A is due to higher silicon content. At the
lower applied static load level of 0.50 σ_f, the A specimens exhibited
approximately the same amount of deformation after static loading for 10
hours at 1100°C, as the B specimens after static loading for 16 hours at
1300°C.

Cavity Formation

Polished tensile surfaces of unbroken static loaded specimens were
examined for microstructural rearrangement by scanning electron microscopy.
This examination revealed formation of cavities in the A and B specimens
static loaded under the higher applied stress of 0.80 σ_f and 0.75 σ_f,
respectively. The cavities were non-uniformly distributed throughout the
tensile plane in regions of locally high SiC content (figures 7a, 7b). The
size and density of cavities in both materials were proportional to the
time and temperature of static loading. There was little evidence of
coalescence or linkage of these cavities for the conditions studied in this
investigation. Material A generally contained more cavities than material
B at correspondingly larger permanent strains. The permanent strains
obtained at these stress levels for both materials is accommodated through
cavity formation and matrix deformation. Cavity formation was not observed
in either the A or B specimens at the lower static load level of 0.50 σ_f.
The absence of cavities suggests a threshold stress or strain for cavity
formation. The permanent strains at the lower applied stress level is
apparently accommodated completely by matrix deformation.

Residual Stresses

In order to examine the possibility of macroscopic stress redistribution
as a cause of strengthening during static loading, specimens were static
loaded at the high applied stress levels, unloaded and fractured at room
temperature. The data collected by fracturing the A and B specimens with

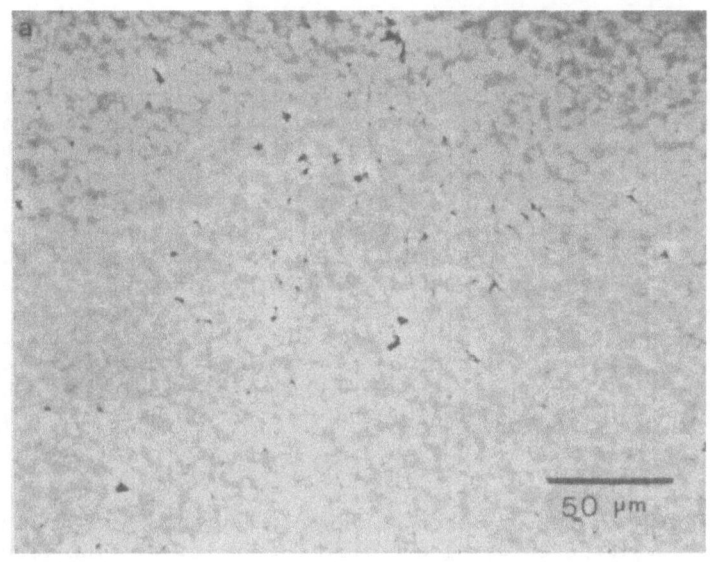

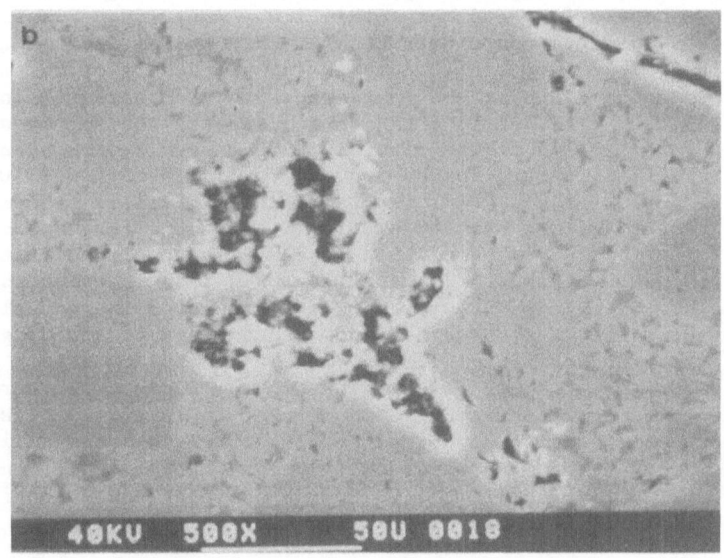

Figure 7. SEM micrographs of polished tensile surfaces of, 7a) material
A, static loaded at 0.80 σ_f for 10 hours at 1100°C, and
7b) material B, static loaded at 0.75 σ_f for 16 hours at
1300°C.

the original tensile surface in tension and the original compressive surface in tension are shown in figures 8 and 9. Material A results are for specimens initially static loaded at 0.80 σ_f for 2 hours at 1000°C. Material B results are for specimens static loaded at 0.75 σ_f for 4 hours at 1300°C. Both materials exhibited a residual compressive stress in the specimens fractured with the original tensile surface in tension. The compressive stress produces an increase in strength from the initial distributions. The residual compressive stress is larger in material A than in material B. For the specimens fractured with the original compressive surfaces in tension, the resulting strengths decreased for both materials as compared to the initial distribution. The strength decrease is attributed to a residual tensile stress. The residual tensile stress is larger in the A specimens than in the B specimens. The existence of these residual stresses after static loading suggests that the type of strengthening described by Cohrt et al. may be operative (8). Since the residual stresses are larger in the A specimens, the stress redistribution process may occur more readily in this material than in material B. This result is expected since the A specimens exhibited more permanent deformation than the B specimens. The larger amount of permanent strain in material A indicates a higher creep rate which would result in a faster stress redistribution process.

Artificial Flaw Tests

Strengthening on the microscopic level through localized flaw blunting processes was investigated using artificial flaw tests. The results for material A are summarized in table 3. The 19.6 N Knoop indentation (followed by surface grinding to remove the crater) decreased the average room temperature strength to 260 MPa. At 1100°C, the strength of the indented specimens with an applied prestress of 0.75 σ_{fI} (195 MPa) increased to 400 MPa. This large increase may be attributed to flaw blunting or healing, although the applied prestress during furnace heatup tends to hold open the flaw and retard healing. Annealing of residual stresses probably did not contribute much to this increase since the specimens were fractured almost immediately upon reaching test temperature. After 10 hours annealing under the same condition, the strength increased slightly to 430 MPa. This small strength increase may be due to annealing of residual stresses or localized flaw blunting caused by the prestress. The indented specimens static loaded at 0.80 σ_{fI}' (320 MPa) for 10 hours at 1100°C were approximately 10% stronger than the annealed specimens. The average fracture stress was 480 MPa. All specimens tested under these conditions fractured at the indentation cracks. The average fracture stress after 10 hours of static loading was still less than the as-received 1100°C strengths. The small 10% strength increase over the annealed specimens suggests that a stress induced microscopic strengthening process has blunted the sharp artificial flaws. Since the applied stress of 320 MPa is well below that needed for strengthening by macroscopic stress redistribution, the strength increase is related to localized blunting processes around the crack tip. Since the 10% strength increase is less than the 27% strength increase observed in proof tested A specimens static loaded at 0.80 σ_f for 10 hours at 1100°C, the dominant strengthening process in this material is macroscopic stress redistribution.

The results for the artificial flaw tests in material B are summarized in table 4. The 88.2 N Knoop indentation (after surface grinding to remove the crater) reduced the average room temperature fracture stress to 180 MPa. At 1300°C, the strength of the indented specimens under an applied prestress of 0.50 σ_{fI} (90 MPa) increased to 240 MPa. All of these specimens fractured at the indentation cracks. The indented specimens static loaded at 0.50 and 0.75 σ_{fI}' exhibited a significant amount of strengthening. Only 2 out of 10 specimens static loaded at 0.50 σ_{fI}' failed at the indentation cracks, while at 0.75 σ_{fI}', none of the specimen

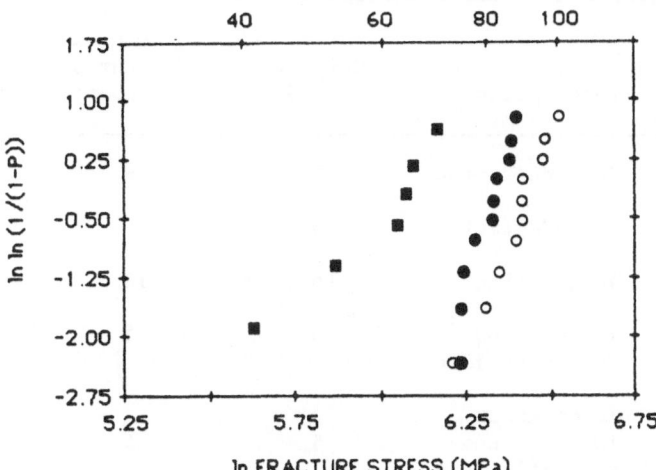

Figure 8. Strength distributions of material A, static loaded at 0.80 σ_f for 2 hours at 1000°C, unloaded and fractured at room temperature.

    ● initial room temperature distribution
    ○ fractured with original tensile surface in tension
    ■ fractured with original compressive surface in tension

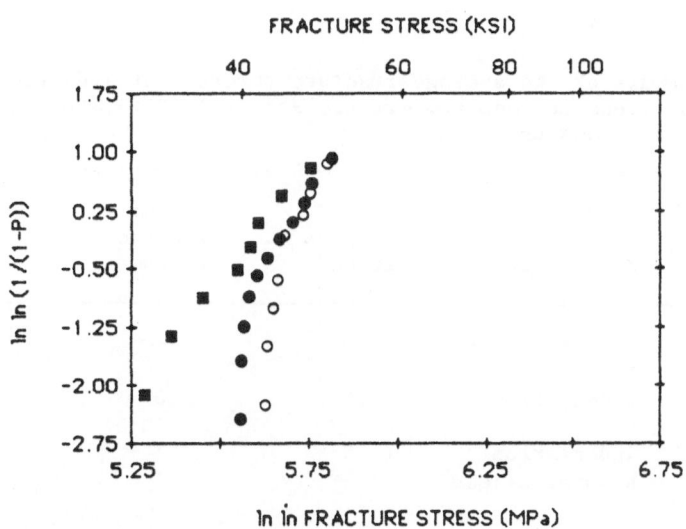

Figure 9. Strength distributions of material B, static loaded at 0.75 σ_f for 4 hours at 1300°C, unloaded and fractured at room temperature

    ● initial room temperature
    ○ fractured with original tensile surface in tension
    ■ fractured with original compressive surface in tension

Table 3. Summary of the average fracture stresses of indented material A specimens at room temperature and at 1100°C after annealing and static loading.

Temperature	Description	Samples Tested	Samples Fractured at Indent	Average Indented Strength (MPa)
Room	No indentation	10	-	570 ± 40
1100°C	No indentation	10	-	600 ± 40
Room	Indentation	10	10	260 ± 15
1100°C	Indentation No static load	5	5	400 ± 30
1100°C	Indentation 10 hour anneal No static load	5	5	430 ± 20
1100°C	Indentation 10 hour static load at $0.80\sigma_{fI}'$	5	5	480 ± 25

Table 4. Summary of the average fracture stresses of indented material B specimens at room temperature and at 1300°C before and after static loading.

Temperature	Description	Samples Tested	Samples Fractured at Indent	Average Indented Strength (MPa)
Room	No indentation	10	-	285 ± 25
Room	Indentation	10	10	180 ± 15
1300°C	Indentation No static load	5	5	240 ± 5
1300°C	Indentation 4 hour static load at $0.50\sigma_{fI}'$	10	2	320 ± 5
1300°C	Indentation 4 hour static load at $0.75\sigma_{fI}'$	10	0	-----

failures occurred at the indentation cracks. It is postulated that even with a prestress of 0.50 σ_{fI} to hold open the flaw, strengthening occurred by a combined effect of flaw blunting and healing. This strength increase may also be complicated by annealing of residual stresses that may have been present after crater removal. Since none of the specimens failed at the indentation cracks under the higher applied stess of 0.75 σ_{fI}', compared to 2 specimens at 0.50 σ_{fI}', a stress induced flaw blunting process must be operative in this material and is apparently the dominant strengthening process under the present experimental conditions.

Summary of the Strengthening Phenomenon

Both materials exhibited a time, temperature and stress dependent strengthening process. This strengthening process occurs on both the macroscopic and microscopic levels. Macroscopic strengthening occurs through stress redistribution in the bend beam. This effect was confirmed by the presence of residual stresses found in specimens unloaded after static loading at the high applied stress levels. The larger residual stresses in the A specimens indicates that the stress redistribution process dominates in material A while playing a lesser role in material B. From the results of the artificial flaw tests, both materials exhibited microscopic flaw blunting behavior. Flaw blunting processes in the indented A specimens produced a 10% strength increase after static loading at 0.80 σ_{fI}' for 10 hours at 1100°C. In comparison, the fracture stress of proof tested A specimens static loaded at 0.80 σ_f for 10 hours at 1100°C increased 27%. The 27% strength increase in the proof tested specimens suggests that macroscopic stress redistribution is the dominant strengthening process in the A material.

The artificial flaw tests with material B have shown a significant amount of strengthening after static loading for four hours at 0.50 and 0.75 σ_{fI}'. Only 2 of the 10 specimens static loaded at 0.50 σ_{fI}' failed at the indentation cracks. At the higher static load level of 0.75 σ_f', no specimen failures occurred at the indentation cracks due to a dominant flaw blunting process. A 14% strength increase was observed in the as-received B specimens after static loading at 0.75 σ_f for four hours. Since the residual stresses were relatively small in similar specimens which were unloaded after static loading and fractured at room temperature, localized flaw blunting processes must be the dominant strengthening process in ths material. This conclusion is based upon the large strengthening increment observed in the static load tests with indented B specimens. In future work, strengthening due to localized flaw blunting processes without the contribution of macroscopic stress redistribution will be determined during static load tests in pure tension.

ACKNOWLEDGEMENTS

The authors thank the Exxon Engineering and Research Company and the Electric Power Research Institute for their support.

REFERENCES

1. B. North and K. E. Kilchrist, "Effect of Impurity Doping on a Reaction-Bonded Silicon Carbide," Am. Ceram. Soc. Bull., 60 (5), 549-52 (1981).
2. G. C. Wei and V. J. Tennery, "Evaluation of Tubular Ceramic Heat Exchanger Materials in Residual Oil Combustion Environments," ORNL/TM-7578, March 1981.

3. C. W. Forrest, P. Kennedy and J. V. Shennan, pp. 99-123 in Special Ceramics Vol. 5, Edited by P. Popper, British Ceramic Research Association, Stoke-on-Trent, England, 1972.

4. R. E. Tressler, and E. J. Minford and D. F. Carroll, pp. 551-563 in Creep and Fracture of Engineering Materials and Structures, Part I. Edited by B. Wilshire and D.R.J. Owen, Pineridge Press, Swansea, U.K., 1984.

5. G. G. Trantina, "Strengthening and Proof Testing of Siliconized Silicon Carbide," J. Mater. Sci., 17 1487-92 (1982).

6. H. Cohrt, G. Grathwohl and F. Thummler, pp. 515-526 in Creep and Fracture of Engineering Materials and Structures, Part I., Edited by B. Wilshire ad D.R.J. Owen, Pineridge Press, Swansea, U.K., 1984.

7. D. F. Carroll and R. E. Tressler, "Time Dependent Strength of Siliconized Silicon Carbide Under Stress at 1000 and 1100°C," J. Am. Ceram. Soc., 68 (3), 143-146 (1985).

8. H. Cohrt, G. Grathwohl and F. Thummler, "Non-Stationary Stress Distribution in a Ceramic Bending Beam During Constant Load Creep," Res. Mechanica, 10 55-71 (1984).

9. M. Srinivason and R. H. Smoak, "Elevated Temperature Strength and Fracture Toughness Determination of sintered Alpha Silicon Carbide," presented at the International Conference on Fracture Mechanics in Engineering Applications, Bangalore, India, March 26-30 (1979).

10. J. J. Petrovic, L. A. Jacobson, P. K. Tally and A. K. Vasudern, "Controlled Surface Flaws in Hot-Pressed Silicon Nitride," J. Am. Ceram. Soc., 58 (3-4), 113-116 (1975).

11. W. Blumenthal and A. G. Evans, "Characterization of Cracks Subjected to Creep," pp. 555-572 in Deformation of Ceramic Materials II., Edited by R. E. Tressler and R. C. Bradt, Plenum Press, New York, NY, 1984.

12. L. Chuck and E. R. Fuller, "Fracture Behavior of a Siliconized Silicon Carbide at Elevated Temperatures," presented at the Fourth International Symposium on the Fracture Mechanics of Ceramics, Blacksburg, VA, June 19-21 (1985).

13. G. Q. Weaver and B. A. Olson, "Process for Fabricating Silicon Carbide Articles," U. S. Patent #4,019,913, April 26, 1977.

14. E. R. Fuller, private communication.

CONTRIBUTORS

Co-Chairmen

G. L. Messing, Associate Professor of Ceramic Science and Engineering, Pennsylvania State University, University Park, PA

R. E. Tressler, Professor and Chairman of Ceramic Science and Engineering, Pennsylvania State University, University Park, PA

C. G. Pantano, Associate Professor of Materials Science and Engineering, Pennsylvania State University, University Park, PA

R. E. Newnham, Professor and Chairman of Solid State Science, Pennsylvania State University, University Park, PA

Conference Session Chairmen

L. E. Cross, Pennsylvania State University, University Park, PA

E. Fitzer, Universitat Karlsruhe, Karlsruhe, W. Germany

W. B. Hillig, General Electric, Schenectady, NY

T. F. Page, University of Cambridge, Cambridge, England

D. W. Readey, Ohio State University, Columbus, OH

L. Toth, National Science Foundation, Washington, DC

H. Yanagida, University of Tokyo, Tokyo, Japan

Conference Coordinator

R. Avillion, J. Orvis Keller Conference Center, Pennsylvania State University, University Park, PA

AUTHORS

J. G. Baldoni, GTE Labs, Waltham, MA

A. S. Bhalla, Pennsylvania State University, University Park, PA

R. T. Bhatt, NASA Lewis Research Center, Cleveland, OH

R. Bordia, Cornell University, Ithaca, NY

R. Bradt, University of Washington, Seattle, WA

J. J. Brennan, United Technologies Research Center, E. Hartford, CT

R. J. Brook, University of Leeds, Leeds, United Kingdom

D. Carroll, Pennsylvania State University, University Park, PA

W. Carlson, Pennsylvania State University, University Park, PA

L. Chuck, National Bureau of Standards, Gaithersburg, MD

N. Claussen, Technische Universitat Hamburg-Harburg, Hamburg, West Germany

D. C. Cranmer, Aerospace Corporation, Los Angeles, CA

S. DaVanzo, Pennsylvania State University, University Park, PA

R. F. Davis, North Carolina State University, Raleigh, NC

R. T. DeHoff, University of Florida, Gainesville, FL

K. T. Faber, Ohio State University, Columbus, OH

E. Fitzer, Universitat Karlsruhe, Karlsruhe, West Germany

S. Freiman, National Bureau of Standards, Gaithersburg, MD

E. R. Fuller, Jr., National Bureau of Standards, Gaithersburg, MD

R. Gadow, Universitat Karlsruhe, Karlsruhe, West Germany

J. Giniewicz, Pennsylvania State University, University Park, PA

A. M. Glaeser, University of California at Berkeley, Berkeley, CA

H. Godard, Ohio State University, Columbus, OH

M. Gomina, University of Caen, Caen, France

D. J. Green, Pennsylvania State University, University Park, PA

T. K. Gupta, Alcoa Labs, Alcoa Center, PA

R. Hannink, CSIRO, Melbourne, Australia

D.P.H. Hasselman, Virginia Polytechnic Institute and State University, Blackburg, VA

R. Hayami, GIRI-Osaka, Osaka, Japan

J. R. Hellmann, Sandia National Labs, Albuquerque, NM

A. H. Heuer, Case Western Reserve University, Cleveland, OH

W. B. Hillig, General Electric Research Center, Schenectady, NY

T. Hirai, Tohoku University, Sendai, Japan

W. Huebner, University of Missouri-Rolla, Rolla, MO

Y. Ikuma, Ikutoku Technical University, Kanagawa, Japan

M. Kahn, National Research Laboratory, Alexandria, VA

K. Kuroda, Waseda University, Tokyo, Japan

O-H. Kwon, Pennsylvania State University, University Park, PA

P. Lagerlof, Case Western Reserve University, Cleveland, OH

W. E. Lee, Case Western Reserve University, Cleveland, OH

M. H. Lewis, University of Warwick, West Midland, United Kingdom

G. L. Messing, Pennsylvania State University, University Park, PA

E. Minford, United Technologies Research Center, East Hartford, CT

K. L. More, North Carolina State University, Raleigh, NC

R. Naslain, de Centre National de La Recherche Scientifique, Talence,
    France

J. N. Ness, University of Cambridge, Cambridge, United Kingdom

R. E. Newnham, Pennsylvania State University, University Park, PA

T. Page, University of Cambridge, Cambridge, United Kingdom

D. A. Payne, University of Illinois, Urbana, IL

H. R. Philipp, General Electric Research Center, Schenectady, NY

S. M. Pilgrim, Pennsylvania State University, University Park, PA

B. Pletka, Michigan Technological University, Houghton, MI

E. J. A. Pope, University of California at Los Angeles, Los Angeles, CA

K. W. Prewo, United Technologies Research Center, East Hartford, CT

A. Revcolevschi, Universite de Paris-Sud, Orsay, France

G. Rossi, Norton Company, Worcester, MA

J. Runt, Pennsylvania State University, University Park, PA

A. Safari, Pennsylvania State University, University Park, PA

A. E. Semple, Pennsylvania State University, University Park, PA

V. S. Stubican, Pennsylvania State University, University Park, PA

W. Y. Sun, University of Newcastle upon Tyne, Newcastle upon Tyne, United
    Kingdom

T. Tiegs, Oak Ridge National Laboratory, Oak Ridge, TN

D. P. Thompson, University of Newcastle upon Tyne, Newcastle upon Tyne,
    United Kingdom

R. Tressler, Pennsylvania State University, University Park, PA

M. Trontelj, University of Ljubljana, Ljubljana, Yugoslavia

Y. Tsai, Pennsylvania State University, University Park, PA

W. M. Waltraud, University of Illinois at Urbana-Champaign, Urbana, IL

S. Weiderhorn, National Bureau of Standards, Gaithersburg, MD

H. Yanagida, University of Tokyo, Tokyo, Japan

J. A. Yeomans, University of Cambridge, Cambridge, United Kingdom

G. Zilberstein, GTE Labs, Waltham, MA

Fracture, 538, 670, 726
Fracture stress (strength), 543–545, 563–567, 611, 642, 644, 654–659, 668, 679–681, 707, 746–752, 764, 768–770, 780
Fracture surface energy, 624, 669
Fracture toughness, 536, 604, 632, 644, 654–659, 695, 727, 770, 779
  94% alumina, 375
  of MgO–ZrO$_2$, 298
  of partially stabilized zirconia, 262
  Si$_3$N$_4$ – TiC composites, 332
  reaction bonded SiC, 359
  of sodium Beta'' -alumina, 282
  of yttria doped zirconia, 66
Glass, 530, 572
  and Al$_2$O$_3$, composites, 41
Grain boundary, 513, 720
  effect of SrO on MgO – partially stabilized zirconia, 71
Grain boundary conductivity, 427, 499
Grain boundary diffusion, 724
Grain boundary phase
  thermodynamic stability of, 16
Grain boundary structure, 487, 490, 495, 511, 666, 720, 738
  reaction bonded SiC, 350
Grain size, 416, 434, 438, 615, 740, 777
Grain sliding, 668
H$_3$O$^+$-B/B'' alumina, 517–528
Hardness, tool materials, 322, 326
  Si$_3$N$_4$ – TiC composites, 332
Hematite (Fe$_2$O$_3$), precipitate morphology, 239
Heterogeneous Ceramics, 408–410, 414, 445
HIP, hot isostatic pressing, 717
  of liquid phase sintered composite, 41
Hot pressed Si$_3$N$_4$, 306, 331
Hot pressing
  of aluminum oxide, 3
  of fiber composites, 11
Impurities, 490, 510, 728, 736
  in Si$_3$N$_4$-TiC composites, 307
Indentation, 291, 314, 318, 330, 352, 358, 375
Indentation crack, 690
Interface characterization electron microscopy, 349, 551, 719

Interface composition, 554–560, 717
Interface reaction, 653, 666
Interfacial energy, 48, 687
  effects of anistropy on phase stability, 23
Interfacial properties, 543, 560, 611, 681, 687–696, 720
Interfacial segregation, 554–560
Interfacial structure, 695
  of ceramic eutectics, 107, 111, 133
  of reaction bonded SiC, 349
Intergranular glass phase, 457, 669
  containing Sr and Si in Mg-partially stabilized zirconia, 71
  in SiAlON ceramics, 96
  in Si$_3$N$_4$ – TiC composites, 312, 336
  in 94% alumina, 373
Internal stresses, 434, 438, 756
  interfacial, during sintering of composites, 37
Ion beam thinning, 351, 510
Ion beam mixing, of SiC surfaces, 198
Laser irradiation, of Ni coated SiC surface, 195
Lattice, 522
Lattice imaging, 719
Lead oxide, 572
Liquid phase bonding, 69, 687–696
Macrovoids, 396, 465–473
Magnesium oxide, 414, 742
  dispersed with ZrO$_2$, 295
Magnetoelastic effects, 387
Mechanical properties, 473, 542, 565–566, 583, 632, 642–644, 671, 678, 695, 725–728, 746–753, 775–787
  of 94% alumina, 373
  of ceramic eutectics, 108, 112
  of partially stabilized zirconia, 272
  of reaction-bonded SiC, 358
  of Si$_3$N$_4$ -TiC composites, 314, 332
  of sodium Beta''-alumina, 282
  of sol-gel composites, 193
  of yttria doped zirconia, 66
Mechanical testing, 633, 640, 689, 756, 778
Metal/ceramic bonding, 367